4 -

AF854566

Transport phenomena:
fundamentals and applications in Metallurgical and Materials Engineering

To

Mr. Gordon Hunter

With best wishes &

Compliments

17th June 2014

Metallurgical, Materials and Mining Books Collection

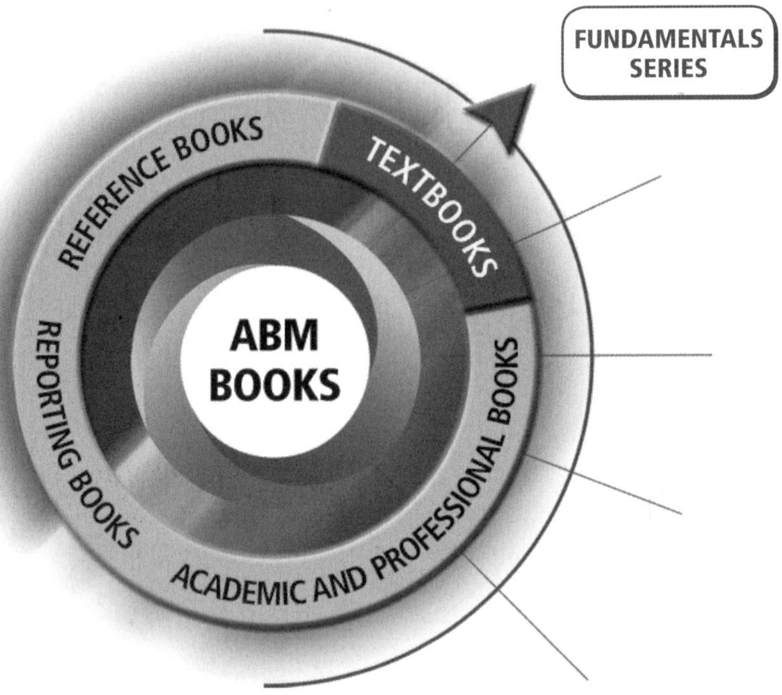

Varadarajan Seshadri
Roberto Parreiras Tavares
Carlos Antonio da Silva
Itavahn Alves da Silva

Transport phenomena: fundamentals and applications in Metallurgical and Materials Engineering

Translated from Portuguese by Varadarajan Seshadri

ABM
ASSOCIAÇÃO BRASILEIRA DE METALURGIA, MATERIAIS E MINERAÇÃO

São Paulo
2011

Cataloguing Card

Transport phenomena: fundamentals and applications in Metallurgical and Materials Engineering / Varadarajan Seshadri ... [et al.]; translated by Varadarajan Seshadri. – São Paulo: Associação Brasileira de Metalurgia, Materiais e Mineração, 2011.
806 p. – (Metallurgical, Materials and Mining Books Collection, Textbook, 2).

ISBN 978-85-7737-040-5

1.Transport Phenomena. 2. Metallurgical Processes.
I. Seshadri, Varadarajan. II. Tavares, Roberto Parreiras.
III. Silva, Carlos Antonio da. IV. Silva, Itavahn Alves. V. Title.

CDD – 536.2

ABM ASSOCIAÇÃO BRASILEIRA DE METALURGIA, MATERIAIS E MINERAÇÃO

ISO 9001 ABM EMPRESA CERTIFICADA

Certificação de Pessoas OPC 0009

Rua Antonio Comparato, 218 - Campo Belo, CEP: 04605-030 - São Paulo - SP - Brasil
Tel.: (55 11) 5534-4333 - Fax: (55 11) 5534-4330
http://www.abmbrasil.com.br

Art:
J.I. Comunicação

Industrial production:
editora cubo
soluções para o universo acadêmico

PRINTED IN BRAZIL

Foreword

TRANSPORT PHENOMENA forms part of the ABM textbooks within the serie "Fundamentals". The objective of ABM in sponsoring this english version is to provide to the international community an updated and modern textbook to the students of undergraduate and post graduate courses in metallurgical and materials engineering in the subject of transport phenomena. This text is in a position to replace others, which are mostly oriented towards mechanical and chemical engineering, used at present in the courses of metallurgical engineering and materials as texts or reference texts. Unlike these, this text focuses on examples and applications in metallurgy which should be more compatible with the courses offered in this field. In addition to emphasis on theory and fundamentals of the subject, the text includes the use of analytical solutions numerical analysis and other techniques for solution of differential equations in steady and unsteady state problems. The students as well as engineers working at the industries should find these aspects of the text quite useful for simulation and optimization of industrial processes. The Associação Brasileira de Metalurgia, Materiais e Mineração - ABM (Brazilian Metallurgical, Materials & Mining Association) congratulates the authors , who are university professors, for writing this excellent text which is a great contribution to the education of metallurgical engineering students The ABM is honoured to publish the same with the conviction that this important contribution should help the students in acquiring not only fundamental knowledge on the subject but also prepare them, while in the university, to solve industrial oriented problems as well as give an opportunity for continuing education in the case of engineers working in the area of metallurgical processes in the industries.

Claudemiro Bolfarini

Editor, Fundamental series - Books ABM

Preface

Rapid advances have been made in recent years in improving the efficiency of metallurgical processes, in order to achieve greater productivity and product quality in the steel industry. It is known that a unified approach to the principles of transport phenomena combining fluid dynamics, heat and mass transfer,for applications in the field of engineering. These aspects are essential to control and improve the processes of metallurgical operations. Books on Transport Phenomena in the english language have been published in recent decades, particularly in the areas of mechanical and chemical engineering and some in the area of metallurgical engineering. This book also deals with the principles of transport phenomena in general, with special emphasis on applications in metallurgical processes and industrial applications. It is useful to the professional engineer as well as to the undergraduate and postgraduate students in metallurgical engineering and materials for acquiring fundamental knowledge of transport processes. The examples and illustrations presented are mainly related to metallurgical processes. Pre-requisites for the effective use of the text are knowledge of mathematics, physics, physical chemistry, thermodynamics, differential calculus and various methods of solving differential equations. Some concepts of transport phenomena, its importance as a field of study, as well as the contents of the various chapters are given in the introductory chapter of the book. This text is based on the lecture notes of the authors for undergraduate and post graduate courses in the area of transport phenomena in the Departments of Metallurgical Engineering, Federal University of Minas Gerais and Federal University of Ouro Preto, in the past decades. The authors wish to thank the Associação Brasileira de Metalurgia, Materiais e Mineração (The Brazilian Association of Metallurgical, Materials & Mining) and Companhia Brasileira de Metalurgia e Mineração - CBMM for their sponsorship which has resulted in the publication of this book.

Varadarajan Seshadri   Roberto Parreiras Tavares

Carlos Antonio da Silva   Itavahn Alves da Silva

Contents

Introduction

During the processing of metals and other materials and the duration of their utility over a period of time, their properties can be changed through transport or movement of the chemical species, thermal cycling and heat fluxes. These flows are caused since thermodynamic equilibrium is not reached. Application of transport phenomena principles has the objective to quantify the fluxes of momentum,energy and species in a system so that it is possible to build models to monitor processes and/or predict the kinetics and duration for attaining equilibrium according to thermodynamic principles. The technical and economical feasibility of a process depends not only on the study of thermodynamics to determine equilibrium conditions but also that of transport phenomena for development of models with the objective of controlling the process variables improving productivity and quality of the product.

This observation is also valid in respect of life cycle after service of material components in different industries and byproducts resulting during processing. Knowing the rate of release of elements harmful to the environment under conditions typical of a landfill may be relevant for the feasibility of a new technology. For example, innovative technologies can be developed for processing rejects from different industries which have accumulated over the years like recuperation of metal values from batteries, and slags from different processes discarded and dumped for decades. The need for specialist knowledge of transport phenomena is even more important in designing a manufacturing process with the objective of getting a material with a specific property. In different areas, there would be different emphasis in the application of these principles:

- *Environmental aspect*, including the interaction of the production process with the environment as well as the product and the environment throughout its lifespan and even after discharge or scrapping after the service period;
- *Product design* involving the various steps in the manufacture of a material with specific properties, quality at low cost;

- *Characterization of the end product* and performance evaluation during service;
- *Processing* steps to get a product with the required properties at competitive cost with minimal environmental impact (sustainable development).

Obviously the aspects mentioned above are arbitrary. However the main objective of the application of transport phenomena principles is summarized in item four, namely manufacture of materials with required properties at competitive cost and minimal environmental impact. The extent of importance of the above four aspects may vary depending on the type of application.

The competition between different materials such as metal and its alloys (the "old" materials) and ceramic, glass, plastics, composite materials, nano materials (the "new" materials) for use in various applications depend on requirements and exigencies in service and environmental conditions, cost/benefits in production and subsequent maintenance costs. Certain class of materials is favored from the environmental point of view and new technology may have to be developed for satisfying the stringent environmental laws that are being enforced these days.

For the development of new technologies and materials which is the trend in the present times, a thorough knowledge of fundamentals of science and engineering as well as technologies of production and application are essential. Knowledge of fundamental sciences such as physics, chemistry and mathematics with an integral approach to the problem applying principles of thermodynamics, chemical kinetics and transport phenomena which includes fluid mechanics, heat transfer and mass transfer would be necessary for this purpose. With this unified treatment based on fundamental concepts, it is possible to handle complex problems in the industry. In this book it is proposed to elaborate the basic concepts and their applications to manufacturing processes of metals, alloys and other materials. Some preliminary examples of complex problems encountered in the metallurgical or any other field where this integral approach can be applied are given below.

Example

The thermal and compositional homogenization of a molten metal bath through gas bubbling, is a common refining operation in extractive metallurgy (Figure below). The bubbles generated in the region of the porous plug, ascending in the liquid, cause turbulence and convection currents, which reduce the temperature and concentration gradients. This is related to momentum transfer in the liquid, coupled with heat and mass transfer. Optimization and control of these phenomena involve control of these flows through process variables.

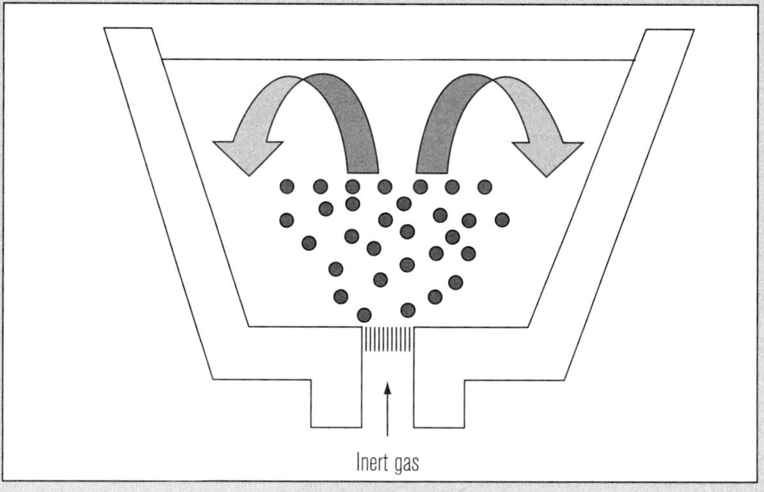

Injection of inert gas into liquid metal.

Example

Several schemes have been proposed for the recycling of household waste, as the one (United States Bureau of Mines-USBM) shown in the Figure below. The main feature includes unit operations with typical mineral processing equipments such as cyclones, screens, magnetic separators etc. The basic scientific and transport phenomena principles applied in the design and operation of these devices are the same as those, applied in mineral processing.

However the aspects of operating parameters may differ. Although household waste recycling is not considered as relevant in the traditional profession of a mining or metallurgical engineer, recycling of this type of waste material use the same principles of transport phenomena, as can be verified from the flow sheet given in the figure below.

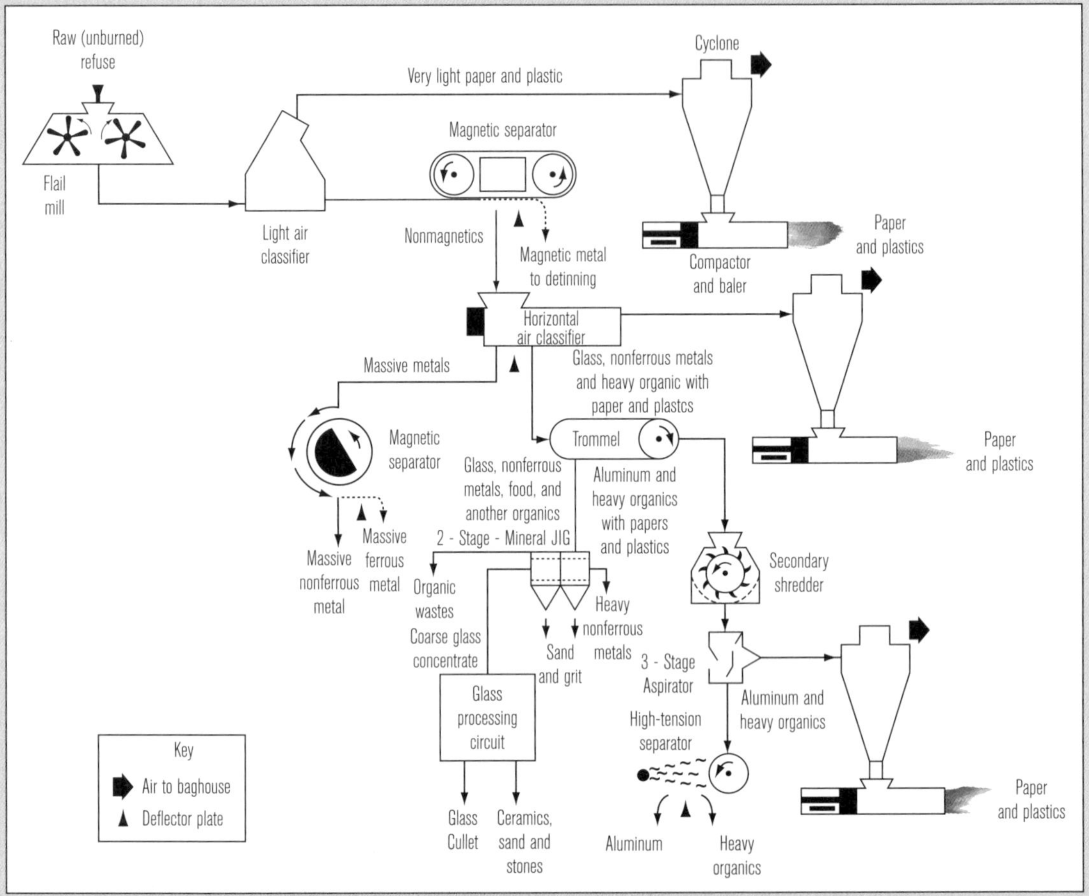

Scheme for recycling of household waste, according to the USBM (GILL, 1991).

Example

Titanium – Aluminium alloys used in the aviation industry have low density, good mechanical properties at high temperatures, resistance to oxidation etc. This is attributed to the formation of Ti-Al intermetallic compounds). A procedure that can be followed for the manufacture of these alloys is to melt a mixture of Al and Ti, 1:1 ratio, to obtain a liquid melt of Ti – Al, followed by casting and mechanical working processes (extrusion, rolling, forging, etc.). The last step may not be successful as Ti – Al alloys are extremely fragile in low temperatures and hence is not suitable for mechanical working. An alternative procedure would be via powder metallurgy techniques. Ti and Al powder obtained separately, with suitable particle sizes are mixed with 1:1 ratio, compacted to desired shape under pressure and finally subjected to sintering at higher temperatures to promote interdiffusion of metals (slightly above the melting temperature of Al, to form the intermetallics). This procedure is also typical in the production of metal ceramics composites. The stages of compaction and sintering etc involve diffusion and mass transfer, which are in general covered by the study of transport phenomena.

These examples seek to highlight the principles that underline the fundamental disciplines such as thermodynamics, chemical kinetics, transport phenomena (and many others) in various engineering applications.

1.1 Unified Treatment

The scope of each of the three branches of the discipline "Transport Phenomena" can be treated separately in a comprehensive manner with examples and applications. As has been previously mentioned, the motivation to assemble together in one text, all the three important areas is due to the fact, that transport of heat, mass and momentum take place in most of the processes simultaneously, with interacting influence. In addition, there are physical and mathematical similarities between these modes of transport, that can lead to a better understanding and easier solution procedures of the equations, if an integral approach is adopted for the study of any process.

For example, denoting by Φ the volume concentration of a given quantity, be it mass, heat and momentum, one can mention at least two contributions to the common transport, namely convective and diffusive.

Convective transport is through a control surface (real or imaginary) by the movement of the medium. For example, in the case of the species (element or compound) A, of concentration C_A (mol/m³), in a medium, that moves with a velocity V_y (m/s), the amount of A going through a static control surface (Figure 1.1) perpendicular to the flow area dS (m²) is given by:

$$V_y(m/s).\ dS(m^2).\ C_A(mols\ A/m^3) \tag{1.1}$$

This corresponds to a relative volumetric flow of the medium times the volume concentration of the species A. Similarly the volume concentration of heat and momentum can be defined, in terms of Φ, being equal to $\rho\ C_p T$ or ρV_i respectively. Here, ρ(kg/m³), stands for the density of the medium; C_p(J/kg.K) the specific heat of the medium, T(K) the temperature of the medium; V_i(m/s), the velocity of the medium in the direction i. Then the equations of transport by convection would be generally defined as V_y(m/s).dS(m²).ϕ:

$$V_y(m/s).\ dS(m^2).\ [\rho\ V_i\ (kg/m^2.s)] \tag{1.2}$$

$$V_y(m/s).\ dS(m^2).\ [\rho\ C_p\ T(J/m^3)] \tag{1.3}$$

$$V_y(m/s).\ dS(m^2).\ C_A(mols\ A/m^3) \tag{1.4}$$

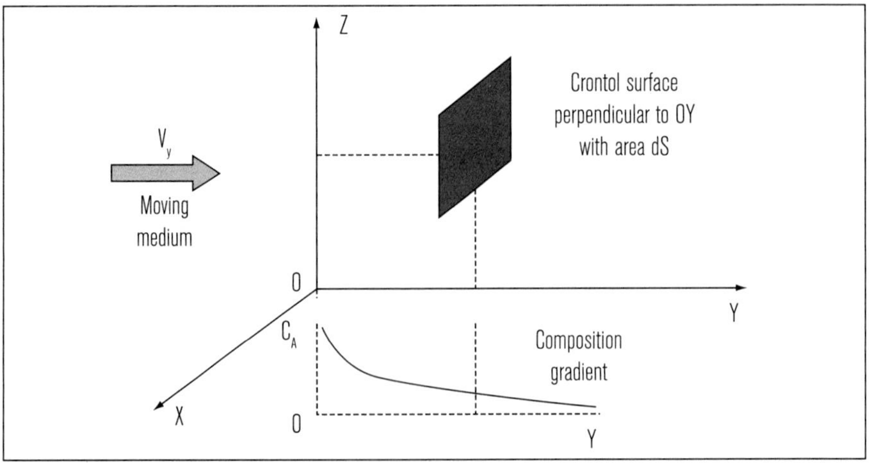

Figure 1.1 - Transport by movement of species (convective).

The driving force for transport processes by diffusion is related to the existence of gradients of a given parameter. For example, there could be transport of a given species under the action of gradients: of temperature, pressure, electrical potential, chemical potential etc. Electrical fields or electrical potential gradients are particularly active in the case of transport of charged species, like ions during electrolysis or in electrorefining. Gradients of chemical potential can be at a given stage, related to gradients of composition, and give rise to ordinary diffusion (most common). Fick's law can be used to calculate the velocity of transport of species by diffusion. Thermodynamics requires that the transport is spontaneous from the point of highest chemical potential (higher concentration) to the point of lowest chemical potential (lower concentration), as shown in Figure 1.2.

$$J_A(\text{mols A}/\text{m}^2.\text{s}) = -D_A(\text{m}^2/\text{s}).\frac{dC_A\ (\text{mol A}/\text{m}^3)}{dy(\text{m})} \qquad (1.5)$$

where D_A represents the diffusion coefficient of species A, usually determined experimentally as a function of properties of the medium and the species A (ie, temperature, pressure, composition, physical state). $\frac{dC_A}{dy}$ represents the magnitude of the gradient or driving force of the process. It can be considered also an indirect measurement of of chemical potential gradient (actual cause of ordinary chemical diffusion).

Expressions corresponding to the diffusive transport of heat and momentum can also be obtained through manipulation of the equations corresponding to the Fourier Law of heat conduction and Newton´s relation for viscosity of fluids.

$$q_y(\text{J}/\text{m}^2.\text{s}) = -\frac{K}{\rho C_p}(\text{m}^2/\text{s}) .\frac{d\,\rho\, C_p\, T(\text{J}/\text{m}^3)}{dy(\text{m})} \qquad (1.6)$$

$$\tau_{yz}(\text{N}/\text{m}^2) = -\frac{\eta}{\rho}(\text{m}^2/\text{s}).\ \frac{d\ \rho\ V_y(\text{kg}/\text{m}^2.\text{s})}{dz(\text{m})} \qquad (1.7)$$

In these expressions η ($Kg.m^{-1.}\ s^{-1}$) is the dynamic viscosity, k (J/m.s.K), the thermal conductivity of the medium. $K/\rho C_p$ is the thermal diffusivity of the medium, η/ρ is the momentum diffusivity Figure 1.2 illustrates the processes of diffusion of species, heat and momentum, in terms of their characteristic gradients. The diffusive and convective contributions per unit area can be identified from the analogies shown in Table 1.1.

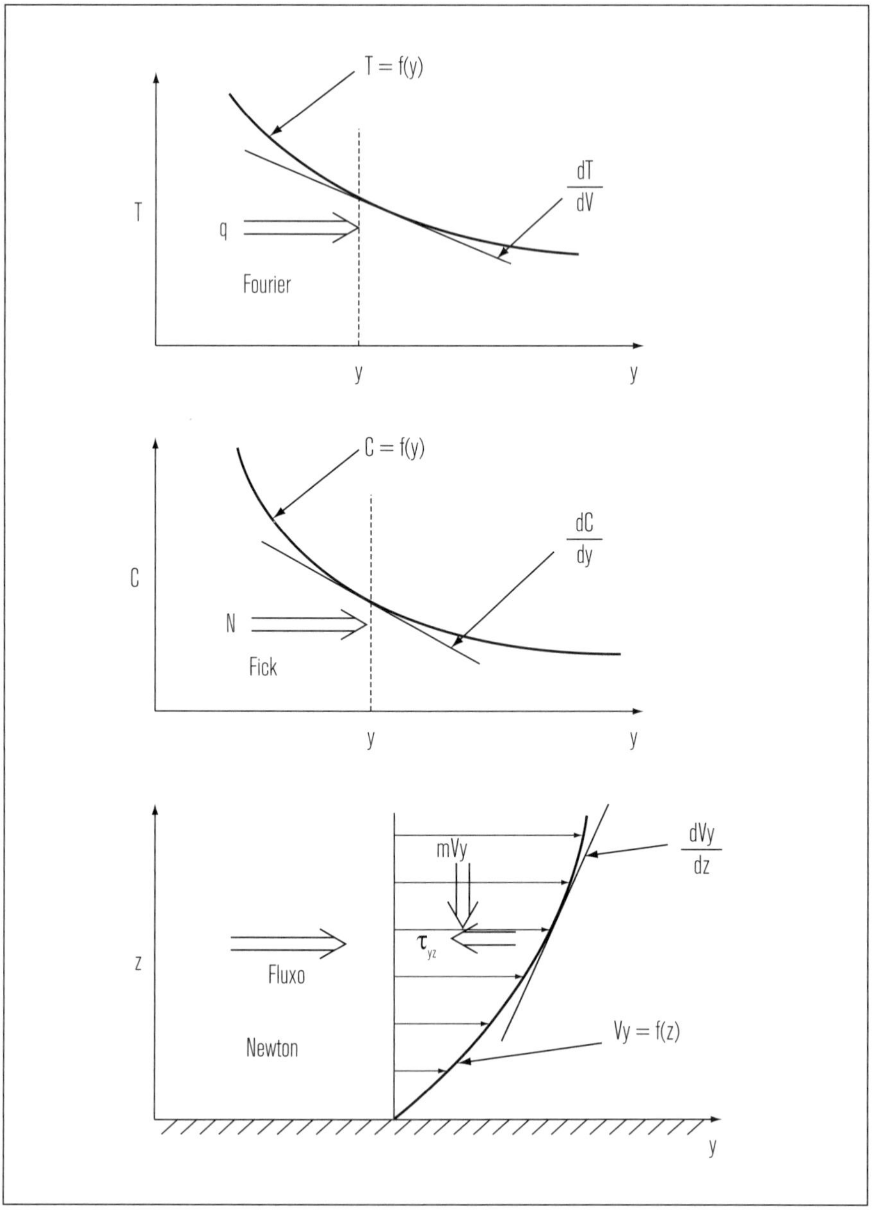

Figure 1.2 - Heat, mass and momentum fluxes based on equations of Fourier, Fick and Newton.

Transport mode	Convective contribution	Diffusive contribution
Mass transfer of Species A	$V_y C_A$	$-D_A \frac{d}{dy} C_A$; Fick's Law
Heat	$V_y (\rho C_p T)$	$-\frac{K}{\rho C_p} \frac{d}{dy} \rho C_p T$; Fourier´s Law
Momentum	$V_y (\rho V_i)$	$-\frac{\eta}{\rho} . \frac{d}{dz} \rho V_Y$, Newton's Law

Table 1.1 - Similarities between expressions for the calculation of diffusive and convective contributions.

Transport Phenomena studies routinely use balances of conservation of heat, mass and momentum for the analysis of problems. In generic terms conservation for any quantity ccan be given as:

"Rate of accumulation within the Volume Control (VC)"

=

"The rate of net entry in Volume Control (VC) through Control Surfaces (SC), through the mechanism of convection"

+

"The rate of net entry in control volume Control Surfaces (SC), through the mechanism of diffusion"

+

"Other Contributions"

The convective and diffusive contributions have been defined previously. Other contributions are those that do not fit in the previous definitions, such as those due to electric and magnetic fields, pressure and gravity, radiation and other mechanisms of heat generation.

Thus one can anticipate that these conservation equations based on Φ regardless of its nature, will be structurally and formally identical, so that analytical procedures and numerical solutions would present common characteristic approaches. This would be an extra attraction of the unified approach of transport phenomena, compared to conventional studies of "Fluid Mechanics", "Heat Transfer" and" Mass Transfer" individually.

1.2 Contents of Various Chapters

In the above, a brief introduction to some industrial applications of transport of momentum, heat and mass, as well as advantages of a unified approach to the solution of problems have been discussed. In the subsequent chapters of this text, momentum, heat and mass transfer are given in that order. Wherever applicable, similarities of approach are highlighted. Much of the treatment described in the section on transport of species is based on the preceding heat transfer section. There is considerable emphasis on fundamentals with a number of examples of applications and solutions to problems.

Chapter 2 deals with a comparison of thermodynamic and transport phenomena approach highlighting the limitations of the former. An overview of application of transport phenomena principles for solution of various aspects of iron making and steelmaking and continuous casting etc are given. In Chapter 3 fundamental concepts of fluids, forces acting on a system, shear stress, energy, units and dimensions of various transport phenomena parameters are discussed. Chapter 4 contains an interpretation of Newton's law of viscosity, definition of viscosity, concepts of viscosity of gases, liquids, metallic melts, slags and development of equations and correlations for the determination of viscosities. Chapter 5 deals with microscopic balance of conservation of mass and momentum for laminar flow and different geometries and different boundary conditions This is continued in Chapter 6 where general equations of conservation of momentum, (Navier-Stokes equations) are developed. Situations are presented highlighting the adaptation of general equations for specific problems with simplifications for easy solution. Turbulent flow is treated in Chapter 7. It is well known that transfer rates (of momentum, heat and species) are more pronounced in this case than in laminar flow. Relevant equations for this case can be derived using the Navier-Stokes equations with time averaged velocity and the concept of turbulent viscosity. More practical approaches to turbulent flow like the use of friction factors for tubes, spheres, packed beds etc with examples have been discussed. In Chapter 8 global balances of mass, momentum and energy, Bernoulli equation and its application in solution of problems involving flow in pipes, flow of liquid metals from ladles, flow measurement techniques etc with many examples from industrial practice have been discussed. In Chapter 9 an introduction to heat transfer is given. Chapter 10 contains basic principles and mechanisms of convection, conduction and radiation modes of transfer of heat energy. The emphasis is on conceptual aspects. In Chapter 11, steady state heat flow is discussed for various geometries with different boundary conditions with a number of examples. It deals with general conservation equations for energy for cartesian, cylindrical and spherical geometries. Analytical Solution methods for steady state equations, numerical methods for complex geometries and examples of solving equations using numerical discretization are given.

Cases of heat transport in the transient regime are presented in Chapter 12, solutions in the form of series, use of spreadsheets to obtain the roots that define the numerical values of the coefficients and the use of numerical discretization of the equations of heat transfer, in the transient regime are explained. Chapter 13 is devoted to various aspects of heat transfer by convection including methods for determining the coefficients of heat transfer and dimensional analysis that helps to find the relevant dimensionless groups in heat transfer problems. Besides the presentation of basic principles a number of applications are discussed.

In Chapter 14, principles of radiation and application to some practical problems are dealt with. Chapter15 is introduction to mass transfer and Chapter 16 discusses some key issues relating to transport of species, establishment of the reference speed, the identification of the driving forces of diffusion, and a brief introduction to the thermodynamics of irreversible processes. Conceptual aspects of the problems are given. The first step in quantifying the transport of species is done in Chapter 17. A review of diffusion coefficients, microscopic structure of solids and its relation to transport of species and other aspects of diffusion are given. Chapter 18 deals primarily with the distribution of species in steady and unsteady states, for simple geometries. The similarities between heat conduction and diffusion are given and it is shown how equations developed for heat transfer can be used for mass transfer with minor modifications. In chapter 19, transport of species to the sites of chemical reactions, that is coupling between chemical reactions and transport of species is covered via model building. Analytical solutions are discussed for specific cases and boundary conditions. Chapter 20 and 21 are dedicated to applications of transport phenomena as a tool for process analysis. Examples are given from the metallurgical industry like blast furnace, steel making and refining processes. Physical modeling techniques are discussed.

1.3 Concluding Remarks

Any study of transport phenomena requires a good understanding of mathematics, especially a deep knowledge of calculus, to be able to handle the differential equations and their solutions. Like many other areas in engineering, the study of transport phenomena and its applications also require knowledge and understanding of thermodynamics and kinetics. Despite the current trend of solution of equations with numerical packages, it is essential that a student should be well versed in fundamentals in order to understand, how the numerical packages have been developed and how they can be used with advantage.for the solution of industrial problems.. Books on Transport Phenomena in general Poirier and Geiger (1994) and Bird, Steward and Lighfood (2002) are quite specialized and usually go into depth in respect of fundamentals. The present text is balanced in respect of all the three areas (momentum, heat and mass transfer) and application of these concepts to metallurgical processes. Hence it is adequate to meet most of the needs of an undergraduate student, without overloading him with too many complex mathematical problems. For this purpose, the text contains several worked out examples and exercises at the end of most of the chapters. The text can also be used by post graduate students in metallurgical engineering and others from various other branches of science and engineering, where transport of fluids, heat and mass transfer are important. However the main objective of the text is meant for specialization in metallurgy and materials, as most of the examples and applications given are drawn from this area of specialization.

References

BIRD, R.B.; STEWART, W.E.; LIGTHFOOT, E.N. *Transport phenomena*. 2. ed. New York: John Wiley & Sons, 2002.

GILL, C.B. *Materials beneficiation*. Berlin: Springer, 1991.

POIRIER, D.H.; GEIGER, G.H. *Transport phenomena in materials processing*. Warrendale, Pennsylvania: The Minerals, Metals and Materials Society, 1994.

Transport Phenomena: Concepts and Applications

Chapter 02

Thermodynamics is a fundamental science in the analysis of the processes involved in the production of metals and other materials. Nevertheless, the thermodynamical approach has a limitation, namely inability to predict the rate of chemical reactions and other phenomena in the processes. The time taken for the completion of a process and the variables affecting it are important parameters, that determine the economical aspects of manufacturing. The following examples illustrate simple situations highlighting the limitations of the thermodynamical approach.

Figure 2.1 shows two bars of a metal, in perfect contact. One of the bars is at 1000°C and the other at 200°C. According to thermodynamics, the heat would be transported from the bar that is at higher temperature to the bar at lower temperature and at equilibrium, the two bars would be at the same temperature. However, it cannot predict how long it will take to achieve this equilibrium or to determine the temperature profiles in the two bars at any given instant.

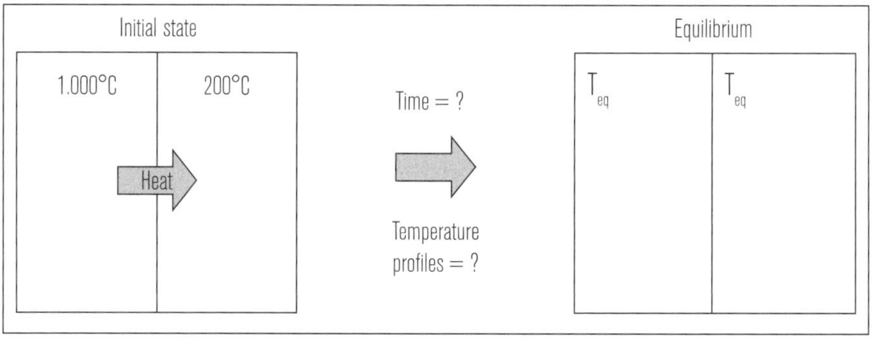

Figure 2.1 - Transport of heat between two metal bars.

A case similar to this can be imagined considering two steel bars at the same temperature, but with different carbon contents, as shown in Figure 2.2. In this case, according to thermodynamical principles there should be transport of carbon from the bar that has a higher concentration to the bar at lower concentration. However, neither the time needed to reach equilibrium, nor the concentration profiles at a certain instant can be determined by applying those principles.

Finally, considering the situation illustrated in Figure 2.3 (showing a ladle with liquid steel), by opening the valve underneath the ladle, steel can be poured into some other reactor. The principles of thermodynamics cannot be applied to estimate the time for emptying the ladle as a function of the amount of steel that it contains, whereas this can assessed applying transport phenomena principles.

These three examples show the three distinct areas that constitute the study of transport phenomena:

- Transport of energy or heat transfer (example of Figure 2.1);
- Mass transport (example of Figure 2.2);
- Transport of momentum (example of Figure 2.3).

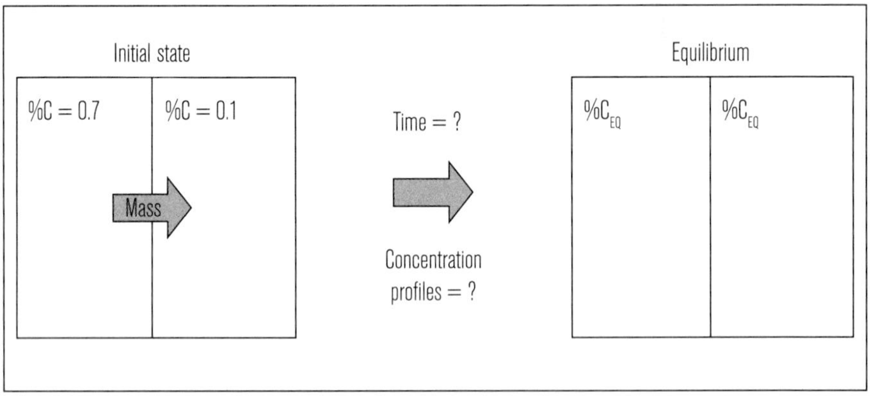

Figure 2.2 - Transport of mass between two steel bars.

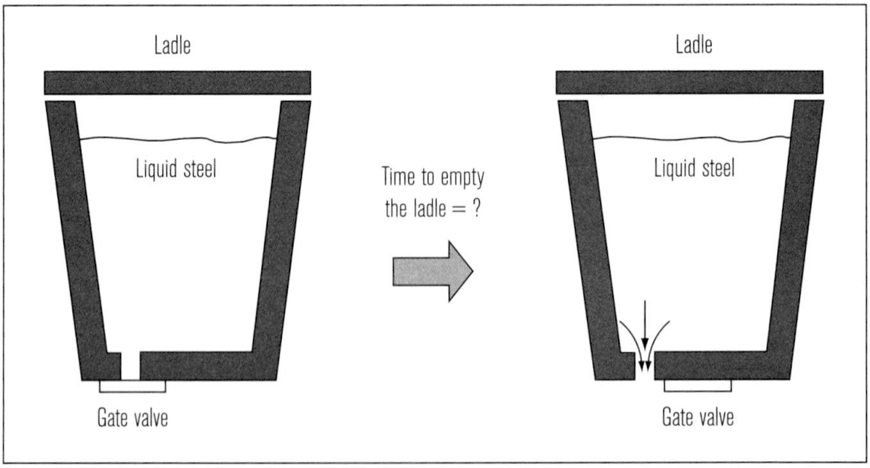

Figure 2.3 - Draining of a steel ladle.

Using the principles of transport phenomena, it should be possible to answer the questions in these three examples regarding time taken for the process to go to completion. The study of transport phenomena finds numerous applications in the various processes in materials and metallurgical engineering and especially in the steel plant, such as sintering, iron making in the blast furnace, steel making and refining, continuous casting and rolling, as illustrated in Figure 2.4, which presents a general flow chart for the production of rolled steel in integrated plants and mini-mills. The important aspects of the study of heat, mass and momentum transfer in the various stages of steelmaking process are discussed below.

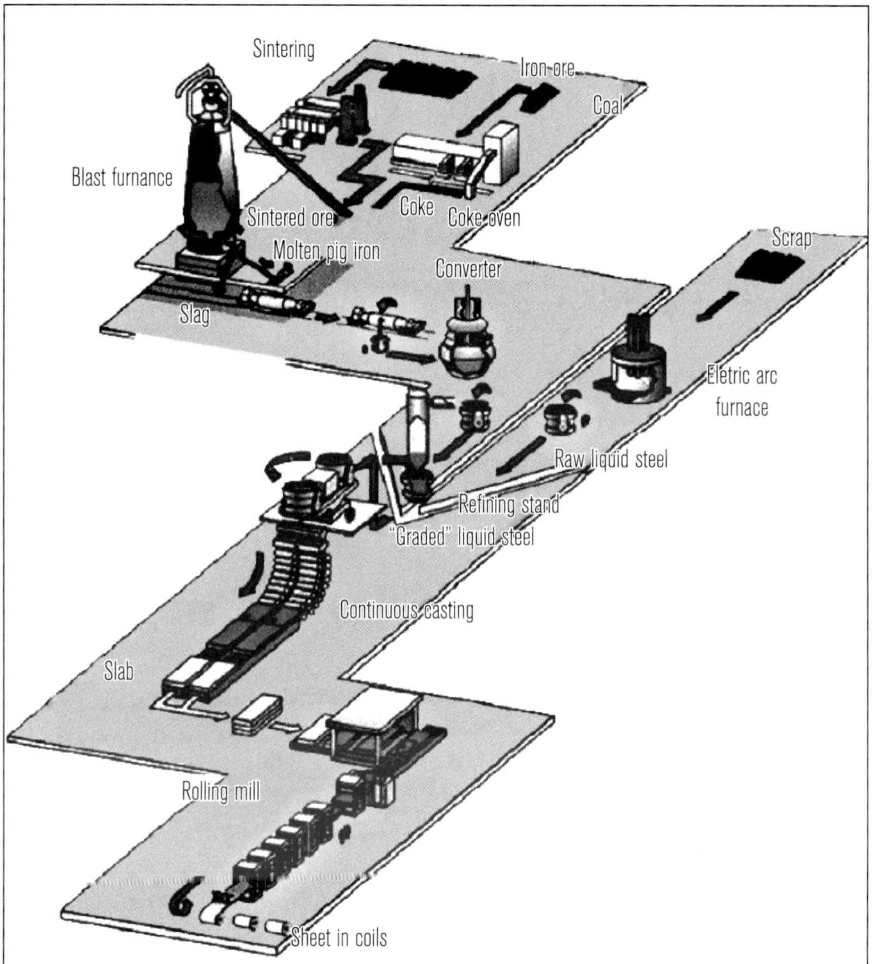

Figure 2.4 - General flow chart for steel production in integrated plants and mini mills (CHO, 2005).

- *Transport of heat*
  - Heat exchange between gases and solids in the sintering and blast furnace processes. With this study, one can determine the rate of heating of solids, and the factors that directly influence the efficiency of heat exchange, which is essential for the energy efficiency, economy and productivity of the respective processes;
  - Heat extraction and solidification in continuous casting machines. Heat transport during solidification is of fundamental importance, since is affects the design, operation and productivity of the casting equipment and also the quality of the products;
  - Heat transfer in reheating furnaces. Reheating furnaces are responsible for delivering steel at the appropriate uniform temperature to the rolling mill. To attain this, it is essential to study the heat transfer between gases and steel inside the furnace. This study can also lead to energy saving and less CO_2 emissions.
- *Mass transport*

All transformations that involve chemical reactions in the various reactors are linked to mass transport, chemical kinetics, coupled with fluid flow and heat transfer:

  - Reduction reactions of iron oxides in blast furnace;
  - Reactions in pig iron desulphurisation and desiliconization in various industrial units;
- Steel making reactions, especially decarburization and dephosphorization;
- Steel refining reactions, such as vacuum degassing, deoxidation, etc.
- *Transport of momentum*

Every step that involves the movement of fluids is linked to the transport of momentum. Some of the examples in iron and steelmaking can be cited:

  - Gas flow through the beds of sintering and blast furnace. In this case, the study of momentum transfer has several applications, including the prediction of the pressure drop of the gas and the evaluation of the horsepower required for the exhaust fan/ air blowing system. Gas distribution in the blast furnace also affects productivity and fuel rate;

- Injection of inert gases in ladles and other reactors for refining of steel. Velocity profiles and mixing times can be predicted and used to select the best location for injection of refining agents;
- Steel flow in the refining process under vacuum, particularly in the RH reactor. In this case, knowledge of the velocity field of steel and how it is affected by the configuration of the equipment can be useful in optimizing the design and operation of the refining plant.

Besides these, numerous other applications can be cited. These applications are becoming more common and important as softwares based on transport phenomena equations are developed (Computational Fluid Dynamics - CFD) with numerical techniques for rapid solution of the conservation equations. Finally, it is important to mention that the science of transport phenomena and its concepts are widely applied in aerospace, chemical, mechanical and civil engineering areas in addition to metallurgical and materials engineering.

References

CHO, J.E. *Some aspects of TRIZ applications in stell making process.* 30 p. Available from: <http://www.realinnovation.com/archives/2005/10/08.pdf>. Access in: 2008 Sep. 25.

Basic Concepts

Chapter 03

Before discussing momentum transfer, in this chapter some of the important concepts and definitions of normal force, shear stress, potential and kinetic energy etc as well as units and dimensions for describing these parameters have been dealt with.

3.1 Concepts

3.1.1 Fluids

One property common to all fluids is that it can not maintain its shape unless it is contained within a receptor or container. A fluid in general assumes the form of the container.

Another property that distinguishes the fluid state is that it deforms continuously under the action of a shear stress, regardless of its value. It is important to note that in the case of solids also, there is deformation with shear stress but not necessarily in a continuous manner as in the case of liquids.

As per this definition, fluids can be liquids or gases. Incompressible fluids are those whose densities remain constant with the variation of pressure. Under normal conditions encountered in engineering problems, liquids are considered incompressible fluids and gases are defined as compressible fluids, as their densities change significantly with pressure. However, under certain situations gases can behave as incompressible fluids.

3.1.2 Force, Momentum and Stress

Another important definition is that of force. The concept of force is derived from Newtons second law which can be defined as follows.

$$\Sigma F_x = m\, a_x \qquad (3.1)$$

where:

- ΣF_x = resultant of forces acting on the body in the x direction;
- m = mass of the body;
- a_x = the acceleration of the body in the x direction.

Another way of expressing this law is:

$$\Sigma F_x = \frac{\partial (m\, v_x)}{\partial t} \qquad (3.2)$$

where:

- v_x = velocity of the body in the x direction;
- t = time.

From equations (3.1) and (3.2) for constant mass, rate of change of velocity can be defined as acceleration:

$$a_x = \frac{\partial v_x}{d\,t} \qquad (3.3)$$

Recalling the definition of momentum:

$$\text{Momentum} = m.v_x \qquad (3.4)$$

One can conclude that force is rate of change of momentum.

The forces acting on a given system can be classified in two categories, volume and surface forces. Volume forces are those caused by gravity or electromagnetic fields in the fluid and act as a whole. These are usually expressed in terms of force per unit volume.

On the other hand, surface forces represent the action of the neighborhood on the fluid element being considered. These are usually given in terms of force per unit area.

Another important concept is that of stress. To define this parameter, the element of fluid volume seen in Figure 3.1 can be considered.

The force applied by the neighborhood on the hatched area, ΔA, that is ΔF, can be resolved into two components: ΔF_n is the normal component and ΔF_t is the tangential component.

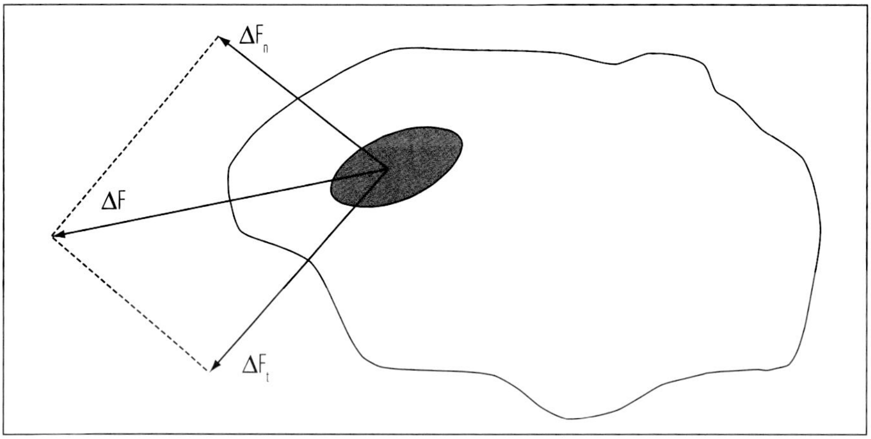

Figure 3.1 - Forces acting on the surface of a volume element.

The quantities ΔF_n and ΔF_t are called normal and shear forces, respectively. Stress is defined as force per unit area. Therefore, these two types of stresses acting on the surface of the fluid element can be defined as:

- Normal stress:

$$\tau_n = \frac{(\Delta F_n)}{(\Delta A)}_{\Delta A \to 0} \tag{3.5}$$

- Shear stress:

$$\tau_t = \frac{(\Delta F_t)}{(\Delta A)}_{\Delta A \to 0} \tag{3.6}$$

More specifically, stress is identified by the direction of the force and the orientation of the area on which it acts. Figure 3.2 shows the volume element in cartesian coordinates indicating nine possibilities of stresses acting on the element in terms of direction and area on which it acts.

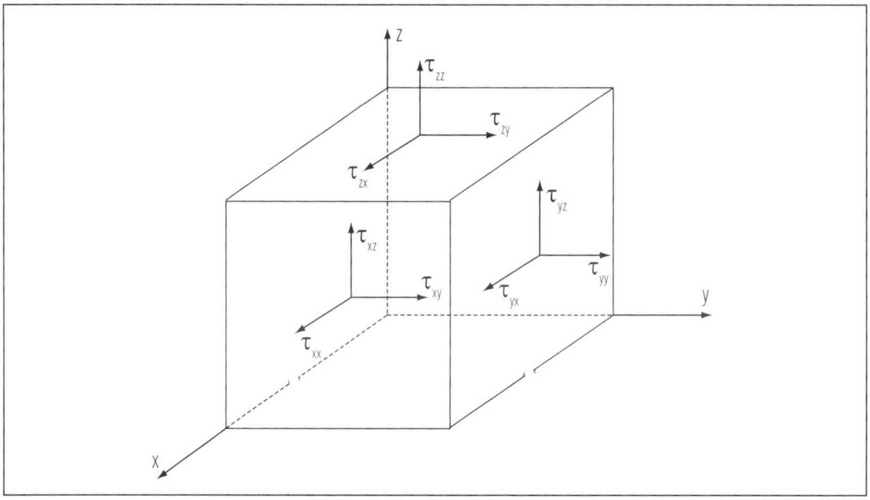

Figure 3.2 - Representation of stresses on a volume element in cartesian coordinates.

The subscripts used in Figure 3.2 have the following meaning:

- First subscript: the direction normal to the surface on which the force is acting;
- Second subscript: direction of the force that produces the stress.

One can easily notice that τ_{xx}, τ_{yy} and τ_{zz} are normal stresses, while τ_{xy}, τ_{xz}, τ_{yx}, τ_{yz}, τ_{zx} and τ_{zy} are shear stresses.

3.1.3 Energy

In the study of fluid flow, two forms of energy are particularly important, the potential energy and kinetic energy.

Potential energy is the energy possessed by the fluid due to its mass, its position and the effect of gravity. Numerically, the potential energy per unit volume of fluid is given by the following relationship:

$$E_p = \rho \, g \, z \tag{3.7}$$

where:

- E_p = potential energy per unit volume of fluid;
- ρ = fluid density (ratio between mass and volume);
- g = acceleration of gravity;
- z = height of fluid in relation to an arbitrary level at which the potential energy is taken as zero.

Kinetic energy is the energy that the fluid possesses by virtue of its macroscopic movement. Its value per unit volume of fluid can be determined by the following relationship:

$$E_c = \frac{1}{2} \rho \, u^2 \tag{3.8}$$

where:

- E_c = kinetic energy per unit volume of fluid;
- u = fluid velocity.

3.1.4 Mechanisms of Transport

Before proceeding to study the units involved in the evaluation of the variables appearing in transport phenomena, an important concept has to be considered. It is related to the mechanisms involved in momentum, heat and mass transfer.

Basically, there are two mechanisms of momentum, heat and mass transfer. These two mechanisms are:

- diffusion;
- convection.

For heat transfer there is an additional mechanism called radiation.

The mechanism of diffusion depends on the existence of a physical medium and occurs due to the presence of a gradient of a given quantity:

- velocity for momentum transfer;
- temperature for heat transfer;
- concentration or chemical potential for mass transfer.

Diffusion does not require a macroscopic movement of the medium.

Convection also depends on the existence of a medium and occurs as a consequence of a macroscopic motion of fluid.

The difference between these two mechanisms is illustrated in Figures 3.3 and 3.4.

In Figure 3.3, inside the metal bar, heat is transferred by diffusion (also known as conduction) due to the temperature gradient. It is observed that there is no macroscopic motion of atoms inside the bar. On the right surface, there is a fan blowing cold air over the bar. In this case, the heat is removed from the bar through the mechanism of convection, since there is a macroscopic movement of the fluid (in this case, air).

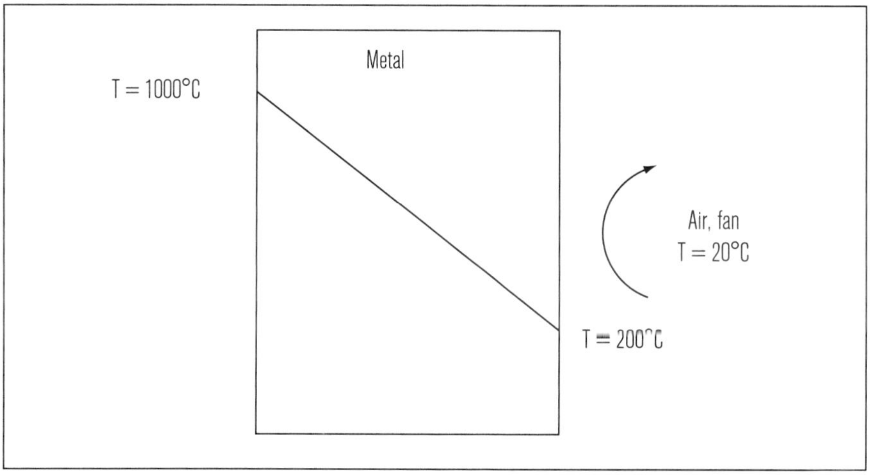

Figure 3.3 - Transport of heat by convection and diffusion.

Figure 3.4a presents a case of mass transport by diffusion. Carbon is transported from one surface to the other due to the concentration gradient. Again, there is no macroscopic motion of the system. In Figure 3.4b, the mass transport occurs by convection. Sugar dissolves in water and is transported to various parts of the system due to the macroscopic movement of water caused by the rotation of the impeller.

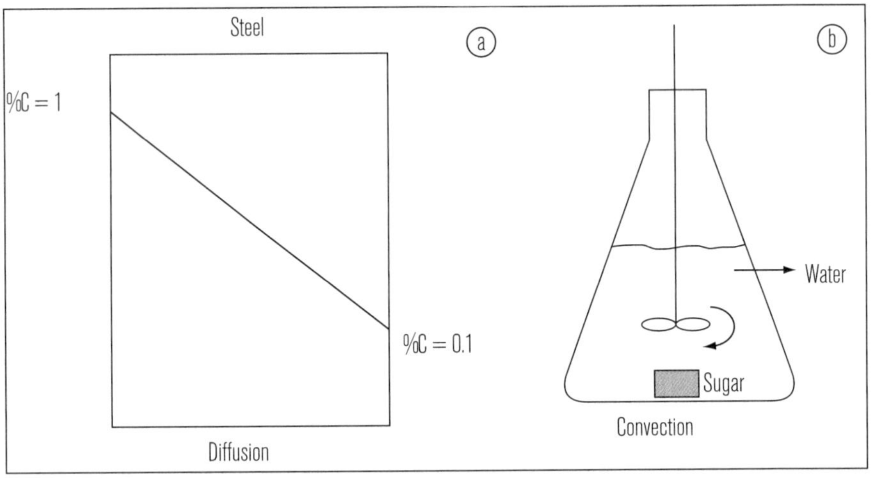

Figure 3.4 - Transport of mass by diffusion and convection.

3.2 Units

The quantitative representation of the phenomena involved in fluid flow requires the use of different types of equations. These equations, describing the physical phenomena, have to be dimensionally homogeneous. In other words, all terms must have the same dimension as expressed in the same units.

Over the years, various systems of units have been adopted by the scientific and engineering communities, for example: the English system, cgs and mks systems.

In 1960, a new and rational system of units has been recommended for international use. It is called the international system of units. In this system, which will be adopted throughout the present text, the unit of mass is the kilogram, the unit of length is the meter and the time unit is second.

Table 3.1 contains a list of units and dimensions of the main quantities involved in transport phenomena, and the nature of these quantities (scalar, vector or tensor).

Despite the recommendation mentioned above, it is still common to find in the literature other systems of units other than the International System being used. Therefore, it is important to familiarize with the necessary conversions.

Table 3.2 shows some useful conversion factors in the study of transport phenomena.

Regarding temperature measurement, two units are used extensively:

- degree Celsius: °C;
- degree Fahrenheit: °F.

Table 3.1 - Units and dimensions of the main quantities involved in transport phenomena.

Quantity	Dimension	Unit (SI)	Nature
Mass	M	kg	scalar
Length	L	m	scalar
Time	t	s	scalar
Temperature	T	K(C)	scalar
Acceleration	$L t^{-2}$	$m.s^{-2}$	vector
Angular velocity	t^{-1}	s^{-1}	vector
Area	L^2	m^2	scalar
Density	ML^{-3}	$kg.m^{-3}$	scalar
Dynamic viscosity	$ML^{-1} t^{-1}$	$kg.m^{-1}.s^{-1}$	scalar
Kinematic viscosity	$L^2 t^{-1}$	$m^2.s^{-1}$	scalar
Energy, work	$ML^2 t^{-2}$	J (Nm)	scalar
Force	$ML t^{-2}$	N (kg $m.s^{-2}$)	vector
Momentum	$ML t^{-1}$	kg.m/s	vector
Pressure	$ML^{-1} t^{-2}$	Pa (Nm^{-2})	scalar
Stress	$ML^{-1} t^{-2}$	Pa (Nm^{-2})	tensor
Power	$ML^2 t^{-3}$	W ($Nm.s^{-1}$)	scalar
Specific Heat	$L^2 t^{-2} T^{-1}$	$J.kg^{-1}.K^{-1}$	scalar
Velocity	$L t^{-1}$	$m.s^{-1}$	vector
Volume	L^3	m^3	scalar

Table 3.2 - Commonly used conversion factors between English system and International system.

Unit	Equivalent in the International System
1 ft (foot)	0.3048 m
1 in (inch)	0.0254 m
1 lb_m (pound mass)	0.45359 kg
1 BTU (British Thermal Unit)	1055 J
1 cal (calorie)	4.184 J
1 lb_f	4.4482 N
1 kg	9.8 N
1 hp	745.7 W

In the absolute scales (Kelvin and Rankine), zero is set as the lowest temperature that is possible. Two units, that are adopted, (Celsius and Fahrenheit) are related to the respective absolute scales in the following way:

- Kelvin, °K (0 °K = –273°C);
- Rankine, °R (0 °R = –460°F).

It is important to note that one degree Celsius is exactly equivalent to one Kelvin and that one degree Fahrenheit is equal to one degree Rankine.

Figure 3.5 shows a diagram relating the different systems of temperature measurements.

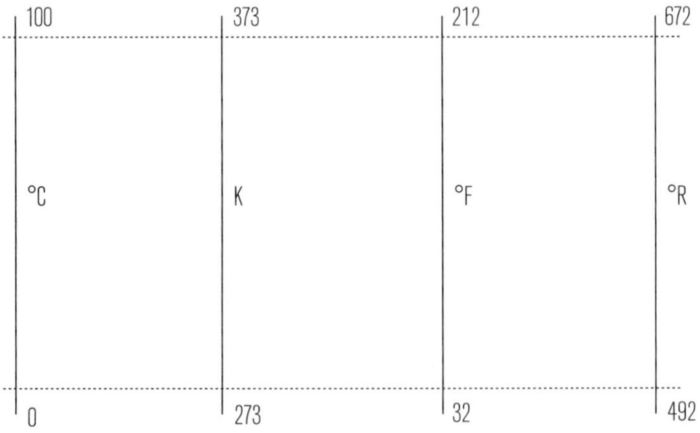

$$\frac{°C}{100} = \frac{K - 273}{100} = \frac{°F - 32}{180} = \frac{°R - 492}{180}$$

Figure 3.5 - Relationships between the different temperature units.

Based on the equations presented in Figure 3.5, the following relationships can be deduced.

$$\Delta°C = 5/9\ \Delta°F \tag{3.9}$$

$$\Delta K = 5/9\ \Delta°F \tag{3.10}$$

The above relationships indicate that the degree Celsius or degree Kelvin is 1.8 times greater than the degree Fahrenheit. The same relationship exists between degree Kelvin and degree Rankine. These relations are useful for conversion of temperatures from one system to the other.

For conversion of temperature from one scale to the other the following relations are useful:

$$T(°R) = T(°F) + 460 \tag{3.11}$$

$$T(K) = T(°C) + 273 \tag{3.12}$$

$$T(°C) = \frac{5}{9}\,[T(°F) - 32] \tag{3.13}$$

Some units have special names and it is important to know these names, as well as their significance,as given below:

- dyne = g.cm/s^2 (force);
- Newton (N) = Kg.m/s^2 (force);
- poundal = lb_m ft/s^2 (force);
- Pascal = N/m^2 (pressure);
- erg = g cm^2/s^2 (energy);
- Joule =Kg.m^2/s^2(energy);
- Poise = g/cm s (viscosity);
- Watt = Joule/s.

Some examples of conversion of units are given below:

Example

An airplane travels at a speed equal to the speed of sound. What is its speed, expressed in units of the international system? Speed of sound = 3.96×10^6 ft/h.

Solution

Using the following conversion factors:

1 ft = 0.3048 m;

1 h = 3600 s.

the speed in units of the international system is given by:

$$\text{velocity (international units)} = 3.96 \times 10^6 \frac{\text{ft}}{\text{h}} = 3.96 \times 10^6 \frac{(0.3048\ \text{m})}{(3.600\ \text{s})} = 335.3\ \text{m/s}$$

Example

100 lb_m of water flow in a pipe with an average velocity of 10 ft/s. What is the kinetic energy of the water in units of international system?

Solution

The kinetic energy is given by:

$$E_c = \frac{1}{2}\, m\, u^2$$

The conversion factors that are necessary are:

1 lb_m = 0.45359 kg;

1 ft = 0.3048 m.

Adopting the same procedure of the previous example,

$$E_c = \frac{1}{2}\, 100\ lb_m\, (10\ \text{ft/s})^2 = \frac{1}{2}\, 100\, (0.45359\ \text{kg})\, [10 \times (0.3048\ \text{m})\ /\text{s}]^2 =$$

$$210.7\ \text{kg}\ \frac{\text{m}^2}{\text{s}^2} = 210.7\ \text{J}$$

Example

What is the potential energy of a body of 30 lb_m situated 10 ft above the level of reference ? Give the result in units of international system.

Solution

The potential energy is given by:

$$E_p = m\,g\,z$$

Applying the conversion factors already used above,

$$E_p = 30\ lb_m\ g\ 10\ ft = 30\ (0.45359\ kg)\left(9.8\ \frac{m}{s^2}\right) 10\ (0.3048\ m)$$

$$E_p = 406.47\ kg\ \frac{m^2}{s^2} = 406.47\ J$$

Example

A parameter widely used in heat transfer studies is the heat transfer coefficient. This parameter is normally estimated by means of empirical correlations. One of these correlations is given below:

$$G = \text{mass flux } (lb_m/(h\ ft^2))$$

where:

- h = heat transfer coefficient (BTU / h ft^2 °F);
- G = mass flow rate (lb_m/h ft^2);
- D = diameter of the tube (ft).

This equation is to be rewritten adopting units of the international system. What should be the new constant instead of 0.026 ?

Solution

In units of the international system, the variables involved are expressed in:

G_{IS} (kg/m^2 s); D_{IS} (m); h_{IS} (J/s m^2 °C)

Using the conversion factors given in Table 3.2:

$$G = G_{SI} \frac{1\ lb_m}{0.45359\ kg} \frac{3600\ s}{h} \frac{(0.3048)^2\ m^2}{ft^2} = 737.34\ G_{SI}$$

$$D = D_{SI} \frac{1\ ft}{0.3048\ m} = 3.28\ D_{SI}$$

$$h = h_{SI} \frac{1\ BTU}{1055\ J} \frac{3600\ s}{h} \frac{(0.3048)^2\ m^2}{ft^2} \frac{1^\circ C}{1.8^\circ F} = 0.176\ h_{SI}$$

Substituting these values in the relation given earlier:

$$0.176\ h_{SI} = 0.026 \frac{(737.34\ G_{SI})^{0,6}}{(3.28\ D_{SI})^{0,4}}$$

$$h_{SI} = 4.8274 \frac{G_{SI^{0.6}}}{D_{SI^{0.4}}}$$

Therefore, the new constant is 4.8274.

Finally, it is important to comment about pressure and the various ways to express this variable. Pressure is usually defined as force per unit area acting in the direction normal to the surface under consideration. Consider, then, the situation presented in Figure 3.6. Inside the glass tube there is a liquid. The force that the liquid exerts on the base of the plate is associated to its weight. Therefore:

$$F = m\ g \tag{3.14}$$

where:

- F = force exerted by the liquid on the base of the plate;
- m = mass of liquid in the tube;
- g = acceleration of gravity.

The mass of liquid in the tube is given by:

$$m = \rho\ V \tag{3.15}$$

where:

- ρ = density of the liquid;
- V = volume of liquid in the tube.

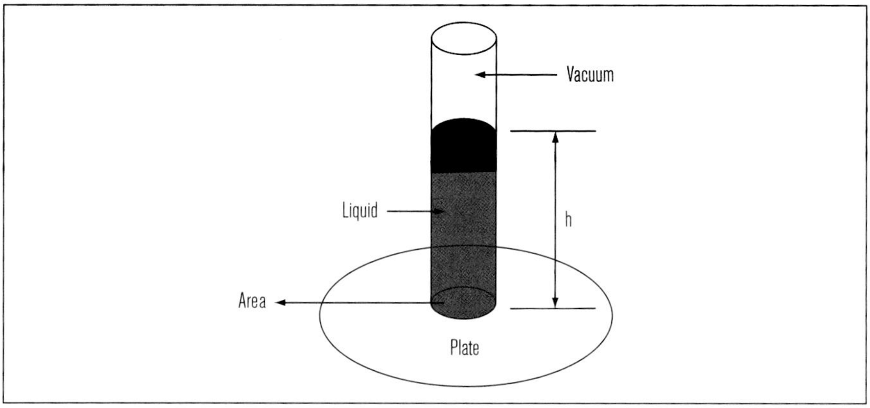

Figure 3.6 - Device for defining pressure.

The volume of liquid in the cylindrical column can be determined by:

$$V = A\,h \tag{3.16}$$

where:

- A = area of the base of the column of liquid;
- h = height of liquid column.

The pressure exerted by the liquid at the base is given by:

$$P = \frac{F}{A} = \frac{m.g}{A} = \frac{\rho.V.g}{A} = \frac{\rho.A.h.g}{A} \tag{3.17}$$

Combining the above relations, one can obtain a generic expression for evaluating the pressure exerted by the liquid column:

$$P = \rho\,g\,h \tag{3.18}$$

The pressure exerted by a column of 0.5 m of mercury (ρ = 13600 kg/m^3) is given by:

$$P = (13600\ \frac{kg}{m^3})\ (9.8\ \frac{m}{s^2})\ (0,5\ m) = 66640\ Pa$$

Sometimes, pressure is expressed in terms of the height of the column of liquid (usually mercury or water). It is common to say a pressure of 20 mm of mercury, referring to the pressure exerted by a column of 20 mm of mercury.Using the above results, one can determine a conversion factor from mm of mercury to Pascal.

500 mm Hg = 66640 Pa

1 mm Hg = 133.33 Pa

Another important point regarding the pressure is related to the way of expressing the values of pressure. Two ways are commonly used, relative pressure and absolute pressure. The difference between them is schematically seen in Figure 3.7.

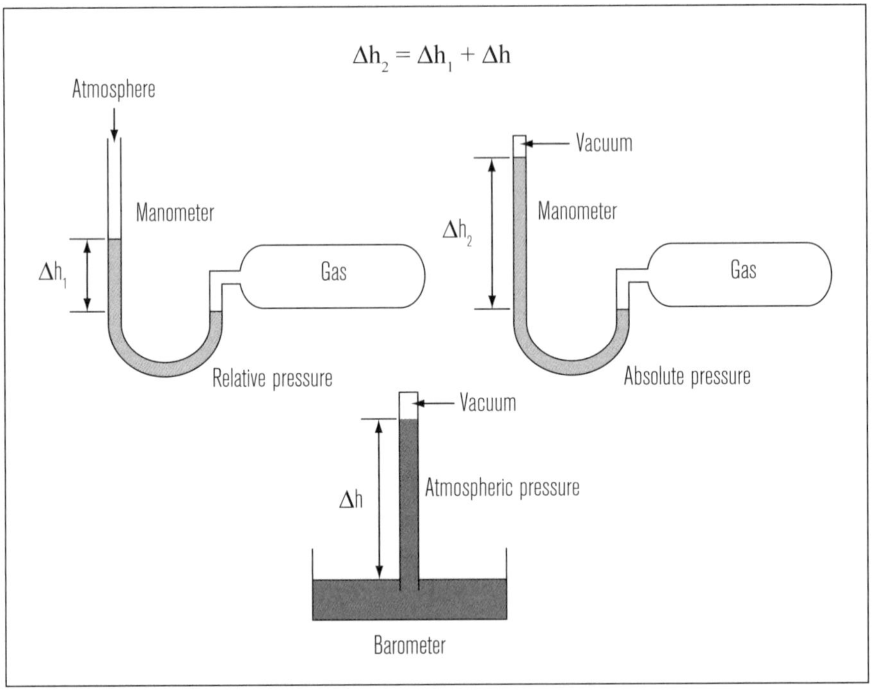

Figure 3.7 - Schematic diagram identifying the difference between absolute and relative pressures.

The absolute pressure is given by:

$$\text{absolute pressure} = \text{relative pressure} + \text{atmospheric pressure} \tag{3.19}$$

or, according to Figure 3.7:

$$\Delta h_2 = \Delta h_1 + \Delta h \tag{3.20}$$

The atmospheric pressure is determined by a device called barometer. In the British system of units, it is common to have pressures given in the following units:

- psia: pounds per square inch absolute;
- psig: pounds per square inch gauge.

Example

Using the same procedure adopted to correlate mm Hg and Pa, determine a conversion factor for meters of water column to Pascal.

Solution

From equation 3.18, we have:

$$P = \rho g h$$

Considering the density of water equal to 1000 kg/m³ :

$$P = (1000 \frac{kg}{m^3})(9.8 \frac{m}{s^2})(1 m) = 9800 Pa$$

Thus, 1 m of water column is equivalent to 9800 Pa.

Example

The atmospheric pressure at sea level is equivalent to 760 mm Hg. Determine this value in psia and Pascal.

Solution

The conversion factor that was determined previously is:

1 mm Hg = 133.33 Pa

Therefore:

760 mm Hg = 760 x 133.33 Pa = 101330 Pa = 101330 N/m²

Using the conversion factors given in Table 3.2,

$1 lb_f = 4.4482 N$

1 in = 0.0254 m

$$101330 Pa = 101330 \left(\frac{1 lb_f}{4.4482 N}\right)\left(\frac{(0.0254)^2 m^2}{in^2}\right) = 14.7 psia$$

Example

A gauge indicates that the pressure inside a tank is 51 psi. The barometric pressure is 28 in Hg. Calculate the absolute pressure of the gas inside the tank in Pa.

Solution

To determine the absolute pressure in Pa, it is necessary to convert the data of barometric and relative pressure to Pa and add the results. The relative pressure is 51.0 psi. From the result of the previous example:

14.7 psi = 101330 Pa

Therefore,

51 psi = 351553 Pa

Since the barometric pressure is 28 in Hg;

1 in = 0.0254 m = 25.4 mm.

The relative pressure in mm Hg is:

28 in = 711.2 mm Hg

From the previous example, it is known that:

1 mm Hg = 133.33 Pa

Hence, 28 in Hg = 948243 Pa.

The absolute pressure is given by:

Absolute pressure = (351553 + 94824.3) = 446377.3 Pa

Example

The density of water at 25°C is 62.4 lb_m/ft^3. Determine the density of water in kg/m^3.

Solution

Using the conversion factors given in Table 3.2:

1 lb_m = 0.45359 kg

1 ft = 0.3048 m

The density of water in units of the international system is then given by:

$$\rho_{H_2O} = 62.4\left(\frac{0.45359\ \text{kg}}{\text{lb}_m}\right)\left(\frac{1\text{ft}^3}{(0.3048)^3\ \text{m}^3}\right) = 999.55\ \frac{\text{kg}}{\text{m}^3}$$

Example

Show that the parameter P/ρ has the dimension of energy per unit mass.

Solution

Referring to Table 3.1:

- pressure: $ML^{-1}\ t^{-2}$;
- density: ML^{-3};
- energy: $ML^2\ t^{-2}$.

Hence:

$$\frac{\text{Pressure}}{\text{Density}} = \frac{P}{\rho} = \frac{ML^{-1}t^{-2}}{ML^{-3}} = \frac{L^2}{t^2}$$

$$\frac{\text{Energy}}{\text{Mass}} = \frac{ML^2t^{-2}}{M} = \frac{L^2}{t^2}$$

Exercises

1 - Calculate the volume of a cylinder (give the answer in m^3) with the following dimensions:

- height or length: 15 in;
- diameter: 3 in.

2 - If a rocket uses 255 ft^3/h of liquid oxygen, what is the oxygen consumption in m^3/s?

3 - Calculate all the temperatures in the table from one of the values given:

Unit	a	b	c	d	e	f	g
°F	140				1000		
°R			500			1000	
K		298					1000
°C				-40			

4 - A pressure gauge indicates that the relative pressure inside a container is 3.53 m of water column. The barometer indicates a pressure of 30.4 in Hg. What is the absolute pressure inside the container in psi and Pa?

5 - Insects and small rodents can live in low pressure (3.0 psia). In one test, a mercury manometer was connected to a container, as pictured below. The reading of the gauge indicates 25.4 in Hg and the barometric pressure is equal to 14.79 psi. Can the insects survive under such conditions?

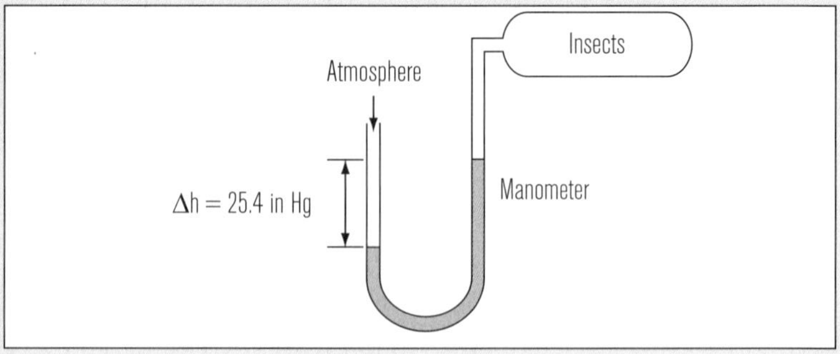

6 - Calculate the kinetic energy of 1 ton of water moving at 60 miles per hour. Give the answer in the following units:

- ft. lb_f;
- ergs;
- Joules;
- hp. sec;
- Watt.sec.

1 mile = 1.6 km.

7 - Densities can be expressed as a linear function of temperature. The general expression is given by:

$$\rho = \rho_0 + A\,T$$

where ρ is given in kg/m^3 and T in °C.

If the equation is dimensionally consistent, what is the unit of A?

8 - In a blast furnace, pressure drop across the bed is given by:

$$\frac{\Delta P}{H} = a\,\mu\,V + b\,\rho\,V^2$$

where:

- ΔP = pressure drop;
- H = height of the bed;
- μ = viscosity of the gas;
- ρ = density of the gas;
- V = velocity of the gas.

Determine the units of the constants "a" and "b" in the international system.

9 - An empirical correlation to determine the heat transfer coefficient from a vertical plate to air can be expressed by:

$$h = 0.29\left(\frac{\Delta T}{L}\right)^{\frac{1}{2}}$$

where:

- h = heat transfer coefficient expressed as BTU/h.ft^2.°F;
- ΔT = temperature difference between the plate surface and air, as °F;
- L = length of the plate as ft.

What should be the new constant instead of 0.29, if the expression above is written in international system of units?

10 - Prove that the following group of parameters are dimensionless:

$$Re = \frac{D\,V\,\rho}{\mu} \qquad Gr = \frac{g\,\beta\,\Delta T\,L^3}{\nu^2}$$

where:

- D = diameter;
- V = velocity;
- ρ = density;
- μ = dynamic viscosity;
- g = acceleration due to gravity;
- ΔT = temperature difference;
- L = length;
- ν = kinematic viscosity;
- β = coefficient of compressibility, defined as:

$$\beta = -\frac{1}{\rho}\left(\frac{\partial \rho}{\partial T}\right)_P$$

Viscosity

Chapter 04

One of the main properties of fluids, which has significant influence on its behavior during flow, is viscosity. This property is very important in liquid metals and slags, as it has a significant influence on aspects of flow and refining reactions. Definition of viscosity and various methods of evaluating viscosity of gases and liquids have been dealt with in this chapter.

4.1 Definition of Viscosity and Newton's Law of Viscosity

Viscosity is a physical property that characterizes the resistance to flow of a simple fluid. To define this property, the situation shown in Figure 4.1 is considered. In this figure, a fluid is contained in between two parallel plates, the upper one is stationary and the lower is moving to the right with velocity V.

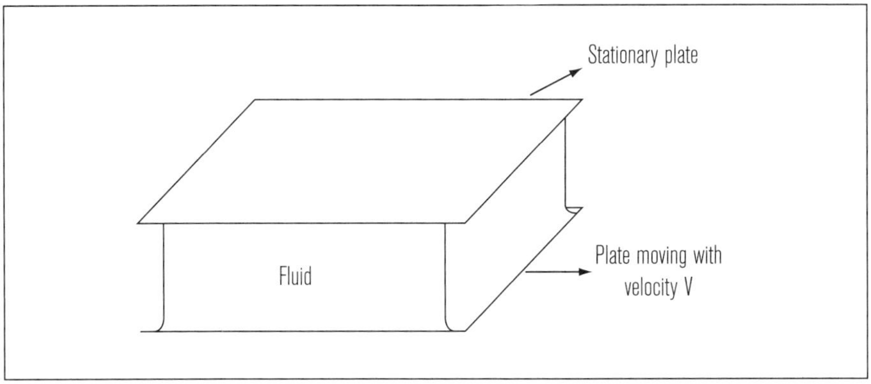

Figure 4.1 - Schematic view of a fluid between two plates.

At time t = 0, the lower plate starts moving with a constant velocity V. From that moment on, the fluid that is above this plate also begins to move. In the course of time, the velocity profile of the fluid reaches steady state, that is to say there is no change in the velocity profile with time. This situation is shown in Figure 4.2, which presents the evolution of the velocity profiles as a function of time.

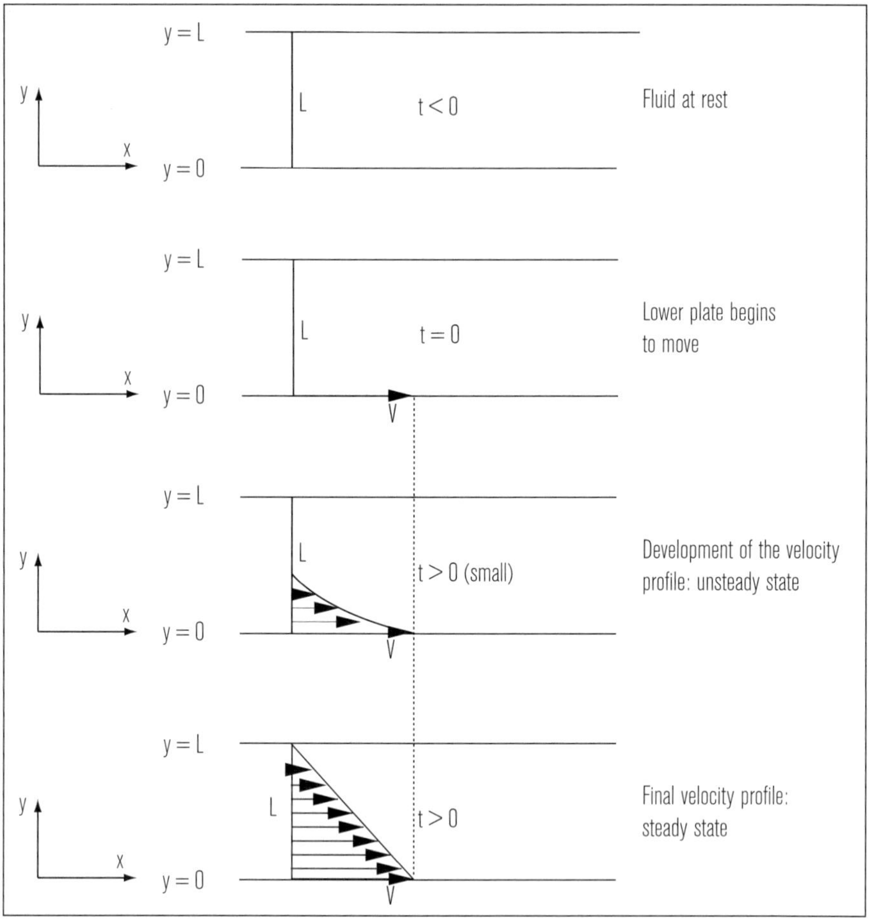

Figure 4.2 - Evolution of the velocity profile in a system of parallel plates.

When steady state is reached, a constant force F is required to maintain the movement of the lower plate. This force can be expressed as follows:

$$\frac{F}{A} = \mu \frac{V}{L} \tag{4.1}$$

where,

- F = force to keep the lower plate in motion;
- A = plate area;
- V = velocity of the lower plate;
- L = distance between the plates;
- μ = constant of proportionality.

This constant depends on a property of the fluid that is between the plates. This property is called dynamic or molecular viscosity.

It is observed that force per unit area is proportional to the velocity and inversely proportional to the distance between the plates. In this case, the force applied is tangential to the surface of the plate and the following relationship can be written:

$$\frac{\text{Force}}{\text{area}} = \tau = \text{Shear stress} \qquad (4.2)$$

At the interfaces between the fluid and the plates, a no-slip condition is assumed. This means that the fluid in contact with the plate assumes the velocity of the plate. Since the plates have different velocities, there is a velocity gradient* in the fluid.

At steady state, when the velocity profile is linear, V/L can be expressed in terms of the velocity gradient. In the case, shown* in Figure 4.2, since the velocity profile is linear, the velocity gradient is given by:

- At $y = 0$, $v_x = V$;
- At $y = L$, $v_x = 0$.

$$\text{Velocity gradient} = \frac{\Delta v_x}{\Delta y} = \frac{dv_x}{dy} = \frac{0 - V}{L - 0} \qquad (4.3)$$

Substituting equations (4.2) and (4.3) in (4.1), a expression for the shear stress in terms of the velocity gradient results:

$$\text{Shear stress} = \tau_{yx} = -\mu \frac{dv_x}{dy} \qquad (4.4)$$

In equation (4.3) given above, there is a direct correspondence between the velocities that appear in the numerator of the fraction and the positions indicated in the denominator. Equation (4.3) for the velocity gradient is valid only for linear profiles.

The relation (4.4) is the mathematical expression of the Newton's law of viscosity applied to cases of one-dimensional flow, in which there is only one component of velocity (v_x) varying only in one direction (y). This law states that the shear stress, τ_{yx}. (y is the direction normal to the surface on which the force acts and x is the direction of the applied force

* Gradient is the change in value of some magnitude with the position within the system.

(Figure 4.2) is proportional to the negative of the velocity gradient . In Chapter 6, a more general form of the Newton´s law of viscosity applied to two or three dimensional flows, will be presented and discussed.

Fluids that obey Newton's law of viscosity are called Newtonian fluids. Many fluids, common in metallurgical processes (water, gases, liquid metals and slags), can be considered as Newtonian fluids. Examples of non-Newtonian fluids are polymers, clay particles suspended in water (used in the process of bonding slurry), pastes and paints. These fluids do not obey equation 4.4. There is a branch of science dedicated to the study of the behavior of non-Newtonian fluids. Equations (similar to equation 4.4), which govern their behavior, have been developed on the basis of theoretical and experimental work. This branch is called rheology. From equation 4.4, it can be observed, that higher the viscosity of the fluid μ, greater is the shear stress or the force required to keep the lower plate moving.

4.1.1 Physical interpretation of the shear stress τ_{yx}

From equation (4.4), it is clear that τ_{yx} is the shear stress due to friction caused by the velocity gradient. Expression (4.4) can be interpreted in another way. In the vicinity of the surface that is moving at $y = 0$, the fluid acquires a certain amount of momentum in the x direction. This fluid transfers a certain amount of momentum to the adjacent layer of fluid, causing it also to move in the direction of x. Hence, one can say that the momentum in the x-direction is transferred in the y-direction. As stated in Chapter 3, momentum is transported by diffusion whenever there is a velocity gradient. τ_{yx} can then be interpreted as flux of momentum** or momentum flux in the x-direction, being transported by diffusion in the y-direction. This interpretation is convenient as it is analogous to the treatment that is used for heat and mass transfer in the later chapters. Moreover, through this interpretation, it becomes easier to understand the sign of τ_{yx}.

Momentum is transported by diffusion from regions of higher to lower velocity. Thus, in Figure 4.2, momentum is transported from $y = 0$ (high speed) to $y = L$ (low velocity). It is important to remember that velocity is a vector. Hence, a velocity of -100 m/s is less than a velocity of 0.01 m/s.

**Flux of momentum, heat or mass represents the amount of momentum, heat or mass that is being transported per unit of time per unit of area normal do the direction of flow. Rate is the amount that is being transported per unit of time.

With this new interpretation for τyx, one can say that Newton's law of viscosity states that the flow of diffusive momentum is proportional to the negative of the velocity gradient. The sign of τyx can be determined based on the direction of the momentum flow. It is positive, if it is in the same direction as the growth of y axis (y is the direction of the velocity gradient). If the flow of momentum is in the opposite direction relative to the increase of the y -axis, it is considered negative.

Finally, the sign of the velocity gradient can be determined in a simple way. If the velocity v_x increases as the y position increases, one can say that the velocity gradient is positive. If it decreases as the position along the y-axis increases, the velocity gradient is negative.

Some examples are given below illustrating the uses of Newton's law of viscosity.

Example

In the figure shown below, determine

- direction and sense of the transport of momentum by diffusion;
- direction and sense of the flow of momentum by convection.

Apply Newton's law of viscosity to the situation shown.

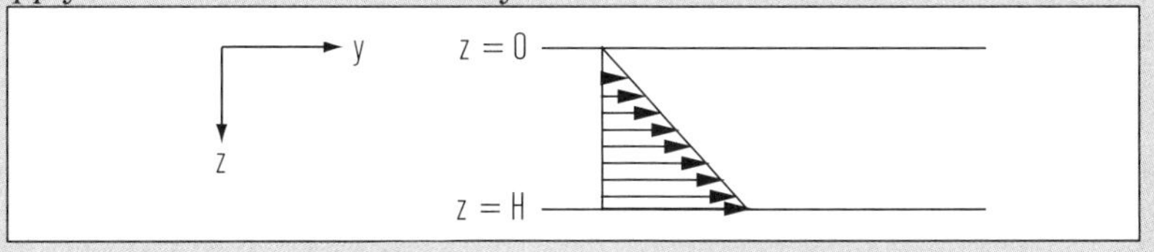

Solution

As seen above, momentum is transported by diffusion in the direction of the velocity gradient. Thus, the direction of transport of diffusive momentum is the direction z.

As momentum is transported by diffusion from the region of high velocity (z = H) to the region of low velocity (z = 0), the sense of transport of diffusive momentum is negative z.

Transport of momentum by convection occurs in the direction of the movement of the fluid, which in this case is the direction y. As the velocities are in the same sense that is the increase of the positions along the y-axis, they are all positive and therefore the sense of flow of momentum by convection is the positive y.

For the system seen above, Newton's law can be written in the following way:

$$\tau_{zy} = -\mu \frac{dv_y}{dz}$$

In this case, the velocity v_y increases with z. Hence the velocity gradient is positive and the diffusive momentum flux τ_{zy}, is negative, as it occurs in the opposite sense to the growth of the z-axis.

Example

Repeat the example above to analyze the situation shown below:

Solution

In this case, the velocities are all negative, except in x = H, because they are in the opposite direction to the growth of axis z. Momentum is transported by diffusion in the x direction (in the direction of the velocity gradient), from higher velocity (x = H) to that of lower velocity (x = 0). Remembering, that velocity is a vector, the direction of transport of diffusive momentum is negative x. Momentum is transported by convection in the direction of macroscopic motion of fluid, that is z-direction in the negative sense of this axis.

For this case, Newton's law of viscosity can be written in the following way:

$$\tau_{xz} = -\mu \frac{dv_z}{dx}$$

The momentum flux by diffusion is negative and the velocity gradient is positive (v_z increases as x increases).

Example

Repeat the same procedure above for the case presented in the following figure.

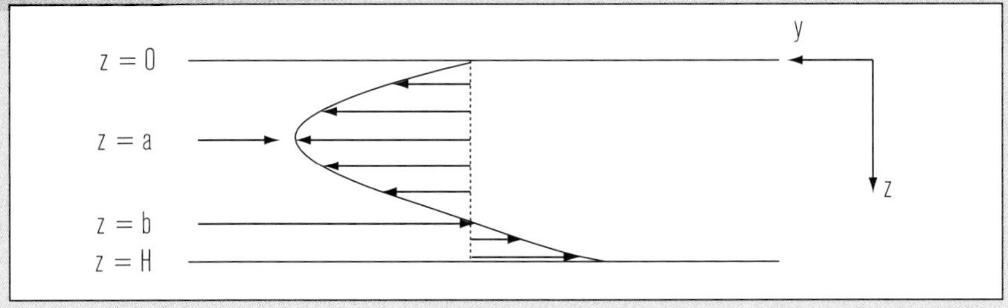

Solution

The present case is a little more complex than the previous ones. In this case, there is a change in the sense of the velocity in the region shown.

Considering the orientation of the y-axis, velocity v_y is positive in the region $0 < z < b$. In the region $b < z < H$, the velocities are negative. The maximum velocity occurs at $z = a$. Therefore, the transport of diffusive momentum occurs in the z direction (direction of the velocity gradient), from $z = a$ to $z = 0$ and from $z = a$ to $z = H$. Hence:

- region: $0 \le z < a$: flux of diffusive momentum is negative - opposite to the growth of axis z;
- region: $a < z \le H$: flux of diffusive momentum is positive - same sense of the growth of axis z.

The transport of momentum by convection occurs in the direction y, that is the direction of the flow of the fluid:

- region: $0 < z < b$: the flux of momentum by convection is positive since the velocities are positive;
- region: $b < z \le H$: the flux of momentum by convection is negative because of the negative velocities.

For the system under study, Newton's law can be written in the following way:

$$\tau_{zy} = -\mu \frac{dv_y}{dz}$$

The signs of the velocity gradient are:

- region: $0 \le z < a$: velocity increases with the increase in z - gradient is positive;
- region: $a < z \le H$: velocity decreases with the increase in z - gradient is negative.

The above examples demonstrate that momentum transport can occur simultaneously by diffusion and convection. Transport of momentum by diffusion does not exclude the possibility of transport by of momentum by convection and vice versa.

4.1.2 Dimensions of viscosity

Using Newton's law of viscosity, expressed by equation (4.4), one can determine the dimension of the molecular or dynamic viscosity, μ. The dimensions of shear stress and velocity gradient are:

$$\text{Shear stress:} \quad \tau_{yx} : \frac{M}{L\,t^2}$$

$$\text{Velocity gradient:} \quad \frac{dv_x}{d_y} : \frac{L}{L\,t} = t^{-1}$$

Substituting these relations in the Newton´s law of viscosity:

$$\frac{M}{L\,t^2} = \mu\,\frac{1}{t}$$

Hence,

$$\mu = \frac{M}{L\,t}$$

In the International System, viscosity is expressed in terms of the following units: $\mu = kg.m^{-1}.s^{-1} = Pa.s$.

The unit of viscosity in the cgs system (centimeter-gram-second) is $g.cm^{-1}.s^{-1}$. This unit is called Poise, P, named after the French scientist Poiseuille, who has done fundamental contributions to the field of fluid mechanics.

The centipoise, cP, is a unit derived from the Poise and is equivalent to one hundredth of that unit:

$$1\ cP = 10^{-2}\ P$$

Another important property in the study of transport of momentum is the kinematic viscosity, defined by the following equation:

$$\nu = \frac{\mu}{\rho} \tag{4.5}$$

where,

- ν = kinematic viscosity;
- ρ = density of fluid.

The kinematic viscosity has dimension of L_2/t (prove this as an exercise) and is also known as diffusivity of momentum. Before dealing with the methods for estimating viscosity of fluids of interest in metallurgical and materials engineering, it is interesting to observe the similarity between Newton's law of viscosity with similar laws governing the transport of heat and mass.

Heat transfer by diffusion is governed by the following equation, known as Fourier's law.

$$q_y = -k\frac{dT}{dy} \tag{4.6}$$

where,

- q = heat flux by diffusion;
- k = thermal conductivity of the material in which heat is transferred;
- T = temperature.

For the mass transport by diffusion, the mass flux is determined using Fick's law, expressed by:

$$j_y = -D\frac{dC}{dy} \tag{4.7}$$

where,

- J_y = mass flux by diffusion;
- D = mass diffusivity of the species;
- C = concentration or chemical potential of the chemical species diffusing.

The similarity between the laws of Newton, Fourier and Fick can easily be seen from the equations. They all state that the flux (of momentum, heat or mass) is proportional to the gradient of a given variable (velocity, temperature and concentration or chemical potential). The constants of proportionality are specific to each situation:

- transport of momentum: μ (viscosity);
- heat transport: k (thermal conductivity);
- mass transport: D (mass diffusivity).

The main difference between Newton's law of viscosity and the laws of Fourier and Fick is related to the nature of the variables involved. In Newton's law, the gradient involves velocity, which is a vector. In the laws of Fourier and Fick, the gradient is of a scalar variable, temperature and concentration (or chemical potential), respectively. On the other hand, the flux of momentum, $\tau_{yx,}$ is a tensor quantity. The first subscript if the direction of the diffusive momentum flux, the second subscript is the direction of the velocity. The first subscript is the direction of the diffusive momentum flux. The second subscript is the direction of the velocity. Fluxes of heat and mass, q_y and j_y are vector magnitudes, and their subscripts indicate the direction of temperature or concentration (or chemical potential) gradients.

4.2 Viscosity of Gases

In the study of diffusive momentum transfer of a fluid, one important property of the fluid that should be known is its viscosity.

A large volume of data on viscosity of fluids can be found in the literature. However, not always all the specific data on viscosities of gases, mixture of gases, liquid metals and slags at high temperatures are available. In such cases, an alternative for experimental determination of viscosity, is to use empirical or semi empirical correlations obtained from theoretical analysis of the gas, metallic or slag systems.

For example evaluation of the viscosities of gases can be carried out based on the kinetic theory of gases. According to this theory, an ideal gas should have the following characteristics:

- gas molecules are rigid like billiard balls, having a diameter "d" and mass "m";
- they do not exert forces on each other, except when they collide;
- the collisions are perfectly elastic and obey the classical laws of conservation of momentum and energy;

- the molecules are evenly distributed in space, continuously moving and are separated by distances that are large compared with their diameters;
- movement of molecules in all the directions are equally possible. The magnitude of the velocity of a molecule can have any value between zero and infinity.

Assuming that the molecules have a velocity distribution given by the Maxwell equation (i.e., the thermal energy of gas is given by the kinetic energy of all molecules that move) and using the concepts of mean free path, the most probable velocity and momentum balance in the microscopic level of molecular layers (Chapman–Enskog, cited by GEIGER and POIRIER, 1980), the following equation to estimate the viscosity of a gas can be obtained:

$$\mu = \frac{2}{3\,\pi^{3/2}} \; \frac{\sqrt{m\,K_B\,T}}{d^2} \tag{4.8}$$

where:

- μ = viscosity of the gas in Poise (g/cm s);
- m = mass of a molecule (g);
- K_B = Boltzmann constant (1.38×10^{16} erg/molecule K);
- T = temperature (K);
- d = diameter of a molecule (cm).

An important conclusion that can be arrived at from the above equation is that the viscosity of a gas is independent of pressure and depends only on the temperature, which is in good agreement with experimental data up to ten atmospheres of pressure. However, the dependence with temperature is only qualitatively correct. The viscosity of a gas increases with temperature. Experimental data obtained for various gases indicate that μ varies T to the power n, where n is between 0.6 and 1.0 instead of 0.5, as shown by equation (4.8). Another theory has replaced the model of billiard balls with a more a more realistic model. This new model considers a force field, encompassing forces of attraction and repulsion between molecules. This theory makes use of the potential energy of interaction between a pair of molecules in the gas. This function, commonly known as Lennard-Jones potential, considers molecular interaction: weak attraction for large separations and strong repulsion for small separations.

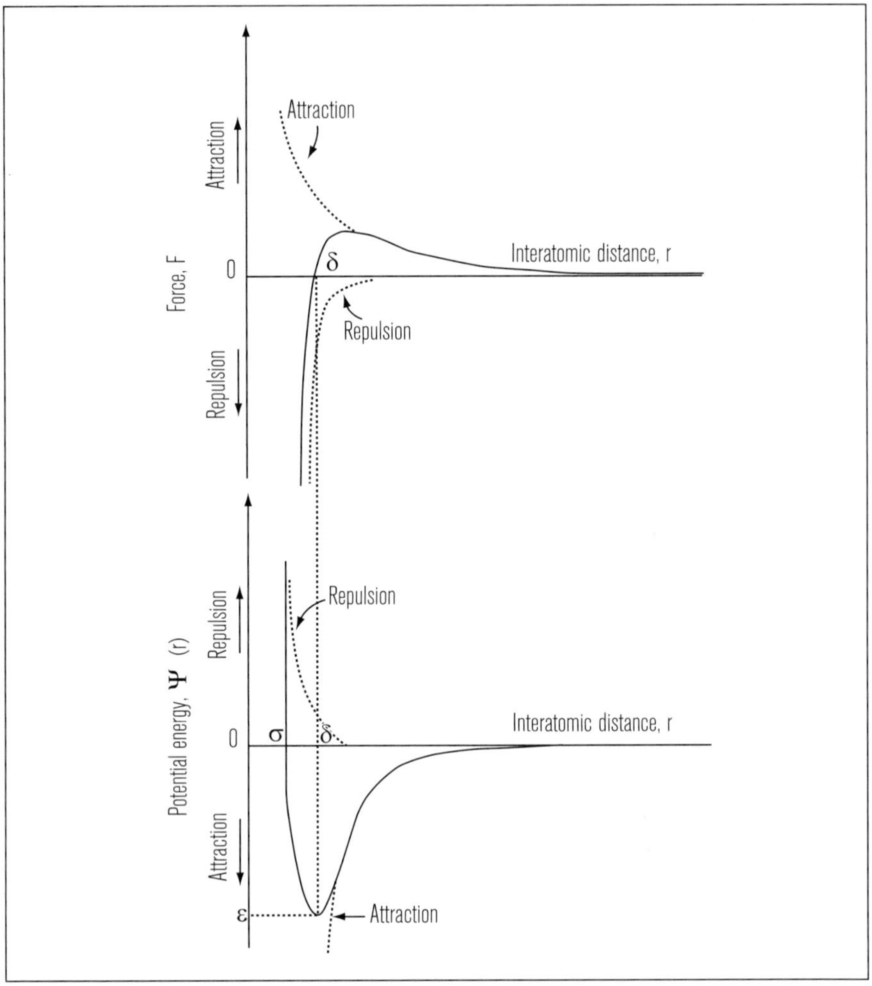

Figure 4.3 - Force and Lennard-Jones potential function, describing the interaction of two nonpolar molecules (JASTRZEBSKI, 1976).

Figure 4.3 shows the variation of the Lennard-Jones potential as a function of the separation between two molecules. The potential function, $\psi(r)$, is described by the following relationship:

$$\Psi(r) = 4\,\varepsilon \left[\left(\frac{\sigma}{r}\right)^{12} - \left(\frac{\sigma}{r}\right)^{6} \right] \tag{4.9}$$

where:

- ε = characteristic energy (erg/molecule);
- σ = collision diameter (Angstrom).

The equilibrium position of the molecules is given by point δ, where the potential energy is minimal and should be equal to - ε. Parameter ε is called the characteristic energy.

Using the Lennard-Jones potential, Chapman and Enskog developed the following equation for calculating the viscosity of nonpolar gases at low pressures:

$$\mu = 2.6693 \times 10^{-5} \frac{\sqrt{M\,T}}{\sigma^2\,\Omega_\mu} \quad (4.10)$$

where:

- μ = viscosity of the gas (Poise);
- σ = collision diameter of the molecule (Angstrom);
- M = molecular weight of gas (g/mol);
- Ω_μ = collision integral of the Chapman-Enskog theory;
- T = temperature (K).

The constant in equation (4.10) already includes conversion factors so that viscosity is obtained in Poise. Values of σ and ε/K_B for a particular gas can be obtained from the Table 4.1.

The collision integral is a function of the dimensionless temperature parameter $K_B.T/\varepsilon$. Values of ε/K_B are also given in Table 4.1. In Table 4.2, the values of Ω_μ are given in terms of $K_B.T/\varepsilon$. If the gas is considered to contain rigid spherical molecules of diameter σ (instead of real molecules with forces of attraction and repulsion), the parameter Ω_μ would be equal to 1.

An alternative to the use of Table 4.2 is the use of mathematical correlations obtained using the data in this table. This way interpolations are avoided since the values of the parameter $K_B.T/\varepsilon$ are not always those indicated in this table. The equations obtained through this mathematical fitting are given below.

for $K_B\,T/\varepsilon < 2$:

$$\log \Omega_\mu = 0.2071 - 0.4662 \log\left(\frac{K_B\,T}{\varepsilon}\right) \quad (4.11)$$

and or $K_B\,T/\varepsilon > 2$:

$$\log \Omega_\mu = 0.0689 - 0.1497 \log\left(\frac{K_B\,T}{\varepsilon}\right) \quad (4.12)$$

Table 4.1 – Parameters of Lennard-Jones (BIRD, STEWART and LIGHTFOOT, 1960; GEIGER and POIRIER, 1980; CUSSLER, 1997).

Substance	σ(Å)	$\frac{\varepsilon}{k_B}$	Substance	σ(Å)	$\frac{\varepsilon}{k_B}$
Argônio	3.418	124.0	CH_3COCH_3	4.600	560.2
He	2.576	10.2	CH_3COOCH_3	4.936	469.8
Kr	3.498	225.0	n-C_4H_{10}	4.687	531.4
Ne	2.789	35.7	Iso-C_4H_{10}	5.278	330.1
Xe	4.055	229.0	$C_2H_5OC_2H_5$	5.678	313.8
Ar	3.617	97.0	$CH_3COOC_2H_5$	5.205	521.3
Br_2	4.268	520.0	n-C_5H_{12}	5.784	341.1
CCl_4	5.947	322.7	$C(CH_3)_4$	6.464	193.4
CF_4	4.662	134.0	C_6H_6	5.349	412.3
$CHCl_3$	5.389	340.2	C_6H_{12}	6.182	297.1
CH_2Cl_2	4.898	356.3	n-C_6H_{14}	5.949	399.3
CH_3Br	4.118	449.2	Cl_2	4.217	316.0
CH_3Cl	4.182	350.0	F_2	3.357	112.6
CH_3oH	3.626	481.8	HBr	3.353	449.0
CH_4	3.822	137.0	HCN	3.630	569.1
CO	3.590	110.0	HCl	3.339	344.7
CO_2	3.996	190.0	HF	3.148	330.0
CS_2	4.483	467.0	HI	4.211	288.7
C_2H_2	4.033	231.8	H_2	2.915	38.0
C_2H_4	4.163	224.7	H_2O	2.641	809.1
C_2H_6	4.443	215.7	H_2O_2	4.196	289.3
C_2H_5CI	4.898	300.0	H_2S	3.623	301.1
C_2H_5OH	4.530	362.6	Hg	2.969	750.0
CH_3OCH_3	4.307	395.0	I_2	5.60	474.2
CH_2CHCH_3	4.678	298.9	NH_3	2.900	558..3
CH_3CCH	4.761	251.8	NO	3.492	116.7
C_3H_6	4.807	248.9	N_2	3.681	91.5
C_3H_8	5.118	237.1	N_2O	3.828	232.4
n-C_3H_7OH	4.549	576.7	O_2	3.433	113.0
CH_3COCH_3	4.600	560.2	SO_2	4.290	252.0

$\frac{k_B T}{\varepsilon}$	Ω_μ	$\frac{k_B T}{\varepsilon}$	Ω_μ	$\frac{k_B T}{\varepsilon}$	Ω_μ	$\frac{k_B T}{\varepsilon}$	Ω_μ
0.30	2.785	1.30	1.399	2.6	1.081	4.6	0.9422
0.35	2.628	1.35	1.375	2.7	1.069	4.7	0.9382
0.40	2.492	1.40	1.353	2.8	1.058	4.8	0.9343
0.45	2.368	1.45	1.333	2.9	1.048	4.9	0.9305
0.50	2.257	1.50	1.314	3.0	1.039	5.0	0.9269
0.55	2.156	1.55	1.296	3.1	1.030	6	0.8963
0.60	2.065	1.60	1.279	3.2	1.022	7	0.8727
0.65	1.982	1.65	1.264	3.3	1.014	8	0.8538
0.70	1.908	1.70	1.248	3.4	1.007	9	0.8379
0.75	1.841	1.75	1.234	3.5	0.9999	10	0.8242
0.80	1.780	1.80	1.221	3.6	0.9932	20	0.7432
0.85	1.725	1.85	1.209	3.7	0.9870	30	0.7005
0.90	1.675	1.90	1.197	3.8	0.9811	40	0.6718
0.95	1.629	1.95	1.186	3.9	0.9755	50	0.6504
1.00	1.587	2.00	1.175	4.0	0.9700	60	0.6335
1.05	1.549	2.1	1.156	4.1	0.9649	70	0.6194
1.10	1.514	2.2	1.138	4.2	0.9600	80	0.6076
1.15	1.482	2.3	1.122	4.3	0.9553	90	0.5973
1.20	1.452	2.4	1.107	4.4	0.9507	100	0.5882
1.25	1.424	2.5	1.093	4.5	0.9464	200	0.5320
						400	0.4811

Table 4.2 - Values of the collision integral, based on the dimensionless temperature (BIRD, STEWART and LIGHTFOOT, 1960; GEIGER and POIRIER, 1980).

Equation (4.10) is then useful for determining the viscosity of nonpolar gases at low densities. However, it cannot be applied with confidence to gases consisting of polar molecules or very large molecules such as H_2O, NH_3, CH_3OH and NOCl.

Another limitation of equation (4.10) is that it provides good results only for temperatures above 100 K. For metallurgical and materials engineers, this is not a significant restriction, because in most cases the temperatures involved are well above this value.

In the field of process metallurgy, gas mixtures are very common in the reactors. Equation (4.13) can be used to estimate the viscosity of gas mixtures, based on the composition of the mixture and the values of viscosities for the individual gases.

$$\mu_{MIXTURE} = \frac{\sum_{i=1}^{n} x_i \mu_i \sqrt{M_i}}{\sum_{i=1}^{n} x_i \sqrt{M_i}} \tag{4.13}$$

where:

- n = number of components of the mixture;
- x_i = mole fraction of component i in the mixture;
- μ_i = viscosity of component i in the mixture;
- M_i = molecular weight of gas i.

The calculations of viscosities of pure gases and gas mixtures can be easily developed using spreadsheets, similar to the file gas-viscosity.xls, included in the CD accompanying this book. Using this file, viscosity of gases and gas mixtures can be calculated as shown in the examples below.

Example

Evaluate the viscosity of hydrogen at 1 atm and 1000 K.

Solution

Using the spreadsheet given in the CD, the value for the collision integral is:

$$\Omega_\mu = 0.7183$$

Therefore,

$$\mu = 1.94 \times 10^{-4} \text{ P.}$$

Example

Calculate the viscosity of CO_2 at 200, 300 and 800 K. Compare with the following experimental data:

$$200 \text{ K: } \mu = 1.015 \times 10^{-4} \text{ P;}$$

$$300 \text{ K: } \mu = 1.495 \times 10^{-4} \text{ P.}$$

Solution

The values for the collision integral at temperatures of 200, 300 and 800 K are:

$$\Omega_\mu\,(200\ K) = 1.573$$

$$\Omega_\mu\,(300\ K) = 1.302$$

$$\Omega_\mu\,(800\ K) = 0.945$$

The calculated viscosities are

$$\text{at } 200\ K: \mu = 9.971 \times 10^{-5}\ P$$

$$\text{at } 300\ K: \mu = 1.475 \times 10^{-4}\ P$$

$$\text{at } 800\ K: \mu = 3.319 \times 10^{-4}\ P$$

Example

Estimate the viscosity of a blast furnace gas with the following composition:

$N_2 = 50\%$ $CO = 24\%$ $CO_2 = 22\%$ $H_2 = 4\%$

at a temperature of 100°C.

Solution

Using the spreadsheet, the viscosities of the individual gases are

$$\mu\,(N_2) = 2.121 \times 10^{-4}\ P$$

$$\mu\,(CO) = 2.169 \times 10^{-4}\ P$$

$$\mu\,(CO_2) = 1.821 \times 10^{-4}\ P$$

$$\mu\,(H_2) = 1.035 \times 10^{-4}\ P$$

The viscosity of the blast furnace gas is then:

$$\mu\,(\text{mixture}) = 2.04 \times 10^{-4}\ P$$

4.3 Viscosity of Liquids

The prediction of the viscosity of liquids is more difficult as the structure of liquids is less well known than that of gases and solids. There is more similarity between solids and liquids than liquids and gases. This is based on the small changes in volume occurring during transformation of a solid to liquid (3 to 5% for metals) (GEIGER and POIRIER, 1980). The value of the heat of fusion in these cases is also significantly smaller than the magnitude of the heat of vaporization.

Several theories have been formulated to enable the prediction of the viscosity of liquids. However, none of these theories is general and can be used with confidence in the prediction of viscosities of different types of liquids. The difficulty in developing a general theory for the viscosity of liquids is understandable considering the wide range of values of viscosities for different liquids, as seen in Table 4.3. There is a variation of approximately five orders of magnitude in the values of viscosities. It is also surprising that liquids having completely different structures show similar values for the viscosity. For gases, the variations of viscosities are relatively small and are usually in the range of 10^{-4} P.

Table 4.3 - Viscosity for different types of liquids (GEIGER and POIRIER, 1980).

Viscosity (Poise)	Materials
1 to100	Slag: CaO - SiO_2 - Al_2O_3
	50% NaOH - 50% H_2O
	Oils
0.1 to 1.0	H_2SO_4
0.01 to 0.1	Molten salts
	Heavy metals (Pb, Au, Zn)
	Alkali metals (Ca, Mg)
	Transition metals (Fe, Ni, Co)
	Water (20°C)
	Kerosene (20°sC)
0.001 to 0.01	Ketones

One simple theory proposes the following relationship for calculation of the viscosity of liquids:

$$\mu = A \exp\left(\frac{\Delta G_{vis}}{R\,T}\right) \tag{4.14}$$

where:

- μ = viscosity of the liquid (Poise);
- A = constant (Poise)
- T = absolute temperature (K);
- R = gas constant (cal/mol.K);
- ΔG_{vis} = activation energy of viscosity (cal/mol).

The constant A has been the subject of many theoretical studies based on the structure of liquids. Application of most of these theories does not provide satisfactory values and cannot be universally applied for all types of liquids. The theory that gives reasonable results is due to Eyring, who proposed the following equation to evaluate the constant A:

$$A = \frac{N_o\,h}{V_m} \tag{4.15}$$

where:

- N_o = Avogadro's number;
- h = Planck's constant (6.624 x 10^{-27} erg.s);
- V_m = molar volume of the liquid (cm^3/mol).

For liquids having predominantly van der Waals type of interaction, the activation energy of viscous flow can be obtained from the energy of vaporization, according to the following relationship:

$$\Delta G_{vis} = 0.41\ \Delta E_{vap} \tag{4.16}$$

in which:

- ΔE_{vap} = energy of vaporization (cal/mol).

The energy of vaporization can be related to the enthalpy of vaporization by the following expression:

$$\Delta E_{vap} = \Delta H_{vap} - R\ T_b \tag{4.17}$$

where:

- ΔH_{vap} = enthalpy of vaporization (cal/mol);
- T_b = boiling temperature (K).

Relations (4.16) and (4.17) are not valid and cannot be used to give accurate results for liquid metals.

4.3.1 Viscosity of liquid metals

Interactions existing in liquid metals are not of van der Waals type and hence relations (4.14) to (4.17) cannot be applied to metallic melts.

Chapman (GEIGER and POIRIER, 1980) developed a theory considering the interactions between atoms in metals and obtained a relationship between three dimensionless parameters:

- μ^*: reduced viscosity;
- T^*: reduced temperature;
- V^*: reduced volume.

These quantities are defined by the following expressions:

$$\mu^* = \frac{\mu\, \delta^2\, N_o}{\sqrt{M\,R\,T}} \tag{4.18}$$

$$T^* = \frac{K_B\, T}{\varepsilon} \tag{4.19}$$

$$V^* = \frac{1}{n\, \delta^3} \tag{4.20}$$

where:

- δ = interatomic distance in the crystal at 0 K (cm);
- ε = energy parameter characteristic of the metal (erg);
- N_o = Avogadro's number;
- M = atomic mass (g/mol);
- R = gas constant (8.314×10^7 g.cm²/s². mol.K);
- T = absolute temperature (K);
- K_B = Boltzmann constant, 1.38×10^{-16} erg/K;
- n = number of atoms per unit volume (atoms/cm^{-3}).

The relationship between the three above parameters is shown in Figure 4.4.

The relationship seen graphically in Figure 4.4 can be expressed by the following equation:

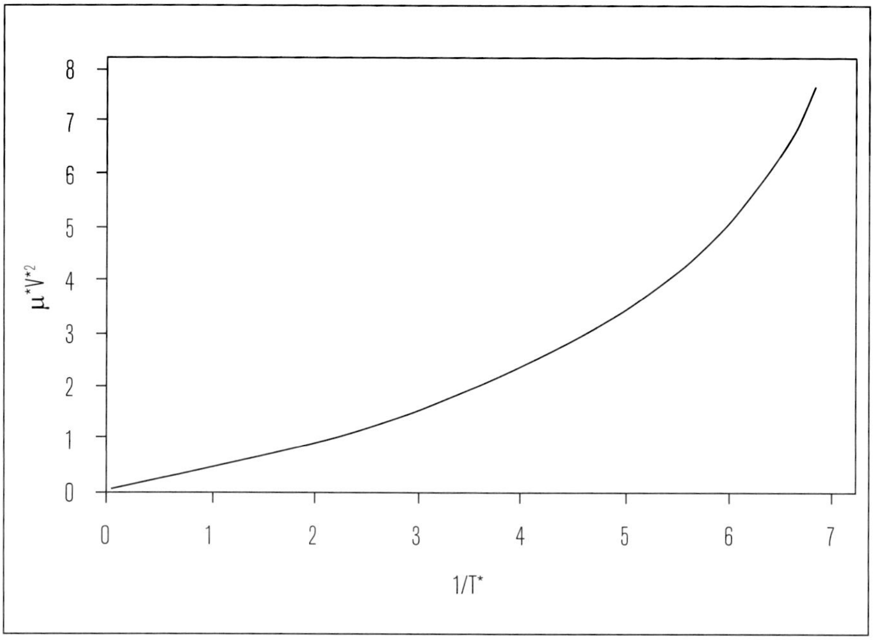

Figure 4.4 – Relation to estimate the viscosity of liquid metals (GEIGER and POIRIER, 1980).

$$\mu^* V^{*2} = -0.0088 + 0.4488\left(\frac{1}{T^*}\right) - 0.0262\left(\frac{1}{T^*}\right)^2 + 0.0156\left(\frac{1}{T^*}\right)^3 \quad (4.21)$$

Table 4.4 provides values of δ and ε/K_B for different metals. The parameter ε/K_B can be expressed as a linear function of the melting temperature of the metal.

$$\frac{\varepsilon}{K_B} = 5.20\ T_f \quad (4.22)$$

where:

- T_f = melting temperature of the metal (K).

The spreadsheet viscosity of metals.xls, contained in the CD that accompanies the book, calculates the viscosity of a particular liquid metal simply by typing its symbol, density and the desired temperature of interest. The following examples demonstrate the use of this spreadsheet and the calculation of viscosity of liquid metals using the relations above.

Table 4.4 - Values of the parameters δ and ε/K_B for various metals (GEIGER and POIRIER, 1980).

Metals	δ (Å)	ε / K_B (K)
In	3.84	1970
K	4.76	1760
Li	3.14	2350
Mg	3.20	4300
Al	2.86	4250
Ca	4.02	5250
Fe	2.52	10900
Co	2.32	9550
Ni	2.50	9750
Cu	2.56	6600
Zn	2.74	4700
Rb	5.04	1600
Ag	2.88	6400
Cd	3.04	3300
In	3.14	2500
Sn	3.16	2650
Au	2.88	6750
Hg	3.10	1250
Pb	3.50	2800
Pu	3.10	5550

Example

Estimate the viscosity of the liquid titanium at 1850°C. The following data are available:

T_f = 1800°C M = 47.9 g/mol

density = 4.5 g/cm^3 δ = 2.89 Å.

Solution

Using equation (4.22), the valor of ε/K_B is determined:

$$\frac{\varepsilon}{K_B} = 5.20\,(1800 + 273) = 10779.6$$

The reduced temperature is given by:

$$T^* = \left(\frac{1850 + 273}{10779.6}\right) = 0.1969 \qquad \frac{1}{T^*} = 5.0787$$

Based on equation (4.21), the following parameter is determined

$$\mu^* V^{*2} = 3.6383$$

The values of n and V* are:

$$n = N_o \frac{\rho}{M} = 6.023 \text{ x } 10^{23} \frac{4.5}{47.9} = 5.6584 \text{ x } 10^{22} \text{ átomos/cm}^3$$

Hence:

$$V^* = \frac{1}{n\,\delta^3} = \frac{1}{(5.6584 \text{ x } 10^{22})\,(2.89 \text{ x } 10^{-8})^3} = 0.7322$$

$$\mu^* = \frac{3.6383}{V^{*2}} = \frac{3.6383}{(0.7322)^2} = 6.7864$$

Using equation (4.18), the viscosity of the metal is finally determined:

$$\mu = \mu^* \frac{\sqrt{M\,R\,T}}{\delta^2\,N_o} = 6.7864 \frac{\sqrt{(47.9)\,(8{,}314 \text{ x } 10^7)\,(2123)}}{(2.89 \text{ x } 10^{-8})^2\,(6.023 \text{ x } 10^{23})} = 0.0392 \text{ P} = 3.92 \text{ cP}$$

Example

Evaluate the viscosity of liquid iron at 1873 K. Density of liquid iron: 7 g/cm^3.

Solution

Using the worksheet, the following values are obtained:

- T* = 0.1718;
- n = 7.549 x 10^{22};
- V* = 0.8278;
- μ* = 6.9795.

And finally:

$$\mu = 5.38 \text{ cP}$$

In the case of alloys, there is no universally applicable model. It is common to represent viscosity data obtained experimentally as isoviscosity lines superimposed to the phase diagram of a particular system. Example of this type (Iron-Carbon diagram) is shown in Figure 4.5.

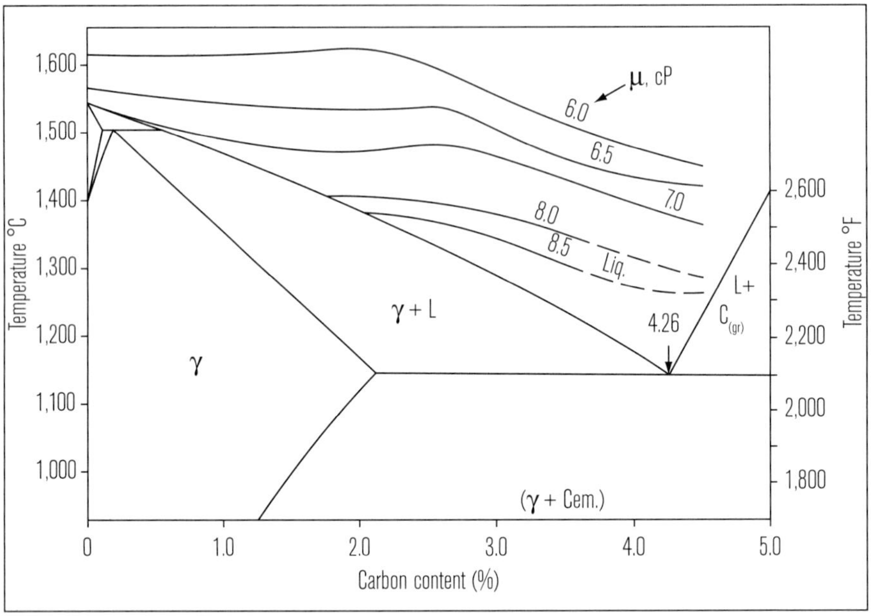

Figure 4.5 - Viscosity of iron-carbon alloys (GEIGER and POIRIER, 1980).

4.3.2 Viscosity of slags

Another type of liquid that is frequently encountered in metallurgical processing is the slag, which is mainly composed by oxides. In metallurgical operations, it is important to have a low viscosity slag. In the blast furnace, for example, a low viscosity slag is essential to attain high productivity and a hot metal with lower content of sulphur. Hence, the study of viscosity of slags and how it is affected by its chemical composition and by the temperature, is very important to optimize operating conditions to achieve increased productivity and efficiency of the processes in terms of the final quality of the product.

Viscosity is an intrinsic property of a given slag. Knowledge of the structure or arrangement of the molecules in a slag helps to understand the factors that affect its viscosity. A liquid slag consists of cations and anions resulting from the ionization of basic and acidic oxides. Acidic oxides are those which, when in the dissolved state in a slag, have the potential to acquire additional oxygen ions to form anionic complexes, and the basic oxides provide the oxygen ions, generating cations that can move freely through the structure of ionic slag. The most common acidic oxides are SiO_2 and Al_2O_3, which have similar behavior, although the latter is sometimes considered amphoteric. The common basic oxides are CaO and MgO.

The structure of liquid silica is similar to that of solid silica, in which each Si^{+4} ion shares an electron with each of the four O^{-2} ions, forming a tetrahedron around the Si^{+4} ion. In solid state, the electro-neutrality is maintained with each oxygen ion sharing the electron with another tetrahedron. The structure is a regular crystalline arrangement of groups SiO_4^{-4}, as shown in Figure 4.6a. When silica is melted, this arrangement continues, as shown in Figure 4.6b. There are still the same Si-O bonds and the viscosity of the liquid SiO_2 is very high (1.5×10^5 Poise at 1940°C) (GEIGER and POIRIER, 1980).

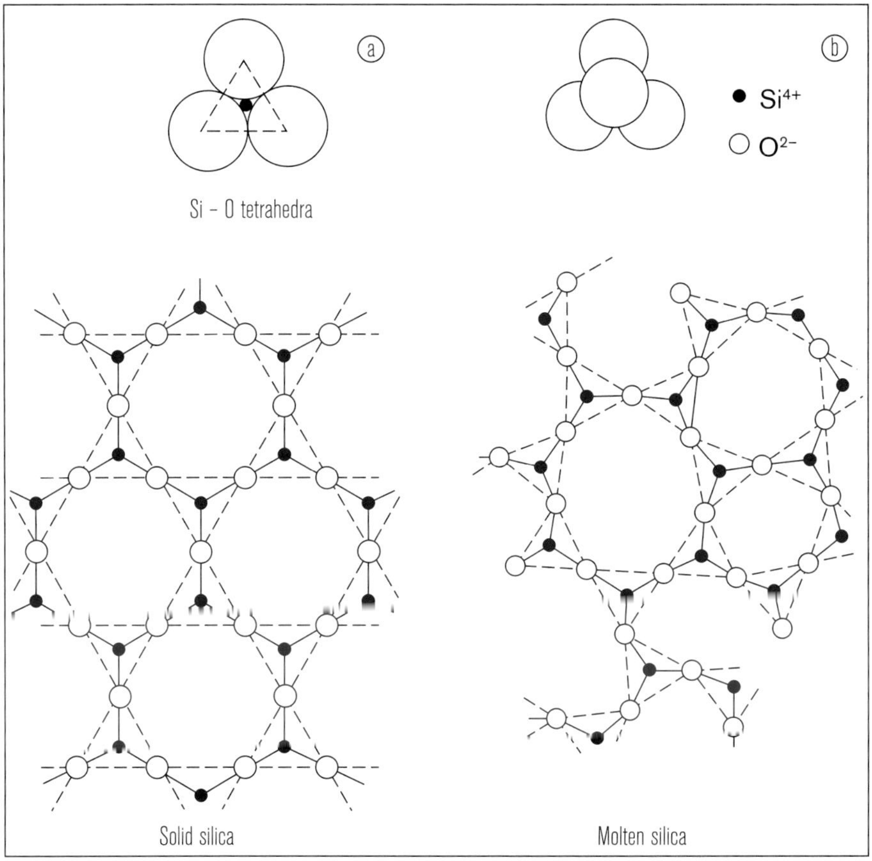

Figure 4.6 - Structure of solid and liquid silica (GEIGER and POIRIER, 1980).

When CaO, or other similar divalent oxide (MgO, for example) is dissolved in liquid silica, it dissociates generating Ca^{+2} ions, which are accommodated in the interstices of the structure of silica and the O^{-2} ion enters into the lattice, as shown in Figure 4.7. Each O^{-2} ion, generated by the dissociation of a basic oxide like CaO, causes the separation of two tetrahedras. Increased the dissolution of CaO, results in further breakage of the original three-dimensional network of silica tetrahedra linked to each other to form various structures of smaller dimensions, leading to a significant reduction in the viscosity of the slag, as shown in Figure 4.8.

This figure also shows that the increase in temperature contributes to an increased bond breaking and thus decreases the viscosity of the melt.

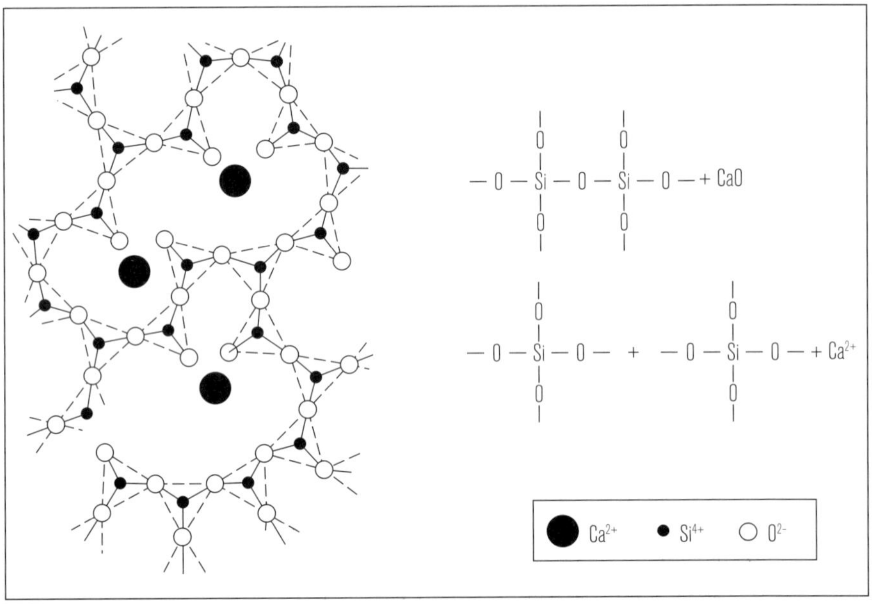

Figure 4.7 - Effect of dissolution of CaO in the structure of liquid silica (GEIGER and POIRIER, 1980).

Figure 4.8 cannot be used in the evaluation of the viscosity of ironmaking and steelmaking slags, since they contain other oxides such as Al_2O_3, MgO, FeO and MnO, in addition to CaO and SiO_2. There has been considerable experimental work in this field, analyzing the effects of various oxides on the viscosity of metallurgical slags. The results of these experimental works are usually presented in the form of isoviscosity diagrams.

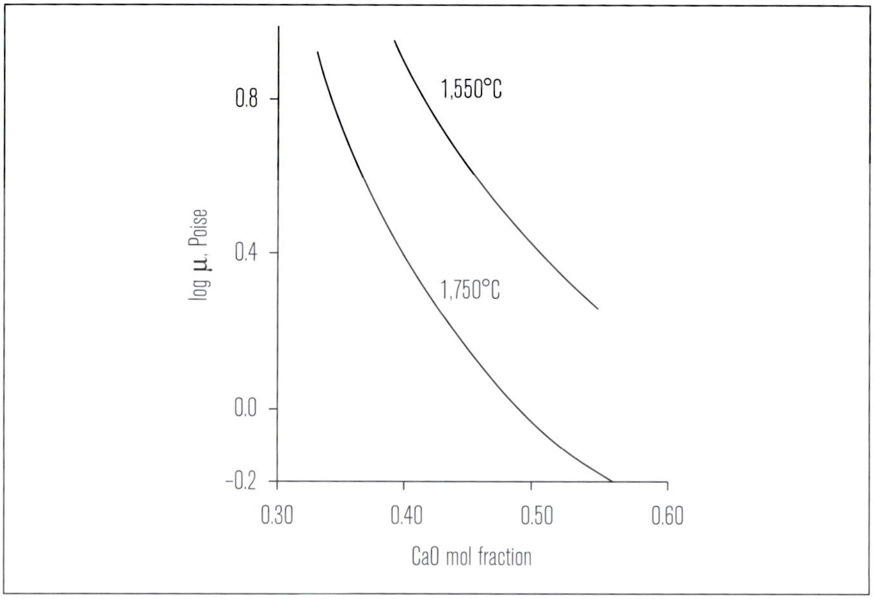

Figure 4.8 - Viscosity of $CaO-SiO_2$ melts (GEIGER and POIRIER, 1980).

4.3.2.1 Isoviscosity diagrams

In the literature (SLAG Atlas, 1995), there are a number of ternary and pseudo-ternary isoviscosity diagrams for various temperatures from which one can obtain viscosity values for different types of slags.

To use these diagrams, one has to first locate the point corresponding to the chemical composition of the slag. Figure 4.9 presents a ternary diagram and illustrates the procedure involved in locating the point associated to composition of the slag. Point A represents the pure component A and any point on the line AC represents a mixture of A and C without component B. The lines parallel to the opposite side of the vertex of component A represent lines of equal concentration of A, and the closer they are of that vertex, the greater the content of A. For example, points 1 and 2 in the diagram represent the following compositions:

- Point 1: A = 40%, B = 20%, C = 40%;
- Point 2: A = 30%, B = 40%, C = 30%.

The sum of the contents of the components of the slag must be equal to 100.

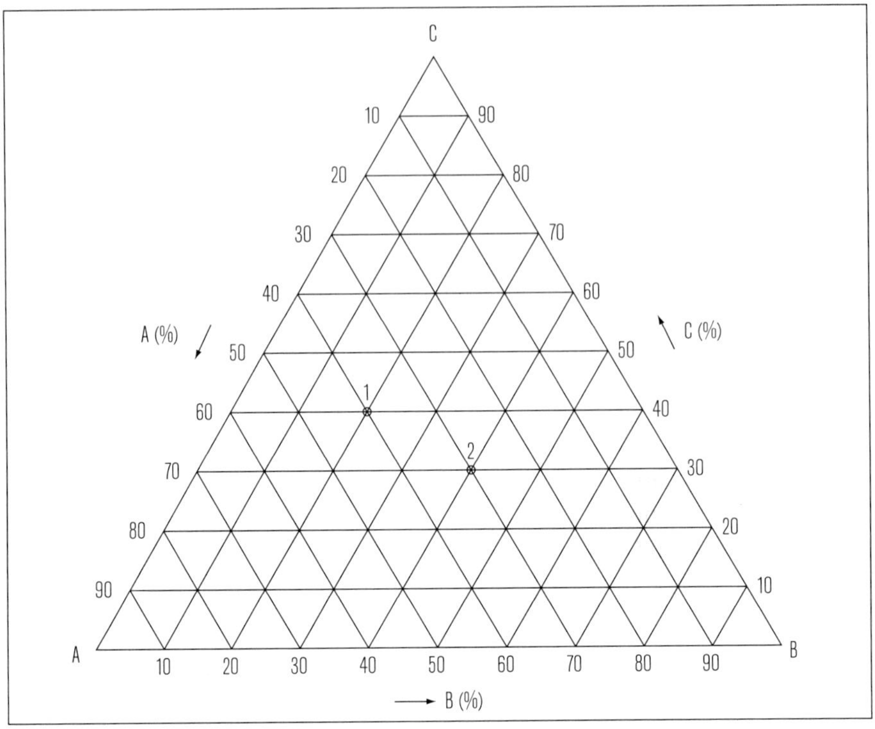

Figure 4.9 - Ternary diagram used in locating points corresponding to the chemical composition of the slags.

Figure 4.10 presents a isoviscosity diagram that can be used in the evaluation of the viscosity of blast furnace slags, containing SiO_2, CaO, Al_2O_3 and MgO. This diagram was obtained for slags containing 35% of SiO_2, with variable proportions of the remaining components.

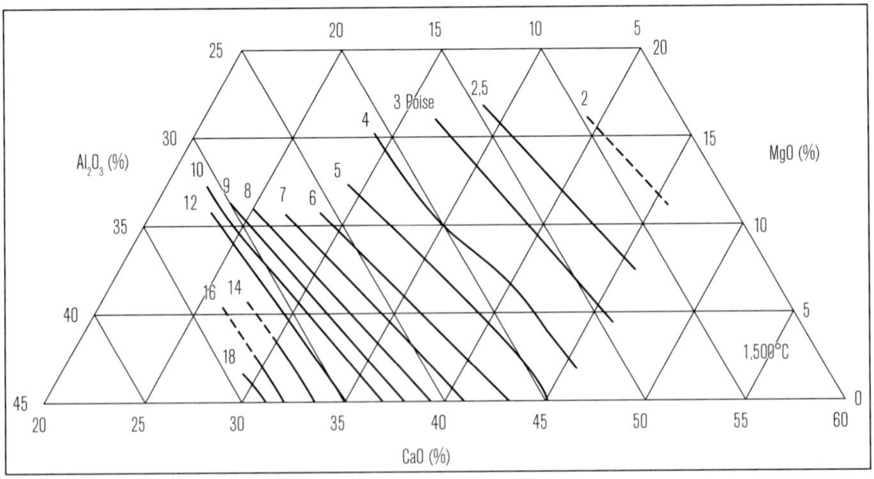

Figure 4.10 - Isoviscosity diagram for the system CaO-SiO_2-Al_2O_3-MgO. Temperature of 1500°C. Silica content: 35% (CARVALHO et al., 1977).

4.3.2.2 Method of silica equivalence and others

Another procedure for estimating the viscosity of slags is called the silica equivalence method. The basis for the development of this method is the silica equivalence of alumina in terms of influence on viscosity.

Alumina (Al_2O_3), when dissolved in slags, forms anions AlO_3^{-3} and its effect on viscosity is similar to that of silica, but not exactly equivalent on a molar basis. Two Al^{+3} ions can replace two Si^{+4} ions only when one Ca^{+2} ion is available to maintain electroneutrality. So with respect to the viscosity, alumina is equivalent to a certain amount of silica (called silica equivalence), which depends on the ratio Al_2O_3/CaO and the total amount of Al_2O_3, as shown in Figure 4.11.

The total mole fraction of silica (mole fraction of silica added to the silica equivalence of alumina, X_a) was correlated to the viscosity of slags in the system CaO - MgO - Al_2O_3 - SiO_2 for various temperatures, as shown in Figure 4.12. To calculate the viscosity of a slag using the silica equivalence method, one must first determine the chemical composition of the slag in terms of mole fraction and then estimate the silica equivalence using Figure 4.11. MgO is considered equivalent to CaO, up to about 10% on molar basis, and their mole fraction should be added to determine X_{CaO}. FeO is also equivalent to CaO up to 16 mole percent (GEIGER and POIRIER, 1980) and its mole fraction should be added to those of MgO and CaO to calculate X_{CaO}.

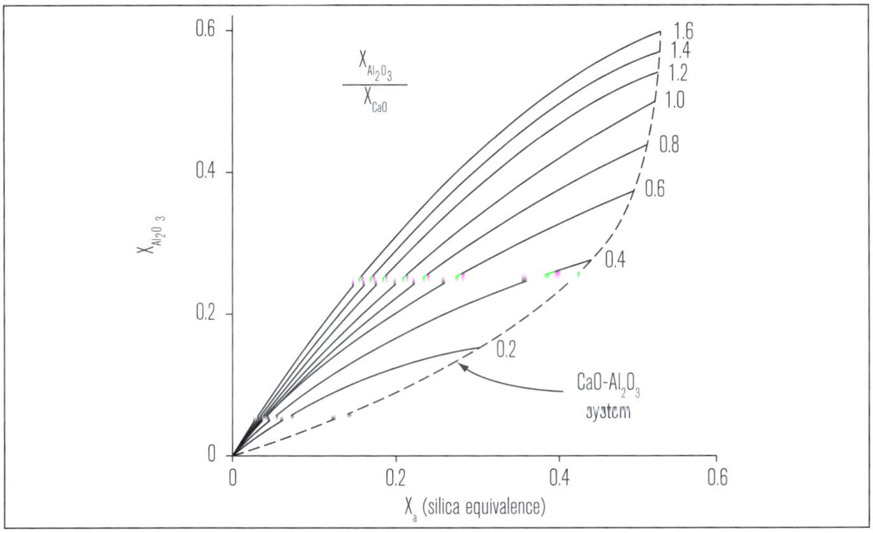

Figure 4.11 - Silica equivalence of alumina as a function of alumina mole fraction and Al_2O_3 to CaO molar ratio (GEIGER and POIRIER, 1980).

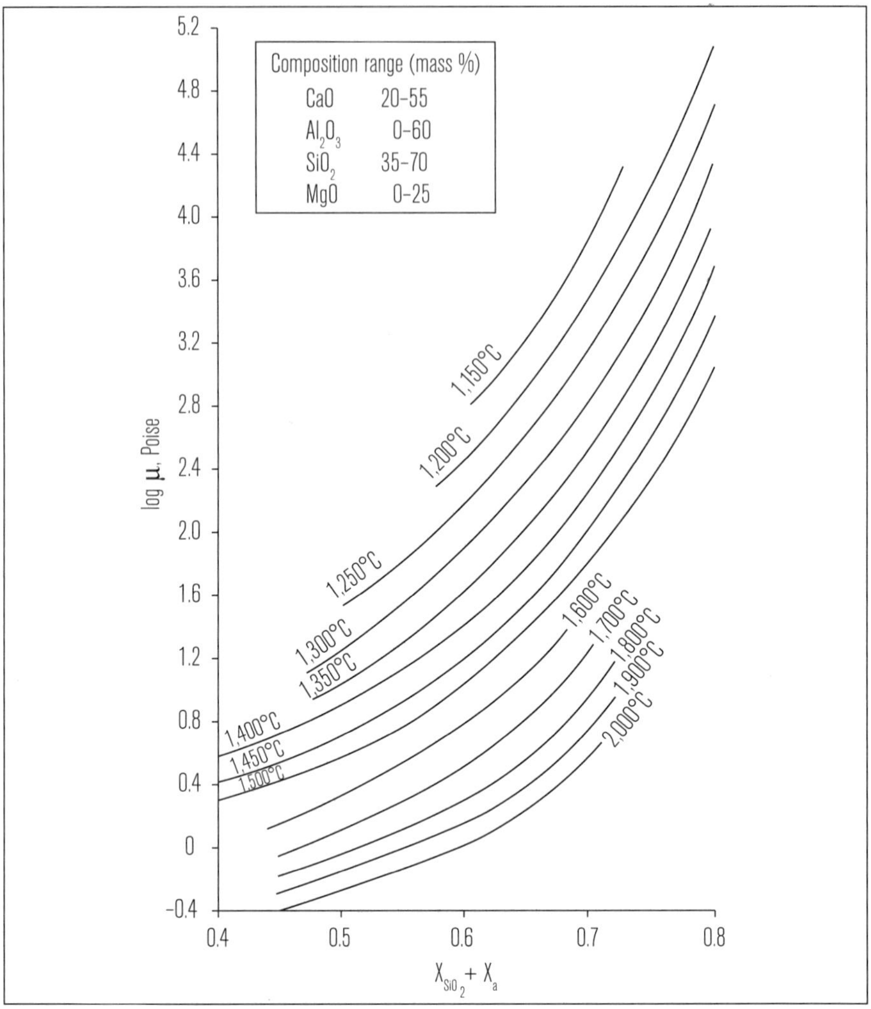

Figure 4.12 - Viscosity of slags in the system CaO - SiO_2 - Al_2O_3 - MgO (GEIGER and POIRIER, 1980).

Another possibility that can be adopted in the calculation of viscosities of slags is to use equations that are arrived at on the basis of regression analysis including the variables that affect the viscosity (chemical composition and temperature) of different types of slags. One example for blast furnace is (CASTRO, TAVARES and FIGUEIRA, 1989):

$$\ln \mu = -10.3469 + \frac{25144}{T} - 0.09633\,(\%\,CaO) - 0.11818\,(\%\,MgO) - 0.008013\,(\%\,Al_2O_3) \qquad (4.23)$$

where:

- μ = viscosity of slag (Pa.s);
- T = temperature (K);
- % i = mass content of the oxide "i" in the slag.

Equation (4.23) is valid for the SiO_2 contents between 35% and 45%.

Example

Estimate the viscosity of a slag with the following composition:

CaO = 41.46% SiO_2 = 35% MgO = 5.62%

Al_2O_3 = 17.92%. Temperature = 1500°C.

Use the three methods presented above.

Solution

Isoviscosity diagram - The temperature and SiO_2 content are within the limits of validity of the diagram presented in Figure 4.10.

The point corresponding to the chemical composition of the slag is located in the diagram below. The viscosity corresponding to this composition can be read from the diagram (usually interpolation between the isoviscosity lines is required). The value obtained in this case is 4 P.

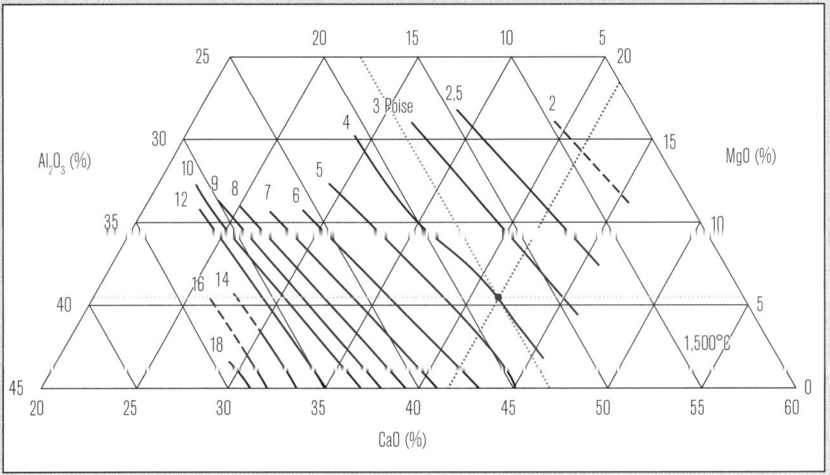

Method of silica equivalence- Initially, the mole fraction of each component of the slag is calculated. The values are:

$$X_{CaO} = 0.4516$$

$$X_{SiO_2} = 0.3558$$

$$X_{Al_2O_3} = 0.1074$$

$$X_{MgO} = 0.0852.$$

Since X_{MgO} is less than 0.1 (10 %), X_{MgO} and X_{CaO} are added together to obtain X_{CaO}. The ratio $X_{Al_2O_3}$ / X_{CaO} is given by:

$$\frac{X_{Al_2O_3}}{X_{CaO} + X_{MgO}} = \frac{0.1074}{0.4516 + 0.0852} = 0.2001$$

Considering the curve corresponding to 0.20, the silica equivalence of alumina is 0.1654 (interpolation from the figure below). This is then added to mole fraction of silica to determine the total mole fraction of silica.

The total mole fraction of silica is then $X_{SiO_2} + X_a$, 0.3558 + 0.1654 = 0.5212. Using this value in Figure 4.12, for the temperature 1500°C, the viscosity is read from the Y-axis and is approximately 4.03 P ($\log \mu = 0.605$) (figure below).

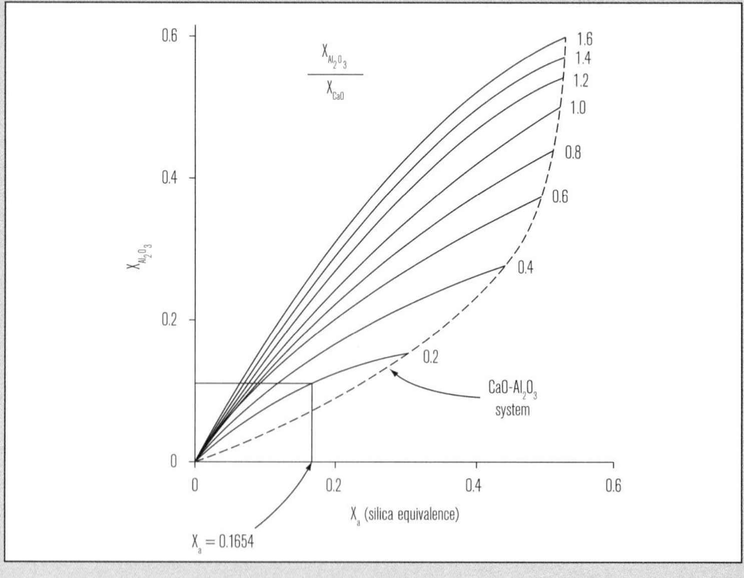

Finally, using the regression equation for viscosity of slags in this range of composition (equation 4.23), the following result is obtained:

$$\ln \mu = -10.3469 + \frac{25144}{1773} - 0.096334\,(41.46) - 0.118176\,(5.62) - 0.0080126\,(17.92) = -0.9670$$

$$\mu = 0.3802\ \frac{kg}{m\,s} = 3.802\ P$$

It is observed that all the three methods give similar results.

Several other models have been developed to predict the viscosity of different types of slags. In view of the importance of this subject, there is extensive published literature on experimental and theoretical work on the influence of different oxides and fluorides on the viscosity of slags used in different stages of iron and steel making.

References

BIRD, R. B.; STEWART, W. E.; LIGHTFOOT, E. N. *Transport phenomena.* New York: John Wiley & Sons, 1960. 780 p.

CARVALHO, J. L. R. et al. *Dados termodinâmicos para metalurgistas.* Belo Horizonte: Edições Engenharia, 1977. 394 p.

CASTRO, L. F. A.; TAVARES, R. P.; FIGUEIRA, R. M. *Tecnologia de fabricação do gusa líquido em altos-fornos.* Belo Horizonte: Fundação Christiano Ottoni, 1989. v. 9: Escórias de alto-forno.

CUSSLER, E. L. *Diffusion*: mass transfer in fluid systems. Cambridge: Cambridge University Press, 1997. p. 105-7.

GEIGER, G. H.; POIRIER, D. R. *Transport phenomena in metallurgy.* Massachussets: Addison-Wesley, 1980.

JASTRZEBSKI, Z. D. *The nature and properties of Engineering Materials.* New York: John Wiley, 1976.

SLAG Atlas. Dusseldorf: StahlEisen, 1995.

Exercises

1 - The velocity profile of a fluid in a given system is expressed by:

$$v_x = -6.25\ y^2 + 6.25\ y - 1 \qquad (v_x\ [=]\ m/s,\ y\ [=]\ m)$$

Make an outline of the velocity profile (y varies between 0 and 0.5 m).

Use the equation of the velocity profile to determine:

- direction and sense of the momentum flux by convection;
- direction and sense of the momentum flux by diffusion;
- flux of momentum by diffusion at y = 0 m;
- shear stress at y = 0 and y = 0.5 m.

The fluid has the following characteristics:

$\rho = 3\ g/cm^3 \qquad \mu = 2 \times 10^{-3}\ lb_f/ft.s.$

2 - For the figure below, indicate the direction and sense of the momentum transport by convection and diffusion. Write the Newton's law of viscosity for the case shown and explain the answers.

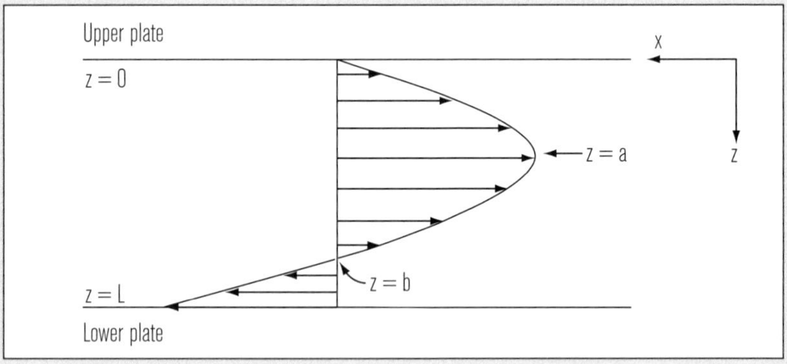

3 - Calculate the viscosity of CO_2 and N_2 in the range of 600 K to 2000 K with 200 K increments. Plot the results in terms of a graph of μ_{CO_2} and μ_{N_2} *versus* temperature (K).

4 - Determine the viscosity of oxygen, nitrogen and methane gas at 20°C and atmospheric pressure, in centipoise. Compare the results with the experimental data below:

- $\mu_{N2} = 0.0175$ cP;
- $\mu_{O2} = 0.0203$ cP;
- $\mu_{CH4} = 0.0109$ cP.

5 - a) Calculate the viscosity of air at 20°C, considering it a mixture of 79% N_2 and 21% O_2 in volume.

b) Compare the result with the experimental value of 0.01813 cP.

6 - Determine the viscosity of liquid chromium at 2000°C. The properties of chromium are:

- melting point: 1898°C;
- atomic mass: 52.01 g/mol;
- density: 7.1 g/cm^3;
- δ: 2.72 Å.

7 - Estimate the viscosity of liquid titanium at 1900°C. The following data are available:

- melting temperature: 1800°C;
- atomic mass: 47.9 g/mol;
- density: 4.50 g/cm^3;
- δ: 2.89 Å.

8 - Evaluate the viscosity of liquid iron at 1800°C. Data:

- melting temperature: 1536°C;
- atomic mass: 55.85 g/mol;
- density: 7 g/cm^3.

9 - Estimate the viscosity of the slag with the following composition at 1500°C.

- % SiO_2 = 45;
- %CaO = 35;
- % Al_2O_3 = 20.

Use the three methods discussed and compare the results.

10 - Determine the viscosity of the blast furnace slag with the following composition at 1400°C.

- % SiO_2 = 40;
- % CaO = 35%;
- % Al_2O_3 = 18;
- % MgO = 7.

Compare the results obtained with the different methods presented in this chapter.

Laminar Flow and Momentum Balance

Chapter 05

In this chapter, the methodology applied in the determination of the velocity distribution of a fluid in laminar flow will be presented and applied to systems of simple geometry. This method is based on the concepts of viscosity, mass and momentum balances. Initially, the general characteristics of laminar and turbulent flows will be discussed.

5.1 Laminar and Turbulent Flows

When a fluid moves through a system, two different regimes of flow can occur. The experiments carried out by Reynolds in 1883 (LIENHARD IV and LIENHARD V, 2005) showed the differences between these two types of flow.

Consider water flowing in a transparent tube. A fine needle is used to inject ink parallel to the direction of flow of water. For low velocities of the fluid, the jet of ink flows in a straight line, without mixing with the adjacent layers of water, as shown schematically in Figure 5.1a. This type of flow is called laminar. In this case, the flow of the fluid consists of streamlines which do not cross each other giving rise to mixing. This happens at low velocities.

As the water velocity increases, the situation shown in Figure 5.1b is reached. Beyond a certain point, the water is all colored by the ink due to intense mixing. This is turbulent flow.

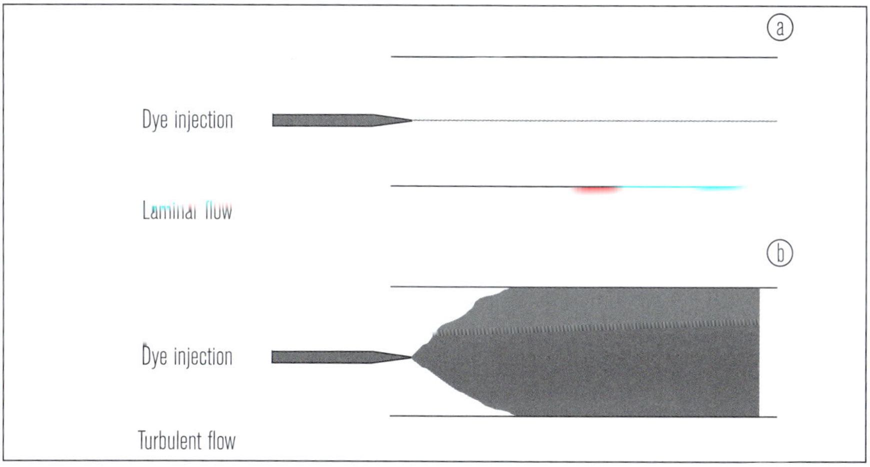

Figure 5.1 - Schematic view of the experience of Reynolds.

In the laminar flow, the trajectory of the fluid is well defined, whereas in the turbulent flow, the motion of the fluid is irregular with intense mixing. The velocity of the fluid varies with time, as shown in Figure 5.2, which presents the variation of velocity at the center of a tube under turbulent conditions. This was obtained experimentally using a laser anemometer.

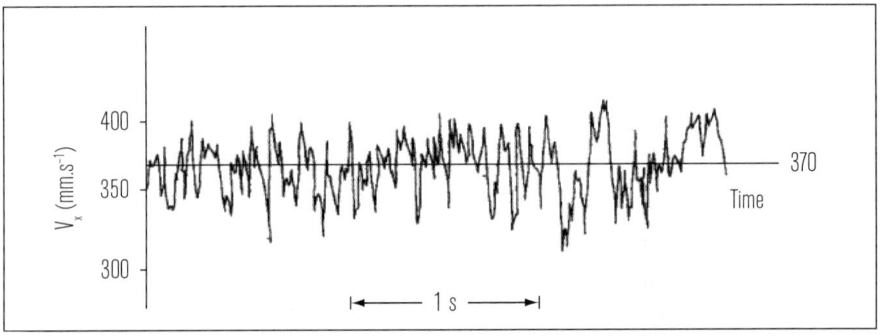

Figure 5.2 - Measurement of velocity at the center of a tube 22 mm internal diameter. Reynolds Number = 6500 (GUTHRIE, 1992).

The conditions for the transition from one flow regime to the other can be determined experimentally and varies according to the configuration of the system. Typically, the criterion for knowing the type of flow prevailing in the fluid is defined in terms of a dimensionless number called the *Reynolds number*, defined by the following relationship:

$$Re = \frac{L \bar{V} \rho}{\mu} \tag{5.1}$$

where:

- L = characteristic length (defined according to the configuration of the system);
- $\bar{V}$ = average velocity of the fluid along the cross section of the duct;
- ρ = density of the fluid;
- μ = dynamic viscosity of the fluid.

(Verify that Re is a dimensionless number).

In the case of tubes, the characteristic length is the diameter. In this case, Reynolds number is defined by:

$$Re = \frac{D \bar{V} \rho}{\mu} \tag{5.2}$$

where

- D is the diameter of the tube.

The value of the Reynolds number for transition from laminar to turbulent flow in pipes, is approximately 2100. This value is determined experimentally and varies according to the configuration of the system. Flows over flat plates, in packed beds and between rotating cylinders have transitions at different Reynolds numbers.

5.2 Mass and Momentum Balances

In this section, the methodology used to obtain velocity profiles of fluids for laminar unidirectional flow in systems of simple geometry is discussed. Initially simple geometries and unidirectional flows are considered to reduce the mathematical complexity and emphasize the concepts of microscopic balance in an infinitesimal control volume typical of the geometry of the system.

Velocity profiles of fluids can be determined through mass and momentum balances and the application of Newton's law of viscosity. The resulting differential equations are solved under specified boundary conditions to obtain the velocity profile.

In engineering problems involving fluid flow, it is important to determine a mathematical relationship between the driving force necessary for the flow such as pressure drop or gravity and the flow rate of the fluid in the system under consideration. Besides the relationship between the driving forces for fluid flow and the flow rate, the procedure to be developed in this chapter can also be used to evaluate the:

- velocity distribution;
- average velocity;
- shear force on the walls.

The mathematical treatment for this kind of problem involves the development of mass and momentum balances applied to the system under study as mentioned earlier. Mass balance is also used in thermodynamics. The main differences between the balances adopted in transport phenomena and in thermodynamics are:

- in transport phenomena, the balances are developed considering the time aspect, i.e., rates of mass and momentum input and output. In thermodynamics, the balances do not consider the rates of mass and energy transport. They are global balances of mass and energy usually linked to a reference amount of product generated or raw materials used.

- in transport phenomena, the balances are applied to infinitesimal parts of the system and not to the system as a whole, as in the case of thermochemical models.

5.2.1 Mass balance

The mass balance is established on the basis of conservation of mass (in nature, whatever exists is transformed one to the other, that is to say nothing is created and nothing is lost). For a system at steady state (parameters do not change with time), mass balance can be expressed by equation (5.3).

$$\left[\text{Rate of mass in}\right] - \left[\text{Rate of mass out}\right] = 0 \tag{5.3}$$

5.2.2 Momentum balance

Momentum balance is actually a balance of the forces acting on the system. Considering a system in steady state, the momentum balance can be expressed in the following generic form:

$$\begin{bmatrix}\text{Rate of} \\ \text{momentum in}\end{bmatrix} - \begin{bmatrix}\text{Rate of} \\ \text{momentum out}\end{bmatrix} + \begin{bmatrix}\text{Sum of forces} \\ \text{acting on the system}\end{bmatrix} = 0 \tag{5.4}$$

This equation is dimensionally correct, since, as has been already mentioned in Chapter 3, the rates of momentum in and out have the same dimension of force. As discussed in Chapters 3 and 4, momentum enters and exits the control volume element by two mechanisms:

- diffusive momentum is due to a velocity gradient. It is estimated by Newton's law of viscosity;
- convective momentum is associated with macroscopic motion of the fluid.

The forces acting on the system are essentially of two types:

- pressure forces that act on surfaces;
- force of gravity, which acts on the control volume.

It is important to clarify that equation (5.4) is another way of writing the well known relationship,

$$\Sigma F_x = m\, a_x \tag{5.5}$$

which is a simplified form of the expression:

$$\Sigma F_x = \frac{d\,(m\, v_x)}{d\,t} \tag{5.6}$$

where,

- ΣF_x = sum of forces acting on the body in the x direction;
- m = mass of the body;
- a_x = acceleration in the x direction;
- v_x = velocity of the fluid in the x direction;
- t = time.

When mass is constant, equations (5.5) and (5.6) are equivalent. Equation (5.6) is actually the Newton's second law of motion, which states that the rate of change of momentum (m.v) is equal to the sum of forces acting on the element, which is similar to equation (5.4).

In general, the procedure to establish mass and momentum balances and solve problems of fluid flow in laminar regime is as follows:

- choice of coordinate axes to represent the geometry of the system being analyzed;
- selection of a volume element in which the balances are to be established;
- development of the mass balance in the form of equation (5.3) for the selected volume element;
- development of the momentum balance in the form of equation (5.4) for the same volume element above;
- division of the terms of the equations above by the volume of the element and evaluation of the limit of these equations, when the volume of the element tends to zero. In this step, two differential equations, representing the principles of conservation of mass and momentum in the system under study, are obtained;
- integration of the equations above. Newton's law of viscosity is incorporated in the momentum conservation equation to obtain an equation for the velocity distribution;

- introduction of boundary conditions to determine the final equation for the velocity profile;
- from the equation for the velocity distribution, average velocity and volumetric flow rate can be evaluated considering the geometry of the system.

Information obtained through the use of the procedure above can be used for determining other parameters of interest such as:

- pressure drop necessary for a certain volumetric flow rate;
- shear forces acting on the surface of the walls in contact with the fluid.

Boundary conditions represent physical phenomena in certain locations of the system. They are necessary for the evaluation of the integration constants, that appear when the momentum conservation is integrated. A list of common boundary conditions is given below:

- at solid-fluid interfaces, a condition referred to as no slip is valid for most fluids and particularly for Newtonian fluids (which obey Newton's law of viscosity), such as gases, water, liquid metals and slags, normally encountered in metallurgical systems. This condition states that at an interface between a solid and a fluid, the fluid assumes the velocity of the solid;
- at gas-liquid interfaces, the shear stress (or the diffusive flux of momentum) in the liquid phase is approximately zero for all practical purposes. This condition arises due to the fact that gases offer little resistance to movement of liquids due to their low viscosity (about 3 to 4 orders of magnitude lower than that of liquids);
- at liquid-liquid interfaces, the flux of diffusive momentum and the velocities are continuous functions.

5.3 Applications of Mass and Momentum Balances

In this section, the procedure described above is used to solve fluid flow problems for laminar flow in steady state conditions (time independent), assuming constant fluid density and viscosity More generalized treatments, without these restrictions, are discussed in Chapters 6 and 7, where more complex situations and also cases involving turbulent flow are analyzed.

5.3.1 Fluid flow between two horizontal flat plates

The system under consideration is schematically seen in Figure 5.3 and involves fluid flow between two horizontal flat plates. Although no system of practical interest is exactly like the situation presented in Figure 5.3, this example is useful for beginners and illustrate the application of mass and momentum balances in Cartesian coordinates to a system with simple geometry to determine the velocity profile and other important parameters.

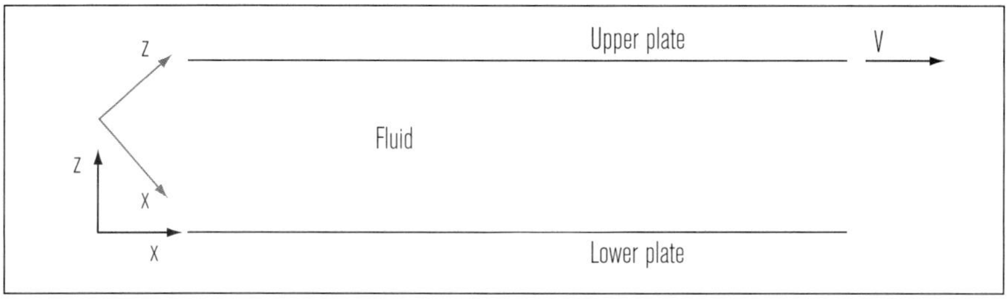

Figure 5.3 - Flow between two horizontal parallel plates.

In the system shown in Figure 5.3, fluid flow is caused by the movement of the upper plate from left the right, with velocity V. The lower plate remains stationary. As mentioned above, for carrying out mass and momentum balances the first step is to choose coordinate axes to represent the geometry of the system. Obviously, any choice of orthogonal axes is correct. However, there is convenient choice, which would simplify the mathematical treatment of the problem. In Figure 5.3, two possible choices of axes are displayed, one in black and one in gray. The fluid flows due to the movement of the upper plate which is the only driving force for flow in the horizontal direction. The choice of axes in gray would mean that there are two velocity components. The vectorial sum of these two components would result in a horizontal velocity. The choice of axes in black leads to the existence of only one component of velocity in the x direction Obviously, this choice is more practical, as there is only one component of velocity. This simplifies the mathematical treatment of the problem. It should be mentioned, however, that it is not always possible to choose coordinate axes to provide only one velocity component, especially in the case of problems involving bidimensional or tridimensional flows.

Having defined the coordinate axes, the volume element to be taken as a reference for the development of mass and momentum balances can be defined. The element is chosen according to the coordinate axes (cartesian) as shown in Figure 5.4.

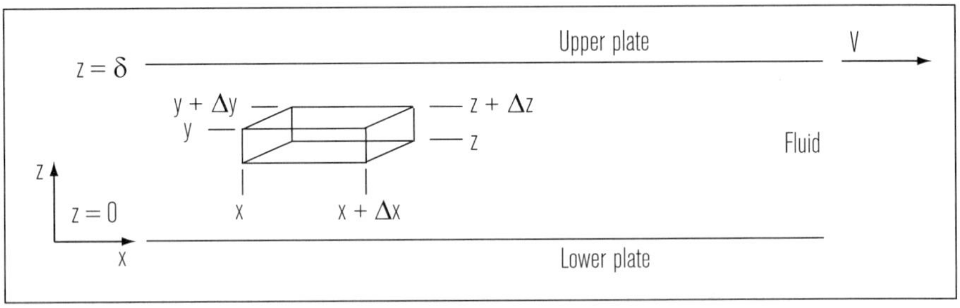

Figure 5.4 - Volume element for mass and momentum balances.

With the choice of axes above, it is seen that the transport of momentum by convection occurs in the x direction, which is the direction of the macroscopic motion of the fluid. The transport of momentum by diffusion occurs in the z direction, which is the direction of the velocity gradient. It is easy to see that there is a velocity gradient in the z direction. At z = 0 (lower plate), the fluid velocity is zero. At any other point along z, the velocity is different from zero. As such, there is a velocity gradient in z direction (the fluid velocity varies with position z). It is important to mention that it is not necessary to know in advance where the velocity is higher or smaller.

5.3.1.1 Mass balance

Considering the direction of the macroscopic motion of fluid, the mass balance is given as follows:

$$\begin{aligned}&[\text{Rate of mass in across surface at } x = x] - \\ &[\text{Rate of mass in across surface at } x = x + \Delta x] = 0\end{aligned} \tag{5.7}$$

The locations of the points for input and output of mass are determined according to the orientation of the axes, and not by the directional sense of the flow. In this case, the fluid is actually flowing from left to right, and inflows and outflows of mass are defined by positions x and x + Δx, respectively. These same locations would have been chosen if the fluid was flowing from right to left. The fact that the points for entry and exit of mass are defined according to the choice of orientation of the coordinate axes, eliminates the need to know beforehand the directional aspect of the flow.

The rates of entry and exit of mass can be evaluated by the following expressions:

$$\left[\text{Rate of mass in across surface at x = x}\right] = (\Delta y\ \Delta z\ v_x \rho)|_{x=x} \tag{5.8}$$

$$\left[\text{Rate of mass in across surface at x = } x + \Delta x\right] = (\Delta y\ \Delta z\ v_x\ \rho)|_{x=x+\Delta x} \tag{5.9}$$

In the equations above, the product $\Delta y.\Delta z$ is the area of the volume element perpendicular to the direction of fluid flow (x direction), v_x is the velocity component and ρ is the density of the fluid, all evaluated at the surface at location x = x. The product of these factors has the dimension of mass per unit time (mass flow rate).The product $\Delta y.\Delta z.v_x$ has the dimension of volume per unit time. (Check the above dimensions as an exercise).

The mass balance can then be written as:

$$(\Delta y\ \Delta z\ v_x\ \rho)|_{x=x} - (\Delta y\ \Delta z\ v_x\ \rho)|_{x=x+\Delta x} = 0 \tag{5.10}$$

Following the sequence of procedures listed in Section 5.2, the next step is to divide the equation above by the volume element, $\Delta x.\Delta y.\Delta z$:

$$\frac{(\Delta y\ \Delta z\ v_x\ \rho)|_{x=x} - (\Delta y\ \Delta z\ v_x\ \rho)|_{x=x+\Delta x}}{\Delta x\ \Delta y\ \Delta z} = 0 \tag{5.11}$$

$$\frac{(v_x\ \rho)|_{x=x} - (v_x\ \rho)|_{x=x+\Delta x}}{\Delta x} = 0 \tag{5.12}$$

Taking the limit when Δx tends to zero and using the concept of first derivative, the following expression is obtained

$$\lim_{\Delta x \to 0}\left[\frac{(v_x\ \rho)|_{x=x} - (v_x\ \rho)|_{x=x+\Delta x}}{\Delta x} = 0\right] = -\frac{\partial(v_x\ \rho)}{\partial x} = 0 \tag{5.13}$$

The negative sign of the derivative at the right hand side of the equation comes from the definition of first derivative, given by:

$$\frac{df}{dx} = \lim_{\Delta x \to 0}\left[\frac{(f)|_{x=x+\Delta x} - (f)|_{x=x}}{\Delta x}\right] \tag{5.14}$$

where f is any function (in equation (5.13), the function is the product $v_x.\rho$). Note that the limits of the evaluation of function f, in the definition of the derivative, are the opposite of those in equation (5.13).

Considering that the density of the fluid is constant and uniform throughout the domain, equation (5.13) can be written as:

$$\rho \frac{\partial v_x}{\partial x} = 0 \tag{5.15}$$

Since the density of the fluid is a positive value, the mass balance is finally written as:

$$\frac{\partial v_x}{\partial x} = 0 \tag{5.16}$$

This equation states that the velocity v_x in the situation being studied does not depend on the position x (v_x is not a function of x). It is common to refer to this as a criterion for fully developed flow. Equation (5.16) represents the result obtained by applying the principle of conservation of mass.

5.3.1.2 Momentum balance

In the momentum balance, the transport of momentum both by diffusion and convection have to be considered. Both contributions have to be included, while evaluating the rate of momentum in and out of the control volume element.

Considering the rates of momentum transported by diffusion in the direction of the velocity gradient (z direction) and that by convection in the direction of the macroscopic motion of the fluid, the following equation can be written:

$$\begin{gathered}
[\text{Rate of momentum in by convection at surface } x = x] - \\
[\text{Rate of momentum out by convection at surface } x = x + \Delta x] + \\
[\text{Rate of momentum in by diffusion at surface } z = z] - \\
[\text{Rate of momentum out by diffusion at surface } z = z + \Delta z] + \\
[\text{Sum of forces acting on the fluid in the x direction}] = 0
\end{gathered} \tag{5.17}$$

As mentioned earlier, the points of entry and exit of momentum by convection and diffusion are determined according to the orientation of the coordinate axes and not through the knowledge of the actual directions of momentum transport by these mechanisms. It is not necessary to know in advance the direction of the momentum fluxes to establish the momentum balance. These directions are naturally determined during the development and analysis of the problem.

The equation of momentum balance contains the following terms:

$$\left[\text{Rate of momentum in by convection across surface at } x = x\right] = (\Delta y\, \Delta z\, v_x\;\rho\; v_x)|_{x=x} \tag{5.18}$$

$$\left[\text{Rate of momentum out by convection across surface at } x = x + \Delta x\right] = (\Delta y\, \Delta z\, v_x\;\rho\; v_x)|_{x=x+\Delta x} \tag{5.19}$$

$$\left[\text{Rate of momentum in by diffusion across surface at } z = z\right] = (\Delta x\, \Delta y\, \tau_{zx})|_{z=z} \tag{5.20}$$

$$\left[\text{Rate of momentum out by diffusion across surface at } z = z+\Delta z\right] = (\Delta x\, \Delta y\, \tau_{zx})|_{z=z+\Delta z} \tag{5.21}$$

As the flow occurs due to displacement of the top plate, there is no other forces (pressure drop or gravity) acting on the system in the direction of flow (x direction). It should be noted that the force of gravity is present, but it acts in the vertical direction (z direction) and has no component in the direction of fluid flow. In equation (5.17), the sum of forces means only those having components in the direction of flow and it is important to remember that momentum is a vectorial quantity. The momentum balance (or balance of forces) in this case refers to the direction x. In equations (5.18) and (5.19), the product $\Delta y.\Delta z.v_x$ represents the mass flow. When this factor is multiplied by the velocity v_x, the rate of momentum being transported by convection is determined. This term can also be interpreted as the inertial force.

In equations (5.20) and (5.21), τ_{zx} represents the diffusive momentum flux (see Chapter 4). This is multiplied by the area normal to its direction, $\Delta x.\Delta y$, to obtain the the rate of diffusive momentum.

Substituting expressions (5.18) to (5.21) in equation (5.17), a general equation expressing the momentum balance for the case under study is obtained.

Dividing by the volume of the element:

$$[\Delta y\, \Delta z\, v_x\, \rho\, v_x)|_{x=x} - (\Delta y\, \Delta z\, v_x\, \rho\, v_x)|_{x=x+\Delta x}] + [(\Delta x\; \Delta y\; \tau_{zx})|_{z=z} - (\Delta x\; \Delta y\; \tau_{zx})|_{z=z+\Delta z}] = 0 \tag{5.22}$$

$$\frac{[\Delta y\, \Delta z\, v_x\ \rho\, v_x)|_{x=x} - (\Delta y\, \Delta z\, v_x\ \rho\ v_x)|_{x=x+\Delta x}]}{\Delta x\ \Delta y\ \Delta z} + \frac{[\Delta x\, \Delta y\, \tau_{zx})|_{z=z} - (\Delta x\ \Delta y\, \tau_{zx})|_{z=z+\Delta z}]}{\Delta x\ \Delta y\ \Delta z} = 0 \tag{5.23}$$

$$\frac{[(v_x\ \rho\ v_x)|_{x=x} - (v_x\ \rho\ v_x)|_{x=x+\Delta x}]}{\Delta x} + \frac{[(\tau_{zx})|_{z=z} - (\tau_{zx})|_{z=z+\Delta z}]}{\Delta z} = 0 \tag{5.24}$$

Taking the limits of equation (5.24) when the volume of the element tends to zero (Δx and $\Delta z \to 0$):

$$\lim_{\Delta x\to 0}\left\{\frac{[(v_x\ \rho\, v_x)|_{x=x} - (v_x\ \rho\ v_x)|_{x=x+\Delta x}]}{\Delta x}\right\} + \lim_{\Delta z\to 0}\left\{\frac{[(\tau_{zx})|_{z=z} - (\tau_{zx})|_{z=z+\Delta z}]}{\Delta z}\right\} = 0 \tag{5.25}$$

Using the concept of first derivative, one can write:

$$-\frac{\partial(v_x\ \rho v_x)}{\partial x} - \frac{\partial(\tau_{zx})}{\partial z} = 0 \tag{5.26}$$

From the mass balance, it is known that:

$$\frac{\partial v_x}{\partial x} = 0 \tag{5.27}$$

As the fluid has constant density, the convection term can be written as:

$$\rho\ \frac{\partial(v_x\ v_x)}{\partial x} = 2\ \rho\, v_x\ \frac{\partial(v_x)}{\partial x} = 0 \tag{5.28}$$

It follows that:

$$\frac{d(\tau_{zx})}{dz} = 0 \tag{5.29}$$

The differential equation above can be integrated by separation of variables, to obtain:

$$\int d(\tau_{zx}) = \int 0\ dz \tag{5.30}$$

$$\tau_{zx} = 0 + C_1 \tag{5.31}$$

where C_1 is an integration constant.

Using Newton's law of viscosity,

$$\tau_{zx} = -\mu \frac{dv_x}{dz} \tag{5.32}$$

and combining equations (5.31) and (5.32), it can be written that:

$$\tau_{zx} = -\mu \frac{dv_x}{dz} = C_1 \tag{5.33}$$

Separating the variables and integrating again, the velocity profile can be obtained:

$$\int dv_x = -\int \frac{C_1}{\mu} dz \tag{5.34}$$

$$v_x = -\frac{C_1}{\mu} z + C_2 \tag{5.35}$$

where C_2 is a new integration constant. The integration above was made assuming that the viscosity of the fluid is constant.

Equation (5.35) is the generic equation for evaluating the velocity profile between two horizontal parallel plates. To have a specific profile for a particular case, the constants C_1 and C_2. should be determined using boundary conditions. From section 5.2.2, it is known that at solid-fluid interfaces, the fluid velocity is equal to that of the solid:

Boundary condition 1 : $v = 0$ at $z = 0$

Boundary condition 2 : $v = V$ at $z = \delta$

Applying the above boundary conditions in the equation for velocity profile one gets:

$$0 = -\frac{C_1}{\mu} 0 + C_2$$

$$V = -\frac{C_1}{\mu} \delta + C_2$$

the values of C_1 and C_2 are determined:

$$C_2 = 0$$

$$C_1 = -\frac{V\mu}{\delta}$$

Finally, the velocity profile for the fluid between the plates can be determined substituting expressions for C_1 and C_2 in equation (5.35), obtaining:

$$v_x = V \frac{z}{\delta} \tag{5.36}$$

(verify that this profile satisfies the boundary conditions).

As can be seen in Figure 4.2, the velocity is a linear function of position z. From the velocity profile, a series of informations about the system can be obtained.

Initially, the profile of shear stress (or flux of diffusive momentum) can be determined. According to Newton's law of viscosity, it is observed that τ_{zx} is constant, not varying with the position between the plates.

$$\tau_{zx} = -\mu \frac{dv_x}{dz} = -\mu \frac{V}{\delta} \tag{5.37}$$

The volume and mass flow rates can also be calculated. Considering that the plates have width W, in the y direction, perpendicular to the plane of the paper, the flow rates are calculated by adding the amounts of fluid that flow in each infinitesimal area W.dz along the distance between the plates. It is known that in each infinitesimal layer, the quantity of fluid that flows is given by:

$$dQ = (W\,dz\,v_x) \tag{5.38}$$

where dQ is the infinitesimal flow rate across a layer of area W.dz.

The volume flow rate can be determined by integrating equation (5.38) in the region between $z = 0$ and $z = \delta$. Hence, integrating equation (5.38) between the specified limits,

$$\int_{z=0}^{z=\delta} dQ = \int_{z=0}^{z=\delta} (W\,dz\,v_x) \tag{5.39}$$

Since the fluid velocity varies with the position, the velocity profile, given by equation (5.36), is introduced in the expression (5.39) before integration, to obtain:

$$Q = \int_{z=0}^{z=\delta} (W\,dz\,v_x) = W\,V\,\frac{\delta}{2} \tag{5.40}$$

The mass flow rate Γ is given by:

$$\Gamma = Q\,\rho = W\,V\,\rho\,\frac{\delta}{2} \tag{5.41}$$

The average velocity is determined by dividing the volume flow rate by the total area effectively used for flow. In this case, this area is given by the product of the width of the plates (W) and the distance between them (δ).

$$\overline{V} = \frac{V}{2} \tag{5.42}$$

This result is expected, since in the case of a linear velocity profile, the average velocity is certainly given by the average of the velocities at the extremes (0 and V).

5.3.2 Flow of a liquid film in an inclined plane

The example discussed in the previous section has the objective of illustrating the basic principles involved in establishing mass and momentum balances to determine velocity profiles of a fluid flowing in a certain system. In the present section, examples which have more practical implications are dealt with. The first system to be analyzed represents the flow of a liquid in an inclined plane, as shown schematically in Figure 5.5. Situations similar to this can be found in casting of metals from various types of furnaces, flow of slags and liquid metals through channels or runners to ladles, flow of water on spillways in hydroelectric power plants, or even water running down from mountains or on the streets in a rainy day. The procedure adopted for analysis of this system is similar to that described in section 5.3.1.

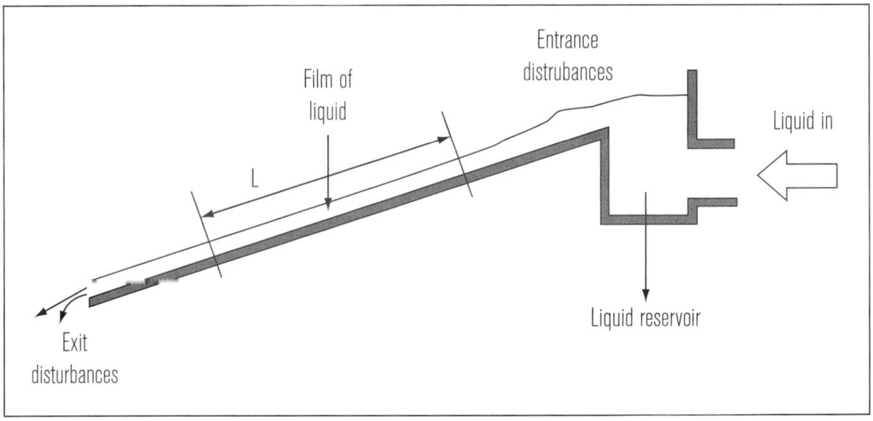

Figure 5.5 - Schematic view of the flow of a liquid in an inclined plane.

In the system shown in Figure 5.5, the flow of the liquid is caused by the gravity component in the direction parallel to the surface of the plane. The liquid layer at the interface with the solid remains stationary and the upper surface is in contact with the atmosphere.

As mentioned earlier, the first step to establish mass and momentum balances is to choose the coordinate axes to represent the geometry of the system. In Figure 5.6, two possible choices of axes are shown, one in black and the other in gray. As the fluid flows due to the component of gravity in the direction parallel to the surface of the plane, there is only driving force to create flow in this direction. Thus, the choice of axes in gray would lead to two velocity components. The vectorial sum of these two components would result in a velocity parallel to the inclined plane.

The choice of axes is based on the existence of only one component of velocity in the y direction. Obviously, this choice is more practical, because there is only one component of velocity, simplifying the mathematical treatment of the problem. It is important to mention that it is not always possible to choose a coordinate axes to provide only one component of velocity.

Having chosen the coordinate axes, the volume element to be taken as a reference for the establishment of mass and momentum balances can be defined. This element is chosen according to the coordinate axes as shown in Figure 5.6.

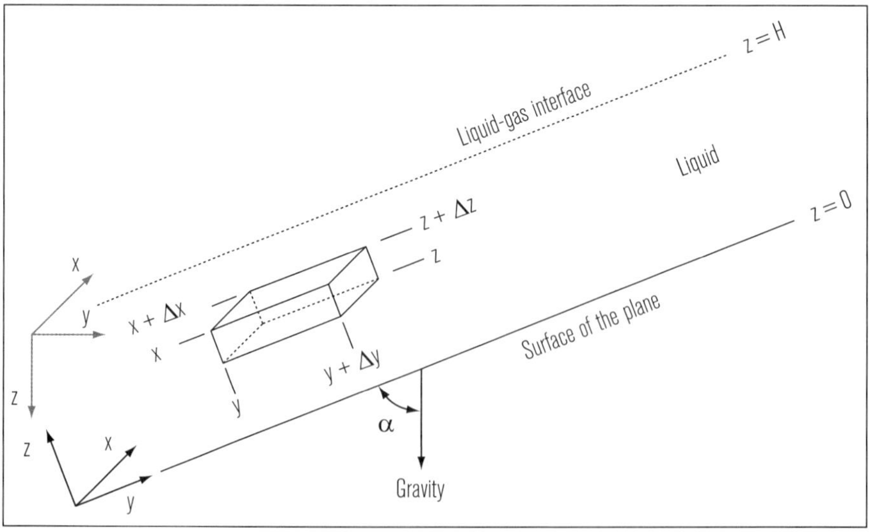

Figure 5.6 - Choice of coordinate axes for the study of flow on an inclined plane and volume element for the establishment of mass and momentum balances.

With the choice of axes above, the transport of momentum by convection occurs in the y direction, which is the direction of the macroscopic motion of the fluid. The transport of momentum by diffusion occurs in the z direction, which is the direction of the velocity gradient. It is easy to see that there is a velocity gradient in the z direction At z = 0 (surface of the plane), the velocity of the fluid is zero. At any other position z, the fluid is moving. Thus, there is a velocity gradient in the z direction (the velocity of the fluid varies with position z). Again, it is important to mention that it is not necessary to know where the velocity is higher or lower or the sense of the flow. It is sufficient to identify the directions of the macroscopic flow and of the velocity gradient. On the basis of the analysis above, one can carry out the mass and momentum balances.

5.3.2.1 Mass balance

Considering the direction of the macroscopic motion of the liquid, the mass balance can be written as follows:

$$\begin{aligned}&\left[\text{Rate of mass in across surface at } y = y\right] - \\ &\left[\text{Rate of mass in across surface at } y = y + \Delta y\right] = 0\end{aligned} \tag{5.43}$$

The surfaces considered in the evaluation of input and output of mass are determined by the orientation of the axes, and not according to the actual direction of the flow. In this case, considering the effect of gravity, the fluid should actually drain from top to downwards, which is the opposite direction to the growth of the y-axis. However, the inflows and outflows of mass are respectively at points y and y + Δy. The fact that the points of entry and exit of mass are defined by the choice of the orientation of the coordinate axes eliminates the need to know in advance the direction of flow. This facilitates the mathematical treatment of the problem in terms of assigning a positive or negative sign to the velocity terms. Velocities are positive when they are in the same direction in respect of the growth of the axis. Negative velocities are in the opposite direction to the growth of the axis.

The rates of entry and exit of mass can be evaluated by the following expressions:

$$\left[\text{Rate of mass in across the surface at } y = y\right] = (\Delta x\, \Delta z\, v_y\, \rho)\big|_{y=y} \tag{5.44}$$

$$\left[\text{Rate of mass in across the surface } y = y + \Delta y\right] = (\Delta x\, \Delta z\, v_y\, \rho)|_{y=y+\Delta y} \quad (5.45)$$

In the above, the product $\Delta x.\Delta z$ is the area of the volume element perpendicular to the direction of fluid flow (y direction), v_y is the velocity component in the y direction and ρ is the density of the fluid.

The product of these factors has the dimension of mass per unit time (mass flow rate).The product $\Delta x.\Delta z.v_y$ is called volumetric flow rate and has the dimension of volume per unit time.

The mass balance gives:

$$(\Delta x\, \Delta z\, v_y\, \rho)|_{y=y} - (\Delta x\, \Delta z\, v_y\, \rho)|_{y=y+\Delta y} = 0 \quad (5.46)$$

Again, following the same sequence of procedures listed in Section 5.2, the next step is to divide the equation above by the volume of the element, $\Delta x.\Delta y.\Delta z$:

$$\frac{(\Delta x\, \Delta z\, v_y\, \rho)|_{y=y} - (\Delta x\, \Delta z\, v_y\, \rho)|_{y=y+\Delta y}}{\Delta x\, \Delta y\, \Delta z} = 0 \quad (5.47)$$

$$\frac{(v_y\, \rho)|_{y=y} - (v_y\, \rho)|_{y=y+\Delta y}}{\Delta y} = 0 \quad (5.48)$$

Taking the limit of Δy tending to zero, the following differential equation is obtained:

$$\lim_{\Delta y \to 0}\left[\frac{(v_y\, \rho)|_{y=y} - (v_y\, \rho)|_{y=y+\Delta y}}{\Delta y} = 0\right] = -\frac{\partial(v_y\, \rho)}{\partial y} = 0 \quad (5.49)$$

It is important to remember that the negative sign in the preceding equation stems from the definition of first derivative.

Since density is constant,

$$\rho\, \frac{\partial v_y}{\partial y} = 0 \quad (5.50)$$

As the density of the fluid is not zero, it can be written that

$$\frac{\partial v_y}{\partial y} = 0 \quad (5.51)$$

The equation shows that the velocity v_y, in the present case, does not depend on the position y (v_y is not a function of y). Equation (5.51) is the main conclusion of the application of the principle of mass conservation.

5.3.2.2 Momentum balance

The momentum balance can be expressed by the following relationship:

$$\begin{aligned}&[\text{Rate of momentum in by convection across surface at } y = y\ ] - \\ &[\text{Rate of momentum out by convection across surface at } y = y + \Delta y\ ] + \\ &[\text{Rate of momentum in by diffusion across surface at } z = z\ ] - \\ &[\text{Rate of momentum out by diffusion across surface at } z = z + \Delta z\ ] + \\ &[\text{Sum of the forces acting on the volume element in the y-direction}] = 0\end{aligned} \quad (5.52)$$

The above equation considers the diffusive momentum in the direction of velocity gradient (z direction) and the transport of momentum by convection in the direction of macroscopic fluid motion.

As before, entry and exit points for the balance are chosen according to orientation of the coordinate axes. Knowledge beforehand of the flow directions is not necessary to establish the momentum balance. This is automatically determined during the course of the analysis of the problem in terms of sign accompanying the parameters. In this case, intuitively, it is known, that the fluid should flow down the slope as the flow is provoked by gravity forces. The terms in the equation (5.52) can be evaluated by:

$$[\text{Rate of momentum in by convection across surface at } y = y] = (\Delta x\, \Delta z\, v_y\, \rho\, v_y)|_{y=y} \quad (5.53)$$

$$[\text{Rate of momentum out by convection across surface at } y = y + \Delta y] = (\Delta x\, \Delta z\, v_y\, \rho\, v_y)|_{y=y+\Delta y} \quad (5.54)$$

$$[\text{Rate of momentum in by diffusion across surface at } z = z] = (\Delta x\, \Delta y\, \tau_{zy})|_{z=z} \quad (5.55)$$

$$[\text{Rate of momentum out by diffusion across surface at } z = z + \Delta z] = (\Delta x\, \Delta y\, \tau_{zy})|_{z=z+\Delta z} \quad (5.56)$$

In this system, the driving force for flow is the component of gravity acting in the direction of flow (y direction). The sum of the forces that have components in the direction of flow would include only the component of gravity in the y direction. The momentum balance (or balance of forces) being established in this case refers to the y direction, which is the direction of fluid flow. The sum of forces acting on the volume element is then given by:

$$[\text{Sum of forces acting on the volume element in the y direction}] = [\Delta x\, \Delta y\, \Delta z\, \rho(-g\cos\alpha)] \tag{5.57}$$

In equations (5.53) and (5.54), the product $\Delta x.\Delta z.v_y.\rho$ represents the mass flow rate. When this factor is multiplied by the velocity v_y, the rate of convective momentum is calculated. In equations (5.55) and (5.56), τ_{zy} represents the flux of diffusive momentum and when this flux is multiplied by the area normal to its direction, $\Delta x.\Delta y$, the rate of diffusive momentum is calculated.

In the equation (5.57), the product $\Delta x.\Delta y.\Delta z$ is the volume element (gravity is a force acting on the volume - see Chapter 3).The term $g\cos\alpha$ represents the component of gravity in the y direction, which is the direction of fluid flow. The negative sign is due to the fact that the component of gravity is in the opposite direction to the growth of y axis.

Substituting the expressions (5.53) to (5.57) in equation (5.52), a general equation for the momentum balance is obtained:

$$[(\Delta x\, \Delta z\, v_y\, \rho\, v_y)|_{y=y} - (\Delta x\, \Delta z\, v_y\, \rho\, v_y)|_{y=y+\Delta y}] + [(\Delta x\, \Delta y\, \tau_{zy})|_{z=z} - (\Delta x\, \Delta y\, \tau_{zy})|_{z=z+\Delta z}] + [\Delta x\, \Delta y\, \Delta z\, \rho(-g\cos\alpha)] = 0 \tag{5.58}$$

Dividing by the volume of the element,

$$\frac{[(\Delta x\, \Delta z\, v_y\, \rho\, v_y)|_{y=y} - (\Delta x\, \Delta z\, v_y\, \rho\, v_y)|_{y=y+\Delta y}]}{\Delta x\, \Delta y\, \Delta z} + \frac{[(\Delta x\, \Delta y\, \tau_{zy})|_{z=z} - (\Delta x\, \Delta y\, \tau_{zy})|_{z=z+\Delta z}] + [\Delta x\, \Delta y\, \Delta z\, \rho(-g\cos\alpha)]}{\Delta x\, \Delta y\, \Delta z} = 0 \tag{5.59}$$

$$\frac{[(v_y\, \rho\, v_y)|_{y=y} - (v_y\, \rho\, v_y)|_{y=y+\Delta y}]}{\Delta y} + \frac{[(\tau_{zy})|_{z=z} - (\tau_{zy})|_{z=z+\Delta z}]}{\Delta z} + [\rho\,(-g\cos\alpha)] = 0 \tag{5.60}$$

Taking the limit of equation (5.60) for the volume of the element tending to zero (Δy and $\Delta z \rightarrow 0$),

$$\lim_{\Delta y, \Delta z \to 0} \left\{ \frac{[(v_y\, \rho\, v_y)|_{y=y} - (v_y\, \rho\, v_y)|_{y=y+\Delta y}]}{\Delta y} + \frac{[(\tau_{zy})|_{z=z} - (\tau_{zy})|_{z=z+\Delta z}]}{\Delta z} + [\rho(-g\cos\alpha)] = 0 \right\} \tag{5.61}$$

and using the concept of the first derivative, the following differential equation results:

$$-\frac{\partial(v_y\ \rho\ v_y)}{\partial y}-\frac{\partial(\tau_{zy})}{\partial z}-\rho\ g\cos\alpha=0 \qquad (5.62)$$

From the mass balance, it is known that:

$$\frac{\partial v_y}{\partial y}=0 \qquad (5.51)$$

As the density of the liquid is constant, the convective terms can be simplified to:

$$\rho\ \frac{\partial(v_y\ v_y)}{\partial y}=2\ \rho\ v_y\ \frac{\partial(v_y)}{\partial y}=0 \qquad (5.63)$$

Combining the equations (5.51), (5.62) and (5.63), the momentum balance can finally be written as:

$$-\frac{d(\tau_{zy})}{dz}-\rho\ g\cos\alpha=\frac{d(\tau_{zy})}{dz}+\rho\ g\cos\alpha=0 \qquad (5.64)$$

This differential equation can be integrated by separation of variables, to obtain:

$$\int d(\tau_{zy})=\int-\rho\ g\cos\alpha\ dz \qquad (5.65)$$

$$\tau_{zy}=-\rho\ g\cos\alpha\ z+C_1 \qquad (5.66)$$

where C_1 is an integration constant.

In the system under consideration, the value of the constant C_1 can be determined using the boundary condition at the interface with air (z = H).

Bondary condition 1 : $\tau_{zy}\cong 0$ in $z=H$ (B.C. 1)

Applying this equation in (5.66), constant C1 can be determined:

$$0=-\rho\ g\cos\alpha\ H+C_1 \qquad \text{(B.C. 1)}$$

$$C_1=\rho\ g\cos\alpha\ H$$

Substituting the value of C_1 above in equation (5.66), it follows that:

$$\tau_{zy}=\rho\ g\cos\alpha\ (H-z) \qquad (5.67)$$

For the situation being studied, Newton's law of viscosity states:

$$\tau_{zy}=-\mu\ \frac{dv_y}{dz} \qquad (5.68)$$

Combining equations (5.67) and (5.68), the momentum flux by diffusion is

$$\tau_{zy} = -\mu \frac{dv_y}{dz} = \rho\, g \cos\alpha\, (H - z) \qquad (5.69)$$

Separating variables and integrating again gives the velocity profile:

$$\int dv_y = -\int \frac{\rho\, g \cos\alpha\, (H - z)}{\mu} dz \qquad (5.70)$$

$$v_y = -\frac{\rho\, g \cos\alpha}{\mu}\left(H\, z - \frac{z^2}{2}\right) + C_2 \qquad (5.71)$$

where C_2 is a new integration constant. The integration above was made assuming that the viscosity of the fluid is constant.

Equation (5.71) is the general equation for the velocity profile of a liquid flowing by gravity on an inclined plane. To have a specific profile as in the present case, the constant C_2 should be determined. This constant is evaluated using a no slip boundary condition. At the solid-liquid interface, for a stationary solid, the fluid velocity is equal to zero. Hence:

Bondary condition 2: $v_y = 0$ in $z = 0$ (B.C. 2)

Applyiing this boundary condition in equation 5.71, one can evaluate the value of the second integration constant gives.

$$0 = -\frac{\rho\, g \cos\alpha}{\mu}\left(H\, 0 - \frac{0^2}{2}\right) + C_2 \qquad \text{(B.C.2)}$$

$$C_2 = 0$$

Hence the velocity profile is given by the following equation:

$$v_y = -\frac{\rho\, g \cos\alpha}{\mu}\left(H\, z - \frac{z^2}{2}\right) \qquad (5.72)$$

(verify that the velocity profile equation satisfies the boundary conditions specified).

Equation (5.72) indicates that the velocity profile is parabolic, with only negative velocities. At $z = 0$, according to boundary condition 2, the velocity is zero. The velocities are negative as the y-axis is directed upwards, and the liquid flows down due to gravity. It is important to note that the flow direction appeared naturally in the development of the balance. Based on the velocity profile, a range of information about the system can be deduced.

The profile of shear stress (or flux of diffusive momentum) is expressed by equation (5.69). τ_{zy} varies linearly with z, and is zero at the interface with air (in accordance with the boundary condition 1). For all other points, τ_{zy} is positive. This diffusive momentum is transported along the z-axis from regions of high to low velocities. In this case, the higher velocity is at z = 0 (velocity is zero). In the region $0 < z \leq H$, the velocities are negative. The smallest velocity is at the liquid-air interface (z = H). (Remember that velocity is a vector).

The volumetric and mass flow rates can also be determined based on the velocity profile. Considering that the inclined plane has width W in the x direction, perpendicular to the plane of the paper, the flow rates are calculated by adding the amounts of fluid that flow in each infinitesimal area W.dz along the thickness of the liquid film. The volume flow rate, in each infinitesimal layer, is given by:

$$dQ = (W\, dz\, v_y) \qquad (5.73)$$

where dQ is the infinitesimal volume flow rate.

The sum of such infinitesimal portions between z = 0 and z = H. is given by integrating (5.73) between the limits specified:

$$\int_{z=0}^{z=H} dQ = \int_{z=0}^{z=H} (W\, dz\, v_y) \qquad (5.74)$$

As the fluid velocity varies with the position (equation (5.72)), the velocity profile is introduced in expression (5.74) to obtain:

$$Q = \int_{z=0}^{z=H} (W\, dz\, v_x) = -\int_{z=0}^{z=H} \left[\frac{W \rho g \cos\alpha}{\mu} \left(Hz - \frac{z^2}{2} \right) dz = -W \frac{\rho g \cos\alpha}{\mu} \frac{H^3}{3} \right] \qquad (5.75)$$

The negative sign in equation (5.75) is the result of the choice of coordinate axes in the system. The mass flow rate is given by:

$$\Gamma = Q \rho = -W \rho \frac{\rho g \cos\alpha}{\mu} \frac{H^3}{3} \qquad (5.76)$$

The equation for the volumetric flow rate can also be written as follows:

$$Q = \frac{(-g \cos\alpha)}{\left(\frac{\mu}{\rho}\right)} \left(W \frac{H^3}{3} \right) = \frac{(-g \cos\alpha)}{(\nu)} \left(W \frac{H^3}{3} \right) \qquad (5.77)$$

In this equation it is easy to identify the driving force for the flow ($-g\cos\alpha$), the term associated with resistance to flow ($\nu = \mu/\rho$), which is the kinematic viscosity of the fluid, and a term associated with the geometry of the system ($WH^3/3$).

Equation (5.77) is similar to the famous Ohm's law of electricity, which states that the current is the ratio of the electromotive force and the resistance. In equation (5.77), the flow rate Q is analogous to the current in electrical circuits, the driving force for the flow ($-g\cos\alpha$) to the electromotive force and the kinematic viscosity to the resistance. It is important to note that the driving force for flow is gravity and the property of the fluid that determines the flow rate is the kinematic viscosity (given by μ/ρ), and not the dynamic viscosity.

The average velocity is determined by dividing the volumetric flow rate by the total area available for flow, (W.H). Hence:

$$\overline{V} = \frac{Q}{W\,H} = -\frac{\rho g \cos\alpha}{\mu}\frac{H^2}{3} \qquad (5.78)$$

Figure 5.7 schematically shows the velocity profile for the system analyzed in this case.

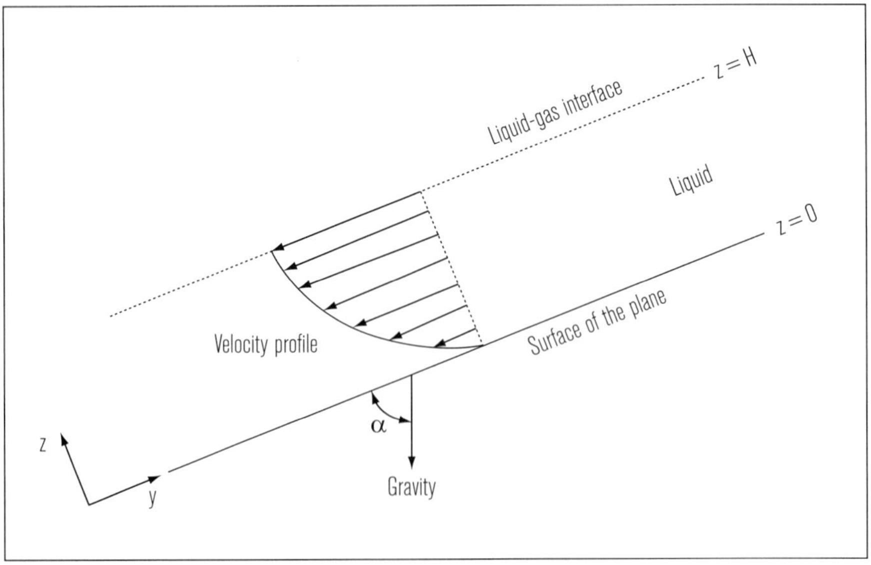

Figure 5.7 - Velocity profile of a liquid flowing in an inclined plane.

The force exerted by the fluid on the surface of the inclined plane in the direction parallel to this surface can be determined. This force is evaluated by integrating $\tau_{zy,}$ evaluated on the surface of the plane (position z = 0) along its width W and length L (double integral)

$$F_z = \int_{y=0}^{y=L} \int_{x=0}^{x=W} \tau_{zy}\big|_{z=0}\, dy\, dx = \int_{y=0}^{y=L} \int_{x=0}^{x=W} \rho\, g \cos\alpha\, H\, dy\, dx \qquad (5.79)$$

$$F_z = \rho\, g \cos\alpha\, H\, L\, W \qquad (5.80)$$

The force value above corresponds exactly to the component of the weight of the fluid in the direction parallel to the inclined plane. This result is compatible with the steady state conditions assumed.

All the relationships obtained are valid only when laminar flow conditions prevail. Low velocity viscous films of small thickness occur in laminar conditions. It was found experimentally that as the film velocity increases, the thickness H increases and the kinematic viscosity decreases, the nature of the flow changes. According to experimental results (BIRD, STEWART and LIGHTFOOT, 1960), flow on an inclined plane is laminar when the condition specified in equation (5.81), expressed in terms of Reynolds number, Re, is satisfied.

$$Re = \frac{H\, \rho\, \overline{V}}{\mu} \leq 6 \qquad (5.81)$$

Example

An oil with kinematic viscosity of 2 x 10^{-4} m^2/s and density of 800 kg/m^3 flows in the form of a film on a vertical wall. What should be the mass flow of the fluid per unit width of the wall, to have a film thickness of 2.5 mm?

Solution

The mass flow rate for this situation is given by

$$\Gamma = W\, \rho\, \frac{\rho g \cos\alpha}{\mu}\, \frac{H^3}{3}$$

For a vertical wall, $\alpha = 0$; W = 1 m (unit width), the equation for the mass flow rate can be written in terms of the kinematic viscosity as

$$\Gamma = \rho \frac{g}{\nu} \frac{H^3}{3}$$

Substituting values, the mass flow rate can be calculated:

- $\rho = 800 \text{ kg/m}^3$;
- $\nu = 2 \times 10^{-4} \text{ m}^2/\text{s}$;
- $H = 2.5 \text{ mm} = 2.5 \times 10^{-3} \text{ m}$.

$$\Gamma = 800 \frac{9.8}{(2 \times 10^{-4})} \frac{(2.5 \times 10^{-3})^3}{3} = 0.2042 \frac{\text{kg}}{\text{m s}}$$

This result is correct only if the flow is laminar. To verify this, the Reynolds number is calculated using equation (5.81):

$$Re = \frac{H \rho \overline{V}}{\mu} = \frac{H \overline{V}}{\nu}$$

The average velocity can be calculated by

$$\overline{V} = \frac{\Gamma}{\rho H W}$$

Therefore, the Reynolds number is given by (for W = 1 m):

$$Re = \frac{\Gamma}{\rho \nu} = \frac{0.2042}{(800)(2 \times 10^{-4})} = 1.276$$

As the Reynolds number is below the threshold for transition from laminar to turbulent, the flow is laminar and the solution is valid.

Example

Liquid slag flows above matte in a copper production unit (figure below) in a furnace 25 m long and 9 m wide. Assuming that the matte is stationary and that the slag flows continuously (flow rate $6.3 \times 10^{-4} \text{ m}^3/\text{s}$), determine:

- equation for the velocity distribution in the slag layer;
- fraction of material remaining in the furnace for a time greater than twice the mean residence time.The thickness of the slag is 0.6 m.

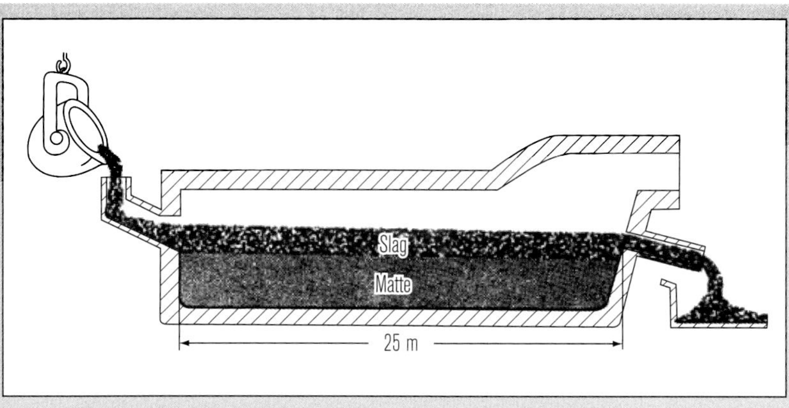

Solution

Assuming that the flow regime is laminar, the volumetric flow of slag is given by

$$Q = W \frac{\rho g \cos\alpha}{\mu} \frac{H^3}{3}$$

Rearranging the equation above, the following expression is obtained

$$\frac{\rho g \cos\alpha}{\mu} = \frac{3\,Q}{H^3\,W}$$

Substituting the expression above in equation (5.72), an expression for the velocity profile is determined (assuming that the flow occurs in the y direction):

$$v_y = \frac{\rho\, g \cos\alpha}{\mu}\left(Hz - \frac{z^2}{2}\right) = \frac{3\,Q}{H^3\,W}\left(Hz - \frac{z^2}{2}\right)$$

Substituting the data given,

$$v_y = \frac{(3)\,(6.3 \times 10^{-4})}{(0.6)^3\,(9)}\left(0.6\,z - \frac{z^2}{2}\right) = 9.72 \times 10^{-4}\,(0.6\,z - 0.5\,z^2)$$

This is the equation for the velocity profile for H = 0.6 m, W = 9 m, Q = 6.3 x 10^{-4} m^3/s;

The mean residence time of the slag in the furnace is given by:

$$\bar{t} = \frac{L}{\bar{\bar{V}}}$$

where L is the length of the inclined plane and $\overline{V}$, the average velocity.

From equation (5.78), the average velocity is given by:

$$\overline{V} = \frac{Q}{W\,H} = \frac{6.3 \times 10^{-4}}{(9)\,(0.6)} = 1.17 \times 10^{-4}\ \mathrm{m/s}$$

Hence, for the material to have at least twice the average residence time in the furnace, its velocity should be less than or equal to half the average velocity. Again from the equation for velocity profile arrived at above:

$$v_y == 9.72 \times 10^{-4}\,(0.6\,z - 0.5\,z^2) = \frac{1.17 \times 10^{-4}}{2}$$

Solution of equation above gives the point z, which is evaluated as

$$z = 0.1102\ \mathrm{m}$$

Since the thickness of the slag layer is 0.6 m, the fraction of material remaining in the furnace at least twice the average residence time is

$$\text{Fraction} = \frac{(0.1102 - 0)}{0.6} = 0.184 \text{ ou } 18.4\%$$

5.3.3 Fluid flow in a cylindrical duct

Fluid flow in a cylindrical pipe or tube is common and is frequently encountered in several engineering applications. For analysis of such systems, it is more convenient to use cylindrical coordinates, instead of cartesian coordinates (x, y and z), applied to the cases studied in previous sections.

The cylindrical coordinate system is shown in Figure 5.8. In this case, the z coordinate corresponds to the axial direction. The coordinate r corresponds to the distance to the origin of the axes and θ is the angle of rotation relative to a reference line. The position of a given point in a cylindrical coordinate system can then be specified in terms of the values of the coordinates r, θ and z. It is important to emphasize that r corresponds to the distance from the origin and never assumes values less than 0. Values of angle θ can vary between 0 and 2π.

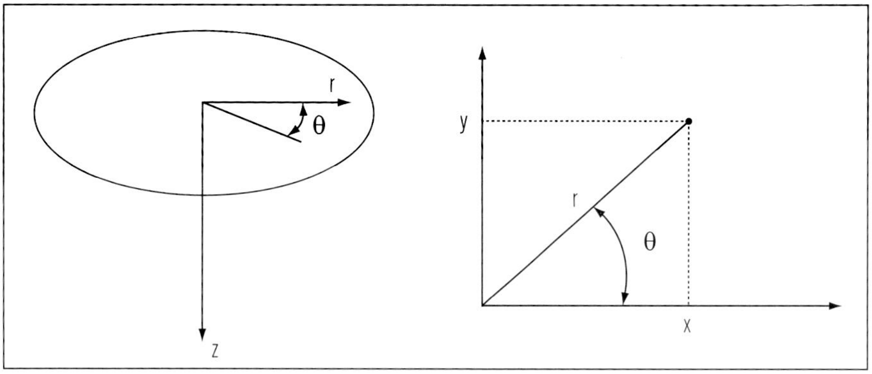

Figure 5.8 - Cylindrical coordinate system.

The use of cylindrical coordinates for the study of flows in tubes is normally preferred. The same problem can be solved using Cartesian coordinates, but the mathematical treatment would be more complicated.

In this section, the case of flow of a fluid in a vertical cylindrical duct under a pressure drop along its length will be analyzed. This system is schematically shown in Figure 5.9. The same sequence of steps adopted in the previous sections will be used here. The control volume to be used for the development of mass and momentum balances is the gray region indicated in Figure 5.9.

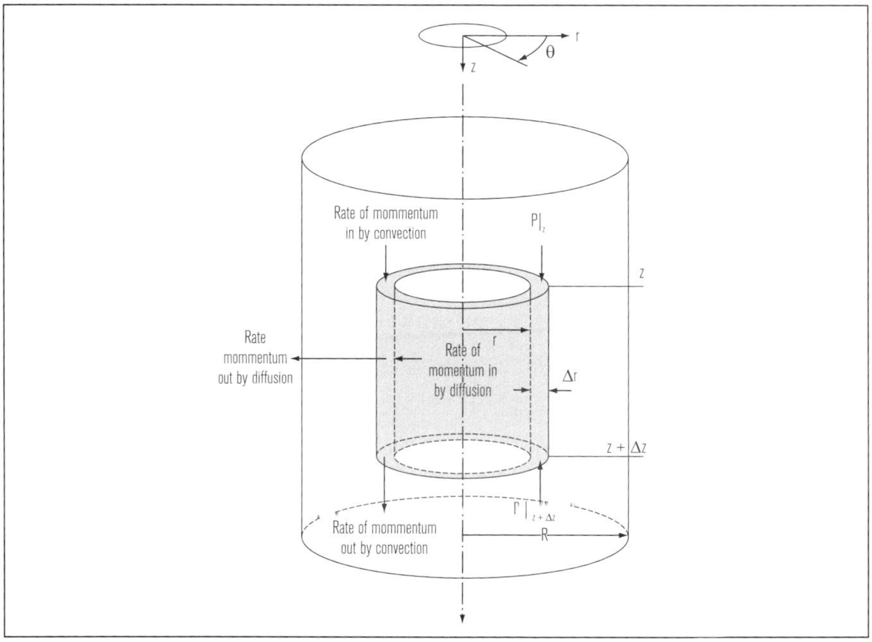

Figure 5.9 - Schematic view of the system to study the flow in cylindrical ducts.

Convective momentum transport occurs in the z direction (axial direction), which is the direction of macroscopic motion of the fluid. There is only driving forces (gravity and pressure difference) for flow in this direction. There is no movement in the radial and angular directions. Diffusive momentum transport occurs in the direction r that is in the direction of the velocity gradient. The existence of this gradient can be demonstrated by noting that at the wall of the tube (position r = R), the velocity of the fluid is zero. At any other point inside the tube ($r \neq R$), the fluid is moving. It is important to mention once again that there is no need to know in advance the sense of the velocity or the location of the maximum velocity, just the knowledge that there is a velocity gradient in the direction r is sufficient.

5.3.3.1 Mass balance

Considering the direction of macroscopic motion of fluid, mass balance can be written as follows:

$$\left[\text{Rate of mass in across the surface at } z = z\right] - \left[\text{Rate of mass out across the surface at } z = z + \Delta z\right] = 0 \tag{5.82}$$

As mentioned before, the points of input and output of mass are determined by the orientation of the axes, and not by the actual flow direction of the fluid. The fact that the points of entry and exit of mass are defined in such a manner eliminates the need to know a priori the direction of flow of the fluid.

The rates of entry and exit of mass can be evaluated by the following expressions:

$$\left[\text{Rate of mass in across the surface at } z = z\right] = (2\,\pi\,r\,\Delta r\,v_z\,\rho)|_{z=z} \tag{5.83}$$

$$\left[\text{Rate of mass out across the surface at } z = z + \Delta z\right] = (2\,\pi\,r\,\Delta r\,v_z\,\rho)|_{z=z+\Delta z} \tag{5.84}$$

In equations (5.83) and (5.84), the product $2\pi\ r\ \Delta r$ is the area of the volume element perpendicular to the direction of flow of the fluid (z direction), vz is the component of velocity and ρ is the density of the fluid. The product of these factors has the dimension of mass per unit time (mass flow rate). The product $2\ \pi r\ \Delta r$ vz is called volumetric flow and has dimension of volume per unit time.

The mass balance is continued with the following steps

$$(2\pi r \Delta r\, v_z\, \rho)|_{z=z} - (2\pi r \Delta r\, v_z\, \rho)|_{z=z+\Delta z} = 0 \tag{5.85}$$

$$\frac{(2\pi r \Delta r\, v_z\, \rho)|_{z=z} - (2\pi r \Delta r\, v_z\, \rho)|_{z=z+\Delta z}}{2\pi r\, \Delta r\, \Delta z} = 0 \tag{5.86}$$

$$\frac{(v_z\, \rho)|_{z=z} - (v_z\, \rho)|_{z=z+\Delta z}}{\Delta z} = 0 \tag{5.87}$$

Following the same sequence of procedures described earlier, (dividing by volume element, $2\ \pi\ r\ \Delta r\ \Delta z$ and applying limit, Δz tends to zero), we get:

$$\lim_{\Delta z \to 0}\left[\frac{(v_z\, \rho)|_{z=z} - (v_z\, \rho)|_{z=z+\Delta z}}{\Delta z} = 0\right] = -\frac{\partial(v_z\, \rho)}{\partial z} = 0 \tag{5.88}$$

The negative sign in the equation above comes from the definition of first derivative.

$$\frac{df}{dz} = \lim_{\Delta z \to 0}\left[\frac{(f)|_{z=z+\Delta z} - (f)|_{z=z}}{\Delta z}\right] \tag{5.89}$$

where f is any function. In equation (5.88), this function is the product $v_z\ \rho$. Note that the limits of the evaluation function f in the definition of derivative are the opposite of that that appears in equation (5.88).

$$\rho\ \frac{\partial v_z}{\partial z} = 0 \tag{5.90}$$

Fluid density is constant, hence:

$$\frac{\partial(v_z)}{\partial z} = 0 \tag{5.91}$$

The above equation states that the velocity v_z in the situation being studied is independent of position z (v_z is not a function of z).

5.3.3.2 Momentum balance

Considering transport of momentum by diffusion in the direction of velocity gradient (direction r) and the transport of momentum by convection in the direction of movement of the fluid (z direction), the balance of transfer of rate of momentum can be written as:

$$\begin{aligned}&\left[\text{Rate of momentum in by convection across the surface at } z = z\right] - \\ &\left[\text{Rate of momentum out by convection across the surface at } z = z + \Delta z\right] + \\ &\left[\text{Rate of momentum in by diffusion across the surface at } r = r\right] - \\ &\left[\text{Rate of momentum out by diffusion across the surface at } r = r + \Delta r\right] + \\ &\left[\text{Sum of forces acting on the volume element in the z direction}\right] = 0\end{aligned} \quad (5.92)$$

Remembering that the points of entry and exit of momentum by convection and diffusion are determined by the choice of orientation of the coordinate axes, and the fact that it is not necessary to know *a priori* the directions of momentum flow, the following equations can be written:

$$\left[\text{Rate of momentum in by convection across the surface at } z = z\right] = (2\pi r \Delta r\, v_z \rho\, v_z)|_{z=z} \quad (5.93)$$

$$\left[\text{Rate of momentum out by convection across the surface at } z = z + \Delta z\right] = (2\pi r \Delta r\, v_z \rho\, v_z)|_{z=z+\Delta z} \quad (5.94)$$

$$\left[\text{Rate of momentum in by diffusion across the surface at } r = r\right] = (2\pi r \Delta z\, \tau_{rz})|_{r=r} \quad (5.95)$$

$$\left[\text{Rate of momentum out by diffusion across the surface at } = r + \Delta r\right] = (2\pi r \Delta z\, \tau_{rz})|_{r=r+\Delta r} \quad (5.96)$$

$$[\text{Sum of forces acting on the volume element in the z direction}] = +(2\pi r \Delta r\, \Delta z\, \rho\, g) + 2\pi r \Delta r\,(P|_z - P|_{z+\Delta z}) = 0 \quad (5.97)$$

In equation (5.97), the sum of forces referred only to the forces that have components in the direction of flow of fluid. It is important to remember that the amount of momentum is a vector. The balance of momentum (or balance of forces) being established in this case refer to the direction z. In equations (5.93) and (5.94), the term $2\ \pi\ r\ \Delta r\ v_z\ \rho$ represents the mass flow rate. When this factor is multiplied by the velocity v_z, momentum per unit time, or force is the result. In equations (5.95) and (5.96), τ_{rz} represents the rate of diffusive momentum. When this flux is multiplied by the area to its normal direction, $2\pi\ r\ \Delta z$, we obtain the total rate of diffusive momentum in the control volume.

In equation (5.97), the first term in the right side refers to the force of gravity (which is force that acts on the volume of the element, $2\ \pi\ r\ \Delta r\ \Delta z$). When this is multiplied by density, ρ, we obtain the mass of the volume element. Finally, multiplying this mass by the acceleration of gravity (g), the force of gravity is determined. The second term of the right side is the force due to pressure difference. This force acts perpendicular (pressure is always associated with a normal force) to the area 2π r. Δ r of the volume element. The product of the area by the pressure difference gives the force term due to pressure difference. Substituting the expressions (5.93) to (5.97) in equation (5.92), one can obtain the general equation of balance of momentum for the problem under consideration:

$$\begin{aligned}&[(2\pi r \Delta r\, v_z\, \rho v_z)|_{z=z} - (2\pi r \Delta r\, v_z\, \rho v_z)|_{z=z+\Delta z}] + [(2\pi r \Delta z\, \tau_{rz})|_{r=r} \\ &- (2\pi r \Delta z\, \tau_{rz})|_{r=r+\Delta r}] + (2\pi r \Delta r \Delta z \rho g) + 2\pi r \Delta r\, (P|_z - P|_{z+\Delta z}) = 0\end{aligned} \tag{5.98}$$

Dividing by the volume element, we obtain:

$$\begin{aligned}&\frac{[(2\pi r \Delta r\, v_z\, \rho v_z)|_{z=z} - (2\pi r \Delta r\, v_z\, \rho v_z)|_{z=z+\Delta z}]}{2\pi r \Delta r \Delta z} + \frac{[(2\pi r \Delta z\, \tau_{rz})|_{r=r} - (2\pi r \Delta z\, \tau_{rz})|_{r=r+\Delta r}]}{2\pi r \Delta r \Delta z} + \\ &\frac{(2\pi r \Delta r \Delta z\, \rho\, g) + 2\pi r \Delta r\, (P|_z - P|_{z+\Delta z})}{2\pi r \Delta r \Delta z} = 0\end{aligned} \tag{5.99}$$

$$\frac{[(v_z\, \rho v_z)|_{z=z} - (v_z\, \rho v_z)|_{z=z+\Delta z}]}{\Delta z} + \frac{[(r\, \tau_{rz})|_{r=r} - (r\, \tau_{rz})|_{r=r+\Delta r}]}{r\, \Delta r} + (\rho g) + \frac{(P|_z - P|_{z+\Delta z})}{\Delta z} = 0 \tag{5.100}$$

In the second part of the equation (5.100), the radius *r* in the numerator and the denominator cannot be cancelled. In the numerator, the radius is measured in two different positions r and r + Δr. The term $(r\, \tau_{rz})\,|_{r+\Delta r}$ is the same as $(r + \Delta r)\, (\tau_{rz}\,|_{r+\Delta r})$. Thus, the simplest way to make the mathematical development, presented below, is to leave the term r in the denominator.

Taking the limits of equation (5.100) when the volume of the element tends to zero (Δr and Δz $\rightarrow$ 0), we have:

$$\lim_{\Delta r, \Delta z \to 0} \left\{ \begin{aligned} &\frac{[(v_z\, \rho v_z)|_{z=z} - (v_z\, \rho v_z)|_{z=z+\Delta z}]}{\Delta z} + \frac{[(r\, \tau_{rz})|_{r=r} - (r\, \tau_{rz})|_{r=r+\Delta r}]}{r\, \Delta r} + \\ &(\rho\, g) + \frac{(P|_z - P|_{z+\Delta z})}{\Delta z} \end{aligned} \right\} = 0 \tag{5.101}$$

Using the concept of first derivative one gets:

$$-\frac{\partial(v_z\, \rho v_z)}{\partial z} - \frac{\partial(r\, \tau_{rz})}{r\, \partial r} + \rho\, g - \frac{\partial P}{\partial z} = 0 \tag{5.102}$$

From the mass balance, it is known that:

$$\frac{\partial(v_z)}{\partial z} = 0 \tag{5.91}$$

As the fluid has constant density, one can write:

$$\rho \frac{\partial(v_z\, v_z)}{\partial z} = 2\rho v_z \frac{\partial(v_z)}{\partial z} = 0 \tag{5.103}$$

Hence,

$$-\frac{\partial(r\, \tau_{rz})}{r\, \partial r} + \rho\, g - \frac{\partial P}{\partial z} = 0 \tag{5.104}$$

In the partial derivatives above, τ_{rz} varies with r and pressure with z. For a length L of the tube, the pressure gradient in equation (5.104) can be written as follows:

$$-\frac{\partial P}{\partial z} = \frac{P_o - P_L}{L} \tag{5.105}$$

where P_0, P_L are the pressure at points $z = z_0$ and $z = z_{L,}$ respectively.

Substituting equation (5.105) in (5.104), the differential equation governing the flow of fluid in a tube is determined:

$$-\frac{\partial(r\, \tau_{rz})}{r\, \partial r} + \rho\, g + \frac{P_o - P_L}{L} = 0 \tag{5.106}$$

This equation can be integrated by separation of variables, to obtain:

$$\int d(r\, \tau_{rz}) = \int \left(\rho\, g + \frac{P_o - P_L}{L}\right) r\, dr \tag{5.107}$$

$$(r\, \tau_{rz}) = \left(\rho\, g + \frac{P_o - P_L}{L}\right) \frac{r^2}{2} + C_1 \tag{5.108}$$

where C_1 is an integration constant.

Equation (5.108) is valid for the following range of values of r:

$$0 \le r \le R$$

For this equation to be valid at r = 0, the value of C_1 must be equal to 0. This is true provided that τ_{rz}, ρ.g and $(P_0 - P_L)/L$ are finite numbers. This condition is usually satisfied and thus C_1 can be assumed to be equal to 0.

With this value of C_1, equation (5.108) can be written as follows:

$$(\tau_{rz}) = \left(\rho\, g + \frac{P_o - P_L}{L}\right)\frac{r}{2} \tag{5.109}$$

For this system, Newton´s law of viscosity is written as

$$\tau_{rz} = -\mu\,\frac{dv_z}{dr} \tag{5.110}$$

Combining equations (5.109) and (5.110), an expression for the velocity gradient is determined:

$$\tau_{rz} = -\mu\,\frac{dv_z}{dr} = \left(\rho\, g + \frac{P_o - P_L}{L}\right)\frac{r}{2} \tag{5.111}$$

Separation of variables and integration yields

$$\int dv_z = \int -\frac{1}{\mu}\left(\rho\, g + \frac{P_o - P_L}{L}\right)\frac{r\,dr}{2} \tag{5.112}$$

$$v_z = -\frac{1}{\mu}\left(\rho\, g + \frac{P_o - P_L}{L}\right)\frac{r^2}{4} + C_2 \tag{5.113}$$

where C_2 is a new integration constant. The integration above is made assuming that the viscosity and density of the fluid are constant.

Equation (5.113) provides a generic equation for the velocity profile within a cylindrical duct. To have a specific profile for the case being studied, the value of constant C_2. must be determined. This constant is determined using the boundary condition below, which implies that at solid-liquid interfaces, fluid velocity equals the velocity of the solid, which in this case is zero.

Boundary condition: $v_z = 0$ at $r = R$

Applying this boundary condition in the equation of velocity profile, we have:

$$0 = -\frac{1}{\mu}\left(\rho\, g + \frac{P_o - P_L}{L}\right)\frac{R^2}{4} + C_2$$

From the equation above, one can determine the value of C_2 and finally obtain the expression for the velocity profile:

$$v_z = \frac{1}{\mu}\left(\rho\, g + \frac{P_o - P_L}{L}\right)\frac{(R^2 - r^2)}{4} \tag{5.114}$$

(verify that the profile meets the above boundary conditions)

Equation (5.114) represents a parabolic profile. As the viscosity of the fluid has a value always greater than zero and the term (R^2 - r^2) is greater than or equal to zero, the sense of the velocity of the fluid, down or up (positive or negative velocities, respectively), depends solely on the sign of the term in square brackets. Note that:

$$\left[\rho\, g + \frac{P_o - P_L}{L}\right] > 0 \text{ positive velocities (fluid goes down)}$$

$$\left[\rho\, g + \frac{P_o - P_L}{L}\right] < 0 \text{ negative velocities (fluid goes up)}$$

$$\left[\rho\, g + \frac{P_o - P_L}{L}\right] = 0 \text{ velocities equal to zero (fluid is stationary)}$$

This term represents the driving force for the occurrence of the flow. When there is no pressure difference, the only driving force is gravity and the velocities are positive. In this case, the velocity profile is given by:

$$v_z = \frac{\rho\, g}{\mu}\frac{(R^2 - r^2)}{4} = \frac{g}{\left(\frac{\mu}{\rho}\right)}\frac{(R^2 - r^2)}{4} = \frac{g}{\nu}\frac{(R^2 - r^2)}{4} \qquad (5.115)$$

For the above equation, it is evident that in the case of a fluid flowing due to gravity, the velocity is inversely proportional to the kinematic viscosity of the fluid. This property determines the resistance to flow. Higher the kinematic viscosity, greater is the resistance to flow resulting in lower velocities. An interesting point to note is that liquid steel and water have almost equal values of kinematic viscosities (verify this using the data from Chapter 4).

Equation (5.114) can be easily modified for the case of inclined pipes. If β is the angle that the tube makes with the vertical, the velocity profile is given by:

$$v_z = \frac{1}{\mu}\left(\rho\, g \cos\beta + \frac{P_o - P_L}{L}\right)\frac{(R^2 - r^2)}{4} \qquad (5.116)$$

When $\beta = 90^o$, the tube is horizontal and the effect of gravity is canceled. The only driving force for the flow in this case would be the difference in pressure.

From the velocity profile above, other parameters of the system can be obtained. From equation (5.114), it is observed that the maximum velocity of the fluid occurs in the center of the tube (r = 0) and is given by:

$$v_z^{max} = \frac{1}{\mu}\left(\rho\, g + \frac{P_o - P_L}{L}\right)\frac{R^2}{4} \tag{5.117}$$

Using this expression for the maximum velocity, the velocity profile can be rewritten in the following way:

$$\frac{v_z}{v_z^{max}} = 1 - \left(\frac{r}{R}\right)^2 \tag{5.118}$$

Equation (5.109) provides the profile of the flux of diffusive momentum or shear stress along the radius of the tube. In the duct wall (r = R), this shear stress is given by:

$$(\tau_{rz})^{wall} = \left(\rho\, g + \frac{P_o - P_L}{L}\right)\frac{R}{2} \tag{5.119}$$

The profile of shear stress can then be expressed as follows:

$$\frac{(\tau_{rz})}{(\tau_{rz})^{wall}} = \frac{r}{R} \tag{5.120}$$

Figure 5.10 schematically shows the velocity and shear stress profiles along the radius of the tube, as indicated by equations (5.118) and (5.120), respectively.

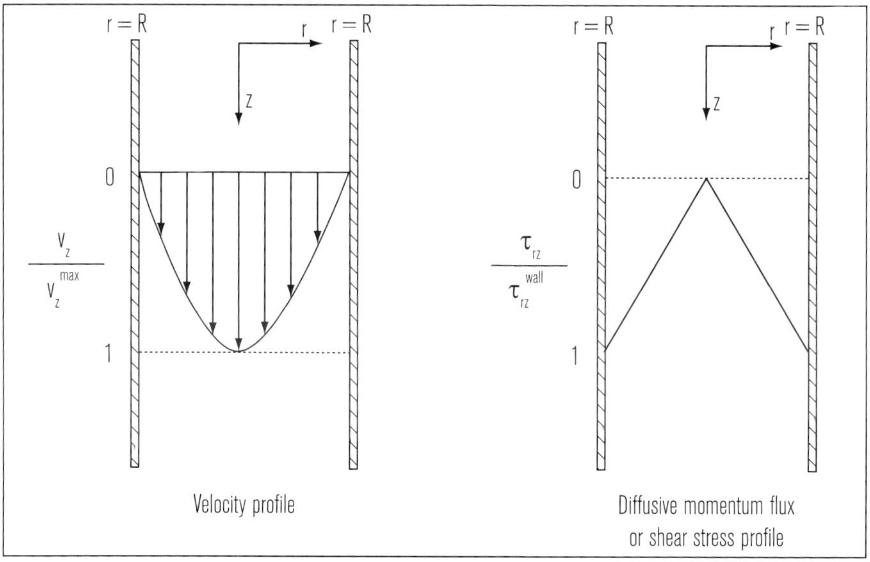

Figure 5.10 - Velocity and shear stress profiles along the radius of a cylindrical duct.

In addition to the velocity and shear stress profiles, the volume and mass flow rates of fluid inside the tube can de determined. The procedure adopted is similar to that used in the cases analyzed before. The flow rates are calculated by adding the amounts of fluid that flow in each infinitesimal portion of the area along the radius of the tube. Since the velocity depends only on the radial position, this infinitesimal element of area would be a small ring of radial thickness dr. For each infinitesimal layer, the amount of fluid that flows is given by:

$$dQ = (2\pi r\, dr\, v_z) \tag{5.121}$$

where dQ is the infinitesimal flow rate over a ring of area equal to $2\pi r\, dr$.

Integrating equation (5.121) between the center and wall of the tube, the volumetric flow of the fluid is determined:

$$\int_{r=0}^{r=R} dQ = \int_{r=0}^{r=R} (2\pi r\, dr\, v_z) \tag{5.122}$$

As the fluid velocity varies with the position (equation (5.114)), the velocity profile is introduced in expression (5.122) to obtain:

$$\int_{r=0}^{r=R} dQ = \int_{r=0}^{r=R} 2\pi r \frac{1}{\mu}\left(\rho g + \frac{P_o - P_L}{L}\right)\frac{(R^2 - r^2)}{4}\, dr \tag{5.123}$$

$$Q = \frac{\pi R^4}{8\mu}\left(\rho g + \frac{P_o - P_L}{L}\right) \tag{5.124}$$

Equation (5.124) is known as Hagen-Poiseuille equation.

The mass flow rate is given by the product of volumetric flow rate and the density of the fluid:

$$\Gamma = \rho Q = \frac{\pi \rho R^4}{8\mu}\left(\rho g + \frac{P_o - P_L}{L}\right) \tag{5.125}$$

The average velocity is determined by dividing the volumetric flow rate by the total area available for flow. In this case, this area is the cross-sectional area of the tube (πR^2). Thus,

$$\overline{V} = \frac{R^2}{8\mu}\left(\rho g + \frac{P_o - P_L}{L}\right) \tag{5.126}$$

Another parameter that can be evaluated from the data above is the vertical component (z direction) of the force that the fluid exerts on the wall of the tube. This force corresponds to the product of the shear stress at the interface fluid-wall surface area of the tube over a length L.

$$F_z = 2\pi R L (\tau_{rz})^{wall} = 2\pi R L \left(\rho g + \frac{P_o - P_L}{L} \right) \frac{R}{2} \quad (5.127)$$

$$F_z = \pi R^2 L \rho g + \pi R^2 (P_o - P_L) \quad (5.128)$$

In equation (5.128), the first term of the right-hand side is the force associated with gravity, and the second term is the force related to the pressure difference. Equation (5.128) shows that the friction force between the fluid and the wall is the sum of the contributions due to the gravity and the force associated with the pressure difference, i.e., the forces that cause the flow are equal to the force that resists the movement of fluid. Overall, the sum of forces acting on the fluid is zero and the fluid moves with a constant velocity (according to Newton's second law).

All the relationships obtained in this section are valid for the case of laminar flow in pipes. This flow regime prevails when the condition below is satisfied.

$$Re = \frac{D \bar{V} \rho}{\mu} < 2100 \quad (5.129)$$

where D is the diameter of the tube.

In this analysis, the effects of inflow and outflow of fluid in the duct have been neglected. In fact, the effects of entry are significant in lengths less than L_e that can be estimated by the following expression:

$$L_e = 0.035\, D\, Re \quad (5.130)$$

Example

Water flows through a horizontal pipe with a diameter of 1.59 x 10^{-3} under a pressure difference of 904.82 Pa/m. Determine the mass flow of water in the tube.

Data:

- density of water = 1,000 kg/m^3;
- viscosity of water = 0.001 Pa.s.

Solution

Using the equation (5.125) for the mass flow rate in a horizontal pipe:

$$\Gamma = \frac{\pi \rho R^4}{8 \mu} \left(\frac{P_o - P_L}{L} \right)$$

And replacing data in the above relation, we obtain:

$$\Gamma = \frac{\pi (1000) \left(\frac{1.59 \times 10^{-3}}{2} \right)^4}{8 (0.001)} (904.82) = 1.42 \times 10^{-4} \text{ kg/s}$$

Finally, the validity of the calculation should be verified. For this, the Reynolds number is calculated. Combining the expressions for the mass flow rate, average velocity and the definition of the Reynolds number for tubes, we have:

$$Re = \frac{4 \Gamma}{\pi D \mu} = \frac{4 (1.42 \times 10^{-4})}{\pi (1.59 \times 10^{-3}) (0.001)} = 113.7$$

Since this value is less than the transition value from laminar to turbulent flow (Re <2100), the calculation method is valid.

Demonstration

Use the concepts of mass and momentum balances to show that, for the system seen in the figure below, the velocity profile and flow rate of fluid flowing between the plates are expressed by:

$$v_x = \left(\frac{P_o - P_L}{L} \right) \frac{\delta^2}{2 \mu} \left[1 - \left(\frac{y}{\delta} \right) \right]^2$$

$$Q = \frac{2}{3} \frac{W \delta^3}{\mu} \left(\frac{P_o - P_L}{L} \right)$$

where W is the width of the plates.

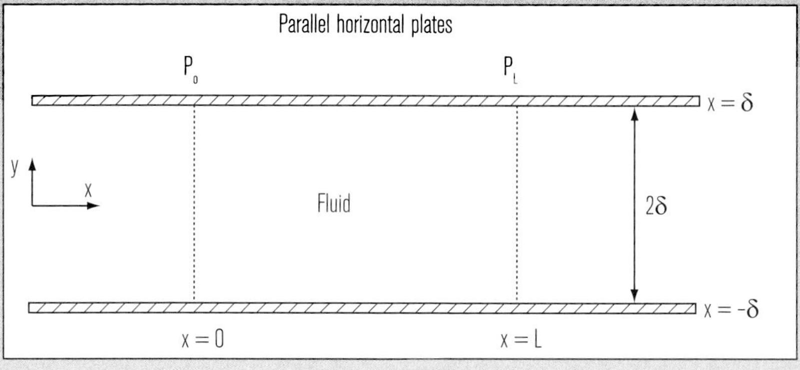

5.3.4 Flow of fluid in concentric duct

In many situations of practical interest, one comes across flow of fluids in concentric tubes. A typical example is the cooling of the lance used for oxygen injection in the BOF steelmaking. In this case, oxygen is transported through a central tube, which is surrounded by other tubes of larger diameter, within which circulates the cooling water.

Figure 5.11 shows a schematic view of the configuration to be studied. The fluid is flowing in the annular area between the tubes of smaller and larger diameters, in the z direction. The driving forces for the flow are the force of gravity and the force due to the pressure difference. The system is similar to that shown earlier in Figure 5.9. The volume element considered in this case is the same as that used in the mass and momentum balance, described in section 5.3.3. This element is highlighted in gray in Figure 5.11.

As the volume element above is analogous to the case studied in the previous section, there is no need to repeat the entire procedure of mass and momentum balances. The starting point for the treatment of this problem

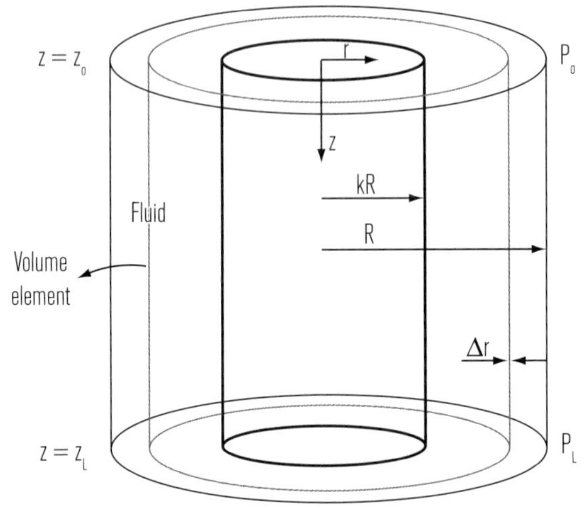

Figure 5.11 - Schematic view of the system of concentric cylindrical tubes and the volume element used in the mass and momentum balances.

is equation (5.108), which provides the profile of shear stress or flux of diffusive momentum in the fluid. This equation is reproduced below:

$$(r\,\tau_{rz}) = \left(\rho\, g + \frac{P_o - P_L}{L}\right)\frac{r^2}{2} + C_1 \tag{5.108}$$

In the case discussed earlier, this equation was valid for the following range values of radial position:

$$0 \le r \le R$$

The condition of finite τ_{rz} at $r = 0$ served as boundary condition to determine the value of C_1.

In the system of concentric tubes, the region of interest for the analysis is defined by the following expression:

$$kR \le r \le R$$

Obviously, this region does not include the point $r = 0$ and thus the boundary condition of the earlier case cannot be applied for the determination of C_1. The value of this constant is obtained by replacing Newton's law of viscosity in the equation (5.108) and integrating to obtain the velocity profile.

Combining equations (5.109) and (5.110), we obtain:

$$\tau_{rz} = -\mu \frac{dv_z}{dr} = \left(\rho g + \frac{P_o - P_L}{L}\right)\frac{r}{2} + \frac{C_1}{r} \tag{5.131}$$

Separating variables and integrating, we have:

$$\int dv_z = \int -\frac{1}{\mu}\left(\rho g + \frac{P_o - P_L}{L}\right)\frac{r\,dr}{2} - \frac{C_1}{\mu\, r}\,dr \tag{5.132}$$

$$v_z = -\frac{1}{\mu}\left(\rho g + \frac{P_o - P_L}{L}\right)\frac{r^2}{4} - \frac{C_1}{\mu}\ln r + C_2 \tag{5.133}$$

where C_2 is a new integration constant. The integration above assumes that viscosity and density of the fluid are constant. The constants C_1 and C_2 are determined from boundary conditions. Considering fluid velocities at the solid-liquid interfaces to be zero, one can define the two boundary conditions:

Boundary condition 1 : $v_z = 0$ at $r = kR$

Boundary condition 2 : $v_z = 0$ at $r = R$

Applying these boundary conditions, the velocity profile equation can be written as:

$$0 = -\frac{1}{\mu}\left[\rho g + \frac{P_o - P_L}{L}\right]\frac{(kR)^2}{4} - \frac{C_1}{\mu}\ln(kR) + C_2$$

$$0 = -\frac{1}{\mu}\left[\rho g + \frac{P_o - P_L}{L}\right]\frac{(R)^2}{4} - \frac{C_1}{\mu}\ln(R) + C_2$$

From the equations above, one can determine the values of C_1 and C_2:

$$C_1 = \left(\rho g + \frac{P_o - P_L}{L}\right)\frac{R^2}{4}\,\frac{(1-k^2)}{\ln k} \tag{5.134}$$

$$C_2 = \left(\rho g + \frac{P_o - P_L}{L}\right)\frac{R^2}{4\mu}\left(1 + \frac{(1-k^2)}{\ln k}\ln R\right) \tag{5.135}$$

Substituting C_1 and C_2 in the equation for the velocity profile and rearranging:

$$v_z = \left(\rho g + \frac{P_o - P_L}{L}\right)\frac{R^2}{4\mu}\left[1 - \left(\frac{r}{R}\right)^2 + \frac{(1-k^2)}{\ln 1/k}\ln\left(\frac{r}{R}\right)\right] \tag{5.136}$$

(verify that this profile satisfies the boundary conditions).

Knowing the velocity profile, other parameters of importance can be determined. Initially, one can determine the position of the maximum (or minimum, depending on the sense of the flow) velocity as well as its value. The first derivative of the velocity with respect to the radial position is zero at the maximum velocity location. This location is determined as:

$$r_{max} = R\sqrt{\frac{(1 - k^2)}{2\ln(1/k)}} \tag{5.137}$$

The value of the maximum (or minimum velocity) is given by:

$$v_{z_{max}} = \left(\rho g + \frac{P_o - P_L}{L}\right)\frac{R^2}{4\mu}\left[1 - \frac{(1-k^2)}{2\ln(1/k)} + \frac{(1-k^2)}{2\ln(1/k)}\ln\left(\frac{(1-k^2)}{2\ln(1/k)}\right)\right] \tag{5.138}$$

Using the procedure similar to that given in the previous sections, one can evaluate the volumetric flow rate of the fluid:

$$Q = \int_{r=kR}^{r=R} 2\pi r v_z\, dr = \left(\rho g + \frac{P_o - P_L}{L}\right)\frac{R^4}{8\mu}\left[1 - k^4 - \frac{(1-k^2)^2}{\ln(1/k)}\right] \tag{5.139}$$

The average velocity can be determined by the following expression:

$$\overline{V_z} = \frac{Q}{\pi R^2(1-k^2)} = \left(\rho g + \frac{P_o - P_L}{L}\right)\frac{R^2}{8\mu}\left[\frac{(1-k^4)}{(1-k^2)} - \frac{(1-k^2)}{\ln(1/k)}\right] \tag{5.140}$$

It is interesting to observe that the equations obtained for the flow in concentric ducts approach those for flow in pipes when k approaches zero. (Verify this as an exercise).

The relationships developed in this item are valid for laminar flow. In this case, the laminar flow prevails when the Reynolds number, defined by the equation below, is less than 2000:

$$Re = \frac{D(1-k)\overline{V_z}\rho}{\mu} < 2000 \tag{5.141}$$

where D is the diameter of the outer tube.

5.3.5 Laminar two-phase flow

Flow of two immiscible fluids of different densities and viscosities occur in many practical applications. This is termed as two-phase flow. A typical example of this type is the flow of a slag along with liquid metal encountered frequently in the metal extraction processes like that of pig iron and slag during tapping of blast furnaces. In this case, the liquid metal flows at the bottom of the channel, and the lighter slag moves over the surface of liquid iron.

What appears new in the analysis of this type of system are the boundary conditions that must be applied in the interface between the two fluids. Other procedures such as determination of general velocity profiles and diffusive and convective momentum fluxes are same as discussed in earlier sections. At the interface between two liquids:

- Velocity of fluid A = Velocity of fluid B;
- Diffusive momentum flux in fluid A = Diffusive momentum flux in fluid B.

These two boundary conditions can be used to determine the velocity profile of the liquids in this two phase system shown in Figure 5.12. Mass

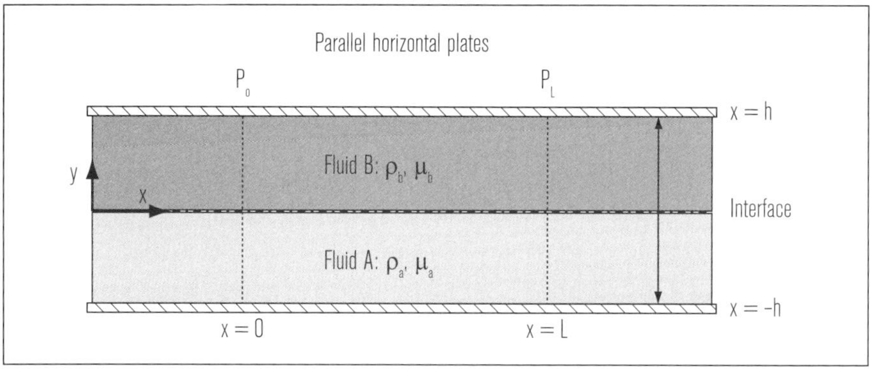

Figure 5.12 - Two phase flow between two flat horizontal parallel plates.

and momentum balances can be developed to determine the velocity profile of liquid.

Flow is occurring in the x direction, and is due only to pressure difference.

For the mass balance, it is known that (see equation (5.16) and others developed previously):

$$\frac{\partial v_x}{\partial x} = 0 \qquad (5.142)$$

From a similar example described earlier, the diffusive momentum fluxes of the two fluids are:

$$\tau_{yx}^{A} = \left(\frac{P_o - P_L}{L}\right) y + C_1^A \tag{5.143}$$

$$\tau_{yx}^{B} = \left(\frac{P_o - P_L}{L}\right) y + C_1^B \tag{5.144}$$

As the profiles of shear stress in the two liquids are not necessarily equal, one should write an equation for each fluid, with their respective constants of integration·. It is known that the interface between the liquids (position y = 0), the two momentum fluxes (or shear stress) are equal. Therefore,

$$y = 0: \ \tau_{yx}^{A} = \tau_{yx}^{B} \qquad \text{(B.C.1)}$$

$$C_1^A = C_1^B = C_1 \ \text{(B.C. 1)}$$

Hence, the two constants of integration are equal and the development can continue considering only constant C_1. Using Newton's law of viscosity, we can write that:

$$\tau_{yx}^{A} = -\mu_a \frac{\partial v_x^A}{\partial y} = \left(\frac{P_o - P_L}{L}\right) y + C_1 \tag{5.145}$$

$$\tau_{yx}^{B} = -\mu_a \frac{\partial v_x^B}{\partial y} = \left(\frac{P_o - P_L}{L}\right) y + C_1 \tag{5.146}$$

Integrating the equations above assuming constant viscosity, we obtain:

$$v_x^A = -\frac{1}{\mu_a}\left(\frac{P_o - P_L}{L}\right)\frac{y^2}{2} - \frac{C_1}{\mu_a} + C_2^A \tag{5.147}$$

$$v_x^B = -\frac{1}{\mu_b}\left(\frac{P_o - P_L}{L}\right)\frac{y^2}{2} - \frac{C_1}{\mu_b} + C_2^B \tag{5.148}$$

The integration constants above are determined with the following boundary conditions:

$$y = 0: \ v_x^A = v_x^B \qquad \text{(B.C.2)}$$

$$y = -h: \ v_x^A = 0 \qquad \text{(B.C.3)}$$

$$y = h: \ v_x^B = 0 \qquad \text{(B.C.4)}$$

From these boundary conditions, the integration constants are evaluated.

$$C_2^A = C_2^B = C_2$$

$$C_1 = -\left(\frac{P_o - P_L}{L}\right) \frac{h}{2} \left(\frac{\mu_a - \mu_b}{\mu_a + \mu_b}\right) \tag{5.149}$$

$$C_2 = \left(\frac{P_o - P_L}{L}\right) \frac{h^2}{2\mu_a} \left(\frac{2\,\mu_a}{\mu_a + \mu_b}\right) \tag{5.150}$$

With these expressions, the equations for the velocity profiles in the two layers of fluids can be finally determined as:

$$v_x^A = \frac{1}{\mu_a}\left(\frac{P_o - P_L}{L}\right) \frac{h^2}{2} \left[\frac{2\,m_a}{\mu_a + \mu_b} + \left(\frac{\mu_a - \mu_b}{\mu_a + \mu_b}\right)\frac{y}{h} - \left(\frac{y}{h}\right)^2\right] \tag{5.151}$$

$$v_x^B = \frac{1}{\mu_b}\left(\frac{P_o - P_L}{L}\right) \frac{h^2}{2} \left[\frac{2\,\mu_b}{\mu_a + \mu_b} + \left(\frac{\mu_a - \mu_b}{\mu_a + \mu_b}\right)\frac{y}{h} - \left(\frac{y}{h}\right)^2\right] \tag{5.152}$$

When the viscosities of the liquids are equal ($\mu_a = \mu_b$), the expressions above are equal to those obtained for flow of a single fluid between horizontal parallel plates.

The expressions for volume and mass flow rates and average velocity are determined using the same procedure adopted in previous sections. It can be shown that the average velocities in the two layers are given by:

$$\bar{v}_x^A = \frac{1}{\mu_a}\left(\frac{P_o - P_L}{L}\right) \frac{h^2}{12} \left(\frac{7\,\mu_a + \mu_b}{\mu_a + \mu_b}\right) \tag{5.153}$$

$$\bar{v}_x^B = \frac{1}{\mu_b}\left(\frac{P_o - P_L}{L}\right) \frac{h^2}{12} \left(\frac{\mu_a + 7\,\mu_b}{\mu_a + \mu_b}\right) \tag{5.154}$$

Finally, it is interesting to determine where the velocity has the maximum (or minimum) value. The location of the maximum velocity can be determined taking the first derivative of the velocity profiles with respect to x and equating it to zero:

$$\frac{\partial v_x^A}{\partial y} = \frac{1}{\mu_a}\left(\frac{P_o - P_L}{L}\right)\left[-y + \frac{h}{2}\left(\frac{\mu_a - \mu_b}{\mu_a + \mu_b}\right)\right] = 0 \tag{5.155}$$

$$\frac{\partial v_x^B}{\partial y} = \frac{1}{\mu_b}\left(\frac{P_o - P_L}{L}\right)\left[-y + \frac{h}{2}\left(\frac{\mu_a - \mu_b}{\mu_a + \mu_b}\right)\right] = 0 \qquad (5.156)$$

In the equations above, the derivative is zero when the term in brackets is zero. Thus, the point of maximum (or minimum) velocity is given by:

$$y_{v_{mx}} = \frac{h}{2}\left(\frac{\mu_a - \mu_b}{\mu_a + \mu_b}\right) \qquad (5.157)$$

An analysis of the above equation enables us to conclude that:

- if $\mu_a > \mu_b$, $y_{vmax} > 0$, which means that the higher velocity occurs in fluid B, which is less viscous fluid;
- f $\mu_a < \mu_b$, $y_{vmax} < 0$, which means that the higher velocity occurs in fluid A, which is less viscous fluid;
- f $\mu_a = \mu_b$, $y_{vmax} = 0$. The maximum velocity occurs at the interface between the two fluids.

The fluid phase with lower viscosity relatively should have a higher average velocity.

References

BIRD, R.B.; STEWART, W.E.; LIGHTFOOT, E.N. *Transport phenomena.* New York: John Wiley & Sons, 1960.

GUTHRIE, R.I.L. *Engineering in process metallurgy.* Oxford: Oxford Science Publications, 1992.

LIENHARD IV, J.H.; LIENHARD V, J.H. *A heat transfer textbook.* Cambridge: Phlogiston, 2005.

Exercises

1 - A method for determining the radius of capillary tubes is to measure the flow of fluid through this tube. Evaluate the diameter of the capillary given the following data

- capillary length: 50.02 cm;
- kinematic viscosity of the fluid: 4.03 x 10^{-5} m^2/s;
- fluid density: 0.9552 x 10^3 kg/m^3;
- pressure drop along the capillary tube (horizontal): 4.829 x 10^5 N/m^2;
- mass flow rate through the tube: 2.997 x 10^{-3} kg/s.

2 - Two horizontal concentric cylinders have a length of 8.23 m. The outer diameter of the inner cylinder is 1.26 cm. The inner diameter of the outer cylinder is 2.79 cm. An aqueous solution is pumped between the two cylinders. The density of this solution is 1286.3 kg/m^3 and its viscosity is 0.0565 kg/m.s. Determine the volumetric flow rate of the fluid when the pressure drop is 3.716 x 10^4 N/m^2.

3 - A liquid is flowing by gravity through a vertical pipe with a length of 1 ft and internal diameter of 0.1 in. The density of the liquid is 1.26 g/cm^3 and the mass flow rate is 0.005 lb/min. Determine the viscosity of the fluid in units of the international system. Test the validity of the results.

4 - Develop expressions for the flow of a fluid between two vertical parallel plates. The plates are separated by a distance 2δ. Consider steady-state and that the density and viscosity of the fluid are constants. Include the difference in pressure. Determine:

- velocity distribution;
- volumetric flow rate of fluid;
- average velocity maximum speed.

5 - Water is placed between two infinite parallel plates, 3 cm distant from each other. If the plates move in opposite directions at velocities of 0.18 m/s and 0.21 m/s, determine the velocity profile and calculate the average velocity of the water. Make a diagram of the system under consideration.

Data:

- density of water: 1 g/cm³;
- viscosity of water: 1 cP.

6 - In experiments of gas absorption, a viscous fluid flows upwards through a circular tube. At the end of the tube is an overflow and fluid flows from top to bottom on the outer surface of the tube. Develop equations for the velocity distribution and volumetric flow rate of the fluid that flows over at the outer surface of the tube (see figure below).

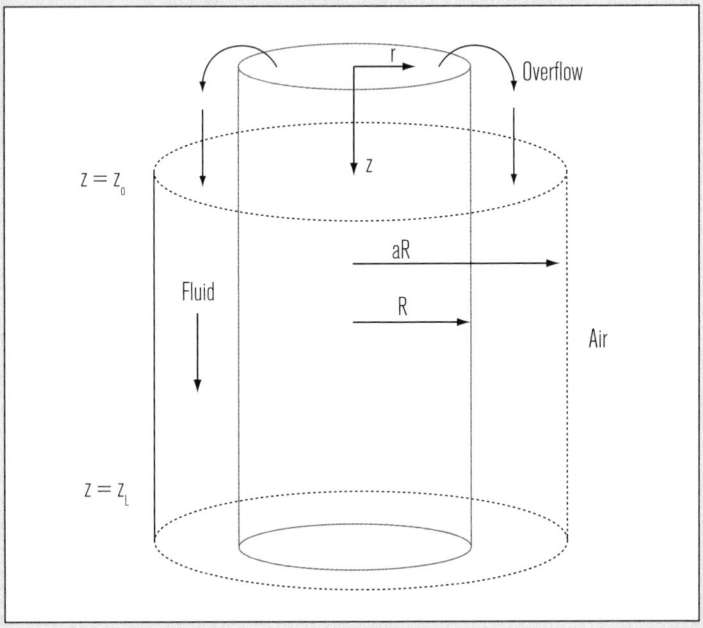

7 - A steel wire of radius R is cooled in an oil tank, as shown in the figure below. This type of cooling occurs in various processes of heat treatment of metals. From the balance equations of mass and momentum, determine an equation for the velocity profile of fluid that flows in the space L in the figure. In this passage, the diameter of the tube inside the tank is KR. Consider steady state, fluid of constant density and viscosity and that the pressure inside the tank is uniform. The velocity of the wire is V.

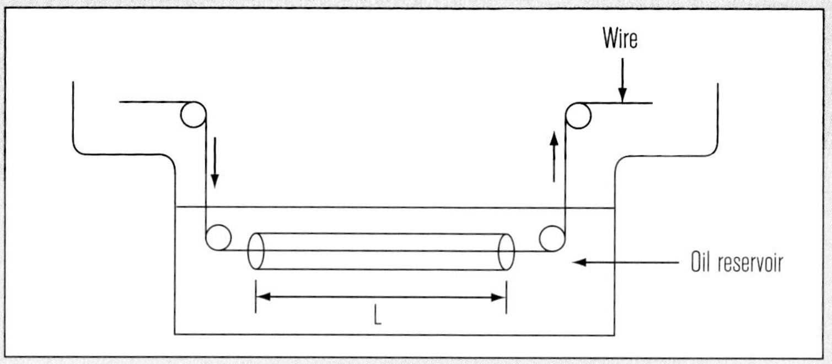

8 - A technique used in determining the flow of fluids in cylindrical ducts is to measure the difference in pressure between two points along the pipe, as seen in the figure below.

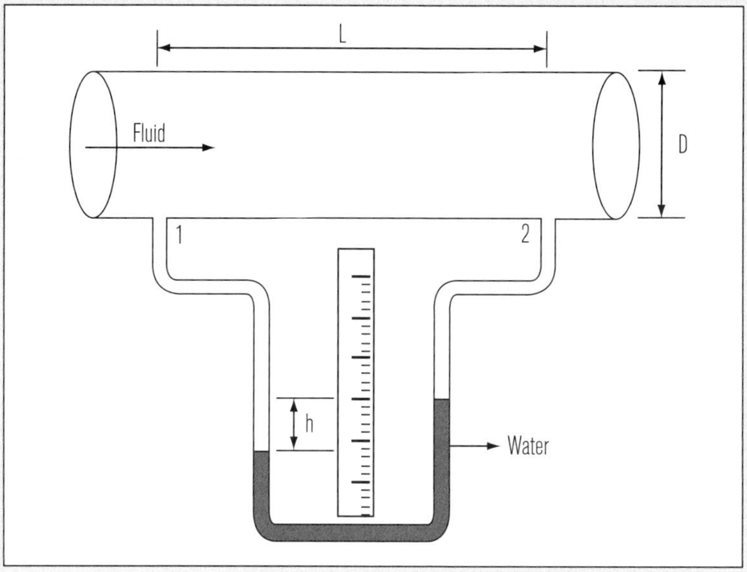

Use mass and momentum balances, to obtain a mathematical equation that expresses the velocity profile of the fluid as a function of radial position in the tube. From this equation, develop a mathematical expression relating the volumetric flow rate of the fluid with the pressure difference between points 1 and 2, expressed in terms of the height of water column in the manometer. Apply the formula developed to determine the flow rate of N_2 in the following conditions:

- tube diameter: 10 inches;
- difference in height of the water column in the gauge: 6 mm;
- density of water: 1 g/cm^3;
- nitrogen density: 1 kg/m^3;
- distance between points 1 and 2: 8 ft ;
- temperature: 298 K;
- properties of nitrogen (to evaluate the viscosity):
  - atomic mass: 28.02 g/mol $\sigma = 3.681$ A $\varepsilon/k_B = 91.5$ K

  Express the results in units of the international system.

9 - The mold of a continuous casting machine has the configuration schematically shown in the picture below. Determine the velocity profile within the lubricant slag layer, as a function of velocities U_{mold} and $U_{casting}$. Assume laminar flow, μ and ρ constants. Consider that the flow reaches steady state instantaneously for each value of the velocity of the mold, which is oscillatory.

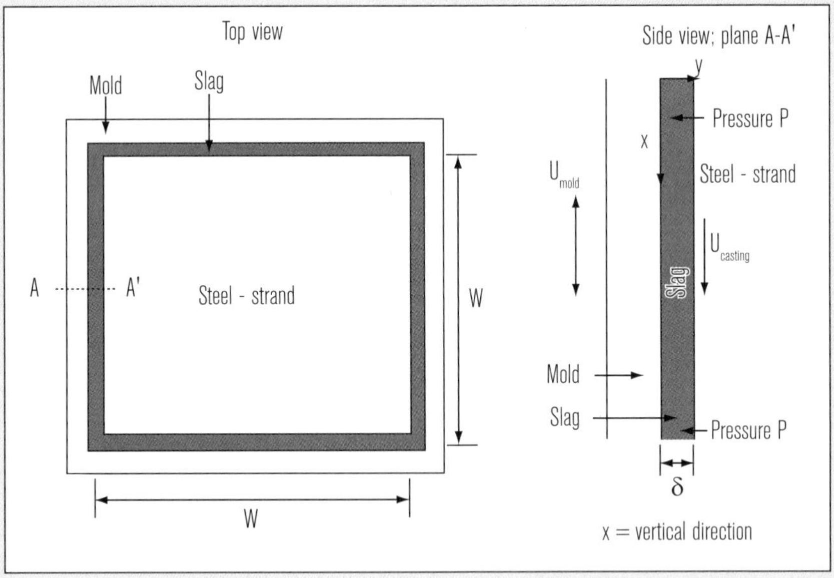

From the result above, how would you evaluate the consumption of slag, if the mold has four sides of equal width W?

Differential Equations of Fluid Flow

Chapter 06

In the previous chapter, the velocity distributions for flow in simple one-dimensional systems were determined from the mass and momentum balance in a given control volume of the system. The velocity distribution can be used to calculate other quantities such as average velocity, flow rate and shear force exerted by the fluid on the surface of the solid. However, it is necessary to develop general mass and momentum balances valid for three dimensions considering pressure drop, gravity and other forces in carteisian cylindrical and spherical coordinates for transient and steady state conditions. These equations will come in handy for the analysis of many practical problems in the industry with reasonable simplification procedures. They also serve as the basis of CFD (computational fluid dynamics) and software packages available now for the solution of sophisticated problems. of fluid flow and heat transfer in the industry.

The equation of continuity for three dimensions can be developed, applying the principle of conservation of mass in a control volume element in the fluid, chosen for particular coordinate system. The equation of motion is just a generalization of momentum balance, for three dimensions similar to the unidimensional cases considered in the previous chapters, for the same control volume. Combining this equation with the continuity equation, many of the problems discussed in the previous chapters can be solved with simplifying assumptions and also other complex problems in two or three dimensions.with appropriate boundary conditions.

Before dealing with the development of the equations of continuity and motion, it is interesting to distinguish, three types of partial derivatives with respect to time. This is illustrated in the following example. Consider fish concentration in a river. As the fish are moving, their concentration "c" should be a function of position (x, y and z) and time t. There can be three types of derivatives with respect to time.

- *Partial derivative with respect to time, $\partial C/\partial t$*

In this case, there is a variation of the concentration of fish over time at a fixed point in space.Thus, $\partial C/\partial t$ means partial derivative of "c" with respect to t, keeping, x, y and z constant.

- *Total derivative with respect to time, dc/dt*

Consider now, instead of analyzing at a fixed point, analysis is done in a moving motor boat in the river. In addition to providing the variation of concentration with time, the analysis should also consider the movement of the boat. Therefore, the total derivative with respect to time is given by:

$$\frac{dc}{dt} = \frac{\partial c}{\partial t} + \frac{\partial c}{\partial x}\frac{\partial x}{\partial t} + \frac{\partial c}{\partial y}\frac{\partial y}{\partial t} + \frac{\partial c}{\partial z}\frac{\partial z}{\partial t} \tag{6.1}$$

where $\partial x/\partial t$, $\partial y/\partial t$ and $\partial z/\partial t$ are components of the boat speed.

- *Substantial time derivative*

Instead of taking a motor boat, supposing one goes on a canoe without paddle and follows the stream in the river. The variation of concentration of fish over time would now depend on the current velocity. This derivative is called the derivative following the motion and is given by:

$$\frac{dc}{dt} = \frac{\partial c}{\partial t} + v_x \frac{\partial c}{\partial x} + v_y \frac{\partial c}{\partial y} + v_z \frac{\partial c}{\partial z} \tag{6.2}$$

and that v_x, v_y and v_z are the components of local velocity of the fluid.

6.1 Equation of Continuity

The general equation of mass balance within the control volume is given by:

$$\begin{aligned}&[\text{Rate of mass input}] - [\text{Rate of mass output}] = \\ &[\text{Rate of mass accumulation within the control volume}]\end{aligned} \tag{6.3}$$

This equation is applied to an infinitismal volume element $\Delta x\, \Delta y\, \Delta z$, along which the fluid moves as shown in Figure 6.1.

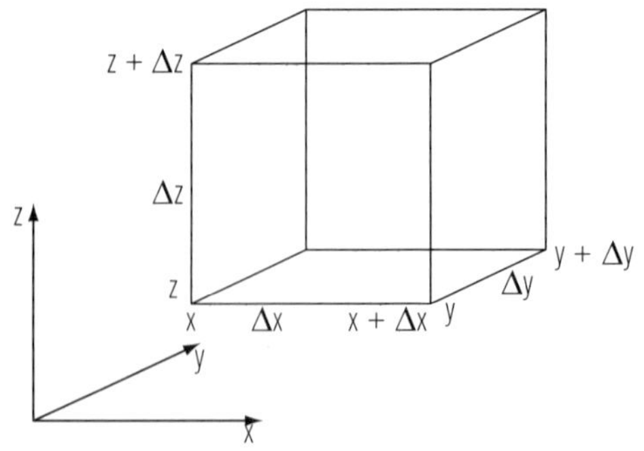

Figure 6.1 – Control volume for mass and momentum balances.

Expressing each of the above terms and considering the three directions:

- Rate of mass accumulation:

$$\Delta x\ \Delta y\ \Delta z\ \frac{\Delta \rho}{\Delta t} \tag{6.4}$$

- Rate of mass input across surface at x:

$$(\Delta y\ \Delta z\ \rho\ v_x)|_x \tag{6.5}$$

- Rate of mass output across surface at x + Δx:

$$(\Delta y\ \Delta z\ \rho\ v_x)|_{x+\Delta x} \tag{6.6}$$

- Rate of mass input across surface at y:

$$(\Delta x\ \Delta z\ \rho\ v_y)|_y \tag{6.7}$$

- Rate of mass output across surface at y + Δy:

$$(\Delta x\ \Delta z\ \rho\ v_y)|_{y+\Delta y} \tag{6.8}$$

- Rate of mass input across surface at z:

$$(\Delta x\ \Delta y\ \rho\ v_z)|_z \tag{6.9}$$

- Rate of mass output across surface at z + Δz:

$$(\Delta x\ \Delta y\ \rho\ v_z)|_{z+\Delta z} \tag{6.10}$$

Combining all these terms according to equation 6.3, we get for mass balance:

$$\begin{aligned}\Delta x\ \Delta y\ \Delta z\ \frac{\Delta \rho}{\Delta t} &= \left[(\Delta y\ \Delta z\ \rho\ v_x)|_x - (\Delta y\ \Delta z\ \rho\ v_x)|_{x+\Delta x}\right] \\ &+ \left[(\Delta x\ \Delta z\ \rho\ v_y)|_y - (\Delta x\ \Delta z\ \rho\ v_y)|_{y+\Delta y}\right] + \left[(\Delta x\ \Delta y\ \rho\ v_z)|_z - (\Delta x\ \Delta y\ \rho\ v_z)|_{z+\Delta z}\right]\end{aligned} \tag{6.11}$$

Dividing this expression by the volume of the element:

$$\frac{\Delta \rho}{\Delta t} = \frac{\left[(\rho\ v_x)|_x - (\rho\ v_x)|_{x+\Delta x}\right]}{\Delta x} + \frac{\left[(\rho\ v_y)|_y - (\rho\ v_y)|_{y+\Delta y}\right]}{\Delta y} + \frac{\left[(\rho\ v_z)|_z - (\rho\ v_z)|_{z+\Delta z}\right]}{\Delta z} \tag{6.12}$$

Taking limits for Δx, Δy and Δz tending to zero and recalling the concept of derivative, we have:

$$\frac{\partial \rho}{\partial t} = -\frac{\partial}{\partial x}(\rho\ v_x) - \frac{\partial}{\partial y}(\rho\ v_y) - \frac{\partial}{\partial z}(\rho\ v_z) \tag{6.13}$$

This is one of the ways to present the equation of continuity. Ther other forms of this equation can be derived as given below:

$$\frac{\partial}{\partial x}(\rho\, v_x) = \rho \frac{\partial v_x}{\partial x} + v_x \frac{\partial \rho}{\partial x} \tag{6.14}$$

$$\frac{\partial}{\partial y}(\rho\, v_y) = \rho \frac{\partial v_y}{\partial y} + v_y \frac{\partial \rho}{\partial y} \tag{6.15}$$

$$\frac{\partial}{\partial z}(\rho\, v_z) = \rho \frac{\partial v_z}{\partial z} + v_z \frac{\partial \rho}{\partial z} \tag{6.16}$$

Substituting relations (6.14) (6.15) and (6.16) in (6.13), we get:

$$\frac{\partial \rho}{\partial t} + \rho \frac{\partial v_x}{\partial x} + v_x \frac{\partial \rho}{\partial x} + \rho \frac{\partial v_y}{\partial y} + v_y \frac{\partial \rho}{\partial y} + \rho \frac{\partial v_z}{\partial z} + v_z \frac{\partial \rho}{\partial z} = 0 \tag{6.17}$$

Rearranging the terms involving derivatives of the density in the left side, we obtain:

$$\frac{\partial \rho}{\partial t} + v_x \frac{\partial \rho}{\partial x} + v_y \frac{\partial \rho}{\partial y} + v_z \frac{\partial \rho}{\partial z} = -\rho \left(\frac{\partial v_x}{\partial x} + \frac{\partial v_y}{\partial y} + \frac{\partial v_z}{\partial z} \right) \tag{6.18}$$

The left side of the above equation represents the derivatives for the variation of the density.with time and x, y and z. In compact notation, we can rewrite the relation (6.18) as follows:

$$\frac{D\rho}{Dt} = -\rho\, (\nabla . v) \tag{6.19}$$

where the gradient operator ∇ means:

$$\nabla = \frac{\partial}{\partial x} + \frac{\partial}{\partial y} + \frac{\partial}{\partial z} \tag{6.20}$$

Equation (6.19) describes the rate of change of density seen by an observer, who moves with the fluid.

It is important to mention that the equation of continuity, in whatever form it appears, is simply the principle of conservation of mass. It should also be noted that the development done is not restricted to rectangular volume elements. The same procedure can be followed with appropriate volume elements in cylindrical and spherical coordinates to arrive at continuity equations.

A special form of the equation of continuity, which is widely used, is the one that is applied to a fluid of constant density. In this case, we have:

$$\left(\frac{\partial v_x}{\partial x} + \frac{\partial v_y}{\partial y} + \frac{\partial v_z}{\partial z}\right) = 0 \qquad (6.21)$$

for a fluid whose density has a value different from zero.

This equation is applicable to incompressible fluids. In reality, no fluid is truly incompressible, but very often in engineering problems this assumption of constant density is assumed for liquids and does not cause significant error in the calculations. However for gases this will not be the case.

6.2 Equation of Motion

For the same volume element Δ x.Δ y.Δ z (Figure 6.1), the general form of momentum balance is given by:

$$\begin{aligned}&[\text{Rate of momentum input}] - [\text{Rate of momentum output}] + \\ &\text{Sum of forces acting on the volume element} = [\text{Rate of accumulation of momentum}]\end{aligned} \qquad (6.22)$$

Equation (6.22) is an extension of equation (5.4), for steady state but applicable to steady and unsteady states. In the case of steady state the accumulation term becomes zero. Otherwise the procedure adopted in the balance is similar to that developed in the previous chapter.

However in this case, unsteady state conditions as well as fluid moving through the six faces of the volume element instead of, in any arbitrary direction are taken into account.

It is important to emphasize that the terms of equation (6.22) are vectorial, with components in all the three directions, x, y and z. For simplicity, one can consider only x direction, the other two (y and z) can be treated similarly. The development of momentum balances in the other two directions (y and z) can be done as an exercise.

Initially, input and output rates of momentum in the direction of x, in the volume element shown in Figure 6.1 are considered. As is known, momentum enters and leaves the volume element by two mechanisms: convection (due to macroscopic movement of the fluid) and diffusion (due to velocity gradients).

The constituent terms of the equation (6.22), for the x direction are given below:

- rate of accumulation of momentum in the x direction:

$$\Delta x\ \Delta y\ \Delta z\ \frac{\Delta(\rho\ v_x)}{\Delta t} \tag{6.23}$$

- input rate of diffusive momentum in x:

$$(\Delta y\ \Delta z\ \tau_{xx})|_x \tag{6.24}$$

- input rate of diffusive momentum in y:

$$(\Delta x\ \Delta z\ \tau_{yx})|_y \tag{6.25}$$

- input rate of diffusive momentum in z:

$$(\Delta x\ \Delta z\ \tau_{yx})|_{y+\Delta y} \tag{6.26}$$

- output rate of diffusive momentum in x + Δx:

$$(\Delta y\ \Delta z\ \tau_{xx})|_{x+\Delta x} \tag{6.27}$$

- output rate of diffusive momentum in y + Δy:

$$(\Delta x\ \Delta z\ \tau_{yx})|_{y+\Delta y} \tag{6.28}$$

- output rate of diffusive momentum in z + Δz:

$$(\Delta x\ \Delta y\ \tau_{zx})|_{z+\Delta z} \tag{6.29}$$

To assess the contribution associated with the mechanism of convection, it is important to remember that momentum is the product of the mass and velocity. Therefore, the rate of momentum by convection can be determined by the product of the mass flow rate (as determined by mass balance) by the velocity (x component of velocity in this case).Thus, we have:

- input rate of momentum by convection in x:

$$(\Delta y\ \Delta z\ v_x\ \rho\ v_x)|_x \tag{6.30}$$

- input rate of momentum by convection in y:

$$(\Delta x\ \Delta z\ v_y\ \rho\ v_x)|_y \tag{6.31}$$

- input rate of momentum by convection in z:

$$(\Delta x\ \Delta y\ v_z\ \rho\ v_x)|_z \tag{6.32}$$

- output rate of momentum by convection at x + Δx:

$$(\Delta y\ \Delta z\ v_x\ \rho\ v_x)|_{x+\Delta x} \tag{6.33}$$

- output rate of momentum by convection at y + Δy:

$$(\Delta x\ \Delta z\ v_y\ \rho\ v_x)|_{y+\Delta y} \tag{6.34}$$

- output rate of momentum by convection at z + Δz:

$$(\Delta x\ \Delta y\ v_z\ \rho\ v_x)|_{z+\Delta z} \tag{6.35}$$

Forces acting on volume element in the x direction:

- force due to pressure difference between x and x+dx:

$$\Delta y\ \Delta z\ (P|_x - P|_{x+\Delta x}) \tag{6.36}$$

- force due to gravity:

$$\Delta x\ \Delta y\ \Delta z\ \rho\ g_x \tag{6.37}$$

Regarding the terms of entry and exit of momentum by convection, an explanation is useful. For example, in the term $\Delta y\ \Delta z\ v_x,\ \rho\ _x,\ v_x$ is the component of velocity in the x direction, $\Delta y\ \Delta z\ v_{x,}\ \rho$ represents the rate of input of mass flow rate across the area $\Delta y\ \Delta z$. The product of these two parameters provides the input rate of convective momentum. Similar explanation applies to the other terms in the directions y and z.

Combining all these terms in the general equation of momentum balance, we obtain the following relationship for the x:

$$\begin{aligned}
\Delta x\ \Delta y\ \Delta z\ \frac{\Delta(\rho\ v_x)}{\Delta t} &= \left[(\Delta y \Delta z\ \tau_{xx})|_x - (\Delta y\ \ \Delta z\ \tau_{xx})|_{x+\Delta x}\right] + t \\
&\left[(\Delta x\ \Delta z\ \tau_{yx})|_y - (\Delta x\ \Delta z\ \tau_{yx})|_{y+\Delta y}\right] + \left[(\Delta x \Delta y\ \tau_{zx})|_z - (\Delta x\ \Delta y\ \tau_{zx})|_{z+\Delta z}\right] + \\
&\left[(\Delta y\ \Delta z\ v_x\ \rho\ v_x)|_x - (\Delta y\ \Delta z\ v_x\ \rho\ v_x)|_{x+\Delta x}\right] + \left[(\Delta x\ \Delta z\ v_y\ \rho\ v_x)|_y\ - (\Delta x\ \Delta z\ v_y\ \rho\ v_x)|_{y+\Delta y}\right] + \\
&\left[(\Delta x\ \Delta y\ v_z\ \rho\ v_x)|_z + (\Delta x\ \Delta y\ v_z\ \rho\ v_x)|_{z+\Delta z}\right] + \\
&\Delta y\ \Delta z\ (P|_x - P|_{x+\Delta x}) + \Delta x\ \Delta y\ \Delta z\ \rho\ g_x
\end{aligned} \tag{6.38}$$

Dividing by the control volume (Δx Δy Δz), we obtain:

$$\begin{aligned}
\frac{\Delta(\rho\ v_x)}{\Delta t} &= \frac{\left[(\tau_{xx})|_x - (\tau_{xx})|_{x+\Delta x}\right]}{\Delta x} + \frac{\left[(\tau_{yx})|_y - (\tau_{yx})|_{y+\Delta y}\right]}{\Delta y} + \frac{\left[(\tau_{zx})|_z - (\tau_{zx})|_{z+\Delta z}\right]}{\Delta z} + \\
&+ \frac{\left[(v_x\ \rho\ v_x)|_x - (v_x\ \rho\ v_x)|_{x+\Delta x}\right]}{\Delta x} + \frac{\left[(v_y\ \rho\ v_x)|_y - (v_y\ \rho\ v_x)|_{y+\Delta y}\right]}{\Delta y} + \\
&+ \frac{\left[(v_z\ \rho\ v_x)|_z + (v_z\ \rho\ v_x)|_{z+\Delta z}\right]}{\Delta z} + \frac{(P|_x - P|_{x+\Delta x})}{\Delta x} + \rho\ g_x
\end{aligned} \tag{6.39}$$

Taking the limits for Δx, Δy and Δz tending to zero and recalling the concept of first derivative, we obtain the following expression:

$$\frac{\partial}{\partial t}(\rho\, v_x) = -\frac{\partial \tau_{xx}}{\partial x} - \frac{\partial \tau_{yx}}{\partial y} - \frac{\partial \tau_{zx}}{\partial z} - \frac{\partial}{\partial x}(v_x\, \rho\, v_x) - \frac{\partial}{\partial y}(v_y\, \rho\, v_x) - \frac{\partial}{\partial z}(v_z\, \rho\, v_x) - \frac{\partial P}{\partial x} + \rho\, g_x \qquad (6.40)$$

For y and z directions, the equations are similar:

- y direction:

$$\frac{\partial}{\partial t}(\rho\, v_y) = -\frac{\partial \tau_{xy}}{\partial x} - \frac{\partial \tau_{yy}}{\partial y} - \frac{\partial \tau_{zy}}{\partial z} - \frac{\partial}{\partial x}(v_x\, \rho\, v_y) - \frac{\partial}{\partial y}(v_y\, \rho\, v_y) - \frac{\partial}{\partial z}(v_z\, \rho\, v_y) - \frac{\partial P}{\partial y} + \rho\, g_y \qquad (6.41)$$

- z direction:

$$\frac{\partial}{\partial t}(\rho\, v_z) = -\frac{\partial \tau_{xz}}{\partial x} - \frac{\partial \tau_{yz}}{\partial y} - \frac{\partial \tau_{zz}}{\partial z} - \frac{\partial}{\partial x}(v_x\, \rho\, v_z) - \frac{\partial}{\partial y}(v_y\, \rho\, v_z) - \frac{\partial}{\partial z}(v_z\, \rho\, v_z) - \frac{\partial P}{\partial z} + \rho\, g_z \qquad (6.42)$$

Considering only the x component of the equation of motion (6.40), one can write the following:

$$v_x \frac{\partial \rho}{\partial t} + \rho \frac{\partial v_x}{\partial t} = -\left[\frac{\partial \tau_{xx}}{\partial x} + \frac{\partial \tau_{yx}}{\partial y} + \frac{\partial \tau_{xz}}{\partial x}\right] - \left\{ v_x \left[v_x \frac{\partial \rho}{\partial x} + \rho \frac{\partial v_x}{\partial x} + v_y \frac{\partial \rho}{\partial y} + \rho \frac{\partial v_y}{\partial y} + v_z \frac{\partial \rho}{\partial z} + \rho \frac{\partial v_z}{\partial z} \right]\right\} + \qquad (6.43)$$
$$-\left\{\rho v_x \frac{\partial v_x}{\partial x} + \rho v_y \frac{\partial v_x}{\partial y} + \rho v_z \frac{\partial v_x}{\partial z}\right\} - \frac{\partial P}{\partial x} + \rho g_x$$

The derivative with respect to time in equation (6.40) actually contains two terms on the left side of equation (6.43). Transposing terms in the above equation, one gets:

$$\rho \left[\frac{\partial v_x}{\partial t} + v_x \frac{\partial v_x}{\partial x} + v_y \frac{\partial v_X}{\partial y} + v_z \frac{\partial v_x}{\partial z}\right] = -\left[\frac{\partial \tau_{xx}}{\partial x} + \frac{\partial \tau_{yx}}{\partial y} + \frac{\partial \tau_{zx}}{\partial z}\right] -$$
$$-\left[v_x \left(\frac{\partial \rho}{\partial t} + v_x \frac{\partial \rho}{\partial x} + \rho \frac{\partial v_x}{\partial x} + v_y \frac{\partial \rho}{\partial y} + \rho \frac{\partial v_y}{\partial y} + v_z \frac{\partial \rho}{\partial z} + \rho \frac{\partial v_z}{\partial z} \right)\right] \qquad (6.44)$$
$$-\frac{\partial P}{\partial x} + \rho\, g_x$$

Recalling continuity equation (6.17) for incompressible fluids, the terms within the rectangular brackets in the equation above vanish. Hence, equation (6.44) can be written as follows:

$$\rho \left(\frac{\partial v_x}{\partial t} + v_x \frac{\partial v_x}{\partial x} + v_y \frac{\partial v_x}{\partial y} + v_z \frac{\partial v_x}{\partial z}\right) = -\left(\frac{\partial \tau_{xx}}{\partial x} + \frac{\partial \tau_{yx}}{\partial y} + \frac{\partial \tau_{zx}}{\partial z}\right) - \frac{\partial P}{\partial x} + \rho\, g_x \qquad (6.45)$$

Recalling the definition of substantial time derivative, equation (6.3), the term in brackets on the left side of the equation (6.45) is equivalent to this. Hence substituting this in the equation of momentumone can write:

$$\rho \frac{Dv_x}{Dt} = -\left(\frac{\partial \tau_{xx}}{\partial x} + \frac{\partial \tau_{yx}}{\partial y} + \frac{\partial \tau_{zx}}{\partial z}\right) - \frac{\partial P}{\partial x} + \rho\, g_x \qquad (6.46)$$

Equation (6.46) represents both the rate of momentum balance as well as force balance. The term in the left represents the product of mass and acceleration (a reference frame moving with the fluid). Terms in the right signify the sum of forces associated with friction (due to viscosity), pressure and gravity (Newton's second law). Equations (6.45) or (6.46) are applicable to any type of fluid, Newtonian or nonnewtonian.

In order that the equations of motion (x component derived above, and the y and z components) can be utilized in an effective way in terms of velocity distribution, one must replace the shear stress or rate of diffusive momentum terms in the equations by velocity components.These expressions are given below for Newtonian fluids with the application of Newton's law of viscosity for three-dimensional systems:

$$\tau_{xx} = -2\,\mu \frac{\partial v_x}{\partial x} + \frac{2}{3}\,\mu \left(\frac{\partial v_x}{\partial x} + \frac{\partial v_y}{\partial y} + \frac{\partial v_z}{\partial z}\right) \qquad (6.47)$$

$$\tau_{yy} = -2\,\mu \frac{\partial v_y}{\partial y} + \frac{2}{3}\,\mu \left(\frac{\partial v_x}{\partial x} + \frac{\partial v_y}{\partial y} + \frac{\partial v_z}{\partial z}\right) \qquad (6.48)$$

$$\tau_{zz} = -2\,\mu \frac{\partial v_z}{\partial z} + \frac{2}{3}\,\mu \left(\frac{\partial v_x}{\partial x} + \frac{\partial v_y}{\partial y} + \frac{\partial v_z}{\partial z}\right) \qquad (6.49)$$

$$\tau_{yx} = \tau_{xy} = -\,\mu \left(\frac{\partial v_x}{\partial y} + \frac{\partial v_y}{\partial x}\right) \qquad (6.50)$$

$$\tau_{yz} = \tau_{zy} = -\,\mu \left(\frac{\partial v_z}{\partial y} + \frac{\partial v_y}{\partial z}\right) \qquad (6.51)$$

$$\tau_{zx} = \tau_{xz} = -\,\mu \left(\frac{\partial v_x}{\partial z} + \frac{\partial v_z}{\partial x}\right) \qquad (6.52)$$

The development of the equations (6.48) to (6.52) has not been dealt with here, as it is quite extensive. However one can refer to Lamb (1945) for details regarding the same. These equations represent the most complete expressions of Newton's law of viscosity, which apply to situations in which the fluid moves in more than one direction. The equations developed for various cases in the earlier chapter through simplified balances can also be arrived at, applying the general equations given above, where shear stress terms have been substituted by Newtons law (6.53) with the same assumptions.

When a fluid moves in the x direction between two parallel plates and perpendicular to the direction y, v_x is a function of y. Hence:

$$\tau_{yx} = -\mu \left(\frac{\partial v_x}{\partial y}\right) \tag{6.53}$$

This is the simplified equation of Newton's law of viscosity used in Chapter 4.

Substituting the expressions (6.47) to (6.52) for shear stress values expessed in terms of viscosity, in equation (6.45), we obtain:

$$\rho\left(\frac{\partial v_x}{\partial t} + v_x\frac{\partial v_x}{\partial x} + v_y\frac{\partial v_x}{\partial y} + v_z\frac{\partial v_x}{\partial z}\right) = -\frac{\partial}{\partial x}\left\{-2\mu\frac{\partial v_x}{\partial x} + \frac{2}{3}\mu\left[\left(\frac{\partial v_x}{\partial x} + \frac{\partial v_y}{\partial y} + \frac{\partial v_z}{\partial z}\right)\right]\right\} + $$
$$-\frac{\partial}{\partial y}\left[-\mu\left(\frac{\partial v_x}{\partial y} + \frac{\partial v_y}{\partial x}\right)\right] - \frac{\partial}{\partial z}\left[-\mu\left(\frac{\partial v_x}{\partial z} + \frac{\partial v_z}{\partial x}\right)\right] - \frac{\partial P}{\partial x} + \rho g_x \tag{6.54}$$

Equation (6.54) is the general equation for momentum balance in the x direction. Similar expressions can be obtained for directions y and z.

For constant density (incompressible fluids), the continuity equation is given by:

$$\left(\frac{\partial v_x}{\partial x} + \frac{\partial v_y}{\partial y} + \frac{\partial v_z}{\partial z}\right) = 0 \tag{6.21}$$

Substituting (6.21) in equation (6.54):

$$\rho\left(\frac{\partial v_x}{\partial t} + v_x\frac{\partial v_x}{\partial x} + v_y\frac{\partial v_x}{\partial y} + v_z\frac{\partial v_x}{\partial z}\right) =$$
$$-\left\{\frac{\partial}{\partial x}\left(-2\mu\frac{\partial v_x}{\partial x}\right) + \frac{\partial}{\partial y}\left[-\mu\left(\frac{\partial v_x}{\partial y} + \frac{\partial v_y}{\partial x}\right)\right] + \frac{\partial}{\partial z}\left[-\mu\left(\frac{\partial v_x}{\partial z} + \frac{\partial v_z}{\partial x}\right)\right]\right\} - \frac{\partial P}{\partial x} + \rho g_x \tag{6.55}$$

Rearranging the above expression and assuming constant viscosity, we obtain:

$$\rho\left[\frac{\partial v_x}{\partial t} + v_x\frac{\partial v_x}{\partial x} + v_y\frac{\partial v_x}{\partial y} + v_z\frac{\partial v_x}{\partial z}\right] =$$
$$+\left[\mu\frac{\partial^2 v_x}{\partial x^2} + \mu\frac{\partial^2 v_x}{\partial y^2} + \mu\frac{\partial^2 v_x}{\partial z^2} + \mu\frac{\partial^2 v_y}{\partial x\,\partial y} + \mu\frac{\partial^2 v_x}{\partial x^2} + \mu\frac{\partial^2 v_z}{\partial x\,\partial z}\right] - \frac{\partial P}{\partial x} + \rho g_x \tag{6.56}$$

Grouping the terms with cross derivatives:

$$\rho\left(\frac{\partial v_x}{\partial t} + v_x \frac{\partial v_x}{\partial x} + v_y \frac{\partial v_x}{\partial y} + v_z \frac{\partial v_x}{\partial z}\right) = + \left[\mu \frac{\partial^2 v_x}{\partial x^2} + \mu \frac{\partial^2 v_x}{\partial y^2} + \mu \frac{\partial^2 v_x}{\partial z^2} + \mu \frac{\partial}{\partial x}\left(\frac{\partial v_x}{\partial x} + \frac{\partial v_y}{\partial y} + \frac{\partial v_z}{\partial z}\right)\right] + - \frac{\partial P}{\partial x} + \rho g_x \qquad (6.57)$$

Again using the continuity equation [equation (6.21)], we obtain:

$$\rho\left(\frac{\partial v_x}{\partial t} + v_x \frac{\partial v_x}{\partial x} + v_y \frac{\partial v_x}{\partial y} + v_z \frac{\partial v_x}{\partial z}\right) = + \left(\mu \frac{\partial^2 v_x}{\partial x^2} + \mu \frac{\partial^2 v_x}{\partial y^2} + \mu \frac{\partial^2 v_x}{\partial z^2}\right) - \frac{\partial P}{\partial x} + \rho g_x \qquad (6.58)$$

This is the equation of motion in the x direction for a Newtonian fluid of constant density and viscosity.

For directions y and z, the equations are:

- y direction:

$$\rho\left(\frac{\partial v_y}{\partial t} + v_x \frac{\partial v_y}{\partial x} + v_y \frac{\partial v_y}{\partial y} + v_z \frac{\partial v_y}{\partial z}\right) = + \left(\mu \frac{\partial^2 v_y}{\partial x^2} + \mu \frac{\partial^2 v_y}{\partial y^2} + \mu \frac{\partial^2 v_y}{\partial z^2}\right) - \frac{\partial P}{\partial y} + \rho g_y \qquad (6.59)$$

- z direction:

$$\rho\left(\frac{\partial v_z}{\partial t} + v_x \frac{\partial v_z}{\partial x} + v_y \frac{\partial v_z}{\partial y} + v_z \frac{\partial v_z}{\partial z}\right) = + \left(\mu \frac{\partial^2 v_z}{\partial x^2} + \mu \frac{\partial^2 v_z}{\partial y^2} + \mu \frac{\partial^2 v_z}{\partial z^2}\right) - \frac{\partial P}{\partial z} + \rho g_z \qquad (6.60)$$

These relationships are shown in Tables 6.1, 6.2 and 6.3 (Annex at the end of chapter), in which the equations of continuity and motion are given in cartesian coordinates. These tables also contain the equations of shear stress for a Newtonian fluid and the momentum conservation equations in terms of shear stresses.

6.3 Equations of Continuity and Motion in Cylindrical and Spherical Coordinates

In the case of cylindrical or spherical systems also, problems can be easily formulated. In these cases, the equations of continuity and motion are different from those derived for rectangular or cartesian coordinates. Similar equations in terms of shear stress and velocity can be derived by defining the control volumes and areas in the respective coordinates for carrying out mass and momentum balances. Alternatively the cylindrical and spherical coordinates can be converted to rectangular coordinates, using the relationships between them as given below. These can be substituted in the relevant equations derived with rectangular coordinates to get the final equations. However this mathematical treatment is quite extensive. and is not presented here.

6.3.1 Cylindrical coordinates

The relationship between the Cartesian and cylindrical coordinates is presented in the equations below The figure is given in the previous chapter (Figure 5.8):

$$x = r \cos \theta \tag{6.61}$$

$$y = r \sin \theta \tag{6.62}$$

$$z = z \tag{6.63}$$

The general equations of continuity and motion, and the expressions for shear stress for a Newtonian fluid in cylindrical coordinates are presented in Tables 6.4, 6.5 and 6.6 in the Annex at the end of this chapter.

6.3.2 Spherical coordinates

The relationship between the rectangular and spherical coordinates is shown in Figure 6.2.The mathematical relationships between these coordinates are given by the following equations:

$$x = r \sin \theta \cos \phi \tag{6.64}$$

$$y = r \sin \theta \sin \phi \tag{6.65}$$

$$z = r \cos \theta \tag{6.66}$$

The general equations of continuity and motion, and the expressions for shear stress for a Newtonian fluid in cylindrical coordinates are presented in Tables 6.7, 6.8 and 6.9 in the annex to this chapter.

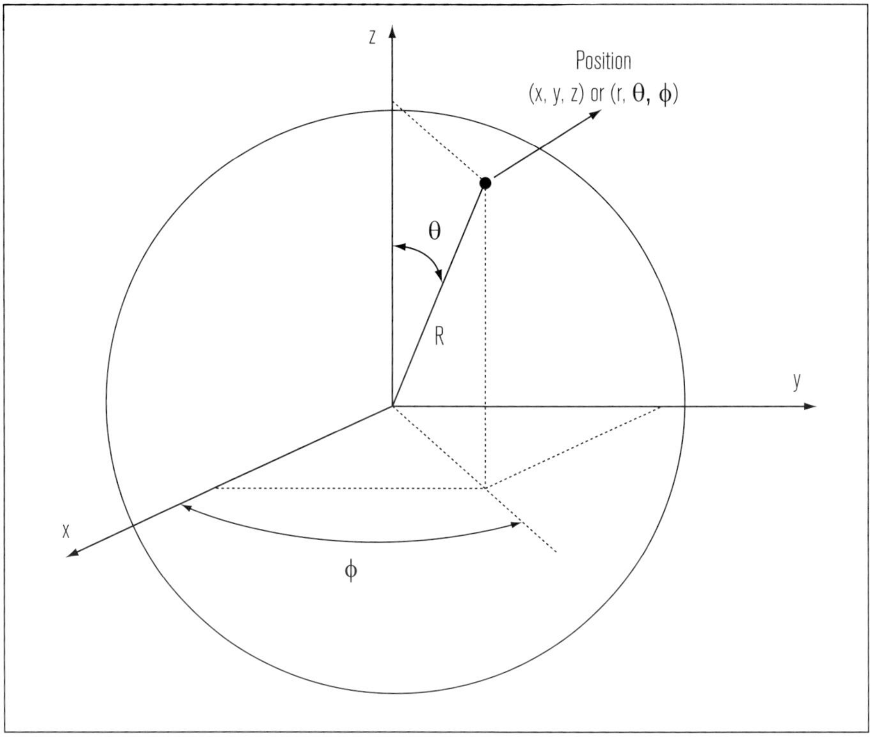

Figure 6.2 - Relationship between cartesian and spherical coordinates.

6.4 Solutions of Differential Equations

In this section, the equations of continuity and motion are used to solve some problems, already discussed in Chapter 5 In the solution procedure, the general equations of mass and momentum balances described earlier and given in the tables at the end, are simplified. Some terms are discarded in accordance with the assumptions. Some other terms are neglected, as they would not alter significantly the end results. To determine the terms to be discarded, one has to carefully analyze the system in terms of flow patterns, pressure distribution, as well as study of process variables. This also helps in identifying the boundary conditions, essential for solution of differential equations. By simplifying the equations, one can also simplify the solution procedures for the resulting differential equations considerably. One advantage of this procedure is that complicated problems in spherical and cylindrical coordinates can be solved analytically or using simpler mathematical procedures.

6.4.1 Flow of a liquid film in an inclined plane

This system is seen schematically in Figure 6.3. According to the orientation given to the axes, there is fluid velocity only in the z direction.

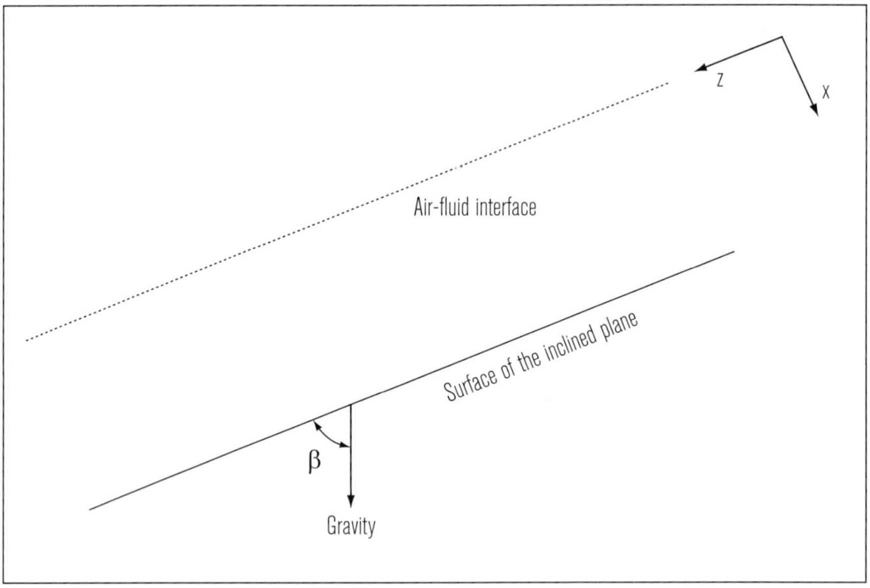

Figure 6.3 - Flow of fluid in an inclined plane.

For a fluid of constant fluid density and viscosity and steady state flow under gravity in the inclined plane in the z direction, the equations of continuity and motion provide:

- equation of continuity:

$$\frac{\partial v_z}{\partial z} = 0 \tag{6.67}$$

- equation of motion (only component z):

$$\mu \frac{\partial^2 v_z}{\partial x^2} + \rho\, g_z = 0 \tag{6.68}$$

$$g_z = g \cos \beta \tag{6.69}$$

Equation (6.68) can be integrated twice to provide the following profile:

$$v_z = -\frac{\rho\, g \cos\beta}{2\,\mu}\, x^2 + C_1\, x + C_2 \qquad (6.70)$$

For the determination of C_1 and $C_{2,}$ the following boundary conditions are used:

$$\text{Boundary condition 1: } x = 0 \quad \tau_{xz} = -\mu\,\frac{\partial v_z}{\partial x} = 0$$

$$\text{Boundary condition 2: } x = \delta \quad v_z = 0$$

Substituting in equation (6.70):

$$C_1 = 0$$

$$C_2 = \frac{\rho\, g \cos\beta}{2\,\mu}\,\delta^2$$

Hence the velocity profile is given by:

$$v_z = \frac{\rho\, g \cos\beta}{2\,\mu}\,(\delta^2 - x^2) \qquad (6.71)$$

This equation is the same obtained through the mass and momentum balances in the volume element discussed in Chapter 5.

6.4.2 Flow in a circular tube

This system has been discussed earlier and is given schematically in Figure 5.9. This is a cylindrical system, and hence the respective equations in cylindrical coordinates should be used. Considering fluid with constant density and viscosity. For steady state conditions and unidimensional flow (fluid velocity in the z direction only) the general equations of continuity and motion get simplified:

- equation of continuity:

$$\frac{\partial v_z}{\partial z} = 0 \qquad (6.72)$$

- equation of motion (only z component) unidirectional flow:

$$\mu\left[\frac{1}{r}\frac{\partial}{\partial r}\left(r\frac{\partial v_z}{\partial r}\right)\right]-\frac{\partial p}{\partial z}+\rho\, g_z=0 \tag{6.73}$$

Forces acting on the system:

$$g_z = g \tag{6.74}$$

$$-\frac{\partial p}{\partial z}=\frac{P_0-P_L}{L} \tag{6.75}$$

Hence:

$$\frac{\mu}{r}\frac{\partial}{\partial r}\left(r\frac{\partial v_z}{\partial r}\right)=-\left(\frac{P_0-P_L}{L}+\rho\, g\right) \tag{6.76}$$

Integrating this equation, we obtain:

$$r\frac{\partial v_z}{\partial r}=-\left(\frac{P_0-P_L}{L}+\rho\, g\right)\frac{r^2}{2\,\mu}+C_1 \tag{6.77}$$

The gradients of velocity and pressure being finite, for the above equation to be valid at r = 0, the value of C_1 should be zero. Using this information, one can integrate equation (6.77) to obtain:

$$v_z=-\left(\frac{P_0-P_L}{L}+\rho\, g\right)\frac{r^2}{4\,\mu}+C_2 \tag{6.78}$$

Boundary condition for the determination of C_2 is:

Boundary condition: $r = R \qquad v_z = 0$

Substituting in equation (6.78):

$$C_2=\left(\frac{P_0-P_L}{L}+\rho\, g\right)\frac{R^2}{4\,\mu} \tag{6.79}$$

The velocity profile is given by:

$$v_z=\left(\frac{P_0-P_L}{L}+\rho\, g\right)\frac{R^2}{4\,\mu}\left[1-\left(\frac{r}{R}\right)^2\right] \tag{6.80}$$

which is the same as equation (5.114), obtained in Chapter 5. Similarly other problems discussed in Chapter 5 can be solved using the general equations.

6.4.3 Tangential annular flow

A schematic view of the equipment for experimentally determining viscosity of liquids using this principle is given in Figure 6.4. This system consists of concentric cylinders with an annular space having the liquid. Viscosity of liquid metals and slags can be determined experimentally, by measuring the torque required to rotate the rod (inner cylinder) at a given speed. The assembly with appropriate material capable of withstanding high temperatures is placed inside a furnace. Viscosity values can be determined at any desired temperature. Commercial viscometers are available for the study of viscosity of liquids at ambient temperatures. These can be adopted for high temperatures with appropriate materials for spindle and crucible. There are specialized equipment manufacturers of viscometers, for the determination of viscosity of liquids at high temperatures.

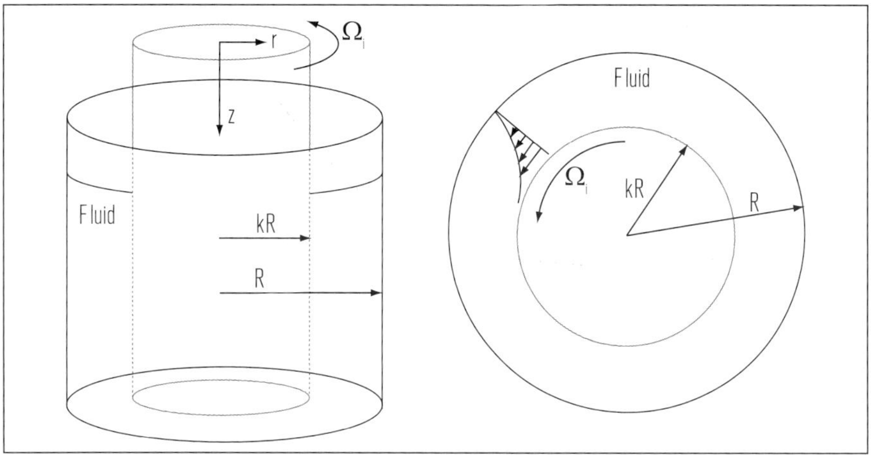

Figure 6.4 - Tangential annular flow between two concentric cylinders.

There are systems with two concentric cylinders (a crucible, a cylindrical rod), where, the internal one can rotate at a speed Ω_i and the outer cylinder is stationary or the external one can rotate and the internal one remains stationary.

Considering laminar flow, fluid density and viscosity constant, one can determine the velocity distribution and shear stress. With this shear stress value, one can relate the torque required to rotate the cylinder and determine the fluid viscosity.

General equations of continuity and motion for components r, θ, and z, given in the tables can be simplified considering the following analysis: $v_z = v_r = 0$ and v_θ is a function of only r; v_θ does not vary with θ (continuity equation). There is no variation of velocity v_θ in the z direction (z component) and hence can be neglected. This approach is reasonably applicable for systems with a high ratio of height to diameter. There is no pressure gradient in the direction θ. The system is in steady state.

(Show that from the general equations of continuity motion in cylindrical coordinates, with the above assumptions, the following equations result)

- equation of continuity:

$$\frac{1}{r}\frac{\partial}{\partial \theta}(\rho\ v_\theta) = 0 \tag{6.81}$$

- equation of motion (component θ) for steady state:

$$\left[\frac{\partial}{\partial r}\left(\frac{1}{r}\ \frac{\partial}{\partial r}(r\ v_\theta)\right)\right] = 0 \tag{6.82}$$

Velocity profile can be determined by integrating equation (6.82).

$$\frac{1}{r}\ \frac{\partial}{\partial r}(r\ v_\theta) = C_1 \tag{6.83}$$

Transposing terms and integrating again, we get:

$$r\ v_\theta = C_1\ \frac{r^2}{2} + C_2 \tag{6.84}$$

$$v_\theta = C_1\ \frac{r}{2} + \frac{C_2}{r} \tag{6.85}$$

The boundary conditions for evaluation of C_1 and C_2 are:

Boundary condition 1: $r = k\,R \quad v_\theta = \Omega_i\ k\,R$

Boundary condition 2: $r = R \quad v_\theta = 0$

Applying the above boundary conditions:

$$\Omega_i\ k\,R = C_1\ \frac{k\,R}{2} + \frac{C_2}{k\,R} \tag{6.86}$$

$$0 = C_1\ \frac{R}{2} + \frac{C_2}{R} \tag{6.87}$$

Combining (6.86) and (6.87):

$$\Omega_i \, k \, R = C_2 \left(\frac{1}{k\,R} - \frac{k}{R} \right) = \frac{C_2}{k\,R} \left(1 - k^2\right) \tag{6.88}$$

Rearranging:

$$C_2 = \frac{\Omega_i \, k^2 \, R^2}{\left(1 - k^2\right)} \tag{6.89}$$

Substituting the value of C_2 in (6.87):

$$0 = C_1 \, \frac{R}{2} + \frac{\Omega_i \, k^2 \, R}{\left(1 - k^2\right)} \tag{6.90}$$

Therefore:

$$C_1 = \frac{-2 \, \Omega_i \, k^2}{\left(1 - k^2\right)} \tag{6.91}$$

Combining these results, the velocity profile is given by:

$$v_\theta = - \frac{\Omega_i \, k^2}{\left(1 - k^2\right)} \, r + \frac{\Omega_i \, k^2 \, R^2}{\left(1 - k^2\right)} \, \frac{1}{r} \tag{6.92}$$

Rearranging, the velocity profile is given by:

$$v_\theta = \frac{\Omega_i \, k^2}{\left(1 - k^2\right)} \left(\frac{R^2}{r} - r \right) \tag{6.93}$$

The shear stress $\tau_{r\theta}$ is given by the following relation taken from Table 6.5:

$$\tau_{r\theta} = - \mu \left[r \, \frac{\partial}{\partial r} \left(\frac{v_\theta}{r} \right) + \frac{1}{r} \, \frac{\partial v_r}{\partial \theta} \right] \tag{6.94}$$

$$\frac{\partial v_r}{\partial \theta} = 0 \tag{6.95}$$

Equation 6.93 can be written as:

$$\frac{v_\theta}{r} = \frac{\Omega_i \, k^2}{\left(1 - k^2\right)} \left(\frac{R^2}{r^2} - 1 \right) \tag{6.96}$$

Substituting 6.96 in 6.94.

$$\tau_{r\theta} = - \mu \left[r \, \frac{\partial}{\partial r} \left(\frac{\Omega_i \, k^2}{\left(1 - k^2\right)} \right) \left(\frac{R^2}{r^2} - 1 \right) \right] \tag{6.97}$$

By differentiating and rearranging:

$$\tau_{r\theta} = -\mu\, r \frac{\Omega_i\, k^2\, R^2}{\left(1-k^2\right)} \left(-\frac{2}{r^3}\right) \qquad (6.98)$$

At r = kR, the shear stress is given by:

$$\tau_{r\theta}\big|_{r=kR} = 2\,\mu \frac{\Omega_i}{\left(1-k^2\right)} \qquad (6.99)$$

The torque Ψ, needed to rotate the inner cylinder, is the product of force (stress x area) and the length of lever arm (kR). This is equal to the frictional force acting on the surface due to the fluid, which is a function of viscosity, as given by the equation below:

$$\Psi = (2\,\pi\,k\,R\,H)\,(\tau_{r\theta}\big|_{r=kR})\,(k\,R) = 4\,\pi\,k^2\,R^2\,H \frac{\Omega_i}{\left(1-k^2\right)}\,\mu \qquad (6.100)$$

where H is the height of the cylinder in contact with the fluid.

Using equation 6.100, it is possible to determine the viscosity of the fluid through the assessment of torque required. There is a linear relationship between these two quantities. This type of viscometer is called Couette-Hatschek viscometer.

6.4.4 Surface profile of a rotating liquid

A fluid density and viscosity constant is contained in a cylindrical container of radius R, as shown in Figure 6.5.

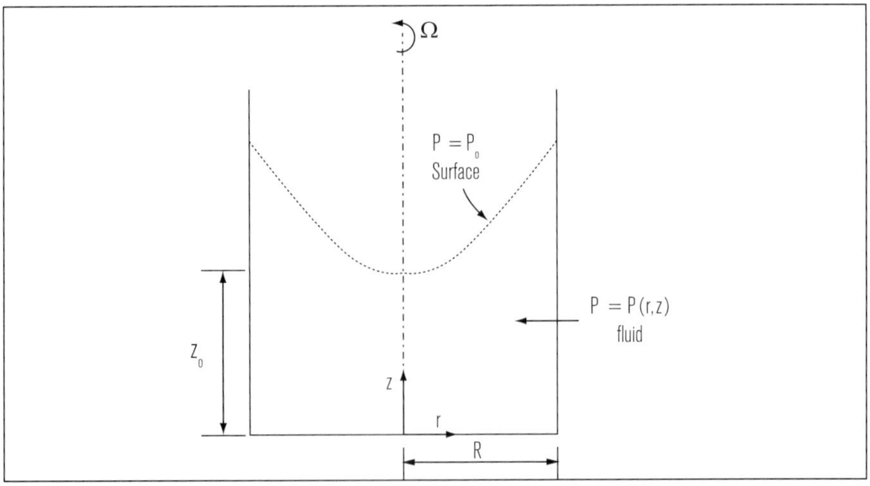

Figure 6.5 -Surface profile of a rotating liquid.

The orientation of the cylinder is such that $g_r = g_\theta = 0$ and g z is (-g), due to the choice of the coordinate system. Equations of motion and continuity determine the shape of the liquid surface in steady state conditions. Obviously, for the system shown in Figure 6.5, cylindrical coordinates have to be used. Considering $v_z = v_r = 0$ and v_θ is a function of r only, the general equations of motion for components r, θ, and z can be simplified. In addition, the derivative of v_θ with θ is zero (continuity equation). There is no pressure gradient in the direction θ. (Show that the simplification of the general equations of motion in cylindrical coordinates, with the above assumptions result in the following equations):

- Component θ:

$$\rho \frac{v_\theta^2}{r} = \frac{\partial P}{\partial r} \qquad (6.101)$$

- Component r:

$$\mu \left[\frac{\partial}{\partial r} \left(\frac{1}{r} \frac{\partial (r v_\theta)}{\partial r} \right) \right] = 0 \qquad (6.102)$$

- Component z:

$$-\frac{\partial P}{\partial z} - \rho g = 0 \qquad (6.103)$$

Integration of the differential equation of θ component gives:

$$v_\theta = C_1 \frac{r}{2} + \frac{C_2}{r} \qquad (6.104)$$

The boundary conditions for evaluation of C_1 and C_2 are:

Boundary condition 1: $r = 0 \quad v_\theta = \text{finite}$

Boundary condition 2: $r = R \quad v_\theta = \Omega R$

Using these boundary conditions, in equation 6.104, one gets:

$$C_1 = 2\Omega$$

$$C_2 = 0$$

Substituting in equation 6.104:

$$v_\theta = \Omega r \qquad (6.105)$$

This expression can be replaced in the equation of component r to determine the pressure profile:

$$\frac{\partial P}{\partial r} = \rho \frac{v_\theta^2}{r} = \rho\, \Omega^2 r \qquad (6.106)$$

$$\frac{\partial P}{\partial z} = -\rho g \qquad (6.107)$$

Assuming,that the pressure is an analytical function of position, it can be written:

$$dP = \frac{\partial P}{\partial r} dr + \frac{\partial P}{\partial z} dz \qquad (6.108)$$

Substituting (6.106) and (6.107) in (6.108), we obtain:

$$dP = \rho \Omega^2 r\, dr - \rho g\, dz \qquad (6.109)$$

Integrating both sides of the equation (6.109), we have:

$$P = \rho\, \Omega^2 \frac{r^2}{2} - \rho g z + C_3 \qquad (6.110)$$

The boundary condition for the evaluating C_3 is:

Boundary condition: $r = 0,\ z = z_o$: $P = P_o$

$$C_3 = P_o + \rho g z_o \qquad (6.111)$$

Substituting 6.111 in 6.110 and rearranging, the pressure distribution is given by:

$$P - P_o = \rho\, \Omega^2 \frac{r^2}{2} - \rho g (z - z_o) \qquad (6.112)$$

The surface is the locus of points where $P = P_o$. Substituting in equation 6.112 one gets the equation that describes the shape of the surface:

$$z - z_0 = \left(\frac{\Omega^2}{2 g}\right) r^2 \qquad (6.113)$$

This is the equation of a parabola, where the point of lowest level occurs exactly in the center of the cylinder.

6.4.5 Laminar flow around a sphere

In this section, the flow of an incompressible fluid around a solid sphere, (Figure 6.6) is analyzed. The fluid approaches the sphere from the bottom upwards along the z axis, with an uniform velocity v∞ (velocity at a point well away from the sphere).

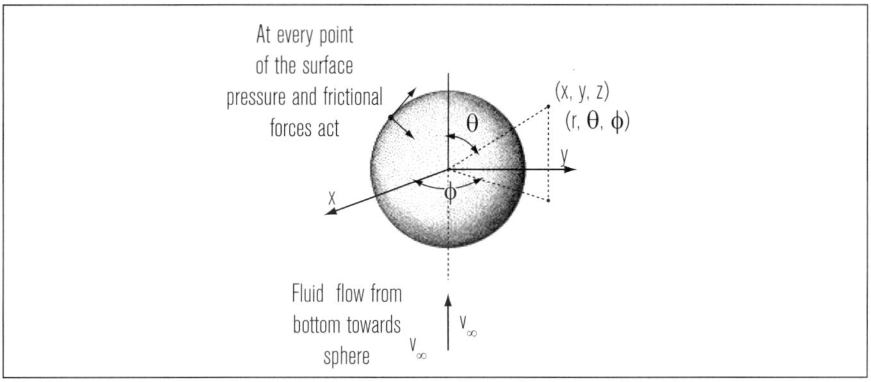

Figure 6.6 - Laminar fluid flow around a sphere.

For evaluating the velocity profile in this case, the assumptions are that the fluid is Newtonian with constant density and viscosity and the system is under steady state conditions. The general equations of continuity and motion in spherical coordinates, given in the tables in the annex, can be simplified. Φ component, as well as all the terms associated with the convective transport of momentum can be neglected for laminar flow conditions in the case of extremely low velocities of the fluid, characterizing creeping flow. (Examine the terms that have been neglected from the general equations given for spherical coordinates and justify the consideration terms given below).

Equations of continuity and motion can be simplified and written as follows:

- Equation of continuity:

$$\rho \left[\frac{1}{r^2} \frac{\partial}{\partial r} (r^2 v_r) + \frac{1}{r \sin \vartheta} \frac{\partial}{\partial \vartheta} (v_\vartheta \sin \vartheta) \right] = 0 \qquad (6.114)$$

- Equations of motion:
  - Component r:

$$\mu \left[\frac{1}{r^2} \frac{\partial}{\partial r} \left(r^2 \frac{\partial v_r}{\partial r} \right) + \frac{1}{r^2 \sin \vartheta} \frac{\partial}{\partial \vartheta} \left(\sin\vartheta \frac{\partial v_r}{\partial \vartheta} \right) - \frac{2}{r^2} v_r - \frac{2}{r^2} \frac{\partial v_\vartheta}{\partial \vartheta} - \frac{2}{r^2} v_\vartheta \cot \vartheta \right] - \frac{\partial P}{\partial r} + \rho g_r \qquad (6.115)$$

  - Component θ:

$$\mu \left[\frac{1}{r^2} \frac{\partial}{\partial r} \left(r^2 \frac{\partial v_\vartheta}{\partial r} \right) + \frac{1}{r^2 \sin \vartheta} \frac{\partial}{\partial \vartheta} \left(\sin\vartheta \frac{\partial v_\vartheta}{\partial \vartheta} \right) + \frac{2}{r^2} \frac{\partial v_r}{\partial \vartheta} - \frac{v_\vartheta}{r^2 \sin^2 \vartheta} \right] - \frac{1}{r} \frac{\partial P}{\partial \vartheta} + \rho g_\vartheta \qquad (6.116)$$

Equations (6.114) (6.115) and (6.116) were solved analytically (Streeter, quoted by Bird, Stewart and Lightfoot (1960)), to obtain the shear stress distribution, pressure and velocity profiles of the fluid flowing around the sphere:

$$\tau_{r\theta} = \frac{3}{2}\frac{\mu\, v_\infty}{R}\left(\frac{R}{r}\right)^4 \sin\vartheta \tag{6.117}$$

$$P = P_o - \rho\, g\, z - \frac{3}{2}\frac{\mu\, v_\infty}{R}\left(\frac{R}{r}\right)^2 \cos\vartheta \tag{6.118}$$

$$v_r = v_\infty\left[1 - \frac{3}{2}\left(\frac{R}{r}\right) + \frac{1}{2}\left(\frac{R}{r}\right)^3\right]\cos\vartheta \tag{6.119}$$

$$v_\theta = -\,v_\infty\left[1 - \frac{3}{4}\left(\frac{R}{r}\right) - \frac{1}{4}\left(\frac{R}{r}\right)^3\right]\sin\vartheta \tag{6.120}$$

where:

- P_o = pressure in tplane z = 0, (far away from the sphere);
- V_∞ = approach velocity of the fluid.

Boundary conditions are:

$$\text{At, } r = R\text{: } v_r = v_\theta = 0$$

$$\text{At, } r = \infty\text{: } v_z = v_\infty$$

The equations (6.117) to (6.120) are valid for Reynolds numbers ($D.v_\infty.\rho/\mu$) less than one. From the above equations, one can evaluate the force exerted by the fluid on the surface of the sphere The total force is determined by integrating the normal and tangential forces acting on the surface of the sphere. The normal force acting on the solid is due to the pressure given by equation (6.118):

- For r = R ; z = R cos θ;
- F_n = normal force < 0 for 0 < θ < $\pi/2$,
- $F_{n>}$ > 0 for $\theta > \pi/2$.

Hence the vertical component of this force is given by:

$$F_n = \int_{\varphi=0}^{\varphi=2\pi}\int_{\vartheta=0}^{\vartheta=\pi}\left(-P|_{r=R}\cos\vartheta\right) R^2 \sin\vartheta\, d\vartheta\, d\varphi \tag{6.121}$$

z component of force acting on area element

The area element is shown in Figure 6.7.

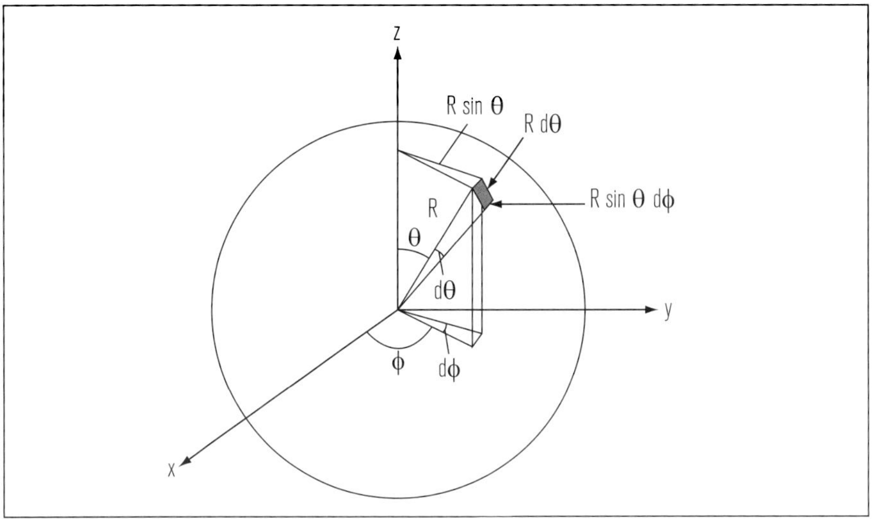

Figure 6.7 - Area element on the surface of a sphere.

Substituting 6.118, the equation for the pressure, in equation 6.121 one gets:

$$F_n = \int_{\varphi=0}^{\varphi=2\pi} \int_{\vartheta=0}^{\vartheta=\pi} \left[-\left(P_o - \rho g z - \frac{3}{2} \frac{\mu v_\infty}{R} \cos\vartheta \right) \cos\vartheta \right] R^2 \sin\vartheta \, d\vartheta \, d\varphi \qquad (6.122)$$

Recalling that $z = R \cos\theta$, and integrating the above equation,we obtain:

$$F_n = \frac{4}{3} \pi R^3 \rho g + 2 \pi \mu R V_\infty \qquad (6.123)$$

(Exercise: Derive the various mathematical steps for arriving at 6.123 from 6.122)

In this equation, the first term in the right side is the buoyancy force and the second term is called form drag

At each point of the surface there is a shear stress acting tangentially.. z component of this force is given by:

$$F_t = \int_{\varphi=0}^{\varphi=2\pi} \int_{\vartheta=0}^{\vartheta=\pi} \left(T_{r\vartheta}\big|_{r=R} \sin\vartheta \right) R^2 \sin\vartheta \, d\vartheta \, d\varphi \tag{6.124}$$

Component Area element

Substituting for $\tau_{r\theta}$ - equation (6.117)-, we obtain:

$$F_t = \int_{\varphi=0}^{\varphi=2\pi} \int_{\vartheta=0}^{\vartheta=\pi} \left[\left(\frac{3}{2} \frac{\mu \, v_\infty}{R} \sin\vartheta \right) \sin\vartheta \right] R^2 \sin\vartheta \, d\vartheta \, d\varphi \tag{6.125}$$

Integrating: (Elaborate the various mathematical steps from 6.125 to get 6.126)

$$F_t = 4 \pi \mu R \, v_\infty \tag{6.126}$$

This force is called *friction drag.*

Thus, the total force exerted by the fluid on the sphere is given by:

$$F_n + F_t = \frac{4}{3} \pi R^3 \rho g + 2 \pi \mu R V_\infty + 4 \pi \mu R V_\infty \tag{6.127}$$

$$F_n + F_t = \frac{4}{3} \pi R^3 \rho g + 6 \pi \mu R V_\infty \tag{6.128}$$

It is common to designate the two terms on the right side of the equation above as follows:

$$F_s = \frac{4}{3} \pi R^3 \rho g \tag{6.129}$$

This is the force that would be exercised even if the fluid is not moving (buoyant force).

$$F_k = 6 \pi \mu R V_\infty \tag{6.130}$$

This force arises due to the movement of fluid. Equation (6.130) is known as Stokes' law and is valid for Reynolds numbers less than 1.

Example

Develop a relationship for assessing the viscosity of a fluid by measuring the speed of a falling sphere in a fluid when it reaches the steady state. Assume laminar flow.

Solution

The sphere inside the liquid initially has zero velocity. lt accelerates due to gravity until it reaches constant speed (terminal velocity). At this stage, the sum of the forces acting on the sphere is zero. The force due to the weight of the solid during the fall acts downwards while the thrust or buoyant force and friction drag act in the opposite direction, as shown in the figure below:

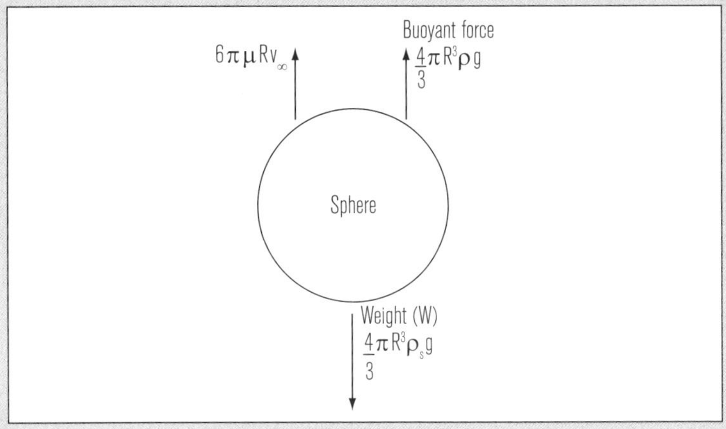

As the sum of forces is zero, we have:

$$W = \frac{4}{3}\pi R^3 \rho_s g = \frac{4}{3}\pi R^3 \rho g + 6\pi\mu R V_t$$

where ρ_s is the density of the solid and V_t is the terminal velocity of the spherel. The viscosity of the fluid is given by:

$$\mu = \frac{2 R^2 (\rho_s - \rho) g}{9 V_t}$$

As mentioned earlier, the above relationship is valid for Re <1. The principle described above is the basis of viscosity measurements using falling sphere viscometers. For example a platinum bob can be used in a slag melt for the determination of viscosity by measuring the the terminal velocity.

6.4.6 Boundary layer

The figure 6.8 shows the velocity profile of a fluid flowing parallel to a flat plate.

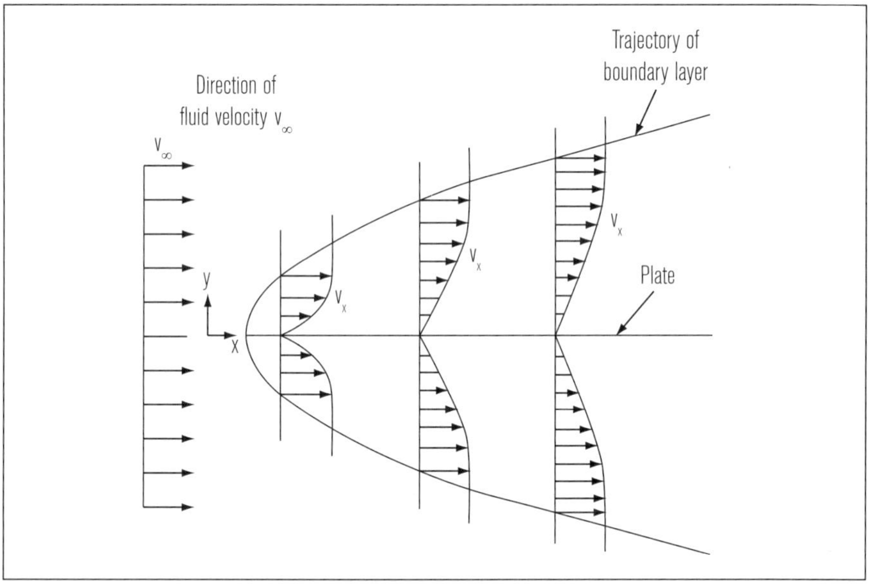

Figure 6.8 - Velocity profile of a fluid flowing parallel to a submerged flat plate.

Before reaching the plate, the fluid has uniform velocity v_∞. When the liquid stream hits the plate, it is observed, that the velocity increases from zero along the surface to a value almost near v_∞ at a distance δ from the wall. This region up to where v_x/v_∞ reaches a value of 0.99 from 0 at the surface, is called the boundary layer.

The location of the point, where $v_x/v_\infty = \delta$ is 0.99, is defined as the boundary layer thickness. At the beginning of the plate (x = 0), δ is equal to zero, increasing gradually with higher values of x, until it reaches δ.

In the case of such problems where fluid flow is in contact with a solid under steady state conditions, viscous effects (friction) are felt in the fluid near the solid, ie at regions where $y < \delta$. It is at this region the behavior of the fluid in respect of momentum transfer and velocity profiles have to considered to understand the characteristics of the boundary layer. The region where $y > \delta$, v_x is essentially constant and where the flow becomes normal with velocity v_∞. Figure 6.8 shows that v_x is a function of y as well as x in the boundary layer. Hence determination of this function is not that simple. There is solid fluid interaction in terms of friction. Which is the cause of variation of velocity in the x direction with x. Hence $\partial v/\partial x$ is not zero.

Equation for continuity and motion can be written for this system from the tables in the appendix for cartesian coordinates. There is no flow in the z direction and hence this equation can be neglected. Neglecting the pressure drop as well as gravity terms and considering steady state, constant viscosity, density and neglecting contributions in the y direction (not significant), the equations of continuity and motion for the system, can be simplified. The resulting differential equations can be solved with boundary conditions. (Starting from the original equations of continuity and motion justify the elimination of various terms to arrive at the final form of differential equations).

- Equation of continuity:

$$\frac{\partial v_x}{\partial x} + \frac{\partial v_y}{\partial y} = 0 \tag{6.131}$$

- Equation of motion:
  - X component:

$$\rho\left(v_x \frac{\partial v_x}{\partial x} + v_y \frac{\partial v_x}{\partial y}\right) = \mu\left(\frac{\partial^2 v_x}{\partial x^2} + \frac{\partial^2 v_x}{\partial y^2}\right) - \frac{\partial P}{\partial x} \tag{6.132}$$

  - Y component:

$$\rho\left(v_x \frac{\partial v_y}{\partial x} + v_y \frac{\partial v_y}{\partial y}\right) = \mu\left(\frac{\partial^2 v_y}{\partial x^2} + \frac{\partial^2 v_y}{\partial y^2}\right) - \frac{\partial P}{\partial y} + \rho\, g_y \tag{6.133}$$

The above equations can be solved considering that v_y is small compared with v_x and the gradient of v_x in the y direction is much larger than in the x direction. In the x direction, it is also assumed, that the component of the convective transport of momentum is much larger than the component associated with the y direction.

The boundary conditions considered are:

a) $x = 0$: $v_x = v_\infty \quad v_y = 0$

b) $y = 0$: $v_x = 0 \quad v_y = 0$

c) $y \rightarrow \infty$: $v_x = v_\infty \quad v_y = 0$

Equations (6.131) to (6.133) can be solved analytically (BIRD, STEWART and LIGHTFOOT, 1960), converting the partial differential equation to a homogeneous 3rd order differential equation using the concept of stream function and other parameters which influence the boundary layer. This equation can be solved and the velocity profile in the boundary layer, its thickness as well as the position for $v_x/v_\infty = 0.99$ can be evaluated. The results are expressed in terms of velocity v_∞, position along the plate and kinematic viscosity of the fluid. The relationship obtained is given below:

$$\delta = 5.0\left(\frac{\mu}{\rho\, v_\infty}\right)^{1/2} x^{1/2} \tag{6.134}$$

Dividing both sides by x, equation (6.134) becomes dimensionless:

$$\frac{\delta}{x} = 5.0\left(\frac{\mu}{\rho\, v_\infty\, x}\right)^{1/2} \tag{6.135}$$

Recalling the definition of Reynolds number,the above equation gives:

$$\frac{\delta}{x} = 5.0\left(\mathrm{Re}_x\right)^{-1/2} \tag{6.136}$$

where the Reynolds number is evaluated for each position x (the x position is used as the characteristic length in defining the Reynolds number).

The thickness of the boundary layer provides a measure of the region of fluid that is affected by the presence of the solid plate. In this region, the effects of viscosity (and friction) are more significant. Outside this region, the fluid velocity is practically uniform and the effects of viscosity are negligible. The relation (6.136) gives the thickness of the boundary layer in terms of Reynolds number. It tends to become smaller with the increase in the Reynolds number. Boundary layer plays an important part in the absorption of gases by liquid metals during flow from ladles and flow in contact with atmospheric gases. Boundary layer in geometries like flow in tubes, flow around sphere etc have also been investigated.

6.4.7 Unsteady flow in a circular tube

This section illustrates a procedure for solving a problem of unsteady flow, in which the velocity varies with time. A fluid of constant density and viscosity is contained within a long horizontal pipe of length L and radius R. Initially the fluid is at rest. At time t = 0, the system is subjected to a pressure gradient : $(P_0 - P_L)/L$. The objective is to determine the variation of velocity profiles of the fluid with time.

For getting the velocity profiles first, the general equations of continuity and motion given in the tables for cylindrical coordinates have to be t simplified considering the characteristics of this system. For the system, $v_r = v_\theta = 0$. Si The flow is in z direction only. Hence $v_z = v_z\,(r, t)$. The simplified equations considering these aspects are given below:

- Equation of continuity:

$$\frac{\partial v_z}{\partial z} = 0 \tag{6.137}$$

- Equation of motion(z component only):

$$\rho\,\frac{\partial v_z}{\partial t} = \frac{P_o - P_L}{L} + \mu\left[\frac{1}{r}\,\frac{\partial}{\partial r}\left(r\,\frac{\partial v_z}{\partial r}\right)\right] \tag{6.138}$$

The initial and boundary conditions are:

$$t = 0 \quad v_z = 0 \qquad 0 \le r \le R$$

$$a)\; t > 0 \quad v_z = 0 \quad r = R$$

$$b)\; t > 0 \quad \frac{\partial v_z}{\partial r} = \text{finite} \;\; r = 0$$

Equation (6.138), can be solved with initial and boundary conditions, using power series. The results are shown in Figure 6.9, where the dimensionless velocity profiles along the radius of the tube are given as a function of time in the form of a dimensionless number involving kinematic viscosity and radius of the tube.

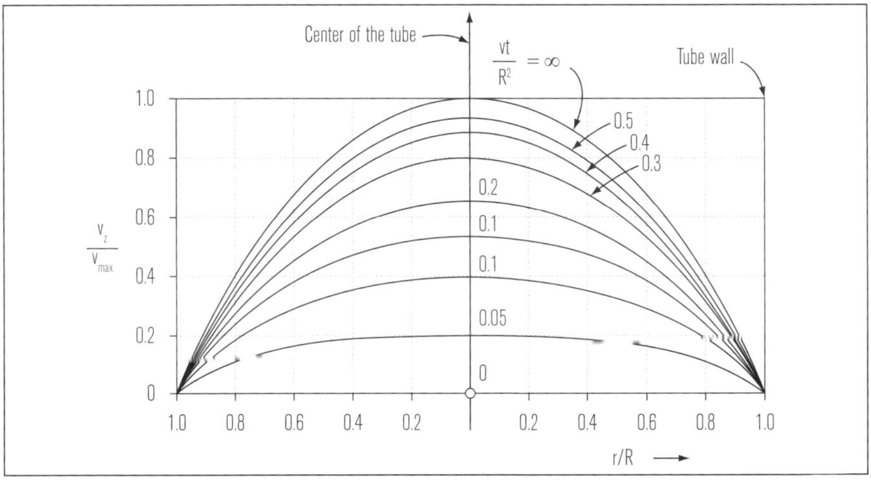

Figure 6.9 – Velocityprofiles for unsteady flow in a circular tube (BIRD, STEWART and LIGHFOOT, 1960).

These types of problems involving steady or unsteady state, two or three dimensions can be solved by numerical methods, since many of them are difficult to solve by analytical methods. A number of computer programs have been developed for the purpose and commercial softwares based on them, are available. These require only the definition of geometry and boundary conditions of the problem and one can easily determine the velocity and pressure profiles as a function of time.

References

BIRD, R.B.; STEWART, W.E.; LIGHTFOOT, E.N. *Transport phenomena.* New York: John Wiley & Sons, 1960.
LAMB, H. *Hydrodynamics.* New York: Dover, 1945.

Appendix

Continuity: $\frac{\partial \rho}{\partial t}+\frac{\partial}{\partial x}(\rho v_x)+\frac{\partial}{\partial y}(\rho v_y)+\frac{\partial}{\partial z}(\rho v_z)=0$
Equation of motion in terms of shear stress
• x component:
$\rho\left(\frac{\partial v_x}{\partial t}+v_x\frac{\partial v_x}{\partial x}+v_y\frac{\partial v_x}{\partial y}+v_z\frac{\partial v_x}{\partial z}\right)=-\left(\frac{\partial \tau_{xx}}{\partial x}+\frac{\partial \tau_{yx}}{\partial y}+\frac{\partial \tau_{zx}}{\partial z}\right)-\frac{\partial P}{\partial x}+\rho g_x$
• y component:
$\rho\left(\frac{\partial v_y}{\partial t}+v_x\frac{\partial v_y}{\partial x}+v_y\frac{\partial v_y}{\partial y}+v_z\frac{\partial v_y}{\partial z}\right)=-\left(\frac{\partial \tau_{xy}}{\partial x}+\frac{\partial \tau_{yy}}{\partial y}+\frac{\partial \tau_{zy}}{\partial z}\right)-\frac{\partial P}{\partial y}+\rho g_y$
• z component:
$\rho\left(\frac{\partial v_z}{\partial t}+v_x\frac{\partial v_z}{\partial x}+v_y\frac{\partial v_z}{\partial y}+v_z\frac{\partial v_z}{\partial z}\right)=-\left(\frac{\partial \tau_{xz}}{\partial x}+\frac{\partial \tau_{yz}}{\partial y}+\frac{\partial \tau_{zz}}{\partial z}\right)-\frac{\partial P}{\partial z}+\rho g_z$

Table 6.1 - Equations of continuity and motion in Cartesian coordinates.

$\tau_{xx}=-2\mu\frac{\partial v_x}{\partial x}+\frac{2}{3}\mu\left(\frac{\partial v_x}{\partial x}+\frac{\partial v_y}{\partial y}+\frac{\partial v_z}{\partial z}\right)$
$\tau_{yy}=-2\mu\frac{\partial v_y}{\partial y}+\frac{2}{3}\mu\left(\frac{\partial v_x}{\partial x}+\frac{\partial v_y}{\partial y}+\frac{\partial v_z}{\partial z}\right)$
$\tau_{zz}=-2\mu\frac{\partial v_z}{\partial z}+\frac{2}{3}\mu\left(\frac{\partial v_x}{\partial x}+\frac{\partial v_y}{\partial y}+\frac{\partial v_z}{\partial z}\right)$
$\tau_{yx}=\tau_{xy}=-\mu\left(\frac{\partial v_x}{\partial y}+\frac{\partial v_y}{\partial x}\right)$
$\tau_{zx}=\tau_{xz}=-\mu\left(\frac{\partial v_x}{\partial z}+\frac{\partial v_z}{\partial x}\right)$
$\tau_{yz}=\tau_{zy}=-\mu\left(\frac{\partial v_z}{\partial y}+\frac{\partial v_y}{\partial z}\right)$

Table 6.2 - Normal and shear forces for a Newtonian fluid. In terms of velocity gradients (Cartesian coordinates).

• x component:

$$\rho\left(\frac{\partial v_x}{\partial t}+v_x\frac{\partial v_x}{\partial x}+v_y\frac{\partial v_x}{\partial y}+v_z\frac{\partial v_x}{\partial z}\right)=\mu\left(\frac{\partial^2 v_x}{\partial x^2}+\frac{\partial^2 v_x}{\partial y^2}+\frac{\partial^2 v_x}{\partial z^2}\right)-\frac{\partial P}{\partial x}+\rho g_x$$

• y component:

$$\rho\left(\frac{\partial v_y}{\partial t}+v_x\frac{\partial v_y}{\partial x}+v_y\frac{\partial v_y}{\partial y}+v_z\frac{\partial v_y}{\partial z}\right)=\mu\left(\frac{\partial^2 v_y}{\partial x^2}+\frac{\partial^2 v_y}{\partial y^2}+\frac{\partial^2 v_y}{\partial z^2}\right)-\frac{\partial P}{\partial y}+\rho g_y$$

• z component:

$$\rho\left(\frac{\partial v_z}{\partial t}+v_x\frac{\partial v_z}{\partial x}+v_y\frac{\partial v_z}{\partial y}+v_z\frac{\partial v_z}{\partial z}\right)=\mu\left(\frac{\partial^2 v_z}{\partial x^2}+\frac{\partial^2 v_z}{\partial y^2}+\frac{\partial^2 v_z}{\partial z^2}\right)-\frac{\partial P}{\partial z}+\rho g_z$$

Table 6.3 - Equations of motion in terms of velocity gradients for a Newtonian fluid of constant density and viscosity (Cartesian coordinates).

Continuity:

$$\frac{\partial \rho}{\partial t}+\frac{1}{r}\frac{\partial}{\partial r}(\rho\, r\, v_r)+\frac{1}{r}\frac{\partial}{\partial \theta}(\rho\, v_\theta)+\frac{\partial}{\partial z}(\rho\, v_z)=0$$

Equations of motion in terms of normal force and shear stress

• r component:

$$\rho\left(\frac{\partial v_r}{\partial t}+v_r\frac{\partial v_r}{\partial r}+\frac{v_\theta}{r}\frac{\partial v_r}{\partial \theta}-\frac{v_\theta^2}{r}+v_z\frac{\partial v_r}{\partial z}\right)=-\left(\frac{1}{r}\frac{\partial(r\tau_{rr})}{\partial r}+\frac{1}{r}\frac{\partial \tau_{\theta r}}{\partial \theta}-\frac{\tau_{\theta\theta}}{r}+\frac{\partial \tau_{zr}}{\partial z}\right)-\frac{\partial P}{\partial r}+\rho g_r$$

• θ component:

$$\rho\left(\frac{\partial v_\theta}{\partial t}+v_r\frac{\partial v_\theta}{\partial r}+\frac{v_\theta}{r}\frac{\partial v_\theta}{\partial \theta}+\frac{v_\theta v_r}{r}+v_z\frac{\partial v_\theta}{\partial z}\right)=-\left(\frac{1}{r^2}\frac{\partial(r^2\tau_{r\theta})}{\partial r}+\frac{1}{r}\frac{\partial \tau_{\theta\theta}}{\partial \theta}+\frac{\partial \tau_{z\theta}}{\partial z}\right)-\frac{1}{r}\frac{\partial P}{\partial \theta}+\rho g_\theta$$

• z component:

$$\rho\left(\frac{\partial v_z}{\partial t}+v_r\frac{\partial v_z}{\partial r}+\frac{v_\theta}{r}\frac{\partial v_z}{\partial \theta}+v_z\frac{\partial v_z}{\partial z}\right)=-\left(\frac{1}{r}\frac{\partial(r\,\tau_{rz})}{\partial r}+\frac{1}{r}\frac{\partial \tau_{\theta z}}{\partial \theta}+\frac{\partial \tau_{zz}}{\partial z}\right)-\frac{\partial P}{\partial z}+\rho g_z$$

Table 6.4 - Equations of continuity and motion in cylindrical coordinates.

$$\tau_{rr} = -\mu\left[2\frac{\partial v_r}{\partial r} - \frac{2}{3}(\Delta \bullet v)\right]$$

$$\tau_{\theta\theta} = -\mu\left[2\left(\frac{1}{r}\frac{\partial v_\theta}{\partial \theta} + \frac{v_r}{r}\right) - \frac{2}{3}(\Delta \bullet v)\right]$$

$$\tau_{zz} = -\mu\left[2\frac{\partial v_z}{\partial z} - \frac{2}{3}(\Delta \bullet v)\right]$$

$$\tau_{rz} = \tau_{zr} = -\mu\left(\frac{\partial v_z}{\partial r} + \frac{\partial v_r}{\partial z}\right)$$

$$\tau_{z\theta} = \tau_{\theta z} = -\mu\left(\frac{\partial v_\theta}{\partial z} + \frac{1}{r}\frac{\partial v_z}{\partial \theta}\right)$$

$$\tau_{r\theta} = \tau_{\theta r} = -\mu\left[r\frac{\partial}{\partial r}\left(\frac{v_\theta}{r}\right) + \frac{1}{r}\frac{\partial v_r}{\partial \theta}\right]$$

$$(\Delta \bullet v) = \frac{1}{r}\frac{\partial}{\partial r}(r\, v_r) + \frac{1}{r}\frac{\partial v_\theta}{\partial \theta} + \frac{\partial v_z}{\partial z}$$

Table 6.5 - Normal and shear forces in terms of velocity gradients for Newtonian fluid (Cylindrical coordinates).

Table 6.6 - Equations of motion in terms of velocity gradients for a Newtonian fluid of constant density and viscosity (Cylindrical coordinates).

• component r:

$$\rho\left(\frac{\partial v_r}{\partial t}+v_r\frac{\partial v_r}{\partial r}+\frac{v_\theta}{r}\frac{\partial v_r}{\partial \theta}-\frac{v_\theta^2}{r}+v_z\frac{\partial v_r}{\partial z}\right)=-\frac{\partial P}{\partial r}+\rho g_r+$$

$$+\mu\left[\frac{\partial}{\partial r}\left(\frac{1}{r}\frac{\partial(r v_r)}{\partial r}\right)+\frac{1}{r^2}\frac{\partial^2 v_r}{\partial \theta^2}-\frac{2}{r^2}\frac{\partial v_\theta}{\partial \theta}+\frac{\partial^2 v_r}{\partial z^2}\right]$$

• component θ:

$$\rho\left(\frac{\partial v_\theta}{\partial t}+v_r\frac{\partial v_\theta}{\partial r}+\frac{v_\theta}{r}\frac{\partial v_\theta}{\partial \theta}+\frac{v_\theta v_r}{r}+v_z\frac{\partial v_\theta}{\partial z}\right)=-\frac{1}{r}\frac{\partial P}{\partial \theta}+\rho g_\theta$$

$$+\mu\left[\frac{\partial}{\partial r}\left(\frac{1}{r}\frac{\partial(r v_\theta)}{\partial r}\right)+\frac{1}{r^2}\frac{\partial^2 v_\theta}{\partial \theta^2}+\frac{2}{r^2}\frac{\partial v_r}{\partial \theta}+\frac{\partial^2 v_\theta}{\partial z^2}\right]$$

• component z:

$$\rho\left(\frac{\partial v_z}{\partial t}+v_r\frac{\partial v_z}{\partial r}+\frac{v_\theta}{r}\frac{\partial v_z}{\partial \theta}+v_z\frac{\partial v_z}{\partial z}\right)=-\frac{\partial P}{\partial z}+\rho g_z+$$

$$+\mu\left[\frac{1}{r}\frac{\partial}{\partial r}\left(r\frac{\partial v_z}{\partial r}\right)+\frac{1}{r^2}\frac{\partial^2 v_z}{\partial \theta^2}+\frac{\partial^2 v_z}{\partial z^2}\right]$$

Continuity:

$$\frac{\partial\rho}{\partial t}+\frac{1}{r^2}\frac{\partial}{\partial r}(\rho r^2 v_r)+\frac{1}{r\sin\theta}\frac{\partial}{\partial\theta}(\rho v_\theta \sin\theta)+\frac{1}{r\sin\theta}\frac{\partial}{\partial\varphi}(\rho v_\varphi)=0$$

Equation of motion in terms of normal and shear stress (spherical coordinates)

• component r:

$$\rho\left(\frac{\partial v_r}{\partial t}+v_r\frac{\partial v_r}{\partial r}+\frac{v_\theta}{r}\frac{\partial v_r}{\partial\theta}-\frac{v_\theta^2+v_\varphi^2}{r}+\frac{v_\varphi}{r\sin\theta}\frac{\partial v_r}{\partial\varphi}\right)=$$

$$-\left[\frac{1}{r^2}\frac{\partial}{\partial r}(r^2\tau_{rr})+\frac{1}{r\sin\theta}\frac{\partial}{\partial\theta}(\tau_{r\theta}\sin\theta)+\frac{1}{r\sin\theta}\frac{\partial\tau_{r\varphi}}{\partial\varphi}-\frac{\tau_{\theta\theta}+\tau_{\varphi\varphi}}{r}\right]-\frac{\partial P}{\partial r}+\rho g_r$$

• component θ:

$$\rho\left(\frac{\partial v_\theta}{\partial t}+v_r\frac{\partial v_\theta}{\partial r}+\frac{v_\theta}{r}\frac{\partial v_\theta}{\partial\theta}-\frac{v_\varphi^2\cot\theta}{r}+\frac{v_\varphi}{r\sin\theta}\frac{\partial v_\theta}{\partial\varphi}+\frac{v_r v_\theta}{r}\right)=$$

$$-\left[\frac{1}{r^2}\frac{\partial}{\partial r}(r^2\tau_{r\theta})+\frac{1}{r\sin\theta}\frac{\partial}{\partial\theta}(\tau_{\theta\theta}\sin\theta)+\frac{1}{r\sin\theta}\frac{\partial\tau_{\theta\varphi}}{\partial\varphi}+\frac{\tau_{r\theta}}{r}-\frac{\cot\theta}{r}\tau_{\varphi\varphi}\right]-\frac{1}{r}\frac{\partial P}{\partial\theta}+\rho g_\theta$$

• component ϕ:

$$\rho\left(\frac{\partial v_\varphi}{\partial t}+v_r\frac{\partial v_\varphi}{\partial r}+\frac{v_\theta}{r}\frac{\partial v_\varphi}{\partial\theta}+\frac{v_\varphi}{r\sin\theta}\frac{\partial v_\varphi}{\partial\varphi}+\frac{v_\varphi v_r}{r}+\frac{v_\theta v_\varphi}{r}\cot\theta\right)=$$

$$-\left[\frac{1}{r^2}\frac{\partial}{\partial r}(r^2\tau_{r\varphi})+\frac{1}{r}\frac{\partial\tau_{\theta\varphi}}{\partial\theta}+\frac{1}{r\sin\theta}\frac{\partial\tau_{\varphi\varphi}}{\partial\varphi}+\frac{\tau_{r\varphi}}{r}+\frac{2\cot\theta}{r}\tau_{\theta\varphi}\right]-\frac{1}{r\sin\theta}\frac{\partial P}{\partial\varphi}+\rho g_\varphi$$

Table 6.7 - Equations of continuity and motion (Spherical coordinates).

$$\tau_{rr} = -\mu\left[2\frac{\partial v_r}{\partial r} - \frac{2}{3}(\Delta \bullet v)\right]$$

$$\tau_{\theta\theta} = -\mu\left[2\left(\frac{1}{r}\frac{\partial v_\theta}{\partial \theta} + \frac{v_r}{r}\right) - \frac{2}{3}(\Delta \bullet v)\right]$$

$$\tau_{\varphi\varphi} = -\mu\left[2\left(\frac{1}{r\sin\theta}\frac{\partial v_\varphi}{\partial \varphi} + \frac{v_r}{r} + \frac{v_\theta \cot\theta}{r}\right) - \frac{2}{3}(\Delta \bullet v)\right]$$

$$\tau_{\varphi\theta} = \tau_{\theta\varphi} = -\mu\left[\frac{\sin\theta}{r}\frac{\partial}{\partial \theta}\left(\frac{v_\varphi}{\sin\theta}\right) + \frac{1}{r\sin\theta}\frac{\partial v_\theta}{\partial \varphi}\right]$$

$$\tau_{r\varphi} = \tau_{\varphi r} = -\mu\left[\frac{1}{r\sin\theta}\frac{\partial v_r}{\partial \varphi} + r\frac{\partial}{\partial r}\left(\frac{v_\varphi}{r}\right)\right]$$

$$\tau_{r\theta} = \tau_{\theta r} = -\mu\left[r\frac{\partial}{\partial r}\left(\frac{v_\theta}{r}\right) + \frac{1}{r}\frac{\partial v_r}{\partial \theta}\right]$$

$$(\Delta \bullet v) = \frac{1}{r^2}\frac{\partial}{\partial r}(r^2 v_r) + \frac{1}{r\sin\theta}\frac{\partial}{\partial \theta}(v_\theta \sin\theta) + \frac{1}{r\sin\theta}\frac{\partial v_\varphi}{\partial \varphi}$$

Table 6.8 - Normal and shear stresses in terms of velocity gradients for Newtonian fluids (Spherical coordinates).

• r component:

$$\rho\left[\frac{\partial v_r}{\partial t}+v_r\frac{\partial v_r}{\partial r}+\frac{v_\theta}{r}\frac{\partial v_r}{\partial \theta}-\frac{v_\theta^{\,2}+v_\varphi^{\,2}}{r}+\frac{v_\varphi}{r\sin\theta}\frac{\partial v_r}{\partial \varphi}\right]=$$

$$-\mu\left[\Delta^2 v_r-\frac{2}{r^2}v_r-\frac{2}{r^2}\frac{\partial v_\theta}{\partial \theta}-\frac{2}{r^2}v_\theta\cot\theta-\frac{2}{r^2\sin\theta}\frac{\partial v_\varphi}{\partial \varphi}\right]-\frac{\partial P}{\partial r}+\rho\, g_r$$

• component θ:

$$\rho\left(\frac{\partial v_r}{\partial t}+v_r\frac{\partial v_r}{\partial r}+\frac{v_\theta}{r}\frac{\partial v_r}{\partial \theta}-\frac{v_\theta^{\,2}+v_\varphi^{\,2}}{r}+\frac{v_\varphi}{r\sin\theta}\frac{\partial v_r}{\partial \varphi}\right)=$$

$$-\mu\left(\Delta^2 v_r-\frac{2}{r^2}v_r-\frac{2}{r^2}\frac{\partial v_\theta}{\partial \theta}-\frac{2}{r^2}v_\theta\cot\theta-\frac{2}{r^2\sin\theta}\frac{\partial v_\varphi}{\partial \varphi}\right)-\frac{\partial P}{\partial r}+\rho\, g_r$$

• component ϕ:

$$\rho\left(\frac{\partial v_\varphi}{\partial t}+v_r\frac{\partial v_\varphi}{\partial r}+v_\theta+\frac{\partial v_\varphi}{\partial \theta}+\frac{v_\varphi}{r\sin\theta}\frac{\partial v_\varphi}{\partial \varphi}+\frac{v_\varphi v_r}{r}+\frac{v_\theta v_\varphi}{r}\cot\theta\right)=$$

$$-\mu\left(\nabla^2 v_\varphi-\frac{v_\varphi}{r^2\sin^2\theta}+\frac{2}{r^2\sin\theta}\frac{\partial v_r}{\partial \varphi}+\frac{2}{r^2}\frac{\cos\theta}{\sin^2\theta}\frac{\partial v_\theta}{\partial \varphi}\right)-\frac{1}{r\sin\theta}\frac{\partial P}{\partial \varphi}+\rho\, g_\varphi$$

In the above equations:

$$\nabla^2=\frac{1}{r^2}\frac{\partial}{\partial r}\left(r^2\frac{\partial}{\partial r}\right)+\frac{1}{r^2\sin\theta}\frac{\partial}{\partial \theta}\left(\sin\theta\frac{\partial}{\partial \theta}\right)+\frac{1}{r^2\sin^2\theta}\left(\frac{\partial^2}{\partial \varphi^2}\right)$$

Table 6.9 - Equations of motion in terms of velocity gradients for a Newtonian fluid of constant density and viscosity (Spherical coordinates).

Exercises

1 - Calculate the torque and power required to rotate the cylinder shown in the figure below. The length of the cylinder is 0.0508 m and it is spinning at 200 rpm. The lubricant which separates the cylinder of the fixed part has a viscosity of 2 P and its density is 800.92 kg/m^3.

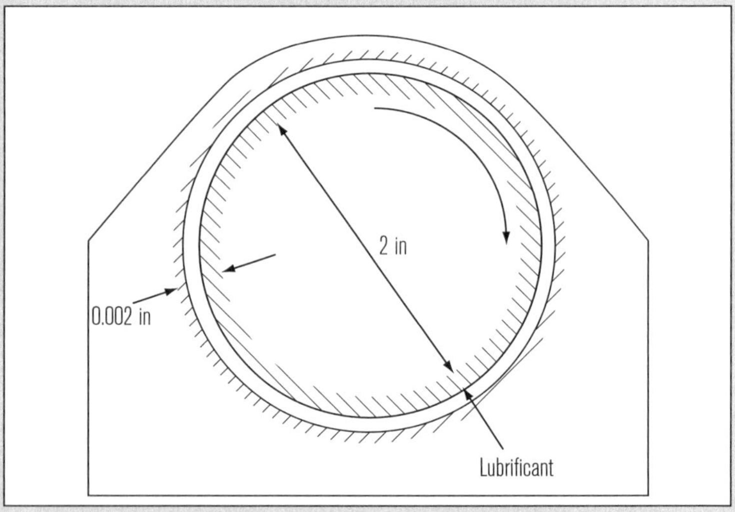

2 - The Stromer Viscometer consists essentially of two concentric cylinders. The external remains stationary and the internal rotates. The viscosity is determined by measuring the speed of the inner cylinder with the application of a certain torque. Develop an expression for the velocity distribution as a function of the applied torque. Consider Newtonian fluid under steady state conditions..

3 - Determine v_θ (r) between two coaxial cylinders of radii R and kR rotating with angular velocities Ω_0 and Ω_1, respectively. Consider that the space between two cylinders is filled with an incompressible fluid in isothermal and laminar flow. Assume steady state.

4 - Liquid steel at 1600°C is deoxidized by adding aluminum to form alumina, (Al_2O_3). For removing oxygen to get a better quality of steel. The alumina particles that are formed float to the surface of the bath. Determine the smallest size particle that reaches the surface, two minutes after the deoxidation. (Height of the bath 1.5 m).

Data:

- $\rho_{steel} = 7100\ kg/m^3$;
- $\rho_{Al_2O_3} = 3000\ kg/m^3$.

Check the validity of the calculation and comment.

5 - A wire is cooled after heat treatment by passing through a tube that is immersed in an oil tank. Get the velocity distribution of oil in the region of the tube, using the equations of continuity and motion. Consider steady state. The system is seen in the figure below. The oil pressure inside the tank is uniform.

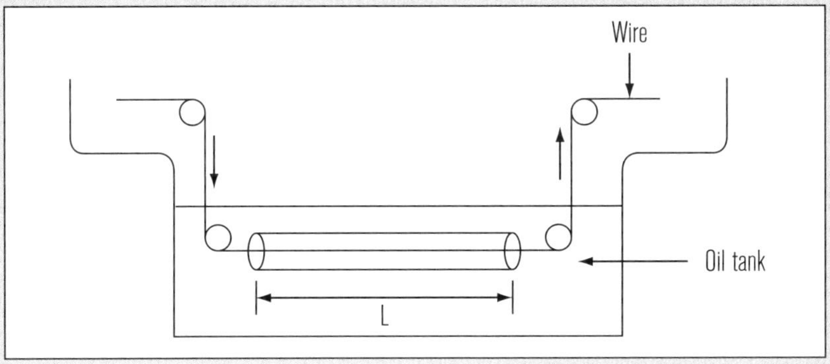

6 - Heavy oil with a kinematic viscosity equal to $3.45 \times 10^{-4} m^2/s$ is at rest in a long vertical tube with a radius of 0.7 cm. Suddenly the fluid is drained out through the bottom due to gravity. How quickly the speed at the center of the tube is equivalent would reach 90% of its final value?

b) What would be the result if the oil was replaced by water at 20°C ($\upsilon = 0.01\ cm^2/s$). Use Figure 6.9 to get the answers.

7 - A fluid is being injected into a reservoir where it undergoes a process of purification. The geometry of the system is shown in the figure below. Using general equations of continuity and motion, obtain the differential equations governing the flow of fluid in the system. Justify the simplifications made. State the boundary conditions necessary for the solution of the equations. Restrict your analysis to the region defined by: $0 < x < L$ and $0 < z < H$. Explain the boundary conditions. Consider steady-state and fluid density and viscosity constant. Velocities in and out of the system are uniform.

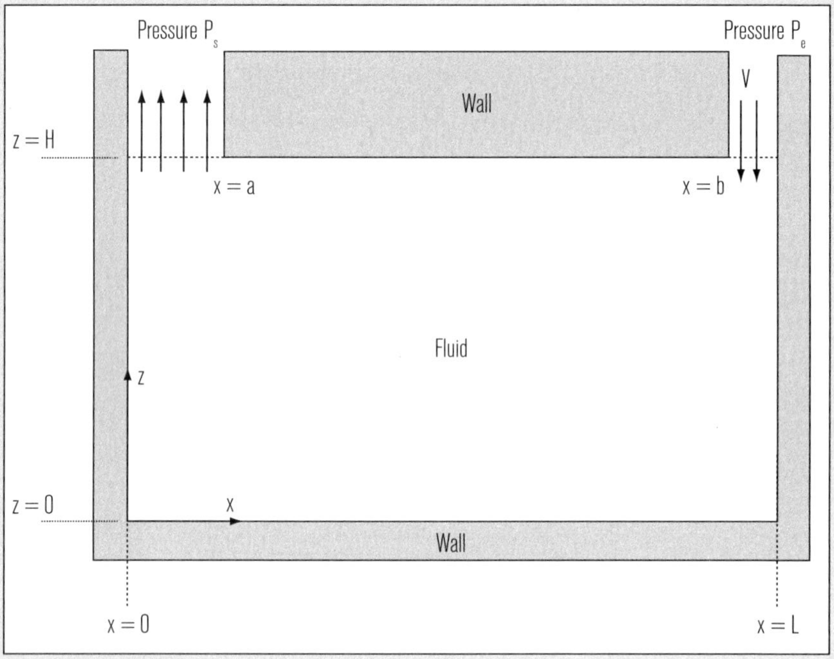

Turbulent Flow and Experimental Results

Chapter 07

In the previous chapters, only laminar flow problems have been dealt with on the basis of solution of differential equations describing the flow. The velocity profiles and other parameters of importance can be determined easily for simple systems. The only limitation that one comes across is related to the mathematical complexity for situations of multiple velocity components and problems in cylindrical and spherical coordinates.

However, there are a large number of engineering processes with turbulent flow and in these cases, laminar flow equations are not directly applicable. Although the general equations of continuity and motion continue to be valid, the existence of velocity fluctuations with high frequencies, as shown in Chapter 5 (Figure 5.2), as well as viscosity variations in turbulent conditions make it difficult to adopt methodologies similar to the those used in Chapters 5 and 6. Hence, for problems involving turbulence, it is more practical to try other types of methods such as empirical approach.

This chapter contains a study of turbulent flow through a much simpler approach of mathematical complexity. It would be helpful to introduce some basic principles of models of turbulent flow that have been proposed for the evaluation of velocity profiles for turbulent flow, as in the case of laminar flow.

7.1 Introduction

In Chapter 5 an introduction was given in respect of transition from laminar to turbulent flow. This transition depends on the system under consideration like tubes, submerged objects like spheres, cylinders and plates. They have different values of Reynolds number characterizing transitions. In the case of flow in pipes, the Reynolds number, evaluated using the following equation, gives the transition Reynolds number amounting to around 2100:

$$Re = \frac{D\,\overline{V}\,\rho}{\mu} \qquad (7.1)$$

where:

- $\overline{V}$ = average velocity of the fluid in the tube;
- D = diameter of the tube;
- ρ = density of fluid;
- μ = dynamic viscosity of the fluid.

The value of Reynolds number for which the transition occurs from laminar to turbulent flow for other systems can be calculated with a characteristic length or parameter for other systems. Some times this number is determined empirically for systems like packed beds etc. In many processes like blast furnace, oxygen steel making or continuous casting, the flow of fluids is turbulent and the treatment needs a different analysis.

In order to get an idea, as to how in the industrial practice turbulent flow is important, consider the example of the process of continuous casting, where liquid steel is fed into a copper mold cooled with water.This feeding is done through a refractory tube, called the submerged valve.

If the equipment produces plates of dimensions of 1.2 x 0.25 m, with a casting speed of 1 m/min, one can evaluate the flow rate of submerged steel valve.This flow rate should be such that it would keep the level of liquid steel in the mold constant. Thus, the flow of liquid through the valve corresponds to actually the rate of production of steel in the form of plates.

Flow rate of steel and steel velocity are given by:

$$\text{Flow rate} = 1\text{ m} \,.\, 0.25\text{ m} \,.\, 1\,\frac{\text{m}}{\text{min}} = 0.25\,\frac{\text{m}^3}{\text{min}} = 0.0042\,\frac{\text{m}^3}{\text{s}}$$

$$\text{Steel velocity} = \frac{Q}{A_{\text{valve}}} = \frac{0.0042}{\left(\dfrac{\pi\, d^2_{\text{valve}}}{4}\right)} = \frac{0.0042}{\left[\dfrac{\pi\,(0{,}070)^2}{4}\right]} = 1.083\text{ m/s}$$

The submerged valve has a diameter of 70 mm and hence one can evaluate the average velocity of liquid steel flowing out of the valve and estimate the Reynolds number using the properties of liquid steel. For the submerged valve, the Reynolds number is given by:

$$Re = \frac{d_{valve} \overline{V} \rho}{\mu}$$

For ρ = 6700 kg/m^3; and μ = 0.0065 Pa.s:

$$Re = \frac{d_{valve} \overline{V} \rho}{\mu} = \frac{0.070 \,.\, 1.083 \,.\, 6700}{0.0065} = 78{,}142$$

From this value, it can be concluded that the flow inside the valve is well above the value of Reynolds number that characterizes the transition from laminar to turbulent flow and hence can be considered turbulent. If this same example is repeated for flow of fluids in other processes of interest like blast furnace, oxygen steel making etc., one can verify, that in most of the cases of these industrial processes, turbulent regimes predominate.

In Chapter 5, for laminar flow in tubes, velocity distribution and the ratio between the average and maximum velocity has been given by:

$$\frac{v_z}{v_{z,max}} = \left[1 - \left(\frac{r}{R}\right)^2\right] \tag{7.2}$$

$$\frac{\overline{v_z}}{v_{z,max}} = \frac{1}{2} \tag{7.3}$$

where $v_{z\ max}$, is the maximum velocity at the center of the tube and R is its radius.

For turbulent flow, it has been shown experimentally, that the velocity profile and the ratio between average and maximum velocities can be given by similar equations:

$$\frac{v_z}{v_{z,max}} = \left[1 - \left(\frac{r}{R}\right)^{\frac{1}{7}}\right] \tag{7.4}$$

$$\frac{\overline{v_z}}{v_{z,max}} = \frac{4}{5} \tag{7.5}$$

However the ratio of average and maximum velocities in the case of turbulent flow is 4/5 as against ½ in the case of laminar flow.

The average velocity is obtained considering the velocity fluctuations over period of time.These equations are valid for Reynolds numbers ranging from 10^4 to 10^5. In this range, the pressure drop is proportional to the volumetric flow rate raised to the power 7/4. A comparison between the velocity profiles for laminar and turbulent flows is shown in Figure 7.1.

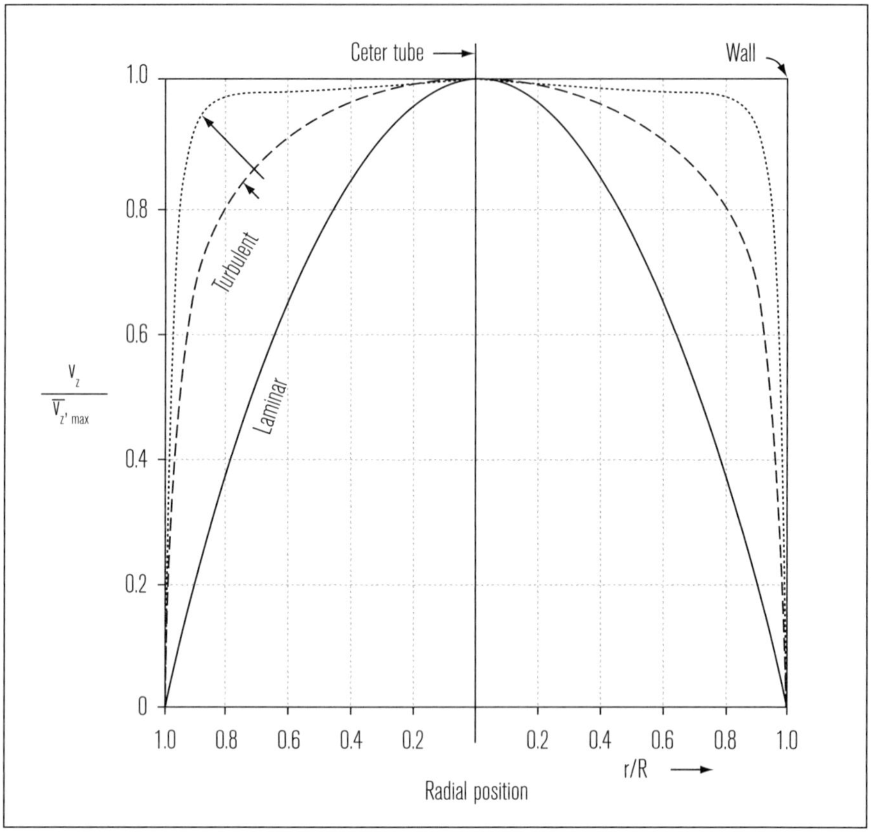

Figure 7.1 - Comparison between the qualitative distributions of velocity in the laminar and turbulent flow (BIRD, STEWART and LIGHTFOOT, 1960).

It is quite evident in Figure 7.1, that there is a transformation of a parabolic profile, characteristic of laminar flow, to a flatter profile in the case of turbulent flow. In the latter, the velocity variations are concentrated in the region near the tube wall. In the central part, the velocities are almost uniform. For turbulent flow, as seen in equation (7.5), the values of average and maximum speeds are very similar and tend to get closer for higher Reynolds numbers as can be concluded from Figure 7.1

In general, problems involving turbulent flow can be resolved through two approaches. One of them, is based on elaborate mathematical treatment for the development of turbulence models to determine the velocity profile of fluid in the system under consideration. From this profile, other parameters of importance can be determined. This type of treatment is extensively used in design problems for new installations, prototypes etc. in the engineering fields and even in the area of weather forecasting.

Another approach is the use of experimental results, where certain parameters of interest are proposed empirically, together with the experimental work to determine others. With these, mathematical relationships that are useful in characterizing the flow and quantifying them, can be developed. This second approach is much simpler compared to the first one and is often called the engineering approach. Most of the problems that are normally encountered in respect of fluid flow in pipes, channels etc, can be solved using this second approach.

In the next section, the first approach,namely mathematical treatment, emphasizing the fundamentals of turbulence models proposed and results obtained, are discussed.

7.2 Turbulence Models

Several turbulence models have been proposed from time to time..A basic feature common to all these models is to work with a velocity averaged over a period of time *(time-averaged velocity or time smoothed velocity)*.That is, averaging the instantaneous velocities, measured experimentally over a period of time. This time interval is large compared to the time associated with fluctuations in velocity, but small relative to changes over time, that may occur due to a change in pressure drop in the system.

The definition of time averaged velocity is illustrated graphically in Figure 7.2 and is expressed mathematically by the following equation:

$$\overline{v_z} = \frac{1}{t_0} \int_{t}^{t+t_0} v_z \, dt \tag{7.6}$$

where t_0 is the time interval used to make the integration and v_z is the instantaneous value of the velocity. The instantaneous values of velocity can then be written as a sum ot time averaged velocity and a term for fluctuation of velocity v_z:

$$v_z = \overline{v_z} + v_z' \tag{7.7}$$

Equations similar to (7.6) and (7.7) can be written for the other components of velocity and pressure, which also have fluctuations in turbulent flow.

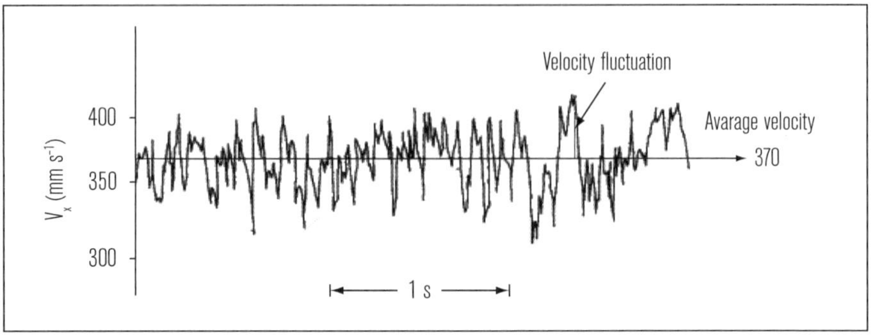

Figure 7.2 - Oscillation of a velocity component relative to a mean value (GUTHRIE, 1992).

The definition of fluctuation of velocity is given by:

$$\overline{v_z^{\prime}} = \frac{1}{t_o} \int_{t}^{t+t_o} v_z^{\prime} \, dt = 0 \tag{7.8}$$

ie, the average velocity fluctuations over a given period of time is zero. However, the mean-square fluctuations will not be zero:

$$\overline{v_z^{\prime 2}} = \frac{1}{t_o} \int_{t}^{t+t_o} (v_z^{\prime})^2 \, dt \neq 0 \tag{7.9}$$

In fact, it is common to use the relationship:

$$\frac{\sqrt{\overline{v_z^{\prime 2}}}}{\overline{v_z}} \tag{7.10}$$

for quantifying the intensity of turbulence. For flow in pipes, the value of the above is usually between 0.01 and 0.10 (BIRD, STEWART and LIGHTFOOT, 1960).

7.2.1 Equations of continuity and motion in terms of time averaged velocities

Using the equation (7.7), we can rewrite the equations of continuity and motion, in terms of time averaged velocities. These new equations can be solved to determine the velocity profiles.

7.2.1.1 Equation of continuity

Considering a fluid with constant density and steady state, we can write the continuity equation as follows:

$$\frac{\partial}{\partial x}(v_x) + \frac{\partial}{\partial y}(v_y) + \frac{\partial}{\partial z}(v_z) = 0 \tag{7.11}$$

Introducing the definition given by equation (7.7) (and its appropriate forms for other components of velocity), we obtain:

$$\frac{\partial}{\partial x}(\overline{v_x} + v_x^{\prime}) + \frac{\partial}{\partial y}(\overline{v_y} + v_y^{\prime}) + \frac{\partial}{\partial z}(\overline{v_z} + v_z^{\prime}) = 0 \qquad (7.12)$$

One can then get the average for the above equation over the interval to, similar to what was done with the velocity components of equation (7.7). This procedure corresponds to time averaging (*time-smoothing*) of the equation of continuity. Through this procedure and using equation (7.8), one gets:

$$\frac{\partial}{\partial x}(\overline{v_x}) + \frac{\partial}{\partial y}(\overline{v_y}) + \frac{\partial}{\partial z}(\overline{v_z}) = 0 \qquad (7.13)$$

This equation is exactly the same as the continuity equation, derived in Chapter 6, but written on the basis of time averaged velocities.

7.2.1.2 Equation of motion

A similar procedure to that used in the previous section can be applied to obtain time averaged equations of motion. The development below is for x component of velocity, but similar procedure can be applied to other components.

Considering a fluid with constant viscosity, we have the following equation of motion for the x component of velocity:

$$\frac{\partial(\rho v_x)}{\partial t} = -\frac{\partial}{\partial x}(\rho v_x v_x) - \frac{\partial}{\partial y}(\rho v_y v_x) - \frac{\partial}{\partial z}(\rho v_z v_x) + \mu \nabla^2 v_x - \frac{\partial P}{\partial x} + \rho g_x \qquad (7.14)$$

Again using the definition of instantaneous velocity,equation (7.7), the above equation can be written in the following way:

$$\frac{\partial[\rho(\overline{v_x} + v_x^{\prime})]}{\partial t} = -\frac{\partial}{\partial x}[\rho(\overline{v_x} + v_x^{\prime})(\overline{v_x} + v_x^{\prime})] - \frac{\partial}{\partial y}[\rho(\overline{v_y} + v_y^{\prime})(\overline{v_x} + v_x^{\prime})] - \\ - \frac{\partial}{\partial z}[\rho(\overline{v_z} + v_z^{\prime})(\overline{v_x} + v_x^{\prime})] + \mu \nabla^2(\overline{v_x} + v_x^{\prime}) - \frac{\partial}{\partial x}(\overline{P} + P^{\prime}) + \rho g_x \qquad (7.15)$$

This can be time averaged over an interval. Using equations (7.8) and (7.9), we obtain:

$$\frac{\partial(\rho \overline{v_x})}{\partial t} = -\frac{\partial}{\partial x}(\rho \overline{v_x}\,\overline{v_x}) - \frac{\partial}{\partial y}(\rho \overline{v_y}\,\overline{v_x}) - \frac{\partial}{\partial z}(\rho \overline{v_z}\,\overline{v_x}) + \mu \nabla^2 \overline{v_x} - \frac{\partial \overline{P}}{\partial x} + \rho g_x - \\ \boxed{- \frac{\partial}{\partial x}(\rho \overline{v_x^{\prime} v_x^{\prime}}) - \frac{\partial}{\partial y}(\rho \overline{v_y^{\prime} v_x^{\prime}}) - \frac{\partial}{\partial z}(\rho \overline{v_z^{\prime} v_x^{\prime}})} \qquad (7.16)$$

The above equation is similar to equation (7.14). However, there are three additional terms, associated with fluctuations in velocity and characteristics of turbulent flow. For convenience, it is common to introduce the following notations:

$$\bar{\tau}_{xx}^{(t)} = (\rho \overline{v_x' \, v_x'}) \tag{7.17}$$

$$\bar{\tau}_{yx}^{(t)} = (\rho \overline{v_y' \, v_x'}) \tag{7.18}$$

$$\bar{\tau}_{zx}^{(t)} = (\rho \overline{v_z' \, v_x'}) \tag{7.19}$$

These terms correspond to the time averaged turbulent flows of momentum, usually called *Reynolds stress.* All terms in equation (7.16) have the dimension of rate of flow of momentum or stress.

Equation (7.16) is difficult to solve, due these time averaged terms. To evaluate these, different models of turbulence have been proposed. To date, there has been no model that is universally applicable. However, with the models already developed there has been substantial progress in the solution of problems of practical interest.

One of the first proposals for evaluation of turbulent flows was made by Boussinesq (BIRD, STEWART and LIGHTFOOT, 1960). Adopting an analogy with Newton's equation of viscosity, the following equation has been suggested:

$$\bar{\tau}_{yx}^{(t)} = -\mu^{(t)} \frac{\partial \overline{v_x}}{\partial y} \tag{7.20}$$

where $\mu^{(t)}$ is the turbulent viscosity and others are time averaged parameters. Similar expressions for other terms can be written. Turbulent viscosity is a property of the fluid and should be evaluated or estimated for each particular system.

Note that this does not solve the problem of evaluating the rate of turbulent flow of momentum. However it enables instead, determination of the turbulent viscosity $\mu^{(t)}$.

The interesting aspect of this proposal is that it makes the equation of motion for turbulent flow, identical to the equation for laminar flow replacing the molecular viscosity, μ, by the effective viscosity, μ_{eff}, expressed as the sum of molecular(or laminar) viscosity and turbulent velocity:

$$\mu_{eff} = \mu + \mu^{(t)} \tag{7.21}$$

A number of other proposals for evaluating the turbulent flow of momentum have been made. Among them, one can cite (BIRD, STEWART and LIGHTFOOT, 1960) the following:

- Proposal of Prandtl (mixing length theory):

$$\bar{\tau}_{yx}^{(t)} = -\rho\, l^2 \left| \frac{d\bar{v}_x}{dy} \right| \frac{d\bar{v}_x}{dy} \tag{7.22}$$

where l is the length of mixing, measured as a function of distance from the point to the wall.

- Proposal of Von Kármán:

$$\bar{\tau}_{yx}^{(t)} = -\rho\, \kappa_2^2 \left| \frac{(d\bar{v}_x/dy)^3}{(d^2\bar{v}_x/dy^2)^2} \right| \frac{d\bar{v}_x}{dy} \tag{7.23}$$

where κ_2 is a constant equal to 0.36 (determined from measurements of velocity profiles in tubes).

- Proposal of Deissler (empirical):

$$\bar{\tau}_{yx}^{(t)} = -\rho\, n^2\, \bar{v}_x\, y \left[1 - \exp\left(- \frac{n^2\, \bar{v}_x\, y}{\nu} \right) \right] \frac{d\bar{v}_x}{dy} \tag{7.24}$$

where y is the distance from the wall and n is a constant empirically evaluated (0.124).

Among these proposals, the widely used one is that of Boussinesq. In this case, a number of approaches have been developed for the assessment of turbulent viscosity. These approaches can be classified into three categories according to the number of additional differential equations that are used to evaluate the viscosity:

- *Zero equation model*: In this case, a constant value for the turbulent viscosity is stipulated for the system. The choice of the value to be used is often arbitrary and seeks agreement between values predicted by the mathematical model and experimental values. This approach was first used in the modeling of turbulence and works reasonably well in systems dominated the transport of momentum by convection (GUTHRIE, 1992);
- *One equation model*: In this type of model, an additional differential equation (in addition to the conservation of mass and momentum) is used for resolution. It is also necessary to specify the value of a parameter called the mixing length, in order to calculate the turbulent viscosity;
- *Two equations models*: have two additional differential equations to estimate the turbulent viscosity. It is not necessary to specify any arbitrary parameter. In this category, there are the popular models κ - ε (in its various

forms), which are used extensively.These models have had a big success in predicting features of turbulent flow in various application areas, including metallurgy. However, no single approach gives quantitatively correct results in a wide range of applications. Generally, this kind of model that works best for a particular type of application.

The model κ - ε proposed by Jones and Launder (1972) is the one that has provided the best results in metallurgical applications. Figures of 7.3 to 7.5 show examples of velocity profiles obtained with the use of these models applied to the RH refining process, tundish and continuous casting mold. Some of these processes have been discussed in chapters 19 and 21. The velocity profiles presented in Figure 7.3 were determined using the commercial software Ansys-CFX. In this case, the mathematical model includes the two phases, liquid and gas. The standard two equation κ-ε model was adopted to model the turbulence.

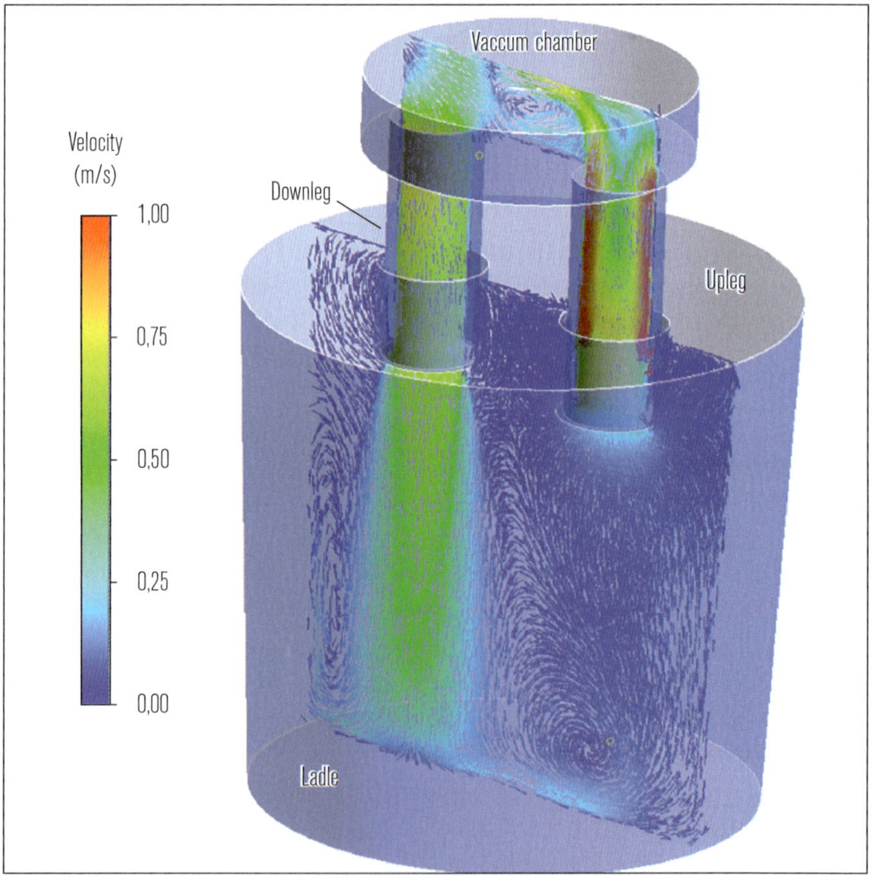

Figure 7.3 - Water velocity profiles in the plane of symmetry of a physical model of RH degasser.

The velocity profiles for the tundish (Figure 7.4) with and without furniture was obtained with the METFLO code, developed at McGill University in Canada. The k-ε model was also adopted to simulate the effects of turbulence on fluid flow. The predictions of these simulations were extensively validated using experimental data obtained in a physical model.

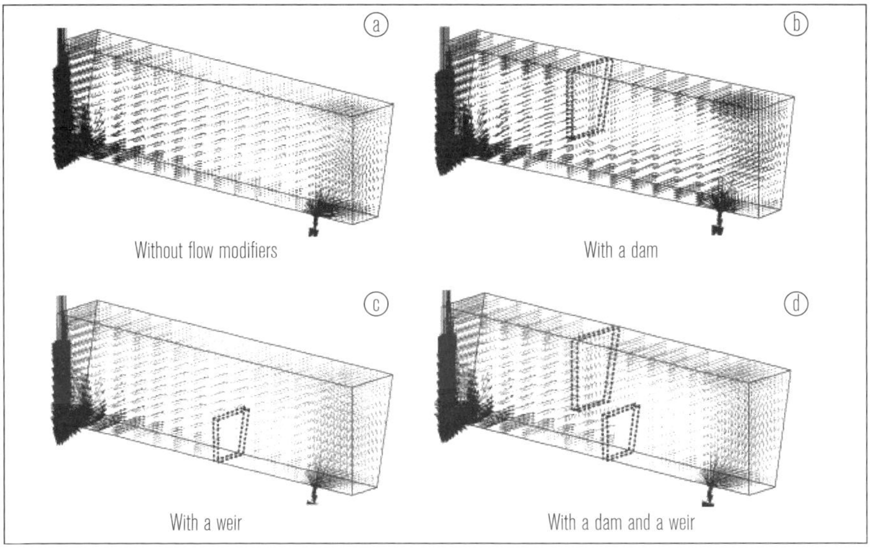

Figure 7. 4 - Velocity profiles in a continuous casting tundish (TAVARES and CASTRO, 1999).

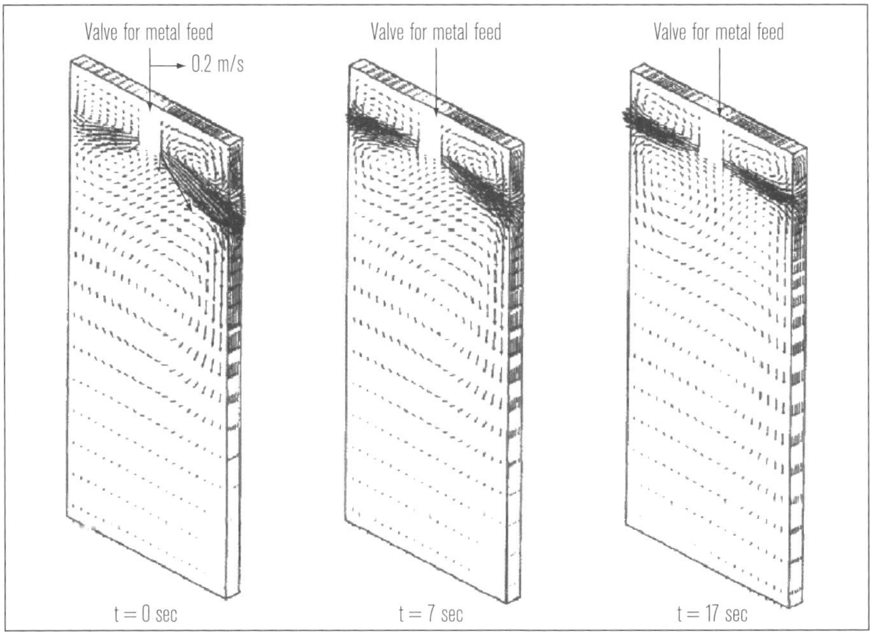

Figure 7.5 - Velocity profiles in a continuous casting mold (HUANG and THOMAS, 1996).

The approach described in respect of turbulent flow above is quite cumbersome and invariably involves the use of complex numerical techniques and computational resources for solution of differential equations of conservation of mass and momentum. As mentioned previously, the use of this approach is generally restricted to specific applications, in which determination of velocity profiles is absolutely essential to solve the problem.

In many problems of practical application in engineering, one can employ simpler techniques (from a mathematical point of view). These provide appropriate responses and satisfactory results.This approach is presented the next section.

7.3 Friction Factors

Many practical problems in the engineering field fall into the categories given below:

- Flow in ducts or channels (internal flow);
- Flow around objects (external flow).

For flow in ducts or channels, one can cite the following examples: pumping oil from pipelines, flow of water in open channels and extrusion of polymer matrix. Examples of flow around objects are: the movement of air around a car, airplane or racket, movement of water around particles undergoing sedimentation (ore treatment and beneficiation operations) or the movement of inclusions in liquid steel.

In problems of flow in channels or ducts, usually one is interested in obtaining a relationship between pressure drop, gravity and flow rate of the fluid. Here friction between the fluid and the walls would influence the pressure drop or rather loss of energy. In respect of problems of flow around submerged objects usually it is necessary to know the relationship between approach speed of the fluid and the drag force of fluid on the particle. As was seen in previous chapters, when we know the distributions of velocity and pressure in a given system, the information mentioned above can be obtained with relative ease. For turbulent regimes, determination of velocity profiles is rather difficult. The empirical treatment given below is relatively simple compared to the complex mathematical treatment of turbulent flow. It has proved very useful in the determination of parameters important in evaluating the variables necessary for appropriate engineering design and equipment specifications in fluid transport systems.

To resolve some of the problems cited in the preceding paragraph in respect of turbulent flow it is necessary to evaluate the frictional force acting on the interface between the fluid and solid, that is the wall of a duct or channel, or the surface of a body immersed in fluid and loss of energy due to the same.

It has been arbitrarily proposed that the frictional force or friction, acting between the fluid and solid is proportional to the area of contact as well as the kinetic energy of the fluid. The proportionality factor f is defined as friction factor, given by the following equation:

$$F_k = A\,K\,f \tag{7.25}$$

where:

- F_k = force of friction between the solid and fluid;
- A = characteristic area;
- K = kinetic energy of fluid per unit volume;
- f = friction factor.

It should be noted that equation (7.25) is an arbitrary definition of the friction factor, which is dimensionless. This is a measure of force exerted on the wall depending on the velocity of fluid and characteristic area of the system. For a given system, f is not defined until the characteristic area, A, is specified. The definition of this area varies depending on the system configuration, internal or external flow, surface smoothness etc.

7.3.1 Flow in ducts (internal)

For pipeline flow, the characteristic area in equation (7.25) is the *wetted* area (area of the region in contact with the fluid). The kinetic energy of the fluid, in turn, is assessed from the average velocity of the fluid.

Similarly for a cylinder of diameter D and length L, the friction force can be estimated by the following equation:

$$F_k = (\pi\,D\,L)\left(\frac{1}{2}\,\rho\,\overline{V^2}\right) f \tag{7.26}$$

where:

- πDL = contact area of fluid-solid;
- $\frac{1}{2}\rho V^2$ = kinetic energy of fluid per unit volume.

The above equation is useful in evaluating the frictional force, if the friction factor term can be determined experimentally. It is quite simple to mount an experimental set up for the determination of friction factor (Figure 7.6) consisting of 2 pressure measuring devices, at different locations to measure the pressure difference.

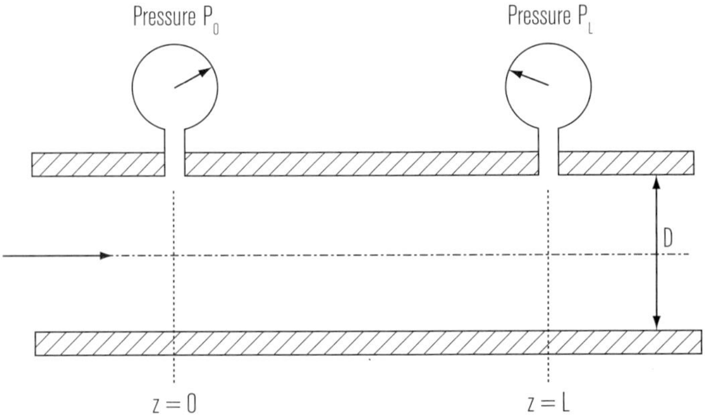

Figure 7.6 - Experimental setup to evaluate the friction factor.

In the system above, the flow of fluid is caused only by the pressure difference, and the fluid moves with constant velocity. If the sum of forces acting on the fluid is zero (Newton's second law), the following expression for the balance of forces can be written:

Force due to pressure difference = Force of friction between solid and fluid

$$\frac{\pi\, D^2}{4}\,(P_0 - P_L) = F_k = (\pi\, D\, L)\left(\frac{1}{2}\,\rho\,\overline{V^2}\right) f \tag{7.27}$$

Assessing experimentally, the pressure difference, $P_0 - P_L$, for a given fluid of known density, flow rate (or average speed), diameter and length of the tube and substituting these values in equation (7.27), one can estimate the friction factor. Equation (7.27) can be written in the following form to facilitate this calculation:

$$\frac{\frac{\pi\, D^2}{4}\,(P_0 - P_L)}{(\pi\, D\, L)\left(\frac{1}{2}\,\rho\,\overline{V^2}\right)} = f \tag{7.28}$$

$$f = \frac{1}{2} \frac{(P_0 - P_L)}{L} \frac{D}{\rho \overline{V}^2} \qquad (7.29)$$

With the equation (7.29), the value of f can be calculated. Variation of f with different experimental conditions such as average velocity, pressure drop and tubes with different surface conditions (smoothness) can be evaluated. Higher the value of friction factor f, greater is the force of friction at the solid-fluid interface.

Certainly a number of factors should affect the value of f.

To determine in a quantitative manner, the effects of these various factors, a very large number of experiments would be necessary. To reduce the number of experiments carried out in the laboratory, a procedure called dimensional analysis of the process is recommended. In fact this procedure would reduce the many variables involved to a few, in the form of dimensionless numbers. This is a very useful tool in the area of physical modeling or cold modeling of simulation of metallurgical industrial processes in the laboratory. There are several ways to conduct this analysis. The technique presented here is based on Buckingham π theorem.This theorem (given here without proof) states that it is possible to group the variables that affect the value of f or any other variable in dimensionless groups, which represent the problem involving the original variables. The number of the dimensionless groups is always less than the number of variables,as from these variables the groups are formed and these new variables are the dimensionless groups. Application of the technique of dimensional analysis has been used various engineering fields like chemical, civil and mechanical engineering and the basis of similarity criteria (geometric and kinematic) between industrial plants and physical models in laboratory scale. Many examples of application of dimensional analysis are given in Szekely and Themelis (1970). The following treatment of dimensional analysis (based on the Buckingham π theorem) is applicable to turbulent flow and determination of friction factors in pipes.

7.3.1.1 Dimensional analysis

The various steps of dimensional analysis are listing of variables that affect the value of the friction factor, their units, dimensions, classification and selection under categories followed finally by dimensional analysis for determining the groups that affect the process.

a) *List of variables*

At the time of listing the variables, knowledge about the system and its operating parameters are helpful in addition to having a broader vision about the process under study. Even if more variables are listed, only those that are actually effective will make a difference during the experimentation. Experiments determine which ones are really significant and which one can be neglected. The following variables have an important influence on the value of the friction factor in pipe:

- Variables: D, $\bar{V}$, L, ρ, μ, and ε

The variable ε, given above corresponds to the roughness of the pipe. This parameter basically depends on the material used for the manufacture of the pipe and gives an idea of its roughness. It represents the average heights of the peaks and valleys, which can be seen on the inner surface of the tube. This can be quantitatively evaluated with a device that gives a magnified view of the surface. This device is know as profiler. In the literature, it is quite common to find values of roughness for pipes of different materials. Figure 7.7 shows schematically the definition of roughness and Table 7.1 gives roughness values of some common materials used for tubes.

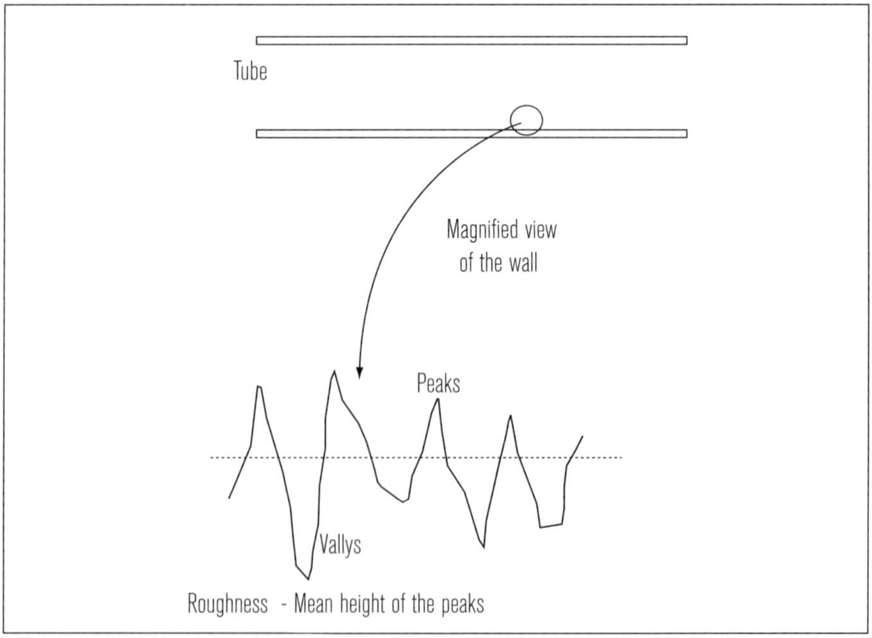

Figure 7.7 - Schematic representation of the roughness of a pipe wall.

Material	Roughness (mm) ε
Commercial steel	0.046
Galvanized iron	0.15
Iron	0.259
Concrete	0.3-3

Table 7.1 - Roughness value for tubes of different materials commonly used.

b) *Dimensions of the variables*

The next step is to to determine its dimensions of the variables. As has been already discussed in Chapter 3 the dimensions of the variables listed above can be given in terms of Mass (M), geometrical dimension (L), time (t) and temperature (T). Hence:

- Diameter D [=] L;
- Length L [=] L;
- Density ρ [=] ML^{-3};
- Average velocity $\overline{V}$ [=] $L\,t^{-1}$;
- Viscosity μ [=] $ML^{-1}\,t^{-1}$;
- ε [=] L.

Using equation (7.29), it is possible to determine the dimension of the friction factor:

$$f = \frac{1}{2}\,\frac{(P_0 - P_L)}{L}\,\frac{D}{\rho\,\overline{V}^2} \qquad (7.29)$$

$$f\,[=]\,\frac{(ML^{-1}t^{-2})}{L}\,\frac{L}{(ML^{-3})\,(L^2\,t^{-2})}\,[=]\text{ dimensionless}$$

As can be seen above, f is a dimensionless parameter.

c) *Classification of variables*

After determining the dimensions, the variables must be classified. This classification is made according to the groups below:

- geometric variables;
- kinematic variables;
- dynamic variables

Table 7.2 provides a list of variables typically involved in problems of transport phenomena and their classification according to the categories above.

The variables that have dimensions of length are called geometric variables. Kinematic variables involve the time element only without mass whereas the dynamic variables have dimensions involving mass and time.

According to this classification criteria, in the present case (friction factor) we have:

- geometric variables: D, L, ε;
- kinematic variables: $\bar{V}$;
- dynamic variables: ρ and μ.

It is interesting to note that, according to the list of variables made, there are 6 independent variables, D, L, ρ, μ, ε and $\bar{V}$ (whose values can be varied at the time of doing the experiment) and one dependent variable, f (whose value is dependent on the values adopted for the independent variables).

Table 7.2 - Classification of different variables.

Geometric			
Variable	Symbol	Dimension	Unit (SI)
Length	L	L	m
Area	A	L^2	m^2
Volume	V	L^3	m^3
Kinematic			
Time	t	t	s
Velocity	V	$L\ t^{-1}$	m/s
Kinematic viscosity	ν	$L^2\ t^{-1}$	m^2/s
Volumetric flow rate	Q	$L^3\ t^{-1}$	m^3/s
Acceleration	a	$L\ t^{-2}$	m/s^2
Dynamic			
Density	ρ	ML^{-3}	kg/m^3
Mass	M	M	kg
Dynamic viscosity	μ	$ML^{-1}\ t^{-1}$	kg/m.s
Mass flow	Γ	$M\ t^{-1}$	kg/s
Momentum	mv	$ML\ t^{-1}$	kg.m/s
Pressure	p	$ML^{-1}\ t^{-2}$	N/m^2
Shear force	τ	$ML^{-1}\ t^{-2}$	N/m^2
Force	F	$ML\ t^{-2}$	N
Energy	E	$ML^2\ t^{-2}$	J
Power	P	$ML^2\ T^{-3}$	W

d) *Selection of variables*

TTo carry out dimensional analysis for this case, initially 3 independent variables, also called basic variables in the process, are selected. They are selected from each of the group of variables, geometric, kinematic and dynamic as given in Table 7.2. The number of basic variables is equal to the number of dimensions needed to express the variables involved in the problem. In this case, the number of dimensions is three (M, L and t). In the selection of variables, one must have one variable for each group, as has been already mentioned (Table 7.2).

An example of selection can be as follows:

- Geometric variable: D;
- Kinematic variable: $\overline{V}$;
- Dynamic variable: ρ.

It is important to emphasize that any other term, which obeys the criterion of a variable from each group would meet the requirements for the development of dimensional analysis.

e) *Development of dimensionless groups*

The number of dimensionless groups that are required to specify the problem is evaluated through the following relationship:

$$\text{Number of dimensionless groups} = \text{number of variables in the process - number of fundamental units} \tag{7.30}$$

There are 7 variables involved (6 independent and 1 dependent) and there are 3 fundamental dimensions (M, L and t). Hence, the number of dimensionless groups are given by:

$$\text{Number of dimensionless groups} = 7 - 3 = 4$$

Of this total, 3 groups are independent and 1 is dependent group it is interesting to comment on the great reduction in the number of experiments required initially and after identifying the dimensionless numbers important for the process through dimensional analysis. Initially, there are 6 independent variables. If experiments have to be done with all these variables for six different values for each of these, the number of experiments necessary to cover all possible combinations of values would be 6^6 (46,656). By dimensional analysis, the number of independent dimensionless groups, has been reduced to 3. Hence for 3 groups of independent variables and 6 different values for each of these groups, it would take 3^6 (729) experiments to cover all possible combinations. There is a reduction of 64 times the number of experiments required! This is one of the great benefits of dimensional analysis.

Now for the case of dimensional analysis for friction factor, diameter, velocity and density have been selected from each group for the formation of groups containing these parameters and other independent variables, as given below:

$$\text{Group } \pi_1 = D^a\ \overline{V}^b\ \rho^c\ \mu \tag{7.31}$$

$$\text{Group } \pi_2 = D^d\ \overline{V}^e\ \rho^f\ L \tag{7.32}$$

$$\text{Group } \pi_3 = D^h\ \overline{V}^i\ \rho^j\ \varepsilon \tag{7.33}$$

$$\text{Group } \pi_4 = D^n\ \overline{V}^o\ \rho^q\ f \tag{7.34}$$

The groups above contain 3 basic variables combined with each of the remaining variables. The basic variables are raised to powers like a, b, c etc, which have to be determined. Each remaining variable in the group is raised to the power 1. Each group is generally termed as π, with subscripts like 1,2,3 etc. The superscripts for the basic variables in the groups namely a, b, c, d, e, f, h, i, j, n, are the exponents to be determined. These exponents are evaluated considering the criteria that would make the groups dimensionless.

Considering initially the first term is dimensionless, one can equate the dimensions of the variables for this criteria namely:

$$\text{Group } \pi_1 = L^a\ (L\,t^{-1})^b\ (ML^{-3})^c\ (ML^{-1}t^{-1}) \tag{7.35}$$

$$\text{Group } \pi_1 = L^0\ M^0\ t^0 \tag{7.36}$$

The group above should be dimensionless. Hence a, b and c must be such that π_1 has no dimension of L, M or t. Equating the equations (7.35) and (7.36), we get 3 equations (for L, M, and t) where the unknowns are the exponents a, b and c:

$$\pi_1 = L^a\ (L\,t^{-1})^b\ (ML^{-3})^c\ (ML^{-1}t^{-1}) = L^0\ M^0\ t^0$$

$$L : a + b - 3c - 1 = 0$$

$$M : c + 1 = 0$$

$$t : -b - 1 = 0$$

The solution of the above system provides:

$$a = -1$$

$$b = -1$$

$$c = -1$$

With these values, we obtain:

$$\text{Group } \pi_1 = D^{-1}\,\overline{V}^{-1}\,\rho^{-1}\,\mu = \frac{\mu}{D\,\overline{V}\,\rho} \tag{7.37}$$

Comparing equations (7.37) and (7.1), it is clear that the group π_1 is the inverse of the Reynolds number. By adopting similar procedures to that used above to determine the exponents a, b and c, one can determine the other exponents that appear in the other dimensionless groups. The results are:

$$d = -1 \quad e = f = 0$$

$$h = -1 \quad i = j = 0$$

$$n = o = q = 0$$

(Demonstrate these results as an exercise).

$$\text{Group } \pi_2 = \frac{L}{D} \tag{7.38}$$

$$\text{Group } \pi_3 = \frac{\varepsilon}{D} \tag{7.39}$$

$$\text{Group } \pi_4 = f \tag{7.40}$$

The group π_3 is commonly known as relative roughness.

The independent groups are π_1, π_2 and π_3. The group π_4 is the dependent group containing f. Thus, one can say that π_4 is a function of π_1, π_2 and π_3.

Hence it can be written:

$$f = \text{func}\left(Re, \frac{L}{D}, \frac{\varepsilon}{D}\right) \tag{7.41}$$

The above relation between friction factor, Reynolds number and roughness factor can be evaluated experimentally. Results are given in Figure 7.8.

Figure 7.8 - Friction factor for tubes; Moody Diagram (GASKELL, 1992).

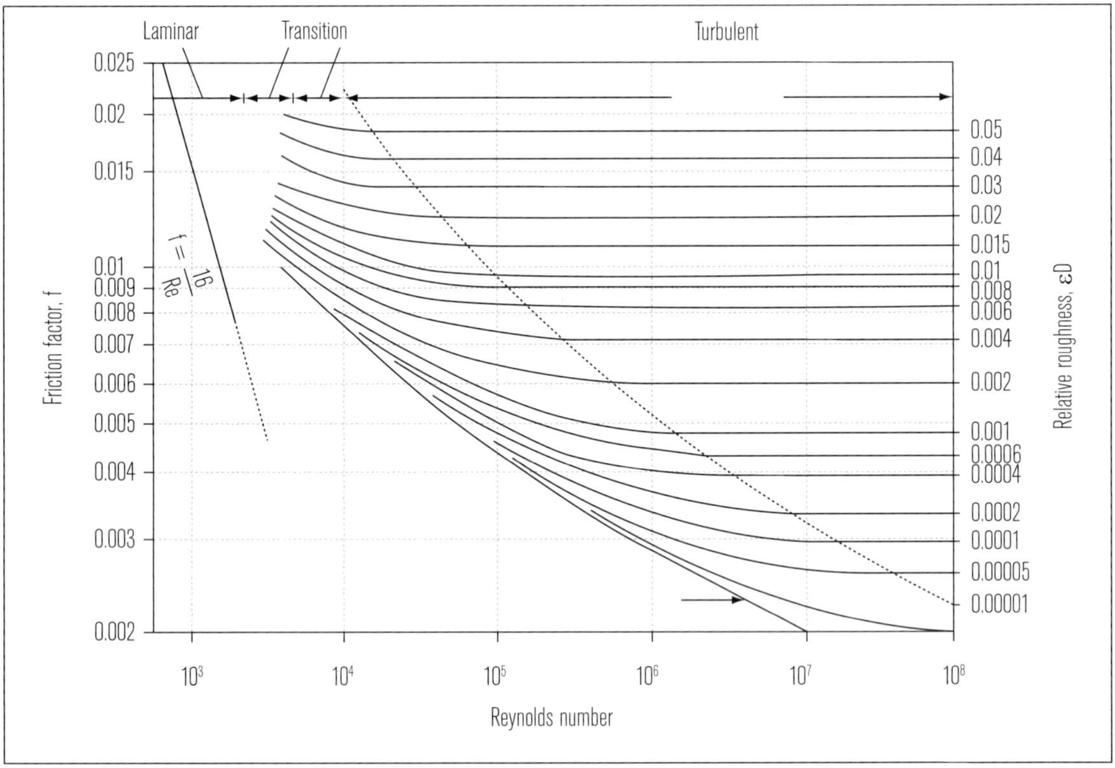

Figure 7.8 shows that the friction factor is a function of Reynolds number and relative roughness.The group L/D has no significant effect on its value even for tube lengths about 50 times larger than the diameter. Hydraulically smooth tubes are those with zero roughness ($\varepsilon = 0$). For these the pressure drop is relatively less compared with tubes with higher roughness. The straight line relationship in the left side of the graph between friction factor and Reynolds number is valid for laminar region. If the velocity of the fluid and tube dimensions are known, Reynolds number can be calculated. Using Figure 7.8 friction factor can be evaluated and the pressure drop can be calculated with equation 7.27.

More recently, Haaland, quoted by Gaskell (1992), developed a mathematical relationship of results presented in Figure 7.8.The function obtained is:

$$\frac{1}{\sqrt{f}} = -3{,}6 \log \left[\left(\frac{\varepsilon/D}{3{,}7} \right)^{1{,}11} + \frac{6{,}9}{Re} \right] \qquad (7.42)$$

The following exercise is resolved as an example of application of the above equation, to evaluate the pressure drop required for a given flow of a fluid in a tube. The same can be used to estimate the flow of fluid for a given pressure drop.

Example

Estimate the pressure drop required to achieve a water flow rate of 0.25 l/s in horizontal pipe of galvanized iron with 1.27 cm in diameter. The length of the tube is 6 m.

Properties of water: ρ = 1000 kg/m ^{3,} μ = 1 cP = 10^{-3} Pa.s.

Solution

This can be solved by balance of forces for the system, discussed earlier resulting in equation 7.27:

$$\frac{\pi D^2}{4} (P_0 - P_L) = (\pi D L) \left(\frac{1}{2} \rho \overline{V}^2 \right) f$$

(Using the above equation prove that for laminar flow f = 16/Re as given in Figure 7.8).

To obtain the value of the difference in pressure, it is necessary to evaluate the average velocity of the fluid in the tube and the friction factor.

$$\overline{V} = \frac{Q}{A} = \frac{Q}{\frac{\pi D^2}{4}}$$

Substituting the values of Q and D,

- D = 1.27 cm = 0.0127 m;
- Q = 0.25 l/s = 2.5 x 10^{-4} m^3/s

Therefore,

$$\overline{V} = \frac{2.5 \cdot 10^{-4}}{\frac{\pi\,(0.0127)^2}{4}} = 1.974 \text{ m/s}$$

To determine the friction factor, Reynolds number and relative roughness have to be evaluated. From table 7.1, for galvanized pipes, the relative roughnessis given by:

$$\varepsilon = 0.15 \text{ mm} = 1.5 \text{ x } 10^{-4} \text{ m}$$

$$Re = \frac{D\,\overline{V}\,\rho}{\mu} = \frac{(0.0127)\,(1.974)\,(1{,}000)}{(0.001)} = 25069.8$$

$$\frac{\varepsilon}{D} = \frac{1.5 \cdot 10^{-4}}{0.0127} = 0.0118$$

With these values, one can determine the value of f using equation (7.42):

$$\frac{1}{\sqrt{f}} = -3.6 \log\left[\left(\frac{0.0118}{3.7}\right)^{1,11} + \frac{6,9}{25069.8}\right]$$

$$f = 0.0105$$

Pressure drop can be calculated from the force balance equation given above:

$$(P_0 - P_L) = 2\left(\frac{L}{D}\right)\rho\,\overline{V}^2 f = 2\left(\frac{6}{0.0127}\right)(1000)\,(1.974)^2\,(0.0105) = 38659.9 \text{ Pa}$$

In some situations, the value of pressure drop measured over a pipe is used to estimate the flow rate of fluid. The following example illustrates this and shows one of the possible approaches that can be adopted in these circumstances.

Example

Estimate the flow of fluid in a vertical pipe, for a measured a differential pressure of 70000 Pa. The fluid flows upwards and the tube length is 5 m.

Data:

- properties of water: $\rho = 1000\ kg/m^3$, $\mu = 10^{-3}$ Pa.s.
- tube diameter: 0.0254 m;
- tube material: cast iron.

The balance of forces for fluid rising in a vertical tube, is given by:

$$\frac{\pi D^2}{4}(P_0 - P_L) = (\pi D L)\left(\frac{1}{2}\rho \overline{V}^2\right) f + \frac{\pi D^2}{4}\rho g L$$

Note that the pressure gradient acts in the opposite direction to the forces of friction and gravity. In this balance of forces, the values of velocity and f are unknown. Therefore, there are two unknowns in the equation. The other equation necessary to solve the problem is equation (7.42), relating the Reynolds number (function of average velocity of the fluid) with friction factor. The above two equations can be solved simultaneously for assessing the average velocity and subsequent calculation of volumetric flow of fluid.

Calculators can be used for the solution of two simultaneous equations given above. The same can be done using a spreadsheet. The solution of the equations can be simplified by transforming the force balance equation into an equation relating the Reynolds number and friction factor. To do this, the average velocity is expressed in terms of Reynolds number:

$$\overline{V} = \frac{Re\,\mu}{D\,\rho}$$

Substituting this expression in the balance of forces, we obtain:

$$\frac{\pi D^2}{4}(P_0 - P_L) = (\pi D L)\left[\frac{1}{2}\rho\left(\frac{Re\,\mu}{D\,\rho}\right)^2\right] f + \frac{\pi D^2}{4}\rho g L$$

After simplifications and transposition of terms, we obtain:

$$Re^2 f = \frac{1}{2}\frac{D^3 \rho}{L \mu^2}\left[(P_0 - P_L) - \rho g L\right]$$

Replacing data in the above expression, we have:

$$Re^2 f = \frac{1}{2}\frac{(0.0254)^3 (1000)}{(6)(0.001)^2}\left[(70000)-(1000)(9.8)(6)\right] = 15294593.07$$

Substituting the data for cast iron (Table 7.1), in equation 7.42 $\varepsilon = 0.259$ mm $= 2.59 \times 10^{-4}$ m we obtain:

$$\frac{1}{\sqrt{f}} = -3.6 \log\left[\left(\frac{(2.59 \,.\, 10^{-4})/(0.0254)}{3.7}\right)^{1,11} + \frac{6.9}{Re}\right]$$

To solve simultaneously the above two nonlinear equations, an iterative procedure can be adopted. Starting with an initial value of f, successive iterations are carried out, to get satisfactory values of f and Re, as illustrated below. For example, consider an initial value of f equal to 0.006. This initial value does not change the final result, but it affects the number of iterations needed to reach an appropriate solution.

$$Re^2 f = 15294593.07$$

$$Re = \sqrt{\frac{15294593.07}{f}} = \sqrt{\frac{15294593.07}{0.006}} = 50488.6$$

Using the initial value of f above, Re is calculated using the above equation of balance of forces. With this value of Re, friction factor is recalculated from the equation of friction factor and a new value of the same is obtained after one itiration:

$$\frac{1}{\sqrt{f}} = -3.6 \log\left[\left(\frac{(2.59 \,.\, 10^{-4})/(0.0254)}{3.7}\right)^{1,11} + \frac{6.9}{50488.6}\right]$$

$$f = 0.0098$$

With this new value of friction factor, using again the equation of balance of forces, an updated value for the Reynolds number is determined. This procedure is repeated until values of friction factor and Re obtained do not have significant variations. The following table summarizes the results for successive iterations.

Iteration	f	Re
1	0.006	50488.6
2	0.009828	39449.0
3	0.009902	39301.4
4	0.009903	39299.4
5	0.009903	39299.4

(Exercise: Do the calculations to verify the results shown in the table above).

From the fourth iteration onwards, values of Reynolds and f begin to repeat and maintain same values. This corresponds to a Reynolds number equal to 39299.4. With this value, the average velocity of the fluid and its flow rate.can be determined:

$$\overline{V} = \frac{Re\ \mu}{D\ \rho} = \frac{(39299.4)\ (0.001)}{(0.0254)\ (1000)} = 1.547\ m/s$$

With this velocity the volumetric flow of fluid is calculated:

$$Q = \frac{\pi\ D^2}{4}\ \overline{V} = \frac{\pi\ (0.0254)^2}{4}\ (1.547) = 7.84\ .\ 10^{-4}\ m^3/s$$

ie, Q = 0.784 L/s.

7.3.1.2 Flow in noncylindrical ducts

All the equations developed for cylindrical systems like ducts and pipes can be used for noncylindrical tubes using the concept of hydraulic equivalent diameter. With this parameter, one an use Figure 7.8 and equation (7.42) developed for cylindrical tubes for the noncylindrical tubes with square or rectangular cross sections. Reynolds number can be calculated using the hydraulic equivalent diameter, as given by the following expression.

$$D_h = \frac{4\ A}{P_M} \quad (7.43)$$

where:

- A = cross-sectional area of the pipeline actually used for the disposal;
- P_M = wetted perimeter (length of the line of contact fluid duct wall).

For a rectangular duct, (Figure 7.9), the hydraulic diameter is defined as:

$$D_h = \frac{4\ W\ H}{(2\ W + 2\ H)}$$

For a cylindrical tube, the hydraulic equivalent diameter equals the diameter of the tube.(Prove this as an exercise).

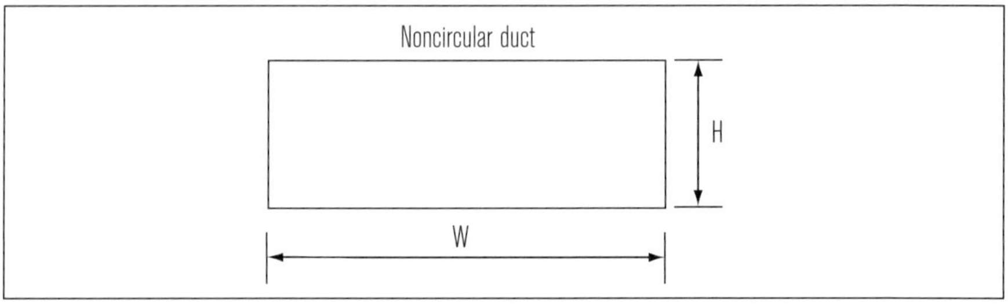

Figure 7.9 - View of the cross section of a noncircular duct for the definition of D_h.

The above approach works quite well for turbulent regime. In laminar flow it is necessary to introduce an additional correction factor of friction, in addition to using the hydraulic diameter eqivalent as a parameter. The friction factor in terms of Reynolds number defined for laminar flow can be used also for fluids flowing in non-cylindrical ducts with a correction factor in the denominator:

$$f = \frac{16}{\varphi\ Re} \tag{7.44}$$

where φ is a parameter that depends on the geometry of the system. For ducts with rectangular cross-section, φ can be evaluated using Figure 7.10.

In Figure 7.10, z_1 corresponds to the length of the lower face and z_2 of the longest side of the rectangle.The balance of forces for noncylindrical ducts can be carried out using the hydraulic equivalent diameter. However, velocity estimation is done using the actual dimensions of the pipe.

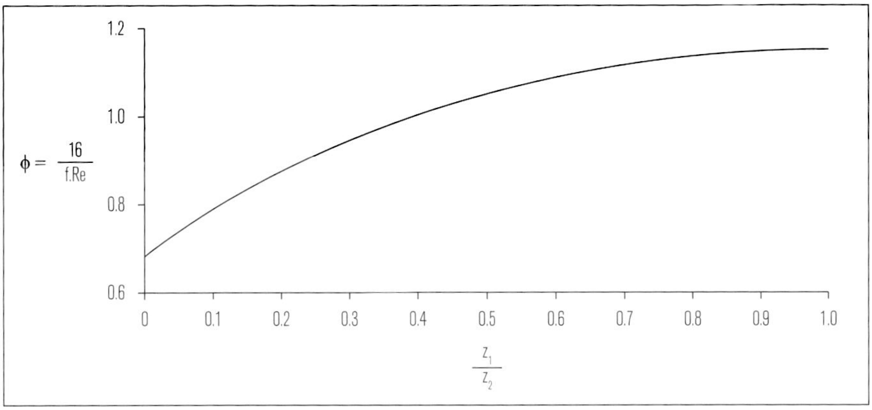

Figure 7.10 - Parameter ϕ - correction for the friction factor for laminar flow in rectangular ducts (GASKELL, 1992).

7.3.2 Flow around submerged objects (external)

As mentioned earlier, in case of external flow, the drag force, the fluid exerts on the object can also be evaluated by equation (7.25):

$$F_k = A\,K\,f \tag{7.25}$$

For this system, the characteristic area, A, is taken as the area obtained by projecting the solid in a plane perpendicular to the approach speed of the fluid. This definition is illustrated schematically in Figure 7.11, for the case of a sphere.

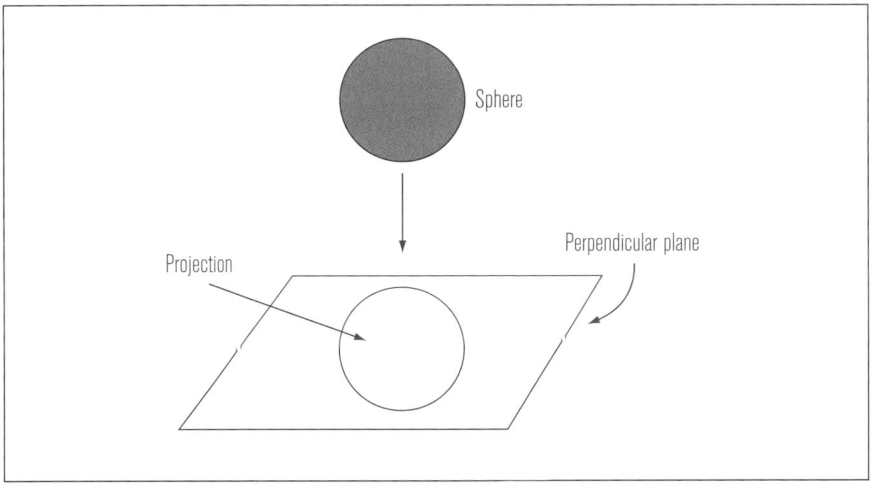

Figure 7.11 - Definition of the area characteristic for the flow around objects.

The kinetic energy per unit volume of fluid is evaluated using the relative velocity between the solid and fluid. For this purpose the fluid velocity sufficiently far from solid, is considered so that the presence of the solid has little influence on the velocity.

Similarly to the case of internal flow, the friction factor can be evaluated experimentally.These experiments show that the value of the friction factor depends on the shape of the object around which the fluid flows. Moreover, the value of f is also affected by the value of the Reynolds number associated with the flow.

7.3.2.1 Flow around a sphere

One of the objectives to study the flow fluid around the a sphere is to determine the value of friction factor through experiments. In these experiments the terminal velocity of a sphere moving in a stagnant fluid. is determined. As has been discussed in Chapter 6, the terminal velocity corresponds to the velocity when the sphere reaches the sum of forces acting on it equals its weight. When a sphere is placed inside a fluid two forces act on it: its weight downwards due to gravity and the buoyancy force acting in the opposite direction.These two forces always exist, regardless whether the sphere is stationary or moving. Buoyancy force is actually is the weight of the fluid that is displaced by the solid body. If the ball moves inside the fluid, there is a frictional force also call Friction drag force, $F_{K,}$ which acts on its surface. It is important to note that the frictional drag force always acts in the opposite direction to the direction of the velocity of the sphere. If the sphere is denser than the fluid, it would fall. In this case, the frictional drag force acts in the same direction upwards as that of buoyancy. When the spherical particle is lighter than the fluid, it rises inside the fluid. The frictional drag force in this case has the same direction of the weight of the sphere that is downwards. These two situations are shown in Figure 7.12.

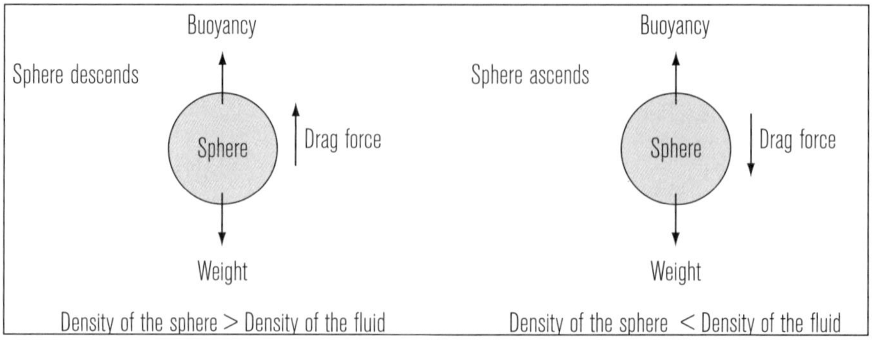

Figure 7.12 - Forces acting on a sphere within a fluid.

The balance of forces for a sphere, moving with constant velocity in the vertical direction in a stagnant fluid (assuming that the density of the sphere is more than that of the fluid)is given by:

Weight = Buoyant force + Frictional drag force (7.45)

$$\frac{\pi D^3}{6} \rho_s g = \frac{\pi D^3}{6} \rho g + \frac{\pi D^2}{4} \left(\frac{1}{2} \rho v_t^2\right) f \qquad (7.46)$$

In the equation (7.46), $\pi D^3/6$ corresponds to the volume of the sphere, ρ_s the density of sphere and v_t the terminal velocity.

Knowing the density of the fluid, the density of the sphere and its diameter, the experimental determination of the terminal velocity can be used to calculate the friction factor:

$$f = \frac{24}{Re} \text{ for } Re \leq 1 \qquad (7.47)$$

$$f = \frac{18,5}{Re^{3/5}} \text{ for } 1 < Re \leq 500 \qquad (7.48)$$

$$f \approx 0.44 \text{ for } Re > 500 \qquad (7.49)$$

The variation of f with the Reynolds number can be expressed mathematically by the above three expressions, valid for specific regimes of Reynolds number. Equation 7.47 can be derived on the basis of stokes equation for laminar flow together with equation 7.46.

(Exercise: Derive and equation of relation of Reynolds number with friction factor for laminar flow)

For the intermediate region of Reynolds numbers equation 7.48 can be used. For high Reynolds numbers the friction factor is practically constant. Figure 7.13 gives the variation of friction factor with Reynolds number for spheres. Similar graphs are available in the literature for other geometries such as cylinders, ellipsoids etc.

The region of Reynolds numbers amounting to less than one corresponds to laminar flow, for which Stokes' law can be applied, as has been demonstrated in Chapter 6.

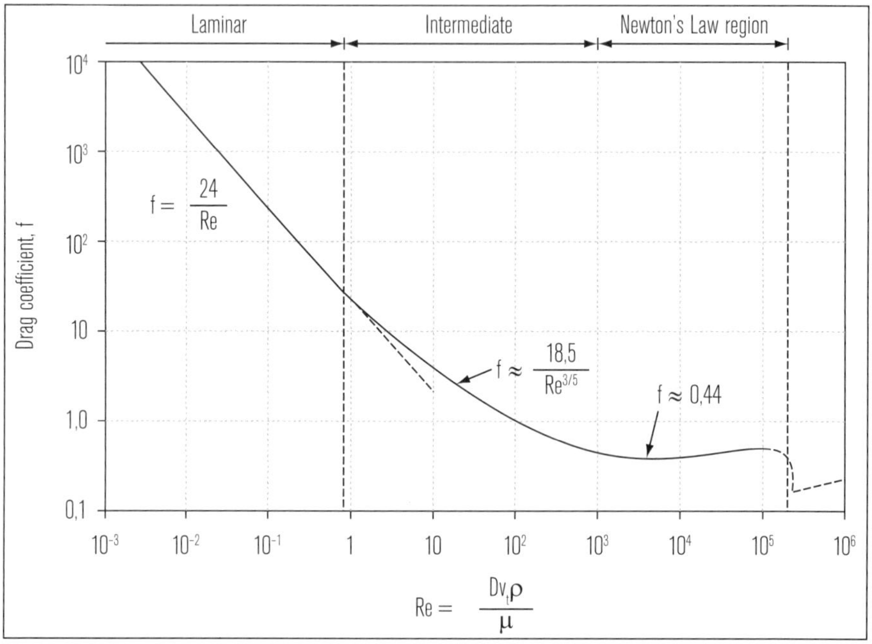

Figure 7.13 - Friction factors for flow around spheres (BIRD, STEWART and LIGHTFOOT, 1960).

Example

Calculate the terminal velocity of an inclusion of alumina in liquid steel during deoxidation when alumina particles rise to the surface.

Data:

- The inclusion diameter: 200 μm;
- Inclusion density: $\rho_s = 2300\ kg/m^3$;
- Density of steel: $\rho = 6700\ kg/m^3$;
- Viscosity of steel: $\mu = 6.5$ cP.

Repeat the calculation for cases of 100 and 50 μm.

Solution

As the inclusion is less dense than steel, the balance of forces can be written as follows:

$$\text{Buoyancy} = \text{Weight} + \text{Drag force}$$

$$\frac{\pi D^3}{6} \rho g = \frac{\pi D^3}{6} \rho_s g + \frac{\pi D^2}{4} \left(\frac{1}{2} \rho v_t^2 \right) f$$

In the above equation, the values of v_t and f are unknown. A similar approach to that adopted in the case of flow in pipes, can be adopted here in order to, obtain a relation between the Reynolds number and friction factor. Expressing the value of v_t in the above equation as a function of Reynolds number:

$$v_t = \frac{Re\,\mu}{D\,\rho}$$

Substituting this value in the previous equation,

$$\frac{\pi\,D^3}{6}\,\rho\,g = \frac{\pi\,D^3}{6}\,\rho_s\,g + \frac{\pi\,D^2}{4}\left[\frac{1}{2}\,\rho\left(\frac{Re\,\mu}{D\,\rho}\right)^2\right] f$$

Making some simplifications and transposing terms, we obtain:

$$Re^2\,f = \frac{4}{3}\,(\rho - \rho_s)\,\rho\,\frac{g\,D^3}{\mu^2}$$

Substituting the data given,

$$Re^2\,f = \frac{4}{3}\,(6700\; - \;2300)\,(6700)\,\frac{(9.8)\,(200\,.\,10^{-6})^3}{(0.0065)^2} = 72.9383$$

As the value of Reynolds number is unknown, one cannot determine f. One is not sure, as to which of the equations for different regimes of Reynolds number (equations (7.47) to (7.49)), is appropriate in this case. Hence a procedure of trial and error.using all the 3 equations can be tried to find the most satisfactory solution. For example, if the equation (7.47) is used for friction factor.

$$Re^2\,f = Re^2\left(\frac{24}{Re}\right) = 72.9383$$

$$Re = 3.039$$

The value of Reynolds number obtained is not compatible with the laminar flow regime characterized by Re = 1, hence the result above is incorrect.

If the second relation (equation 7.48) is used,

$$Re^2\,f = Re^2\left(\frac{18.5}{Re^{3/5}}\right) = 72.9383$$

$$Re = 2.664$$

This value of Reynolds number obtained is within the range of validity of the equation used and hence can be used for calculating the terminal velocity of the inclusion:

$$v_t = \frac{Re\ \mu}{D\ \rho} = \frac{(2.664)\ (0.0065)}{(200\ .\ 10^{-6})\ (6700)} = 0.0129 \text{ m/s}$$

Similarly one can calculate the velocities of inclusions of other particles sizes namely 100 and 50 μm.

- Inclusion of 100 μm: v_t = 0.00369 m/s;
- Inclusion of 50 μm: v_t = 0.00092 m/s.

(Calculate and show that these are the values obtained in the above case)

7.4 Friction Factors for Packed Beds

Flow through beds of particles is also a system of interest to the metallurgist. Fixed and moving beds consisting of solid granules or clusters of fine particles, are encountered frequently in various metallurgical processes, like agglomeration processes of sintering and pelletizing of iron ores, blast furnace, process, direct reduction, drying, heating and cooling of beds with gases etc. In many systems, hot gases pass through a bed of particles imparting heat and causing chemical reactions like reduction. Permeability of beds is an important parameter for ensuring good solid gas contact and this is related to the pressure drop across the bed for a given flow rate of gas. This information is vital for getting design details of equipment for injection (or suction) of gas through these beds.

Although in practice one deals with beds of non-uniform particles and their distribution in the reactor giving rise to preferential flow or channeling etc., the following treatment presented below assumes bed of uniform particles and no preferential flow It is also assumed that the diameter of the particles that make up the bed is small compared with the diameter of the reactor and there is uniform flow of gas through the bed.

7.4.1 Ergun equation

The flow of gas through the beds of particles and the pressure drop are influenced by the characteristics of the bed which in turn are defined by the particle size shape,the void fraction uniformity of particle size etc. Figure 6.14 shows a schematic view of particles in a packed bed. Ergun considered that the voids in the packed bed can be considered as a bundle of tubes running through from one end to the other through tortuous paths. Chracterization of the bed in terms of particle sizes, void fraction etc. is necessary for the calculation of a hydraulic diameter. Using this parameter, equations are developed for laminar and turbulent flows in the bed on the basis of analogy of flow in tubes. The treatment is sometimes called tube bundle theory, since the continuous pores in the bed are analogical to a bundle of tortuous tubes.

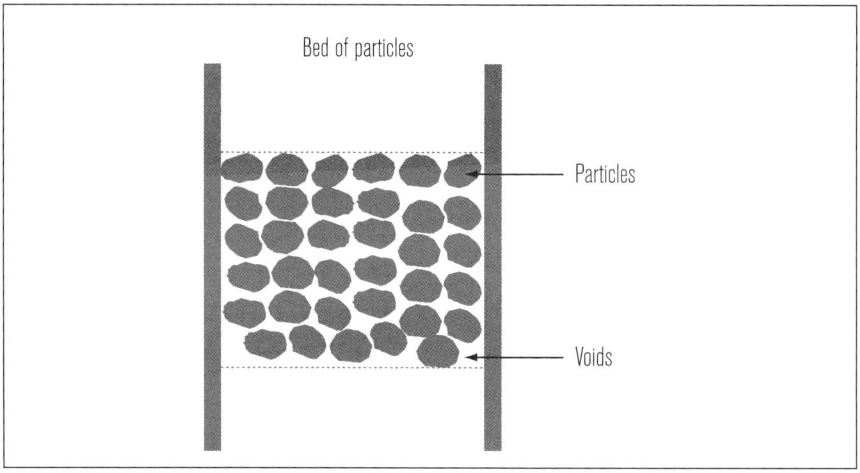

Figure 7.14 - Schematic view of a bed of particles.

It is observed that the bed is composed by the particles and voids that form between them.Thus, one can write that:

$$\text{Volume of the bed} = \text{Volume of the particles} + \text{Volume of voids} \quad (7.50)$$

An important parameter in the characterization of a bed is the void fraction. Dividing the two sides of the equation above by the volume of the bed, we get:

$$1 = \frac{\text{volume of particles}}{\text{volume of bed}} + \frac{\text{volume of voids}}{\text{volume of bed}} \quad (7.51)$$

Void fraction is defined by the following equation:

$$\text{Void fraction} = \omega = \frac{\text{volume of voids}}{\text{volume of bed}} \qquad (7.52)$$

Hence,

$$1 = \frac{\text{volume of particles}}{\text{volume of bed}} + \omega \qquad (7.53)$$

$$\frac{\text{volume of particles}}{\text{volume of bed}} = 1 - \omega \qquad (7.54)$$

A number of factors interfere with the value of the fraction of a bed. Among them, the most important is the particle size distribution and average size of particles that compose it. Another variable of importance in their beds is the surface area. This area is defined by the equation below:

$$a = \frac{\text{Surface area of particles}}{\text{Volume of bed}} \qquad (7.55)$$

This can be rewritten in the following manner:

$$a = \left(\frac{\text{Surface area of particles}}{\text{Volume of particles}}\right)\left(\frac{\text{Volume of particles}}{\text{Volume of bed}}\right) \qquad (7.56)$$

Spherical particles of uniform size (diameter d). Hence,

$$\left(\frac{\text{Surface area of particles}}{\text{volume of particles}}\right) = \frac{\pi d^2}{\frac{\pi d^3}{6}} = \frac{6}{d} \qquad (7.57)$$

Combining the equations (7.54) (7.56) and (7.57), we obtain:

$$a = \frac{6}{d}(1 - \omega) \qquad (7.58)$$

The above relationship is valid only for spherical particles. It is common to have spherical particles in countercurrent beds like in direct reduction processes. To deal with non-spherical particles is common to use the concept of sphericity. The sphericity is a measure as to to what degree the shape of a particle approaches the shape of a sphere. Its definition is given in Figure 7.15.

Hence sphericity can be defined as the ratio of surface area of the sphere and the particle, both having equal volume:

$$\text{spherificity} = \varphi = \frac{\text{area of sphere}}{\text{area of particle}} \qquad (7.59)$$

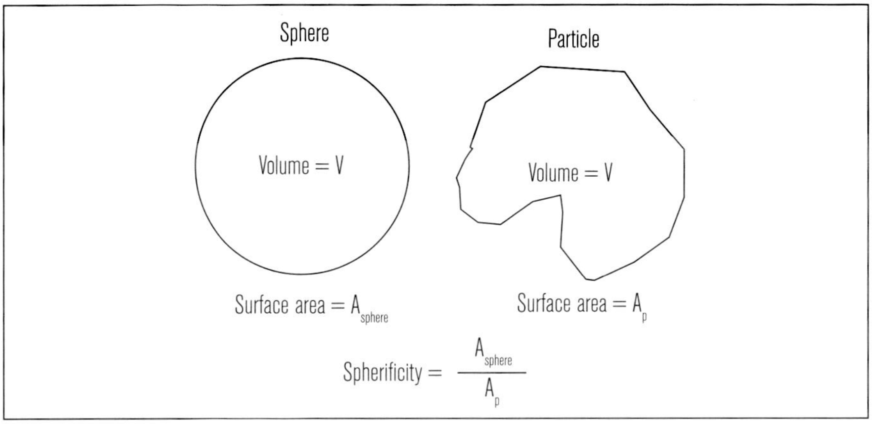

Figure 7.15 - Definition of sphericity of a particle.

As the solid sphere has least surface area per unit volume, sphericity values are always smaller than one. The spherificity of a sphere is = 1.

Equation (6.59) can be written in the following way:

$$\text{area of particle} = \frac{\text{area of sphere}}{\varphi} \quad (7.60)$$

Combining (7.60) and (7.58), we obtain an expression for evaluating the surface area of a bed composed of non-spherical particles.

Have:

$$a = \frac{6}{d\,\varphi}(1 - \omega) \quad (7.61)$$

Example

Estimate the sphericity of lamellar type of particles of iron ore with approximate dimensions as given in the figure below. The shape of the particle has been simplified to facilitate the calculations.

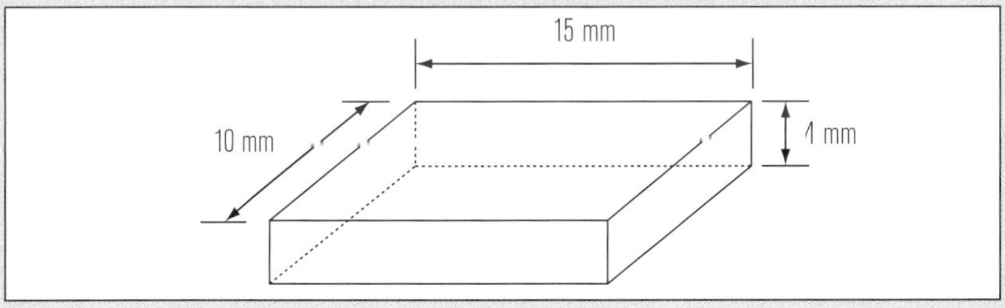

Solution

Volume of the particle of iron ore:

$$V_p = 15 \,.\, 10 \,.\, 4 = 600 \text{ mm}^3$$

Surface area of the particle:

$$A_p = (15 \,.\, 10 + 10 \,.\, 4 + 15 \,.\, 4) \,.\, 2 = 500 \text{ mm}^2$$

The area above is the area of the six surfaces of the particle. The surface area of sphere of same volume of the particle can be calculated by evaluating the value for radius of the sphere of same volume:

$$V = \frac{4}{3} \pi R^3 = 600 \text{ mm}^3$$

$$R = \left(\frac{3 \,.\, 600}{4 \pi} \right)^{\frac{1}{3}} = 5.23 \text{ mm}$$

Surface area of the sphere with a radius of 5.23 mm:

$$A_{sphere} = 4 \pi R^2 = 344 \text{ mm}^2$$

Therefore, the sphericity of the particle of iron ore is given by:

$$\frac{A_{sphere}}{A_p} = \frac{344}{500} = 0.688$$

In the equation (7.61), d is the diameter of the sphere of the same volume of the particle. Instead of determination of this diameter exactly, it is customary to work with the particle sizes defined in terms of openings of the sieves used for screening the particles. In this case the particle size is not uniform. One can define an average size of the particles from the size analysis during screening. This approach is the same, used in mineral processing. In the case of nonspherical particles with a certain particle size distribution, the value of the average size of particles is determined by the following relationship:

$$d = \frac{1}{\sum_{i=1}^{i=n} \frac{(\%\, i)/100}{d_i}} \tag{7.62}$$

where:

- n = number of sieves used in the screening and where material was retained;
- d = diameter of particles;
- d_i = average diameter of the material retained on sieve i;
- (%.i) = percentage of material retained on sieve i.

The average diameter of the material retained on sieve is determined using the geometric mean of the sieve opening in which the material is retained and immediately above the sieve through which the material passed through.The geometric mean is calculated by the square root of the product of the openings of sieves. The example below illustrates the calculation of the average particle size from the sieve analysis of particles of different sizes

Example

The table below shows the size analysis of iron ore. From these data, determine the average size of the ore.

Opening of sieves (mm)		Percent
Top	Bottom	retained (%)
25.4	19.1	0.44
19.1	15.9	2.27
15.9	12.7	18.37
12.7	6.3	68.66
6.3	4.8	5.95
4.8	1	4.31

Solution with the data from the table above:

Opening of sieves (mm)		Average size of material retained (mm)	Percent retained (%)	$\frac{(\%i)/100}{d_i}$
Top	Bottom			
25.4	19.1	22.03	0.44	0.0002
19.1	15.9	17.43	2.27	0.0013
15.9	12.7	14.21	18.37	0.0129
12.7	6.3	8.94	68.66	0.0768
6.3	4.8	5.5	5.95	0.0108
4.8	1	2.19	4.31	0.0197
				Σ = 0.1217

Average diameter = d = 8.215 mm

With the help of discussions and definitions above, we can finally determine relationships to estimate the pressure drop in beds traversed by gases.

The concept of equivalent hydraulic diameter. can be used as in this case, the bed contains particles of irregular shape, through which the gas flows. Recalling the definition of the hydraulic equivalent diameter:

$$D_h = \frac{4\ A}{P_M} \qquad (7.43)$$

where A represents cross-sectional area through which the fluid flows and P_M is the wetted perimeter.

For calculating the hydraulic diameter, first the numerator and denominator of the above equation are multiplied by bed height, L, to get:

$$D_h = \frac{4\ A\,L}{P_M\ L} \qquad (7.63)$$

Analyzing the above equation, AL is the volume available for gas flow through the bed, equal to the volume of voids. In the denominator, the product P_M L is the wetted area, which is practically equivalent to the gas-particle contact area (the area of contact with the walls of the vessel containing the bed is very small compared with the surface area of particles). The contact area of gas-particles is actually the surface area of these particles (ignoring the the contact areas between the particles). On the basis of this discussion one can write the equation (7.63) in the following way:

$$D_h = 4\,\frac{\text{volume of voids}}{\text{surface area of particles}} \qquad (7.64)$$

Dividing equation (7.64) by the volume of the bed:

$$D_h = \frac{4\left(\frac{\text{volume of voids}}{\text{volume of bed}}\right)}{\left(\frac{\text{surface area of particles}}{\text{volume of bed}}\right)} \qquad (7.65)$$

Combining 7.65 with equations (7.52) and (7.61), one can write:

$$D_h = \frac{4\ \omega}{\frac{6}{d\ \varphi}\ (1\ -\omega)} = \frac{2\ \omega\ d\ \varphi}{3\ (1\ -\omega)} \qquad (7.66)$$

Equation (7.66) expresses the equivalent hydraulic diameter of the bed as a function of particle and bed characteristics. This equation can be used to calculate the pressure drop and flow for laminar and turbulent regimes, in packed beds as well as to evaluate permeability.

7.4.1.1 Laminar flow in packed beds

The analogy of laminar flow in tubes can be applied to beds as the voids can be considered tubes through wich the gas passes through. Equation (5.126) was used to estimate the pressure drop of a fluid in terms o average velocity of the gas and neglecting the force of gravity (for gases, it is a reasonable assumption due to their low densities), one can write equation (5.126), in terms of equivalent hydraulic diameter:

$$\frac{P_0 - P_L}{L} = \frac{8\,\mu\,\overline{V}}{R_h^2} = \frac{32\,\mu\,\overline{V}}{D_h^2} \tag{7.67}$$

Substituting the definition of the hydraulic equivalent diameter equation (7.66)

$$\frac{P_0 - P_L}{L} = \frac{32\,\mu\,\overline{V}}{D_h^2} = \frac{32\,\mu\,\overline{V}}{\left(\dfrac{2\,\omega\,d\,\varphi}{3\,(1-\omega)}\right)^2} \tag{7.68}$$

$$\frac{P_0 - P_L}{L} = \frac{72\,\mu\,\overline{V}\,(1-\omega)^2}{\omega^2\,d^2\,\varphi^2} \tag{7.69}$$

The influence of variables on pressure drop predicted by equation 7.69, are qualitatively compatible with the experimental data. However, the constant in the equation, that gives the best fit for the experimental results, is 150 instead of 72. This is due to the fact that the distance traveled by the gas is longer than the bed height, L, considered in the evaluation of pressure drop. The gas actually traverses a tortuous path entering the voids in between the particles. Hence the equation that is used to estimate pressure drop in the bed of particles for laminar flow is:

$$\frac{P_0 - P_L}{L} = \frac{150\,\mu\,\overline{V}\,(1-\omega)^2}{\omega^2\,d^2\,\varphi^2} \tag{7.70}$$

The above expression is known as Blake-Kozeny equation. It is usual to replace the gas velocity through the voids of the bed, by V_0, the superficial velocity. Both of them are related through the void fraction of the bed:

$$\overline{V} = \frac{V_o}{\omega} \tag{7.71}$$

The velocity of gas in the voids would be much higher than the superficial velocity at the the cross section of the bed. Substituting (7.71) in (7.70), we get Blake-Kozeny equation in terms of superficial velocity.

$$\frac{P_0 - P_L}{L} = \frac{150\,\mu\,V_o\,(1-\omega)^2}{\omega^3\,d^2\,\varphi^2} \tag{7.72}$$

7.4.1.2 Turbulent flow in packed beds

Again following the analogy of fluid flow in tubes, equation (7.27) can be used to estimate the pressure drop of gas in turbulent flow in a tube. This equation can be used for packed beds substituting D with equivalent hydraulic diameter:

$$\frac{(P_0 - P_L)}{L} = 2\,\rho\,\overline{V^2}\,f\,\frac{1}{D_h} \tag{7.73}$$

Substituting for the hydraulic equivalent diameter (equation (7.68)):

$$\frac{(P_0 - P_L)}{L} = \frac{2\,\rho\,\overline{V^2}\,f}{\left(\dfrac{2\,\omega\,d\,\varphi}{3\,(1-\omega)}\right)} = \frac{3\,\rho\,\overline{V^2}\,(1-\omega)}{\omega\,d\,\varphi}\,f \tag{7.74}$$

The friction factor for beds was evaluated experimentally and the value amounted to:

$$f = \frac{1.75}{3} \tag{7.75}$$

Substituting this value in (7.76) and using the superficial velocity, we get:

$$\frac{(P_0 - P_L)}{L} = \frac{1.75\,\rho\,V_o^2\,(1-\omega)}{\omega^3\,d\,\varphi} \tag{7.76}$$

The above expression is known as Burke Plummer equation and can be used to estimate the pressure drop of gas passing though the bed under turbulent flow conditions. Knowing the pressure drop and bed characteristics one can determine the flow rate.

Ergun unified the equations of Blake-Kozeny and Burke-Plummer, considering thereby the contributions of laminar and turbulent flow to the pressure drop. This takes into account the pressure drop over the entire range of flow regimes and is given by:

$$\frac{P_0 - P_L}{L} = \frac{150\,\mu\,V_o\,(1-\omega)^2}{\omega^3\,d^2\,\varphi^2} + \frac{1.75\,\rho\,V_o^2\,(1-\omega)}{\omega^3\,d\,\varphi} \qquad (7.77)$$

This equation is known as the Ergun equation and can be used to determine the pressure drop in beds both for laminar and turbulent regimes.From the above equation, it is observed that the parameters that favor a decrease in pressure drop of gas in the bed (or improvement of bed permeability) are:

- Greater void fraction, ω;
- Larger diameter of the particle, d;
- Increased sphericity, ϕ;
- Lower viscosity μ, density ρ, and superficial gas velocity V_0.

References

BIRD, R.B.; STEWART, W.E.; LIGHTFOOT, E.N. *Transport phenomena.* New York: John Wiley & Sons, 1960. 780 p.

GASKELL, D.R. *An introduction to transport phenomena in materials engineering.* New York: Macmillan, 1992. 637 p.

GUTHRIE, R.I.L. *Engineering in process metallurgy.* Oxford: Oxford Science Publications, 1992. 528 p.

HUANG, X.; THOMAS, B.G. Modeling of transient flow phenomena in continuous casting of steel. Computational fluid dynamics and heat/mass transfer modelingin the metallurgical industry. In: ANNUAL CONFERENCE OF METALURGISTS OF CIM, 35., 1996, Montreal, Canadá. *Annals...* [S.n.t.]. p. 129-45.

JONES, W.P.; LAUNDER, B.E. The prediction of laminarization with a two-equation model of turbulence. *International Journal of Heat and Mass Transfer*, v. 15, n. 2, p. 301-14, Feb. 1972.

SZEKELY, J.; THEMELIS, N.J. *Rate phenomena in process metallurgy.* New York: Wiley Interscience, 1970. 784 p.

TAVARES, R.P.; CASTRO, L.F.A. Mathematical modeling of fluid flow and heat transfer in a continuous casting distributor. In: ANNUAL CONGRESS OF ABM, 54., São Paulo, 1999. *Annals...* São Paulo: ABM, 1999. p. 544-54.

Exercises

1 - Calculate the terminal velocity of rise of a solid inclusion with 20 μm in diameter, rising in stagnant liquid steel.

Data:

- $\rho_{inclusion} = 2.7 \times 10^3$ kg/m³;
- $\rho_{steel} = 7.1 \times 10^3$ kg/m³;
- $\mu_{steel} = 5.5 \times 10^{-3}$ kg/ms.

The inclusion can be considered spherical. Check the validity of the calculations.

2 - A technique used to determine the viscosity of fluids is to measure the terminal velocity of a sphere that falls within the fluid. Determine the viscosity of the fluid Data for the sphere and the liquid are given below:

- $D_{sphere} = 1$ cm;
- $\rho = 1.261$ g/cm³;
- $\rho_{sphere} = 7.1$ g/cm³.

We know also that the constant rate period, the sphere travels 2 m in 7 seconds.

3 - A hollow steel sphere with 5 mm in diameter and weighing 0.05 g, is dropped from the surface of a liquid column and reaches a terminal velocity of 0.5 cm/s, The density of the liquid is 0.9 g/cm³ and the acceleration due to gravity is 980.7 cm/s². The sphere is well away from the walls of the duct. Determine the following:

- Drag force;
- Friction factor;
- Viscosity of the fluid.

4 - Calculate the sphericity of a cube with two inches square on each side.

5 - Calculate the average diameter of the material that has the following sieve analysis:

Opening of sieves (mm)		Percent retained (%)
Top	Bottom	
25.4	19.1	0.22
19.1	15.9	4.3
15.9	12.7	15.36
12.7	6.4	72.95
6.4	4.8	3.89
4.8	1	3.28

6 - A steel ball (radius = 8.87 cm) is thrown into molten slag to determine the viscosity of fluid.The density of steel is twice that of the slag and the terminal velocity of the sphere is 1.524 m/s (experimentally determined).

Calculate the kinematic viscosity of the slag.

7 - Gas is passed through a bed of square section of 3.048 m side and 14.11 m in length.The pressures at the entry and exit of gas are respectively, 104109.97 and 103 420.5 N/m^2, respectively. The mass flow of gas is 90.72 km/h. Evaluate the void fraction of bed (between 0 and 0.6) for the following conditions:

- Particle diameter = 3.048 cm;
- Viscosity of gas = 2.067 x 10^{-5} kg/ms;
- Density of gas = 0.12 kg/m^3 (medium density).

8 - Calculate the pressure difference needed to make water rise in a vertical tube 10 m long at a flow rate of 0.5 l/s. The tube diameter is 1.5 cm and roughness of 0.1 mm.

9 - Evaluate the water flow in a horizontal pipe of 1 inch in diameter, over which the measured differential pressure is 50,000 Pa. The roughness of the tube is 0.5 mm. The length of the tube is 5 m.

Overall Energy Balances in Isothermal Flow of Fluids

Chapter 08

In most engineering problems involving fluid flows, the important objective is to obtain a relationship between flow rate of the fluid and the variables that cause the flow such as differential pressure, gravity, electromagnetic forces etc.

For this, microscopic as well as macroscopic, balances are used. The microscopic balance, illustrated in earlier chapters is given schematically in Figure 8.1a. Here the control volume is infinitesimal and is located far from the boundaries of the system. Micreoscopic balance in this control volume is carried out with input and output terms resulting in differential equations and these can be solved with boundary conditions to get velocity profile etc. This was treated with extensive examples in Chapters 5 and 6.

In the case of macroscopic approach, illustrated in Figure 8.1b, the control volume is the total system and therefore the conditions of input and output are at the entrance and exit of the reactor.

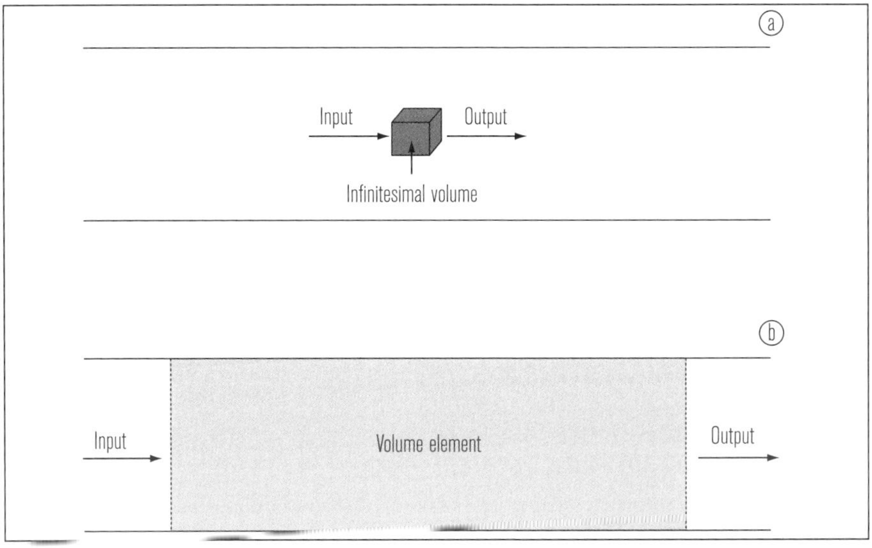

Figure 8.1 - Control volume elements in: a) microscopic and b) macroscopic fluid flow.

In general, the establishment of global balance (macroscopic treatment) results in algebraic equations for steady state systems and differential equations of first order for unsteady state. This method simplifies the mathematical manipulations necessary, but the resulting solutions provide less information about the system. The macroscopic method has been used in Chapter 7, in terms of global balance of forces for flow in pipelines (internal flow) and around objects (external flow). In this chapter, macroscopic approach, is adopted for establishing global balances of mass and energy applied to flow in pipelines, flow from ladles,flow measurement techniques and other practical applications particularly in metallurgical engineering.

8.1 Global Mass Balance

For global mass balance system shown in Figure 8.2. is considered:

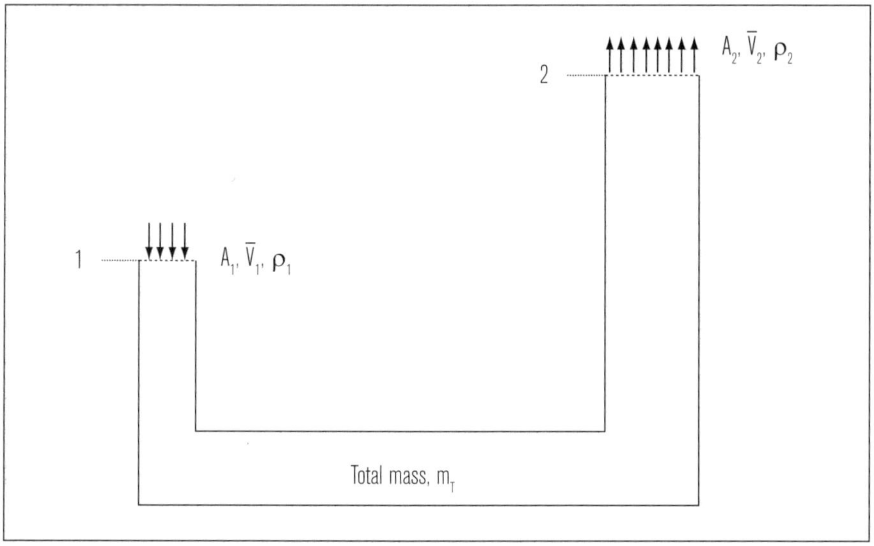

Figure 8.2 - System for the development of the global mass balance.

Development of global mass balance can be made with the following assumptions:

- average speeds in planes 1 and 2 are parallel to the walls of the duct;
- density and other physical properties do not vary over the cross sections 1 and 2.

The equation of conservation of mass states that:

$$[\text{Rate of mass input}] - [\text{Rate of mass output}] = [\text{Rate of mass accumulation}] \quad (8.1)$$

That is to say:

$$A_1 \rho_1 \overline{V_1} - A_2 \rho_2 \overline{V_2} = \frac{d m_T}{d t} \tag{8.2}$$

where,

- A_1, A_2 = area of cross sections in planes 1 and 2;
- ρ_1, ρ_2 = density of the fluid in planes 1 and 2;
- $\overline{V_1}, \overline{V_2}$ = average speed of the fluid in planes 1 and 2;
- m_T = total mass of fluid in the system;
- t = time.

One can also define the following variable:

$$\dot{m} = A \rho \overline{V} \tag{8.3}$$

representing the mass flow of fluid in a given plane. With this variable, equation (8.2) becomes:

$$\dot{m}_1 - \dot{m}_2 = \frac{d m_T}{d t} \tag{8.4}$$

For steady state: conditions:

$$\frac{d m_T}{d t} = 0 \tag{8.5}$$

Hence,

$$\dot{m}_1 - \dot{m}_2 = 0 \tag{8.6}$$

Example

Liquid steel is poured from a ladle through a nozzle placed at the bottom. The diameter of this nozzle is 7.62 cm. Calculate the time required to empty the ladle. Consider, as a first approximation, that the velocity of liquid steel at the nozzle is a function of the height of the steel ladle, according to the following equation:

$$\overline{V}_{nozzle} = C_D \sqrt{2 g h}$$

where:

- h = height of the metal in the ladle;
- C_D is the discharge coefficient (in this case, considered equivalent to 0.9).

other data:

- Ladle diameter: 3.0 m;
- Initial height of liquid: 3.3 m;
- Density of steel: 7000 kg/m³.

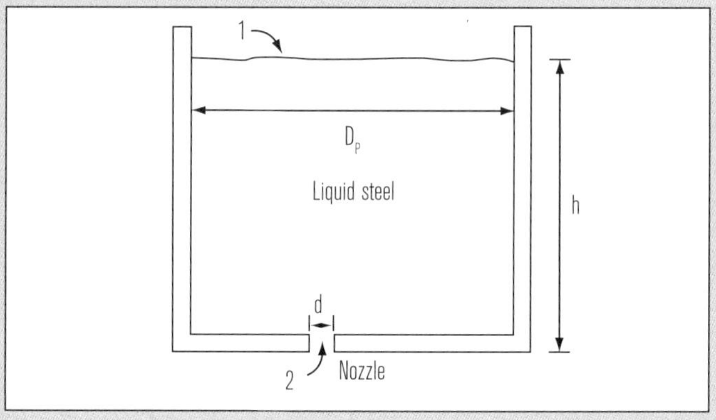

Solution

The above figure shows schematically the ladle containing steel melt.

Level 1 is the surface of liquid steel in the ladle and level 2 is the outlet of the nozzle for steel flow.

In the equation (8.6):

$$\dot{m}_1 = 0$$

Hence the equation of mass balance reduces to:

$$-\dot{m}_2 = \frac{d\, m_T}{d\, t}$$

$$\dot{m}_2 = A_2\, \rho_2\, \overline{V}_2$$

Using the given relation for average velocity in the nozzle, we have:

$$\dot{m}_2 = A_2\, \rho_2\, C_D \sqrt{2\, g\, h}$$

Density of steel being constant the total mass of steel in the ladle is:

$$m_T = A_P \rho h$$

Differentiating this equation (A_p is the cross-sectional area of the ladle; considered constant over the height of the ladle).

$$d\,m_T = A_P \rho\, d h$$

$$\frac{d\,m_T}{d t} = A_P \rho \frac{d h}{d t} = -\dot{m}_2 = -A_2 \rho_2 C_D \sqrt{2 g h t}$$

Separating variables in the equation above, we have:

$$\frac{dh}{h^{1/2}} = -\frac{A_2}{A_p} C_D \sqrt{2g}dt$$

The above equation can be integrated, with boundary conditions:

$h = h_i$ for $t = 0$

$h = 0$ for $t = t_e$

where:

- h_i is the initial height of the steel ladle;
- t_e is the time of emptying the ladle.

$$\left(2\, h^{1/2}\right)_{h_0}^{0} = -\frac{A_2}{A_P} C_D \sqrt{2\, g}\, \left(t\right)_0^{t_e}$$

Substituting the limits of integration, we have:

$$-2\, h_0^{\ 1/2} = -\frac{A_2}{A_P} C_D \sqrt{2\, g}\ \ t_e$$

Finally, the time necessary for emptying the ladle is given by:

$$t_e = \frac{A_P}{A_2 C_D} \left(\frac{2\ h_0}{g}\right)^{1/2}$$

$$A_2 = \pi \frac{d^2}{4}$$

$$A_P = \pi \frac{D_P^2}{4}$$

Incorporating the areas of cross sections of the nozzle and the ladle t (considered circular) given above we get:

$$t_e = \frac{D_P^2}{d^2 C_D}\left(\frac{2\ h_0}{g}\right)^{1/2}$$

Using the data given:

$$t_e = \frac{D_P^2}{C_D d^2}\left(\frac{2\ h_0}{g}\right)^{1/2} = \frac{(3)^2}{(0.9)(0.0762)^2}\left(\frac{2 \times 3.3}{9.8}\right)^{1/2} = 1413\ \text{s}$$

This time necessary for emptying out the ladle is approximately equal to 24 minutes.

8.2 Global Energy Balance

To develop an overall or global energy balance consider the system shown in Figure 8.3 for application of the principle of conservation of energy:

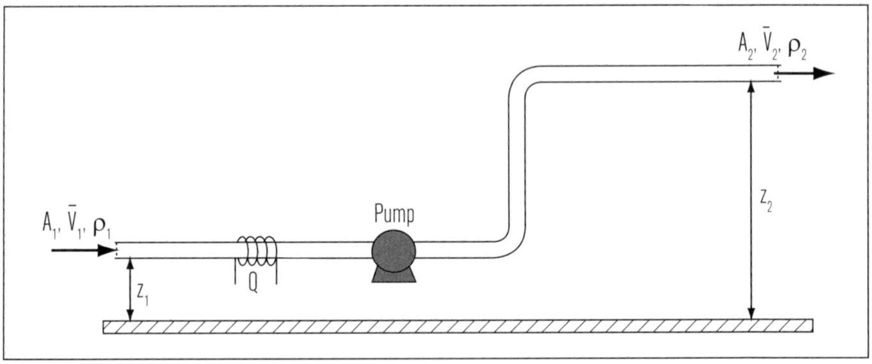

Figure 8.3 - System for global energy balance.

[Rate of input of energy] - [Rat of output of energy] = [Rate of energy accumulation] (8.7)

The various types of energy entering and leaving the system and the accumulation of energy can be represented by the following equation:

$$\frac{d}{dt}(E_{total}) = -\Delta\left[(H + E_P + E_c)\ \dot{m}\right] + Q + S_R - M \tag{8.8}$$

- E_{total} = total energy of the fluid, (internal energy, potential and kinetic energies etc.);
- H = enthalpy of the fluid per unit mass;
- E_P = potential energy of fluid per unit mass;
- E_c = kinetic energy of fluid per unit mass;
- $\dot{m}$ = mass flow of fluid in the system;
- Q = net rate of heat input into the system;
- M = mechanical work done by the fluid on the pump (or other device for fluid handling);
- S_R = net generation of energy in the system due to chemical reactions or other sources.

In the above equation, the operator Δ means (output - input) and -Δ (input - output).

Assuming steady state conditions:

$$\frac{d}{dt}(E_{total}) = 0 \tag{8.9}$$

Considering systems without chemical reactions and where there are no other sources of energy, we have:

$$S_R = 0 \tag{8.10}$$

The equation (8.8) becomes:

$$\Delta\left[(H + E_P + E_c)\ \dot{m}\right] - Q + M = 0 \tag{8.11}$$

In the following section evaluation of these terms, as measurable parameters, are discussed.

8.2.1 Evaluation of the kinetic energy term

The input by kinetic energy in the system through the area A_I (normal to the flow) can be evaluated using the following equation.

$$\dot{m}_I\ E_{c_I} = \int_0^{A_I}\left[\frac{1}{2}(\rho_I\ v_I\ dA_I)\ v_I^2\right] \tag{8.12}$$

For a cylindrical duct, the differential element area is given by:

$$d\,A_1 = 2\,\pi\,r\,d\,r \tag{8.13}$$

Combining the equations (8.12) and (8.13):

$$\dot{m}_1\;E_{c_1} = \int_0^{R_1}\left[\frac{1}{2}\,(\rho_1\;v_1\;2\,\pi\,r\,dr)\;v_1^2\right] \tag{8.14}$$

where R_1 is the radius of the pipe section 1 (Figure 8.3)

To integrate the above equation, it is important to remember that the velocity of the fluid varies along the cross section of the pipeline. Taking into account this fact, two limiting cases namely laminar flow and highly turbulent flow can be analyzed.

a) *Laminar flow*

The following equation for the velocity profile along the cross section of an inclined pipe (Chapter 5) was obtained for laminar flow:

$$v_z = \frac{1}{\mu}\left(\rho\,g\cos\beta + \frac{P_o - P_L}{L}\right)\frac{(R^2 - r^2)}{4} \tag{5.116}$$

The average velocity along the cross section of the pipe is given by:

$$\overline{V} = \frac{R^2}{8\,\mu}\left(\rho\,g\cos\beta + \frac{P_o - P_L}{L}\right) \tag{5.126}$$

Combining the two equations above, one can obtain an expression relating he velocity profile with the average speed:

$$v_z = 2\,\overline{V}\,\frac{(R^2 - r^2)}{R^2} \tag{8.15}$$

Substituting equation (8.15) the expression for profile velocity in the kinetic energy equation (8.14):

$$\dot{m}_1\,E_{c_1} = \int_0^{R_1}\left\{\frac{1}{2}\,\rho_1\left[2\,\overline{V}_1\,\frac{(R^2 - r^2)}{R^2}\right]2\,\pi\,r\,d\,r\left[2\,\overline{V}_1\,\frac{(R^2 - r^2)}{R^2}\right]^2\right\} \tag{8.16}$$

Integration of the above equation gives:

$$\dot{m}_1\;E_{c_1} = \pi\,\rho_1\,\overline{V}_1^{\,3}\,R_1^2 \tag{8.17}$$

(integration can be done as an exercise).

Flow at level 1 equals (Figure 8.3)

$$\dot{m}_1 = \pi R_1^2 \rho_1 \overline{V}_1 \tag{8.18}$$

Comparing 8.17 and 8.18, one can write:

$$E_{c_1} = \overline{V}_1^2 \tag{8.19}$$

The above equation is valid for laminar flow and using the same determination of kinetic energy of fluid per unit mass in terms of average speed can be assessed.

b) *Turbulent flow*

In the turbulent regime, the velocity profile of fluid in a given cross section of the pipe, is quite different from the parabolic profile, which prevails in laminar flow. This can be seen in Figure 8.4.

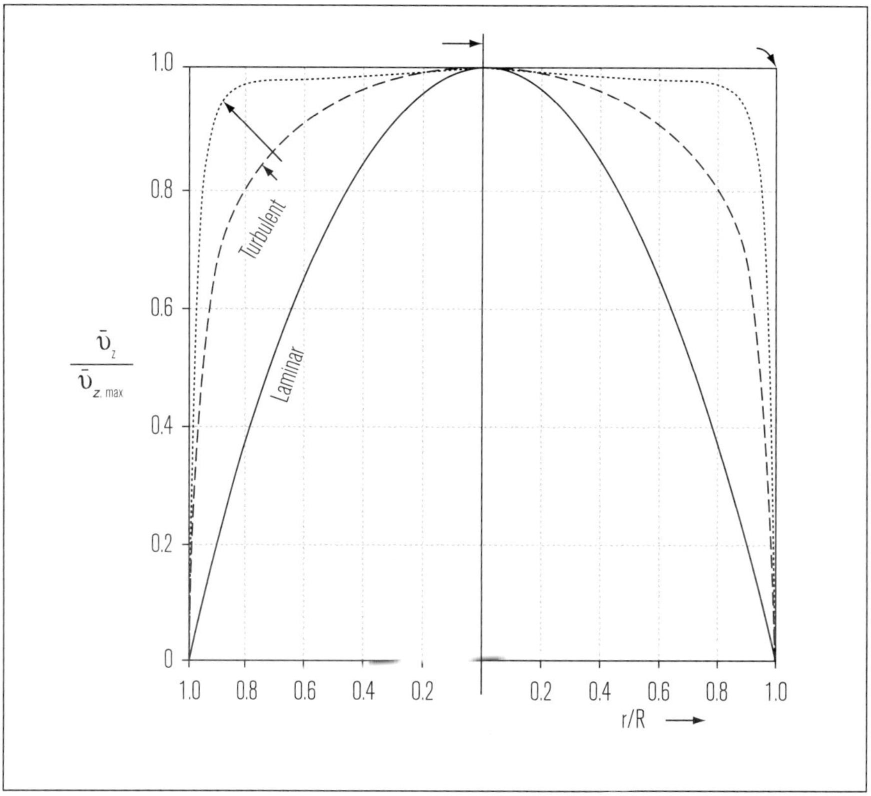

Figure 8.4 - Comparison between the qualitative distributions of velocity in laminar and turbulent flows (BIRD, STEWART and LIGHTFOOT, 1960).

For highly turbulent regime, it is observed that the velocities are approximately constant in the central region of the duct. Velocity gradients are confined to a very narrow region near the walls of the duct. This means that that the value of average speed can be approximately consider as the velocity profile of the fluid.

Combining equations (8.14) and (8.20), we get:

$$v_z = \overline{V} \tag{8.20}$$

$$\dot{m}_1 \, E_{c_1} = \int_0^{R_1} \left[\frac{1}{2} \rho_1 \left(\overline{V_1} \right) 2 \pi r \, d r \left(\overline{V_1} \right)^2 \right] \tag{8.21}$$

Integrating the above equation:

$$\dot{m}_1 E_{c_1} = \frac{1}{2} \pi \rho_1 \overline{V}_1^3 R_1^2 \tag{8.22}$$

(prove this as an exercise).

Comparing equations 8.22 and 8.18,

$$E_{c_1} = \frac{1}{2} \overline{V}_1^2 \tag{8.23}$$

The above equation can be used for the determination of the kinetic energy of fluid per unit mass in terms of their average speed for turbulent flow.

The equations for laminar and turbulent regime can be written in the same general way as shown below:

$$E_{c_1} = \frac{1}{2\beta_1} \overline{V}_1^2 \tag{8.24}$$

- β_1 = 1/2 for laminar;
- β_1 = 1 for turbulent regime.

8.2.2 Evaluation of the potential energy term

The potential energy is defined in relation to a given arbitrary plane.

The rate of entry of potential energy in plane 1 can be estimated by the following equation:

$$\dot{m}_1 \, E_{p_1} = \dot{m}_1 \, g \, z_1 \tag{8.25}$$

where z_1 is the height of the midpoint of the cross section of the duct in the plane 1 with respect to the reference plane.

Eliminating the mass flow on both sides of equation (8.25), one gets the following equation to estimate the potential energy per unit mass of fluid:

$$E_{p_1} = g\, z_1 \quad (8.26)$$

8.2.3 Bernoulli's equation

Dividing the general equation of energy balance for steady state conditions, 8.11 by the mass flow of fluid (which is constant throughout the system - conservation of mass), we can write that:

$$\Delta H + \Delta E_P + \Delta E_c - Q^* + M^* = 0 \quad (8.27)$$

$$Q^* = \frac{Q}{\dot{m}_1} \quad (8.28)$$

$$M^* = \frac{M}{\dot{m}_1} \quad (8.29)$$

The above 2 terms represent the net rate of heat input and mechanical work done by the fluid, both per unit mass of fluid that flowing through the system.

Recalling now the definitions of thermodynamics, we have:

$$H = E + \frac{P}{\rho} \quad (8.30)$$

where:

- E = internal energy per unit mass of fluid;
- P = pressure of the fluid.

Combining equations (8.27) and (8.30), we obtain:

$$\Delta E + \Delta\left(\frac{P}{\rho}\right) + \Delta E_P + \Delta E_c - Q^* + M^* = 0 \quad (8.31)$$

Assuming an infinitesimal length of the system, the equation above can be converted to the following differential form:

$$dE + d\left(\frac{P}{\rho}\right) + dE_P + dE_C - \delta Q^* = 0 \quad (8.32)$$

It is worth noting, that the term M* disappears in the equation, because it is normally associated with pumps or any other equipment to transport the fluid. These devices are not included in an infinitesimal volume element.

The most common form of the energy balance applied to the flow of fluids is known as the balance of mechanical energy (which is a form of Bernoulli's theorem).This is developed below.

The variation of internal energy per unit mass of fluid as it passes through a small segment of the pipeline, is given by:

$$dE = \delta Q^{*} - P\, d\left(\frac{1}{\rho}\right) + \delta E_f \tag{8.33}$$

where δE_f is the mechanical energy per unit mass of fluid that is converted into heat due to friction. Equation (8.33) is actually is first law of thermodynamics.

Remembering:

$$d\left(\frac{P}{\rho}\right) = Pd\left(\frac{1}{\rho}\right) + \frac{1}{\rho}\, dP \tag{8.34}$$

Combining (8.32) (8.33) and (8.34), we obtain:

$$\delta Q^{*} - Pd\left(\frac{1}{\rho}\right) + \delta E_f + Pd\left(\frac{1}{\rho}\right) + \frac{1}{\rho}\, dp + dE_P + dE_c - \delta Q^{*} = 0 \tag{8.35}$$

Simplifying the terms in equation (8.35):

$$\frac{1}{\rho}\, dp + dE_P + dE_c + \delta E_f = 0 \tag{8.36}$$

Integration of this equation over the entire system (with the term M* appearing again) gives one of the versions of Bernoulli's equation which can be applied to solve most problems of fluid flow:

$$\int_1^2 \frac{1}{\rho}\, dp + g\,(z_2 - z_1) + \left(\frac{\overline{V}_2^{\,2}}{2\,\beta_2} - \frac{\overline{V}_1^{\,2}}{2\,\beta_1}\right) + M^{*} + E_f = 0 \tag{8.37}$$

Each of the terms in the above equation is per unit mass of fluid flow.

The term E_f above is represents losses by friction along the pipe.

Equation (8.37) can be rewritten in two basic forms, depending on the fluid that is flowing. One of them applicable to incompressible fluids.

Where, ρ is constant throughout the system and can be taken out of the integral resulting in:

$$\frac{P_2 - P_1}{\rho} + g\,(z_2 - z_1) + \left(\frac{\overline{V}_2^{\,2}}{2\,\beta_2} - \frac{\overline{V}_1^{\,2}}{2\,\beta_1}\right) + M^* + E_f = 0 \qquad (8.38)$$

The other version is applied to compressible fluids. Like gases. Considering the case of an isothermal ideal gas, one can obtain the following equation for density as a function of pressure:

$$\rho = \frac{P\,M_M}{R\,T} \qquad (8.39)$$

where M_M is the molecular weight of gas (Derive this equation from ideal gas law).

Substituting equation (8.39) in (8.37) and integrating, we get:

$$\frac{R\ T}{M_M}\ \ln\left(\frac{P_2}{P_1}\right) + g\,(z_2 - z_1) + \left(\frac{\overline{V}_2^{\,2}}{2\,\beta_2} - \frac{\overline{V}_1^{\,2}}{2\,\beta_1}\right) + M^* + E_f = 0 \qquad (8.40)$$

The equations (8.38) and (8.40) are the most common forms of the Bernoulli equation.

8.2.4 Evaluation of friction losses

For practical application of the equations (8.38) and (8.40), it is necessary to develop methods for estimating losses due to friction, during flow in various parts of a system like bends, valves,varying cross sectional areas, of the pipes.

Frictional loss can be determined experimentally by measuring all other terms which appear in equations (8.38) or (8.40), and leaving only its value as unknown in the equations. However, what is usual is to estimate E_f from the characteristics of the system and use the above equations to determine the energy required to pump the fluid at a given speed along the pipe. Once this is done the capacity of pump necessary for the system can be designed. Hence the evaluation of friction losses that occur in different parts of fluid flow system is a very important aspect of project design for transport of fluids, encountered in every field of engineering.

8.2.4.1 Friction losses in straight pipes

Let us consider the case of a constant density fluid flowing at a given velocity in a horizontal duct, as shown in Figure 8.5.

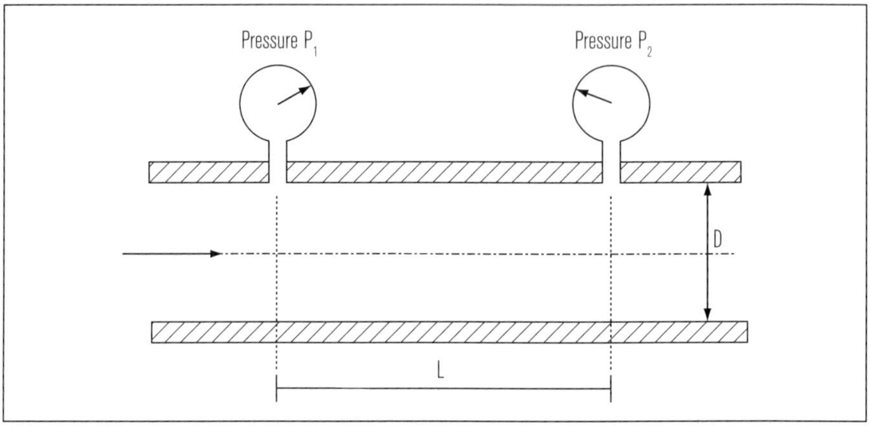

Figure 8.5 - Fluid flowing in a horizontal duct with constant cross section under a pressure drop.

Assuming that the fluid flows due to a difference in pressure and steady state, applying balance of forces:

$$F_K = (P_1 - P_2)\ A \qquad (8.41)$$

where:

- F_K = force of friction between the fluid and the duct wall;
- P_1 - P_2 = pressure difference between points 1 and 2 (the fluid leaks from point 1 to 2);
- A = cross-sectional area of the duct.

The above equation is established considering that when the fluid flows with constant velocity, the sum of forces acting on it is zero.

Now applying energy balance for the system considering energy loss due to friction associated with the walls of the tube, shown in Figure 8.5, we obtain:

$$\frac{P_2 - P_1}{\rho} + E_f = 0 \qquad (8.42)$$

$\overline{V_1}$, $\overline{V_2}$, as the duct cross section is constant. Potential energy difference is 0, between points 1 and 2 in the the horizontal tube ($Z_1 = Z_2$) and there is no equipment for pumping fluid between points 1 and 2 ($M^* = 0$).

Combining equations (8.41) and (8.42), we have:

$$E_f = \frac{F_K}{\rho A} \tag{8.43}$$

In Chapter 7, the relation between frictional force between the fluid and the walls of the duct in terms of friction factor has been given by the following equation:

$$F_k = (\pi\ D\ L)\left(\frac{1}{2}\ \rho\ \overline{V^2}\right) f \tag{7.26}$$

For a duct of circular cross section, we have:

$$A = \pi\ \frac{D^2}{4} \tag{8.44}$$

Combining equations (8.43) (7.26) and (8.44), one can obtain an expression for estimating energy losses by friction in straight section of pipe:

$$E_f = \frac{(\pi\ D\ L)\left(\frac{1}{2}\ \rho\ \overline{V^2}\right) f}{\rho\ \pi\ \frac{D^2}{4}} = 2\ f \left(\frac{L}{D}\right) \overline{V}^2 \tag{8.45}$$

(Demonstrate that an equation identical to the above is obtained for a vertical duct).

Equation (8.45) above can also be used for non-circular ducts, simply replacing the diameter D by the equivalent hydraulic diameter, defined by equation (7.43).

Example

A fan kept at one end of a duct blows air through the duct. Given: rectangular duct dimensions: section: 0.20 m x 0.30 m and length of 50.0 m Air enters at 20°C and 750 mm Hg pressure. The air flow in the pipe is 0.5 m^3/s. Consider that the duct is hydraulically smooth (roughness = 0) and horizontal. What should be the fan power for the flow above, assuming that output air has the same temperature and pressure as the input air?

Solution

The system under consideration is schematically given in the figure below:

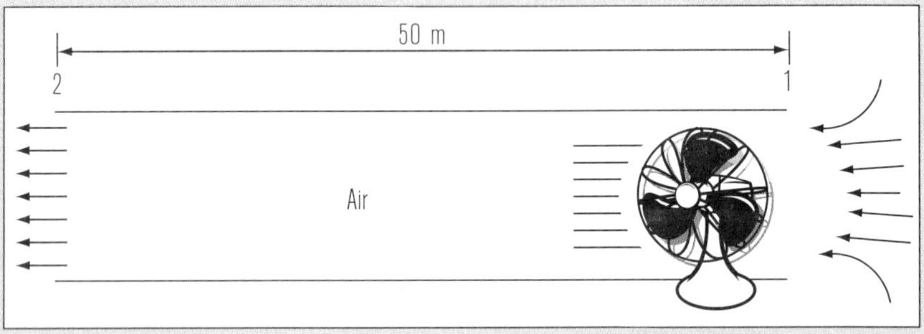

In this case, the temperature and pressure of gas blown by the fan does not vary. One can consider Bernoulli's equation applied to an incompressible fluid:

$$\frac{P_2 - P_1}{\rho} + g(z_2 - z_1) + \left(\frac{\overline{V}_2^2}{2\beta_2} - \frac{\overline{V}_1^2}{2\beta_1}\right) + M^* + E_f = 0$$

As the pressures are the same in 1 and 2, we have that:

$$P_2 = P_1$$

Since the tube is horizontal:

$$z_2 = z_1$$

The fan captures the air at rest from one end of the tube.

Hence:

$$\overline{V}_1 = 0$$

With these considerations, the Bernoulli equation reduces to:

$$\left(\frac{\overline{V}_2^2}{2\beta_2}\right) + M^* + E_f = 0$$

To determine M* it is necessary to evaluate the velocity of air at point 2 and also evaluate E_f, the frictional energy. For the calculation of velocity at level 2 and also E_f it is necessary to know the air density, cross-sectional area of the duct and Reynolds number.

The air density can be calculated by the following relationship and data:

$$\rho = \frac{P\, M_M}{R\; T}$$

$P = 750\ mmHg = 0.9868\ atm = 0.9868 \times 101330\ Pa = 99992.44\ Pa$

$T = 20 + 273 = 293\ K$

$P_M = 0.02884\ kg / mol$ (considerando 21% de 02 e 79% de N_2)

$R = 8.31\ J / mol$

Substituting the values, we obtain:

$$\rho = \frac{P\; M_M}{R\;\; T} = \frac{99992.44\ .\ 0.02884}{8.31\ .\ 293} = 1.1844\ kg/m^3$$

The cross-sectional area of the duct is:

$$A = 0.2\ .\ 0.3 = 0.06\ m^2$$

Hence the velocity at point 2 is:

$$\overline{V}_2 = \frac{Q}{A} = \frac{0.5}{0.06} = 8.33\,m/s$$

To determine the Reynolds number, it is still necessary to know the viscosity of air and hydraulic equivalent diameter of the pipe.

The viscosity of air at that temperature is:

$$\mu = 1.8\ .\ 10^{-5}\ kg/m.s$$

(Recall the calculation of viscosity of gases in Chapter 4 - Equation (4.10)).

The hydraulic equivalent diameter is calculated using the following equation:

$$D_h = \frac{2\ .\ 0.2\ .\ 0.3}{(0.2 + 0.3)} = 0.24\ m$$

With these values, we can calculate the Reynolds number

$$Re = \frac{D_h \cdot \overline{V}_2 \cdot \rho}{\mu} = \frac{0.24\ .\ 8.33\ .\ 1.1844}{1.18\ .\ 10^{-5}} = 131547.36$$

As the pipe is hydraulically smooth, one can determine the friction factor from the following equation with $\varepsilon = 0$:

$$\frac{1}{\sqrt{f}} = -3.6 \log \left[\left(\frac{\varepsilon / D}{3.7} \right)^{1.11} + \frac{6.9}{Re} \right] \qquad f = 0.0042$$

Substituting f = 0.0039 in the above equation in the, frictional energy equation:

$$E_f = 2 f \left(\frac{L}{D_h} \right) \overline{V}^2 = 2 \,.\, 0.0042 \,.\, \left(\frac{50}{0.24} \right) . (8.33)^2 = 121.77 \ m^2/s^2$$

Returning to Bernoulli equation and transposing terms, we obtain:

$$M^* = -\left(\frac{\overline{V}_2^2}{2 \beta_2} \right) - E_f$$

As the flow is highly turbulent (Re >> 2100), $\beta_2 = 1$, then:

$$M^* = -\left[\frac{(8.33)^2}{2 \times 1} \right] - 112.757 = -156.46 \ m^2/s^2$$

The value of M* is negative because it represents the work done by the fan per unit mass of fluid being transported. Therefore:

$$|M| = |M^*| Q \rho = 147.45 \,.\, 0.5 \,.\, 1.1844 = 92.66 \ W$$

Using the conversion factor (Chapter 3 - Table 3.2), we obtain:

$$|M| = \frac{87.32}{745.7} = 0.124 \ hp$$

8.2.4.2 Friction losses in expansion and contraction

The friction losses associated with the presence of expansions or contractions along the pipes are calculated using empirical correlations, using a parameter called the friction loss factor (e_f).

These losses are estimated through the following relationship:

$$E_f = \frac{1}{2} e_f \overline{V}^2 \tag{8.46}$$

The parameter e_f is determined by experimental correlations, expressing its value on the basis of type of expansion or contraction (sudden or gradual), the area ratio before and after the contraction/expansion and the value of the Reynolds number.

a) Contractions

Figure 8.6 shows schematically a sudden contraction in a pipe.

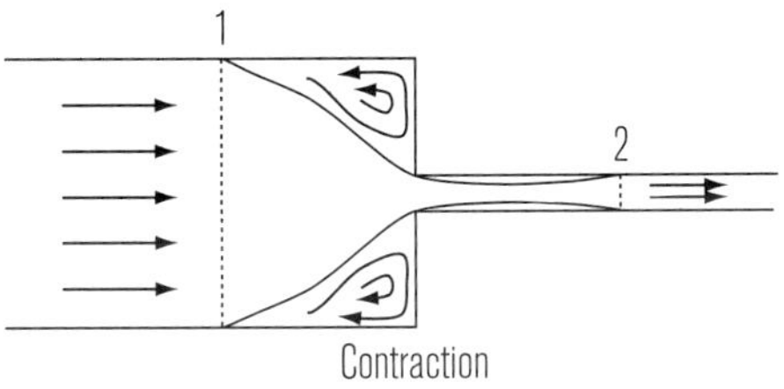

Figure 8.6 - Schematic view of a contraction.

In the case of sudden contractions (such as that shown in Figure 8.6) and highly turbulent regime, the value of friction loss factor (e_f) can be evaluated using the following equation:

$$e_f = 0.45\,(1 - \alpha) \tag{8.47}$$

$$\alpha = \frac{\text{Cross sectional area of smaller region}}{\text{Cross sectional area of larger region}} \tag{8.48}$$

When equation (8.47) is used compute friction loss factor the velocity term in equation (8.46) should be estimated using the cross-sectional area after contraction (smaller area).

In the case of contractions, not exactly defined by parameter α, having different geometries at the region of transition from larger to smaller sections also affect the value of friction loss factor as shown in Figure 8.7.

As shown in the figure, the rounding of input region's contraction results in a friction loss factor, which is 1/3 of that predicted by equation 8.47, for sudden contraction.

In other texts (WHITE, 1979; GASKELL, 1992) there are correlations to predict the values of ef for various configurations of contraction and for several range of Reynolds numbers.

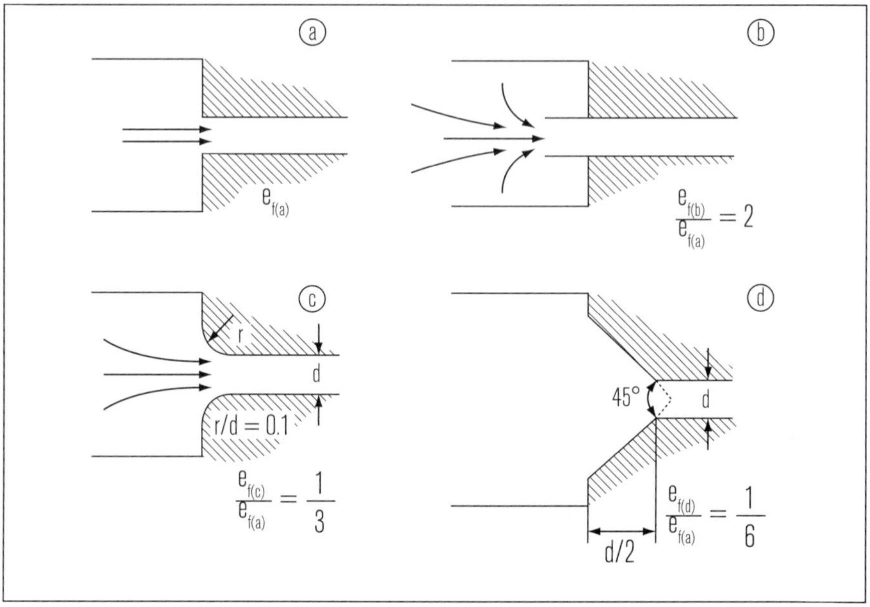

Figure 8.7 - Influence of contraction geometry on friction loss factor (GEIGER and POIRIER, 1980).

b) Expansion

For a sudden expansion and highly turbulent flow, the friction loss factor, can be estimated from the following correlation:

$$e_f = (1 - \alpha)^2 \tag{8.49}$$

While using the above equation to evaluate the friction loss factor, the velocity that appears in equation (8.46) should be estimated using the cross-sectional area before the expansion (smaller area).The values of e_f in the case of expansions apply equally well to all types of angles in the transition region from smallest to largest section (except for gradual expansion, as discussed below), since the formation of vortices after the expansion does not change whether the corners are rounded or not.

For flow through gradual expansion, the friction losses are significantly reduced due to the elimination of vortices. In accordance with the experimental results shown in Figure 8.8, friction loss factor is a function of opening angle and area ratio A_1/A_2.

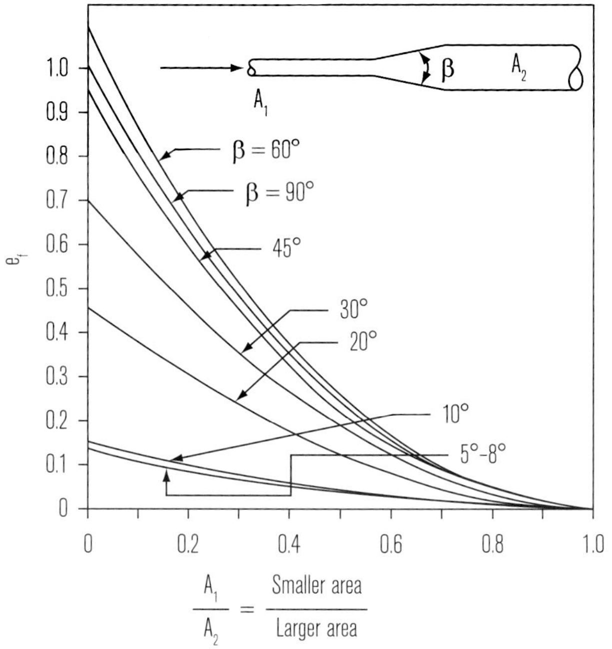

Figure 8.8 - Values of friction loss factor for gradual expansion in the case of turbulent flow (GEIGER and POIRIER, 1980).

8.2.4.3 Friction losses in valves and connections

To evaluate the friction losses for flow through valves and connections, one can use the technique of equivalent length.The friction losses are given by the following equation:

$$E_f = 2\,f\left(\frac{L_e}{D}\right)\overline{V}^2 \tag{8.50}$$

where:

- f = friction factor evaluated for a Reynolds number of a tube with a certain diameter having the the valve or connection;
- L_e = equivalent length of the valve or connection. It is the length of the tube (the same diameter of the connection or valve) that would cause the same loss by friction as the valve or connection.

It is interesting to note, that the equation (8.50) is similar to (8.45), used to predict friction losses in straight sections of pipes. The ratio values L_e/D for some types of connections and valves are provided in Table 8.1.

The data shown in Table 8.1 are valid for turbulent flow. When a system has several valves and connections, the equivalent lengths (L_e /D) of all of them are added and the equation (8.50) is used for estimating energy of friction loss.

Table 8.1 - L_e/D ratio for some types of valves and connections (GEIGER and POIRIER, 1980).

Component	L_e/D
Elbow 45	15
Elbow 90°, default radius	31
Elbow 90°, radius	26
Elbow 90°, square	65
Return 180°	75
Gate valve, open	7
Gate Valve, ¼ closed	40
Gate Valve, ½ closed	190
Gate Valve, ¾ closed	840
Unions	Negligible

Example

What is the power required to pump the water through the system shown in the following figure? Water (ρ = 1000 kg/m^3 and μ = 1 cP) discharges into the upper tank with a flow rate of 6 x 10^{-3} m^3/s. Every part of the pipe has an internal diameter of 10.16 cm (4 inches). The roughness of the pipe is 0.1 mm.

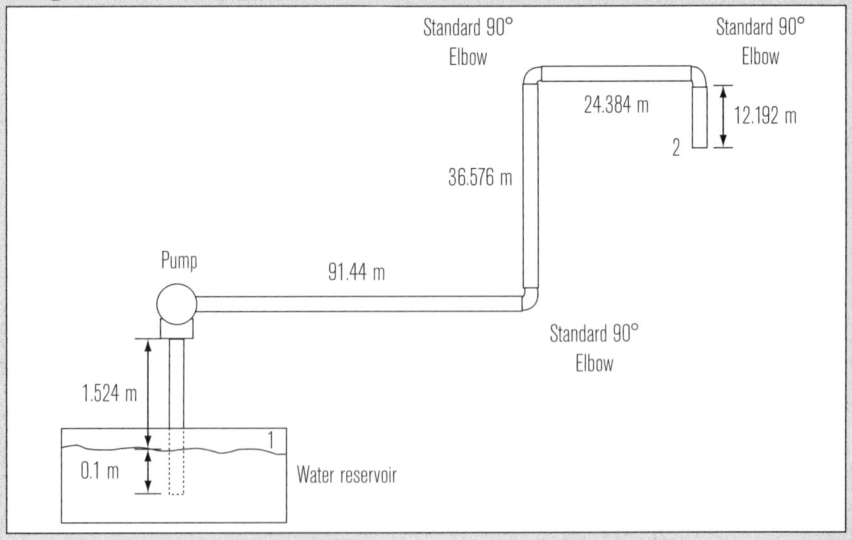

For the solution, points 1 and 2 are marked in the figure are taken as the basis for the balance of energy. As water is an incompressible fluid, Bernoulli equation can be used:

$$\frac{P_2 - P_1}{\rho} + g\,(z_2 - z_1) + \left(\frac{\overline{V}_2^2}{2\,\beta_2} - \frac{\overline{V}_1^2}{2\,\beta_1}\right) + M^* + E_f = 0$$

Considering levels 1 and 2,

$$P_2 = P_1 = P_{atmosphere}$$

$$z_2 - z_1 = (1.524 + 36.576 - 12.192) = 25.908 \text{ m}$$

To calculate the variation of kinetic energy and friction loss, one must determine the velocities at points 1 and 2 along the pipe. Flow rate can be evaluated from overall mass balance:

$$A_1\,\rho_1\,\overline{V}_1 = A_2\,\rho_2\,\overline{V}_2 = A_{duct}\,\rho_{duct}\,\overline{V}_{duct}$$

As the density of water is constant, we have:

$$Q = A_1\,\overline{V}_1 = A_2\,\overline{V}_2 = A_{duct}\,\overline{V}_{duct}$$

Cross-sections of pipe in the levels 1 and 2 are the same, so the velocity of water in these two regions are equal. Considering also that the reservoir area (level 1) is much larger compared to level 2 (outlet pipe), for purposes of estimating the kinetic energy variation, it can be assumed that:

$$\overline{V}_1 <<< \overline{V}_2$$

and hence negligible.

Using the diameter of the pipe, one can calculate the area in level 2 and along the pipeline:

$$A_2 = A_{duct} = \pi\,\frac{D_{duct}^2}{4} = \pi\,\frac{(0.1016)^2}{4} = 0.0081 \text{ m}^2$$

Therefore, the velocities along the pipe and level 2 are:

$$\overline{V}_2 = \overline{V}_{duct} = \frac{Q}{A_{duct}} = 0.74 \text{ m/s}$$

We now determine the value of β_2. This is done by evaluating the Reynolds number in level 2, to see if the flow is laminar or turbulent:

$$Re = \frac{D_{duct}\ \overline{V}_2\ \rho}{\mu} = \frac{0.1016\ .\ 0.74\ .\ 1{,}000}{0.001} = 75184$$

Since Re > 2,100, the flow is turbulent and β_2 is then equal to 1. One can now evaluate the variation of kinetic energy between points 1 and 2:

$$\left(\frac{\overline{V}_2^2}{2\,\beta_2} - \frac{\overline{V}_1^2}{2\,\beta_1}\right) = \left(\frac{(0.74)^2}{2\ .\ 1} - 0\right) = 0.274\ m^2/s^2$$

To determine the pump power, it is necessary to estimate the energy losses by friction between points 1 and 2, considering the following:

- contraction at the entrance of the duct;
- friction along the straight sections of pipe;
- friction loss in 3 Elbows of 90°.

a) Friction loss associated with contraction

$$E_f = \frac{1}{2}\ e_f\ \overline{V}_{duct}^2$$

This can be estimated from the following equation for sudden contraction:

$$e_f = 0.45\ (1 - \alpha)$$

For the configuration being studied:

$$\alpha\ \approx\ 0$$

Considering the configuration of the region where the fluid enters the duct (Figure 7.6):

$$e_f = 0.45\ .\ (1 - 0)\ .\ 2 = 0.90$$

Fraction losses by contractions are given by:

$$E_f = \frac{1}{2}\ e_f\ \overline{V}_{duct}^2 = \frac{1}{2}\ (0.9)\ (0.74)^2 = 0.246\ m^2/s^2$$

b) Friction losses through pipe,elbows etc.

The losses in the straight sections are evaluated by the equation:

$$E_f = 2\,f\left(\frac{L}{D_{duct}}\right)\overline{V}^2_{duct}$$

The friction factor for the flow inside the pipe is evaluated using the equation:

$$\frac{1}{\sqrt{f}} = -3.6\log\left[\left(\frac{\varepsilon/D}{3.7}\right)^{1.11} + \frac{6.9}{Re}\right]$$

Substituting values:

$$\frac{1}{\sqrt{f}} = -3.6\log\left[\left(\frac{(0.0001)/(0.1016)}{3.7}\right)^{1,11} + \frac{6.9}{75184}\right] \quad \therefore\ f = 0.0056$$

The total length of the straight sections is:

$$L = 0.1 + 1.524 + 91.44 + 36.576 + 24.384 + 12.192 = 166.216\ m$$

Friction losses in the straight sections are given by:

$$E_f = 2\,f\left(\frac{L}{D_{duct}}\right)\overline{V}^2_{duct} = 2\,(0.0056)\left(\frac{166.216}{0.1016}\right)(0.74)^2 = 10.034\ m^2/s^2$$

c) Losses in standard elbows of 90°.

The friction losses in connections are determined by the following relationship:

$$E_f = 2\,f\left(\frac{L_e}{D_{duct}}\right)\overline{V}^2_{duct}$$

For each elbow of 90°:

$$\left(\frac{L_e}{D}\right) = 31$$

Hence:

$$E_f = 2\,f\left(\frac{L_e}{D}\right)\overline{V}^2_{duct} = 2\,(0.0056)\,(3)\,(31)\,(0.74)^2 = 0.570\ m^2/s^2$$

With these values, we can use Bernoulli equation, to evaluate the power of the:

$$\frac{P_2 - P_1}{\rho} + g(z_2 - z_1) + \left(\frac{\overline{V}_2^2}{2\,\beta_2} - \frac{\overline{V}_1^2}{2\,\beta_1}\right) + M^* + E_f =$$

$$0 + 9.8\,(25.908) + 0.274 + M^* + (0.246 + 10.034 + 0.570)$$

$$M^* = -265.02\ m^2/s^2$$

The negative value of M * indicates that for transport of fluid the energy has to come form pump. The power of the pumping equipment can be determined by multiplying the above value by the mass flow of water in the pipe:

$$\text{Power of pump} = |M^*|\,Q\,\rho = 265.02\ .\ 6\ .\ 10^{-3}\ .\ 1000 = 1{,}590.12\ W = 2.13\ hp$$

The above is the amount of energy that the pump effectively transfers to the fluid. The pump power must be greater than the above value due to electrical and mechanical losses. The negative value of M* indicates that for transport of fluid the energy has to come from pump. The power of the pumping equipment can be determined by multiplying the above value by the mass flow of water in the pipe:

$$\text{Power of pump} = \frac{2.13}{(50/100)} = 4.16\ hp$$

8.3 Flow from Ladle and Tundish

In many situations of practical interest, the metal contained in ladles and tundishes flows into other refining reactors or molds. The metal is normally discharged at the bottom of the ladle through a nozzle. The time taken for emptying the ladle is important as the metal should have sufficient superheat during pouring. Determining pouring rate of metal contained in a ladle can be ascertained through the use of Bernoulli equation. Also parameters, such as time necessary for the liquid metal to fill in a mold can be calculated. Hence the study of emptying ladles through nozzles at the bottom is a very important field metallurgical engineering.

8.3.1 Flow from ladle through nozzle

The configuration of the system under study is shown in Figure 8.9.

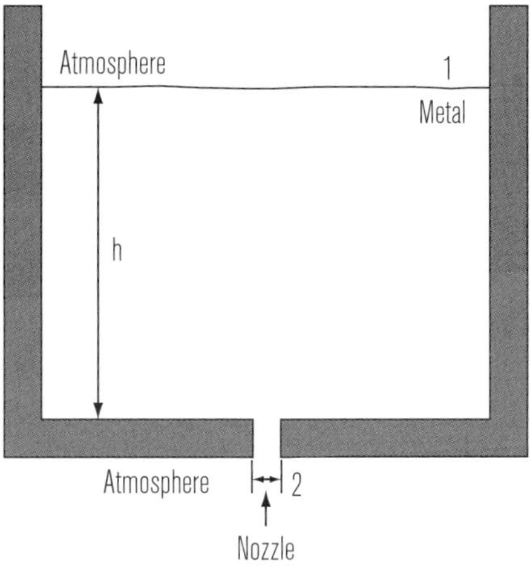

Figure 8.9 - Schematic view of a ladle containing metal provided with a nozzle.

To establish an equation relating the velocity of metal flow through the nozzle and the height of metal in the ladle one can apply Bernoulli's equation to levels 1 and 2, as shown in the figure:

$$\frac{P_2 - P_1}{\rho} + g\,(z_2 - z_1) + \left(\frac{\overline{V}_2^{\,2}}{2\,\beta_2} - \frac{\overline{V}_1^{\,2}}{2\,\beta_1} \right) + M^* + E_f = 0 \qquad (8.38)$$

The pressure at both levels is equal of atmospheric pressure:

$$P_2 = P_1 = P_{atmosphere} \qquad (8.51)$$

$$z_2 - z_1 = -h \qquad (8.52)$$

$$A_1\, \rho_1\, \overline{V_1} = A_2\, \rho_2\, \overline{V_2} \qquad (8.53)$$

To calculate the variation of kinetic energy, one must relate the velocities at levels 1 and 2. For this, a mass balance can be carried out between these levels, assuming constant density of metal:

$$A_1 \overline{V_1} = A_2 \overline{V_2} \qquad (8.54)$$

Cross-sections in levels 1 and 2 are evaluated using the following equations:

$$A_1 = \pi \frac{D_{Ladle}^2}{4} \quad (8.55)$$

$$A_2 = \pi \frac{D_{nozzle}^2}{4} \quad (8.56)$$

Substituting the above relations in equation (8.54), we obtain:

$$\overline{V}_1 = \frac{D_{nozzle}^2}{D_{Ladle}^2} \overline{V}_2 \quad (8.57)$$

As the nozzle diameter is much smaller than that of ladle, it can be stated that:

$$\overline{V}_1 <<<< \overline{V}_2 \quad (8.58)$$

Hence, the variation of kinetic energy between levels 1 and 2 can be estimated by the following expression:

$$\left(\frac{\overline{V}_2^2}{2\,\beta_2} - \frac{\overline{V}_1^2}{2\,\beta_1} \right) \approx \frac{\overline{V}_2^2}{2\,\beta_2} \quad (8.59)$$

Since no mechanical work is done:

$$M^* = 0 \quad (8.60)$$

Friction losses have to be assessed levels 1 and 2. These losses are associated with:

- friction in cross section inside the ladle;
- friction due to contraction at the nozzle hole during flow of metal.

Friction loss inside the ladle can be evaluated using the following equation:

$$E_f = 2\, f_{Ladle} \left(\frac{h}{D_{Ladle}} \right) \overline{V}_1^2 \quad (8.61)$$

Friction loss in the nozzle is given by:

$$E_f = \frac{1}{2}\, e_f\, \overline{V}_2^2 \quad (8.62)$$

For contraction, the loss factor for friction is given by:

$$e_f = 0.45\,(1 - \alpha) \tag{8.63}$$

$$\alpha = \frac{\text{Cross sectional area of nozzle}}{\text{Cross sectional area of ladle}} \tag{8.64}$$

where:

$$\alpha \approx 0 \tag{8.65}$$

$$e_f = 0.45\,.\,(1 - 0) = 0.45 \tag{8.66}$$

The total friction loss is given by:

$$E_f = 2\, f_{ladle} \left(\frac{h}{D_{ladle}} \right) \overline{V}_1^{\,2} + \frac{1}{2}\,(0.45)\, \overline{V}_2^{\,2} \tag{8.67}$$

As seen above, the velocity at level 1 is much smaller than the velocity at level 2. Thus, the term associated with losses within the ladle can be neglected when compared with the loss due to contraction. Incorporating the above in equation 8.38 (Bernoulli equation), we get:

$$g\,(-h) + \left(\frac{\overline{V}_2^{\,2}}{2\,\beta_2} \right) + \frac{1}{2}\,(0.45)\, \overline{V}_2^{\,2} = 0 \tag{8.68}$$

Rearranging terms, one can obtain an equation for the velocity of metal flowing through the nozzle from the ladle:

$$\overline{V}_2 = \left[\frac{2\,g\,h}{\frac{1}{\beta_2} + 0.45} \right]^{\frac{1}{2}} \tag{8.69}$$

Equation (8.69) is commonly written in the following way:

$$\overline{V}_2 = C_D\,(2\,g\,h)^{\frac{1}{2}} \tag{8.70}$$

where C_D is known as discharge coefficient and is evaluated by:

$$C_D = \left[\frac{1}{\frac{1}{\beta_2} + 0.45} \right]^{\frac{1}{2}} \tag{8.71}$$

Considering that flow from ladle is turbulent ($\beta_2 = 1$) and neglecting the friction losses, the value of C_D becomes unity and the maximum velocity of metal flowing out of nozzle is given by:

$$\overline{V}_2 = (2\ g\ h)^{\frac{1}{2}} \tag{8.72}$$

Example

Apply equation (8.70) gisven above, to the situation shown in the figure below, which has a refractory duct coupled to the nozzle of the ladle. with a gate valve, whose opening can be controlled. For assessing the velocity of metal stream from the ladle.

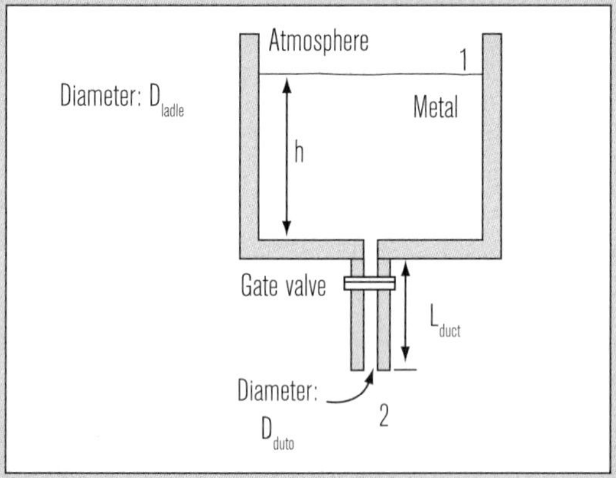

Solution

To establish a relationship for assessing the velocity of the metal stream from the ladle, Bernoulli equation can be applied for levels 1 and 2 indicated in the figure above. The change in location of the level 2 is made for convenience, since at this location, one can assume that the pressure is equal to atmospheric pressure. Using the location selected for previous example, this is not valid. Hence:

$$P_2 = P_1 = P_{atmosphere}$$

$$z_2 - z_1 = -(h + L_{duct})$$

where L_{duct} the length of the refractory duct coupled to the ladle.

To calculate the variation of kinetic energy, one must relate the velocities at points 1 and 2. From mass balance between levels 1 and 2. and considering constant density for the metal, we have:

$$A_1 \overline{V_1} = A_2 \overline{V_2}$$

Cross-sections in levels 1 and 2 are evaluated using the following equations:

$$A_1 = \pi \frac{D^2_{ladle}}{4}$$

$$A_2 = \pi \frac{D^2_{duct}}{4}$$

Substituting the above in the equation of mass balance:

$$\overline{V_1} = \frac{D^2_{duct}}{D^2_{ladle}} \overline{V_2}$$

As the diameter of the duct is much smaller than that of ladle, change of kinetic energy between levels 1 and 2 can be estimated by the following expression:

$$\left(\frac{\overline{V_2}^2}{2\,\beta_2} - \frac{\overline{V_1}^2}{2\,\beta_1} \right) \approx \frac{\overline{V_2}^2}{2\,\beta_2}$$

For the system under consideration, no mechanical work is done on the same. Hence:

$$M^* = 0$$

For this case, the friction losses are associated with:

- friction in cross section inside the ladle;
- friction due to contraction at the entrance orifice;
- friction in cross section of the refractory tube at the bottom of the ladle;
- friction due to the gate valve.

The loss by friction in the cross section inside the pot is evaluated using the following equation:

$$E_f = 2\, f_{ladle} \left(\frac{h}{D_{ladle}} \right) \overline{V_1}^2$$

The friction loss in the contraction is evaluated by the following expression:

$$E_f = \frac{1}{2} e_f \overline{V}_2^2$$

For contraction, the friction loss factor is given by:

$$e_f = 0.45\,(1 - \alpha)$$

$$\alpha = \frac{\text{Cross sectional area of nozzle}}{\text{Cross sectional area of ladle}}$$

where:

$$\alpha \approx 0$$

Considering that the contraction has no special finish in the input:

$$e_f = 0.45\,.\,(1 - 0) = 0.45$$

The loss by friction in cross section in the duct is given by:

$$E_f = 2\, f_{duct} \left(\frac{L_{duct}}{D_{duct}}\right) \overline{V}_2^2$$

The friction loss in the gate valve is estimated by:

$$E_f = 2\, f_{duct} \left(\frac{L_e}{D}\right)_{gate\ valve} \overline{V}_2^2$$

where the parameter value of L/D depends on the valve opening.

The total friction losses are given by:

$$E_f = 2\, f_{ladle} \left(\frac{h}{D_{ladle}}\right) \overline{V}_1^2 + \frac{1}{2}(0.45)\,\overline{V}_2^2 + 2\, f_{duct} \left(\frac{L_{duct}}{D_{duct}}\right) \overline{V}_2^2 + 2\, f_{duct} \left(\frac{L_e}{D}\right)_{gate\ valve} \overline{V}_2^2$$

As seen above, the the velocity at level 1 smaller than that at level 2. Thus, the term associated with tfriction loss inside the ladle can be neglected compared to other losses. Returning to the Bernoulli equation, incorporating the above, we have:

$$-g\,(h + L_{duct}) + \left(\frac{\overline{V}_2^2}{2\,\beta_2}\right) + \frac{1}{2}(0.45)\,\overline{V}_2^2 + 2\, f_{duct} \left(\frac{L_{duct}}{D_{duct}}\right) \overline{V}_2^2 + 2\, f_{duct} \left(\frac{L_e}{D}\right)_{gate\ valve} \overline{V}_2^2 = 0$$

Rearranging terms, one can obtain an equation for the velocity of metal of metal stream from the ladle:

$$\overline{V}_2 = \left[\frac{2\,g\,(h + L_{duct})}{\dfrac{1}{\beta_2} + 0.45 + 4\,f_{duct}\left(\dfrac{L_{duct}}{D_{duct}}\right) + 4\,f_{duct}\left(\dfrac{L_e}{D}\right)_{gate\ valve}} \right]^{\frac{1}{2}}$$

The difficulty with the above equation is friction factor depends on Reynolds number which is also a function of velocity. Hence the friction factor and velocity are delpendent on each other and it is necessary to know one to fix the other.This difficulty is circumvented by using an iterative method by assigning arbitrarily an initial value for f_{duct} (which may not be exact as the velocity is not known) to determine the velocity of metal,discharged from the ladle. This value of velocity calculated in this manner is also approximate, since the calculation is based on an unknown value of friction factor. With this new value the Reynolds number is estimated and an updated value for the friction factor is calculated. With this value, restart the process and carry out one more iteration. Usually, this iterative process converges and the values of velocity and friction factor begin to repeat after successive iterations. An example of application of this method is shown below. This will be used for the following data:

- height of the steel ladle, h = 3 m;
- length of duct, L = 1 m;
- roughness of the duct, ε = 0.0002 m;
- diameter of duct, D_{duct} = 0.075 m;
- gate valve half open, (L_e/D) = 190.

For liquid steel, we have:

- density, ρ = 7000 kg/m^3;
- viscosity, μ = 0.007 Pa.s.

Substituting the above values in the equation for velocity,

$$\overline{V}_2 = \left[\frac{2\,.\,9.8\,.\,(3 + 1)}{\dfrac{1}{\beta_2} + 0.45 + 4\,f_{duct}\left(\dfrac{1}{0.075}\right) + 4\,f_{duct}\,(190)} \right]^{\frac{1}{2}} \qquad \text{(A)}$$

The friction factor is estimated by the following expression (assuming turbulent flow):

$$\frac{1}{\sqrt{f_{duct}}} = -3.6 \log\left[\left(\frac{\varepsilon_{duct}/D_{duct}}{3.7}\right)^{1.11} + \frac{6.9}{\frac{D_{duct}\,\overline{V}_2\,\rho}{\mu}}\right]$$

Substituting the data given above in this equation:

$$\frac{1}{\sqrt{f_{duct}}} = -3.6 \log\left[\left(\frac{0.0002/0.075}{3.7}\right)^{1.11} + \frac{6.9}{\frac{(0.075)(\overline{V}_2)(7000)}{0.007}}\right] \quad \text{(B)}$$

The iterative process can be started with an arbitrary value of f_{duct} = 0.005. Substituting this value in the equation for the velocity (equation (A)), we have:

$$\overline{V}_2 = 3.7698 \text{ m/s}$$

One can substitute the velocity value in equation B, and carry out the iteration.

These two calculations can be done for one iteration. Repeating this process, we obtain the values of friction facor and velocity as given in the table below:

Friction factor values for variation in velocities by iteration process:

Iteration	Velocity (m/s)	Friction factor
1	3.7698	0.006459
2	3.4200	0.006470
3	3.4176	0.006470
4	3.4176	0.006470

The table shows that after the fourth iteration the values of velocity and friction factor begin to repeat. Thus, the correct svelocity is 3.4176 m/s. Another way to solve the problem is through the use of *spreadsheets*. In this case, the equation for $f_{pipeline}$ would be replaced in the equation for the velocity and seek *zero* of the following function:

$$\text{function} = \overline{V}_2 - \left[\frac{2\,.\,9.8\,.\,(3+1)}{\frac{1}{\beta_2} + 0.45 + 4\, f_{duct}^{f(\overline{V}_2)} \left(\frac{1}{0.075}\right) + 4\, f_{duct}^{f(\overline{V}_2)}\,(190)} \right]^{\frac{1}{2}} = 0$$

The result obtained is identical to that obtained by iterative method.

Exercise

Using the spreadsheet, analyse the effect of opening the valve and the length of the refractory duct on the velocity of metal stream from the ladle. Construct graphs showing the results obtained.

8.3.2 Metal transfer from tundish to continuous casting mold

The process of transfer of metal from tundish to the continuous casting mold can be studied on the basis of Bernoulli equation. The system is shown in Figure 8.10. For this configuration, the objective is to develop an equation for the flow rate of liquid steel between the two reactors, relating the height of steel melt in the tundish and the opening position of the gate valve, using Bernoulli equation:

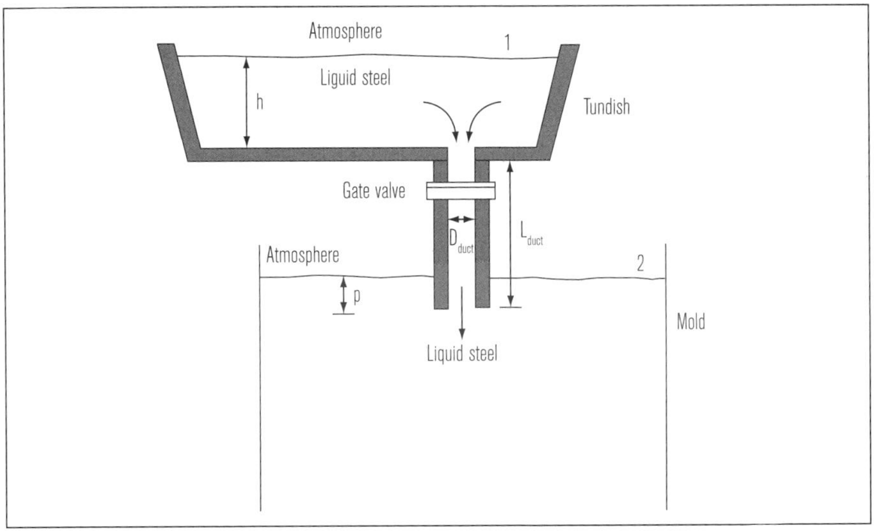

Figure 8.10 - Schematic view of the system of transfer of metal from tundish to the continuous casting mold.

As there is no external work done on the system, Bernoulli equation is given below without that term:

$$\frac{P_2 - P_1}{\rho} + g\,(z_2 - z_1) + \left(\frac{\overline{V}_2^{\,2}}{2\,\beta_2} - \frac{\overline{V}_1^{\,2}}{2\,\beta_1}\right) + E_f = 0 \tag{8.73}$$

The choice of location of levels 1 and 2 should take into consideration that these levels represent the parameters that appear in Bernoulli's equation, such as pressure,velocity and height above a given plane. In the case of pressure, one should try to choose locations, where this is equal to the atmospheric pressure. In this case, the most convenient choice is the one shown in Figure 8.10, with levels 1 and 2 located on the surfaces of metal in the tundish and the mold, respectively.

For these we have:

$$P_2 = P_1 = P_{atmosphere} \tag{8.74}$$

For these relative positions, we can write the following equation:

$$z_2 - z_1 = -(\,h + L_{duct} - p\,) \tag{8.75}$$

p in the above equation, is defined in the Figure 8.10 (It has to be included because plane 2 is at the level of steel in the mold and not at the exit of the submerged nozzle).

To calculate the variation of kinetic energy and friction loss, one should know the velocities in different parts of the system. The relationship between these velocities is expressed by the equations (obtained through mass balance) as given below:

$$A_1\,\rho_1\,\overline{V_1} = A_2\,\rho_2\,\overline{V_2} = A_{duct}\,\rho_{duct}\,\overline{V}_{duct} \tag{8.76}$$

As the density of metal is constant:

$$A_1\,\overline{V_1} = A_2\,\overline{V_2} = A_{duct}\,\overline{V}_{duct} \tag{8.77}$$

Cross-sections in 1, 2 are given by:

$$A_1 = W_{dist} \cdot T_{dist} \tag{8.78}$$

$$A_2 = W_{mold} \cdot T_{mold} \tag{8.79}$$

$$A_{duct} = \pi\,\frac{D_{duct}^2}{4} \tag{8.80}$$

where:

- W_{dist} = width of the tundish;
- T_{dist} = thickness of the tundish;
- W_{mold} = width of the mold;
- T_{mold} = thickness of the mold.

Substituting the relation 8.78, 8.79 and 8.80 in the equation (8.77), we obtain:

$$\overline{V}_1 = \frac{\pi\ D^2_{duct}}{4\ W_{dist}\ T_{dist}}\ \overline{V}_{duct} \tag{8.81}$$

$$\overline{V}_2 = \frac{\pi\ D^2_{duct}}{4\ W_{mold}\ T_{mold}}\ \overline{V}_{duct} \tag{8.82}$$

For the usual sizes of tundishes and molds, one can can write:

$$\overline{V}_1 <<<< \overline{V}_{duct} \tag{8.83}$$

$$\overline{V}_2 << \overline{V}_{duct} \tag{8.84}$$

Thus, the terms associated with kinetic energy are negligible:

$$\left(\frac{\overline{V}_2^{\,2}}{2\ \beta_2} - \frac{\overline{V}_1^{\,2}}{2\ \beta_1}\right) \approx 0 \tag{8.85}$$

For the evaluation of friction losses between levels 1 and 2. One considers:

- (a) friction in the walls of the tundish;
- (b) friction due to the contraction at the exit of the tundish;
- (c) friction in the walls of the duct;
- (d) friction due to the gate valve;
- (e) friction due to the expansion in the outlet of the duct and entry into the mold;
- (f) friction in the cross section of the mold.

Friction losses are proportional to the square of the velocity in any system under consideration (equation (8.62)). As the velocities inside the tundish and the mold are small (especially when compared with the velocity of metal in the pipeline), one can neglect the losses relating to items "a" and "f" listed above. Friciton losses associated with other items listed above are as follows:

b - Friction loss due to contraction at the tundish outlet:

$$E_f = \frac{1}{2} e_f \overline{V}^2_{duct} \tag{8.86}$$

Friction loss factor for contraction, is given by:

$$e_f = 0.45\,(1 - \alpha) \tag{8.87}$$

where:

$$\alpha = \frac{\text{Cross sectional area of duct}}{\text{Cross sectional area of tundish}} \tag{8.88}$$

As mentioned before, for this case also:

$$\alpha \approx 0 \tag{8.89}$$

because the contraction has no special finish in the region of entry:

$$e_f = 0.45 \,.\, (1 - 0) = 0.45 \tag{8.90}$$

$$E_{f_{contract}} = \frac{1}{2} \,.\, 0.45 \,.\, \overline{V}^2_{duct} \tag{8.91}$$

c - Friction loss at the straight duct connecting tundish and mold.

These losses are evaluated using the following equation:

$$E_{f_{duct}} = 2\, f_{duct} \left(\frac{L_{duct}}{D_{duct}}\right) \overline{V}^2_{duct} \tag{8.92}$$

d - Friction loss due to the gate valve:

$$E_{f_{valve}} = 2\, f_{duct} \left(\frac{L_e}{D}\right)_{valve} \overline{V}^2_{duct} \tag{8.93}$$

where $(L_e/D)_{valve}$ value depends on the opening of the gate valve.

e - Friction loss due to expansion in outlet from duct to mold:

$$E_f = \frac{1}{2} e_f \overline{V}^2_{duct} \tag{8.94}$$

For expansion, the friction loss factor is given by:

$$e_f = (1 - \alpha)^2 \tag{8.95}$$

$$\alpha = \frac{\text{Cross sectional area of duct}}{\text{Cross sectional area of mold}} \tag{8.96}$$

where

$$\alpha \approx 0 \tag{8.97}$$

Hence:

$$E_{f_{expansion}} = \frac{1}{2} . 1 . \overline{V}^2_{duct} \tag{8.98}$$

Adding up all the losses by friction (b,c,d,e), we have:

$$E_f = \frac{1}{2} . 0.45 . \overline{V}^2_{duct} + 2 f_{duct} \left(\frac{L_{duct}}{D_{duct}}\right) \overline{V}^2_{duct} + 2 f_{duct} \left(\frac{L_e}{D}\right)_{valve} \overline{V}^2_{duct} + \frac{1}{2} . 1 . \overline{V}^2_{duct} \tag{8.99}$$

$$E_f = \frac{1}{2} \overline{V}^2_{duct} \left[0.45 + 4 f_{duct} \left(\frac{L_{duct}}{D_{duct}}\right) + 4 f_{duct} \left(\frac{L_e}{D}\right)_{valve} + 1 \right] \tag{8.100}$$

Substituting these terms in Bernoulli's equation:

$$-g (h + L_{duct} - p) + \frac{1}{2} \overline{V}^2_{duct} \left[0.45 + 4 f_{duct} \left(\frac{L_{duct}}{D_{duct}}\right) + 4 f_{duct} \left(\frac{L_e}{D}\right)_{valve} + 1 \right] = 0 \tag{8.101}$$

Rearranging, we obtain an expression for the velocity of metal in the pipe:

$$\overline{V}_{duct} = \left\{ \frac{2 g (h + L_{duct} - p)}{\left[0.45 + 4 f_{duct} \left(\frac{L_{duct}}{D_{duct}}\right) + 4 f_{duct} \left(\frac{L_e}{D}\right)_{valve} + 1 \right]} \right\}^{\frac{1}{2}} \tag{8.102}$$

Example

Using the equation (8.98) determine the velocity of steel in the duct that connects the Tundish to the continuous casting mold. Consider the following data:

- duct length, L = 1 m
- Tundish, h = 0.80 m;
- diameter of duct, D_{duct} = 0.075 m;
- penetration of the product inside the mold, p = 0.20 m;
- roughness of the duct, ε = 0.0002 m;
- position of the gate valve: half open.

Solution

Substituting the data given in equation (8.102), we have:

$$\overline{V}_{duct} = \left\{ \frac{2 \,.\, 9.8 \,.\, (0.8 + 1 - 0.20)}{\left[0.45 + 4 \, f_{duct} \left(\frac{1}{0.075} \right) + 4 \, f_{duct} \, (190) + 1 \right]} \right\}^{\frac{1}{2}} \quad \text{(A)}$$

To determine the velocity in the duct, the friction factor is calculated by:

$$\frac{1}{\sqrt{f_{duct}}} = -3.6 \log \left[\left(\frac{\varepsilon_{duct} / D_{duct}}{3.7} \right)^{1,11} + \frac{6.9}{\frac{D_{duct} \, \overline{V}_{duct} \, \rho}{\mu}} \right]$$

Substituting data:

$$\frac{1}{\sqrt{f_{duct}}} = -3.6 \log \left[\left(\frac{0.0002/0.075}{3.7} \right)^{1,11} + \frac{6.9}{\frac{0.075 \, \overline{V}_{duct} \, 7000}{0.007}} \right] \quad \text{(B)}$$

The equations (A) and (B) must be solved simultaneously in order to determine the velocity in the duct. Using iterative methods, as before, we obtain:

$$\overline{V}_{duct} = 2.152 \text{ m/s} \quad e \quad f_{duct} = 0.006542$$

From the velocity, volumetric and mass flow rate can be determined:

$$Q = \pi \frac{D^2_{duct}}{4} \overline{V}_{duct} = \pi \frac{(0.075)^2}{4} . 2.152 = 0.0095 \text{ m}^3/\text{s}$$

$$\Gamma = Q . \rho = (0.0095) .(7000) = 66.55 \text{ kg/s} = 239.6 \text{ ton/h}$$

Exercise: Analyze the effect of opening of the gate valve on the velocity and flow of steel.

8.4 Techniques of measuring fluid flow

In many situations, efficient operation and control of metallurgical processes and experimental set-ups require quantitative information on fluid flow like flow rates and pressure at different locations. To measure flow in pipelines, there are a wide variety of equipments based on differential pressure at two locations like orifice meters, venturi meters, pitot tubes, variable area meters, rotameters etc. Many of these equipments, have been designed and operated on the basis of principles of fluid flow discussed earlier and the Bernoulli equation.

8.4.1 Flow meters based on differential pressure measurements

A group of devices for measuring the flow rate of fluids are based on determination of pressure differences between two locations. In this group, one can include orifice meter,venturi meter and pitot tube.

8.4.1.1 Orifice and Venturi meters

Figures 8.11 and 8.12 show schematic diagrams of orifice meter and venturi meters. Both have the same operating principle. In the case of orifice meter an orifice plate containing a small hole is introduced at the cross section of the pipe line through which the fluid flows. This sudden reduction causes an increase in the local velocity of the fluid with a corresponding decrease in the pressure which can be measured using a manometer. From the pressure difference the flow rate can be determined with the application of Bernoulli equation. In the case of venturi meter, the change in the area of cross section is gradual and pressure difference between the two locations can be measured as in the case of orifice meter and related to the flow rate of the fluid.

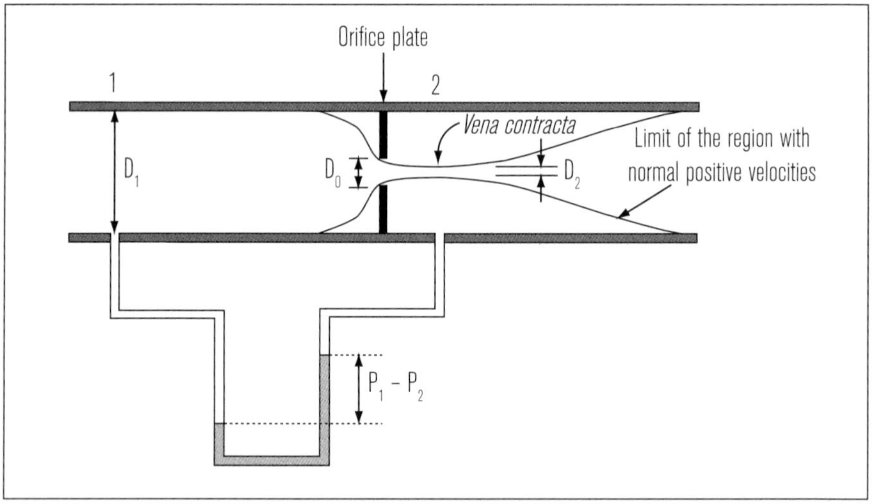

Figure 8.11 - Schematic view of an orifice plate.

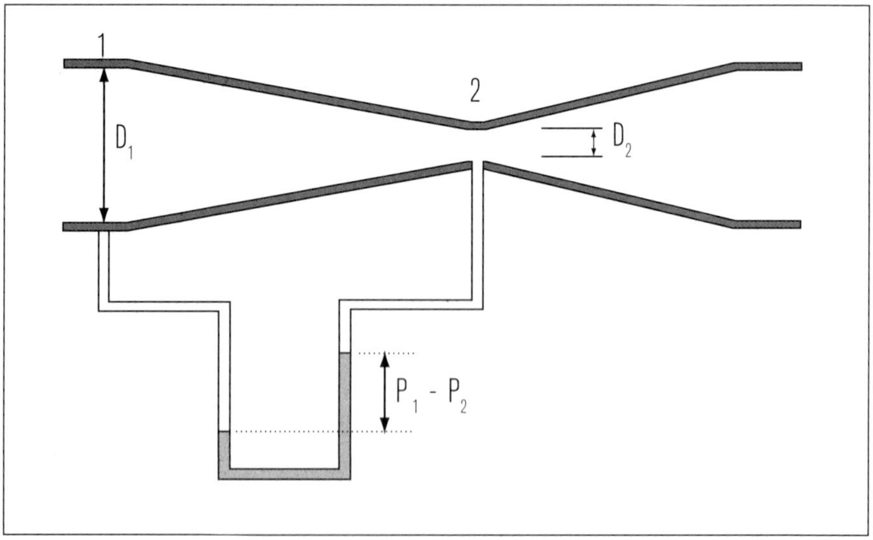

Figure 8.12 - Schematic view of a Venturi meter.

As shown in Figure 8.11, the flow contracts before the hole and continue to contract for a short distance from the position of the orifice plate, forming a region where the flow area is minimal. The position where this happens is called the "vena contracta".

To understand the principle of operation of this equipment, let us apply Bernoulli's equation with reference to locations 1 and 2 in Figure 8.10 (Points of measurement of differential pressure). If the friction loss is neglected and eliminating the potential and mechanical energy terms (horizontal tube, no energy input) Bernoulli equation for points 1 and 2 can be written as:

$$\frac{P_2 - P_1}{\rho} + \left(\frac{\overline{V}_2^2}{2\,\beta_2} - \frac{\overline{V}_1^2}{2\,\beta_1} \right) = 0 \tag{8.103}$$

For turbulent flow on both points, we have: $\beta_1 = \beta_2 = 1$, then:

$$\frac{P_2 - P_1}{\rho} + \frac{\overline{V}_2^2}{2} - \frac{\overline{V}_1^2}{2} = 0 \tag{8.104}$$

Fluid has constant density and applying a mass balance between points 1 and 2:

$$A_1\overline{V_1} = A_2\overline{V_2} \tag{8.105}$$

$$A_1 = \pi \, \frac{D_1^2}{4} \tag{8.106}$$

$$A_2 = \pi \, \frac{D_2^2}{4} \tag{8.107}$$

Combining equations (8.106) (8.107) and (8.108):

$$D_1^2 \, \overline{V_1} = D_2^2 \, \overline{V_2} \tag{8.108}$$

$$\overline{V_1} = \left(\frac{D_2}{D_1} \right)^2 \overline{V_2} \tag{8.109}$$

Substituting (8.105) in Bernoulli's equation:

$$\frac{P_2 - P_1}{\rho} + \frac{\overline{V}_2^2}{2} \left[1 - \left(\frac{D_2}{D_1} \right)^4 \right] = 0 \tag{8.110}$$

Velocity at location 2 is given by:

$$\overline{V}_2 = \sqrt{\frac{2\,(P_1 - P_2)}{\rho\left[1 - \left(\frac{D_2}{D_1}\right)^4\right]}} \qquad (8.111)$$

This is the theoretical velocity at the point of *vena contracta*. This expression does not consider losses due to friction and the velocity calculated is not achieved in practice. Furthermore, the above equation is not useful for determining the flow rate of fluid, since diameter D_2, is not known. Hence it would be better to to have an equation to assess the velocity as a function of the diameter of the opening of the orifice plate and also take into account the effects of friction.To take into account these, a discharge coefficient, C_D, determined empirically, is introduced into the equation (8.112). With the introduction of this coefficient, the equation can be written as follows:

$$\overline{V}_o = C_D\sqrt{\frac{2\,(P_1 - P_2)}{\rho\left[1 - \left(\frac{D_o}{D_1}\right)^4\right]}} \qquad (8.112)$$

Another way to write the above equation is:

$$\overline{V}_o = K\sqrt{\frac{2\,(P_1 - P_2)}{\rho}} \qquad (8.113)$$

$$K = \frac{C_D}{\sqrt{\left[1 - \left(\frac{D_o}{D_1}\right)^4\right]}} = \frac{C_D}{\sqrt{1 - B^4}} \qquad (8.114)$$

K in the above equation is is called runoff coefficient and B is the ratio between the diameter of the orifice and the duct:

$$B = \frac{D_o}{D_1} \qquad (8.115)$$

Figure 8.13, determined experimentally, shows the values of K according to the parameter B, and the position of the pressure measurement after the orifice plate. Hence K is a function of position of measurement and diameter of orifice.

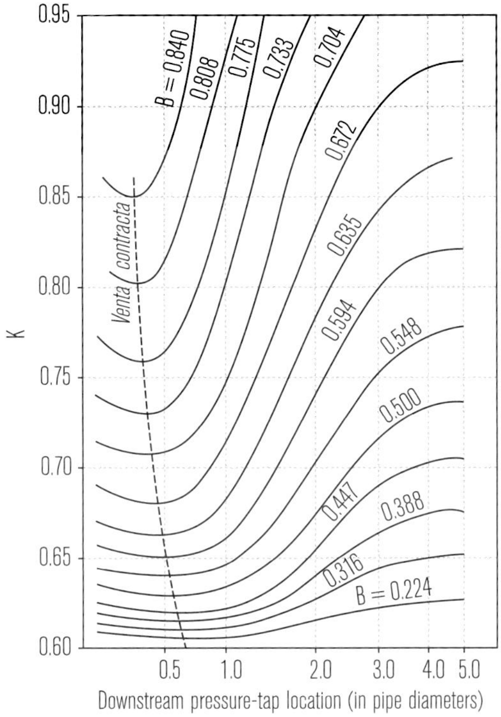

Figure 8.13 - Values of the runoff coefficient for the orifice plate (GEIGER and POIRIER, 1980).

Example

Determine the value of the runoff coefficient for an orifice plate with apertures of 20 cm, installed in a duct with a diameter of 40 cm. The second pressure gauge is installed 60 cm after the orifice plate.

$$D_o = 20 \text{ cm} \qquad D_1 = 40 \text{ cm}$$

$$B = \frac{D_o}{D_1} = 0.5$$

Solution

Based on the data, we have that:

The position of the second pressure gauge, expressed in terms of the ratio of its distance to the orifice and the diameter of the duct, is:

$$\text{Position of the second pressure gauge} =$$

$$= \frac{\text{Distance between pressure gauge and orifice}}{\text{Diameter of the duct}} = \frac{60 \text{ cm}}{40 \text{ cm}} = 1.5$$

Using the above data, one can use Figure 8.13 to determine the value of K.

For the configuration of the orifice plate and position of the pressure gauge, the value of K is about 0.65 (See figure below). Hence, the average velocity of the fluid in the orifice is given by:

$$\overline{V}_o = 0.65\sqrt{\frac{2(P_1 - P_2)}{\rho}}$$

The flow rate is given by:

$$Q = \pi\frac{D_o^2}{4}\overline{V}_o = 0.65\,\pi\frac{D_o^2}{4}\sqrt{\frac{2(P_1 - P_2)}{\rho}}$$

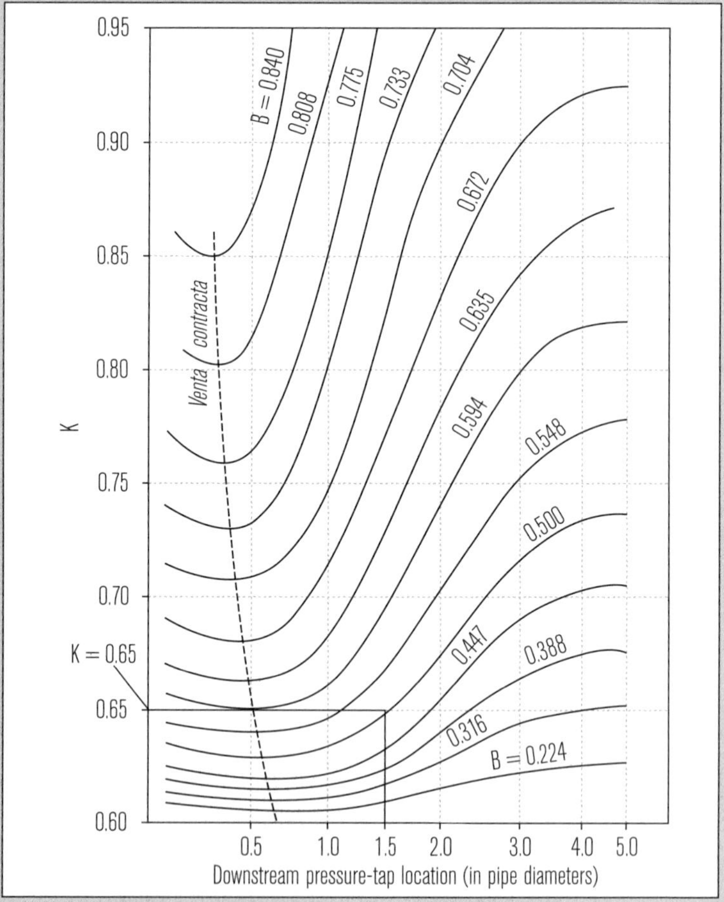

Figure for evaluating K

Equation (8.113) can be used also for the Venturi meter. However, the value of the discharge coefficient, C_D, is close to one for this type of flow meter. The maximum contraction corresponds exactly to the position where P_2 is measured.

As seen in Figure 8.13, in the case of an orifice plate also, it is important to choose the position of pressure gauges. Usually, a pressure gauge is placed one to two pipe diameters ahead of the orifice plate, while the other pressure gauge is placed in the middle diameter of the tube after the orifice plate or else at the *vena contracta*, whose position can be determined experimentally.

8.4.1.2 Pitot tube

The Pitot tube can also be used for evaluating fluid velocities. In this case the velocity is determined by measuring the difference between the static pressure and impact pressure (called stagnant pressure) at a given point. The opening for impact is positioned perpendicular to the flow, while the static holes are parallel to the direction of flow. The Figure 8.14 shows schematically a Pitot tube.

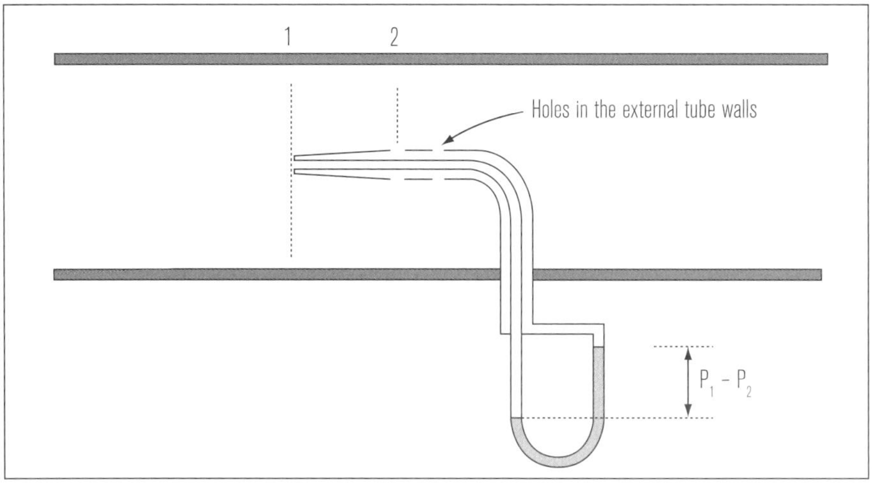

Figure 8.14 - Schematic view of a Pitot tube.

To obtain a relationship between the measured pressure difference and fluid velocity at a given point along the cross section of a duct, one should establish an energy balance (Bernoulli equation) between points 1, the opening for mpact, and point 2, as shown in Figure 8.14. Applying the

Bernoulli equation, assuming incompressible fluid and neglecting the losses by friction, and other criteria like before we have:

$$\frac{P_2 - P_1}{\rho} + \left(\frac{\overline{V}_2^{\,2}}{2\,\beta_2} - \frac{\overline{V}_1^{\,2}}{2\,\beta_1} \right) = 0 \qquad (8.116)$$

For turbulent flows ($\beta_2 = 1$) and at 1 point the fluid velocity drops to zero. Hence one can rewrite the equation (8.116) in the following way:

$$\frac{P_2 - P_1}{\rho} + \frac{\overline{V}_2^{\,2}}{2} = 0 \qquad (8.117)$$

or:

$$\overline{V}_2 = \sqrt{\frac{2\,(P_1 - P_2)}{\rho}} \qquad (8.118)$$

To correct the effect of approximations made in developing the above relationship (incompressibility of the fluid and no friction losses), a coefficient is incorporated into the equation (8.118) Incorporating C_P coefficient of Pitot tube, in the above equation:

$$\overline{V}_2 = C_P \sqrt{\frac{2\,(P_1 - P_2)}{\rho}} \qquad (8.119)$$

Generally, the value of this coefficient ranges from 0.98 to 1.00. Caution should be exercised in the use of Pitot tubes with reference to the location and form of static openings, for getting a real measure of static pressure along the same line of flow in which the impact pressure is measured. Locations of these openings not parallel to the line of flow may introduce errors in measurements.

As the Pitot tube measures only local rates, in order to determine the average velocity, velocity data should be collected at different points along the cross section of the pipeline. The value of density ρ, is determined for the temperature of the fluid stream. at the vicinity of the Pitot tube. In some situations, the maximum speed, (measured at the center of the duct) can be related to the average speed. This avoids determination of velocities at various points. As discussed in Chapter 5 from (equations (5.117) and (5.126)) for maximum and average respectively for laminar flow in circular ducts, we have:

$$\frac{\overline{V}}{V_{max}} = \frac{1}{2} \qquad (8.120)$$

For turbulent flow in circular ducts (for Reynolds numbers between 10^4 and 10^7) the following equation is recommended:

$$\frac{\bar{V}}{V_{max}} = 0.62 + 0.04 \cdot \log\left(\frac{D\, V_{max}\, \rho}{\mu}\right) \qquad (8.121)$$

For Reynolds numbers between 2100 and 10^4 there seems to be no expression relating V_{av} and V_{max}

Example

A Pitot tube is installed in a tube with its impact opening along the centerline. Diameter of the tube is 0.3048 m. Air at 65.5 °C and 82,736.4 Pa pressure flows through the tube. Barometric pressure is 99,323.4 Pa. Pressure difference measured by the Pitot tube is 104.55 Pa. Viscosity of air is: 2 x 10^{-5} kg/ms. Estimate the mass flow rate of air.

Solution

First, determine the air density under the conditions of measurement.

Absolute pressure in the region of interest is:

$$P_2 = 82736.4 + 99323.4 = 182059.8 \text{ Pa}$$

The air density is calculated as follows:

$$\rho = \frac{P\, M_M}{R\, T}$$

where:

- M_M = molecular weight of air = 0.02884 kg/mol;
- R = gas constant = 8.31 J/mol.K;
- T = temperature (K).

Substituting values, we have:

$$\rho = \frac{P\, M_M}{R\, T} = \frac{(182059.81) \cdot (0.02884)}{(8.31) \cdot (65.5 + 273)} = 1.867 \text{ kg/m}^3$$

Assuming $C_p = 0.99$, one can calculate the velocity of air (in the center of the tube):

$$\overline{V}_2 = C_P \sqrt{\frac{2\ \Delta P}{\rho}} = 0.99 \sqrt{\frac{2\ .\ 104.55}{1.867}} = 10.477 \text{ m/s}$$

Hence the Reynolds number:

$$Re_{max} = \frac{D\ V_{max}\ \rho}{\mu} = \frac{0.3048\ .\ 10.477\ .\ 1.886}{2\ .\ 10^{-5}} = 301424$$

Substituting Reynolds Number in equation 8.122:

$$\frac{\overline{V}}{V_{max}} = 0.62 + 0.04\ .\ \log\left(301424\right) = 0.839$$

$$\overline{V} = 0.839\ .\ 10.477 = 8.79 \text{ m/s}$$

Mass flow is given by:

$$\Gamma = \pi\ \frac{D^2}{4}\ .\ \overline{V}\ .\ \rho = \pi\ \frac{(0.3048)^2}{4}\ .\ 8.79\ .\ 1.866 = 1.197 \text{ kg/s}$$

8.4.2 Rotameters

This type of flow meter is a device suitable for measurement of relatively small flows of liquids and gases.

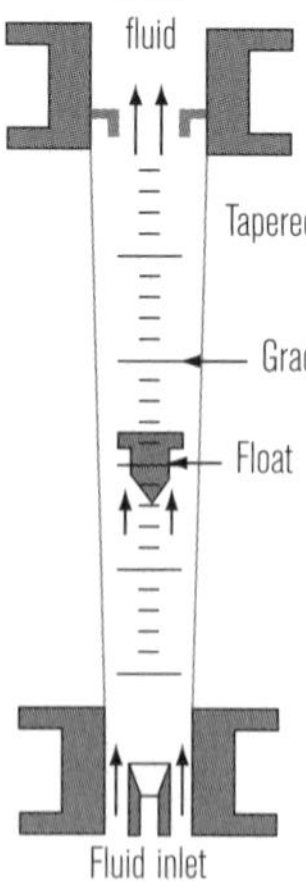

Figure 8.15 - Schematic view of a rotameter.

This operation is based on the principle of introducing a restriction to the flow of the fluid, inside the tube in the form of a float, which changes the position along the height of the tapered tube with a corresponding change in velocity of the fluid. This also creates a pressure drop in this region with the reduced area for flow. However, in this case, the pressure drop remains constant and the area of flow changes as the air velocity changes. This type of flow meter is illustrated in Figure 8.15. The flow rate of the fluid corresponds to the height indicated by the float along a slightly tapered section. This type of flow meter has to be calibrated previously for different known flow rates for a particular fluid.

A balance of forces can be applied to float at its equilibrium position. For a float density of ρ_f and a fluid density of ρ kept in suspension, the forces acting on the float is balanced.

The forces acting on the float are:

- F_G: weight, acting down;
- F_E: buoyancy, acting in order to suspend the float;
- F_A: drag force resulting from friction between the fluid and the float. Operates in the same direction of the fluid velocity.

In the state of balance of forces:

$$F_G = F_E + F_A \tag{8.122}$$

Transposing terms and expressing the weight and buoyancy in terms of volume of the float, V_f, and the fluid density, ρ, and float density, ρ_f, we have:

$$V_f \, (\rho_f - \rho) \, g = F_A \tag{8.123}$$

But the volume of the float is given by:

$$V_f = \frac{m_f}{\rho_f} \tag{8.124}$$

Where m_f is the mass of the float:

Combining equations (8.124) and (8.125), we have:

$$\frac{m_f}{\rho_f} \, (\rho_f - \rho) \, g = F_A \tag{8.125}$$

For a given flow meter, through which a fluid flows, the left side of the equation (8.125) is a constant. Hence, F_A is constant when the float is in equilibrium and when the flow rate of the fluid changes, the float counteracts this effect by assuming a new position of equilibrium. For example, if the float is in a position of equilibrium corresponding to a given mass flow, then the mass flow rate becomes larger, the float rises inside the tube. However, as the float rises, the cross-sectional area of the tube increases and the velocity of the fluid between the float and the tube wall decreases, so as to reach a value of F_A which satisfies the equation (8.125). These type of flow meters are calibrated for known flows for giving accurate results.

References

BIRD, R.B; STEWART, W.E.; LIGHTFOOT, E.N. *Transport phenomena.* New York: John Wiley & Sons, 1960. 780 p.

GASKELL, D.R. *An introduction to transport phenomena in materials engineering.* New York: Macmillan, 1992. 637 p.

GEIGER, G.H.; POIRIER, D.R. *Transport phenomena in metallurgy.* Massachusetts: Addison-Wesley, 1980. 616 p.

WHITE, F.M. *Fluid mechanics.* New York: McGraw-Hill, 1979. 701 p.

Exercises

1 - Consider the continuous casting unit according to the following figure:

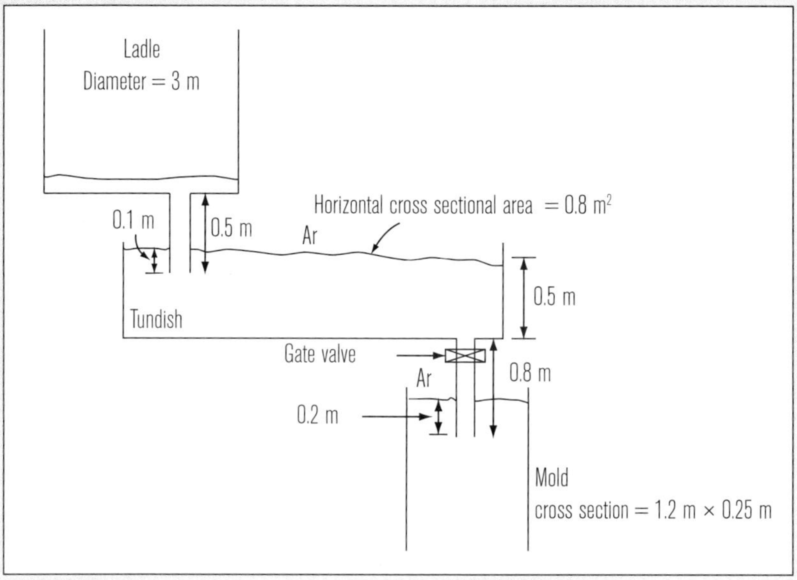

The diameters of the pipelines feeding the tundish and the mold are, respectively, 70 and 60 mm. The roughness of the refractory is 0.1 mm.The steel ladle is emptied and and gets filled with liquid metal until it is full. This operation takes 1 minute. In this period there is no supply of steel in the distributor, but the supply of steel in the mold is kept constant and is equal to 108 tons/hour. Estimate the decrease in the level of steel in the distributor during this period.

2 - A ladle is feeding pouring liquid steel into the ingot, mold as seen in the figure below. Determine the time taken to fill the mold with a capacity of 2 tons of steel. Given Steel: - density: 6.7 g/cm³ - viscosity: 0.07 P. The nozzle opening at the bottom of the ladle has a diameter of 70 mm. Disregard the thickness of the refractory.

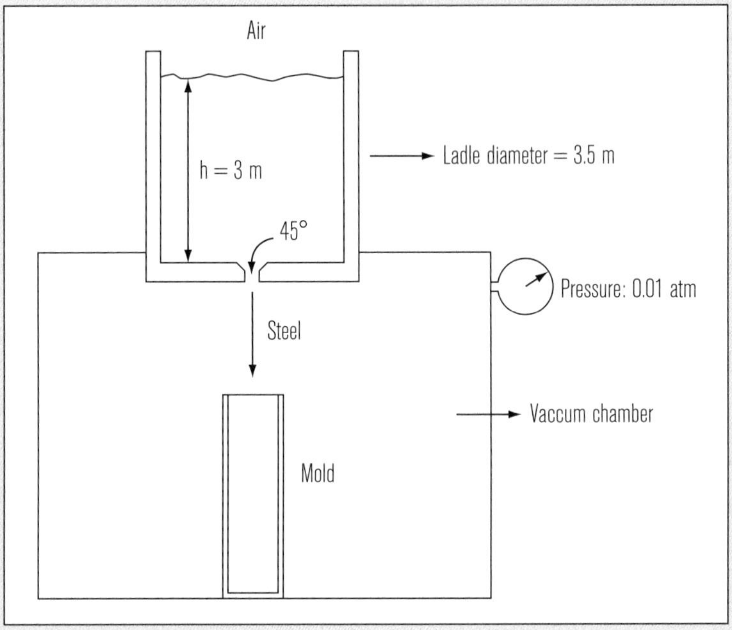

3 - In a continuous casting plant, a ladle is used to feed liquid steel into a tundish, which supplies two strands for casting of slabs, as shown below:

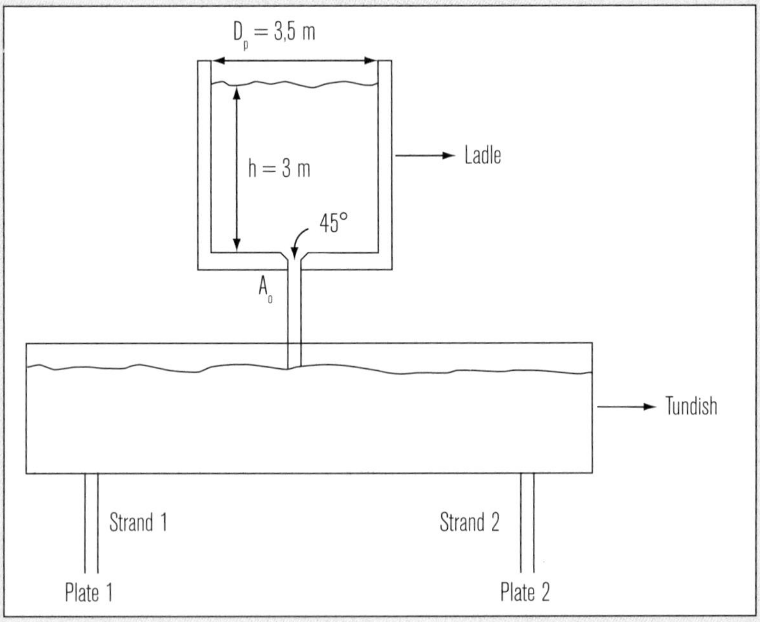

If the level of steel in the tundish is to be maintained as constant as possible ,the flow rate of metal draining from it through the nozzle can be controlled by the extent of opening of the nozzle, that is by changing its diameter. Get a mathematical relationship between the opening area of the nozzle and the height of liquid steel inside the ladle, to ensure a constant height of steel melt in the tundish, until the steel height of 100 mm of steel in the ladle. Using the relationship developed, to calculate, what should be the areas of the nozzle opening for an initial height of 3 m and final height of 100 mm of steel in the ladle.

Assume turbulent flow.

Dimensions of slab: 2 m in width - 0.25 m thick

Strand speed: 1.5 m/min.

4 - Determine the internal pressure, P, to be taken in the pressure chamber in order to obtain a flow of liquid steel compatible with that shown in the figure below:

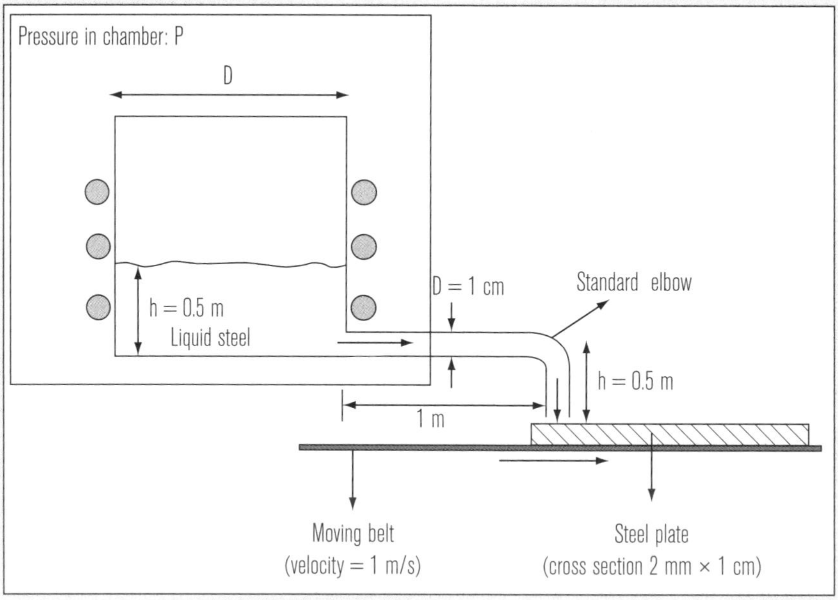

Assume turbulent flow, - Tubes are hydraulically smooth.

- Density of steel: 7000 kg/m^3;
- Viscosity of steel: 7 cP.

5 - Determine the power of the the pump for the system represented in the figure below, which requires a *spray* of water for accelerated cooling of a steel strip after rolling. Data:

- Duct diameter: 2.54 cm;
- Roughness on the pipeline, $\alpha v \delta / D = 0.004$;
- Elbows: 90° standard;
- Flow of water: 1 l/s;
- Water:
  - density: 1000 kg/m^3;
  - viscosity: 1cP. - 1 hp = 745.7 W.

Consider that the pressure drop in the spray nozzle is 1.7 atm. 1 atm = 101,330 Pa.

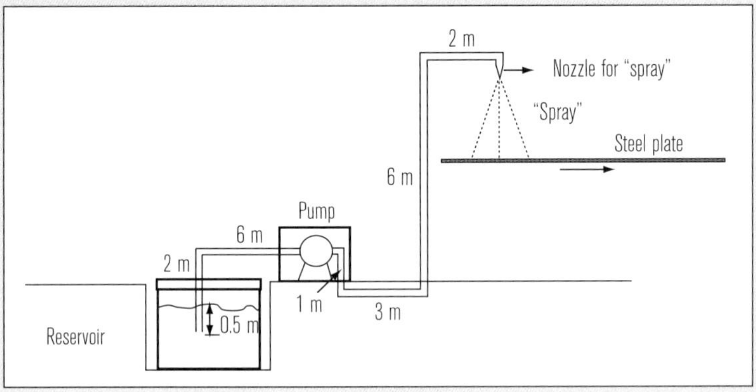

6 - For the system shown below in the figure, estimate the time required to empty the reservoir 1 to the level of inlet. The inlet in the reservoir 2 is closed.

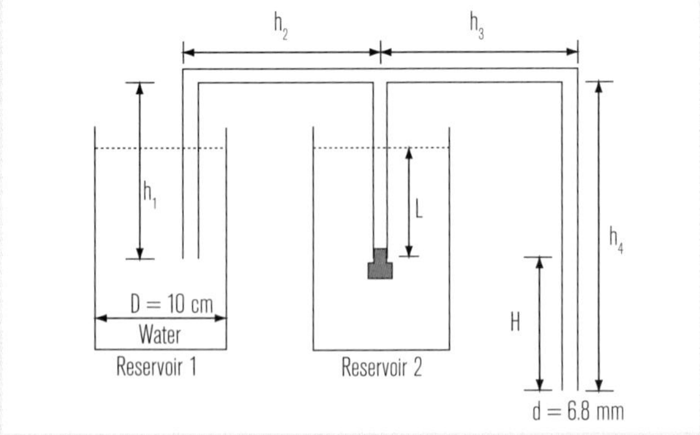

If the stopper of the tube in the tank 2 is removed, will the time for emptying the tanks be the same? Justify the answer.

- $h = 10$ cm;
- $h_2 = 15$ cm;
- $h_3 = 15$ cm;
- $h_4 = 20$ cm;
- $L = 8$ cm;
- $H = 10$ cm.

Consider that the glass tube is hydraulically smooth and the friction factor is given by:

$$f = \frac{8}{Re} + \frac{0.0385}{(0.839 - \log Re)^2}$$

7 - A ladle with an internal diameter of 1 m containis liquid aluminum. The initial height of liquid metal is 1.5 m and the nozzle diameter is 0.1 m. Determine:

- time required to empty the ladle through the nossle at the bottom;
- initial rate of discharge of metal in kg/s;
- Discharge rate of metal (kg/s) when the ladle is 50% full.

Develop all relations used in the calculations. Assume turbulent flow and consider friction losses.

Properties of aluminum:

- density: 2410 kg/m^3;
- viscosity: 2.75×10^{-3} Pa.s.

8 - Water is being siphoned from the tank seen schematically in the figure below.

Determine the average velocity of water in the siphon outlet for the situation seen in the figure below.

Fluid properties:

- density: 1 g/cm^3;
- viscosity: 1 cP (1 P = 1 g/cm.s).

Tubing:

- diameter: 0.0254 m;
- relative roughness: $\varepsilon/D = 0.001$.

9 - Considering the physical model shown in the figure below, determine the minimum diameter of the duct, which guarantees a flow rate of water supply of 50 liters per minute in the tundish, only by gravity. The roughness of the pipe is 0.1 mm. Properties of water: density: $1 g/cm^3$, viscosity 1 cP.

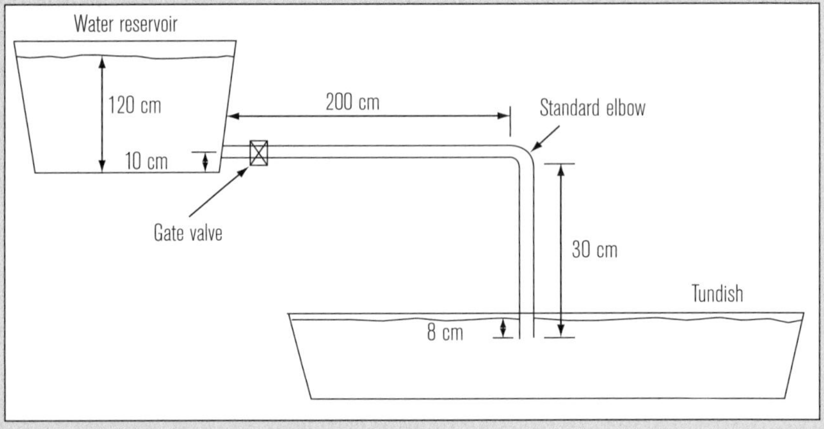

10 - Determine if there is risk of overflow of water in the reservoir below, if the feed flow is 400 l/min. Roughness of duct = 0.3 mm.

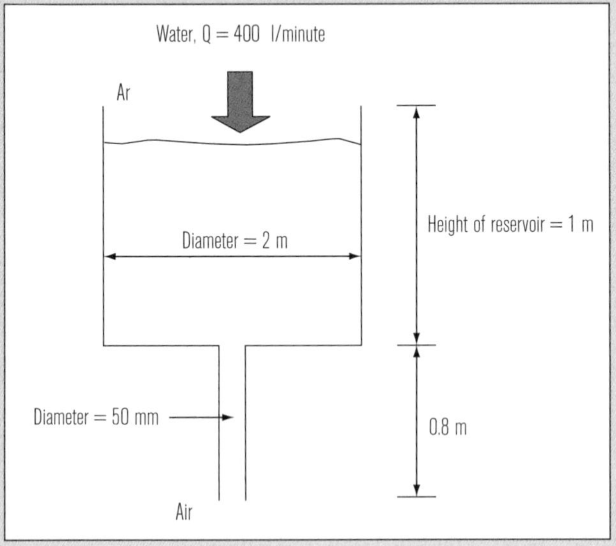

Introduction to Heat Transfer

Chapter 09

The study of thermodynamics generally deals with the concepts of equilibrium, enthalpy, internal energy, work, free energy, etc. The second law of thermodynamics shows that heat is transferred from regions of high to low temperature. However, the concepts of thermodynamics do not through any light on the evaluation of the rate at which energy or mass is transferred. In many practical applications in the engineering field, the knowledge of the rate of heat transfer is important. In the area of process metallurgy, heat treatment and deformation of metals at high temperatures, the rate of heat transfer is very vital as efficient transfer of heat under controlled conditions determines the productivity and quality of the products. As has already been discussed in Chapter 2, the subject of heat transfer is very important in various stages of steel industry as illustrated in Figure 2.4 that presents the general flow diagram for the production of end products like coils in steel mills. Throughout this sequence of steps, one can identify a number of applications involving the study of heat transfer.

In the regenerators, heat is recovered from the hot gases resulting from combustion of blast furnace gas and coke oven gas with air. This is accomplished by heat transfer to the refractories in the Cowper stoves and subsequent heat transfer to air for blowing into the blast furnace. Heat transfer study of regenerators is very important for the design of these units. Similarly, in the coke ovens, there is transfer of heat to the coal mixture in the chambers for converting it to coke. Thermal efficiency and productivity of these processes depend on the rate of heat transfer.

In the sintering of iron ores, heat exchange occurs between the hot gases and the sinter mixture, when these gases flow through the sinter bed. The rate of heat transfer between the solid and the gas determines the velocity of descent of the combustion front from top to bottom of the sinter bed. This rate of descent of the combustion front determines the time required for sintering process to complete and hence the productivity of the sinter plant. Heat transfer also plays an important role in drying and firing of iron ore pellets.

In the production of pig iron in blast furnaces, heat exchange between the burden material and ascending hot gases from the raceway determines the location and the extent of the thermal reserve zone, as well as the geometry and position of softening and melting zone. Variables in the heat exchange process have to be carefully controlled in order to minimize the fuel consumption and to increase the productivity of the reactor. The heat loss through the walls of various furnace units in the steel plants is minimized by use of proper refractories and appropriate cooling for reduction of energy consumption. These aspects can be studied based on the principles of heat transfer.

In the steel industry, the thermal state of ladles and torpedo cars, for the transport of liquid metal is also of great importance. Temperature and composition control of the liquid metal is essential at every stage until the process of continuous casting and subsequent rolling for obtaining products with good quality. The characterization of the thermal state of the various reactors and ladles involves the application of knowledge of heat transfer.

Throughout the continuous casting process, heat is extracted from the liquid steel to promote its solidification. This extraction is performed in different stages. The rate of heat removal in these different stages affects the speed of solidification of steel, which determines the productivity of the casting machine. In addition, the defects in the cast products have their origins associated to heat extraction rates at different stages. Heat transfer study is also important in the conventional foundry metallurgy involving solidification in sand molds and permanent molds.

After casting and before mechanical working to produce the desired end products, like plates, bars and wires, reheating in appropriate furnaces is usually necessary. In subsequent processing, like heat treatment or temperature homogenization, the residence time of these materials in the reheating furnaces depends on the rate of heating at appropriate temperatures.

Finally, heat treatments are performed in the semi-finished products involving controlled heating and cooling in order to have desired phase transformations and to obtain appropriate structures for desired properties. In these operations, the rates of heating and cooling are decisive in defining the structures and properties of the products obtained.

All the examples presented above illustrate the great importance of heat transfer in the sequence of steps involved in the production of metals. Similar examples can be given in the extraction and processing of other metals and also in other branches of engineering and even in day-to-day life of people.

In the next chapter, the different mechanisms of heat transport and the equations governing each one of them will be discussed. Subsequently, the concept of energy balance, steady and unsteady state heat transfer as well as heat transfer by radiation will be discussed.

Exercise

1 - Besides the cases mentioned in this chapter, identify and analyze other systems and situations in the field of metallurgical engineering and materials, where heat transfer study is relevant.

Energy Balances and Mechanisms of Heat Transport

Chapter 10

In this chapter there is a brief introduction to mechanisms of heat transfer and the equations that are used for each of these to evaluate the rate of heat transfer followed by concepts of thermal energy balance and simple solutions of equations that result from them. Applications of the same will be discussed in the subsequent chapters of heat transfer.

10.1 Mechanisms of Heat Transfer

Heat is normally transferred by three mechanisms:

- conduction (or diffusion of heat);
- convection;
- radiation.

The application and relevance of each mechanism depend on the system, phases involved (solid, liquid or gas) and temperature ranges.

10.1.1 Conduction

The transfer of heat by conduction occurs when there is a temperature gradient in a material. The temperature gradient can be defined as a temperature variation with position. In this mechanism, heat is transferred from regions of high to low temperature. Figure 10.1 illustrates the heat transfer by conduction in a solid medium. Conduction can also considered as diffusive transfer of heat through the medium.

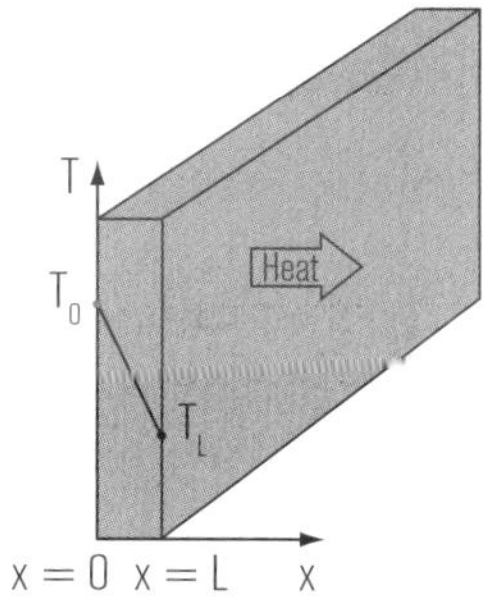

Figure 10.1 - Transfer of heat by conduction with temperature gradient along the x direction in a solid plate.

For the situation shown in Figure 10.1, the heat flux in x direction (direction of the gradient of temperature) can be calculated by the Fourier equation:

$$q_x = -k\frac{\partial T}{\partial x} \tag{10.1}$$

where:

- q_x = heat flux in x direction (W/m²);
- k = thermal conductivity (W/mK);
- $\frac{\partial T}{\partial x}$ = temperature gradient in the x direction (K/m or °C/m);
- T = temperature (K or °C);
- x = position (m).

As the temperature profile illustrated in Figure 10.1 is linear, the temperature gradient can be estimated by the following equation:

$$\frac{\partial T}{\partial x} = \frac{T_L - T_O}{L - 0} = \frac{T_L - T_O}{L} \tag{10.2}$$

where:

- T_L = temperature at x = L (K or °C);
- T_O = temperature at x = 0 (K or °C).

T_L greater than T_O, and hence the temperature gradient is negative (the temperature decreases as the position x increases).

As the thermal conductivity is always positive, the heat flow shown in Figure 10.1 is also positive. The convention for the sign of the heat flow is:

- positive, if it is in the same direction of growth in respect of x;
- negative if it is in the opposite direction of growth in respect of x.

In the case, shown in Figure 10.1, the heat is transferred from T_O to T_L (high to low temperature). Thus, the flow of heat is in the same direction of growth as x-axis and is therefore positive.

The thermal conductivity is a physical property of the medium in which heat is being conducted. Its value tends to be higher for solids, particularly for metals, and lower for gases.

The rate of heat transfer can be obtained by multiplying the flow area perpendicular to the direction of heat transfer:

$$Q_x = A.q_x = -A.k.\frac{\partial T}{\partial x} \quad (10.3)$$

where:

- Q_x = rate of heat transfer in the x direction (W);
- A = area perpendicular to the direction of heat transfer (m^2).

10.1.2 Convection

In the case of the conduction, there's no macroscopic displacement of material,only heat energy is transported by diffusive mechanism. This type of mechanism is common in the case of transport of heat through solid bodies.

In a system, apart from conduction, there can be transport of heat energy through macroscopic motion of the medium. This mode of heat transfer is called convection. Some authors use the term to indicate the convective transport due only to the macroscopic motion of the medium. However, one can say that convection is the overlap of conduction as well as convection involving macroscopic motion.

An important issue associated with convective heat transfer is the identification of the cause of the movement of macroscopic medium. When the flow is caused by an external agent, such as as a ventilator, exhaust fan or a compressor, the convection is termed as forced convection. In the case, where the movement is caused by the buoyant force associated with the variation of fluid density due to change in temperature, it is called natural convection. There can be mixed convection, when both these are present simultaneously in a system.

In some cases of interest in engineering particularly in metallurgy, convection is associated with phase changes, such as during boiling and condensation. These phase changes usually have a significant effect on heat transfer.

A very common situation that involves the mechanism of convection is the heat exchange between a solid surface and a fluid in contact with it. The heat flow in this case can be estimated from the following equation:

$$q_x = h.(T_s - T_\infty) \quad (10.4)$$

where:

- q_x = heat flow between the surface of the solid and fluid in the x direction (W/m^2);
- h = coefficient of heat transfer ($W/m^2.K$ or $W/m^2.°C$);
- T_s = surface temperature of the solid (K or °C);
- T_∞ = temperature of fluid at a point away from the surface of the solid (K or °C).

Equation (10.4) refers to a situation like the one given in Figure 10.2, which shows the convective heat transfer between a .solid and a fluid, very frequently encountered in various metallurgical operations.

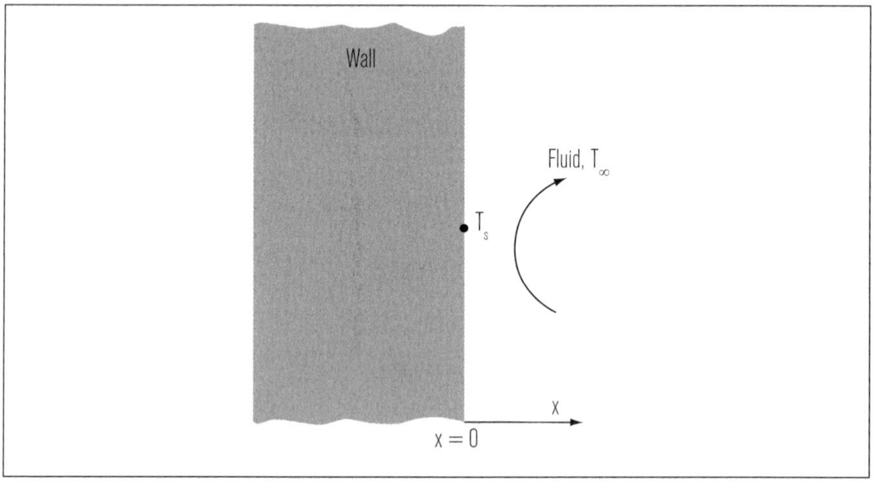

Figure 10.2 - Transfer of heat by convection.

Some relevant points related to the equation (10.4) can be highlighted. Initially it is important to mention that in the equation (10.4), the term, heat transfer coefficient (h) depends on a number of parameters such as physical properties of the fluid as well as solid geometry and actually represents the resistance to transfer of heat through the boundary layer between them. The heat transfer coefficient is usually determined experimentally and expressed in the form of correlations arrived at through dimensional analysis of a particular system. Expressions of such type for estimation of this coefficient are given in Chapter 13.

Another important aspect concerning the equation (10.4) is the order in which the temperatures T_s and T_∞ appear. The choice of this order is linked to the direction that is given to the x-axis. In the situation shown in Figure 10.2, if T_s is greater than T_∞, heat should flow from surface of the solid to the fluid. Thiat is the same direction of growth of x-axis, so the heat flow is positive, consistent with what is shown in equation (10.4).

It should be noted that the heat transfer coefficient is always a positive number.

If the direction of growth of the x-axis in Figure 10.2 is reversed, the order of temperatures T_s and T_∞ in equation (10.4) should also be reversed to maintain consistency with the convention for the sign of heat flux (positive when this occurs in the same direction of growth of position along the x-axis).

The table 10.1 gives the magnitude of values of the heat transfer coefficient (INCROPERA and DEWITT, 2003). For natural and forced convection as well as those involving phase changes for gases and liquids.

Type of convection	h ($W/m^2.K$)
Natural	
Gases	2-25
Liquids	50-1000
Forced	
Gases	25-250
Liquids	100-20000
Convection with phase change	
Boiling/condensation	2500-100000

Table 10.1 - Values for the coefficient of heat transfer (INCROPERA and DEWITT, 2003).

10.1.3 Thermal radiation

The third mechanism for heat transport is thermal radiation. The thermal radiation has very different characteristics compared to conduction and convection.The first difference is associated to the transport of energy. In the thermal radiation, heat is transported by electromagnetic waves. Hence, unlike the mechanisms cited for conduction and convection, radiation does not need a material medium to be transported. The heat transfer by radiation occurs more efficiently in a vacuum.

The other difference is associated with the dependence on temperature in the case of conduction and convection. In these cases, the heat flow is a linear function of the temperature difference between the regions or areas concerned. In thermal radiation, this flow depends on the difference of the fourth power of temperature.

According to the Stefan-Boltzmann law, the maximum power output of energy radiated by a body or surface can be determined by equation (10.5):

$$E_n = \sigma . T_s^4 \tag{10.5}$$

where:

- E_n = maximum emitting power of energy by radiation (W/m^2);
- σ = Stefan-Boltzmann constant = 5.67 x 10^{-8} $W/m^2.K^4$;
- T_s = absolute temperature of the body surface (K).

The surface that has a transmitting power equal to the maximum indicated by equation (10.5) is called an ideal radiator or black body.

Normal surfaces have transmitting power less than the black body. This is given by the following equation:

$$E_n = \varepsilon . \sigma . T_s^4 \tag{10.6}$$

where:

- ε = emissivity of the surface.

Emissivity is a property of the surface. Its value is in the range 0 to 1 and depends on the type of material and its surface finish. The emissivity may also vary with the temperature of the surface. In addition to reflection of radiation energy, a surface can also receive energy by radiation. The total amount of energy that falls on the surface by radiation is generally designated by the letter G.

The total energy that falls on a surface by radiation, a portion is absorbed, another reflected and a third portion is transmitted. Considering these, one can write the following relationship:

$$G = \alpha . G + \rho . G + \tau . G \tag{10.7}$$

where:

- G = radiation (W/m^2);
- α = absorbivity or fraction of incident energy absorbed the surface;
- ρ = reflectivity or fraction of incident energy reflected from the surface;

- τ = transmissivity or fraction of incident energy transmitted through the surface.

Dividing the equation (10.7) by G, we get:

$$\alpha + \rho + \tau = 1 \qquad (10.8)$$

Many of the solids are opaque, ie, they do not transmit energy by radiation. Therefore, for these bodies it can be written:

$$\alpha + \rho = 1 \qquad (10.9)$$

$$\tau = 0 \qquad (10.10)$$

Of the three parts mentioned above, only that which is absorbed by the body is taken into account for energy balance.

Energy exchange by radiation is of considerable importance. It is quite common, that a body is surrounded completely by a neighborhood having a different temperature. In this case, the net heat exchange by radiation between the body and the neighborhood can be expressed by:

$$q_{rad} = \varepsilon.\sigma.(T_s^4 - T_{viz}^4) \qquad (10.11)$$

where:

- q_{rad} = net heat exchange by radiation between the body and the neighborhood (W/m^2);
- T_{viz} = temperature of the neighborhood (K).

Again, the order of temperatures given in the equation (10.11) can be reversed depending on the convention adopted for the coordinate axes.

Equation (10.11) can be written in a manner similar to heat transfer by convection:

$$q_{rad} = h_{rad}.(T_s - T_{viz}) \qquad (10.12)$$

where h_{rad} is defined by equation (10.13) and significantly depends on the temperatures involved:

$$h_{rad} = \varepsilon.\sigma.(T_s^2 + T_{viz}^2).(T_s + T_{viz}) \qquad (10.13)$$

10.2 Energy Balance

The first law of thermodynamics is essesntially the principle of conservation of energy. Based on this, one can carry out an energy balance of a system, as discussed in the previous chapters. The resulting equations from these balances constitute the basis for solving the problems in both the areas of thermodynamics and transport phenomena. As has already been stated, the fundamental difference between the two approaches is that thermodynamics deals with equilibrium states and does not consider the rates of energy transfer, whereas transport phenomena deals with the rates, mechanisms and process dynamics. The final objective is the solution of engineering problems using the principles of both the areas.

In establishing tenergy balances various forms of energy should be considered. However in this treatment of heat transfer, one deals with situations, where only the thermal energy is relevant.

Two types of energy balances, equally important, will be considered:

- energy balance in a control volume;
- energy balance based on surface area.

10.2.1 Energy balance in a control volume

The equation for thermal energy balance can be written in the following general form:

$$\begin{aligned}&[\text{Rate of energy input}]-[\text{Rate of energy output}]+\\&[\text{Rate of generation of energy}]=[\text{Rate of energy accumulation}]\end{aligned} \qquad (10.14)$$

The first step in implementing the energy balance is to select a control volume, which is different for different coordinate systems. The input as well as output of energy occurs at the boundary of the control volume and are functions of the surface area, which again are different for rectangular, cylindrical and spherical coordinates. .These energy balances are usually associated with transport of heat by one or more of the mechanisms presented earlier (conduction, convection and radiation). The input and output of mass along with the associated energy through the relevant area of the contour of the control volume has also to be considered in respect of the transport of a particular specie.

The contribution of heat generation due to chemical reactions in a system or the conversion of a given form of energy into heat energy has to be also considered. The usual forms of heat generation are associated with:

- conversion of electrical energy into heat energy. This can be estimated by multiplying the electrical resistance of the material (R) by electric current, (I) squared (This rate is always positive and contributes to the generation of energy in the control volume);
- conversion of chemical energy into heat energy. This term is associated with the occurrence of chemical reactions in the control volume, which may be positive or negative, depending on whether the reaction is exothermic or endothermic. This contribution is assessed by multiplying the enthalpy of the reaction by the rate of reaction determined from studies of reaction kinetics. This item includes absorption or release of heat energy due to phase changes (latent heat). Heat generation and the accumulation terms in the balance equations are usually evaluated.on a volumetric basis in the control volume.In the case of heat balances, the main feature is the temperature and the variation of temperature with time is an important parameter in many operations or transient situations. In problems involving heat transport in the steady state, the accumulation term becomes zero. An example of application of energy balance is given below:

Example

An electric current passes through a wire whose diameter is D and electrical resistance R per unit length. The wire is in contact with a fluid at a temperature T_∞ and is surrounded by a neighborhood at a temperature T_{viz}. Considering that the temperature inside the wire is uniform, develop an equation to predict the temperature variation with time, from the moment , when the electrical current starts passing. Given: h is the heat transfer coefficient for the wire in contact with the fluid and ε is the emissivity of the wire.

Solution

The temperature varies in the wire as a function of time and this variation can be determined starting with an energy balance. From the equation (10.14), we have:

- Rate of energy input, Q_{ent}.

There is no energy input on the wire. Thus:

$$\left[\text{Rate of energy input}\right] = Q_{ent} = 0$$

- Rate of energy output, Q_{out}.

Considering the heat exchanges by convection and radiation (assuming that the neighborhood completely encloses the wire), we have:

$$\left[\text{Rate of output of energy}\right] = Q_{out} = \pi.D.L.h.(T_a - T_\infty) + \pi.D.L.\varepsilon.\sigma.(T_a^4 - T_{viz}^4)$$

In the above equation, the product π.DL corresponds to the surface area of the wire, where L is the length and D its diameter. T_a is the temperature of the wire.

- Rate of energy generation, Q_{ger}.

This generation is due to the conversion of electrical energy into heat by Joule effect. This value is calculated by the following equation:

$$\left[\text{Rate of energy generation}\right] = Q_{ger} = R^*.L.I^2$$

The product R^*.L gives the total electrical resistance of the wire of length L.

- Rate of accumulation of energy, Q_c.

The rate of accumulation of energy is associated with variation in temperature of the wire over time. This term can be evaluated using the following equation:

$$\left[\text{Rate of acummulation of energy}\right] = Q_{ac} = \frac{\pi D^2}{4}.L.\rho.C_p.\frac{dT_a}{dt}$$

In this equation, the product $\pi D^2/4$. L corresponds to the volume of the wire.

Combining all parts of the energy balance, we have:

$$0 - \pi.D.L.h.(T_a - T_\infty) - \pi.D.L.\varepsilon.\sigma.(T_a^4 - T_{viz}^4) + R^*.L.I^2 = \frac{\pi D^2}{4}.L.\rho.C_p.\frac{dT_a}{dt}$$

The rate of change of temperature is given by:

$$\frac{dT_a}{dt} = \frac{-\pi.D.h.(T_a - T_\infty) - \pi.D.\varepsilon.\sigma.(T_a^4 - T_{viz}^4) + R^*.I^2}{\frac{\pi D^2}{4}.\rho.C_p}$$

The equation above, with the terms of radiation and heat generation, cannot be solved analytically. Hence a simple numerical technique is used to evaluate the temperature variation with time. The principle of this technique is based on the concept of first derivative, shown by the equation below:

$$\frac{dT_a}{dt} = \lim_{\Delta t \to 0}\left[\frac{T_a(t+\Delta t) - T_a(t)}{\Delta t}\right]$$

For Δt small enough, one can make the following approximation:

$$\frac{dT_a}{dt} \approx \left[\frac{T_a(t+\Delta t) - T_a(t)}{\Delta t}\right]$$

With this approach, the equation for the rate of change of temperature of the wire can be written as follows:

$$\frac{T_a(t+\Delta t) - T_a(t)}{\Delta t} = \frac{-\pi.D.h.\ [T_a(t) - T_f)] - \pi.D.\sigma.\varepsilon.[(T_a^4(t) - T_{viz}^4)] + R_e.I^2}{\frac{\pi.D^2}{4}.L.\rho.C_p.}$$

or:

$$T_a(t+\Delta t) = T_a(t) + \Delta t.\left\{\frac{-\pi.D.h.\ [(T_a(t) - T_f)] - \pi.D.\sigma.\varepsilon.[(T_a^4(t) - T_{viz}^4)] + R_e.I^2}{\frac{\pi.D^2}{4}.L.\rho.C_p.}\right\}$$

Through this numerical technique it is possible to calculate the temperatures at time instants separated by an interval Δt, and no longer continuously as is usual in analytical solutions (T as a function time).

In the first step, the initial temperature ($t = 0$) is used to calculate the temperature at time $t = 0 + \Delta t$. This new temperature is then used to calculate the temperature at $t = 0 + 2.\Delta t$. This procedure is repeated successively until one reaches the desired simulation time.

To get precision in the calculations, it is essential that the values of Δt are quite small. However for values of Δt extremely small, time spent in the calculations become extremely high. To determine the appropriate value of Δt in order to provide accurate results without excessive computation time, one should perform calculations with successively smaller Δt. From a certain point, reduction in the values of Δt does not cause significant changes in the values of temperature, with the increase in time consumed in the calculations. This Δt is chosen for the calculations.

The *wire*.xls spreadsheet (in the CD,that accompanies this book), presents the calculations of temperature of the wire, specifying the values of R * D, h, ε, I, ρ and Cp.The figure below gives the heating curves of the wire for different values of Δt, in accordance with the discussion above, regarding the proper choice of value of Δt. In the numerical solution, the figure shows that for the values Δt of 1 s and 0.1 s, the deviation of temperature is very little compared to larger times. Hence this should be the region of choice.

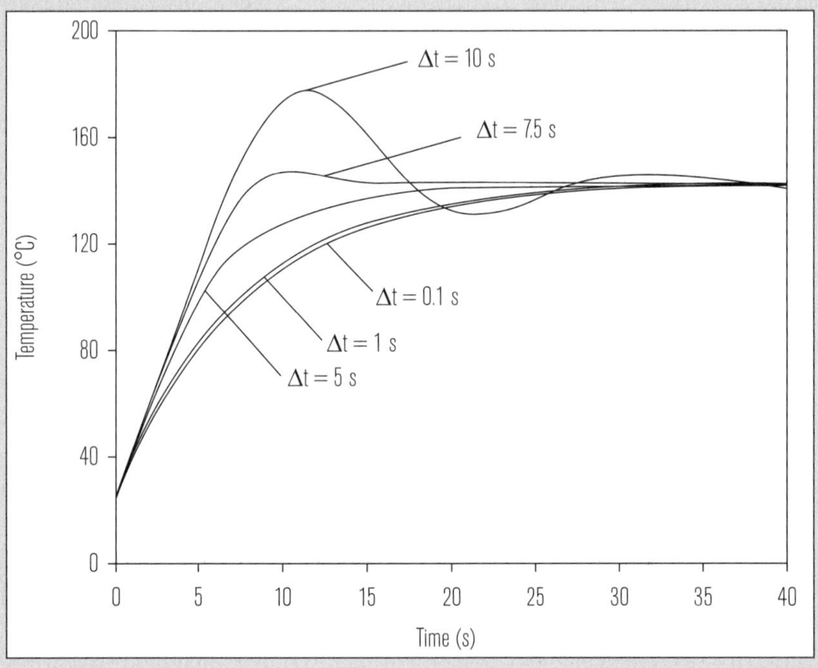

It is also observed, that for large times, the temperature of the wire stabilizes. This is the temperature corresponding to the steady state (no more accumulation of energy). The value of this temperature can also be determined by solving the equation of energy balance, but assuming that the accumulation term is zero:

$$R^*.I^2 = \pi.D.h.(T_a - T_\infty) + \pi.D.\varepsilon.\sigma.(T_a^4 - T_{viz}^4)$$

This equation can be solved using the SOLVER command in the spreadsheet. This feature is also illustrated in the worksheet *wire.xls*.

(Can the energy balance developed in the previous example be used to simulate the cooling of the wire after heating is stopped and also for heating of the wire if it is placed inside a furnace.? Explain you answers)

Example

Estimate the power consumption of a reheating furnace of steel plates for following situation. Plates come into the furnace at 300 deg.K and leave at 1400 deg.K. The speed of displacement of the plates is 1 m/min; cross section of the plates is 1.5 m in width (l) and 0.20 m thick (e). The temperature of the outer wall of the furnace , T_a = measured value of 400 K. The heat transfer oefficient for the walls is 8 W/m 2.K and emissivity is 0.7. The dimensions of the furnace are:

- length = 30 m;
- width = 5 m;
- height = 3 m.

Consider that the density of steel is 7500 kg/m^3 and its average specific heat in the specified temperature range is 750 J/kg.K. Assume that heat losses through the bottom of the furnace are negligible. Air temperatures in the vicinity of the furnace is 298 K.

Solution

The power consumption can be estimated from an energy balance for the reheating furnace.

Since the furnace is operating in steady state, we have:

$$[\text{Rate of energy input}] - [\text{Rate of energy output}] + [\text{Rate of generation of energy}] = 0$$

The first 2 terms can be evaluated from the data given and the unknown is the rate of generation of energy associated with the conversion of electrical energy into heat.

The rate of energy input, Q_{ent}, is associated with the steel that goes into the furnace temperature T_{ent} (300 deg.K). Assuming that T_{ref}

is the reference temperature for the assessment of energy contained in the plates, we have:

$$[\text{Rate of input of energy}] = Q_{ent} = \Gamma.C_p.(T_{ent} - T_{ref})$$

where:

- Γ = mass of steel that enters the furnace per unit time (kg/s);
- C_p = average specific heat of steel (J/kg.K).

The mass of steel that goes into the furnace per unit time can be assessed by the following equation:

$$\Gamma = v_p.\ l.\ e.\ \rho$$

where:

- v_p = velocity of displacement of the plates in the oven (m/s);
- l = width of plates (m);
- e = thickness of plate (m);
- ρ = density of steel (kg/m^3).

Combining the two equations above, we obtain:

$$Q_{ent} = v_p.\ l.\ e.\ \rho\ .C_p.(T_{ent} - T_{ref})$$

The rate of heat output has two parts:

- energy entering the steel that comes from inside the furnace;
- thermal losses through the walls of the furnace, including convection to air and radiation to the neighborhood.

This rate of output can be estimated by the following equation:

$$[\text{Rate of output of energy}] = Q_{out} = v_p.l.e.\rho.C_p.(T_{out} - T_{ref}) + A_t.[h.(T_a - T_\infty) + \varepsilon.\sigma.(T_a^4 - T_{viz}^4)]$$

where:

- A_t = total area of the outer surface of the furnace;
- T_{out} = outlet temperature of the steel from the furnace;
- $T_{viz} = T_\infty$ = ambient temperature;
- T_a = temperature of the outer wall of the furnace;
- T_{ref} = reference temperature for the assessment of energy contained in the plates.

The total area of the outer surface of the furnace can be determined by the following equation:

$$A_t = 2.[C_f.H_f + L_f.H_f] + C_f.L_f$$

where:

- C_f = length of the furnace (m);
- H_F = height of the furnace (m);
- L_f = width of the furnace (m).

Replacing these in the equation of energy balance, we have:

$$v_p.l.e.\rho.C_p.(T_{ent} - T_{ref}) - v_p.l.e.\rho.C_p.(T_{out} - T_{ref}) -$$
$$A_t.[h.(T_a - T_\infty) + \varepsilon.\sigma.(T_a^4 - T_{viz}^4)] + Q_{elet} = 0$$

where: Q_{elet} is the power consumption of the furnace.

Transposing terms in the equation above, we have:

$$Q_{elet} = v_p.l.e.\rho.C_p.(T_{out} - T_{ent}) + A_t.[h.(T_a - T_\infty) + \varepsilon.\sigma.(T_a^4 - T_{viz}^4)]$$

For the above equation, the choice of reference temperature does not affect the final result.

The total area of the outer surface of the furnace is:

$$A_t = 2.[(30).(3) + (5).(3)] + (30).(5) = 360 \text{ m}^2$$

Returning to the equation of energy balance, we obtain:

$$Q_{elet} = (1/60).(1.5).(0.2).(7{,}500).(750).(1400\text{-}300) +$$
$$(360).[(8).(400 - 298) + (0.7).(5.67 \,.\, 10^{-8}).[(400)^4 - (298)^4)]$$

$$Q_{elet} = 30{,}937{,}500 + 546.862{,}6 \quad W = 31{,}484{,}362.6 \quad W = 31.5 \text{ MW}$$

10.2.2 Energy balance based on surface areas

In many situations involving heat transfer becomes necessary to establish the energy balance for surfaces.

$$\text{Rate of energy input} - \text{Rate of energy output} = 0 \qquad (10.15)$$

As mentioned in the item of energy balance for control volumes, the generation and accumulation of energy are volumetric parameters. For surface areas, the terms in the equation are similar to the energy balance for volumes, but with the exclusion of terms of heat generation and energy storage. It is important to note that the above equation is valid for steady and transient regimes. The example below illustrates this type of energy balance

Example

A manufacturer of refractory want to evaluate the working conditions of bricks that are being used in microwave reheating of billets in a steel mill. The temperature of the external face of the brick, which is in contact with air, is 150°C. under steady state conditions. The thickness of the brick is 0.23 m and its thermal conductivity is 1 W/K. The coefficient of heat transfer from the brick to the air is $h = 15\ W/m^2.K$ and emissivity of the brick is 0.75. The temperature of air and the neighborhood is 25°C, Estimate the temperature of the inner surface of the firebrick.

Solution

The figure below illustrates the data given for temperature conditions of the brick.

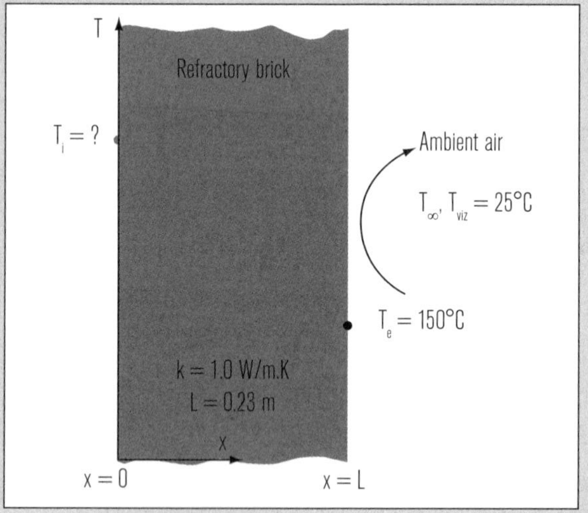

To calculate the temperature T_i, one can make an energy balance for the outer surface of the brick. For this surface, one can write:

$$[\text{Rate of heat energy entering the surface}] - [\text{Rate of heat energy leaving the surface}] = 0$$

Heat reaches the surface by conduction through the brick. Assuming a linear temperature profile, the flow of heat that reaches the surface is given by:

$$[\text{Rate of heat entry at the surface}] = q_e = -k\frac{dT}{dx} = k\frac{(T_i - T_e)}{L}$$

The heat flow coming out of the surface is the sum of the flows by convection to air and radiation to the neighborhood:

$$[\text{Rate of heat exit at the surface}] = q_s = h.(T_e - T_\infty) + \sigma.\varepsilon.(T_e^4 - T_{viz}^4)$$

Replacing data in the expressions and putting all temperatures in K, we have:

$$q_s = (15).(423 - 298) + (5.67 \times 10^{-8}).(0.75).[(423)^4 - (298)^4] = 2901.1 \text{ W/m}^2$$

Equating inflows and outflows of heat:

$$(1).\frac{(T_i - 423)}{(0.23)} = 2901.1$$

to obtain:

$$T_i = 1{,}090.3 \text{ K} = 817.3°\text{C}$$

Temperature in the firebrick varies from 817.3°C at the hot face to 150°C at the cold surface.

References

INCROPERA, F.P.; DEWITT, D.P. *Fundamentos da transferência de calor e massa*. 5. ed. Rio de Janeiro: LTC, 2003.

Exercises

1 - One of the surfaces of a wall of steel, 10 mm thick (k = 15 W/m.K) is maintained at 90°C by condensation of water vapor, the opposite face is exposed to a stream of air at 20°C and h = 25 W/m^2.K. What is the temperature of the surface in contact with air? What is the heat flow in this area?

2 - The heat transfer coefficient by natural convection in the case of a vertical plate suspended in air can be measured by determining the temperature variation of the plate as a function of time during cooling. Assuming that the temperature inside the plate is uniform and that radiation is negligible, evaluate the heat transfer coefficient at the moment the plate temperature reaches 225°C The rate of change of temperature (dT/dt) is -0.022 deg.K/s. The air temperature is 25°C. The height and width of the plate amount to 0.3 m. The thickness is 1 mm. and the mass of the plate is 3.75 kg .Its specific heat is 2770 J/kg deg.K.

3 - A ladle containing liquid steel is transported from the steel refining unit to the continuous casting unit. This takes 5 minutes. Estimate the temperature drop of steel during this time. The ladle is approximately cylindrical shape with a diameter of 3 m and height of 3.3 m. Consider also the following data:

- steel temperature at the exit of the installation of refinement: 1,580°C;
- emissivity of liquid steel: 0.28;
- heat transfer oefficient at the upper surface of the steel: 10 W/m^2.deg.K;
- thermal conductivity of liquid steel: 35 W/m deg.K;
- specific heat of liquid steel: 680 J/kg.deg.K;
- density of liquid steel: 7020 kg/m^3;
- refractory lining: thickness of 30 cm and of extremely low thermal conductivity. The lining is in thermal equilibrium with the steel.

List the assumptions made in the calculations.

4 - The five walls (ceiling and the four side faces) of a microwave reheating of ingots were built using the configuration shown in the figure below:

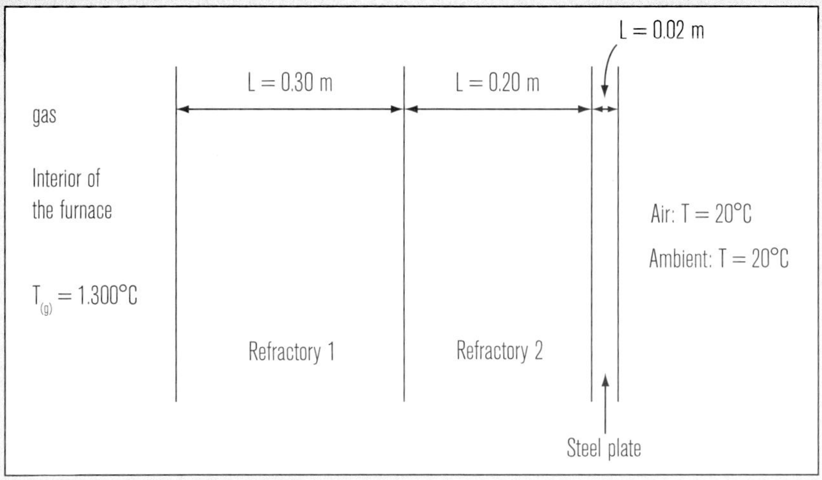

The temperature of the outer surface of the steel sheet is uniform and is equal to 300°C. Estimate the rate of heat loss from the furnace (including radiation). Assume that the bottom of the furnace is adiabatic. The temperature inside the furnace is 1,300°C. The emissivity of the steel sheet is 0.8 and the furnace has length of 10 m, width 5 m and height of 3 m. Calculate the temperature of the inner side of the furnace in the vertical walls.

The coefficient of heat transfer in all external vertical surfaces 5 W/m^2.K and the ceiling is 12 W/m^2·deg.K.

Data for the refractory and steel sheet:

- Refractory 1: density = 2500 kg/m^3, specific heat = 700 J/kg deg.K, thermal conductivity = 5 W/m deg.K;
- Refractory 2: density = 2800 kg/m^3, specific heat = 800 J/kg.deg.K, thermal conductivity = 2 W/m. deg.K;
- Steel: density = 7800 kg/m^3, specific heat = 670 J/kg.deg.K, thermal conductivity = 35 W/m. deg.K.

5 - In the hot rolling of steel plates the thickness is reduced by passage through rolls. The steel in the form of strip go out through series of rolls and is cooled before being wound. The cooling of the strip is done in several steps, including cooling with jets of water. For the production of strip steel (density = 7800 kg/m^3, specific heat = 950 J/kg.deg.K), the operating conditions of the process correspond to an outlet temperature of the rolling rolls at 1000°C, water temperature in the jets of 25°C and a speed of the strip of 12 m/s. The width and thickness of the strips are respectively 1.8 m 2 mm. What is the average rate with which heat must be extracted from the strip so that its temperature to be wound should be 500°C the extent of cooling line is 100 m. Identify all the mechanisms of heat transfer that contribute to the cooling of the strip.

6 - A crucible is used in the transport of liquid aluminum from the furnace until the unit of strip casting. The transport time is 2 minutes. Estimate the temperature drop of aluminum on the way. The crucible is rectangular with a width of 20 cm and length 18 cm. The height of aluminum is 10 cm. Consider also the following data:

- temperature of aluminum in the furnace exit: 720°C;
- air temperature and the neighborhood 25 deg.C;
- emissivity of liquid aluminum: 0.12;
- heat transfer coefficient of the upper surface of aluminum: 20 W/m^2.deg.K;
- thermal conductivity of liquid aluminum: 200 W/m deg.K;
- specific heat of liquid aluminum: 1200 J/kg.deg.K;
- density of liquid aluminum: 2388 kg/m^3;
- crucible has a lining thickness of 1.5 cm and of extremely low thermal conductivity.The lining is in thermal equilibrium with the aluminum.

List the assumptions made in the calculations.

Steady State Conduction

Chapter 11

11.1 Fourier's Law of Conduction

As mentioned in the previous chapter, the transfer of heat by conduction occurs when there is a temperature gradient in a given environment. The flow of heat by conduction is calculated by Fourier's law. As this flow is a vector, it can be calculated for the three orthogonal directions. In cartesian coordinates, we can write:

$$q_x = -k_x \frac{\partial T}{\partial x} \tag{11.1}$$

$$q_y = -k_y \frac{\partial T}{\partial y} \tag{11.2}$$

$$q_z = -k_z \frac{\partial T}{\partial z} \tag{11.3}$$

where:

- q_x, q_y, q_z = heat flows in orthogonal directions x, y and z, respectively (W/m^2);
- k_x, k_y, k_z = thermal conductivity in directions x, y and z (W/m.K);
- T = temperature (K or °C).

Anisotropic materials have different thermal conductivities along the three orthogonal directions. For isotropic materials, the thermal conductivity values in all the three directions are equal. The treatment in this chapter is restricted to only isotropic materials. Thermal conductivity is an important property of materials, which together with the temperature gradient determines the rate of heat transfer by conduction.

11.1.1 Thermal conductivity

Thermal conductivity of a material depends on a number of factors. such as the state of a material. Solids have higher thermal conductivity values than liquids, which in turn, have higher conductivities than gases.

Pure metals have higher thermal conductivity compared to the alloys. In general, thermal conductivities of metals and alloys are higher than those of non-metallic materials.

For the same material, the temperature can also have a significant effect on the value of thermal conductivity. This effect varies depending on the nature of the material. For metals, there is a general trend of decreasing conductivity with increasing temperature. However, changes in crystal structure can change this trend. For gases, the thermal conductivity tends to increase with increasing temperature.

Solid materials having crystalline structures, conduct heat more rapidly than amorphous materials. The pores in the material tend to diminish the value of thermal conductivity. The filling of pores with gas also affects the value of conductivity. This effect depends on the type of gas that fills the pores. Gases, that are better conductors, tend to mitigate the effect of the presence of pores. Porous solids are generally poor conductors of heat, making them suitable for thermal insulation.

The Figures of 11.1 to 11.3 show the values of thermal conductivity as a function of temperature for different materials. Some theories have been proposed for the estimation of thermal conductivity of materials (GEIGER and POIRIER, 1980). Theories for solids and liquids are still limited and provide only estimates of the order of magnitude. For gases, there is an equation for relating the viscosity to the thermal conductivity (GEIGER and POIRIER, 1980):

$$k = \mu . \left(C_p + \frac{1.25.R}{M} \right) \tag{11.4}$$

where:

- k = thermal conductivity of the gas (W/m.K);
- μ = viscosity of gas (Pa.s);
- C_p = specific heat of gas at constant pressure (J/kg.K);
- R = gas constant (J/mol.K);
- M = molecular weight of gas (kg/mol).

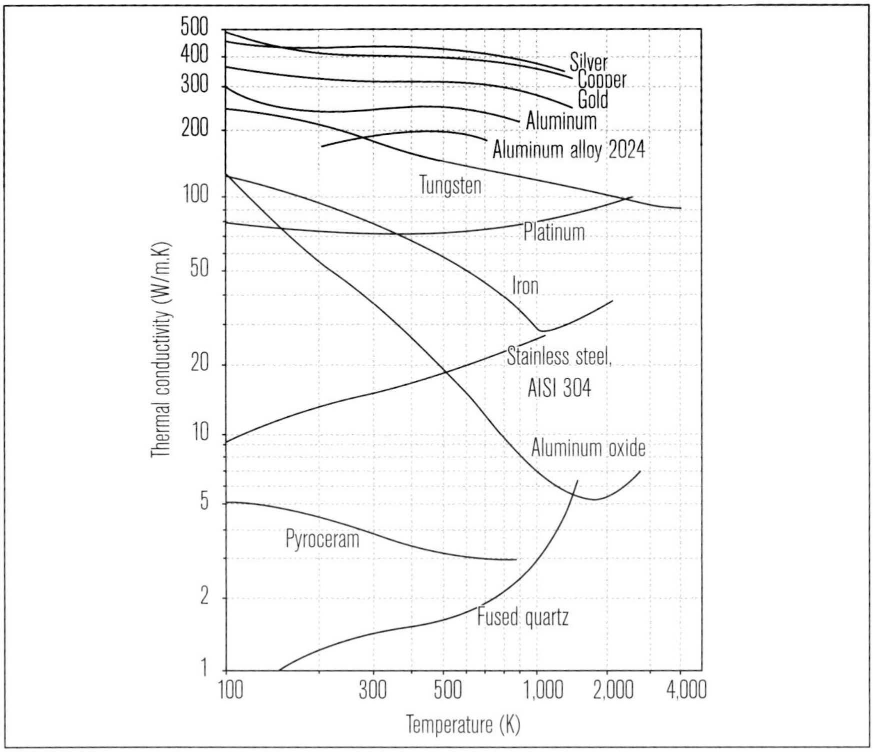

Figure 11.1 - Thermal conductivity of some solid materials (INCROPERA and DEWITT, 2003)

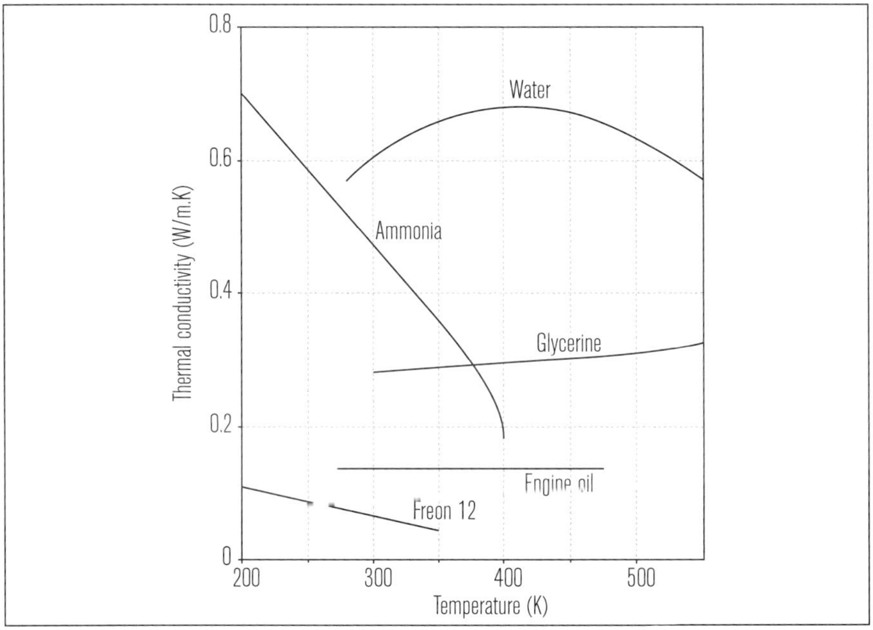

Figure 11.2 - Thermal conductivity of some liquids (INCROPERA and DEWITT, 2003).

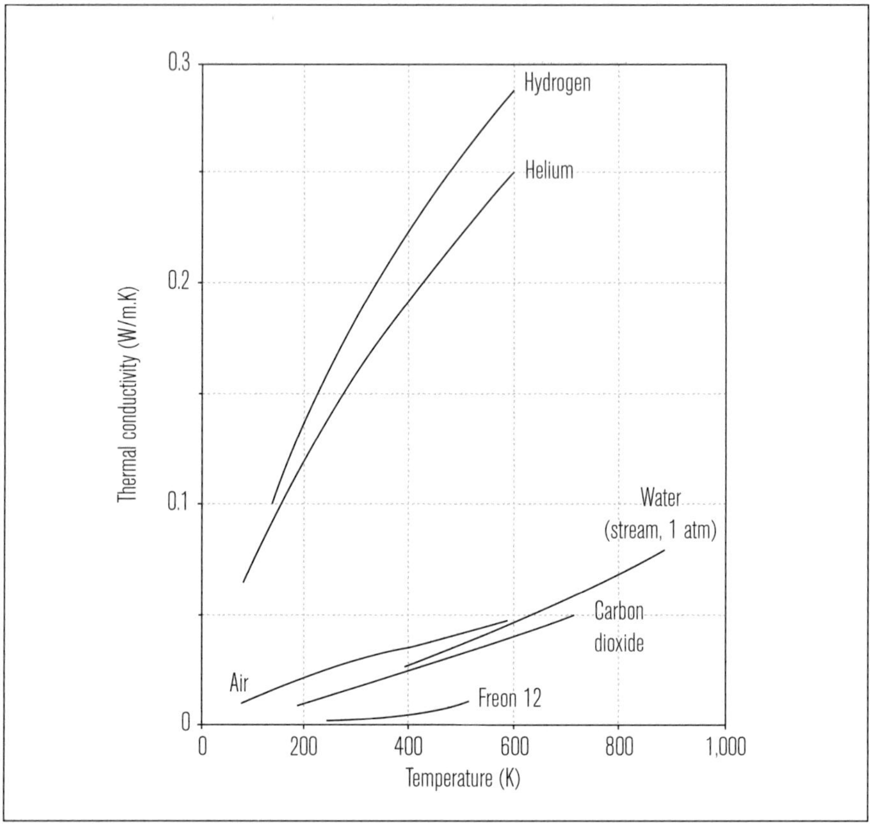

Figure 11.3 - Thermal conductivity of some gases (INCROPERA and DEWITT, 2003).

For the mixture of gases, thermal conductivity can be estimated from the chemical composition of the mixture and the and the individual conductivity of the gases that make up the mixture like in the case of viscosity of gases:

$$k_m = \frac{\sum_{i=1}^{n} x_i.k_i.M_i^{1/3}}{\sum_{i=1}^{n} x_i.M_i^{1/3}} \quad (11.5)$$

where:

- n = number of components in the gas mixture;
- x_i = mole fraction of gas i;
- M_i = molecular weight of gas i;
- k_i = thermal conductivity of gas i.

Estimation of the thermal conductivity of gases by the above equations deviate by less than 5% from the values determined experimentally. Figure 11.4 presents a summary of the ranges of conductivity for different materials.

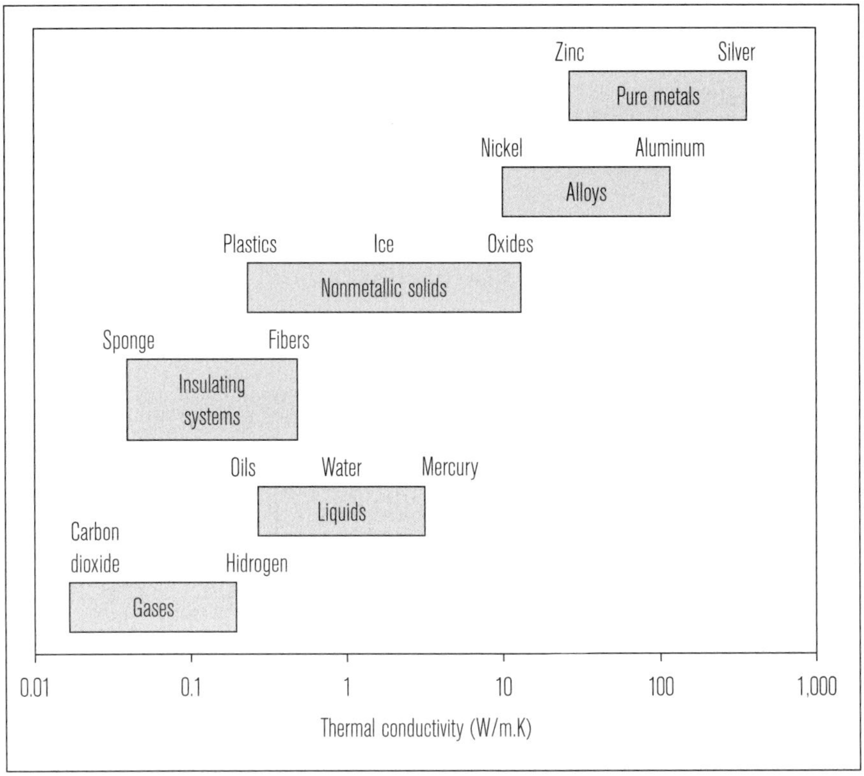

Figure 11.4 - Values of thermal conductivity for various materials (INCROPERA and DEWITT, 2003).

Another parameter of the materials, important in heat transfer is the thermal diffusivity, defined by:

$$\alpha = \frac{k}{\rho . C_p} \tag{11.6}$$

where:

- α = thermal diffusivity (m^2/s).

The thermal diffusivity is a property of the material that is related to its ability to transfer and store energy.

11.2 General Equations of Conduction

Determination of temperature distribution within a solid body, is of great relevance and practical importance especially in applications such as heat treatment of metals etc. The starting point is to obtain a mathematical equation that describes this variation as a function of position and time. This can be obtained by developing an energy balance in an infinitesimal volume element within the solid. As in the previous cases discussed, , the volume element is chosen depending on the geometry of system (cartesian, cylindrical or spherical coordinates) and one can go ahead with the energy balance.

11.2.1 Energy balance in cartesian cordinates

The energy balance is developed for the infinitesimal volume element shown in Figure 11.5.

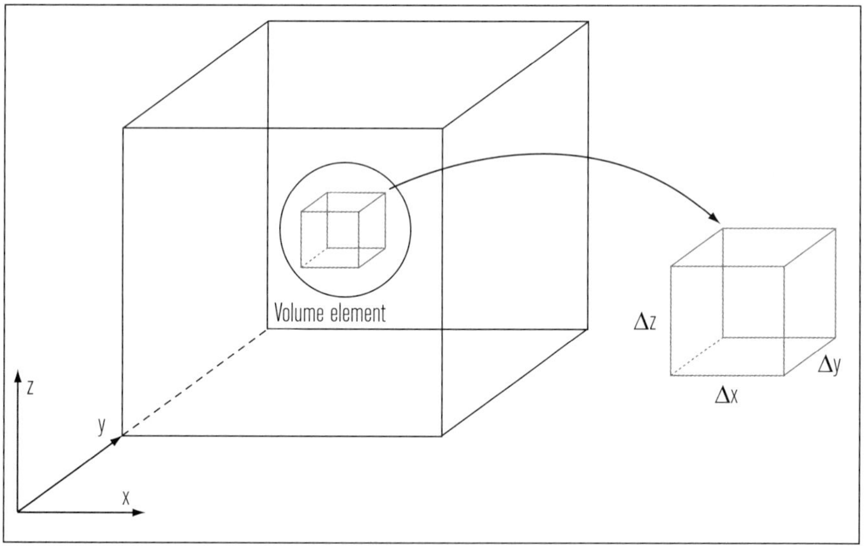

Figure 11.5 - Volume element in Cartesian coordinates.

To develop the energy balance one considers the case of an isotropic material twith temperature gradients in three orthogonal directions and the temperature can also vary withr time.

The energy balance for the volume element shown in Figure 11.5 can be expressed by the following equation:

$$[\text{Rate of energy input}] - [\text{Rate of energy output}] + [\text{Rate of energy generation}] = [\text{Rate of accumulation of energy}] \quad (11.7)$$

To assess the rates of input and output of energy, the heat flow is broken down into its three orthogonal components, q_x, q_y and q_z. The faces of the cubical volume element are considered for inputs and outputs of energy and these are chosen according to the orientation of the coordinate axes. Inputs are evaluated on the faces in position x, y and z and outputs the faces located at, $x + \Delta x$,$y+ \Delta y$ and $z + \Delta z$.

- Rate of energy input:

$$[\text{Rate of input of energy}] = (\Delta y.\Delta z.q_x)|_x + (\Delta x.\Delta z.q_y)|_y + (\Delta x.\Delta y.q_z)|_z \quad (11.8)$$

- Rate of energy output:

$$[\text{Rate of output of energy}] = (\Delta y.\Delta z.q_x)|_{x+\Delta x} + (\Delta x.\Delta z.q_y)|_{y+\Delta y} + (\Delta x.\Delta y.q_z)|_{z+\Delta z} \quad (11.9)$$

- Rate of energy generation:

$$[\text{Rate of generation of energy}] = \Delta x.\Delta y.\Delta z.\dot{q} \quad (11.10)$$

- Rate of accumulation of energy:

$$[\text{Rate of accumulation of energy}] = \Delta x.\Delta y.\Delta z.\rho.C_p.\frac{T(t+\Delta t) - T(t)}{\Delta t} \quad (11.11)$$

where:

- q = rate of energy generation per unit volume (W/m^3);
- Δt = time interval (s).

Combining the different contributions above, we get:

$$\begin{aligned}&[(\Delta y.\Delta z.q_x)|_x + (\Delta x.\Delta z.q_y)|_y + (\Delta x.\Delta y.q_z)|_z] - \\ &[(\Delta y.\Delta z.q_x)|_{x+\Delta x} + (\Delta x.\Delta z.q_y)|_{y+\Delta y} + (\Delta x.\Delta y.q_z)|_{z+\Delta z}] + \\ &\Delta x.\Delta y.\Delta z.\dot{q} = \Delta x.\Delta y.\Delta z.\rho.C_p.\frac{T(t+\Delta t) - T(t)}{\Delta t}\end{aligned} \quad (11.12)$$

The mathematical treatment developed in order to obtain the differential equations is similar to the case of rate of momentum in fluid flow discussed in earlier chapters.

Dividing equation (11.12) by the volume element:

$$\frac{q_x|_x - q_x|_{x+\Delta x}}{\Delta x} + \frac{q_y|_y - q_y|_{y+\Delta y}}{\Delta y} + \frac{q_z|_z - q_z|_{z+\Delta z}}{\Delta z} + \dot{q} = \rho.C_p.\frac{T(t+\Delta t) - T(t)}{\Delta t} \qquad (11.13)$$

Taking the limit of equation (11.13) for Δx, Δy, Δz and Δt tending to zero, we get:

$$-\frac{\partial q_x}{\partial x} - \frac{\partial q_y}{\partial y} - \frac{\partial q_z}{\partial z} + \dot{q} = \rho.C_p.\frac{\partial T}{\partial t} \qquad (11.14)$$

To get the equation (11.14) one should remember the definition of first derivative, as per the following equation:

$$\lim_{\Delta x \to 0} \frac{\left[(f)|_{x+\Delta x} - (f)|_x\right]}{\Delta x} = \frac{df}{dx} \qquad (11.15)$$

Substituting for heat flux terms with equations of Fourier's law, given earlier we have:

$$q_x = -k_x \frac{\partial T}{\partial x} \qquad (11.1)$$

$$q_y = -k_y \frac{\partial T}{\partial y} \qquad (11.2)$$

$$q_z = -k_z \frac{\partial T}{\partial z} \qquad (11.3)$$

Hence:

$$\frac{\partial}{\partial x}\left(k_x \frac{\partial T}{\partial x}\right) + \frac{\partial}{\partial y}\left(k_y \frac{\partial T}{\partial y}\right) + \frac{\partial}{\partial z}\left(k_z \frac{\partial T}{\partial z}\right) + \dot{q} = \rho.C_p.\frac{\partial T}{\partial t} \qquad (11.16)$$

This differential equation represents the energy balance for a volume element in a general form. Some simplified versions are given below.

For an isotropic material:

$$k_x = k_y = k_z = k \qquad (11.17)$$

Hence:

$$\frac{\partial}{\partial x}\left(k \frac{\partial T}{\partial x}\right) + \frac{\partial}{\partial y}\left(k \frac{\partial T}{\partial y}\right) + \frac{\partial}{\partial z}\left(k \frac{\partial T}{\partial z}\right) + \dot{q} = \rho.C_p.\frac{\partial T}{\partial t} \qquad (11.18)$$

When thermal conductivity is constant, this equation of conservation of energy can be written in the following way:

$$\alpha\left(\frac{\partial^2 T}{\partial x^2}+\frac{\partial^2 T}{\partial y^2}+\frac{\partial^2 T}{\partial z^2}\right)+\frac{\dot{q}}{\rho.C_p}=\frac{\partial T}{\partial t} \qquad (11.19)$$

where α is the thermal diffusivity of the material, defined by equation (11.6).

For conditions of steady state, the temperature is not a function of time and the equation of conservation of energy is simplified to:

$$\frac{\partial}{\partial x}\left(k\frac{\partial T}{\partial x}\right)+\frac{\partial}{\partial y}\left(k\frac{\partial T}{\partial y}\right)+\frac{\partial}{\partial z}\left(k\frac{\partial T}{\partial z}\right)+\dot{q}=0 \qquad (11.20)$$

When heat transfer occurs only in one direction (y) and there is no heat generation (q = 0) and for steady state conditions:

$$\frac{\partial}{\partial y}\left(k\frac{\partial T}{\partial y}\right)=0 \qquad (11.21)$$

Integration of the equation above, provides:

$$k\frac{\partial T}{\partial y}=C \qquad (11.22)$$

where C is a constant.

In this case, it is observed that the heat flux in the y direction is a constant.

Similar equations obtained for rectangular coordinates can be obtained for cylindrical and spherical coordinates following similar procedures.

11.2.2 Energy balance cylindrical coordinates

Figure 11.6 shows the volume element in cylindrical coordinates.

Fourier equations for cylindrical coordinates are expressed by:

$$q_r=-k\frac{\partial T}{\partial r} \qquad (11.23)$$

$$q_\theta=-\frac{k}{r}\frac{\partial T}{\partial \theta} \qquad (11.24)$$

$$q_z=-k\frac{\partial T}{\partial z} \qquad (11.25)$$

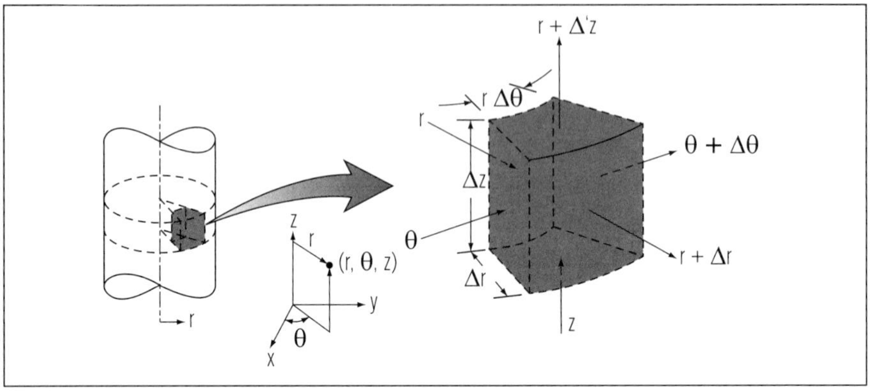

Figure 11.6 - Volume element in cylindrical coordinates (INCROPERA and DEWITT, 2003).

Following the procedure similar to that used for rectangular coordinates, one can obtain the equation of energy conservation in cylindrical coordinates in the volume element shown above:

$$\frac{1}{r}\frac{\partial}{\partial r}\left(k.r.\frac{\partial T}{\partial r}\right)+\frac{1}{r^2}\frac{\partial}{\partial \theta}\left(k.\frac{\partial T}{\partial \theta}\right)+\frac{\partial}{\partial z}\left(k\frac{\partial T}{\partial z}\right)+\dot{q}=\rho.C_p\frac{\partial T}{\partial t} \tag{11.26}$$

This equation can also be obtained substituting in the equation developed for in rectangular coordinates. The mathematical relationships between the two coordinate systems as given below.

$$x = r.\cos\theta \tag{11.27}$$

$$y = r.\sin\theta \tag{11.28}$$

$$z = z \tag{11.29}$$

11.2.3 Energy balance in spherical coordinates

Figure 11.7 shows a volume element in spherical coordinates.

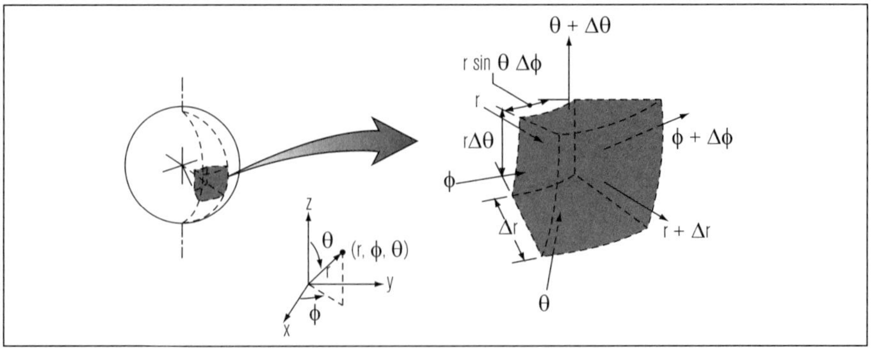

Figure 11.7 - Volume element in spherical coordinates (INCROPERA and DEWITT, 2003).

The equations of Fourier for spherical coordinates are:

$$q_r = -k\frac{\partial T}{\partial r} \quad (11.30)$$

$$q_\theta = -\frac{k}{r}\frac{\partial T}{\partial \theta} \quad (11.31)$$

$$q_\varphi = -\frac{k}{r\sin\theta}\frac{\partial T}{\partial \varphi} \quad (11.32)$$

The differential equation of energy conservation for the volume element in spherical geometry can be obtained from the energy balance (as in the case of rectangular geometry) or incorporating the following definitions in the equationdeveloped for cartesian coordinates namely equation (11.16) and the equations below relating both the coordinates.

$$x = r.\sin\theta.\cos\varphi \quad (11.33)$$

$$y = r.\sin\theta.\text{sen}\varphi \quad (11.34)$$

$$z = r.\cos\theta \quad (11.35)$$

The result is:

$$\frac{1}{r^2}\frac{\partial}{\partial r}\left(k.r^2.\frac{\partial T}{\partial r}\right)+\frac{1}{r^2\sin^2\theta}\frac{\partial}{\partial \varphi}\left(k.\frac{\partial T}{\partial \varphi}\right)+\frac{1}{r^2\sin\theta}\frac{\partial}{\partial \theta}\left(k.\sin\theta.\frac{\partial T}{\partial \theta}\right)+\dot{q}=\rho.C_p\frac{\partial T}{\partial t} \quad (11.36)$$

To determine the temperature distribution in solids it is necessary to solve the differential equation of energy conservation with boundary conditions. In the case of unidimensional transient problems one requires boundary conditions and an initial condition. Solution of differential equations with different boundary conditions are discussed in the next section.

11.3 Boundary Conditions and Initial Condition

In heat transfer, boundary conditions can usually be found by analyzing the problem. The types are illustrated in Figure 11.8.

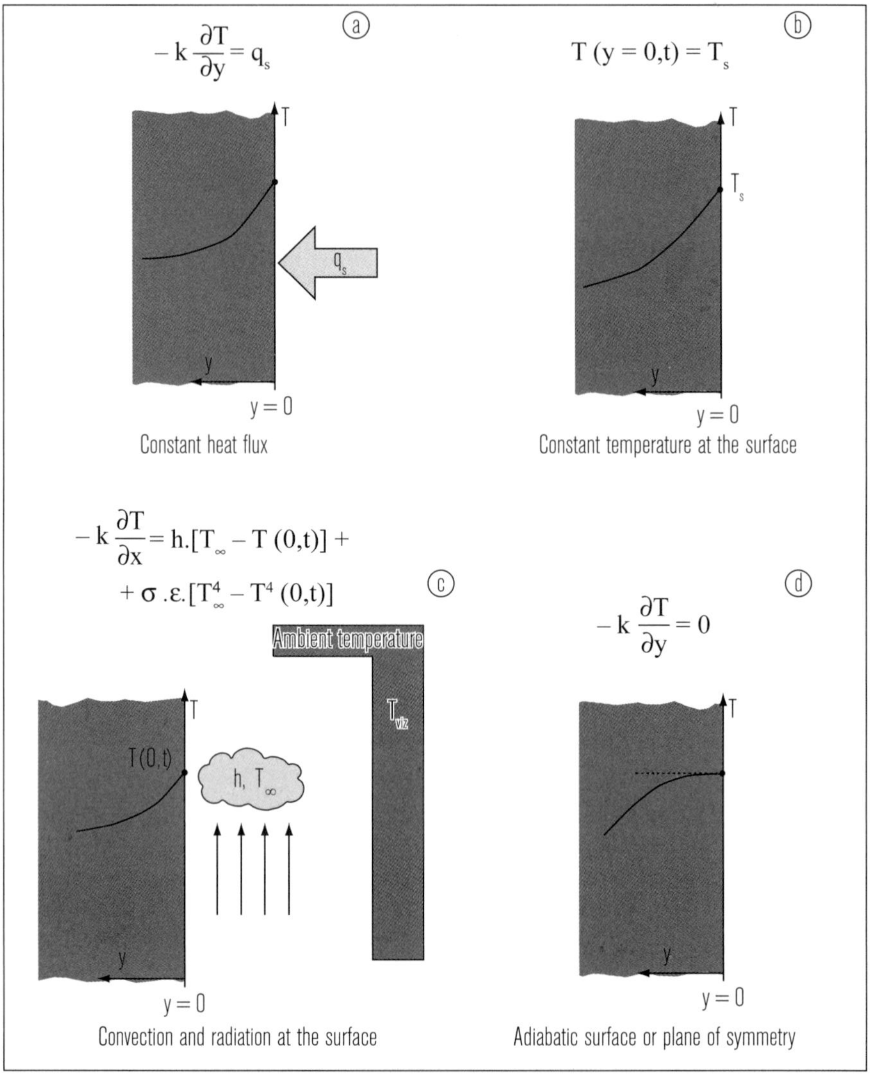

Figure 11.8 - Usual boundary conditions in simple problems of heat transfer.

The boundary condition of type (a) can occur when the surface is in perfect contact with a pure material that undergoes phase change. In this case, the temperature remains constant at the value of the temperature of phase change and heat flow will be associated with the release or absorption of latent heat of phase change.

The situation seen in the boundary condition of type (b) is related to the placement of an electric heater near the surface.This heater, with a given power, provides a constant flow of heat to the surface.

The condition of type (c) is commonly found in practical cases. In this situation, there is heat exchange between the surface and a fluid in contact with it and a neighborhood that surrounds it completely. There can be contribution from both convection and radiation. If the contribution by radiation is negligible, the second term of the right side of the equation in Figure 11.8c ,the radiation contribution can be neglected.

A particular case of boundary condition occurs when the heat flow is zero (type d). This situation occurs in isolated areas as well. Another possibility of heat flow is zero for symmetry planes within the material undergoing heating or cooling.

It is important to mention that the equations for the boundary conditions (b) to (d) can be established by energy balance at the surface.

For transient problems, besides these boundary conditions, it is necessary to specify the condition of the system at the moment when we begin to count the time.This condition is usually called the initial condition. The most common initial condition is a uniform temperature throughout the system under study. This condition is seen schematically in Figure 11.9.

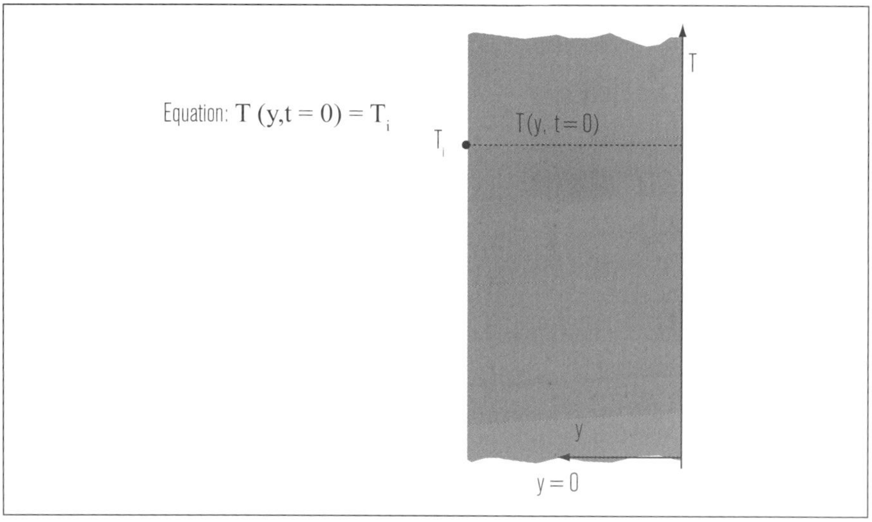

Figure 11.9 - Schematic view of an initial condition of uniform temperature.

Following are several examples of determination of temperature distribution and other quantities of relevant to various systems with different geometries. under steady state conditions. Transient situations will be addressed in the next chapter.

11.4 One-Dimensional Profiles of Temperature

Under this general title initially let us consider cases involving plane geometry with varying boundary conditions for infinite flat plates and composite walls.

11.4.1. Infinite flat plate composite walls

The system under consideration is seen schematically in Figure 11.10:

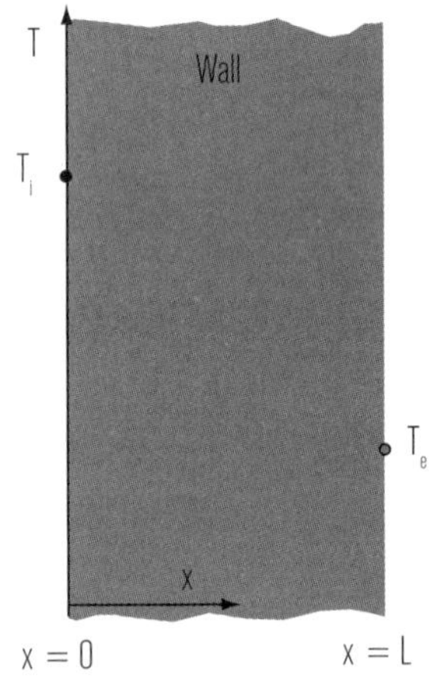

Figure 11.10 - Unidimensional system of heat transfer in plane geometry.

The heat transfer in this case is only in the x direction (direction of the gradient of temperature).

The differential equation that determines the temperature distribution can be obtained by simplifying the general equation (equation 11.18). The terms associated with heat transfer in directions y and z, the generation and accumulation terms can be ignored (unidimensional, steady state conditions with no heat generation) and the equation obtained is:

$$\frac{d}{dx}\left(k\frac{dT}{dx}\right)=0 \tag{11.37}$$

Integrating the equation (11.37), we obtain:

$$k\frac{dT}{dx}=C_1' \tag{11.38}$$

where C_1 is a constant of integration.

Assuming that the thermal conductivity is constant, we can write that:

$$\frac{dT}{dx} = \frac{C_1^{'}}{k} = C_1 \quad (11.39)$$

Integrating the equation (11.40), we have:

$$T = C_1.x + C_2 \quad (11.40)$$

where C_2 is a constant of integration.

It is observed that, the temperature profile is linear. The constants of integration C_1 and C_2 can be determined using the boundary conditions.The boundary conditions that are adopted are the type (a), in the Figure 11.8:

Boundary conditions 1: $x = 0 \quad T = T_i$;

Boundary conditions 2: $x = L \quad T = T_e$

Applying these conditions the equation of the temperature profile is obtained:

$$T_i = C_1.(0) + C_2 \quad (11.41)$$

$$T_e = C_1.(L) + C_2 \quad (11.42)$$

Solving the equations above, we have:

$$C_2 = T_i \quad (11.43)$$

$$C_1 = \frac{T_e - T_i}{L} \quad (11.44)$$

Substituting the values of C_1 and C_2 in equation (11.40), we obtain:

$$T = (T_e - T_i).\frac{x}{L} + T_i \quad (11.45)$$

From the equation (11.45), one can calculate the heat flow through the material.Using Fourier's law, we have:

$$q_x = -k\frac{dT}{dx} = -k.\frac{(T_e - T_i)}{L} = k\frac{(T_i - T_e)}{L} \quad (11.46)$$

The rate of heat transfer is given by:

$$Q_x = A.q_x = A.k\frac{(T_i - T_e)}{L} \tag{11.47}$$

where A is the area of the material perpendicular to the direction of heat transfer.

The equation (11.47) can be rewritten in the following way:

$$Q_x = \frac{(T_i - T_e)}{\left(\frac{L}{k.A}\right)} \tag{11.48}$$

The equation is analogical with with Ohm's law of electricity, given by:

$$i = \frac{V}{R} \tag{11.49}$$

where i, is the current, V is the electromotive force and R the resistance.

The heat flux is analogous to current. Temperature difference is the analogue of electromotive force, representing the driving force for heat transfer. The term L/Ak can be interpreted as a thermal resistance to heat transfer by conduction.

Instead of a specific temperature at x = L, another boundary condition of convection can be specified like heat transfer to a a fluid at a temperature $T_{\infty,e}$ with a heat transfer coefficient. The temperature profile can be evaluated with this data and this second boundary condition:

$$\text{Boundary condition 2: } x = L: -k\frac{dT}{dx} = h_e[T(x = L) - T_{\infty,e}]$$

This condition is of type (b) in Figure 11.8. It is important to note the order of terms in this equation for heat flow by convection. It is chosen this way to maintain consistency with the x-axis The temperature profile can be determined from the relationship obtained from the integration of the equation given earlier

$$T = C_1.x + C_2 \tag{11.40}$$

Applying the Boundary condition 1 (x = 0; T = Ti):

$$C_2 = T_i \tag{11.43}$$

Applying the second boundary condition, we have:

$$C_1 = \frac{-h_e(T_i - T_{\infty,e})}{h_e.L + k} \tag{11.50}$$

Returning to the expression for the temperature profile and substituting the equations for the constants of integration, we have:

$$T = T_i - \frac{h_e(T_i - T_{\infty,e})}{h_e.L + k}.x \tag{11.51}$$

This profile can be placed in the following dimensionless form:

$$\frac{T_i - T}{T_i - T_{\infty,e}} = \frac{h_e.x}{h_e.L + k} = \frac{x/L}{1 + k/h_e.L} \tag{11.52}$$

The parameter $k/h_e.L$ is a dimensionless number. Its inverse is a dimensionless group called the Biot number, Bi:

$$\text{Biot} = \text{Bi} = \frac{h_e.L}{k} \tag{11.53}$$

This number is very important in heat transfer, especially in respect of transient problems, as discussed in Chapter 12. Rewriting the temperature profile in accordance with this number, we have:

$$\frac{T_i - T}{T_i - T_{\infty,e}} = \frac{x/L}{1 + 1/Bi} \tag{11.54}$$

The temperature at position x = L is given by:

$$\frac{T_i - T|_{x=L}}{T_i - T_{\infty,e}} = \frac{1}{1 + 1/Bi} \tag{11.55}$$

If the Biot number tends to zero, the temperature at x = L tends to T_i. If Bi tends to infinity, the temperature at x = L approaches T_∞. These two situations are illustrated schematically in Figure 11.11.

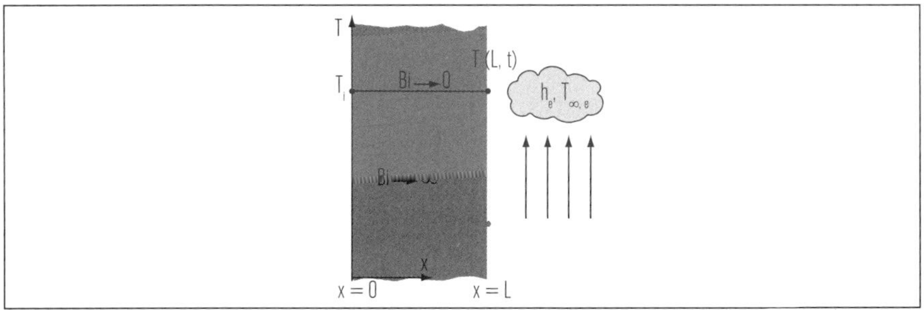

Figure 11.11 - Schematic temperature profiles for two limits of Biot Number.

According to Figure 11.11, increase in Biot number causes temperature gradients within the material more pronounced.

The flow of heat through the material can be expressed following equation:

$$q_x = -k\frac{dT}{dx} = \frac{k.h_e.(T_i - T_{\infty,e})}{h_e.L + k} \quad (11.56)$$

and the rate of heat transfer:

$$Q_x = A.q_x = A.\frac{k.h_e.(T_i - T_{\infty,e})}{h_e.L + k} \quad (11.57)$$

Dividing the numerator and denominator of the right side of equation (11.57) by ($A.h_e.K$):

$$Q_x = \frac{(T_i - T_{\infty,e})}{\left(\frac{L}{k.A} + \frac{1}{h_e.A}\right)} \quad (11.58)$$

Again one can consider the above equation analogical to Ohm's law. In this case, there is resistance to heat transfer by convection ($1/h_e.A$). The analogy is with an electrical circuit in series, the total resistance is the sum of individual resistances (terms in the denominator). The temperature difference in the numerator is analogical to the voltage. The situation is illustrated schematically in Figure 11.12.

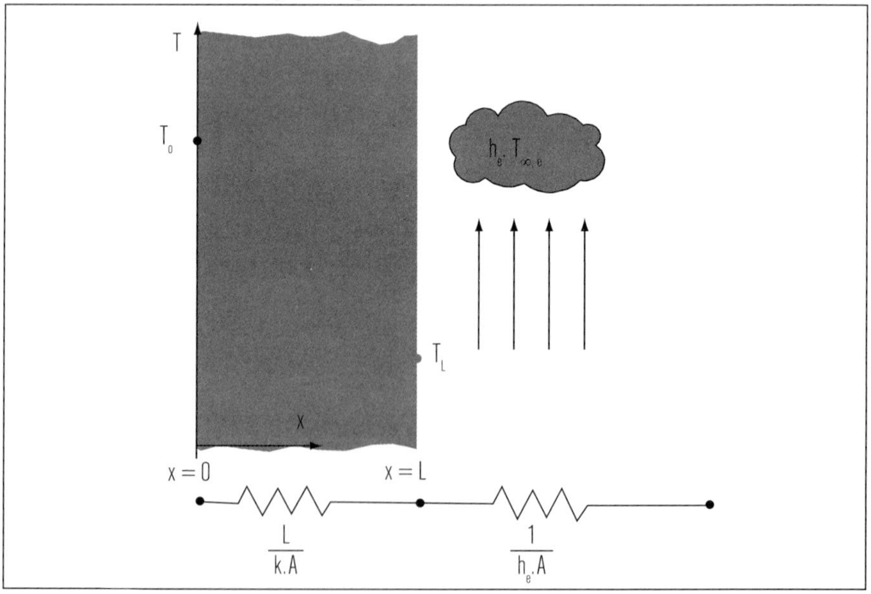

Figure 11.12 - Analogy of heat transfer in a flat plate with an electrical circuit in series.

By analogy with the expression (11.58), one can deduce that if $x = 0$ in the boundary condition, transfer of heat by convection to a fluid at a temperature T_∞ and h_i. is the heat transfer coefficient$_i$, the rate of heat transfer is given by:

$$Q_x = \frac{(T_{\infty,i} - T_{\infty,e})}{\left(\frac{1}{h_i.A} + \frac{L}{k.A} + \frac{1}{h_e.A}\right)} \tag{11.59}$$

This equation can also be obtained considering that the rate of heat transfer is constant over the plate, (steady state):

$$Q_x = A.h_i.(T_{\infty,i} - T_i) \tag{11.60}$$

$$Q_x = \frac{A.k}{L}.(T_i - T_e) \tag{11.61}$$

$$Q_x = A.h_e.(T_e - T_{\infty,e}) \tag{11.62}$$

$$(T_{\infty,i} - T_i) = \frac{Q_x}{A.h_i} \tag{11.63}$$

$$(T_i - T_e) = \frac{Q_x}{\frac{A.k}{L}} \tag{11.64}$$

$$(T_e - T_{\infty,e}) = \frac{Q_x}{A.h_e} \tag{11.65}$$

Adding the three equations above, we have:

$$(T_{\infty,i} - T_{\infty,e}) = Q_x.\left(\frac{1}{h_i.A} + \frac{L}{k.A} + \frac{1}{h_e.A}\right) \tag{11.66}$$

Hence:

$$Q_x = \frac{(T_{\infty,i} - T_{\infty,e})}{\left(\frac{1}{h_i.A} + \frac{L}{k.A} + \frac{1}{h_e.A}\right)} \tag{11.59}$$

This is the same equation obtained previously.

From the equations deduced above, one can obtain a general equation for rate of heat transfer in the case , shown schematically in Figure 11.13. In this case, the wall has multiple layers with different thicknesses and thermal conductivities. On either side of the surface of the external walls there is convective heat transfer. Interfaces between these layers have different temperatures.

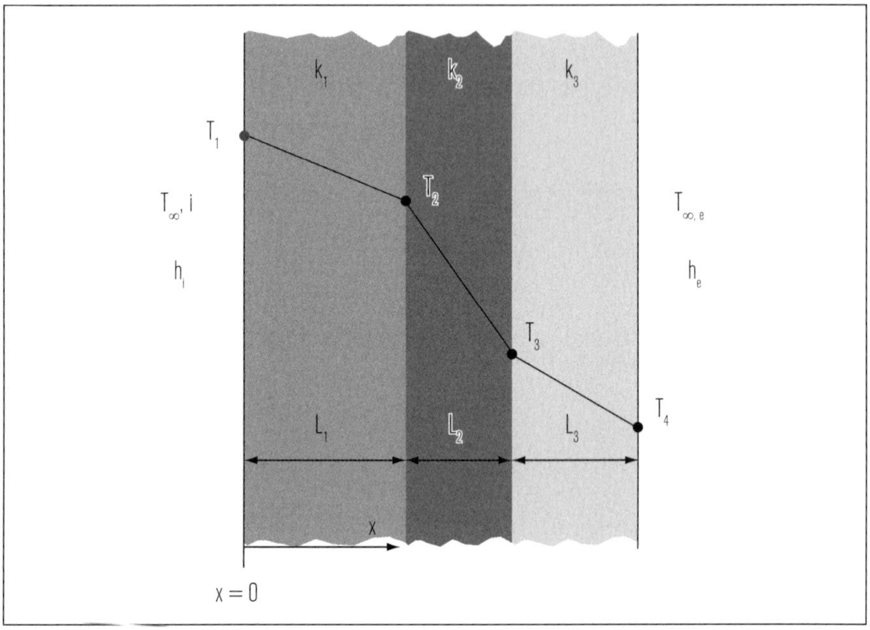

Figure 11.13 - Flat wall with several layers of material.

The equation for the rate of heat transfer in this case is:

$$Q_x = \frac{(T_{\infty,i} - T_{\infty,e})}{\left(\frac{1}{h_i.A} + \frac{L_1}{k_1.A} + \frac{L_2}{k_2.A} + \frac{L_3}{k_3.A} + \frac{1}{h_e.A}\right)} \tag{11.67}$$

If there should be more layers of material, then one has to add more terms corresponding to their thermal resistances in the form of L/k.A, in the denominator of equation (11.67).

Example

A surface whose temperature is kept at 400°C, is separated from the ambient temperature by a layer of insulating material with a thickness of 30 mm and thermal conductivity of 0.2 W/m.K. If the air temperature is 30°C and the heat transfer coefficient between air and outer surfaces of the insulation is equal to 400 W/m^2.K, what is the temperature of the outer surface of the insulation? What is the heat flux through the insulation layer?

Solution

The system under consideration is seen in the figure below:

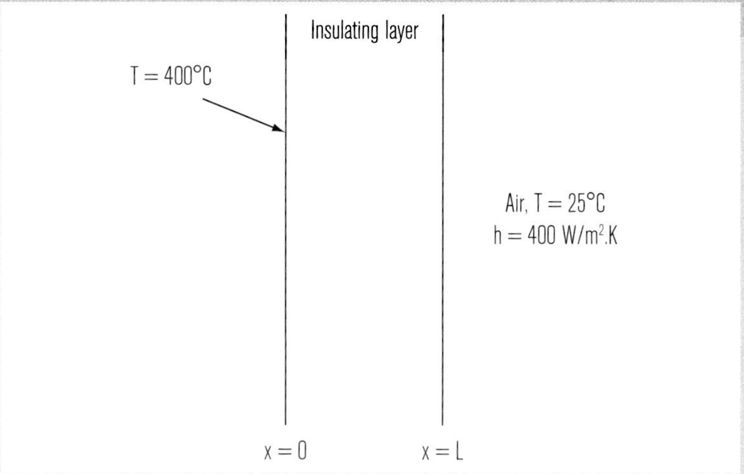

An energy balance for the surface at x = L states:

[Rate of energy input] – [Rate of energy output] = 0

(steady state unidimensional heat transfer)

Applying energy balance, we can write:

$$-k\frac{dT}{dx} = k.\frac{T(x=0) - T(x=L)}{L} = h.[T(x=L) - T_\infty]$$

Replacing data, we obtain:

$$(0.2).\frac{400 - T(x=L)}{0.03} = (400).[T(x=L) - 25]$$

Solving for T (x = L), we have:

$$T(x=L) = 31.1°C$$

The heat flux can be calculated as shown below:

$$q_x = h.[T(x=L) - T_\infty] = (400).(31.1 - 25) = 2440\ W/m^2$$

11.4.2 Cylindrical geometry

The system under consideration is seen schematically in Figure 11.14.

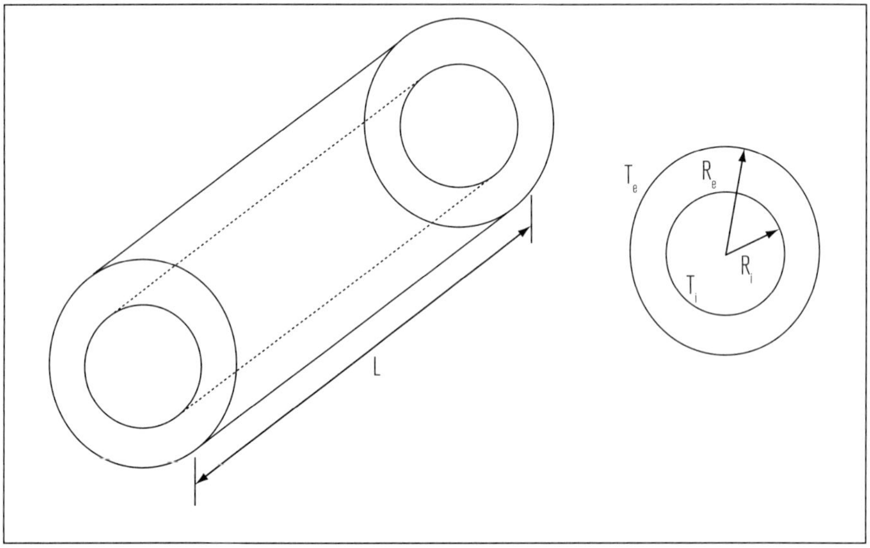

Figure 11.14 - Cylindrical duct with heat transfer in the radial direction.

For steady state, without generation of heat and temperature gradient in the radial direction only, we can simplify the general equation of energy conservation in cylindrical coordinates (equation 11.26) to:

$$\frac{1}{r}\frac{d}{dr}\left(k.r.\frac{dT}{dr}\right) = 0 \qquad (11.68)$$

The heat flow in the radial direction is given by Fourier's law:

$$q_r = -k.\frac{dT}{dr} \qquad (11.69)$$

The rate of heat transfer is given by:

$$Q_r = -k.A.\frac{dT}{dr} = -k.2\pi.r.L.\frac{dT}{dr} \qquad (11.70)$$

Considering a cylinder of length L.

Integrating the equation (11.70), we obtain:

$$k.r.\frac{dT}{dr} = C_1^{'} \qquad (11.71)$$

Multiplying both sides of the equation (11.70) by (-2.π.L), we have:

$$-2.\pi.r.L.k\frac{dT}{dr} = Q_r = -2.\pi.L.C_1' \qquad (11.72)$$

Since 2.π.LC$_1$ is a constant, it can be said that the rate of heat transfer in the radial direction is also constant in this case. From the equation (11.69) it can be concluded that the heat flux, q_r, is not constant along the radial direction and its value is inversely proportional to r.

Integrating the equation (11.71), we have:

$$T = C_1.\ln r + C_2 \qquad (11.73)$$

where $C_1 = C_1'/k$.

The boundary conditions are:

Boundary conditions 1: $r = R_i \quad T = T_i$;

Boundary conditions 2: $r = R_e \quad T = T_e$.

Applying these boundary conditions in equation (11.73), we have:

$$T_i = C_1.\ln R_i + C_2 \qquad (11.74)$$

$$T_e = C_1.\ln R_e + C_2 \qquad (11.75)$$

Solving the system of equations above, we get:

$$C_1 = \frac{T_i - T_e}{\ln\frac{R_i}{R_e}} \qquad (11.76)$$

$$C_2 = T_i - \frac{T_i - T_e}{\ln\frac{R_i}{R_e}}.\ln R_i \qquad (11.77)$$

The temperature profile is then expressed by the following equation:

$$T = T_i - \frac{T_i - T_e}{\ln\left(\frac{R_e}{R_i}\right)}.\ln\left(\frac{r}{R_i}\right) \qquad (11.78)$$

$$q_r = -k.\frac{dT}{dr} = k.\frac{C_1}{r} = k.\frac{(T_i - T_e)}{\ln\left(\frac{R_e}{R_i}\right)}.\frac{1}{r} \qquad (11.79)$$

The het flux is given by:

$$Q_r = 2.\pi.r.L.k.\frac{(T_i - T_e)}{\ln\left(\frac{R_e}{R_i}\right)}.\frac{1}{r} = 2.\pi.L.k.\frac{(T_i - T_e)}{\ln\left(\frac{R_e}{R_i}\right)} \quad (11.80)$$

This rate can also be expressed as follows:

$$Q_r = \frac{(T_i - T_e)}{\frac{\ln\left(\frac{R_e}{R_i}\right)}{2.\pi.L.k}} \quad (11.81)$$

Using the analogy with Ohm's law, the equation for the thermal resistance for conduction in cylindrical coordinates can be given as:

$$R_{T,cond} = \frac{\ln\left(\frac{R_e}{R_i}\right)}{2.\pi.L.k} \quad (11.82)$$

Now considering a system as shown in Figure 11.15, one can obtain a general equation for the rate of heat transfer for multiple layers of material having different K values and thicknesses and convection in both internal and external areas of the duct.

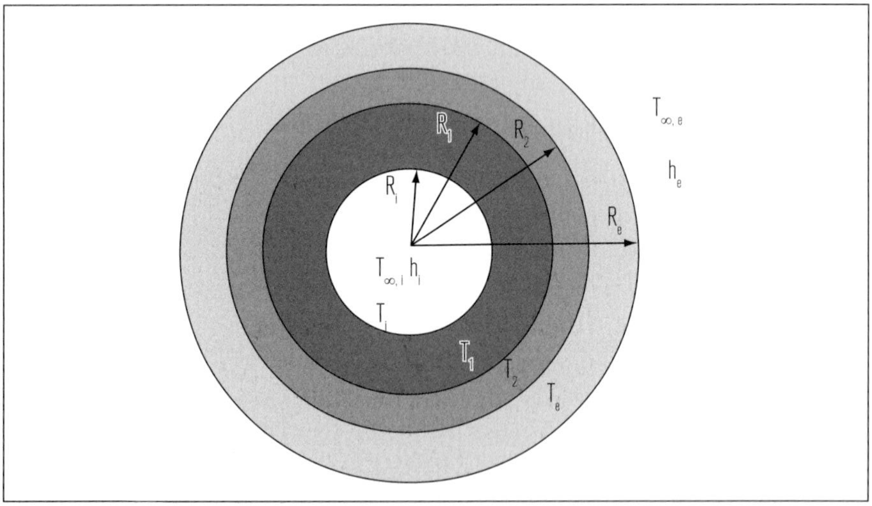

Figure 11.15 - Cylindrical duct with several concentric layers of material (composite wall) and convection in the outer and inner surfaces.

As the rate of heat transfer through the different layers is constant, we can write:

$$Q_r = 2.\pi.R_i.L.h_i.(T_{\infty,i} - T_i) \quad (11.83)$$

$$Q_r = \frac{(T_i - T_1)}{\frac{\ln\left(\frac{R_1}{R_i}\right)}{2.\pi.L.k_1}} \quad (11.84)$$

$$Q_r = \frac{(T_1 - T_2)}{\frac{\ln\left(\frac{R_2}{R_1}\right)}{2.\pi.L.k_2}} \quad (11.85)$$

$$Q_r = \frac{(T_2 - T_e)}{\frac{\ln\left(\frac{R_e}{R_2}\right)}{2.\pi.L.k_3}} \quad (11.86)$$

$$Q_r = 2.\pi.R_e.L.h_e.(T_e - T_{\infty,e}) \quad (11.87)$$

Since only the temperatures T_∞ external and internal are known, one can develop an equation similar to that for plane geometry (equation 11.67) to obtain the following expression:

$$Q_r = \frac{(T_{\infty,i} - T_{\infty,e})}{\frac{1}{2.\pi.R_i.L.h_i} + \frac{\ln\left(\frac{R_1}{R_i}\right)}{2.\pi.L.k_1} + \frac{\ln\left(\frac{R_2}{R_1}\right)}{2.\pi.L.k_2} + \frac{\ln\left(\frac{R_e}{R_2}\right)}{2.\pi.L.k_3} + \frac{1}{2.\pi.R_e.L.h_e}} \quad (11.88)$$

In the equation (11.88), the terms that appear in the denominator on the right represent the thermal resistance for convection and conduction.

An interesting analysis can be developed to check what happens to the total resistance to heat transfer like increasing the thickness of the outer wall.or thickness of the layers, leading to an increase in the total thermal resistance.

For the case shown in Figure 11.15, the total resistance is given by:

$$R_T = \frac{1}{2.\pi.R_i.L.h_i} + \frac{\ln\left(\frac{R_1}{R_i}\right)}{2.\pi.L.k_1} + \frac{\ln\left(\frac{R_2}{R_1}\right)}{2.\pi.L.k_2} + \frac{\ln\left(\frac{R_e}{R_2}\right)}{2.\pi.L.k_3} + \frac{1}{2.\pi.R_e.L.h_e} \quad (11.89)$$

Changes in the thickness of the last layer would affect only the last two contributions of the equation (11.89). To determine how R_T varies with changes in the thickness of the outer layer one can take the derivative of R_T with respect to R_e then:

$$\frac{dR_T}{dR_e} = \frac{1}{2.\pi.L.k_3.R_e} - \frac{1}{2.\pi.L.h_e.R_e^2} \tag{11.90}$$

Setting this derivative equal to zero, one can verify that R_T has a maximum or minimum when R_e varies:

$$\frac{1}{2.\pi.L.k_3.R_e} - \frac{1}{2.\pi.L.h_e.R_e^2} = 0 \tag{11.91}$$

Therefore:

$$R_e^{crit} = \frac{k_3}{h_e} \tag{11.92}$$

Hence this is the point of critical radius.

To determine whether maximum or minimum, one can evaluate the second derivative of R_T with respect to R_e at the the point where the first derivative vanishes:

$$\frac{d^2R_T}{d^2R_e} = -\frac{1}{2.\pi.L.k_3.R_e^2} + \frac{2}{2.\pi.L.h_e.R_e^3} \tag{11.93}$$

$$\frac{d^2R_T}{d^2R_e} = \frac{1}{2.\pi.L.R_e^2}\left(-\frac{1}{k_3} + \frac{2}{h_e.R_e}\right) \tag{11.94}$$

The term within the brackets determines the sign of the second derivative. For Re = R_e^{crit} (equation 11.92), we have:

$$\frac{d^2R_T}{d^2R_e} = \frac{1}{2.\pi.L.R_e^2}\left(-\frac{1}{k_3} + \frac{2}{k_3}\right) > 0 \tag{11.95}$$

As the second derivative is greater than zero, this is a minimum point.

Further analysis of this result can be made by evaluating the pattern of variation of resistance for conduction in the outer layer and by convection between this layer and the fluid in contact with it, depending on the thickness of this layer. Assuming:

$$R_e = R_2 + e \tag{11.96}$$

where e represents the thickness of the outer wall, these two resistances can be expressed as follows:

$$R_{conduction} = \frac{\ln\left(1 + \frac{e}{R_2}\right)}{2.\pi.L.k_3} \quad (11.97)$$

$$R_{convection} = \frac{1}{2.\pi.(R_2 + e).L.h_e} \quad (11.98)$$

The resistance to conduction increases with the thickness (e) , the convective resistance decreases with increasing (e) (as the area for heat transfer is increased). Figure 11.16 summarizes the variation of these two resistances with thickness (e) . The figure has an additional line representing the sum of two resistances, enabling one to observe the existence of a minimum point in total resistance and consequently a maximum in terms of rate of heat transfer.

In the design of a thermal insulation for a cylindrical pipe, the goal is to reduce heat losses, and hence thicknesses near the critical value should be avoided. On the other hand, in systems which require quick dissipation of e heat one should prefer thicknesses equal to the critical value as shown in Figure 11.16.

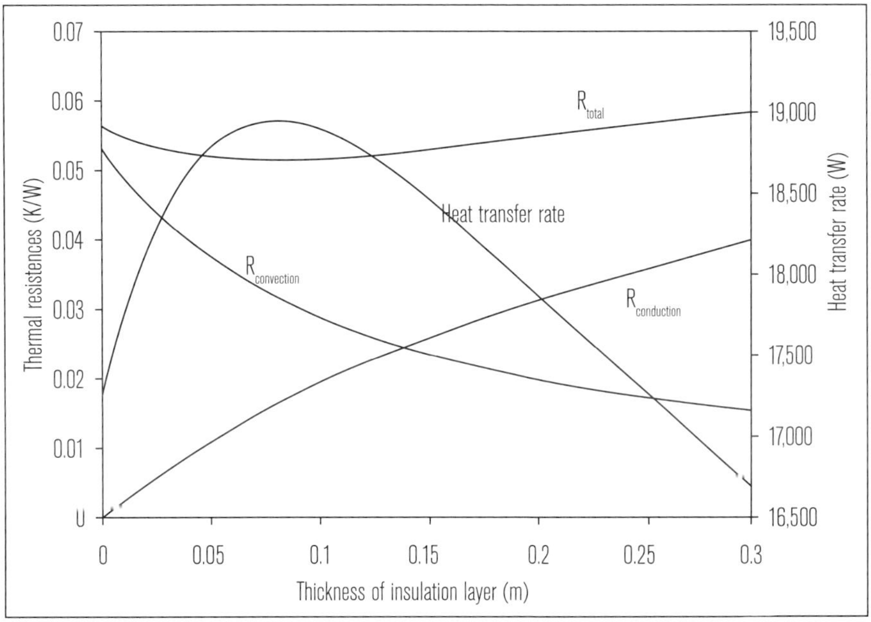

Figuro 11.16 Variation of thermal resistance as a function of thickness of the outer wall for a cylindrical duct.

Example

The cross section of a copper wire, which is good conductor of electricity with a plastic covering outside is given below:

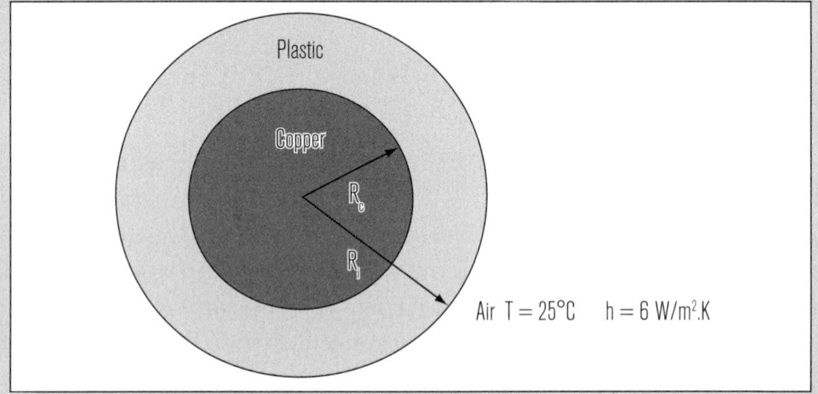

The electrical resistance of copper wire is 0.5 Ω/m. Determine the maximum current that can pass through the copper wire, without risk of melting the insulating plastic. The melting temperature of the plastic is 200°C

Data: $R_C = 1.25$ mm $R_i = 1.75$ mm

Thermal conductivity: copper: 400 W/m.K; plastic: 0.16 W/m.K.

Solution

Energy balance for the copper conductor for the situation is given below:

$$[\text{Rate of energy input}] - [\text{Rate of energy output}] + [\text{Rate of energy generation}] = 0$$

The rate of heat generation by Joule effect occurs and is given by:

$$[\text{Rate of energy generation}] = R_e.L.I^2$$

The rate of heat output is given by:

$$Q_r = \frac{[T(r = R_c) - T_\infty)}{\left[\frac{\ln\left(\frac{R_i}{R_c}\right)}{2\pi LK} + \frac{1}{2\pi\ R_i\ L\ h}\right]}$$

Replacing data, we obtain:

$$Q_r = \frac{L.(200-25)}{\left[\frac{\ln\left(\frac{1.75}{1.25}\right)}{2\pi(0.16)} + \frac{1}{2\pi(0.00175)(6)}\right]} = 11.30.L \text{ W}$$

In the above equation, the maximum surface temperature of the wire corresponds to the plastic temperature at $r = R_c$.

Hence the maximum current permissible is 4.75 amperes:

$$R_e.L.I^2 = (0.5).L.I^2 = 11.30.L$$
$$I = 4.75 \text{ A}$$

11.4.3 Spherical geometry

Figure 11.17 gives the profile of a sphere with temperatures defined at Ri = T_i and R_e = Te at the external surface.

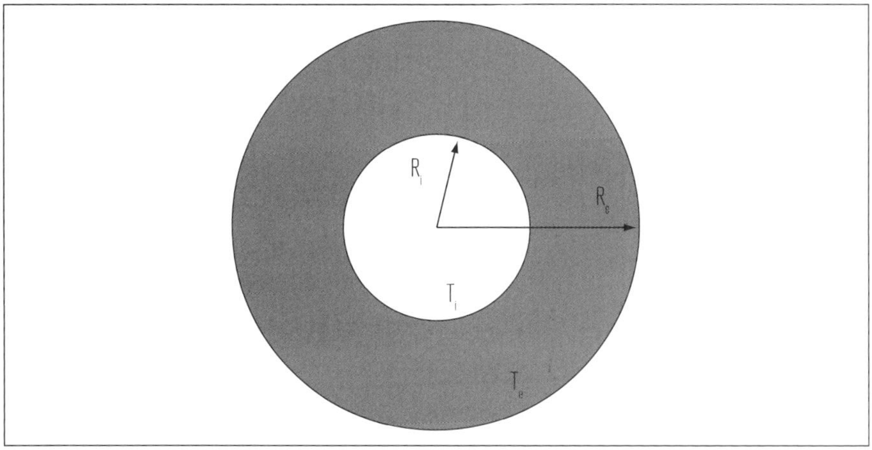

Figure 11.17 - Heat transfer in the radial direction in a sphere.

Assuming steady state, without generation of heat and temperature gradient in the radial direction only, and for steady state conditions one can simplify the general equation of energy in spherical coordinates (equation 11.36) to:

$$\frac{1}{r^2}\frac{d}{dr}\left(k.r^2.\frac{dT}{dr}\right) = 0 \qquad (11.99)$$

For spherical geometry, the flow of heat in the radial direction is given by:

$$q_r = -k.\frac{dT}{dr} \quad (11.100)$$

The rate of heat transfer is the product of theheat flow given above multiplied by area normal to the directionof flow (direction r)

$$Q_r = -k.A.\frac{dT}{dr} = -k.4\pi.r^2.\frac{dT}{dr} \quad (11.101)$$

Integrating the equation (11.99), we have:

$$k.r^2.\frac{dT}{dr} = C'_1 \quad (11.102)$$

Multiplying the equation (11.102) by - 4 π, we get:

$$-k.4.\pi.r^2.\frac{dT}{dr} = -4.\pi.C'_1 \quad (11.103)$$

Comparing equations (11.101) and (11.103) it is clear that rate of heat transfer in the radial direction is constant and the heat flow is inversely proportional to the radial position squared.

Integrating the equation (11.102), we obtain:

$$T = \frac{C'_1}{k}\frac{1}{r} + C_2 = \frac{C_1}{r} + C_2 \quad (11.104)$$

where $C_1 = C'_1/k$.

The boundary conditions are:

Boundary conditions 1: $r = R_i \quad T = T_i$;

Boundary conditions 2: $r = R_e \quad T = T_e$.

Substituting these boundary conditions in equation (11.104):

$$T_i = \frac{C_1}{R_i} + C_2 \quad (11.105)$$

$$T_e = \frac{C_1}{R_e} + C_2 \quad (11.106)$$

Solving the system of equations, one can determine the constants C_1 and C_2:

$$C_1 = \frac{T_i - T_e}{\frac{1}{R_i} - \frac{1}{R_e}} \tag{11.107}$$

$$C_2 = T_i - \frac{T_i - T_e}{\frac{1}{R_i} - \frac{1}{R_e}} \cdot \frac{1}{R_i} \tag{11.108}$$

Returning to the equation of the temperature profile, we have:

$$T = T_i + \frac{T_e - T_i}{\frac{1}{R_i} - \frac{1}{R_e}} \cdot \left(\frac{1}{R_i} - \frac{1}{r} \right) \tag{11.109}$$

The rate of heat transfer is given by:

$$Q_r = \frac{4.\pi.k.(T_e - T_i)}{\frac{1}{R_i} - \frac{1}{R_e}} \tag{11.110}$$

Rearranging:

$$Q_r = \frac{(T_e - T_i)}{\frac{\left(\frac{1}{R_i} - \frac{1}{R_e} \right)}{4.\pi.k}} \tag{11.111}$$

The term in the denominator of the right side of the equation (11.111) can be considered as the thermal resistance to heat transfer by conduction in spherical coordinates:as in the case a similar equation in cylindrical coordinates.

$$R_{T,cond} = \frac{\left(\frac{1}{R_i} - \frac{1}{R_e} \right)}{4.\pi.k} \tag{11.112}$$

Another relevant situation of heat transfer in spherical coordinates is shown in Figure 11.18, in which a sphere, whose surface temperature is maintained constant, is in contact with a stagnant fluid. The temperature of the fluid at a point well away from the sphere is equal to T_∞. The region of interest in this is the layer of fluid surrounding the sphere, defined by the following.

$$R \leq r < \infty$$

where R is the radius of the sphere.

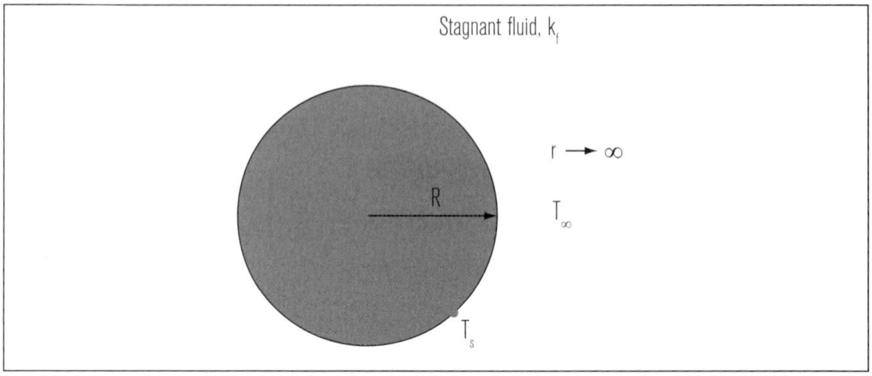

Figure 11.18 - Sphere surrounded by stagnant fluid.

As the fluid is stagnant (no macroscopic motion), the heat transfer should occur only by conduction. Considering again steady state without heat generation and heat transfer occurring only in the radial direction, we have the following differential equation that governs the distribution of temperature:

$$\frac{1}{r^2}\frac{d}{dr}\left(k.r^2.\frac{dT}{dr}\right) = 0 \tag{11.113}$$

Assuming constant k, the above equation can be integrated to provide the following temperature distribution:

$$T = \frac{C_1}{r} + C_2 \tag{11.114}$$

The boundary conditions are:

Boundary condition 1: $r = R$ $\quad T = T_s$;

Boundary condition 2: $r \rightarrow \infty$ $\quad T = T_\infty$.

Applying these boundary conditions:

$$T_S = \frac{C_1}{R} + C_2 \tag{11.115}$$

$$T_\infty = C_2 \tag{11.116}$$

The constant C_1 is given by:

$$C_1 = R.(T_S - T_\infty) \tag{11.117}$$

The temperature profile is then given by:

$$T = T_\infty + (T_S - T_\infty).\frac{R}{r} \qquad (11.118)$$

The heat flux on the surface of the sphere is:

$$q_r \big|_{r=R} = \frac{k}{R}(T_S - T_\infty) \qquad (11.119)$$

As mentioned earlier, the flow of heat between a solid and a fluid in contact with it can be estimated through the following relationship:

$$q = h.(T_S - T_\infty) \qquad (11.120)$$

From equations (11.119) and (11.120), the heat transfer coefficient h is:

$$h = \frac{k}{R} \qquad (11.121)$$

This value of h represents its lower limit, as it was assumed that the fluid is stagnant. Any movement of the fluid, due to variations in density with temperature lead to an increase in the value of h.

Equation (11.121) can also be written in the following way:

$$\frac{h.D}{k} = 2 \qquad (11.122)$$

The dimensionless number expressed by h.D/k is called the Nusselt number. As it will be seen later, this number is is a function of other dimensionless numbers and these correlations are used to evaluate the heat transfer coeeficient.

The Nusselt number is a dimensionless number like the Biot number. However k in Nusselt number, refers to the thermal conductivity of the fluid, whereas in the Biot number k the thermal conductivity refers to that of the solid,with which the fluid is in contact. It is very important to remember this.

11.4.4 Contact resistances

In situations involving more than one layer of solid material, it is normally considered that the contact between these layers is perfect, that is to say, the thermal contact resistance is neglected.

The thermal contact resistance is due to the fact that the solid surfaces are not perfectly smooth and and there is variation in the roughness. Hence, when two materials are placed to form adjacent layers, in reality, there are a number of points of actual contact, interspersed with empty spaces filled with gas or air due to roughness.

The unit of contact resistance is the inverse that of heat transfer coefficient ($m^2 \cdot K/W$ in the international system of units). The magnitude of this resistance depends on a number of factors. Among them are:

- contact pressure between the materials;
- hardness of the materials involved;
- roughness of the surfaces;
- presence or absence of a fluid (paste, liquid or gas) in the interface.

The value of contact resistance is usually determined through experiments. Incropera and DeWitt (2003) provide some resistance values for contact materials in different conditions. Cho et al. (1998a,b) obtained values of contact resistance between the flux powder and wall of the mold in continuous slab casting. In this system, the contact resistance can be the main resistance to the process of heat transfer. The authors discuss also the techniques for determining this contact resistance.

11.4.5 Temperature profile with heat generation inside a system

Systems considered here are in steady state, but with heat generation, which may be due to Joule effect due to the passage of electric current or chemical reactions (endothermic or exothermic).

Consider the situation shown in Figure 11.19.

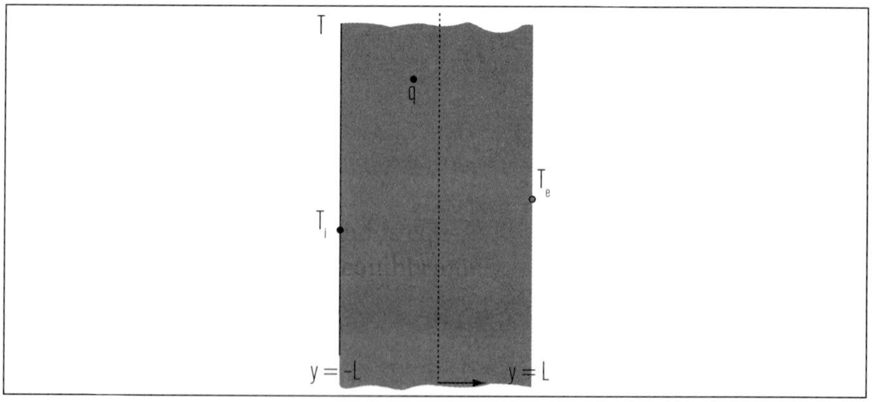

Figure 11.19 - Flat plate or wall with heat generation.

There is heat transfer only in the y direction, The general equation of conservation of energy in cartesian coordinates (equation 11.18), for steady state and heat generation inside the material, simplifies to:

$$\frac{d}{dy}\left(k\frac{dT}{dy}\right) + \dot{q} = 0 \qquad (11.123)$$

$\dot{q}$ is the rate of heat generation per unit volume.

Integrating the above equation, we obtain:

$$k\frac{dT}{dy} = -\dot{q}.y + C_1' \qquad (11.124)$$

Assuming constant k, the above equation can be integrated to provide:

$$T = -\frac{\dot{q}.y^2}{2.k} + C_1.y + C_2 \qquad (11.125)$$

where:

$$C_1 = C_1'/k$$

The constants C_1 and C_2 can be determined with the use of boundary conditions:

Boundary condition 1: $y = -L \quad T = T_i$;

Boundary condition 2: $y = L \quad T = T_e$.

Applying these boundary conditions in equation (11.125), we obtain:

$$C_1 = \frac{T_e - T_i}{2.L} \qquad (11.126)$$

$$C_2 = \frac{\dot{q}.L^2}{2.k} + \frac{T_e + T_i}{2} \qquad (11.127)$$

Substituting the values of these constants in equation (11.125), determine the profile of temperatures:

$$T = -\frac{\dot{q}.(L^2 - y^2)}{2.k} + \frac{T_e - T_i}{2}.\frac{y}{L} + \frac{T_e + T_i}{2} \qquad (11.128)$$

The profile obtained shows a quadratic dependence with position y. If $\dot{q}$ is zero, then a linear temperature profile , like the one obtained previously, results. If the temperatures Ti and Te are equal, the second term on the right side of the equation (11.128) vanishes and a symmetric profile results for the position y = 0.

Taking the derivative of the equation (11.128) and equating the result to zero, one can determine, whether there will be point of maximum or minimum temperature along the wall:

$$\frac{dT}{dy} = -\dot{q}.y + \frac{T_e - T_i}{2.L} = 0 \qquad (11.129)$$

Point of minimum or maximum temperature is evaluated by making the first derivative zero:

$$y = \frac{k.(T_e - T_i)}{2.\dot{q}.L} \qquad (11.130)$$

To determine if this is a point of maximum or minimum, the second derivative of temperature with respect to y is evaluated at the point where the first derivative is zero.

$$\frac{d^2T}{dy^2} = -\frac{\dot{q}}{k} \qquad (11.131)$$

The sign of the second derivative depends on the sign of the term power generation. If it is positive (heat generation by Joule effect or exothermic reaction), the second derivative is negative and there will be a maximum temperature. If it is negative (consumption of heat due to an exothermic reaction, for example), the second derivative is positive and there will be a minimum point.

It is also interesting to note that when the temperatures T_i and T_e are equal, the point of maximum or minimum should l occur at the point y = 0, ie the center of the piece.

Considering the case, where the temperatures on the surfaces are equal and that the surfaces are in contact with a fluid at T_∞ with a heat transfer coefficient h, one can establish a relationship between the surface temperature and the temperature of the fluid. For this first step is develop an energy balance for each of the surfaces. For the surface at y = L, we have:

$$-k.\left.\frac{dT}{dy}\right|_{y=L} = h.(T_{sup} - T_\infty) \qquad (11.132)$$

where T_{sup} is the temperature of the surface at y = L ($T_{sup} = T_i = T_e$).

$$\dot{q}.L = h.(T_{sup} - T_{\infty}) \quad (11.133)$$

Hence:

$$T_{sup} = \frac{\dot{q}.L}{h} + T_{\infty} \quad (11.134)$$

It can be seen that higher values of heat transfer coefficient h result in T_{sup} becoming closer to T_{∞}.

For cylindrical systems, we can adopt a similar procedure to determine the temperature profile, when there is a term for heat generation. The differential equation, that governs heat transfer in this case is given by (considering the radial direction and steady state):

$$\frac{1}{r}\frac{d}{dr}\left(k.r.\frac{dT}{dr}\right) + \dot{q} = 0 \quad (11.135)$$

Integrating the above equation, we obtain:

$$k.r.\frac{dT}{dr} = -\frac{\dot{q}.r^2}{2} + C_1^{'} \quad (11.136)$$

For constant k, integration gives:

$$T = -\frac{\dot{q}.r^2}{4.k} + C_1.\ln r + C_2 \quad (11.137)$$

where:

$$C_1 = C_1^{'}/k$$

For a solid cylinder, the equations (11.136) and (11,137) are valid for values of radial position in the range $0 \le r \le R$, where R is the radius of the cylinder. Substituting the radial position r = 0 in equation (11.136), the constant $C_1^{'}$ (and hence C_1) should be zero, if quantities k, $\dot{q}$, dT/dr are finite.

Hence:

$$T = -\frac{\dot{q}.r^2}{4.k} + C_2 \quad (11.138)$$

The constant C_2 can be determined by considering the temperature on the surface of the cylinder is known:

$$r = R \qquad T = T_{sup}$$

For this condition, we get:

$$C_2 = T_{sup} + \frac{\dot{q}.R^2}{4.k} \quad (11.139)$$

and the temperature profile is then expressed by:

$$T = T_{sup} + \frac{\dot{q}}{4.k}.(R^2 - r^2) \quad (11.140)$$

If the surface of the cylinder is in contact with a fluid at T_∞, one can relate T_{sup} and T_∞ through an energy balance at the surface:

$$-k.\left.\frac{dT}{dr}\right|_{r=R} = h.(T_{sup} - T_\infty) \quad (11.141)$$

The temperature profile is given by:

$$T_{sup} = T_\infty + \frac{\dot{q}.R}{2.h} \quad (11.142)$$

11.5 Temperature Profiles in Two and Three Dimensions

In some situations of practical interest to the metallurgists, the one-dimensional approach presented above provides adequate representation for the study of heat transfer. However, not all cases can be adequately treated assuming one-dimensional heat transfer. In these cases,two or three-dimensional approaches should be adopted.

The difficulty involved in the study of heat transfer in two and three dimensions is linked to obtaining a temperature profile from the analytical solution of differential equations governing the process. This is the methodology that has been applied for quite some time in the past.

The analytical solutions to obtain temperature profiles in two and three dimensions exist only for special situations involving simple boundary conditions and material properties with physical constants. Even in these cases, obtaining these solutions are labor intensive and require specialized knowledge of mathematics.

Hence, it is interesting to explore an alternative way to obtain the temperature profiles without the difficulties and limitations of analytical solutions. This involves the use of numerical techniques for obtaining solutions.

There are several methods for numerical solutions of differential equations resulting from energy balance. The method presented here is the *finite volume technique*. This involves the following steps:

- discretization of the domain through its division into volume elements;
- obtain equations of conservation of energy for each element of volume as a function of temperature (in terms of temperature and volume of neighboring elements). In this step, one gets a system of linear equations, where the number of equations equals the number of volume elements in which the unknowns are the temperatures of these elements;
- solution of the system of linear equations obtained for determination of temperatures.

In the following each of the above steps are discussed in detail.

11.5.1 Discretization

Several methodologies can be adopted in the discretization step. The choice of the technique to be used depends on the complexity of the geometry of the domain to be discretized. In this section, one of these techniques, normally used in the discretization of domains with simple geometries, is discussed. Consideration is given to cases of discretization in two dimensions for easy understanding. However, the concept used is easily generalized to heat transfer in three dimensions.

Figure 11.20 shows the wall along which occurs heat transfer. To discretize this area, lines are drawn parallel to its boundary in two orthogonal directions, as shown in Figure 11.20(b). The spacing between the lines need not be uniform. One can use smaller spacing in the regions where it is expected that the temperature gradients are higher. The points of intersection between the lines drawn are called *nodes*.

Regarding the dotted lines shown in Figure 11.21, these are drawn exactly in the middle of the opening lines used in the discretization of the domain to establish the limits of the elements of volume. It should be noted that each element of volume contains a node inside.

Figures 11.20 and 11.21 present a discretization procedure in two directions (x and y).The procedure of discretization for the three dimensions (including the z-direction) is similar.

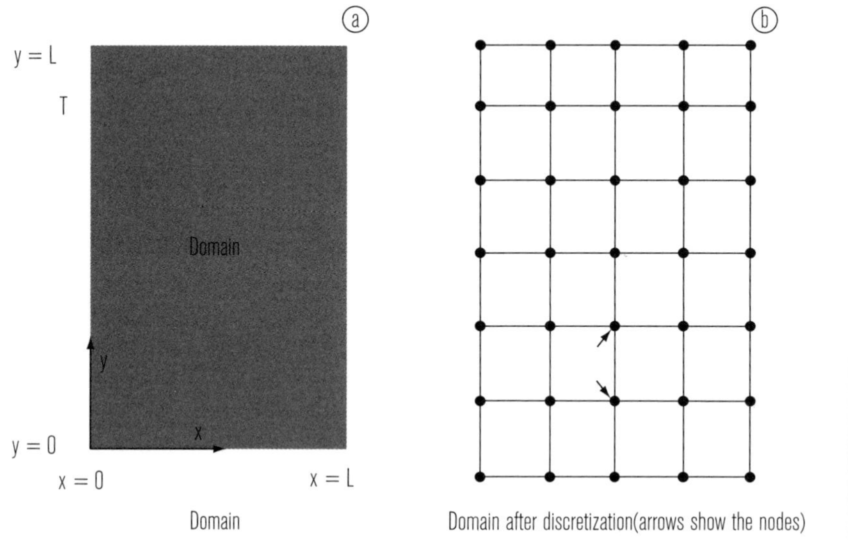

Figure 11.20 - Bidimensional domain before (a) and after (b) discretization.

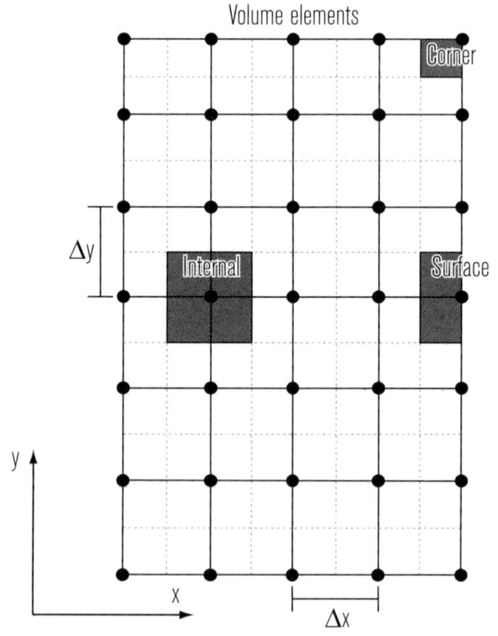

Figure 11.21 - Bidimensional domain discretized with defining of the volume elements.

As can be seen n Figure 11.21 that there are three types of volume elements in the discretized domain. One is inside the area, the other on one side and the third one one in the corner. These elements are highlighted.

The discretization process is completely controlled by what determines the temperature profiles. As will be seen later, use of a smaller spacing between lines (smaller volume elements) results in more accurate results. However, the computational effort (time) in determining the temperature profiles increases. Hence one must seek a compromise between these two aspects to arrive at optimum conditions.

It is important to note that in such numerical solutions, essentially the temperatures of the nodes are determined. The nodes represent the temperatures in every volume element. Hence, regardless of, how the discretization has been done in the numerical solution, one can get the temperatures for a finite number of points. In analytical solutions based on equations of conservation, the result is a mathematical function describing the temperature distribution. In this case, one can evaluate the temperature at any position within the field.

11.5.2 Obtaining the equations of conservation

The differential equations from which the temperature profiles were obtained in the previous chapters were based on development of energy balances in infinitesimal volume elements. In the method of *finite volume*, the equations necessary for evaluating the temperature distribution are also determined from energy balances in each volume element.

In the following section, some examples of getting energy balances in volume elements at different positions within the domain, are discussed. Consider a situation, in which the contours of the domain has heat exchange with a fluid at a temperature, T_∞ and heat transfer coefficient, *h*.

Assuming that the element of internal volume represents an area of unit size, in the direction perpendicular to the sheet of paper, the volume element including its closest neighbours can be represented as shown in Figure 11.22

Now the balance of energy can be carried out in the volume element considering the inputs and outputs of energy according to the orientation of the coordinate axes assuming steady state and absence of heat generation.

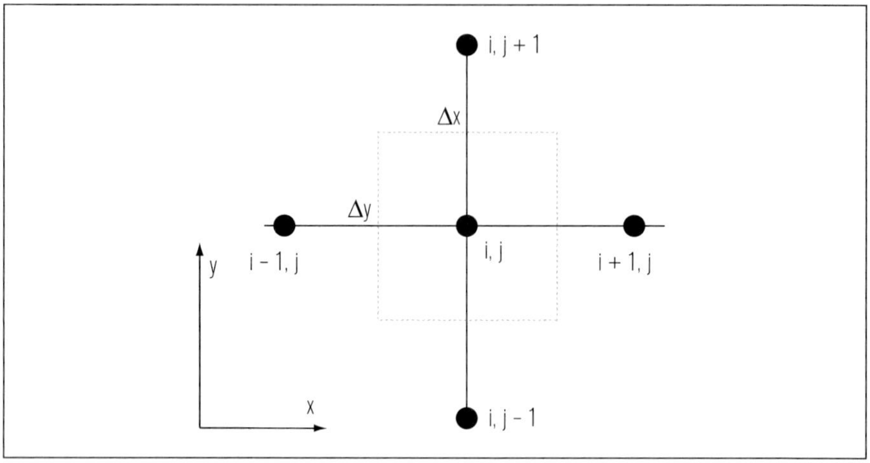

Figure 11.22 - Details of internal volume element and its nearest neighbors.

$$\left[\text{Rate of energy input}\right] - \left[\text{Rate of energy output}\right] = 0 \tag{11.143}$$

Assuming that between two neighboring volume elements, temperature variation is linear, one can express the heat flow between two volume elements as follows (considering the elements i-1, j and i, j, as an example):

$$-k\frac{\partial T}{\partial x} \approx -k\frac{T_{i,j} - T_{i-1,j}}{\Delta x} = k\frac{T_{i-1,j} - T_{i,j}}{\Delta x} \tag{11.144}$$

Using similar procedure for all faces of the volume element i, j, the following terms for the balance of energy can be written:

- rate of energy input:
    - x direction: $\Delta y.1.k\frac{(T_{i-1,j} - T_{i,j})}{\Delta x}$
    - y direction: $\Delta x.1.k\frac{(T_{i,j-1} - T_{i,j})}{\Delta y}$
- rate of energy output:
    - x direction: $\Delta y.1.k\frac{(T_{i,j} - T_{i+1,j})}{\Delta x}$
    - y direction: $\Delta x.1.k\frac{(T_{i,j} - T_{i,j+1})}{\Delta y}$

The values Δx.1 and Δy.1 represent areas perpendicular to the direction of heat flowr. Although the thermal conductivity is considered same in all elements, it (*k*) can have different values on each side, for materials whose thermal conductivity is a function of temperature. In this case, the thermal conductivity on each side can be estimated using harmonic or arithmetic mean values obtained for each volume element, using their respective temperatures.

Replacing the terms in the equation of energy balance, we have:

$$\Delta y.1.k\frac{T_{i-1,j}-T_{i,j}}{\Delta x}+\Delta x.1.k\frac{T_{i,j-1}-T_{i,j}}{\Delta y}-\Delta y.1.k\frac{T_{i,j}-T_{i+1,j}}{\Delta x}-\Delta x.1.k\frac{T_{i,j}-T_{i,j+1}}{\Delta y}=0 \qquad (11.145)$$

Considering a case where Δx and Δy are equal and thermal conductivity constant and uniform throughout the domain, we obtain the following equation:

$$T_{i-1,j}+T_{i,j-1}+T_{i+1,j}+T_{i,j+1}-4.T_{i,j}=0 \qquad (11.146)$$

Any element of volume within the domain has an equation of conservation of energy similar to those given in equations (11.145) or (11.146).

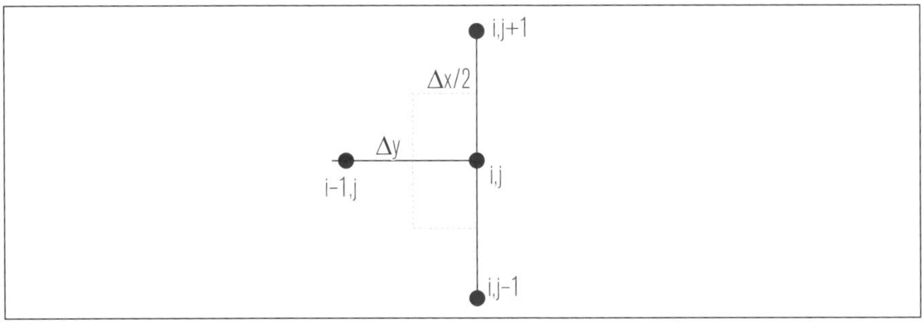

Figure 11.23 - Details of a volume element located on a lateral face.

For a volume element located in a lateral face, (Figure 11.21) and seen in detail in Figure 11.23, the input and output of heat energy are given by:

- rate of energy input:
  - x direction: $\Delta y.1.k\frac{(T_{i-1,j}-T_{i,j})}{\Delta x}$
  - y direction: $\frac{\Delta x}{2}.1.k\frac{(T_{i,j-1}-T_{i,j})}{\Delta y}$

- rate of energy output:
  - x direction: $\Delta y.1.h.(T_{i,j} - T_{\infty})$
  - y direction: $\frac{\Delta x}{2}.1.k\frac{(T_{i,j} - T_{i,j+1})}{\Delta y}$

Replacing these terms in the equation of energy balance, we get:

$$\Delta y.1.k.\frac{(T_{i-1,j} - T_{i,j})}{\Delta x} + \frac{\Delta x}{2}.1.k.\frac{(T_{i,j-1} - T_{i,j})}{\Delta y} - h.\Delta y.(T_{i,j} - T_{\infty}) - \frac{\Delta x}{2}.1.k.\frac{(T_{i,j} - T_{i,j+1})}{\Delta y} = 0 \quad (11.147)$$

For a case where k is constant and uniform and $\Delta x = \Delta y$, we obtain the following simplified version of equation (11.147):

$$2\frac{\Delta x.h}{k}T_{\infty} + 2.T_{i-1,j} + T_{i,j-1} + T_{i,j+1} - (2\frac{\Delta x.h}{k} + 4).T_{i,j} = 0 \quad (11.148)$$

The factor $\Delta x.h/k$ can be defined as Biot number where the characteristic length is the value of Δx, defined in the discretization.

Expressions for the other elements of volume positioned on the faces can be obtained by analogy with equation (11.147).

For the volume element located in a corner (Figure 11.24), the energy balance gives, (assuming k constant and uniform in the domain and $\Delta x = \Delta y$) the following relationship: (prove this relationship as an exercise).

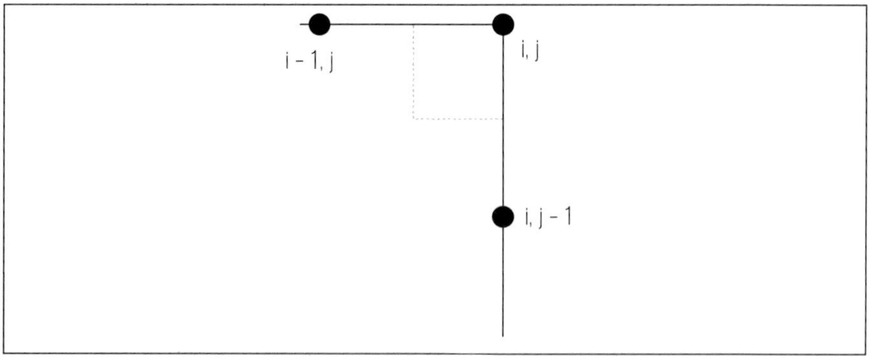

Figure 11.24 - Detail of a volume element located in a corner.

$$2\frac{\Delta x.h}{k}T_{\infty} + T_{i-1,j} + T_{i,j-1} - (2\frac{\Delta x.h}{k} + 2).T_{i,j} = 0 \quad (11.149)$$

Expressions for other volume elements located in the corners can also be obtained by analogy with equation (11.149).

The conservation equations obtained for the different volume elements are all linear equations, where the unknowns are the temperatures. If a given domain is discretized into 100 x 100 volume elements, 10,000 balance equations of energy are obtained. However, in the case of a conservation equation for a given volume element, only the temperatures of neighboring volume elements appear. All temperatures of volume elements farther away will have zero coefficients.To determine the values of temperatures, the system of linear equations must be solved.

11.5.3 Solution of system of linear equations

In matrix notation, the system of linear equations can be written as follows:

$$[A].[T] = [B] \qquad (11.150)$$

where:

- [A] is the coefficient matrix;
- [T] is the vector of temperatures (unknowns);
- [B] the vector of constant terms.

Several conventional techniques, including matrix inversion, can be employed for the solution of the system of linear equations obtained. However, in the specific case of the systems being treated here, these techniques are not very efficient in terms of numbers and usually require large computer memory, compared to the Gauss-Seidel method presented below.

The method of Gauss-Seidel is an iterative method. The first step of this method is to rewrite all the equations of energy balance as follows:

$$T_{i,j} = -\frac{\sum a_{viz}.T_{viz}}{a_{i,j}} + \frac{b_{i,j}}{a_{i,j}} \qquad (11.151)$$

where:

- a_{viz}: are the coefficients of temperatures of neighboring volume elements;
- $a_{i,j}$: the coefficient of temperature $T_{i,j}$;
- $b_{i,j}$: the term independent of the equation of energy balance for the volume element i, j.

The above procedure corresponds to finding the temperature $T_{i,j}$ in the equation of energy balance for the volume element *i, j*. This should be done for all the equations of energy balance.

The second step of the method is to propose an initial *solution* for the temperatures to be calculated.This is necessary to initiate the iterative process, not affecting the final result. However, the initial proposition is important, as it affects the number of iterations required to achieve this result.

From the initial *solution* proposed, one can start the iterative process of substitution in the equations of the form shown in equation (11.151) to correct the initial *solution* and to get a new temperature value. The process is repeated gradually approaching the final solution. Every time one goes through the entire domain, correcting the values of temperature, it is considered one iteration. This process must be repeated until the variations in temperature between two successive iterations fall below a predetermined value, called the convergence criteria.

For the Gauss-Seidel convergence (ie, the corrections applied are getting smaller as it performs the iterations), it is necessary that the following conditions are met:

$$\left|a_{i,j}\right| \geq \sum \left|a_{viz}\right| \tag{11.152}$$

for all equations of the system. For at least one of the equations, one must have:

$$\left|a_{i,j}\right| > \sum \left|a_{viz}\right| \tag{11.153}$$

For problems of heat transfer, these conditions are always met and the method of Gauss-Seidel converges, enabling solution of the system of equations.

The following is an example of numerical solution to determine the profile of temperatures for a situation for which there is an analytical solution. In addition to illustrating the application of the method, the existence of analytical solution enables to evaluate the accuracy of the numerical solution as well as the effect of mesh spacing on the accuracy of results obtained.

The system under consideration is given schematically in Figure 11.25.

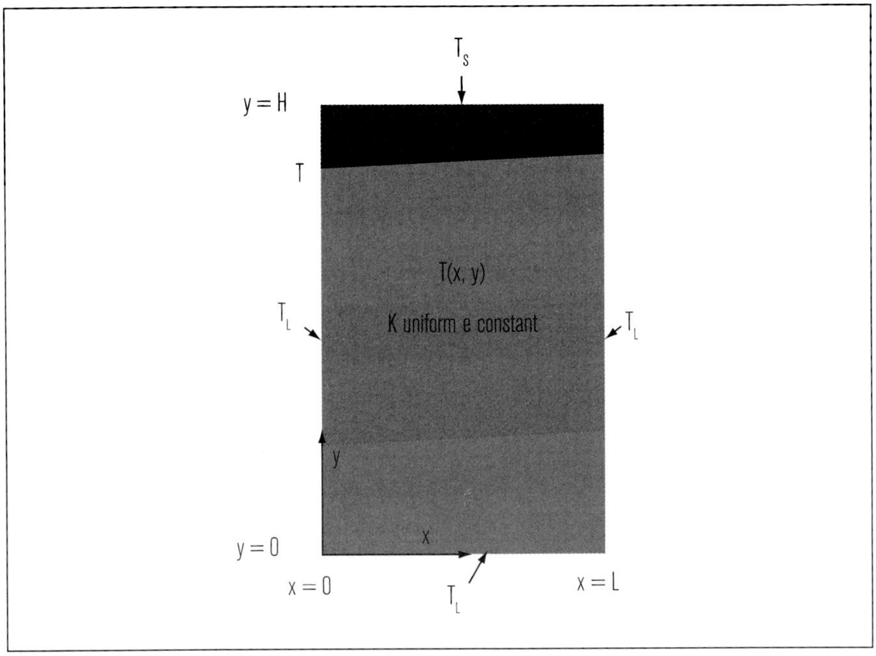

Figure 11.25 - Bidimensional system of heat transfer.

For this system, the analytical solution for the distribution of temperatures is as follows (INCROPERA and DEWITT, 2003):

$$\frac{T(x,y)-T_L}{T_S-T_L}=\frac{2}{\pi}\sum_{n=1}^{\infty}\frac{(-1)^{n+1}+1}{n}\sin\left(\frac{n\pi x}{L}\right)\frac{\sinh(n\pi y/L)}{\sinh(n\pi H/L)} \tag{11.154}$$

Considering the following data:

$T_S = 1000°C \qquad T_L = 100°C \qquad L = H = 0.10$ m.

The details for the system is given in Figure 11.25 and . the discretization of the domain in Figure 11.26.

In the case of numerical methods it is interesting to note that the existence of symmetry planes. significantly simplify the calculation. As can be observed in Figure 11.26, there is a plane of symmetry. Hence, we have the following equalities between temperatures:

$$T_{2,2} = T_{4,2}$$
$$T_{2,3} = T_{4,3}$$
$$T_{2,4} = T_{4,4}$$

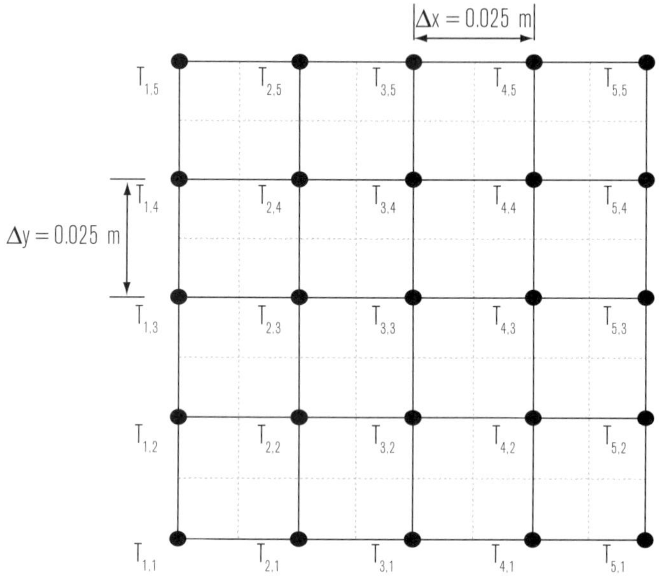

Figure 11.26 - Discretization of the system.

Thus, the unknowns to be determined are the temperatures $T_{2,2}$, $T_{3,2}$, $T_{2,3}$, $T_{3,3}$, $T_{2,4}$ and $T_{3,4}$. Since all these are temperatures of internal volume elements, the energy balances for the volume elements are given using theses in the following expressions:

$$T_{1,2} + T_{2,1} + T_{3,2} + T_{2,3} - 4.T_{2,2} = 0 \qquad \text{(volume element 2,2)}$$
$$T_{2,2} + T_{3,1} + T_{4,2} + T_{3,3} - 4.T_{3,2} = 0 \qquad \text{(volume element 3,2)}$$
$$T_{1,3} + T_{2,2} + T_{3,3} + T_{2,4} - 4.T_{2,3} = 0 \qquad \text{(volume element 2,3)}$$

$$T_{2,3} + T_{3,2} + T_{4,3} + T_{3,4} - 4.T_{3,3} = 0 \qquad \text{(volume element 3,3)}$$
$$T_{1,4} + T_{2,3} + T_{3,4} + T_{2,5} - 4.T_{2,4} = 0 \qquad \text{(volume element 2,4)}$$
$$T_{2,4} + T_{3,3} + T_{4,4} + T_{3,5} - 4.T_{3,4} = 0 \qquad \text{(volume element 3,4)}$$

To solve these system of equations, Gauss-Seidel method is used. The first step in applying this method is to rearrange the above equations in terms of unknown temperatures for the volume elements where the balance of energy have been developed. Thus, we get:

$$T_{2,2} = \frac{T_{1,2} + T_{2,1} + T_{3,2} + T_{2,3}}{4}$$

$$T_{3,2} = \frac{T_{2,2} + T_{3,1} + T_{4,2} + T_{3,3}}{4}$$

$$T_{2,3} = \frac{T_{1,3} + T_{2,2} + T_{3,3} + T_{2,4}}{4}$$

$$T_{3,3} = \frac{T_{2,3} + T_{3,2} + T_{4,3} + T_{3,4}}{4}$$

$$T_{2,4} = \frac{T_{1,4} + T_{2,3} + T_{3,4} + T_{2,5}}{4}$$

$$T_{3,4} = \frac{T_{2,4} + T_{3,3} + T_{4,4} + T_{3,5}}{4}$$

The temperatures $T_{1,j}$, $T_{i,1}$, $T_{5,j}$ and $T_{i,5}$ are obtained through the boundary conditions. For the Gauss-Seidel method, one has to start with an initial solution. The following values are considered for the initial solution:

$$T_{2,2} = T_{3,2} = 200°C$$
$$T_{2,3} = T_{3,3} = 400°C$$
$$T_{2,4} = T_{3,3} = 700°C$$

Applying successively the equations for calculating the temperatures one can reach the solution for the whole system. Table 11.1 shows the results of successive iterations, made with the aid of a spreadsheet. As can be seen, the method converges. The solution obtained for discretization is given below:

$$T_{2,2} = 164.29°C$$
$$T_{3,2} = 188.39°C$$
$$T_{2,3} = 268.75°C$$
$$T_{3,3} = 325.00°C$$
$$T_{2,4} = 485.71°C$$
$$T_{3,4} = 574.11°C$$

Table 11.1 - Results obtained using the method of Gauss-Seidel to a discretization with $\Delta x = \Delta y = 0.025$ m.

T11	T12	T13	T21	T 22	T32	T31	T23	T33	T14	T24	T34	T15	T25	T35
100.00	100.00	100.00	100.00	200.00	200.00	100.00	400.00	400.00	100.00	700.00	700.00	1000.00	1000.00	1000.00
100.00	100.00	100.00	100.00	200.00	225.00	100.00	350.00	425.00	100.00	550.00	700.00	1000.00	1000.00	1000.00
100.00	100.00	100.00	100.00	193.75	231.25	100.00	318.75	406.25	100.00	537.50	631.25	1000.00	1000.00	1000.00
100.00	100.00	100.00	100.00	187.50	223.44	100.00	309.38	375.00	100.00	512.50	620.31	1000.00	1000.00	1000.00
100.00	100.00	100.00	100.00	183.20	212.50	100.00	293.75	365.63	100.00	507.42	600.00	1000.00	1000.00	1000.00
100.00	100.00	100.00	100.00	176.56	208.01	100.00	289.06	350.00	100.00	498.44	595.12	1000.00	1000.00	1000.00
100.00	100.00	100.00	100.00	174.27	200.78	100.00	281.25	345.31	100.00	496.04	586.72	1000.00	1000.00	1000.00
100.00	100.00	100.00	100.00	170.51	198.46	100.00	278.91	337.50	100.00	491.99	584.35	1000.00	1000.00	1000.00
100.00	100.00	100.00	100.00	169.34	194.63	100.00	275.00	335.16	100.00	490.81	580.37	1000.00	1000.00	1000.00
100.00	100.00	100.00	100.00	167.41	193.46	100.00	273.83	331.25	100.00	488.84	579.20	1000.00	1000.00	1000.00
100.00	100.00	100.00	100.00	166.82	191.52	100.00	271.88	330.08	100.00	488.26	577.23	1000.00	1000.00	1000.00
100.00	100.00	100.00	100.00	165.85	190.93	100.00	271.29	328.13	100.00	487.28	576.65	1000.00	1000.00	1000.00
100.00	100.00	100.00	100.00	165.55	189.96	100.00	270.31	327.54	100.00	486.98	575.67	1000.00	1000.00	1000.00
100.00	100.00	100.00	100.00	165.07	189.66	100.00	270.02	326.56	100.00	486.50	575.38	1000.00	1000.00	1000.00
100.00	100.00	100.00	100.00	164.92	189.17	100.00	269.53	326.27	100.00	486.35	574.89	1000.00	1000.00	1000.00
100.00	100.00	100.00	100.00	164.68	189.03	100.00	269.38	325.78	100.00	486.10	574.74	1000.00	1000.00	1000.00
100.00	100.00	100.00	100.00	164.60	188.78	100.00	269.14	325.63	100.00	486.03	574.50	1000.00	1000.00	1000.00
100.00	100.00	100.00	100.00	164.48	188.71	100.00	269.07	325.39	100.00	485.91	574.42	1000.00	1000.00	1000.00
100.00	100.00	100.00	100.00	164.44	188.59	100.00	268.95	325.32	100.00	485.87	574.30	1000.00	1000.00	1000.00
100.00	100.00	100.00	100.00	164.38	188.55	100.00	268.91	325.20	100.00	485.81	574.27	1000.00	1000.00	1000.00
100.00	100.00	100.00	100.00	164.37	188.49	100.00	268.85	325.16	100.00	485.79	574.20	1000.00	1000.00	1000.00
100.00	100.00	100.00	100.00	164.33	188.47	100.00	268.83	325.10	100.00	485.76	574.19	1000.00	1000.00	1000.00
100.00	100.00	100.00	100.00	164.33	188.44	100.00	268.80	325.08	100.00	485.75	574.16	1000.00	1000.00	1000.00
100.00	100.00	100.00	100.00	164.31	188.43	100.00	268.79	325.05	100.00	485.74	574.15	1000.00	1000.00	1000.00
100.00	100.00	100.00	100.00	164.31	188.42	100.00	268.77	325.04	100.00	485.73	574.13	1000.00	1000.00	1000.00
100.00	100.00	100.00	100.00	164.30	188.41	100.00	268.77	325.02	100.00	485.73	574.13	1000.00	1000.00	1000.00
100.00	100.00	100.00	100.00	164.30	188.41	100.00	268.76	325.02	100.00	485.72	574.12	1000.00	1000.00	1000.00
100.00	100.00	100.00	100.00	164.29	188.40	100.00	268.76	325.01	100.00	485.72	574.12	1000.00	1000.00	1000.00
100.00	100.00	100.00	100.00	164.29	188.40	100.00	268.76	325.01	100.00	485.72	574.11	1000.00	1000.00	1000.00
100.00	100.00	100.00	100.00	164.29	188.40	100.00	268.75	325.01	100.00	485.72	574.11	1000.00	1000.00	1000.00
100.00	100.00	100.00	100.00	164.29	188.40	100.00	268.75	325.00	100.00	485.72	574.11	1000.00	1000.00	1000.00
100.00	100.00	100.00	100.00	164.29	188.40	100.00	268.75	325.00	100.00	485.72	574.11	1000.00	1000.00	1000.00
100.00	100.00	100.00	100.00	164.29	188.39	100.00	268.75	325.00	100.00	485.72	574.11	1000.00	1000.00	1000.00
100.00	100.00	100.00	100.00	164.29	188.39	100.00	268.75	325.00	100.00	485.72	574.11	1000.00	1000.00	1000.00
100.00	100.00	100.00	100.00	164.29	188.39	100.00	268.75	325.00	100.00	485.71	574.11	1000.00	1000.00	1000.00
100.00	100.00	100.00	100.00	164.29	188.39	100.00	268.75	325.00	100.00	485.71	574.11	1000.00	1000.00	1000.00
100.00	100.00	100.00	100.00	164.29	188.39	100.00	268.75	325.00	100.00	485.71	574.11	1000.00	1000.00	1000.00
100.00	100.00	100.00	100.00	164.29	188.39	100.00	268.75	325.00	100.00	485.71	574.11	1000.00	1000.00	1000.00

Using the expression for analytical solution , equation (11.154) one can get the values for the corresponding positions of x and y:

$T_{2,2}$ = 161.17°C (x = 0.025 m, y = 0.025 m)

$T_{3,2}$ = 185.87°C (x = 0.050 m, y = 0.025 m)

$T_{2,3}$ = 263.83°C (x = 0.025 m, y = 0.050 m)

$T_{3,3}$ = 325.00°C (x = 0.050 m, y = 0.050 m)

$T_{2,4}$ = 488.83°C (x = 0.025 m, y = 0.075 m)

$T_{3,4}$ = 586.48°C (x = 0.050 m, y = 0.075 m)

The differences observed between the numerical solution and analytical are due to coarse discretization.With refinement in the mesh discretization, one can obtain numerical solutions closer to the analytical. The values obtained with values of $\Delta x = \Delta y = 0.005$ m are closer to those predicted by analytical solution (for comparison, the temperature values found in the positions are shown along with results of the analytical solution in the table above):

T = 161.31°C (x = 0.025 m, y = 0.025 m)

T = 186.00°C (x = 0.050 m, y = 0.025 m)

T = 264.10°C (x = 0.025 m, y = 0.050 m)

T = 325.00°C (x = 0.050 m, y = 0.050 m)

T = 488.68°C (x = 0.025 m, y = 0.075 m)

T = 585.77°C (x = 0.050 m, y = 0.075 m)

As can be seen, these values are closer to the analytical solution than those obtained with the coarser discretization.

References

CHO, J. et al. Heat transfer across mold flux film in mold during initial solidification in continuous casting of steel. *ISIJ International*, v. 38, n. 8, p. 834-42, Aug. 1998a.

CHO, J. et al. Thermal resistance at the interface between mold flux film and mold for continuous casting of steels. *ISIJ International*, v. 38, n. 5, p. 440-6, May 1998b.

GEIGER, G.H.; POIRIER, D.R. *Transport phenomena in metallurgy*. Massachussetts: Addison-Wesley, 1980.

INCROPERA, F.P.; DEWITT, D.P. *Fundamentos da transferência de calor e massa*. 5. ed. Rio de Janeiro:LTC, 2003.

Exercises

1 - Determine which of the following figures represents the temperature profile at steady state in a plate of a material whose thermal conductivity decreases with increasing temperature.Explain the answer.

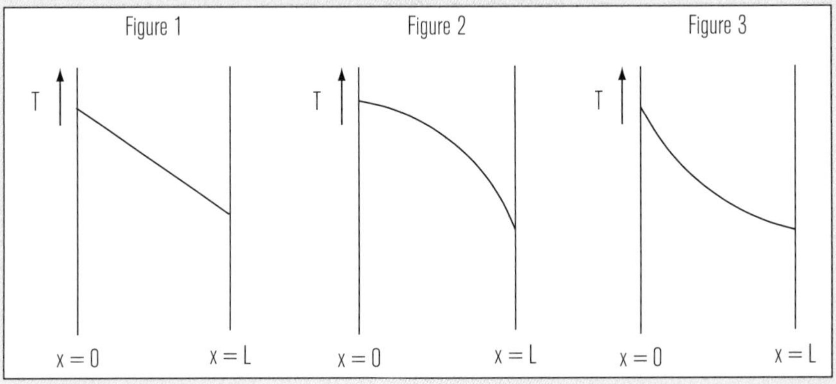

2 - Determine, which of the following figures represents the temperature profile of a plate that has constant thermal conductivity and undergoing cooling. Justify the answer.

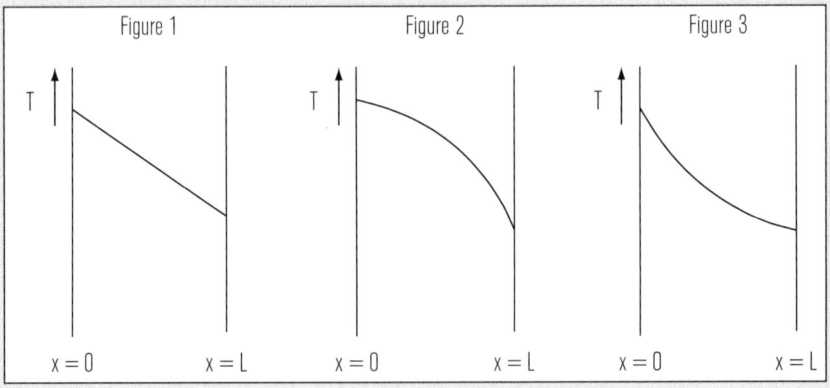

3 - For the system given in the figure below, obtain an equation for the temperature profile within the refractory lining through an energy balance. Consider steady state and that the material has constant physical properties.

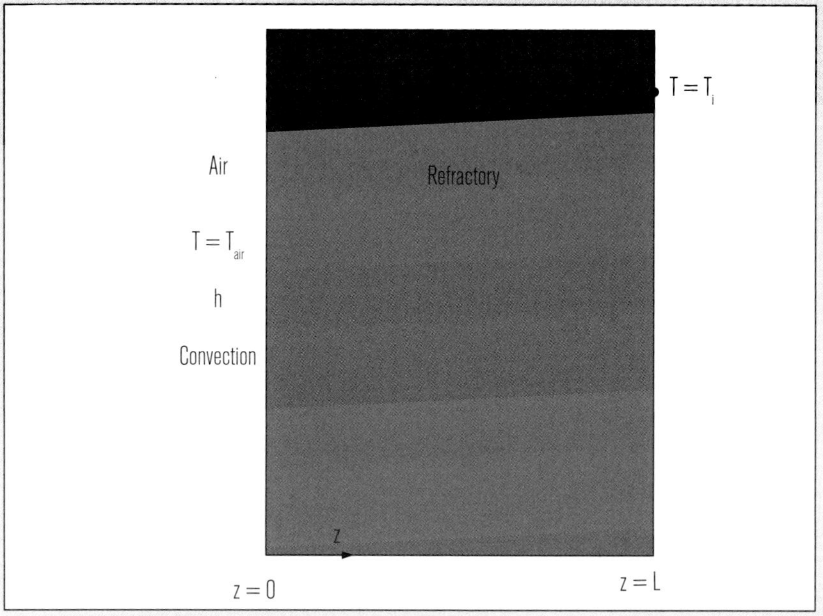

Determine the value of the temperature of the refractory at z = 0, considering the following data:

- $T_i = 1200°F$;
- $T_{air} = 25°C$;
- thermal conductivity of refractory, k = 0.60 BTU/h.ft.°F;
- specific heat of the refractory, $C_p = 0.27$ BTU/lb_m.°F;
- density of the refractory, ρ = 171 lb_m/ft^3;
- coefficient of heat transfer, h = 15 W/m^2.K;
- thickness, L = 30 cm.

Analyze the effect of heat transfer coefficient, h, on the surface temperature of the refractory at z = 0.

- 1 lb_m = 0.45359 kg;
- 1 ft = 0.3048 m;
- 1 lb_f = 4.4482 N;
- 1 in = 0.0254 m;
- 1 BTU = 1055 J;
- T(°C) = 5/9 [T(°F) – 32].

4 - For the system seen in the figure below.temperatures at points 1 and 2 were measured, obtaining values of 150 and 300°C, respectively, when steady state was reached. From the data provided in the following figure, evaluate the thermal conductivity of refractory. Determine the heat flux at z = 0.1 m.

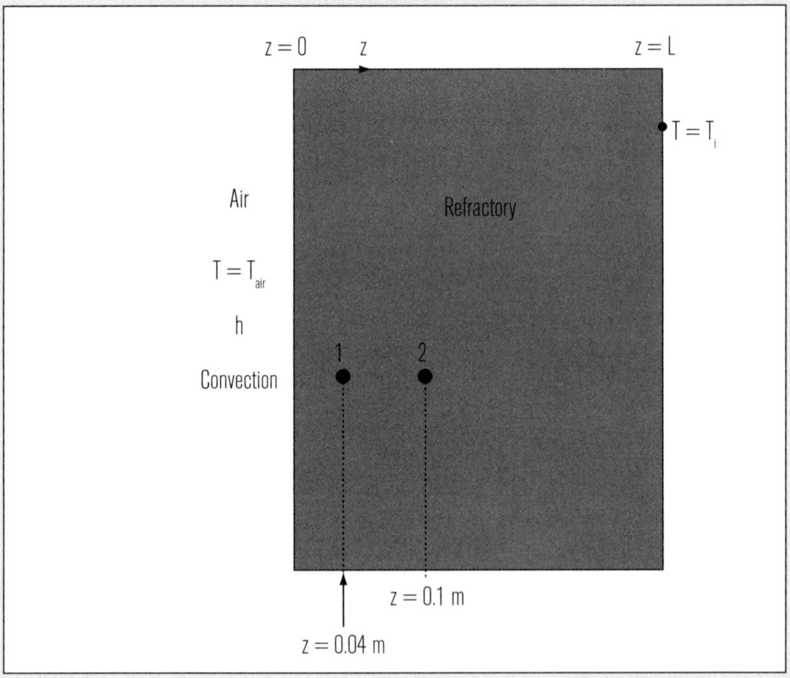

Determine the temperature of the refractory at z = 0 and z = L. Consider:

- T_{air} = 25°C;
- Specific heat of the refractory, C p = 0.27 BTU/lb_m. deg F
- Density of the refractory, r = 171lb_m/ft^3;
- Heat transfer coefficient , h = 15 W/m^2.K;
- Thickness, L = 30 cm.

Provide all answers in international unit system.

5 - For the system seen in the figure below. In x = 0, the temperature is 300°C. At, x = L, the surface of the plate is in contact with water, whose temperature is 25°C. The coefficient of heat transfer between the plate and the water is 200 W/m².K. There is also heat generation due to the passage of current through the plate. This generation is 27,000 W/m³. Knowing that the plate is aluminum (density = 2700 kg/m³, specific heat = 900 J/kg.K and thermal conductivity = 220 W/mK), determine the location and value of the maximum temperature inside the plate. L = 10 cm.

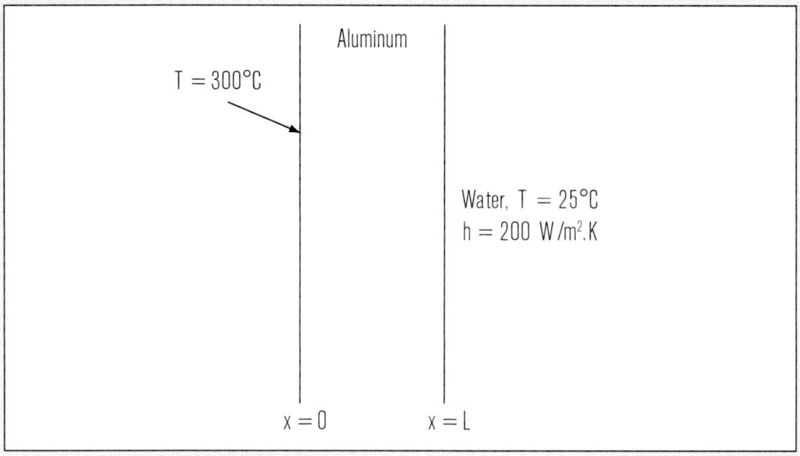

6 - For the system seen in the figure below, determine the value of temperature T_1. Consider that the material is copper (density = 8900 kg/m³, specific heat = 385 J/kg.K and thermal conductivity = 400 W/m.K). All temperatures are given in celsius. The distance between nodes is 5 mm, horizontally and vertically.

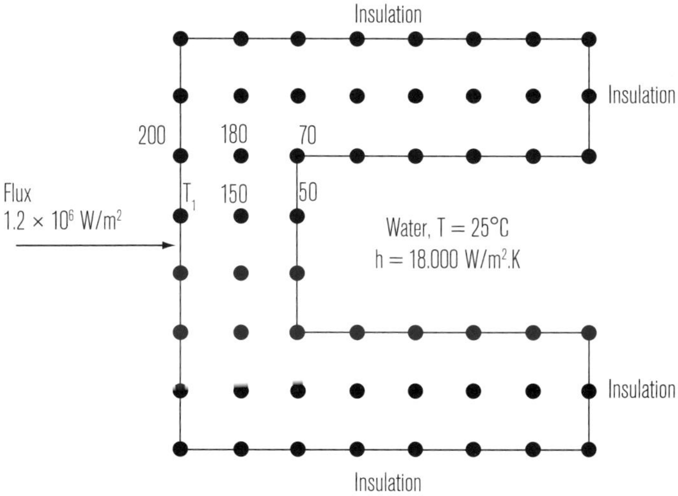

7 - A surface whose temperature is kept at 400°C has a layer of insulating material with a thickness of 30 mm and thermal conductivity of 0.2 W/mK. If the air temperature is 30°C and the coefficient of heat transfer between air and outer surfaces of the insulation is equal to 400 W/m^2.K, what is the temperature of the outer surface of the insulation? What is the flow of heat through the insulation layer?

8 - The temperature profile through the wall of furnace consisting of two layers of insulation is shown in the figure below. Which of the two insulation materials has the lowest thermal conductivity? Explain your answer.

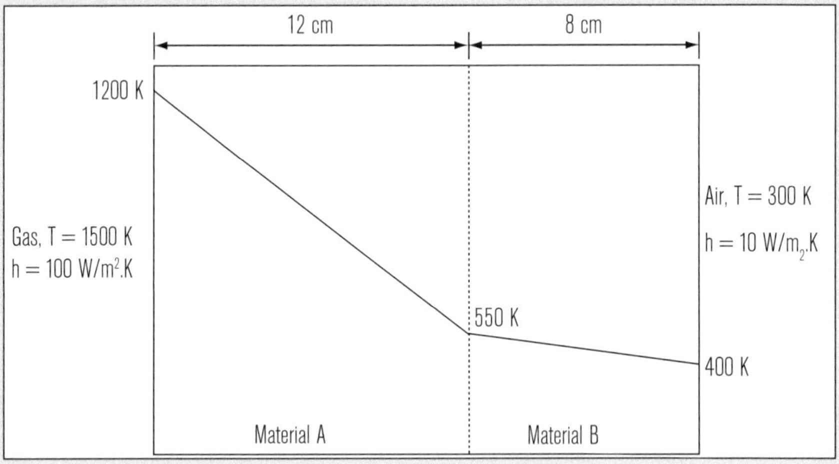

9 - A wire,conductor of electricity,has 1.5 mm in diameter and ithe external surface has a plastic insulation of 1 mm thickness and thermal conductivity of 0.25 W/mK. The interface between the wire and the plastic has a contact resistance of 1 x 10^{-3}m^2.K/W. The coefficient of heat transfer on the outer surface of the coating is 15 W/m^2.K. If the temperature of the insulation should not exceed 320 K, what maximum power that can be dissipated by the wire? Consider that its length is 5 m and the room temperature is 293 K. Determine the temperature of the interface of the wire with the insulating layer.

Unsteady State Heat Conduction

Chapter 12

In the previous chapter general equations for heat conduction for steady and unsteady states for cartesian cylindrical and spherical coordinates were derived. However only steady state situations were treated with examples.

In this chapter unsteady state or transient problems, involving temperature variation with position and time will be dealt with. There are many unsteady state situations of interest to the metallurgist and materials engineer, like solidification of metals, heat losses in ladles and furnaces, heating and cooling rate of metals and materials etc. In the heat treatment of metals, for example, temperature variation with time (cooling rate) has a decisive role in defining the structure of a metal or alloy and hence its properties. The reheating of metal for mechanical working processes, and residence time of metallic parts inside the furnace for reaching desired temperatures are other examples. For making these processes efficient and productive, a basic understanding of unsteady state heat transfer is necessary.

In transient problems, Biot number, defined in the previous chapter, has great importance Figure 12.1 shows schematically the effect of Biot number on heat transfer in the transient regime. Values of Biot number well

Figure 12.1 - Effect of Biot number temperature profiles in transient conditions.

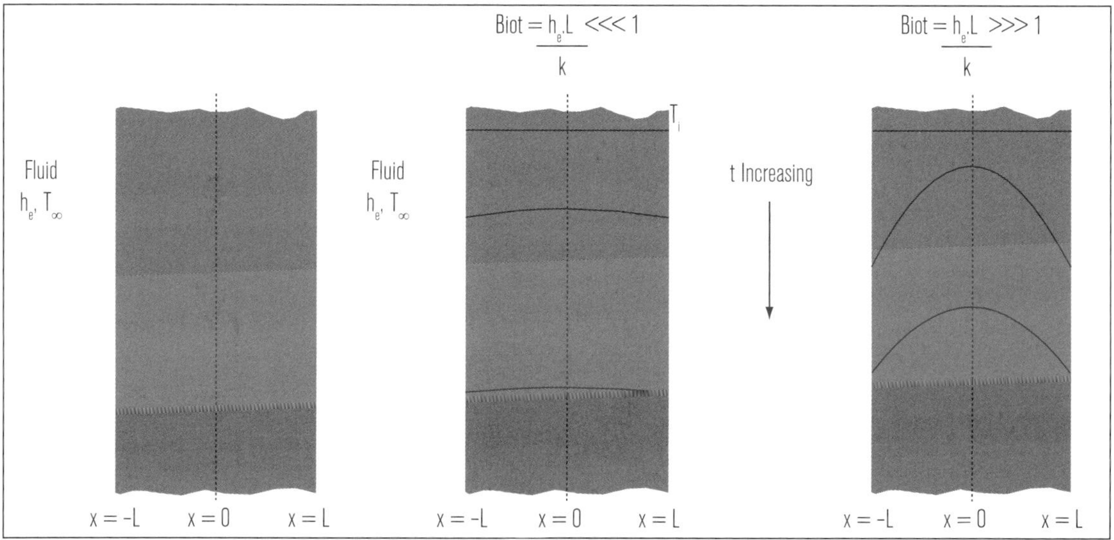

below 1, indicate that cooling or heating tends to occur in isothermal mode, that is to say with negligible temperature gradients inside the material. On the other hand higher Biot numbers signify higher temperature gradients in the material.

Typically, in the treatment of transient problems, for Biot numbers less than 0.1, it is reasonable to assume that the piece is heated or cooled isothermally, that is only one temperature is sufficient to describe the temperature variation of material. This approach has given better results, lower the value of Biot number.

Figure 12.1, shows that the characteristic length that appears in the definition of the Biot number is the semi-thickness of the flat plate. For cylindrical and spherical parts, the characteristic length is the radius. Some authors (INCROPERA and DEWITT, 2003) have suggested that the characteristic length can be evaluated from the relationship:

$$L = \frac{V}{A_{sup}} \tag{12.1}$$

where:

- V = volume;
- A_{sup} = surface area of the part.

For plane geometry, the characteristic length defined by equation (12.1) becomes half thickness of the piece.For cylindrical and spherical parts, equation (12.1) provides characteristic lengths equivalent to R/2 and R/3 (R = radius), respectively. (Prove this as an exercise). However, the criterion cited earlier (characteristic length equal to the radius) is normally preferred to that proposed by equation (12.1) Isothermal heating or cooling is also known as Newtonian cooling and heating.

12.1 Isothermal Heating or Cooling

When the heating or cooling occurs isothermally, temperature variation in the part as a function of time can be evaluated from simple energy balance.

Consider a metal of high thermal conductivity with volume V and surface area $A_{sup,}$ exchanging heat with a fluid at temperature T_∞ having a heat transfer coefficient h.

$$\begin{bmatrix}\text{Rate of input}\\ \text{of energy}\end{bmatrix} - \begin{bmatrix}\text{Rate of output}\\ \text{of energy}\end{bmatrix} = \begin{bmatrix}\text{Rate of accumulation}\\ \text{of energy}\end{bmatrix} \tag{12.2}$$

$$0 - h.A_{sup}.(T - T_{\infty}) = V.\rho.C_p.\frac{dT}{dt} \qquad (12.3)$$

where:

- ρ, C_p and T are density, specific heat at constant pressure and temperature of the piece, respectively.

The equation (12.3) can be applied for both heating and cooling. In the case of heating, T_{∞} is greater than T and the heat output is negative, that is to say, it becomes heat input. On cooling, T is greater than T_{∞} and signifies heat output.

Rearranging equation (12.3), we obtain an equation for the rate of heating or cooling:

$$\frac{dT}{dt} = -\frac{h.A_{sup}.(T - T_{\infty})}{V.\rho.C_p} \qquad (12.4)$$

Using the equation above, it is possible to identify the main factors affecting dT/dt:

- A_{sup}/V: Larger the surface area per unit volume, higher is the rate of heating/cooling of the body;
- $V\rho.C_p$: This product is the heat capacity. Higher is this value lower is the rate of change of temperature of the piece;
- h: Higher the values of heat transfer coefficient higher are the rates of heating/cooling.The parameters that affect the value of h will be discussed in the next chapter;
- $(T\text{-}T_{\infty})$ Greater the temperature difference between the body and the fluid with which it exchanges heat, higher is the rate of heating/cooling.This indicates that early in the process of heat transfer when there is maximum difference between the temperatures of the body and the fluid, the rate of change of temperature is higher.

Considering the physical properties and heat transfer coefficient constant, one can integrate equation (12.4):

$$\ln(T - T_{\infty}) = -\frac{h.A_{sup}}{V.\rho.C_p}.t + C_1 \qquad (12.5)$$

In this case, the integration constant C_1 can be determined using an initial condition, which establishes the temperature value at time t = 0.

- For the Initial condition: $t = 0$, $T = T_i$ we obtain:

$$C_1 = \ln(T_i - T_\infty) \tag{12.6}$$

Substituting 12.6 in 12.5 we get:

$$\ln\frac{(T - T_\infty)}{(T_i - T_\infty)} = -\frac{h.A_{sup}}{V.\rho.C_p}.t \tag{12.7}$$

$$\frac{(T - T_\infty)}{(T_i - T_\infty)} = \exp\left(-\frac{h.A_{sup}}{V.\rho.C_p}.t\right) \tag{12.8}$$

The left side is the dimensionless temperature, ranging between 0 and 1. It has the value 1 at $t = 0$ and tends to zero as time increases. This is valid, regardless of whether the process is heating or cooling.

Figure 12.2 illustrates schematically the variation of dimensionless temperature with time. Two curves can be constructed for different values of $h.A_{sup}/V.\rho.C_p$. Lower the value of this parameter, more pronounced are the variations in temperature.

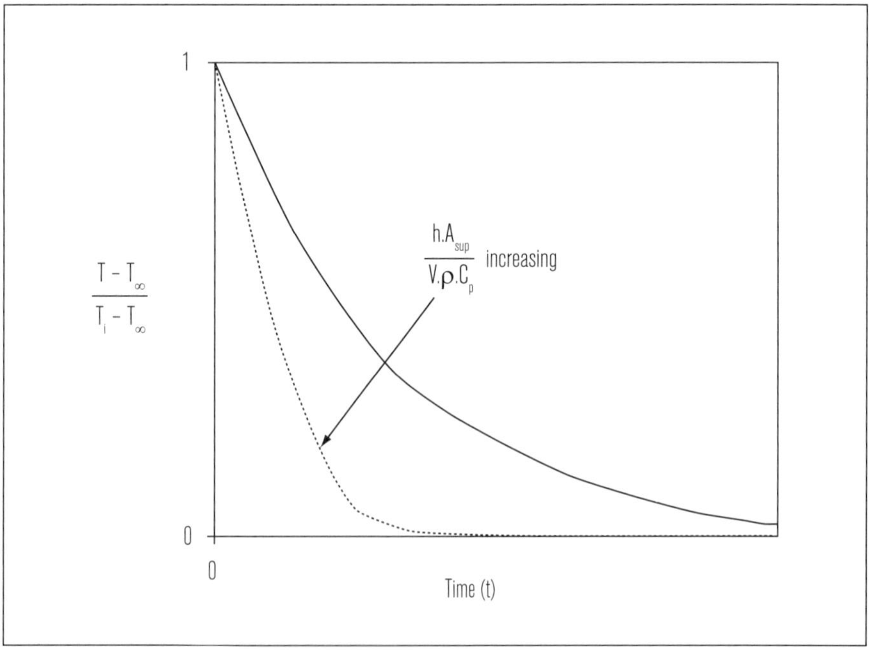

Figure 12.2 - Variation of dimensionless temperature with time, for the case of isothermal heating/cooling.

Example

Cylindrical parts of carbon steel with 5 cm in diameter are heated in a furnace. The gases inside the furnace are maintained at 1400°C and heat transfer coefficient is 80 W/m².K. These parts are fed into the furnace at 25°C. Determine the time that they should remain in the furnace to reach a minimum temperature of 1000°C.

Solution

Initially Biot number should be evaluated. Considering the properties of steel at an average temperature of 500°C, we have:

$$\rho = 7832 \text{ kg/m}^3 \quad \overline{C_p} = 541 \text{ J/kg.K} \quad \overline{k} = 51.3 \text{ W/m.K}$$

The Biot number is given by:

$$Bi = \frac{h.R}{\overline{k}} = \frac{(80).(0.025)}{51.3} = 0.039 < 0.1$$

As the value obtained, is less than 0.1, it is isothermal heating. In this case, the heating time can be estimated through the following relationship:

$$\frac{T - T_\infty}{T_i - T_\infty} = \exp\left(-\frac{h.A_{sup}.t}{V.\rho.C_p}\right)$$

Rearranging the equation:

$$t = -\ln\left(\frac{T - T_\infty}{T_i - T_\infty}\right)\frac{V.\rho.C_p}{h.A_{sup}}$$

From the data provided, we have:

$$T = 1000°C \qquad T_i = 25°C \qquad T_\infty = 1400°C$$

$$h = 80 \text{ W/m}^2.K \qquad V = \frac{\pi.D^2}{4}.L \qquad A_{sup} = \pi.D.L$$

$$\frac{V}{A_{sup}} = \frac{D}{4}$$

Substituting the values in the equation:

$$t = -\ln\left(\frac{1000 - 1400}{25 - 1400}\right)\frac{(0.05).(7832).(541)}{(80).(4)} = 817.5 \text{ s}$$

When the physical properties and the heat transfer coefficient vary with temperature and/or there is a role of radiation, equation (12.4) can no longer be integrated analytically. In this case, the numerical integration has to be adopted similar to the procedure described in Chapter 10, for the case of wire with Joule heating.

When the Biot number is greater than 0.1, the treatment presented here is no longer valid. In this case, the temperature gradients become more significant and temperature variation within the piece also should be taken into account. This is considered in the next section (transient one dimensional cases). Subsequently, the treatment can be extended to situations of two and three-dimensional cases.

12.2 One Dimensional Unsteady Heat Transfer in a Flat Plate

The situation considered is given schematically in Figure 12.3.

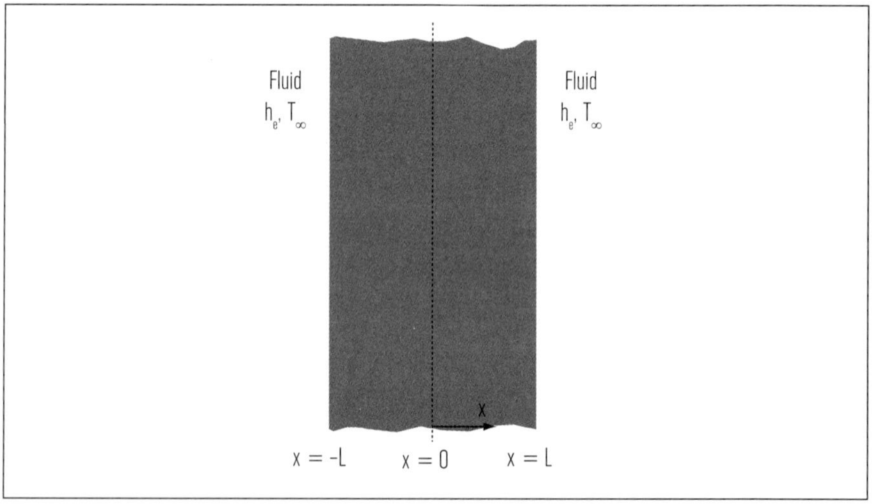

Figure 12.3 - Unsteady state heat transfer in infinite flat plate.

Assuming, constant physical properties (k, ρ and C_p), with no heat generation as well as neglecting heat transfer by radiation on surfaces at x = –L x = L and for unidimensional heat transfer in the x direction only, we can simplify the general equation of energy balance in cartesian coordinates (11.19):

- Equation of energy balance:

$$k\frac{\partial^2 T}{\partial x^2} = \rho.C_p.\frac{\partial T}{\partial t} \quad (12.9)$$

$$\alpha \frac{\partial^2 T}{\partial x^2} = \frac{\partial T}{\partial t} \tag{12.10}$$

- boundary conditions:

$$x = -L \ : \ -k\frac{\partial T}{\partial x}\Big|_{x=-L} = h[T_\infty - T(x = -L,t)] \tag{12.11}$$

$$x = L \ : \ -k\frac{\partial T}{\partial x}\Big|_{x=L} = h[T(x = L,t) - T_\infty] \tag{12.12}$$

- initial condition:

$$T(x, t = 0) = T_i \quad \text{(uniform)} \tag{12.13}$$

As the boundary conditions at x = -L and x = L are similar, the temperature profile will be symmetrical to the origin (position x = 0). Hence, one of the boundary conditions above can be replaced by:

$$x = 0 \ : \quad -k.\frac{\partial T}{\partial x}\Big|_{x=0} = 0 \tag{12.14}$$

Analyzing the conservation equation and boundary conditions, the temperature distribution inside the body is a function of several variables, as given below:

$$T(x,t) = f(x, L, T_i, T_\infty, k, \alpha, h) \tag{12.15}$$

These variables can be grouped into dimensionless numbers (as discussed in Chapter 7 for the case of fluid flow) The temperature profile can be expressed as a function of of the following dimensionless groups:

- Dimensionless temperature, θ:

$$\theta = \frac{T(x,t) - T_\infty}{T_i - T_\infty} \tag{12.16}$$

- Dimensionless position, x*:

$$x^* = \frac{x}{L} \tag{12.17}$$

- Dimensionless time, Fourier number, (Fo):

$$t^* = Fo = \frac{\alpha.t}{L^2} \tag{12.18}$$

- Biot number:

$$Bi = \frac{h.L}{k} \tag{12.19}$$

With these variables, one can rewrite the equation (12.15) in the form:

$$\theta = f(x^*, Fo, Bi) \tag{12.20}$$

Considering the above, the equation (12.10) can be solved analytically. The analytical solution, is given as follows (INCROPERA and DEWITT, 2003):

$$\theta = \sum_{n=1}^{\infty} C_n.\exp(-\varsigma_n^2.Fo).\cos(\varsigma_n.x^*) \tag{12.21}$$

where C_n is given by:

$$C_n = \frac{4.\sin(\varsigma_n)}{2.\varsigma_n + \sin(2.\varsigma_n)} \tag{12.22}$$

The parameter ζ_n represents the positive roots of the following equation:

$$\varsigma_n.\tan(\varsigma_n) - Bi = 0 \tag{12.23}$$

A procedure that determines the first four roots of equation (12.23) for a given biot number has been implemented in the *worksheet-roots plane.xls* contained in the CD, accompanying this book. The roots that can be determined using the *SOLVER* function of *Excel*, specifying the range of values where the root must be found.

Table 12.1 presents values of ζ_n for the first four roots, for selected values of Biot number.

The solution of equation (12.21) is an infinite series. As the Fourier number (Fo) increases, the number of terms needed to achieve convergence diminishes. This can be seen in the spread sheet giving *transiente1D-plane.xls*. When the Fourier number is greater than 0.2, just one term of the series is enough to calculate the temperatures. In this case, the temperature θ is given by:

$$\theta = C_1.\exp(-\varsigma_1^2.Fo).\cos(\varsigma_1.x^*) \tag{12.24}$$

In the center of the plate, $x^* = 0$, the temperature is given by:

$$\theta_0 = C_1.\exp(-\varsigma_1^2.Fo) \tag{12.25}$$

$Bi = \frac{h.L}{k}$	ζ_1	ζ_2	ζ_3	ζ_4
0	0	3.1416	6.2832	9.4248
0.001	0.0316	3.1419	6.2833	9.4249
0.002	0.0447	3.1422	6.2835	9.4250
0.004	0.0632	3.1429	6.2838	9.4252
0.006	0.0774	3.1435	6.2841	9.4254
0.008	0.0893	3.1441	6.2845	9.4256
0.01	0.0998	3.1448	6.2848	9.4258
0.02	0.1410	3.1479	6.2864	9.4269
0.04	0.1987	3.1543	6.2895	9.4290
0.06	0.2425	3.1606	6.2927	9.4311
0.08	0.2791	3.1668	6.2959	9.4333
0.1	0.3111	3.1731	6.2991	9.4354
0.2	0.4328	3.2039	6.3148	9.4459
0.3	0.5218	3.2341	6.3305	9.4565
0.4	0.5932	3.2636	6.3461	9.4670
0.5	0.6533	3.2923	6.3616	9.4775
0.6	0.7051	3.3204	6.3770	9.4879
0.7	0.7506	3.3477	6.3923	9.4983
0.8	0.7910	3.3744	6.4074	9.5087
0.9	0.8274	3.4003	6.4224	9.5190
1.0	0.8603	3.4256	6.4373	9.5293
1.5	0.9882	3.5422	6.5097	9.5801
2.0	1.0769	3.6436	6.5783	9.6296
3.0	1.1925	3.8088	6.7040	9.7240
4.0	1.2646	3.9352	6.8140	9.8119
5.0	1.3138	4.0336	6.9096	9.8928
6.0	1.3496	4.1116	6.9924	9.9667
7.0	1.3766	4.1746	7.0640	10.0339
8.0	1.3978	4.2264	7.1263	10.0949
9.0	1.4149	4.2694	7.1806	10.1502
10.0	1.4289	4.3058	7.2281	10.2003

Table 12.1 - First four roots of equation (12.23) for different values of Biot number (INCROPERA and DEWITT, 2003).

$Bi = \frac{h.L}{k}$	ζ_1	ζ_2	ζ_3	ζ_4
15.0	1.4729	4.4255	7.3959	10.3898
20.0	1.4961	4.4915	7.4954	10.5117
30.0	1.5202	4.5615	7.6057	10.6543
40.0	1.5325	4.5979	7.6647	10.7334
50.0	1.5400	4.6202	7.7012	10.7832
60.0	1.5451	4.6353	7.7259	10.8172
80.0	1.5514	4.6543	7.7573	10.8606
100.0	1.5552	4.6658	7.7764	10.8871
∞	1.5708	4.7124	7.8540	10.9956

Example

A steel plate with 5 m Length, 20 cm thick and 1.0 m wide, initially at 400 K, is placed in a reheating furnace. The temperature inside is maintained at 1500 K. Assuming a residence time in the furnace of 50 minutes, determine the maximum temperature difference in the plate, when it is removed from the furnace. What is the minimum residence time in the furnace to ensure that the plate comes out at a temperature of at least 1,280 deg.K? Given: Heat transfer coefficient: 200 W/m².K. Consider heat transfer only along the thickness of the plate.

Solution

Start by calculating the Biot number to check the heating rate, with the following values for the properties of steel:

$$\rho = 7830 \text{ kg/m}^3;\ C_p = 550 \text{ J/kg.K; e k} = 48 \text{ W/m.K.}$$

Considering heat transfer along the thickness of the plate, Biot number is given by:

$$Bi = \frac{h.L}{k} = \frac{(200).(0,10)}{48} = 0,42 > 0,1 \text{ (nonisothermal)}$$

The Fourier number is given by:

$$Fo = \frac{\alpha.t}{L^2} = \frac{k.t}{\rho.C_p.L^2} = \frac{(48).(50 \times 60)}{(7830).(550).(0.1)^2} = 3.34$$

Using the worksheet that accompanies this text, one can determine:

$$Bi = 0.42: \ \varsigma_1 = 0.604 \quad C_1 = 1.06$$

As Fo > 0.2, only the first term of the series is sufficient for calculations of temperatures in the center and the plate surface (the difference between these two temperatures, determines the maximum temperature difference in the plate). From the results provided by the spreadsheet, we get:

$$\theta_0\,(\text{center}) = 0.313;$$
$$\theta_s\,(\text{surface}) = 0.258.$$

Using the definition of dimensionless temperature, one can determine the difference in temperature at the center and the surface of the plate. The values obtained are:

$$T(\text{center}) = 1155.7 \text{ K};$$
$$T(\text{surface}) = 1216.2 \text{ K}.$$

The maximum temperature is 60.5 K.

The time required for heating the center part reaches the temperature of 12.80 K (temperature of the plate) is obtained from the relationship:

$$\theta_0 = C_1.\exp\left(-\varsigma_1^2 \frac{k.t}{\rho.C_p.L^2}\right)$$

Substituting the we obtain:

$$\frac{1280-1500}{400-1500} = (1.06).\exp\left[-(0,604)^2 \frac{(48).t}{(7830).(550).(0.10)^2}\right]$$

$$t = 4,101 \text{ s} \qquad Fo = 4.57$$

As Fo> 0.2, only the first term of the series is sufficient for the calculation of the temperature.

When the number of Fo is greater than 0.2, it is possible to obtain an expression for the rate of change of temperature, dT/dt.

Differentiating θ in the Fourier number (Fo) and applying the chain rule, we get:

$$\frac{dT}{dt} = (T_i - T_\infty)\frac{\alpha}{L^2}.C_1.(-\varsigma_1^2).\exp(-\varsigma_1^2.Fo).\cos(\varsigma_1.x^*) \quad (12.26)$$

(Demonstrate this relationship as an exercise from equations (12.24) and (12.25).

Another relevant parameter is the amount of energy supplied to or removed from the plate during a given time.This value is estimated by the following equation:

$$E = \int_0^V \rho.C_p.[T(x,t) - T_i].dV \quad (12.27)$$

where:

- dV is a differential volume of the piece.

  In the case of plane geometry, the volume differential is given by:

$$dV = A.dx \quad (12.28)$$

where:

- A is the cross-sectional area part in the direction perpendicular to the direction of heat transfer.

  In the evaluation of energy, by convention energy stored in the piece is positive (heating) and energy removed is negative (cooling).

The maximum amount of energy that can be removed or supplied to the piece is:

$$E_{max} = \rho.C_p.(T_i - T_\infty).V \quad (12.29)$$

This result is obtained considering constant physical properties of the material. When Fo is greater than 0.2, the integral expressed by equation (12.27) can be solved, yielding (INCROPERA and DEWITT, 2003):

$$\frac{E}{E_{max}} = 1 - \frac{\sin(\varsigma_1)}{\varsigma_1}.\theta_0 \quad (12.30)$$

12.3 Heat Transfer in Transient One Dimensional Cylindrical Geometry

The system considered in this case is shown schematically in Figure 12.4. Heat transfer is considered only in the radial direction. This approach is reasonable for long cylinders, which have lengths of at least ten times the radius R.

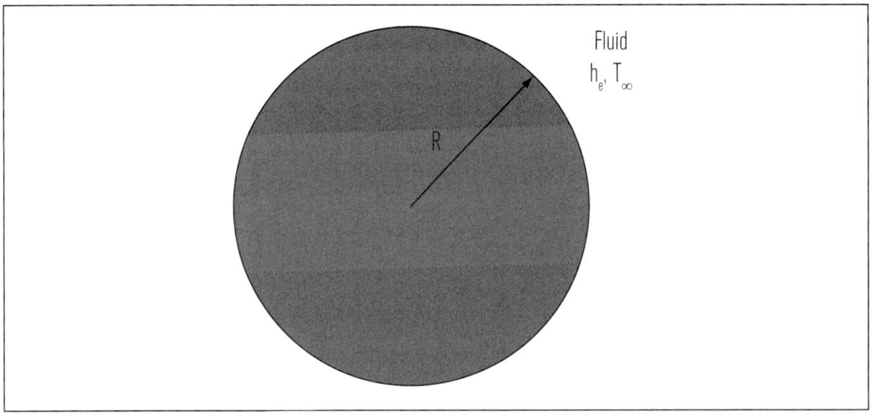

Figure 12.4 - Transient heat transfer in the radial direction for infinite cylinder.

For this system, the definition of dimensionless variables becomes:

- dimensionless position, r*:

$$r^* = \frac{r}{R} \tag{12.31}$$

- dimensionless time, Fourier number, Fo:

$$t^* = Fo = \frac{\alpha.t}{R^2} \tag{12.32}$$

- biot number:

$$Bi = \frac{h.R}{k} \tag{12.33}$$

For boundary conditions analogous to the case of heat transfer in plane geometry, the dimensionless temperature profile is given by (INCROPERA and DEWITT, 2003):

$$\theta = \sum_{n=1}^{\infty} C_n.\exp(-\varsigma_n^2.Fo).J_0(\varsigma_n.r^*) \tag{12.34}$$

where C_n is determined from the following equation:

$$C_n = \frac{2}{\varsigma_n} \cdot \frac{J_1(\varsigma_n)}{J_0^2(\varsigma_n) + J_1^2(\varsigma_n)} \qquad (12.35)$$

The parameter ζ_n represents the positive roots of the following equation:

$$\varsigma_n \cdot \frac{J_1(\varsigma_n)}{J_0(\varsigma_n)} - Bi = 0 \qquad (12.36)$$

J_0 and J_1 are Bessel functions of first orders 0 and 1, respectively. Their values can be obtained in *Excel* by typing the function argument, x, and the order of the function.

Using the *spreadsheet-roots cylindrical xls* it is possible to determine the first four roots of equation (12.36), depending on the value of the Biot number.The other roots can be determined using procedures similar to that implemented in the spreadsheet above.

The series expressed by equation (12.34) is an infinite series. Again the number of terms needed to obtain convergence decreases with the increase in the Fourier number. For Fourier values greater than 0.2, only one term of the series is sufficient for the calculation of the temperature. The spreadsheet *transient-1D-cylindrical allows* the calculation of temperatures along the radial positions in a cylinder, including the first four terms in the series.

When the Fourier number is greater than 0.2, the temperature can be calculated by the equation:

$$\theta = C_1.\exp(-\varsigma_1^2.Fo).J_0(\varsigma_1.r^*) \qquad (12.37)$$

In the center of the cylinder, r* = 0, the temperature is given by:

$$\theta_0 = C_1.\exp(-\varsigma_1^2.Fo) \qquad (12.38)$$

When Fo is greater than 0.2, the amount of energy supplied or removed during a certain time interval is given by:

$$\frac{E}{E_{máx}} = 1 - \frac{2.\theta_0}{\varsigma_1}.J_1(\varsigma_1) \qquad (12.39)$$

Example

Long cylindrical bars of plain carbon steel with a diameter of 190 mm are heated , during passage through a furnace, whose temperature is maintained at 1,600°C. The furnace is 5 m long and the bars come inside at 50°C. What should be the speed of passage of the bars through the furnace, so that they come out at a temperature of at least 100°C? The heat transfer coefficient is 150 W/m².K. What is the maximum temperature difference in the bar when it gets out of the furnace under the above conditions?

Solution

The speed of displacement of the plates can be assessed by determining the heating time to reach a temperature of 1,100°C in the center of the piece considering the following values for the properties of steel:

$$\rho = 7830 \text{ kg/m}^3;\ C_p = 550 \text{ J/kg.K};\ e\ k = 48 \text{ W/m.K.}$$

Considering heat transfer along the radial direction of the cylinder, the value for the Biot number is given by:

$$Bi = \frac{h.R}{k} = \frac{(150).(0.095)}{48} = 0.30 > 0.1 \text{ (nonisothermal)}$$

Using the worksheet that accompanies this text, one can determine the parameters.

$$Bi = 0.30 : \varsigma_1 = 0.743\ C_1 = 1.056$$

Assuming that the Fourier number is greater than 0.2, only the first term of the series is sufficient for calculations of temperatures in the center and the plate surface (the difference between these two temperatures, is the maximum temperature difference).

The dimensionless temperature at the center of the cylinder is:

$$\theta_0(\text{center}) = \frac{1100 - 1600}{50 - 1600} = 0.32$$

Time required to achieve this temperature can be evaluated using:

$$\theta_0 = C_1.\exp\left(-\zeta_1^2.\frac{k.t}{\rho.C_p.R^2}\right)$$

Substituting data given:

$$0.32 = (1.057).\exp\left[-(0.743)^2 \frac{(48).t}{(7830).(550).(0.095)^2}\right]$$

$$t = 1,752.6 \text{ s} \qquad Fo = 2.16$$

As Fo > 0.2, only the first term of the series is sufficient for the calculation of the temperature.

The speed of passage of the bars through the furnace is given by:

$$v = \frac{1}{t} = \frac{5}{1752.6} \; 0.0029 \text{ m/s} = 0.17 \text{ m/min}$$

For the time obtained above, the surface temperature of the cylinder is:

$$\theta(\text{surface}) = 0.236$$

$$T(\text{surface}) = 1234.2°C$$

The difference in temperature between the center and the surface is 134.2 °C.

12.4 Heat Transfer in Transient One Dimensional Spherical Geometry

In this case also heat transfer in the radial direction is considered within a sphere. This system is represented in the Figure 12.4. Definitions of dimensionless variables are the same presented for cylindrical geometry.

For boundary conditions analogous to the case of heat transfer in cylindrical geometry, the dimensionless temperature profile is given by (INCROPERA and DEWITT, 2003):

$$\theta = \sum_{n=1}^{\infty} C_n.\exp(-\varsigma_n^2.Fo).\frac{1}{\varsigma_n.r^*}.\sin(\varsigma_n.r^*) \qquad (12.40)$$

where C_n is given by:

$$C_n = \frac{4.[\sin(\varsigma_n) - \varsigma_n.\cos(\varsigma_n)]}{2.(\varsigma_n) - \sin(2.\varsigma_n)} \qquad (12.41)$$

In this equation, the values of ζ_n are positive roots of the equation:

$$1-\varsigma_n.\cot(\varsigma_n)-Bi=0 \tag{12.42}$$

The *spreadsheet-roots spherical.xls* enables evaluation of the first roots of the equation (12.42), for known Biot. Numbers the other roots can be calculated using a methodology similar to that implemented in the spreadsheet.

With the spreadsheet *transient-1D-spherical*, one can calculate the temperatures along the radial positions using the first four terms in the series for the spherical geometry also the contribution of successive terms of the series becomes less significant. For Fourier values greater than 0.2, only one term of the series is sufficient for the calculation of the temperature. In this case, one can write:

$$\theta=C_1.\exp(-\varsigma_1^2.Fo).\frac{1}{\varsigma_1.r^*}.\sin(\varsigma_1.r^*) \tag{12.43}$$

In the center of the sphere, r* = 0, the temperature is given by:

$$\theta_0=C_1.\exp(-\varsigma_1^2.Fo) \tag{12.44}$$

When Fo is greater than 0.2, the amount of energy supplied or removed during a certain time interval can be calculated by:

$$\frac{E}{E_{máx}}=1-\frac{3.\theta_0}{\varsigma_1^3}.[\sin(\varsigma_1)-(\varsigma_1).\cos(\varsigma_1)] \tag{12.45}$$

Example

Iron ore pellets are loaded into a furnace at a temperature of 1100 K (heated by gas). Determine the temperatures at the center and the surface of the pellet after a heating time of 20 seconds. The initial temperature of the pellet is 300 K and the coefficient of heat transfer is 300 W/m^2.K. The characteristics of pellets are: diameter: 18 mm, specific heat: 800 J/kg.K, density = 3500 kg/m^3 and thermal conductivity = 6.75 W/m.K.

Solution

Considering heat transfer in the radial direction, one can get the following value for the Biot number:

$$Bi = \frac{h.R}{k} = \frac{(300).(0.009)}{6.75} = 0.40 > 0.1 \text{ (nonisothermal)}$$

Using the worksheet that accompanies this text, one can determine:

$$Bi = 0.40 : \varsigma_1 = 1.053 \quad C_1 = 1.116$$

With the given time, one can determine the Fourier number:

$$Fo = \frac{\alpha.t}{L^2} = \frac{k.t}{\rho.C_p.R^2} = \frac{(6.75).(20)}{(3500).(800).(0.009)^2} = 0.595$$

With the help of the worksheet, that accompanies this text, one can determine the temperatures at the center and the surface of the pellet. The values obtained are:

$$\theta_0(\text{center}) = 0.577 \qquad T(\text{center}) = 638.4°C$$

$$\theta(\text{surface}) = 0.476 \qquad T(\text{surface}) = 719.2°C$$

The worksheets provided can also be used to evaluate temperature profiles in a body depending on its geometry. As mentioned earlier for values of Bi less than 0.1, the temperature inside the body tends to be uniform. Figure 12.5 shows the profiles of dimensionless temperature θ for a flat plate for three different values of Biot number.

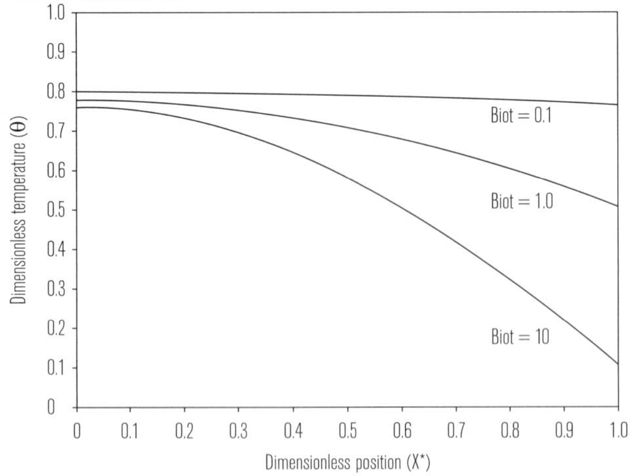

Figure 12.5 - Temperature profiles in a flat wall for different Biot numbers.

Through this figure, one can confirm, that in cases where Biot number is 0.1 or less (isothermal heating or cooing), the temperature profile is flat, implying insignificant temperature gradients. This justifies the mathematical treatment proposed for these cases.

Another approach that can be verified is that for values of Fo greater than 0.2, just the first term of the series (equations 12:24, 12.37 and 12.43) is important for the calculation of temperature profiles of a system undergoing heating or cooling. Table 12.2 shows the results provided by the worksheet, in respect of the contribution of the first four terms in the series, for Fo = 0.2. It can be seen clearly that the change in temperature caused by the inclusion of the second to fourth terms is negligible, which confirms the proposed approach.

Table 12.2 - Results provided by the spreadsheet indicating the contribution of the first four terms of the series in the calculation of temperatures in the transient regime.

Fo = 0.20 Position x*	Terms of the series				Dimensionless temperature
	1	2	3	4	
0	0.839	-9.63E-03	6.07E-06	-1.19E-10	0.829
0.1	0.830	-8.75E-03	4.55E-06	-6.24E-11	0.821
0.2	0.805	-6.27E-03	7.56E-07	5.40E-11	0.798
0.3	0.763	-2.65E-03	-3.42E-06	1.19E-10	0.760
0.4	0.705	1.46E-03	-5.88E-06	7.04E-11	0.707
0.5	0.633	5.30E-03	-5.40E-06	-4.52E-11	0.639
0.6	0.549	8.17E-03	-2.22E-06	-1.18E-10	0.557
0.7	0.453	9.55E-03	2.07E-06	-7.80E-11	0.462
0.8	0.348	9.19E-03	5.33E-06	3.61E-11	0.357
0.9	0.235	7.14E-03	5.92E-06	1.16E-10	0.242
1	0.118	3.80E-03	3.55E-06	8.50E-11	0.122

12.5 Heat Transfer in Semi-Infinite Solid

In some situations of practical importance involving heat transfer, we can define a region within the body, that is not affected by the process of heat transfer. This region practically remains at the initial temperature. This situation is commonly referred to as heat transfer in a semi-infinite solid and is represented schematically in Figure 12.6, showing three distinct cases.

Figure 12.6 - Situations heat transfer involving the semi-infinite solid.

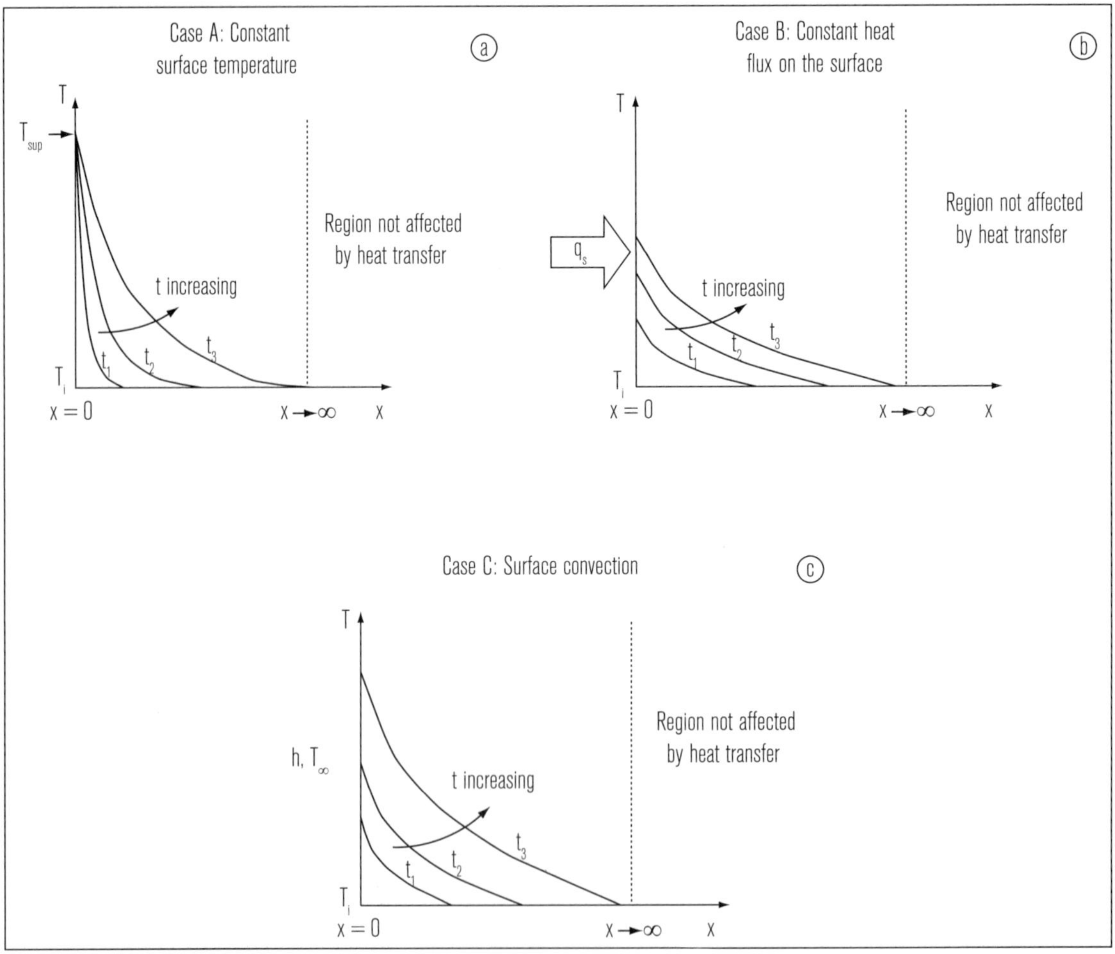

For the situations shown in Figure 12.6, like in the previous cases,the equation of conservation of energy can be simplified to:

$$\alpha\frac{\partial^2 T}{\partial x^2}=\frac{\partial T}{\partial t} \tag{12.46}$$

The initial conditions and boundary for the three cases shown in Figure 12.6 are:

- Case A: Constant surface temperature:

  Initial:

  $t=0 \qquad T(x,t)=T_i \qquad 0\le x<\infty$

  Boundary conditions:

  $x=0 \;:\; T=T_{sup} \qquad t>0$

  $x\to\infty \;:\; T=T_i \qquad t>0$

- Case B: Constant heat flux at surface:

  Initial:

  $t=0 \qquad T(x,t)=T_i \qquad 0\le x<\infty$

  Boundary conditions:

  $x=0 \;:\; -k\frac{\partial T}{\partial x}\Big|_{x=0}=q_s \qquad t>0$

  $x\to\infty \;:\; T=T_i \qquad t>0$

- Case C: Surface convection:

  Initial:

  $t=0 \qquad T(x,t)=T_i \qquad 0\le x<\infty$

  Boundary conditions:

  $x=0 \;:\; -k\frac{\partial T}{\partial x}\Big|_{x=0}=h\left(T_\infty-T\big|_{x=0}\right) \qquad t>0$

  $x\to\infty \;:\; T=T_i \qquad t>0$

A parameter η ucan be defined in terms of x,t, and thermal diffusivity as follows:

$$\eta=\left(\frac{x}{2\sqrt{\alpha.t}}\right) \tag{12.47}$$

This parameter can be used to transform the partial differential equation arrived at though the equation of conservation of energy (equation 12.46) to an ordinary differential equation for which analytical solutions can be arrived at for each of the cases illustrated in Figure 12.6 as given below:

- Case A: Constant surface temperature:

Temperature profile: $$\frac{T(x,t)-T_{sup}}{T_i - T_{sup}} = \text{erf}\left(\frac{x}{2\sqrt{\alpha.t}}\right) \quad (12.48)$$

Heat flux x = 0: $$q\big|_{x=0} = \frac{k(T_{sup} - T_i)}{\sqrt{\pi.\alpha.t}} \quad (12.49)$$

- Case B: Constant heat flux at the surface:

Temperature profile:

$$T(x,t)-T_i = \frac{2.q_s.(\alpha.t/\pi)^{1/2}}{k}\exp\left(-\frac{x^2}{4.\alpha.t}\right) - \frac{q_s.x}{k}.\text{erfc}\left(\frac{x}{2\sqrt{\alpha.t}}\right) \quad (12.50)$$

Heat flux x = 0: $$q\big|_{x=0} = q_s \quad (12.51)$$

- Case C: Surface convection:

Temperature profile:

$$\frac{T(x,t)-T_i}{T_\infty - T_i} = \text{erfc}\left(\frac{x}{2\sqrt{\alpha.t}}\right) - \left[\exp\left(\frac{h.x}{k}+\frac{h^2.\alpha.t}{k^2}\right)\right] . \left[\text{erfc}\left(\frac{x}{2\sqrt{\alpha.t}}+\frac{h.(\alpha.t)^{1/2}}{k}\right)\right] \quad (12.52)$$

Heat flux x = 0: $$q\big|_{x=0} = -k\frac{\partial T}{\partial x} = h.\left(T_\infty - T\big|_{x=0}\right) \quad (12.53)$$

In the above, *erf* and *erfc* functions represent the error function and complementary error function.These functions are defined through the following relations:

$$\text{erf}(\eta) = \frac{2}{\sqrt{\pi}}\int_0^\eta e^{-x^2}.dx \quad (12.54)$$

$$\text{erfc}(\eta) = 1 - \text{erf}(\eta) \quad (12.55)$$

Table 12.3 presents the values of error function for η ranging from 0 to 3. These and other values of the error function can also be obtained through the *ERF* function of Excel (Spread sheet).

The equations of the temperature profiles above can be applied to finite solid also provided the following condition is met:

$$\frac{L}{2\sqrt{\alpha.t}} \geq 2 \qquad (12.56)$$

where:

- L is the thickness or the semi-thickness of the piece, or characteristic parameter depending on the situation, as illustrated in Figure 12.7.

Table 12.3 - Values of error function of Gauss (INCROPERA and DEWITT, 2003).

η	erf η	η	erf η	η	erf η
0.00	0.00000	0.36	0.38933	1.04	0.85865
0.02	0.02256	0.38	0.40901	1.08	0.87333
0.04	0.04511	0.40	0.42839	1.12	0.88679
0.06	0.06762	0.44	0.46622	1.16	0.89910
0.08	0.09008	0.48	0.50275	1.20	0.91031
0.10	0.11246	0.52	0.53790	1.30	0.93401
0.12	0.13476	0.56	0.57162	1.40	0.95228
0.14	0.15695	0.60	0.60386	1.50	0.96611
0.16	0.17901	0.64	0.63459	1.60	0.97635
0.18	0.20094	0.68	0.66378	1.70	0.98379
0.20	0.22270	0.72	0.69143	1.80	0.98909
0.22	0.24430	0.76	0.71754	1.90	0.99279
0.24	0.26570	0.80	0.74210	2.00	0.99532
0.26	0.28690	0.84	0.76514	2.20	0.99814
0.28	0.30788	0.88	0.78669	2.40	0.99931
0.30	0.32863	0.92	0.80677	2.60	0.99976
0.32	0.34913	0.96	0.82542	2.80	0.99992
0.34	0.36936	1.00	0.84270	3.00	0.99998

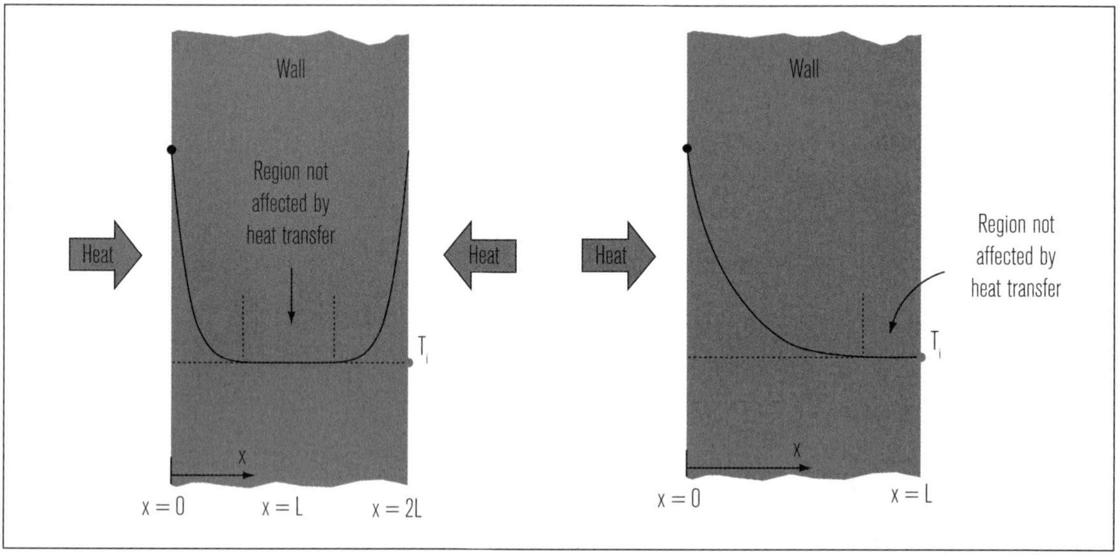

Figure 12.7 - Approximation of semi-infinite solid to cases of finite solid.

From equation (12.56), it can be inferred that a thick finite (large L) body with low thermal conductivity and for short times can be treated as semi-infinite and the relevant equations can be applied.

It is interesting to observe, that equation (12.52) can be used as an alternative to equation (12.21) for low values of Fourier number and in this case equation (12.21), the solution in the form of series would require a very large number of terms for reaching convergence and hence its application would be impractical.The equation (12.51) on the other hand determines the profile of temperature with less effort, subject to satisfying the condition specified in equation (12.56). However, it should be noted that in equation (12.21), the origin is at the center of the piece, while in equation (12.53), the origin is on the surface.

Example

Estimate the temperature drop of the liquid steel contained in a ladle over a period of 10 minutes. The ladle has an internal diameter of 2.3 m and height of liquid steel is 2.5 m. The lining material is silica-alumina refractory (thermal conductivity = 1.5 W/m.K, specific heat = 960 J/kg.K and density = 2325 kg/m^3) and has a uniform thickness equal to 30 cm. Iinitial temperature of the steel and refractory lining are 1600 and 1200°C, respectively. Consider that the top surface of the steel is covered by a layer of slag, 50 mm thick. Heat losses through this steel and slag layer amount to 120 kW/m^2. Properties of liquid steel are: thermal conductivity = 20 W/mK, specific heat = 680 J/kg.K and density = 7000 kg/m^3.

Solution

The system under consideration is seen schematically in the figure below.

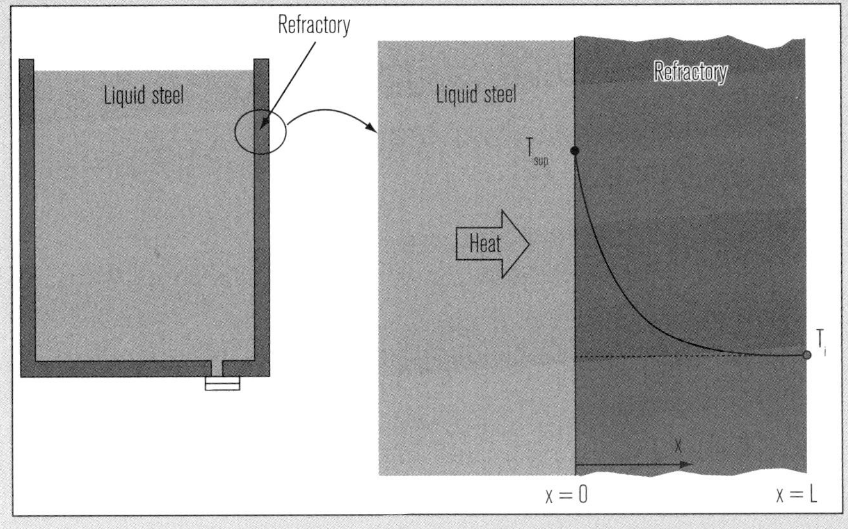

Initially, one should assess, whether the refractory in the period of time given can be considered as a semi-infinite solid. For this:

$$\alpha = \frac{k}{\rho . C_p} = \frac{1.5}{(2325).(960)} = 6.72 \,.\, 10^{-7} \ m^2/s$$

$$\frac{L}{2\sqrt{\alpha . t}} = \frac{(0.30)}{2\sqrt{(6.72 \,.\, 10^{-7}).(10 \,.\, 60)}} = 7.47 > 2$$

Hence, the refractory can be assumed to behave as a semi-infinite solid during the period of 10 minutes. For the purpose of evaluating the heat flux at the refractory-steel interface, it is assumed that this interface between refractory and steel melt instantly assumes the temperature of the steel (this assumption tends to overestimate the flow but is a reasonable approximation). The flow of heat in the refractory-steel interface can then be estimated from the relationship:

$$q\big|_{x=0} = \frac{k(T_{sup} - T_i)}{\sqrt{\pi.\alpha.t}}$$

Integrating this flow during the time (10 minutes) and multiplying by the the area of steel-refractory interface, the total amount of heat transferred from steel melt to the refractory lining can be estimated:

$$Q_{slag\text{-}ref} = A_{ref\text{-}steel}.\frac{k(T_{sup} - T_i)}{\sqrt{\pi.\alpha}}.2.t^{1/2}$$

$$A_{ref\text{-}slag} = \pi.D.H + \frac{\pi.D^2}{4} = \pi.(2.3).(2.5) + \frac{\pi.(2.3)^2}{4} = 22.22\ m^2$$

$$Q_{steel \to ref} = (22,22).\frac{(1.5).(1600 - 1200)}{\sqrt{\pi.(6.72 \times 10^{-7})}}.2.(10 \times 60)^{1/2}$$

$$Q_{steel \to ref} = 4.495 \times 10^8\ W$$

Heat loss through the slag can be estimated by the following equation:

$$A_{slag\text{-}steel} = \frac{\pi.D^2}{4} = \frac{\pi.(2.3)^2}{4} = 4.15\ m^2$$

$$Q_{steel \to slag} = (4.15).(120000).(10.60) = 2.99 \times 10^8\ W$$

To estimate the temperature variation of liquid steel, it is assumed that its temperature is uniform. Thus, we can write that:

$$Q_{steel \to ref} + Q_{steel} = m_{steel}.C_{p_{steel.}}\Delta T_{steel}$$

$$m_{steel} = \frac{\pi.D^2}{4}.H.\rho_{steel} = \frac{\pi.(2.3)^2}{4}.(2.5).(7,000) = 72708.2\ kg$$

$$C_{p_{steel}} = 680\ J/kg.K$$

$$\Delta T_{steel} = \frac{(4.495 + 2.99).10^8}{(72\ x\ 708.2).(680)} = 15.1°C$$

12.6 Two and Three Dimensional Heat Transfer

When the physical properties of the solid are constant and heat transfer by radiation can be neglected, temperature profiles for cases of heat transfer in two and three dimensions, can be obtained from combinations of one-dimensional solutions presented above, through the method of separation of variables (CARSLAW and JAEGER, 1959). Recalling the dimensionless expressions discussed earlier :

- flat wall:

$$\theta_P(x,t) = \frac{T(x,t) - T_\infty}{T_i - T_\infty} \qquad (12.57)$$

- infinite cylinder:

$$\theta_C(r,t) = \frac{T(r,t) - T_\infty}{T_i - T_\infty} \qquad (12.58)$$

- semi-infinite solid:

$$\theta_S(x,t) = \frac{T(x,t) - T_\infty}{T_i - T_\infty} \qquad (12.59)$$

For this case, x is distance to the surface that is exchanging heat.

12.6.1 Bidimensional heat transfer

Figure 12.8 shows a situation involving heat transfer in two dimensions in plane geometry.

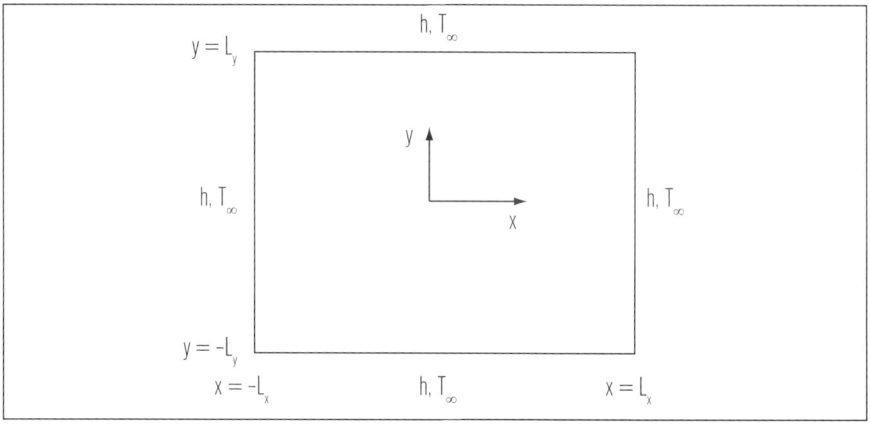

Figure 12.8 - Two dimensional system in plane geometry.

For this case, the temperature profile can be obtained by the following equation:

$$\theta(x,y,t)=\theta_P(x,t).\theta_P(y,t)=\left[\frac{T(x,t)-T_\infty}{T_i-T_\infty}\right].\left[\frac{T(y,t)-T_\infty}{T_i-T_\infty}\right] \quad (12.60)$$

Figure 12.9 represents a two dimensional system involving plane and cylindrical geometries.

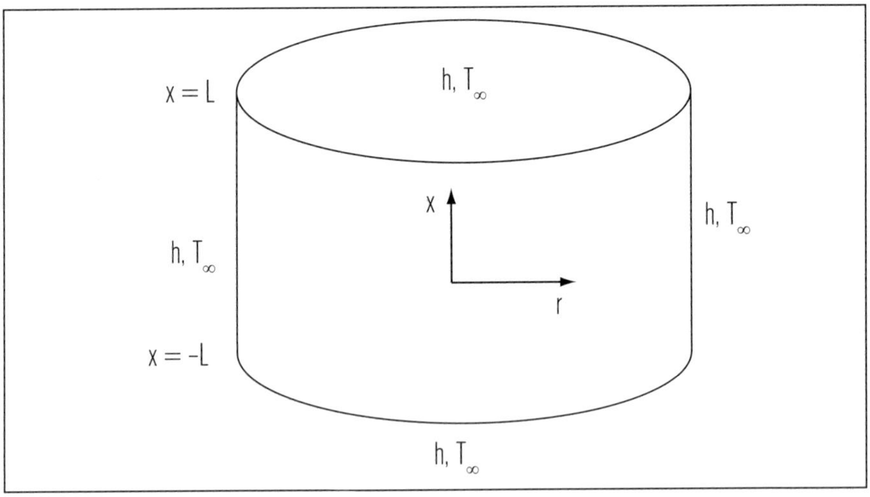

Figure 12.9 - Two dimensional system in flat and cylindrical geometries.

The temperature profile in this case is determined by the following equation:

$$\theta(x,r,t)=\theta_P(x,t).\theta_C(r,t)=\left[\frac{T(x,t)-T_\infty}{T_i-T_\infty}\right].\left[\frac{T(r,t)-T_\infty}{T_i-T_\infty}\right] \quad (12.61)$$

Other situations of two-dimensional heat transfer are shown in Figure 12.10, with the respective equations to calculate the temperature profiles.

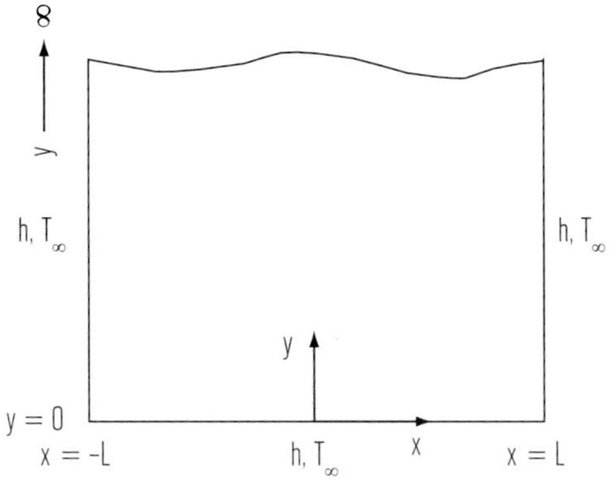

$$\theta(x,y,t) = \theta_P(x,t).\theta_S(y,t) = \left[\frac{T(x,t)-T_\infty}{T_i - T_\infty}\right].\left[\frac{T(y,t)-T_\infty}{T_i - T_\infty}\right] \quad (12.62)$$

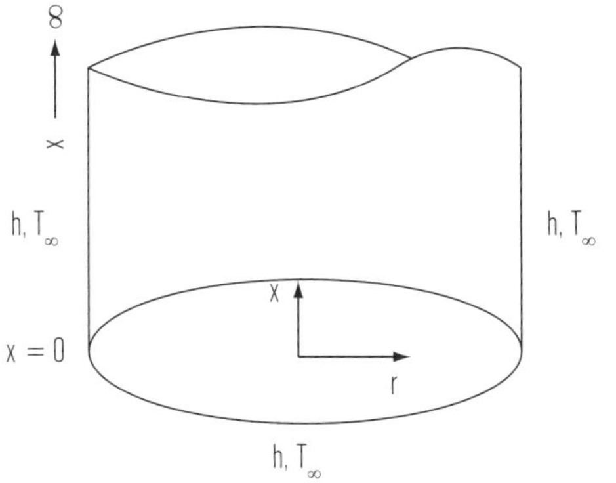

$$\theta(x,r,t) = \theta_S(x,t).\theta_C(r,t) = \left[\frac{T(x,t)-T_\infty}{T_i - T_\infty}\right].\left[\frac{T(r,t)-T_\infty}{T_i - T_\infty}\right] \quad (12.63)$$

Figure 12.10 - Examples of heat transfer and their two-dimensional equations to calculate the temperature profiles.

Example

A long block of carbon steel not alloyed, with a square cross section (0.4 m by 0.4 m), initially at a temperature of 300 deg K, is placed inside furnace at a temperature of 1500 deg.K. The heat transfer coefficient by convection is 150 W/m².K. How long the steel should remain in the furnace until the temperature in its center reaches 1300 deg K? What is the temperature in the corner of the block at this time? Consider that the heat transfer by radiation is included in the heat transfer coefficient value provided.

Solution

First the heating rate is determined by calculating the Biot number. The properties of the steel are:

$$\rho = 7830 \text{ kg/m}^3; \ C_p = 550 \text{ J/kg.K; e } k = 48 \text{ W/m.K.}$$

The Biot number is given by:

$$Bi = \frac{h.L}{k} = \frac{(150).(0.2)}{48} = 0.63 > 0,1 \text{ (nonisothermal)}$$

Heat transfer in two dimensions for this situation is represented figure below, where $L_x = L_y = 0.2$ m.

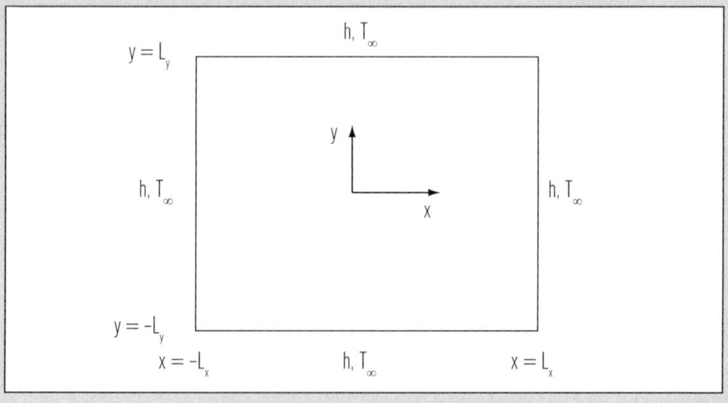

For this case, the temperature can be calculated using the following equation:

$$\theta(x,y,t) = \theta_P(x,t).\theta_P(y,t) = \left[\frac{T(x,t) - T_\infty}{T_i - T_\infty}\right].\left[\frac{T(y,t) - T_\infty}{T_i - T_\infty}\right]$$

The temperature at the center of the block is 1300 K. The corresponding dimensionless temperature is:

$$\theta(x=0,y=0,t)=\theta_P(x=0,t).\theta_P(y=0,t)=\left(\frac{1300-1500}{300-1500}\right)=0.166$$

As the block has square section, one can write:

$$\theta_P(x=0,t)=\theta_P(y=0,t)=(0.166)^{1/2}=0.41$$

The values of the following parameters were evaluated using the worksheet provided:

$$C_1 = 1.084 \qquad \varsigma_1 = 0.717$$

From the equation below, one can estimate the heating time, initially assuming that Fo greater than 0.2:

$$\theta_p(x=0,t)=C_1.\exp\left(-\varsigma_1^2.\frac{k.t}{\rho.C_p.L^2}\right)$$

Substituting the values:

$$0.41=(1.084).\exp\left[-(0.717)^2.\frac{(48).t}{(7830).(550).(0.20)^2}\right]$$

$$t=6787.1\text{ s} \qquad Fo=1.89$$

At this time, the dimensionless temperature in the block will be:

$$\theta_P(x=L_x,t)=\theta_P(y=L_y,t)=0.309$$

$$\theta(x=L_x,y=L_y,t)=\theta_P(x=L_x,t).\theta_P(y=L_y,t)=(0.309)^2=0.095$$

$$T(x=L_x,y=L_y,t)=1500-1200.(0.095)=1386°C$$

Example

Steel discs are preheated to temperatures of at least 1300 K before being forged. The preheating is performed in a furnace whose temperature is 1700 K. The heat transfer coefficient (including the contribution of radiation) is 500 W/m².K the discs are initially at 300 K. Determine the temperatures at the points indicated in the following figure after a heating time of 300 s. Assess whether this time is appropriate, considering the desired temperature for forging. If not, what should be done?

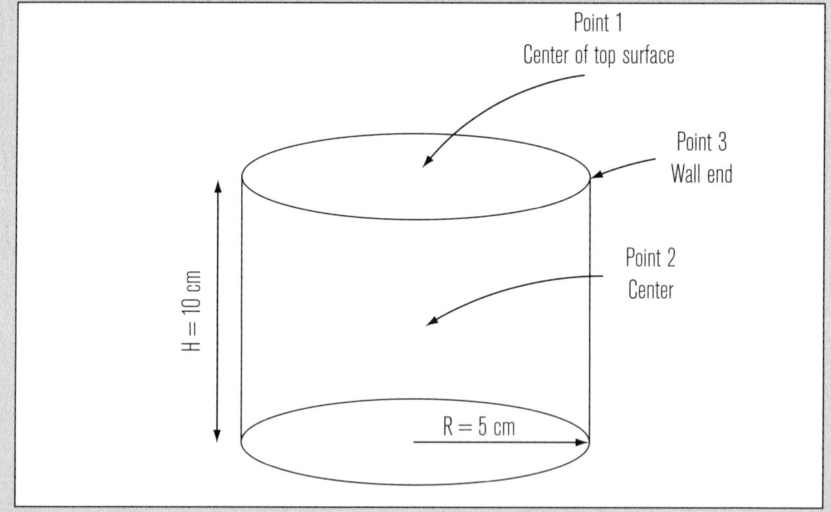

Solution

Using the properties of steel given below, one can evaluate the Biot number for the axial and radial directions of the cylinder.

$$\rho = 7830\ \text{kg/m}^3;\ C_p = 550\ \text{J/kg.K};\ \text{e}\ k = 48\ \text{W/m.K}$$

$$Bi_L = \frac{h.L}{k} = \frac{(500).(0.05)}{48} = 0.52 > 0.1\ \text{(nonisothermal)}$$

$$Bi_R = \frac{h.R}{k} = \frac{(500).(0.05)}{48} = 0.52 > 0.1\ \text{(nonisothermal)}$$

For, two-dimensional heat transfer in the axial and radial directions,with the help of the worksheet provided, one determine :

- Axial direction:

$$C_1 = 1.073 \qquad \varsigma_1 = 0.665$$

- Radial direction:

$$C_1 = 1.096 \qquad \varsigma_1 = 0.958$$

The Fourier number in the axial and radial directions are the same and is given by:

$$Fo = \frac{\alpha.t}{L^2} = \frac{\alpha.t}{R^2} = \frac{(48).(300)}{(7830).(550).(0.05)^2} = 1.338$$

Temperatures in the center and one-dimensional surface in both directions are considered (calculated with the aid of the worksheet provided):

$$\theta_P(x = 0,\ t = 300\ s) = 0.594$$

$$\theta_P(x = L,\ t = 300\ s) = 0.468$$

$$\theta_C(r = 0,\ t = 300\ s) = 0.321$$

$$\theta_C(r = R,\ t = 300\ s) = 0.185$$

Temperatures can be determined from combinations of the dimensionless temperatures:

- Point 1:

$$\theta(x = L, r = 0, t = 300\ s) = \theta_P(x = L, t).\theta_C(r = 0, t) = (0.468).(0.321) = 0.150$$

- Point 2:

$$\theta(x = 0, r = 0, t = 300\ s) = \theta_P(x = 0, t).\theta_C(r = 0, t) = (0.594).(0.321) = 0.191$$

- Point 3:

$$\theta(x = L, r = R, t = 300\ s) = \theta_P(x = L, t).\theta_C(r = R, t) = (0.468).(0.185) = 0.087$$

Temperatures at these points are given by:

- Point 1: 1490 K;
- Point 2: 1432 K;
- Point 3: 1578.2 K.

The requirement of minimum temperature of 1300 K is being met. There is the possibility to reduce the time of residence in the furnace and and still meet the specification, with possible gains in productivity.

12.6.2 Tridimensional heat transfer

Figure 12.11 shows a case where heat transfer occurs in three directions in plane geometry.

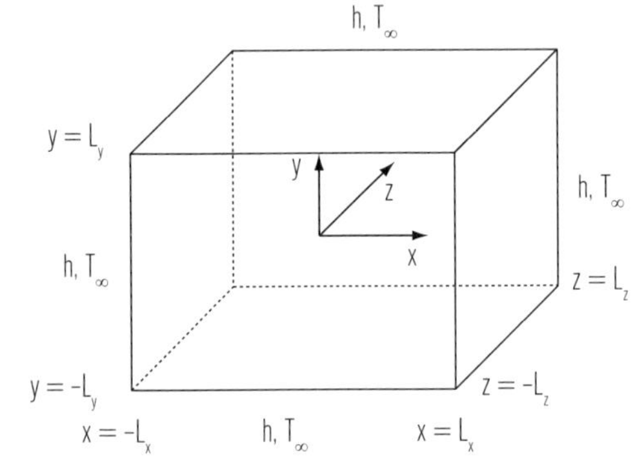

Figure 12.11 Three-dimensional heat transfer in Cartesian coordinates.

For the case shown in Figure 11.11, the temperature profile can be determined by a combination of three-dimensional solutions, as seen in equation (12.64).

$$\theta(x,y,t)=\theta_P(x,t).\theta_P(y,t).\theta_P(z,t)=\left[\frac{T(x,t)-T_\infty}{T_i-T_\infty}\right].\left[\frac{T(y,t)-T_\infty}{T_i-T_\infty}\right].\left[\frac{T(z,t)-T_\infty}{T_i-T_\infty}\right] \quad (12.64)$$

Figure 12.12 is shown another case with three-dimensional heat transfer, which can also be calculated temperatures by combination of one-dimensional solutions, as shown in equation (12.65).

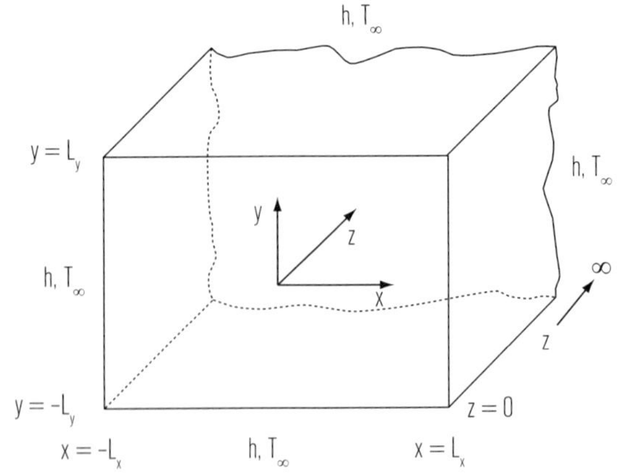

Figure 12.12 - Three dimensional system with plane geometry and semi-infinite solid.

$$\theta(x,y,t) = \theta_P(x,t).\theta_P(y,t).\theta_S(z,t) = \left[\frac{T(x,t)-T_\infty}{T_i - T_\infty}\right].\left[\frac{T(y,t)-T_\infty}{T_i - T_\infty}\right].\left[\frac{T(z,t)-T_\infty}{T_i - T_\infty}\right] \quad (12.65)$$

Other options discussed here can have their temperature profiles determined by adopting criteria similar to those presented.

Example

A cubic billet aluminum with 20 cm side is heated prior to forging. The minimum temperature required for forging is 673 K. Determine the heating time necessary to meet this requirement. Consider that the billet is initially at 300 K and the heat transfer coefficient is 300 W/m².K. The temperature of the gases inside the furnace is 1,000 K.

Solution

First the heating rate is determined by calculating the Biot number. The properties of aluminum are:

$$\rho = 2700 \text{ kg/m}^3;\ C_p = 900 \text{ J/kg.K};\ k = 240 \text{ W/m.K}$$

The Biot number is given by:

$$Bi = \frac{h.L}{k} = \frac{(300).(0.1)}{240} = 0.125 > 0.1 \text{ (nonisothermal)}$$

Heat transfer in three dimensions is considered as seen in the figure below, where $L_x = L_y = L_z = 0.1$ m.

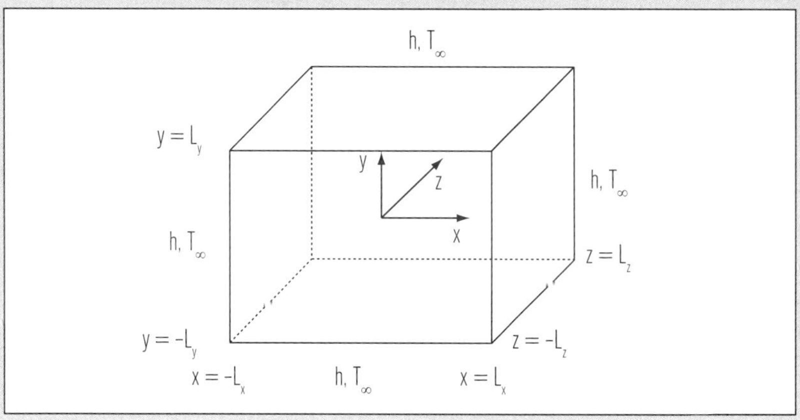

For this case, the temperature can be calculated using the following equation:

$$\theta(x,y,t) = \theta_P(x,t).\theta_P(y,t).\theta_P(z,t) = \left[\frac{T(x,t)-T_\infty}{T_i - T_\infty}\right].\left[\frac{T(y,t)-T_\infty}{T_i - T_\infty}\right].\left[\frac{T(z,t)-T_\infty}{T_i - T_\infty}\right]$$

The temperature at the center of the billet is 673 K. The corresponding dimensionless temperature is:

$$\theta(x=0,y=0,z=0,t) = \theta_P(x=0,t).\theta_P(y=0,t).\theta_P(z=0,t) = \left(\frac{673-1000}{300-1000}\right) = 0.467$$

As the billet is cubic, we can write that:

$$\theta_P(x=0,t) = \theta_P(y=0,t) = \theta_P(z=0,t) = (0.467)^{1/3} = 0.776$$

With the help of the worksheet provided, one can determine:

$$C_1 = 1.02 \qquad \varsigma_1 = 0.346$$

From the equation below, one can estimate the heating time, initially assuming that Fo is greater than 0.2.

$$\theta_p(x=0,t) = C_1.\exp\left(-\varsigma_1^2.\frac{k.t}{\rho.C_p.L^2}\right)$$

Substituting data:

$$0.776 = (1.02).\exp\left[-(0.346)^2.\frac{(240).t}{(2700).(900).(0.10)^2}\right]$$

$$t = 231.4 \text{ s} \qquad Fo = 2.29$$

As mentioned earlier, obtaining three dimensional solutions from one-dimensional solutions, is limited to simple situations in which the physical properties of the material are constant and heat transfer by radiation can be disregarded. In addition the heat transfer coefficient, h, and the temperature of the fluid that heats or cools the body, T_∞, should also be constant and equal across the various areas of the solid for using the above approach.

When one or more of the above restrictions are not met, the alternative is to obtain the temperature profiles with numerical methods.

These methods also enable determination of temperature profiles for systems with more complex geometries than those dealt with in previous sections.

Numerical methods, that can be used for the solution of the equations of heat transfer by conduction in the transient regime will be discussed in the next section These methods are an extension of the methods presented in Chapter 11, considering also the accumulation term.

12.6.3 Numerical solutions

The steps involved in obtaining numerical solutions for the determination of temperature profiles of transient problems are the same as those presented for the problems in steady state:

- discretization of the domain elements of volume and space-time simulation;
- obtaining equations of conservation of energy for each volume element. These equations, when solved, enable the calculation of temperatures in the volume elements in different moments of time;
- solution of the conservation equations for determining the temperature profiles.

12.6.3.1 Discretization

The discretization of the domain is done in a similar way tas discussed in item 11.5.1 in the previous chapter. However in the unsteady state or transient problems,discretization of time also should be included in the discretization domain. The solution is obtained in certain moments of time, separated by an interval Δt. Lower the value of Δt, more precise is the solution. However, the time spent in the calculations increases. Similar to the methodology of discretization of the domain, one should seek a value of Δt to optimize aspects of accuracy and computation time.This optimization can be found obtaining solutions for values of Δt successively smaller. From a certain point, reductions in the value of Δt does not cause significant changes in the temperatures. Only the time consumed in the calculations increases. Hence one fixes the optimum value of Δt for the calculations.

12.6.3.2 Equations of conservation for volume elements

The equations of conservation of energy for the various elements of volume obtained in step discretization are determined by establishing energy balances for these elements.

The method presented is called explicit mehod and as it enables obtaining the equations in which one can define the temperature in a given internal volume element at any instant of time, depending on the temperatures of neighboring elements and the temperature of the volume element at a previous instant of time. The main characteristics of this method are its simplicity and ease of implementation in the calculations with a computer program. The limitation is linked to the fact that one cannot establish independently the size of the volume element (values of Δx, Δy, Δz) and the value of Δt, as these values should meet the criteria for stability of the method.This is discussed in the following.

Similar to the procedure adopted in Chapter 11, equations of energy balance in the elemental volume in different positions within the domain of calculation are established. Cases of two-dimensional heat transfer (the method is easily implemented in three-dimensional ones), in which a body exchanges heat with a fluid and a neighborhood at a temperature of T_∞. The heat transfer coefficient is h and e is emissivity of the solid.

In the energy balance for an internal volume element, shown in Figure 12.22. the superscript m to denotes the temperature at time t. The superscript m +1 indicates the temperature at time t + Δt.

Neglecting heat generation term, the energy balance is given by:

$$[\text{Rate of energy input}]-[\text{Rate of energy output}]=[\text{Rate of accumulation}] \quad (12.66)$$

The various terms of the above equation above can be defined as follows:

- Rate of energy input:
  - x direction:

$$\Delta y.1.k\frac{T^m_{i-1,j}-T^m_{i,j}}{\Delta x}$$

  - y direction:

$$\Delta x.1.k\frac{T^m_{i,j-1}-T^m_{i,j}}{\Delta y}$$

- Rate of energy output:
  - x direction:

$$\Delta y.1.k\frac{T_{i,j}^{m}-T_{i+1,j}^{m}}{\Delta x}$$

  - y direction:

$$\Delta x.1.k\frac{T_{i,j}^{m}-T_{i,j+1}^{m}}{\Delta y}$$

- Rate of accumulation of energy:

$$\Delta x.\Delta y.1.\rho.C_{p}.\frac{T_{i,j}^{m+1}-T_{i,j}^{m}}{\Delta t}$$

Substituting the terms in energy balance equation:

$$\Delta y.1.k\frac{T_{i-1,j}^{m}-T_{i,j}^{m}}{\Delta x}+\Delta x.1.k\frac{T_{i,j-1}^{m}-T_{i,j}^{m}}{\Delta y}-\Delta y.1.k\frac{T_{i,j}^{m}-T_{i+1,j}^{m}}{\Delta x}-\Delta x.1.k\frac{T_{i,j}^{m}-T_{i,j+1}^{m}}{\Delta y}=\Delta x.\Delta y.1.\rho.C_{p}.\frac{T_{i,j}^{m+1}-T_{i,j}^{m}}{\Delta t} \quad (12.67)$$

From the equation above, it is possible to evaluate temperature $T_{i,j}^{+1m}$ (temperature at point i, j at time t + Δt), which is the only unknown:

$$T_{i,j}^{m+1}=\frac{\alpha.\Delta t}{\Delta x^{2}}(T_{i-1,j}^{m}+T_{i+1,j}^{m})+\frac{\alpha.\Delta t}{\Delta y^{2}}(T_{i,j-1}^{m}+T_{i,j+1}^{m})+\left(1-\frac{\alpha.\Delta t}{\Delta x^{2}}-\frac{\alpha.\Delta t}{\Delta y^{2}}\right).T_{i,j}^{m} \quad (12.68)$$

Thermal diffusivity of the medium being $\alpha = k/\rho.C_p$)

One can define the following dimensionless parameters:

$$Fo_{x}=\frac{\alpha.\Delta t}{\Delta x^{2}} \quad (12.69)$$

$$Fo_{y}=\frac{\alpha.\Delta t}{\Delta y^{2}} \quad (12.70)$$

They have the same form of the Fourier number, defined previously.

With these definitions, we can rewrite the equation (12.68) in the following way:

$$T_{i,j}^{m+1}=Fo_{x}(T_{i-1,j}^{m}+T_{i+1,j}^{m})+Fo_{y}(T_{i,j-1}^{m}+T_{i,j+1}^{m})+\left(1-Fo_{x}-Fo_{y}\right).T_{i,j}^{m} \quad (12.71)$$

It should be noted that the term which multiplies the temperature $T_{i,j}^{m}$ in equation (12.71) if negative, then an increase of temperature $T_{i,j}^{m}$ would result in a reduction in $T_{i,j}^{m+1}$ and vice versa. This type of behavior in the numerical solution has no physical meaning and leads to fluctuations in temperatures calculated for each time interval. Hence, we can establish a stability criterion for the explicit method:

$$\left(1 - Fo_x - Fo_y\right) \geq 0 \tag{12.72}$$

Substituting the definitions of Fourier number Fo_x and Fo_y we have:

$$1 - 2.\alpha.\Delta t\left(\frac{1}{\Delta x^2} + \frac{1}{\Delta y^2}\right) \geq 0 \tag{12.73}$$

Transposing terms, we obtain:

$$\Delta t \leq \frac{1}{2.\alpha.\left(\dfrac{1}{\Delta x^2} + \dfrac{1}{\Delta y^2}\right)} \tag{12.74}$$

In the above equation, the values of Δt, Δx and Δy cannot all be set arbitrarily. For specified values of Δx and Δy, maximum value of Δt can be estimated from the equation (12.74). If the values of Δx and Δy are simultaneously reduced by half, the value of Δt must be divided by four, to continue to meet the criterion of stability.This relationship between the spacing of the mesh (Δx Δy) and time interval (Δt) represents the main limitation of the explicit method.There are significant increases in computation time with increase in the accuracy of the solutions through reductions in Δx and Δy.

Energy balances similar to those presented for the internal volume element can be developed for other locations of the volume element. Energy balance of an element at a lateral surface in a location similar to the one shown in the previous chapter (Figure 11.23) is given below.

For this element in a lateral surface location, input, output and energy accumulation terms are:

- Rate of energy input:
  - x direction:

$$\Delta y.1.k\frac{T_{i-1,j}^{m} - T_{i,j}^{m}}{\Delta x}$$

- y direction:

$$\frac{\Delta x}{2}.1.k\frac{T_{i,j-1}^{m}-T_{i,j}^{m}}{\Delta y}$$

- Rate of energy output:
  - x direction:

$$\Delta y.1.h.(T_{i,j}^{m}-T_{\infty})$$

  - y direction

$$\frac{\Delta x}{2}.1.k\frac{T_{i,j}^{m}-T_{i,j+1}^{m}}{\Delta y}$$

- Rate of accumulation of energy:

$$\frac{\Delta x}{2}.\Delta y.1.\rho.C_{p}.\frac{T_{i,j}^{m+1}-T_{i,j}^{m}}{\Delta t}$$

Substituting these terms in the energy balance equation for the lateral face element, and rearranging one obtains for $T_{i,j}^{m+1}$:

$$\begin{aligned}T_{i,j}^{m+1} &= 2.Bi_{x}.Fo_{x}.T_{\infty}+2.Bi_{y}.Fo_{y}.T_{\infty}+2.Fo_{x}.T_{i-1,j}^{m}+T_{i,j-1}^{m})+ \\ &\left(1-2.Bi_{x}.Fo_{x}-2.Bi_{y}.Fo_{y}-2.Fo_{x}-2.Fo_{y}\right).T_{i,j}^{m}\end{aligned} \qquad (12.75)$$

The stability criterion is established considering that the term which multiplies $T_{i,j}^{m}$ in equation (12.75) cannot be negative.Therefore:

$$1-2.Bi_{x}.Fo_{x}-2.Fo_{x}-2.Fo_{y}\geq 0 \qquad (12.76)$$

In these equations, Bi_x is given by:

$$Bi_{x}=\frac{h.\Delta x}{k} \qquad (12.77)$$

For a volume element in a corner, as shown in Figure 12.24, adopting the same procedure of substitution in the energy balance equation for the corner element temperature $T_{i,j}^{m+1}$.

$$\begin{aligned}T_{i,j}^{m+1} &= 2.Bi_{x}.Fo_{x}.T_{\infty}+2.Bi_{y}.Fo_{y}.T_{\infty}+2.Fo_{x}.T_{i-1,j}^{m}+2.Fo_{y}.T_{i,j-1}^{m})+ \\ &\left(1-2.Bi_{x}.Fo_{x}-2.Bi_{y}.Fo_{y}-2.Fo_{x}-2.Fo_{y}\right).T_{i,j}^{m}\end{aligned} \qquad (12.78)$$

The stability criterion is expressed by the following equation:

$$1-2.Bi_{x}.Fo_{x}-2.Bi_{y}.Fo_{y}-2.Fo_{x}-2.Fo_{y}\geq 0 \qquad (12.79)$$

For other settings of the volume, similar equations can be obtained by making the necessary adjustments considering the locations of neighboring volume elements.

Applying the method presented here, one can get equations for the calculation of temperatures in the entire field at a given moment t + Δt, knowing the temperature at time t.

12.6.3.3 Solution of the conservation equations

The calculation of the temperature profile in a transient simulation adopting the explicit method is simpler than that presented in the case of steady state.

In a transient simulation, the temperature profile at the initial time, t = 0, must be known. The first step in calculating the profile is to determine the values of physical properties using the temperatures in the initial time. The heat transfer coefficient can be estimated using surface temperatures of the field at the initial time. If radiation is considered, it is possible to incorporate its effect by using a coefficient of heat transfer by radiation,as in the a following equation:

$$h_{rad} = \sigma.\varepsilon.(T_{i,j}^{m^2} + T_{viz}^2).(T_{i,j}^m + T_{viz}) \qquad (12.80)$$

where e is the emissivity of the surface volume elements around a temperature $T_{i,j}^{m}$.

In the conservation equations one can use a global coefficient of heat transfer, incorporating the contributions of convection and radiation:

$$h_g = h_{rad} + h \qquad (12.81)$$

With these values substituted in the conservation equations for all volume elements in the field, one can calculate the temperature at time t + Δt.

After determining all the temperatures within the domain at time t + Δt,these temperatures can be used for the calculation of temperatures at time t + 2.Δt, repeating the steps above. This procedure is repeated successively until one reaches total simulation time desired.This procedure is illustrated in the flow chart shown in Figure 12.13.

The following example illustrates the application of numerical method for one-dimensional case. The results can be compared with the values obtained by analytical solution to evaluate the effect of Δ t on the accuracy of the results.

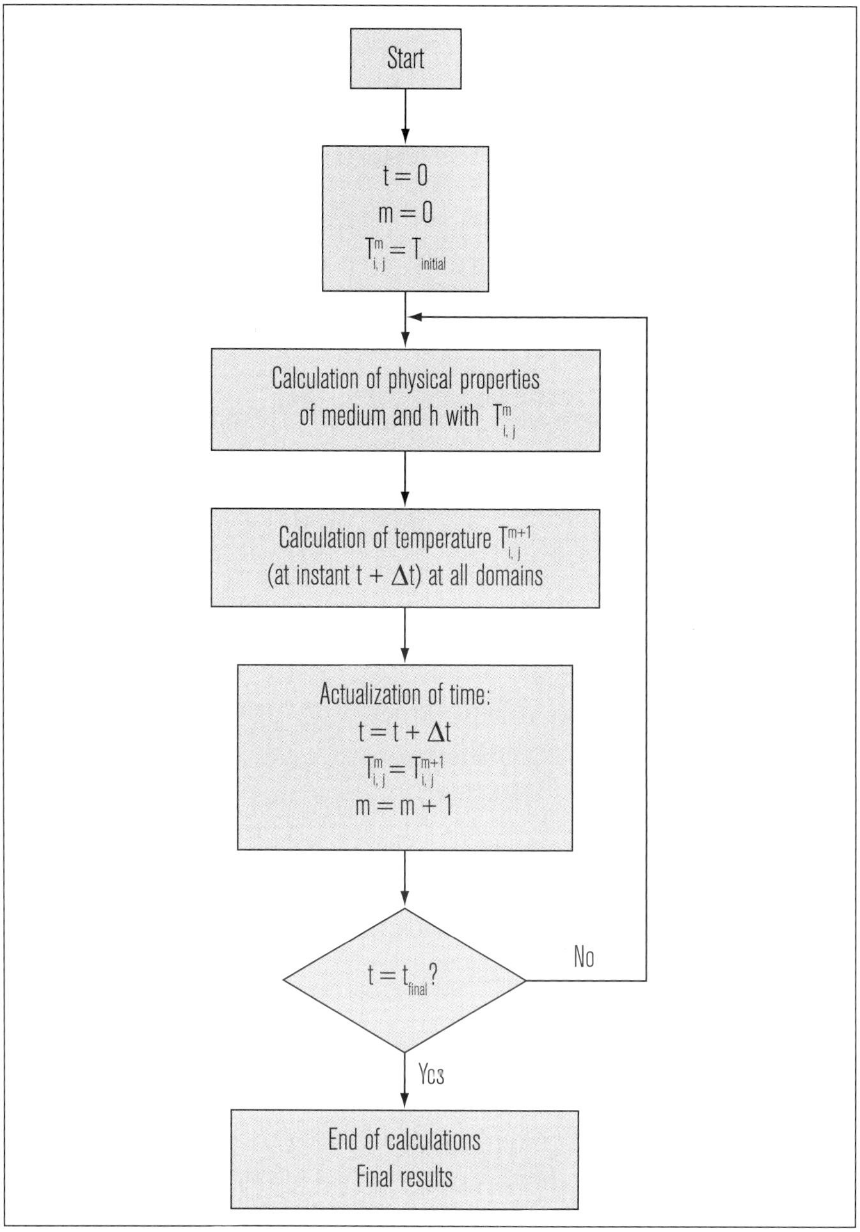

Figure 12.13 - Flow chart illustrating the procedure for evaluating temperature profiles in a transient simulation.

Example

Using the spreadsheet, determine the one dimensional temperature profile of carbon steel plate with 0.1% C (AISI 1010) with a thickness of 20 cm.The piece is initially at 25°C and is placed in a furnace where the gases and the neighborhood are at 1500°C. The properties of steel are:

- density: 7020 kg/m^3;
- specific heat: 600 J/kg.K;
- thermal conductivity: 35 W/m.K;
- emissivity: 0.8.

The heat transfer coefficient is 200 W/m^2.K.

Assuming that these properties are constant and the two surfaces at both sides are in contact with gas.calculate the temperature profiles the following moments of time: t = 100 s, t = 500 s e t = 1000 s. Analyze the effect of Δt on the results.

Get the temperature profiles disregarding the radiation. Compare these with the results obtained through analytical solutions. In all cases, use Δx equal to 1 cm.

Solution

The system under consideration has symmetry and hence it is sufficient to calculate the temperatures in only half the thickness of the piece. After discretization, we have the following arrangement of volume elements and nodes.

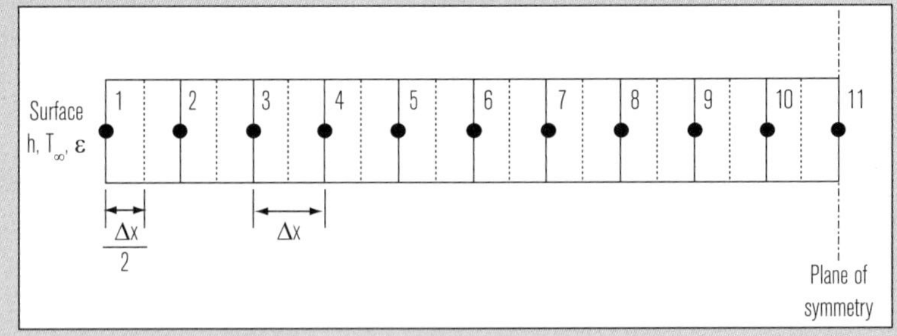

Just like in the two-dimensional case discussed earlier one can establish energy balances for the different volume elements. For the surface volume element 1:

$$\left[\text{Rate of input of energy}\right] - \left[\text{Rate of output of energy}\right] = \left[\text{Rate of accumulation}\right]$$

Taking height and depth of volume element to be unity and substituting the terms:

$$1.1.h.(T_\infty - T_1^m) - 1.1.k\frac{T_1^m - T_2^m}{\Delta x} = \frac{\Delta x}{2}.1.1.\rho.C_p.\frac{T_1^{m+1} - T_1^m}{\Delta t}$$

And explicating the value of T_1^{m+1}:

$$T_1^{m+1} = 2.Bi_x.Fo_x.T_\infty + 2.Fo_x.T_2^m + (1 - 2.Bi_x.Fo_x - 2.Fo_x).T_1^m$$

This equation can also be obtained through simplification of equation (12.75) removing the terms associated with the y direction, including the temperatures of neighboring elements in this direction.

For the volume elements 2 to 10, the energy balance yields the following equation:

$$T_i^{m+1} = Fo_x.(T_{i-1}^m + T_{i+1}^m) + (1 - Fo_x).T_1^m$$

For the volume element 11, we obtain the following equation for the energy balance:

$$T_{11}^{m+1} = 2.Fo_x.T_{10}^m + (1 - Fo_x).T_{11}^m$$

In these equations:

$$Bi_x = \frac{h_G.\Delta x}{k}$$

$$Fo_x = \frac{\alpha.\Delta t}{\Delta x^2}$$

A more rigorous stability criterion is given by:

$$Fo_x \leq \frac{1}{2.(1 + Bi_x)}$$

$$\frac{\alpha.\Delta t}{\Delta x^2} \leq \frac{1}{2.(1 + Bi_x)}$$

$$\Delta t \leq \frac{\Delta x^2}{2.\alpha.(1 + Bi_x)}$$

To assess the maximum Δt that can be used without causing numerical instability, first one has to make an estimate of Bi_x. For this, the global heat transfer coefficient has to be estimated:

$$h_G = h + \sigma.\varepsilon.(T_1^{m^2} + T_\infty^2).(T_1^m + T_\infty)$$

This value can be estimated assuming a maximum value for $T^m{}_1$ equal to 1500°C (1773 deg. K). This consideration leads to obtaining the highest value for h_G and Bi which results in a lower value for Δt. Substituting the data in the above equation:

$$h_G = 200 + (5.67\ .\ 10^{-8}).(0.8).[(1773)^2 + (1773)^2].(1773 + 1773)$$
$$h_G = 1211.25\ \text{W/m}^2.\text{K}$$

From the data of physical properties, we have:

$$\alpha = \frac{k}{\rho.C_p} = \frac{35}{(7020).(600)} = 8.31\ .\ 10^{-6}\ \text{m}^2/\text{s}$$

$$Bi = \frac{h_G.\Delta x}{k} = \frac{(1211.25).(0.01)}{35} = 0.35$$

$$\Delta t \le \frac{\Delta x^2}{2.\alpha.(1 + Bi_x)} = \frac{(0.01)^2}{2.(8.31x10^{-6}).(1 + 0.35)} = 4.47\ \text{s}$$

During the simulation, the values of h_G are updated every time step using the value of temperature $T^m{}_1$.

The following table presents a view of the spreadsheet showing the results for the first time intervals.

Figure for temperature as a function of position for the three times(t) in the problem.

View of worksheet transient-1d.xls

Transient simulation of unidimensional heat transfer - Explicit method

Δx (m) = 0.01

Δt (s) = 5

h (W/m².K) = 200

k (W/m.K) = 35

ρ (kg/m³) = 7020

Cp (J/kg.K) = 600

α (m²/s) = 8.31E-06

Fox = 4.15E-01

ε = 0.80

t (s)	hG (W/m².K)	Bi x	T1 (K)	T2 (K)	T3 (K)	T4 (K)	T5 (K)	T6 (K)	T7 (K)	T8 (K)	T9 (K)	T10 (K)	T11 (K)
0	503.65	0.14	298.00	298.00	298.00	298.00	298.00	298.00	298.00	298.00	298.00	298.00	298.00
5	543.39	0.16	474.372	298.00	298.00	298.00	298.00	298.00	298.00	298.00	298.00	298.00	298.00
10	548.69	0.16	495.351	371.28	298.00	298.00	298.00	298.00	298.00	298.00	298.00	298.00	298.00
15	565.49	0.16	558.690	392.38	328.45	298.00	298.00	298.00	298.00	298.00	298.00	298.00	298.00
20	572.41	0.16	583.525	434.92	342.36	310.65	298.00	298.00	298.00	298.00	298.00	298.00	298.00
25	583.44	0.17	621.687	458.20	367.64	318.57	303.26	298.00	298.00	298.00	298.00	298.00	298.00
30	590.51	0.17	645.319	488.50	384.88	332.59	307.43	300.18	298.00	298.00	298.00	298.00	298.00
35	599.06	0.17	673.107	510.60	406.21	343.86	314.88	302.29	298.91	298.00	298.00	298.00	298.00
40	605.83	0.17	694.508	534.75	423.68	357.72	321.69	306.11	299.94	298.38	298.00	298.00	298.00
45	613.07	0.18	716.877	554.98	442.42	370.16	330.19	310.02	301.85	298.87	298.16	298.00	298.00
50	619.43	0.18	736.069	575.48	459.16	383.58	338.41	315.01	304.01	299.81	298.39	298.07	298.00
55	625.87	0.18	755.120	593.87	476.08	396.22	347.45	320.16	306.83	300.96	298.85	298.17	298.05
60	631.83	0.18	772.381	611.93	491.84	409.14	356.37	325.96	309.93	302.52	299.45	298.40	298.15
65	637.71	0.18	789.151	628.70	507.37	421.58	365.66	331.94	313.51	304.32	300.29	298.73	298.36
70	643.30	0.18	804.781	644.96	522.14	433.99	374.88	338.29	317.35	306.47	301.32	299.23	298.67
75	648.77	0.19	819.849	660.33	536.54	446.05	384.24	344.79	321.53	308.85	302.59	299.86	[illegible]
80	654.03	0.19	834.108	675.18	[illegible]	457.97	393.53	351.52	325.93	311.52	304.06	300.69	299.74
85	659.16	0.19	847.830	689.36	563.83	469.59	402.85	358.34	330.57	311.40	305.76	301.69	300.53
90	664.13	0.19	860.931	703.05	576.83	481.02	412.09	365.30	335.39	317.53	307.66	302.90	301.50
95	668.98	0.19	873.547	716.20	589.46	492.19	421.29	372.31	340.39	320.85	309.78	304.30	302.66
100	673.69	0.19	885.659	728.92	601.71	503.14	430.40	379.40	345.54	324.37	312.10	305.90	304.02

The figure below shows the temperature profiles for the three times desired, including radiation.

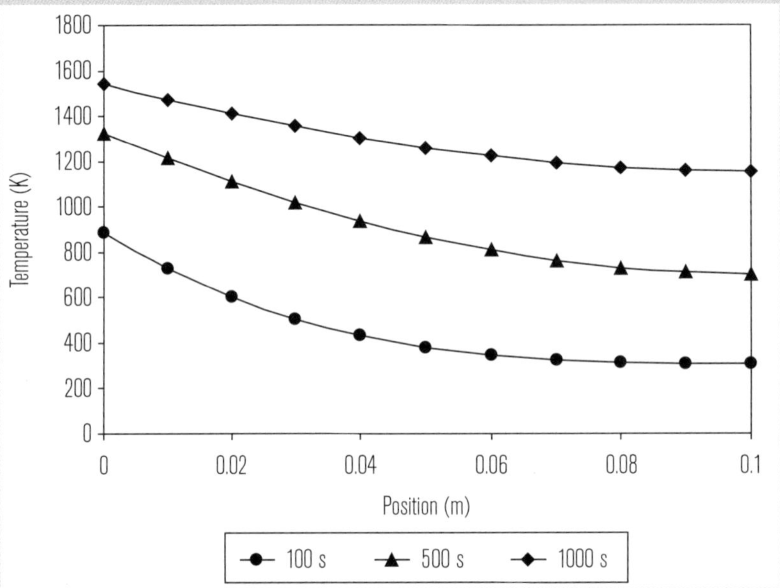

In the table below, the radiation was excluded assuming $\varepsilon = 0$ and the values obtained are compared with analytical solutions. As can be seen, there is an excellent agreement between the two types of solution, even using a coarse one ($\Delta x = 1$ cm).

	100	536.7912	472.3138	420.4163	380.4764	351.1632	330.6996	317.1538	308.6973	303.7946	301.3110	300.5596
Numerical Times (s)	500	766.5864	711.5544	661.7405	617.3667	578.6147	545.6280	518.5161	497.3588	482.2116	473.1102	470.0743
	1000	949.0029	903.8857	862.9207	826.3033	794.2082	766.7882	744.1738	726.4724	713.7683	706.1216	703.5686
	Posição x (m)	0.00	0.01	0.02	0.03	0.04	0.05	0.06	0.07	0.08	0.09	0.1
	100	536.6007	472.2577	420.5325	380.7698	351.6027	331.2279	317.7067	309.2242	304.2710	301.7422	300.9730
Analytical Times (s)	500	766.4347	711.4432	661.6731	617.3452	578.6396	545.6977	518.6266	497.5040	482.3833	473.2984	470.2682
	1000	948.7555	903.661	862.7171	826.1192	794.0415	766.6366	744.0349	726.3436	713.6468	706.0045	703.4530

The advantage of the numerical technique, besides its simplicity, is the great flexibility on boundary conditions and the possibility of incorporating physical property variables as a function of temperature.These properties can be updated for each time step, using the temperatures for each of the volume elements.

The spreadsheet Transient-1D.xls is useful in carrying out transient simulations in one dimension, enabling changes in physical properties and including radiation.

References

CARSLAW, H.S.; JAEGER, J.C. *Conduction of heat in solids*. 2. ed. New York: Clarendon, 1959.

INCROPERA, F.P.; DEWITT, D.P. *Fundamentos da transferência de calor e massa*. 5. ed. Rio de Janeiro: LTC, 2003.

Exercises

1 - A long wire with a diameter of 1 mm is submerged in a bath of oil which is at 25°C. The wire has an electrical resistance per unit length of 0.01 $\Omega/$ m . A current of 100 A passes through the wire. The heat transfer coefficient by convection is 500 W/m^2.K. What is the temperature of the wire in steady state? From the moment, the electric current starts passing , how much time is necessary for the wire reach a temperature of 1°C lower than that in steady state? The properties of the wire are: thermal conductivity = 20 W/mK; specific heat = 500 J/kg.K and density = 8000 kg/m^3.

2 - A procedure to determine the thermal conductivity of a solid material is to introduce a thermocouple into a thick sheet of material and measuring the change in temperature after a given change in surface temperature.Consider a situation where the thermocouple is 10 mm from the surface, which has its temperature suddenly raised and maintained at 373 K. The initial temperature of the plate is 300 K. Two minutes after the surface reaches a temperature of 373 K, the thermocouple indicates a temperature of 330 K. From these data, determine the thermal conductivity of the material. Its density and specific heat are, respectively, 2200 kg/m^3 and 700 J/kg.K.

3 - Metal spheres with 12 mm in diameter are heat treated in order to use in grinding mills for processing of ores. This treatment consists of heating them up to 1200 deg.K and subsequent cooling to 350 deg.K in air, at 300 deg.K, with a heat transfer coefficient equal to 22 W/m^2.K. Consider the following properties of the steel:

- Thermal conductivity = 42 W/m.K;
- Specific heat = 580 J/kg.K;
- Density: 7500 kg/m^3.

Estimate the time required for cooling. Evaluate the rate of cooling of the surface of the sphere, when its temperature is 600 deg. K.

4 - A long ingot of carbon steel (not alloyed), with square cross section of 0.4 m by 0.4 m, initially at a uniform temperature of 300 K, is placed inside a furnace at a temperature of 1,100 K. If the heat transfer coefficient convection is 100 W/m².K, how long the steel should remain in the furnace until the temperature in the center reaches 900 K? What is the temperature in the corner of the block at this time? It is not necessary to consider heat transfer by radiation.

5 - The following figure shows a section of a metal part made of pure copper. The data shown in the figure corresponds to the values of temperatures in the nodal points at a certain moment of time. Considering point 1, as reference, determine if the system is in steady state. If not, estimate the temperature value at that point when there is an elapsed time interval of 5 seconds. All temperatures are given in Celsius. The distance between the nodes is 6 cm horizontally and vertically. Deduce all the equations required for calculations. The emissivity of copper is 0.8.

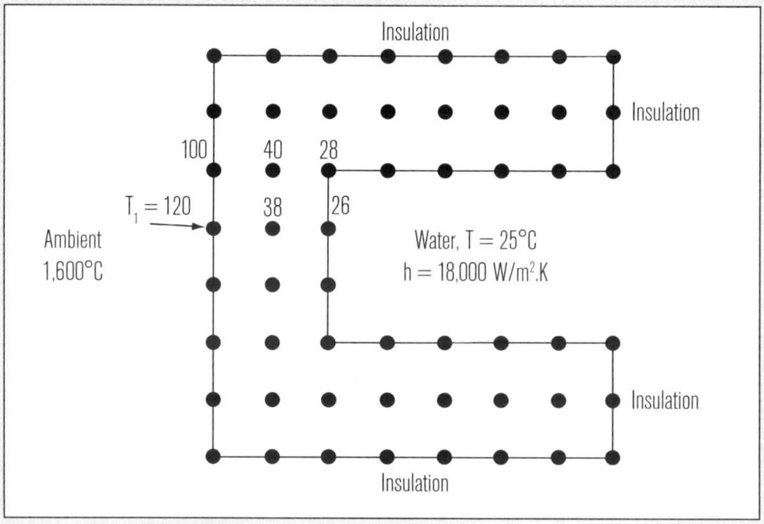

6 - A strip of steel coming out of a rolling unit undergoes cooling of its two large surfaces by convection and radiation to the environment, which is at 300 deg.K. The strip is 5 mm thick, and its physical properties are: thermal conductivity: 30 W/mK, density: 7900 kg/m³; specific heat: 640 J/kg.K and emissivity: 0.7. Determine the following:

- time to cool the strip at 940°C (ex-mill) to 540°C (temperature of coiling). The heat transfer coefficient for convection to air is 25 W/m².K. As an approximation incorporate the effect of radiation by defining a global heat transfer coefficient, estimated from the average temperature of the steel on the way to the winding unit;

- the length of the cooing region for a strip speed of 10 m/s.

Comment on the efficiency of this method of cooling.

7 - Cylindrical steel bars with diameter of 5 cm undergo heat treatment during passage through a furnace maintained at 700°C. The furnace is 5 m long and the bars come inside at 50 deg. C. What should be the speed of passage of the bars through the furnace, so that they come out at a temperature of 600°C? The heat transfer coefficient 135 W/m^2.K. Properties of steel are as follows:

- Thermal conductivity: 50 W/m.K;
- Density: 7800 kg/m^3;
- Specific heat: 434 J/kg.K.

8 - The refractory wall of a furnace reheating plates is 200 mm thick. The refractory has the following characteristics: k = 1 W/mK, C_p = 1200 J/kg.K and ρ = 2500 kg/m^3. The surface of the outer jacket is well insulated. At the beginning of the operation of the furnace, the refractory is at 300 K and its inner surface is exposed to gases at 1273 deg.K. The heat transfer coefficient between the gases and the refractory is 150 W/m^2.C. Estimate the time required for the outer surface of the refractory to reach a temperature of 1000 deg.K. Give all the equations used in the calculations and indicate how the data necessary to solve the problem can be obtained.

9 - A crucible is used for the transport of liquid aluminum from the melting furnace to the strip casting unit. The transport time is 2 minutes. Determine the amount of heat transferred by the aluminum to the refractory during transport. Estimate the temperature drop of aluminum on the way. The crucible is rectangular in shape with a width of 20 cm and length of 18 cm (internal dimensions). The height of aluminum is 10 cm. Consider also the following data:

- temperature of aluminum in the furnace exit: 720°C;
- air temperature and the neighborhood: 25°C;
- emissivity of liquid aluminum: 0.12;

- heat transfer coefficent for the upper surface of aluminum: 20 $W/m^2.K$;
- thermal conductivity of liquid aluminum: 200 W/m.K;
- specific heat of liquid aluminum: 1200 J/kg.K;
- density of liquid aluminum: 2388 kg/m^3;
- crucible refractory: thickness of 4.0 cm, thermal conductivity equal to 0.1 W/m.K, C_p = 960 J/kg.K and ρ = 1800 kg/m^3. The initial temperature of refractory is 298 K.

Assume that the interface between the refractory and Al liquid instantly reaches the temperature of the liquid Al.

10 - A long cylinder of 30 mm in diameter, initially at a uniform temperature of 1000 K, is immersed in a large oil bath maintained at a temperature of 350 K. The properties of the cylinder are:

- Thermal conductivity: 1.7 W/m.K;
- Specific heat: 1700 J/kg.K;
- Density: 1200 kg/m^3.

The heat transfer coefficient is 100 $W/m^2.K$.

Calculate the time to reach the surface temperature of 450 K. What is the temperature of the center of the cylinder at that instant?

11 - In the heat treatment for hardening steel ball bearings (thermal conductivity: 50 W/mK, specific heat: 500 J/kg.K, density: 7800 kg/m^3), it is desirable to raise the temperature of the surface in a short time without heating up the interior of the spheres. This type of heating can be obtained by immersing the sphere in a liquid salt bath at a temperature of 1300 K . The heat transfer coefficient for this stage is 5000 $W/m^2.K$. Any location within the sphere, where the temperature exceeds 1000.K, gets the desired heat treatment Estimate the time needed for this type of heat treatment to reach a depth of 1 mm in a ball with 20 mm in diameter, initially at 300 K.

12 - Steel plates AISI 1010, 5 m long, 1.5 m wide and 25 cm thick, are placed in a furnace to reheat before being rolled. The temperature of the gases and the walls of the furnace is 1700 K. The global heat transfer coefficient inside the furnace is 250 $W/m^2.K$. The initial temperature of the plates is 300 K, Calculate the temperature at the center, on the surface and at the midpoint between the center and surface of the plate after a time of 1 hour. Assume heat transfer only in the direction of the thickness of the plates.

13 - Axles of carbon steel AISI 1010 with 0.2 m in diameter are undergoing treatment in a gas fired furnace. The temperature of the gases inside the furnace is 1800 K and the heat transfer coefficient is 500 $W/m^2.K$. If the axles enter the furnace at 300 K, how long they should be kept for the temperature at the center line reaching up to 1500 K? What is the temperature at the surface at that moment ?

14 - A steel with carbon-silicon-manganese undergoes heating before being subjected to heat treatment.The steel plate is 100 mm thick, 2 m wide and 5 m long and is initially at 300 K. It is heated inside a gas fired furnace. The temperature of the gas inside the furnace is 1200 K and the heat transfer coefficient is 500 $W/m^2.K$. How long should it take for the minimum temperature of the plate to reach 1000 K? What is the maximum temperature at that moment?

15 - Axles of carbon steel with 0.1 m in diameter are heat treated by heating in a furnace, kept at 1200 K. The heat transfer coefficient inside the furnace is 90 $W/m^2.K$. If the axles enter the oven at 298 K, how long they should stay inside so that the center reaches a temperature of 800 K? What is the temperature of the surface at that moment?

Convection

Chapter 13

The systems treated so far mostly were about heat transfer in a medium that had no macroscopic motion. The mechanism of heat transfer was essentially conduction.

In many processes of interest to the metallurgist the study of heat transfer with macroscopic movement, involving liquids and gases has become important.In these situations, two approaches can be adopted. The choice of approach depends on the objectives of the study.

It is important to know, the temperature distribution in a fluid in motion. For this, the equation of energy conservation equations, including the transport of energy associated with the movement, have to be resolved along with the equations of continuity (mass conservation) and conservation of momentum, as are related to the macroscopic movement. The solution process becomes very laborious particularly in the case of two and three-dimensional flows. In such cases, computational techniques have be employed to solve the equations of conservation, since they do not have analytical solutions.

In situations, where one wants to quantify the rate of heat transfer between a fluid and a solid surface in contact with the fluid, the approach mentioned above can also be employed. However, these cases are simpler and there are easier alternatives that enable evaluation of the rate of heat transfer without the need to solve complex equations of conservation of momentum, even in turbulent flow conditions. In this case, the rate of heat transfer is calculated using the using the heat transfer coefficient which in turn can be determined experimentally and expressed as correlations between dimensionless numbers. This is widely applied in most engineering problems.

The methodology is very similar to that described in the earlier chapter dealing with balance of momentum. For example the force of friction between a solid and a moving fluid in contact with it, can be evaluated through determination of velocity profiles of the fluid, from conservation equations for force balance and then application of Newton's law of viscosity to estimate shear stress at the solid-fluid. interface. The other possibility of evaluating this frictional force is to use experimental results to estimate friction factor, (f) (Chapter 7). The friction factor is used to to calculate the frictional force similar to the use of heat transfer coefficient to calculate the rate of heat transfer between a solid and a fluid in contact with it. The heat transfer coefficient is analogous to the mass transfer coefficient used in most of the problems of mass flow.

In the next section some fundamentals of using the first approach, both to assess the temperature profile and to estimate the rate of heat transfer will be discussed. In the remainder of the chapter, the focus will be on the use of experimental results to estimate the heat transfer coefficient and the rate of heat transfer between a solid and fluid.

13.1 Energy Balance

In Chapter 11, formulation of equations of energy balance in rectangular, cylindrical and spherical, coordinates has been discussed. In these, it was assumed that the medium had no macroscopic motion. When the medium moves, the input and output of heat due to the macroscopic motion should be incorporated. The existence of gradients of velocity also cause the conversion of mechanical energy into heat.

The equations (13.1) (13.2) and (13.3) are the general equations of conservation of energy for all the three coordinate systems, cartesian, cylindrical and spherical, respectively. These equations include the terms associated with macroscopic motion of fluid and were obtained considering Newtonian fluids, with constant density, viscosity and thermal conductivity (BIRD, STEWART and LIGHTFOOT, 1960).

$$\rho.C_p.\left(\frac{\partial T}{\partial t}+v_x\frac{\partial T}{\partial x}+v_y\frac{\partial T}{\partial y}+v_z\frac{\partial T}{\partial z}\right)=k\left[\frac{\partial^2 T}{\partial x^2}+\frac{\partial^2 T}{\partial y^2}+\frac{\partial^2 T}{\partial z^2}\right]$$
$$+2.\mu\left\{\left(\frac{\partial v_x}{\partial x}\right)^2+\left(\frac{\partial v_y}{\partial y}\right)^2+\left(\frac{\partial v_z}{\partial z}\right)^2\right\}+\mu\left\{\left(\frac{\partial v_x}{\partial y}+\frac{\partial v_y}{\partial x}\right)^2+\right. \qquad (13.1)$$
$$\left.\left(\frac{\partial v_x}{\partial z}+\frac{\partial v_z}{\partial x}\right)^2+\left(\frac{\partial v_y}{\partial z}+\frac{\partial v_z}{vy}\right)^2\right\}$$

$$\rho.C_p.\left(\frac{\partial T}{\partial t}+v_r\frac{\partial T}{\partial r}+\frac{v_\theta}{r}\frac{\partial T}{\partial \theta}+v_z\frac{\partial T}{\partial z}\right)=k\left[\frac{1}{r}\frac{\partial}{\partial r}\left(r\frac{\partial T}{\partial r}\right)+\frac{1}{r^2}\frac{\partial^2 T}{\partial \theta^2}+\frac{\partial^2 T}{\partial z^2}\right]+$$

$$+2.\mu\left\{\left(\frac{\partial v_r}{\partial r}\right)^2+\left[\frac{1}{r}\left(\frac{\partial v_\theta}{\partial \theta}+v_r\right)\right]^2+\left(\frac{\partial v_z}{\partial z}\right)^2\right\}+ \qquad (13.2)$$

$$\mu\left\{\left(\frac{\partial v_\theta}{\partial z}+\frac{1}{r}\frac{\partial v_z}{\partial \theta}\right)^2+\left(\frac{\partial v_z}{\partial r}+\frac{\partial v_r}{\partial z}\right)^2+\left[\frac{1}{r}\frac{\partial v_r}{\partial \theta}+r\frac{\partial}{\partial r}\left(\frac{v_\theta}{r}\right)\right]^2\right\}$$

$$\rho.C_p.\left(\frac{\partial T}{\partial t}+v_r\frac{\partial T}{\partial r}+\frac{v_\theta}{r}\frac{\partial T}{\partial \theta}+\frac{v_\varphi}{r\sin\theta}\frac{\partial T}{\partial \phi}\right)=k\left[\frac{1}{r^2}\frac{\partial}{\partial r}\left(r^2\frac{\partial T}{\partial r}\right)+\right.$$

$$\left.\frac{1}{r^2\sin\theta}\frac{\partial}{\partial \theta}\left(\sin\theta\frac{\partial T}{\partial \theta}\right)+\frac{1}{r^2\sin^2\theta}\frac{\partial^2 T}{\partial \phi^2}\right]+2.\mu\left[\left(\frac{\partial v_r}{\partial r}\right)^2\right.$$

$$\left.+\left(\frac{1}{r}\frac{\partial v_\theta}{\partial \theta}+\frac{v_r}{r}\right)^2+\left(\frac{1}{r\sin\theta}\frac{\partial v_\phi}{\partial \phi}+\frac{v_r}{r}+\frac{v_\theta\cot\theta}{r}\right)^2\right]+ \qquad (13.3)$$

$$\mu\left\{\left[r\frac{\partial}{\partial r}\left(\frac{v_\theta}{r}\right)+\frac{1}{r}\frac{\partial v_r}{\partial \theta}\right]^2+\left[\frac{1}{r\sin\theta}\frac{\partial v_r}{\partial \phi}+r\frac{\partial}{\partial r}\left(\frac{v_\phi}{r}\right)\right]^2+\right.$$

$$\left.\left[\frac{\sin\theta}{r}\frac{\partial}{\partial \theta}\left(\frac{v_\phi}{\sin\theta}\right)+\frac{1}{r\sin\theta}\frac{\partial v_è}{\partial \phi}\right]^2\right\}$$

To use the equations 13.1 to 13.3 for determining the temperature profiles in a system, it is necessary to know the profiles of fluid velocity. Hence, the conservation equations of mass and momentum should also be considered. In this solution, two distinct cases can be identified. In a situation, where the movement of the fluid is not significantly affected by temperature distribution the conservation equations of mass and momentum can be resolved before the solution of equations of conservation of energy. The velocity profiles can be determined first and then these can be used to calculate the temperature profiles in problems of forced convection, where the fluid motion is caused mainly by some external force (like a pump or a fan) that causes the fluid flow.

In another situation, the fluid motion can be caused by changes in fluid temperature. Temperature variations in a fluid cause changes in its density, which in turn, induces movement. The fluid material that is at higher temperatures is less dense and tends to go to up whereas the cooler fluid being denser, moves to the lower regions. In this case, profiles of temperature and velocity are coupled and the equations of conservation of momentum and energy must be solved simultaneously. In this case, the problem is natural or free convection. The movement of the fluid is caused by the buoyant force associated with changes in fluid density. Besides the temperature, density variations can also be caused by variation in the concentration of solutes in problems involving mass transfer.

Figure 13.1 shows the temperature distribution in a strand of continuous slab casting. In this case, the temperature profile can be used to analyze the effect of setting of the submerged valve on the progress of solidification front. In this case of practical application, it is essential to solve the equations of conservation of momentum and energy simultaneously.

This temperature distribution was obtained by solving the equations of continuity and for momentum and energy conservation simultaneously. The effect of solidification on fluid flow was introduced using the enthalpy-porosity technique. These results were attained using the Metflo code developed at McGill University.

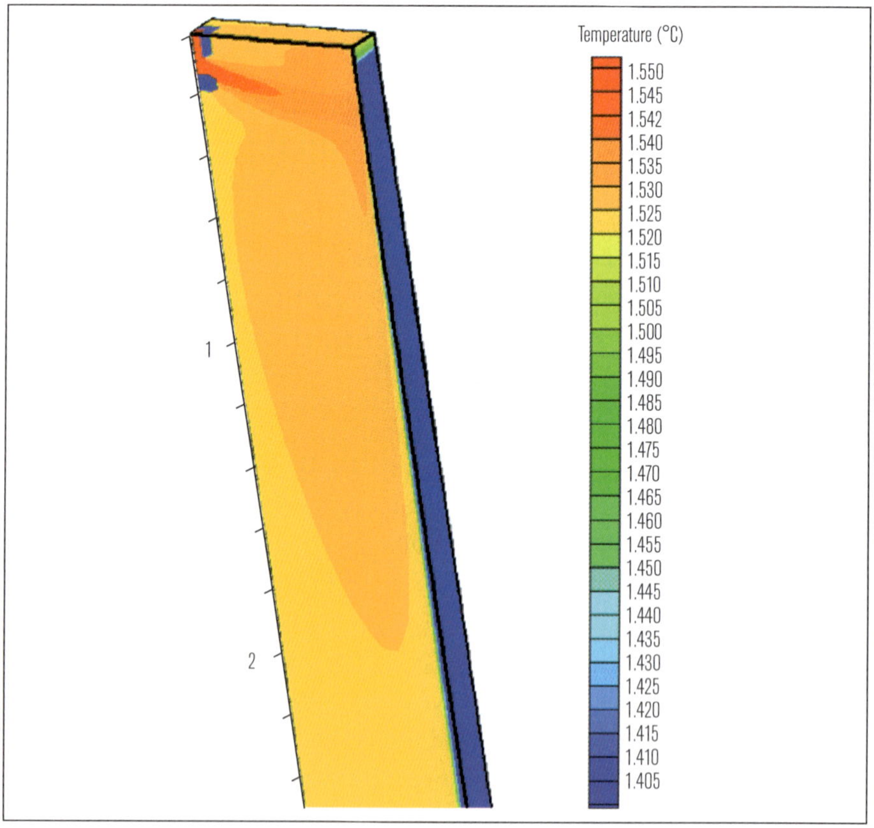

Figure 13.1 - Temperature profile in a strand of continuous slab casting (PEREIRA et al., 2001).

When the temperature changes in the system are significant, the effect of density variation with temperature can be included through the Boussinesq approximation (JOO, HAN and GUTHRIE, 1993). In this approach, the density is expanded using a Taylor series in terms of temperature, using a reference temperature T_{ref} one can write:

$$\rho = \rho\Big|_{T_{ref}} + \frac{\partial\rho}{\partial T}\Big|_{T_{ref}} .(T - T_{ref}) + \frac{1}{2!}\frac{\partial^2\rho}{\partial T^2}\Big|_{T_{ref}} .(T - T_{ref})^2 + ... \quad (13.4)$$

Truncating the series at the end of the first order, we have:

$$\rho = \rho_{ref} + \frac{\partial\rho}{\partial T}\Big|_{T_{ref}} .(T - T_{ref}) \quad (13.5)$$

where:

- ρ_{ref} is the density at reference temperature.

Substituting this relation in the term associated with the force of gravity in the equation of conservation of momentum, we obtain:

$$\rho.g = \rho_{ref}.g + g.\frac{\partial\rho}{\partial T}\Big|_{T_{ref}} .(T - T_{ref}) \quad (13.6)$$

or:

$$\rho.g = \rho_{ref}.g - \rho_{ref}.g.\beta.(T - T_{ref}) \quad (13.7)$$

where:

- β is the coefficient of volume expansion, defined by the following equation:

$$\beta = -\frac{1}{\rho_{ref}}.\frac{\partial\rho}{\partial T}\Big|_{T_{ref}} \quad (13.8)$$

The term $-\rho_{ref}.g.\beta.(T - T_{ref})$ above represents the buoyant force per unit volume of fluid. As β is usually a positive number (density tends to decrease with increasing temperature), when T is greater than T_{ref}, the term associated with the buoyant force is negative, indicating an upward force, contrary to the direction of acceleration due to gravity.

When the temperature distribution in the fluid is known, it is possible to determine the flow of heat between the fluid and solid in contact with the fluid. In a region sufficiently close to the solid, there will be no movement of fluid in the direction perpendicular to the surface of the solid. Hence, the heat flux can be calculated using the following equation:

$$q = -k_f \cdot \left.\frac{\partial T}{\partial n}\right|_{sup} \tag{13.9}$$

where:

- k_f = thermal conductivity of the fluid;
- $\left.\frac{\partial T}{\partial n}\right|_{sup}$ = temperature gradient normal to the surface of the solid.

The temperature gradient that appears in equation (13.9) depends on the velocity profile of the fluid, especially in the region near the wall. Hence, using this approach to estimate the rate of heat exchange between fluid and solid is extremely complex, even with the computing resources available today, as in the case of determination of velocity profiles.

When the goal is simply to estimate the rate of heat exchange between fluid and solid in contact with him, there is an alternative approach, simpler, which is adequate for most engineering problems. This approach is discussed in the next section.

13.2 Heat Transfer Coefficient

Figure 13.2 illustrates a simple situation and heat transfer between a solid and a fluid in motion.In this figure, the fluid motion is predominantly parallel to the solid surface. This system is similar to what was presented in Chapter 6.

Figure 13.2 - Fluid flow and heat transfer on a flat plate submerged in a fluid.

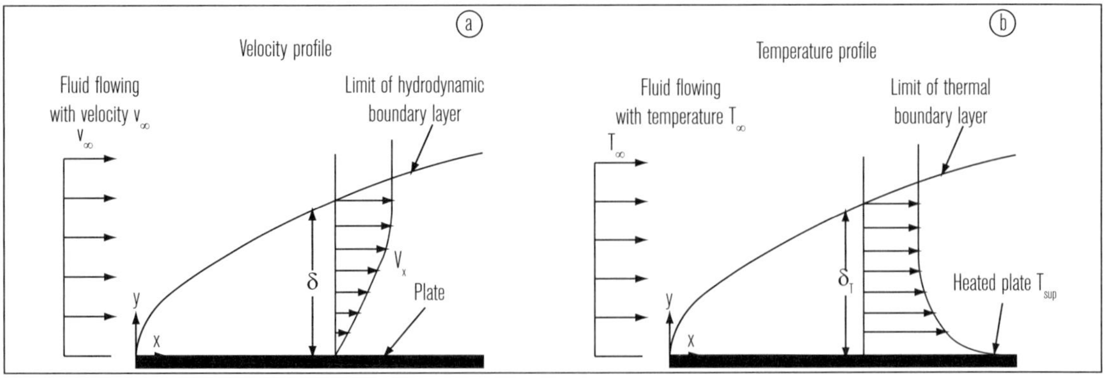

Figure 13.2a illustrates the development of boundary layer and flow of fluid inside the same.The thickness δ is called the boundary layer thickness and corresponds to the region where 99% of the variation in velocity (from 0 at the solid surface, to v_∞, at a point well away from the plate). As mentioned earlier, the thickness of the boundary layer grows with the increase of kinematic viscosity of the fluid and the reduction of velocity v_∞.

Figure 13.2 (b) is shows the development of thermal boundary layer. In this case, the thickness δ_T is defined for the region where there is 99% of the variation of temperature (T_{sup} between the surface and T_∞ ,a point, well away from the plate). The thickness of the thermal boundary layer for submerged plate in a fluid stream can be evaluated using the energy equation coupled with Navier stokes equation after simplifying them and converting the partial differential equations into homogeneous ones using parameters based on stream function (GEIGER and POIRIER, 1973) The results indicate that its thickness tends into increase with the increase of thermal diffusivity of fluid, ($\alpha = k/\rho.C_p$) and decrease with increasing velocity v_∞.

The relationship between the thickness of hydrodynamic and thermal boundary layers, depends on the values of thermal diffusivity and kinematic viscosity of the fluid:

$$\frac{\delta_T}{\delta} \cong \left(\frac{\alpha}{\nu}\right)^n \tag{13.10}$$

where:

- n is a positive exponent.

The ratio between kinematic viscosity and thermal diffusivity is a dimensionless number known as Prandtl number, Pr. The relation (13.10) can then be rewritten in the following way:

$$\frac{\delta_T}{\delta} \cong \frac{1}{Pr^n} \tag{13.11}$$

The Prandtl number varies with the nature of the fluid. Gases have Prandtl numbers near 0.7 and these are little affected by temperature. Water and oil have Prandtl numbers in the range of $1\text{-}10^3$, with a tendency to decrease with increasing temperature. Liquid metals have Prandtl numbers in the range 0.001 to 0.03 (GEIGER and POIRIER, 1973).

As seen in Figure 13.2 b, for a given temperature difference between fluid and solid surface, the thickness of thermal boundary layer affects the temperature gradient in the fluid near the surface. This temperature gradient, together with the thermal conductivity of the fluid determines the flow of heat between the solid and fluid (equation 13.9).

As seen in previous chapters, the flow of heat between a solid surface and a fluid in contact with it is estimated by the following relationship:

$$q = h.(T_{sup} - T_{\infty}) \tag{13.12}$$

Equation (13.12) is not a fundamental law of transport phenomena. It can be interpreted as a definition of heat transfer coefficient, h. Combining equations (13.9) and (13.12), we have:

$$h = -\frac{k_f . \left.\frac{\partial T}{\partial n}\right|_{sup}}{(T_{sup} - T_{\infty})} \tag{13.13}$$

The simple definition of the heat transfer coefficient given by equation (13.12) does not solve the problem of evaluating the heat flux at the solid-fluid, unless the coefficient of heat transfer is known. It can be seen by equation (13.13) that the value of h depends on the temperature gradient near the solid/liquid interface. This temperature gradient is a function of thickness of the thermal boundary layer, which, in turn, depends on the velocity profile of fluid along the wall (boundary layer hydrodynamics). The definition of heat transfer coefficient, as shown in equation (13.13), is suitable for external flows (around objects), because then one can always specify a temperature T_{∞} outside the thermal boundary layer. In the case of flow in ducts (inside the duct), the boundary layer grows from the entry of fluid in the pipeline and eventually reaches its center. From this point, it is possible to specify the temperature of the fluid outside the boundary layer. Hence , for both internal and external flows, the heat transfer coefficient is usually is defined by the following equation:

$$h = -\frac{k_f . \left.\frac{\partial T}{\partial n}\right|_{sup}}{(T_{sup} - T_m)} \tag{13.14}$$

where:

- T_m is the average temperature of the fluid along the cross section of the pipeline.

With this definition, the heat flow between the duct wall and the fluid inside is expressed by:

$$q = h.(T_{sup} - T_m) \qquad (13.15)$$

The evaluation of the coefficient of heat transfer through the fundamental equations of fluid flow and heat transfer is a laborious task. Moreover, as the thickness of the boundary layer and the temperature gradient vary along the surface of the solid, the coefficient of heat transfer is also a function of position.

An approach that is commonly used in the engineering fields is to evaluate the coefficient of heat transfer through experiments.These experiments are conducted in different conditions and for systems with different geometries and flow conditions. The results are usually expressed as mathematical correlations, whose validity is limited to conditions under which the experiments had been done. These correlations enable the estimation of an average value for the heat transfer coefficient along the solid-fluid interface. However, there are correlations also for local values of this coefficient. In the next section, aspects of experimental evaluation of the coefficient of heat transfer and development of correlations will be discussed.

13.3 Experimental Evaluation of Heat Transfer Coefficient

There are several ways to carry out experimental evaluations of the heat transfer coefficient The choice of technique depends on, the configuration and the type of the system, being studied. Figure 13.3 shows a schematic view of an equipment that can be used to determine the average value of the coefficient of heat transfer between a flat plate(with internal heating through passage of current) under steady state conditions and a fluid flowing parallel to its surface.

In the experimental set up above,the heat transfer coefficient can be determined by measurements of voltage, (V) current (I) and temperatures of solid and fluid in steady state conditions. Assuming that all heat is dissipated by Joule effect in the resistance and is transferred to the fluid,

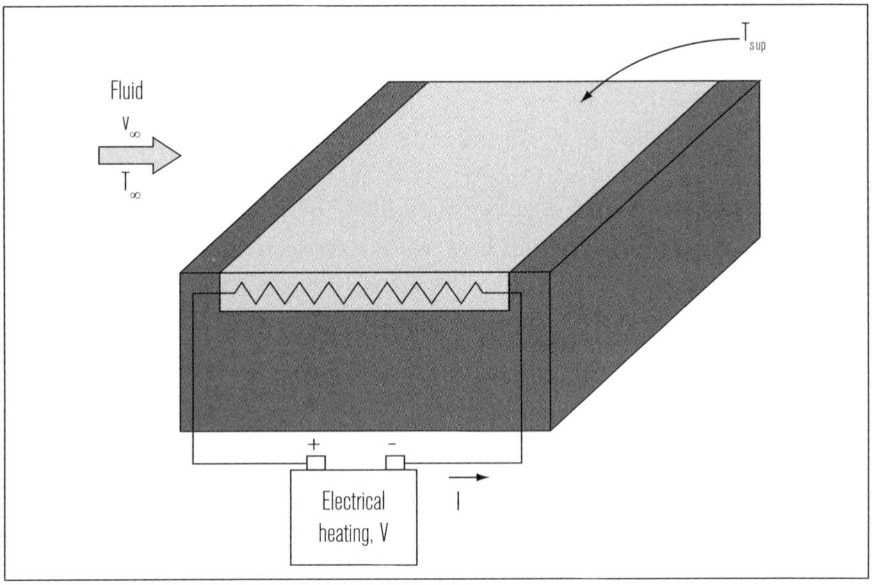

Figure 13.3 - Experimental setup for determining heat transfer coefficient.

one can calculate the average value of heat transfer coefficient along the surface of the solid by the following equation:

$$V.I = \bar{h}.A.(T_{sup} - T_{\infty}) \tag{13.16}$$

where:

- A is the surface area in contact with the fluid.

Another possibility for experimental evaluation of the coefficient of heat transfer is to arrange temperature sensors in various locations inside a solid material in contact with the fluid. This can be considered an inverse problem of heat transfer. For determining the coefficient of heat transfer, the measured temperatures from the sensors should be the same predicted by a mathematical model involving convection in the surface and conduction through the solid material. When one gets this setting correct, the value of the coefficient of heat transfer should be appropriate. These experiments, usually have the objective to determine the effect of different variables on the coefficient of heat transfer. To reduce the number of experiments it is necessary to characterize these effects. For this it is desirable to develop a dimensional analysis similar to that used in evaluating the friction factor (Chapter 7). The same methodology adopted for the friction factor can be applied to the heat transfer coefficient. Applying this technique, the experimental results are provided in the form of correlations involving dimensionless numbers.

In the case of forced convection, the average heat transfer coefficient depends on a number of physical properties of the fluid (density ρ, viscosity μ, specific heat at constant pressure, C_p and thermal conductivity, k) ,velocity of the fluid V, and a characteristic length of the system under consideration, L. In all, seven variables (one dependent variable namely heat transfer coefficient h, and six independent variables mentioned above) and four basic dimensions (mass, length, time and temperature). According to the Buckingham π theorem, one can obtain three (7 variables – 4 dimensions) dimensionless groups that characterize this system.The dimensionless numbers that are important in heat transfer are:

$$\text{Número de Nusselt: } Nu = \frac{\bar{h}.L}{k} \tag{13.17}$$

$$\text{Número de Reynolds: } Re = \frac{L.V.\rho}{\mu} \tag{13.18}$$

$$\text{Número de Prandtl: } Pr = \frac{C_p.\mu}{k} \tag{13.19}$$

From the experimental results, it is possible to develop correlations for the Nusselt number as a function of Reynolds and Prandtl Numbers. These correlations are usually expressed in the following general form:

$$Nu = C.Re^a.Pr^b \tag{13.20}$$

where:

- C, a and b are constants specific for each system configuration occurs where the heat transfer. The values of C, a and b also vary with the regime of fluid flow, laminar or turbulent.

The Nusselt number can be interpreted as a dimensionless temperature gradient measured along the surface of the solid. Defining a dimensionless temperature as:

$$\theta = \frac{T - T_{sup}}{T_\infty - T_{sup}} \tag{13.21}$$

and a dimensionless distance in the direction normal to the surface:

$$n^* = \frac{n}{L} \tag{13.22}$$

With these definitions, equation (13.13) can be written in the following way:

$$h = -\frac{k_f.\left.\frac{\partial T}{\partial n}\right|_{sup}}{(T_{sup} - T_\infty)} = -\frac{k_f.\left.\frac{\partial \theta}{\partial n^*}\right|_{sup}(T_\infty - T_{sup})}{L.(T_{sup} - T_\infty)} \tag{13.23}$$

or:

$$\frac{h.L}{k_f} = \left.\frac{\partial \theta}{\partial n^*}\right|_{sup} \tag{13.24}$$

As stated earlier, the Reynolds number represents a ratio of inertial forces and viscous forces. Reynolds number affects the thickness of the hydrodynamic boundary layer. Higher Reynolds number results in thinner boundary layer (Chapter 6).

Prandtl number is a ratio between the diffusivities of momentum and heat. This ratio between diffusivities determines the relationship between the thicknesses of thermal and velocity boundary layers:

$$\frac{\delta}{\delta_T} = Pr^n \tag{13.25}$$

where:

- n is a positive number.

In the case of natural convection, as the fluid velocity is no longer determined by an external agent, but by the change in fluid density with temperature, the Reynolds number is replaced by a dimensionless number, called Grashof number and is defined by:

$$Gr = \frac{g.\beta.(T_{sup} - T_\infty).L^3}{\nu^2} \tag{13.26}$$

where:

- β is the coefficient of volume expansion, defined by equation (13.8).

Grashof number represents a ratio between the viscous force and the buoyancy forces In this case, the empirical correlation for determining heat transfer coefficient through Nusselt number is given by:

$$Nu = C.Gr^a.Pr^b \tag{13.27}$$

Sometimes this correlation is given in terms of Rayleigh number, which is the product of the Grashof and Prandtl:numbers.

$$Ra = Gr.Pr = \frac{g.\beta.(T_{sup} - T_{\infty}).L^3}{\nu.\alpha} \qquad (13.28)$$

The Grashof and Rayleigh numbers are always positive. Hence when T_{∞} is more than $T_{sup,}$ the temperature difference that appears in the definition of these numbers must be reversed.

In the case of empirical correlations to evaluate the coefficient of heat transfer, two aspects should be paid attention. One of them is related to the temperature that is used to evaluate the physical properties of the fluid. In the case of flow around objects (external flow), it is conventional to adopt the average temperature of the solid surface and the fluid outside the boundary layer. This temperature is called the temperature of the film and is calculated by the following relationship:

$$T_f = \frac{T_{sup} + T_{\infty}}{2} \qquad (13.29)$$

In some correlations, the fluid properties are evaluated at temperature T_{∞}. In these cases, it is usual to incorporate a correction factor in the correlation for compensating the effect of variations in the properties. This factor is usually a ratio in the form (Pr_{∞}/Pr_{sup}) or (μ_{∞}/μ_{sup}) raised to a positive power. The subscripts ∞ and "sup" refer to temperatures T_{∞} (Temperature of the fluid in the bulk) and T_{sup}, (at the surface) respectively.

In the case of flow inside the ducts (internal flow), the properties of the fluid are determined at the average temperature along the cross section of the pipeline.In some cases, a correction factor is introduce to take care of the the change in the temperature of the fluid, especially when there is a large difference between the temperature of the fluid and the duct wall.

Another aspect concerning the use of empirical correlations is associated with the conditions of validity of these expressions. As these correlations are proposed from experimental data, not from fundamental concepts of transport phenomena, their validity is limited to experimental conditions that are adopted in obtaining the same. These limit the validity of correlations which are usually expressed in terms of ranges of values of Reynolds, Prandtl and Grashoff (or Rayleigh), numbers depending on the correlations proposed.

In the later sections, correlations to evaluate the heat transfer coefficient for forced and natural convection, external and internal flows and for different systems are given. The objective is to illustrate the use of these correlations that have been developed especially for metallurgical applications and for high temperatures. In the literature on heat transfer, one can find a large number of number of correlations, for situations relevant to application in the respective fields of engineering. These correlations are derived from the results of experiments done with specific experimental set ups and conditions relevant to the particular application. It is also important to stress that the correlations presented below, may have errors even up to about 20%. Whenever possible, it is recommended that the values of heat transfer coefficient obtained in different correlations for the same situation should be compared in order to decide o which one is more reliable for a particular application.

In the correlations presented in this text, the indices (subscripts) used for Reynolds number, Nusselt, Rayleigh and Grashof, match the characteristic dimensions, used in the definition of these numbers.

13.4 Forced Convection and External Flow

As mentioned earlier, external flow refers to situations where the fluid flows around an object. Three geometries will be discussed: flat plate, cylinder and sphere.

13.4.1 Forced convection over submerged plate

The system considered in this case is seen schematically in Figure 13.4. In this case, the flow is predominantly parallel to the surface of the plate.

For this system the transition from laminar to turbulent flow occurs for Reynolds number equal to 5×10^5:

$$Re_{transition} = \frac{x.v_{\infty}.\rho}{\mu} = 5 \,.\, 10^5 \qquad (13.30)$$

If the Reynolds number is less than 5×10^5, the flow around the plate is laminar (using L as the characteristic dimension in the Reynolds number).

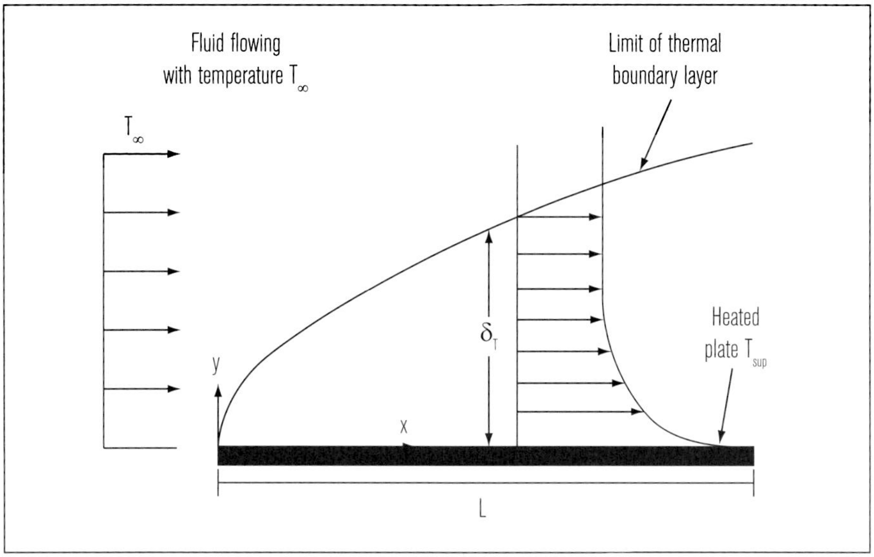

Figure 13.4 - External fluid flow parallel to a flat plate.

In this case, the average Nusselt number along the length of the plate can be determined by the following equation (INCROPERA and DEWITT, 2003):

$$Nu_L = \frac{h.L}{k} = 0.664.Re_L^{1/2}.Pr^{1/3} \qquad (13.31)$$
$$Pr \geq 0.6$$

When the Reynolds number is greater than 5 x 10^5, the transition from laminar to turbulent over the surface of the plate takes place.The Nusselt number is then given by (INCROPERA and DEWITT, 2003):

$$Nu_L = (0.037.Re_L^{4/5} - 871).Pr^{1/3}$$
$$Pr < 60 \qquad (13.32)$$
$$5 \,.\, 10^5 < Re_L < 10^8$$

The correlations presented above are valid irrespective of the position of the plate (horizontal, vertical or inclined), provided that the fluid flows parallel to the surface. These correlations are valid,when the temperature of the slab is uniform.

Example

During the process of cold rolling, a strip of steel at 800°C moves with a speed of 10 m/s. Knowing that the air is 30°C, estimate the rate of heat loss from the strip by forced convection to the air. The width of the strip is 1.5 m and the length is 100 m. Compare this rate with that associated with radiation, if the emissivity of the steel is 0.78.

Solution:

Initially, one should evaluate the heat transfer coefficient for forced convection. The properties of air should be evaluated at the temperature of the film, given by:

$$T_f = \frac{T_{sup} + T_\infty}{2} = \frac{800 + 30}{2} = 415°C = 688\ K$$

At this temperature, we have:

$$\rho = 0.4975\ kg/m^3 \quad \mu = 338.8\ .\ 10^{-7}\ Pa.s \quad Pr = 0.695$$
$$k = 52.4\ .\ 10^{-3}\ W/m.K$$

The Reynolds number is:

$$Re_L = \frac{L.v.\rho}{\mu} = \frac{(100).(10).(0.4975)}{(338.8\ .\ 10^{-7})} = 1.47\ .\ 10^7$$

The Nusselt number is calculated by the following correlation:

$$Nu_L = (0.037.Re_L^{4/5} - 871).Pr^{1/3}$$
$$Nu_L = [0.037.(1.47\ .\ 10^7)^{4/5} - 871].(0.695)^{1/3} = 16970.77$$

The coefficient of heat transfer is:

$$h = \frac{Nu_L.k}{L} = \frac{(16970.77).(52.4\ .\ 10^{-3})}{(100)} = 8.89\ W/m^2.K$$

The rate of heat loss by forced convection is:

$$Q = A.h.(T_{sup} - T_\infty) = (100\ .\ 1.5).(8.89).(800 - 30) = 1.03\ .\ 10^6\ W$$

Loss due to radiation, is:

$$Q = A.\sigma.\varepsilon.(T_{sup}^4 - T_\infty^4) = (100\ .\ 1.5).(5.67\ .\ 10^{-8}).(0.78).[(1{,}073)^4 - (303)^4] = 8.74\ .\ 10^6\ W$$

For the conditions under consideration, the loss by radiation is about 8 times higher than that by convection.

13.4.2 Forced convection around cylinder

Figure 13.5 illustrates the case of transversal flow around a cylinder.

In this type of flow, the heat transfer coefficient varies with the angle θ.The local values of this coefficient can be integrated over the surface of the cylinder to obtain average values.

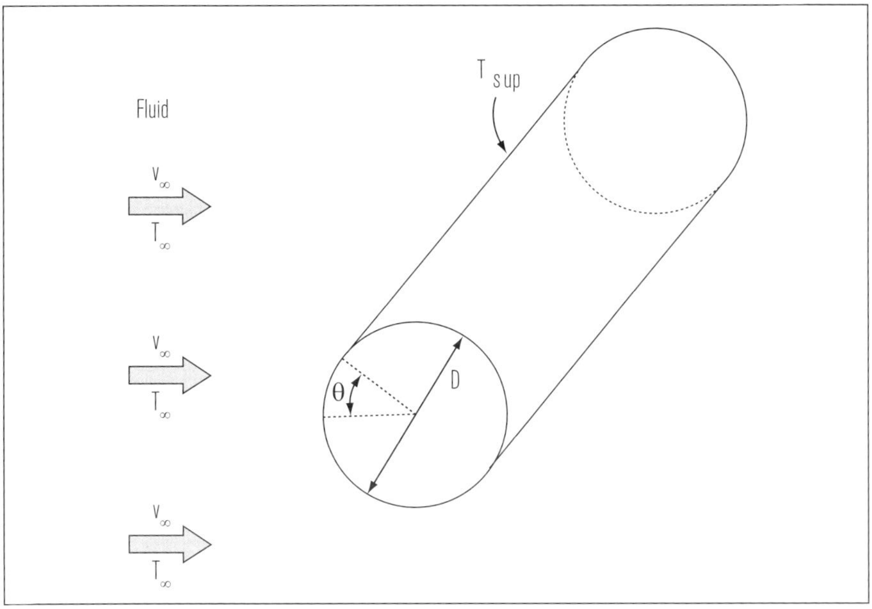

Figure 13.5 - Cross-flow around a cylinder.

An equation that can be used to evaluate the heat transfer coefficient in this case is given by (INCROPERA and DEWITT, 2003):

$$Nu_D = 0.3 + \frac{0.62.Re_D^{1/2}.Pr^{1/3}}{\left[1+\left(\frac{0.4}{Pr}\right)^{2/3}\right]^{1/4}}.\left[1+\left(\frac{Re_D}{282000}\right)^{5/8}\right]^{4/5} \qquad (13.33)$$

$Re_D.Pr > 0.2$

13.4.3 Forced convection around sphere

For flow around a sphere, the average Nusselt number along the surface can be estimated from the following correlation (INCROPERA and DEWITT, 2003):

$$Nu_D = 2 + (0.4.Re_D^{1/2} + 0.06.Re_D^{2/3}).Pr^{0,4}.\left(\frac{\mu}{\mu_s}\right)^{1/4} \quad (13.34)$$

$0.71 < Pr < 380$

$3.5 < Re_D < 7.6 \,.\, 10^4$

In using this correlation, all the properties of the fluid should be evaluated at temperature T_∞. The viscosity μ_s is evaluated at the temperature of the surface of the sphere.

Another widely used correlation is the equation of Ranz-Marshall (GEIGER and POIRIER, 1973):

$$Nu_D = 2 + 0.6.Re_D^{1/2}.Pr^{1/3.} \quad (13.35)$$

$1 < Re_D < 10^5$

It is worth noting, that in these two correlations, when the Reynolds number is zero, Nusselt number becomes 2. This value, corresponding to a situation, where the fluid is completely stagnant, was derived in Chapter 11, considering the heat transfer by conduction only around a sphere.

Example

Determine the initial rate of heat transfer in heating a pellet of iron ore at 25°C when placed in contact with blast furnace gas from the furnace at 130°C. The diameter of the pellet is 15 mm. Consider the properties of the blast furnace gas as being equivalent to those of nitrogen. The gas velocity is 2 m/s.

Solution

To assess the rate of heat transfer, the heat transfer coefficient ils evaluated using the following correlation.

$$Nu_D = 2 + (0.4.Re_D^{1/2} + 0.06.Re_D^{2/3}).Pr^{0.4.}.\left(\frac{\mu}{\mu_s}\right)^{1/4}$$

The properties of the fluid should be evaluated at an average temperature between T_∞ (130°C) and T_s the surface temperature of the pellet (25°C). Equation (13.40) can be used to evaluate the average temperature of the fluid, T_m, for each position z along the pipe. Figure 13.7 shows, schematically, the variation of fluid temperature along the length of the pipe.

$\rho = 0.8425\ kg/m^3$ $\mu = 220.4 . 10^{-7}$ Pa.s $\mu_s = 178.2 . 10^{-7}$ Pa.s

$Pr = 0.704$ $k = 32.7 . 10^{-3}$ W/m.K

Reynolds number:

$$Re_D = \frac{D.v.\rho}{\mu} = \frac{(0.015).(2).(0.8425)}{(220.4 . 10^{-7})} = 1{,}146.78$$

Applying the correlation, we have:

$$Nu_D = 2 + [0.4.(1146.78)^{1/2} + 0.06.(1146.78)^{2/3}].(0.704)^{0.4.}.\left(\frac{220.4}{178.2}\right)^{1/4}$$

$$Nu_D = 20.44$$

The heat transfer coefficient is:

$$h = \frac{Nu_D.k}{D} = \frac{(20.44).(32.7 . 10^{-3})}{(0.015)} = 44.56\ W/m^2.K$$

The rate of heat transfer is:

$$Q = A.h.(T_\infty - T_{sup}) = \pi.D^2.h.(T_\infty - T_{sup}) = \pi.(0.015)^2.(44.56).(130 - 25) = 3.31\ W$$

13.5 Forced Convection and Internal Flows

Internal flow refers to situations of fluid flow within pipes. When the tube wall temperature is different from that of the fluid, there is heat transfer between them. In this case, assessments of changes in fluid temperature along the pipe and the total quantity of heat exchanged between the pipe and the fluid are important.

For these assessments, one has to develop an energy balance for the fluid flowing through the duct. To determine the rate of heat exchange between the fluid and the duct, it is necessary to know the coefficient of heat transfer between them. This can be evaluated from appropriate correlations presented in the following sections.

13.5.1 Energy balance for the fluid flowing inside a duct

In the case of a fluid flowing inside a duct, when the the walls are at different temperature relative to the fluid temperature, the first step is to determine the temperature variation of the fluid along the pipe. For this, one can establish an energy balance considering an average temperature of the fluid, T_m, along the cross section of the pipe.

Figure 13.6 illustrates the system being considered with the choice of an infinitesimal volume element for developing the energy balance.

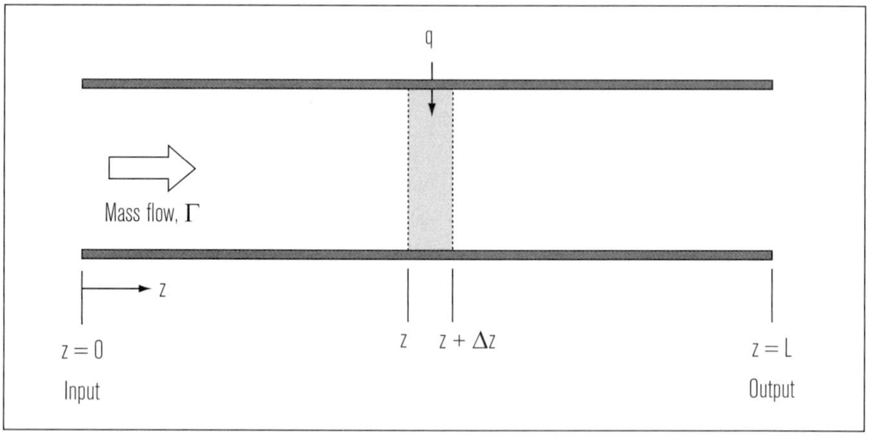

Figure 13.6 - Control volume for energy balance for the fluid.

The energy balance for the volume element of fluid in Figure 13.6 is given by:

$$\left[\text{Rate of input at z}\right]+\left[\begin{array}{l}\text{Rate of heat exchange}\\ \text{through duct wall}\end{array}\right]=\left[\text{Rat of output at z}+\Delta z\right] \quad (13.36)$$

In this equation, the rate of heat exchange between the fluid and the duct wall is considered positive when the heat flows from the wall to the fluid. Substituting for each of the terms above, we get:

$$\left[\Gamma.C_p.(T_m - T_{ref})\right]\Big|_z + q.P.\Delta z = \left[\Gamma.C_p.(T_m - T_{ref})\right]\Big|_{z+\Delta z} \quad (13.37)$$

where:

- Γ is mass flow of fluid;
- P is the wetted perimeter of the pipe;
- q is the heat flow from duct to the fluid.

Considering C_p constant, the above equation can be written:

$$\Gamma.C_p.(T_m|_{z+\Delta z} - T_m|_z) = q.P.\Delta z \tag{13.38}$$

Dividing both sides of the equation by Δz and making the limit Δz tends to zero, we have:

$$\Gamma.C_p.\frac{dT_m}{dz} = q.P \tag{13.39}$$

For the integration of equation (13.39), three cases are considered:

a) constant temperature at the surface of the duct;

b) constant heat flux;

c) heat exchange with a fluid flowing on the outside of the duct.

The case «a» can occur when the external face of the product is a pure material undergoing phase change.The second situation (case "b") occurs when the heat is supplied through an electrical resistance dissipating a certain energy. The case "c" is the most common and involves the exchange of heat with a fluid that flows in the outer duct.

Case (a): In this case, the heat flow between the duct and the fluid is provided by the following equation:

$$q = h.(T_{sup} - T_m) \tag{13.40}$$

where:

- T_{sup} is the temperature of the surface of the duct.

Substituting (13.40) in (13.39), we obtain:

$$\Gamma.C_p.\frac{dT_m}{dz} = h.(T_{sup} - T_m).P \tag{13.41}$$

Separating variables and integrating equation (13.41), assuming constant h, we get:

$$-\ln(T_{sup} - T_m) = \frac{P.h}{\Gamma.C_p}.z + C_1 \tag{13.42}$$

The integration constant C_1 can be determined by the boundary condition:

$$z = 0 \qquad T_m = T_m^e$$

where:

- T_m^e, is the input temperature at the entry of the pipe. Hence:

$$C_1 = -\ln(T_{sup} - T_m^e) \quad (13.43)$$

and finally:

$$\frac{(T_{sup} - T_m)}{(T_{sup} - T_m^e)} = \exp\left(-\frac{P.h}{\Gamma.C_p}.z\right) \quad (13.44)$$

Equation (13.44) can be used to evaluate the average temperature of the fluid, T_m, for each position z along the pipe. Figure 13.7 shows, schematically, the variation of fluid temperature along the length of the pipe.

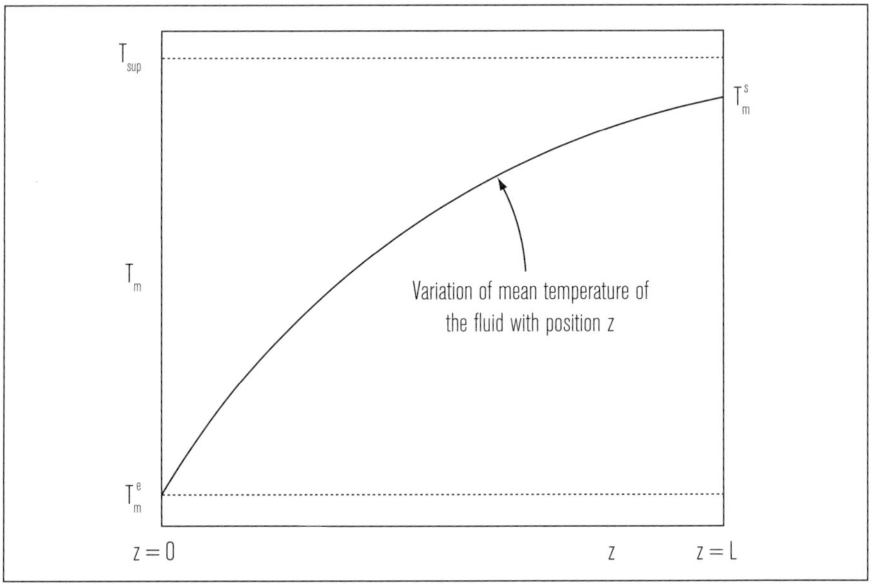

Figure 13.7 - Variation of fluid temperature along the length of the duct wall at constant temperature T_{sup} and higher than T_m.

Equation (13.44) is valid for both situations: wall warmer or colder than the fluid. Case (b): In the case of constant heat flux, equation (13.41) can integrated to give:

$$T_m = \frac{q.P}{\Gamma.C_p}.z + T_m^e \quad (13.45)$$

The convention adopted is that the heat flux is positive when it occurs from duct wall to the fluid.In this case, the fluid temperature rises as it traverses the pipe. Equation (13.45) provides that in the case of constant flux, the fluid temperature varies linearly along the duct. When the heat flow is constant, the temperature of the duct wall varies depending on the position z. In a given position, this temperature can be estimated by the following equation:

$$T_{sup} = \frac{q}{h} + T_m \qquad (13.46)$$

Case (c): This is illustrated schematically in Figure 13.8.

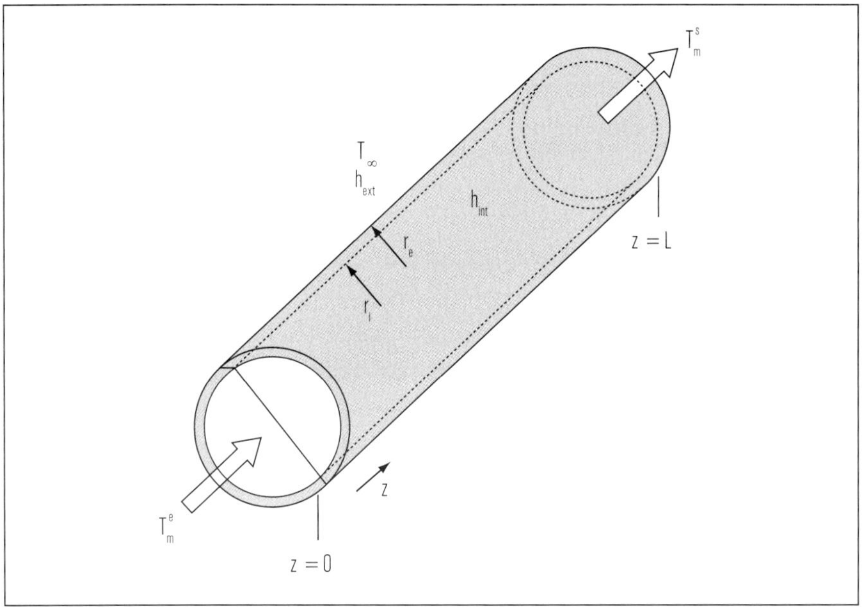

Figure 13.8 - Schematic view of the case "c": - for heat transfer in the case of fluid flowing outside of the duct.

Adopting a similar procedure, given in Chapter 11 for cylindrial geometry (equation 11.89) the rate of heat transfer between the external and the internal fluid can be estimated by the following equation:

$$Q = \left[\frac{T_\infty - T_m}{\dfrac{1}{2.\pi.r_i.h_{int}} + \dfrac{\ln(r_e/r_i)}{2.\pi.k_d} + \dfrac{1}{2.\pi.r_e.h_{ext}}} \right].\Delta z \qquad (13.47)$$

where:

- k_d is the thermal conductivity of the duct wall.

Equation (13.47) can be written:

$$Q = \left(\frac{T_\infty - T_m}{R_T} \right).\Delta z \qquad (13.48)$$

where:

$$R_T = \frac{1}{2.\pi.r_i.h_{int}} + \frac{\ln(r_e/r_i)}{2.\pi.k_d} + \frac{1}{2.\pi.r_e.h_{ext}} \qquad (13.49)$$

This term represents the sum of thermal resistances to heat transfer.

The term in brackets in equation (13.48) can replace the product q.P in equation (13.39). Hence:

$$\Gamma.C_p.\frac{dT_m}{dz} = \frac{T_\infty - T_m}{R_T} \qquad (13.50)$$

Separating variables and integrating, one obtains the following expression for the fluid temperature along the pipe:

$$\frac{(T_\infty - T_m)}{(T_\infty - T_m^e)} = \exp\left(-\frac{z}{R_T.\Gamma.C_p}\right) \qquad (13.51)$$

The evaluation of the coefficient of heat transfer to flow inside the pipeline is discussed in the next section.

13.5.2 Heat transfer coefficient for cylindrical ducts

Figure 13.9 shows, schematically, the region of entry of a fluid in a duct, illustrating the development of hydrodynamic and thermal boundary layers. In this case, it is considered that the fluid and the duct are at different temperatures.

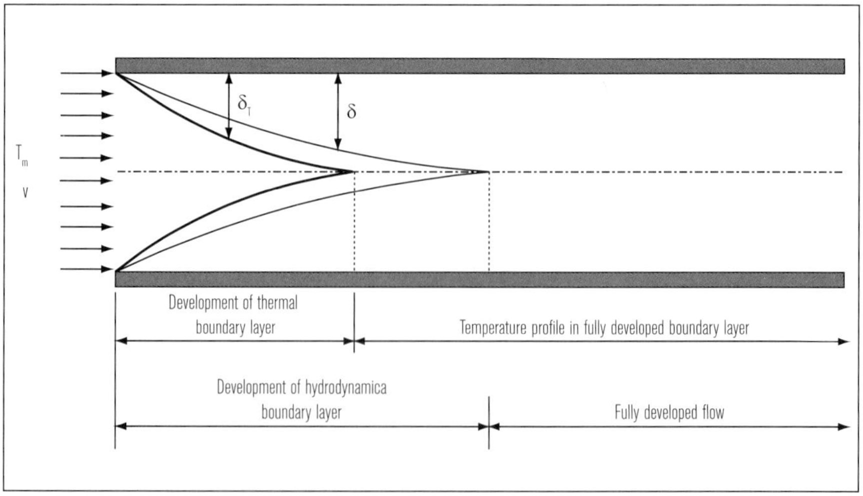

Figure 13.9 - Development of hydrodynamic and thermal boundary layers, inside a pipeline.

From the points where the boundary layers reach the center of the duct,the velocity and temperature profiles become fully developed.

In the region of the boundary layer, Nusselt number varies along the length of the pipe. After the development of boundary layers, there is no variation along the length. However it can vary due to change in physical properties of fluid with temperature. Correlations for calculating the Nusselt number, presented below, apply to the region after the development of thermal boundary layer. For turbulent flow, it is considered, that it is fully developed for positions z along the length of the pipeline, more than 10 times the diameter of the pipe. The position $z = 0$ corresponds to the entry point of the fluid in the duct.

For laminar flow ($Re < 2100$), Nusselt number (INCROPERA and DEWITT, 2003) amounts to:

- For constant heat flux on the duct wall:

$$Nu_D = \frac{h.D}{k} = 4.36 \qquad (13.52)$$

- For constant temperature of the duct wall:

$$Nu_D = \frac{h.D}{k} = 3.66 \qquad (13.53)$$

For turbulent flow, the same correlations apply as in conditions of constant heat flux or constant temperature of the duct wall as given above. One of the other correlations proposed for evaluation of the Nusselt number for flow in cylindrical ducts is given by:

$$\begin{aligned} & Nu_D = \frac{h.D}{k} = 0.023.Re_D^{4/5}.Pr^n \\ & 0.7 \le Pr \le 160 \\ & Re_D \ge 10000 \\ & \frac{L}{D} \ge 10 \end{aligned} \qquad (13.54)$$

The exponent n is 0.3, for the fluid undergoing cooling and 0.4, if it is being heated.

A more recent correlation includes the effect of roughness of the duct wall (through friction factor, f) and expands the application range in terms of Reynolds number and Prandtl. This correlation is given by:

$$Nu_D = \frac{h.D}{k} = \frac{(f/2).(Re_D - 1000).Pr}{1 + 12.7.(f/2).(Pr^{2/3} - 1)}$$
$$0.5 \leq Pr \leq 2000 \qquad (13.55)$$
$$3000 \leq Re_D \leq 5 \,.\, 10^6$$

The friction factor is determined using Reynolds number and roughness of the duct, with the empirical equation presented in Chapter 7 (Equation 7.42).

Example

Motor oil flows at a rate of 1.5 kg/min through a tube 4 mm in diameter and 25 m in length. The inlet temperature of oil is 30°C, while the wall temperature is kept at 100°C by vapor condensation on the outer surface. Calculate:

a) average coefficient of heat transfer between the tube and oil;
b) temperature of the oil output;
c) total rate of heat transfer between the tube and oil.

Solution:

To determine the heat transfer coefficient, it necessary to know the average temperature of the oil flowing at the exit of the pipe. However, to evaluate this temperature one has to know the value of the coefficient of heat transfer. Since both are unknown, an initial value for the temperature of oil at the outlet can be assumed (80°C) in order to estimate the physical properties of the oil. With these data the heat transfer coefficient h can be be evaluated for estimating the outlet temperature of oil This temperature is compared with the initially assumed value. If they are very different, the procedure is repeated until the calculated outlet temperatures have little variation. Assuming the outlet temperature to 80°C for the outlet temperature of the oil:

Properties evaluated at: T_m : $T_m = \dfrac{T_{m,e} + T_{m,s}}{2}$

Initial temperature $T_{m,s}$: $T_{m,s}$ =80°C

T_m = 55°C = 328 K

$\rho = 853.9$ kg/m^3	$\mu = 3.56 \,.\, 10^{-2}$ Pa.s	Pr = 546
$k = 138 \,.\, 10^{-3}$ W/m.K	$\mu_s = 1.725 \,.\, 10^{-2}$ Pa.s	C_p = 2118 J/kg.K

With these data, we can estimate the relevant dimensionless numbers, as seen below:

$$Re_D = \frac{D.\overline{V}.\rho}{\mu} = \frac{4.\Gamma}{\pi.D.\mu} = \frac{4.(1.5/60)}{\pi.(0.004).(3.56\ .\ 10^{-2})} = 223.53$$

$$\frac{L}{D} = \frac{25}{0.004} = 6{,}250$$

From the data, one should use the following correlation for evaluation of the Nusselt number (assuming fully developed laminar flow and surface temperature of the tube constant: Equation 13.53):

$$Nu_D = \frac{h.D}{k} = 3.66$$

The heat transfer coefficient is then given by:

$$h = \frac{Nu_D.k}{D} = \frac{(3.66).(138\ .\ 10^{-3})}{0.004} = 126.27\ W/m^2.K$$

The outlet temperature of oil is estimated by the relationship:

$$\frac{(T_{sup} - T_{m,s})}{(T_{sup} - T_{m,e})} = \exp\left(-\frac{P.h}{\Gamma.C_p}.L\right)$$

Substituting data, we have:

$$\frac{(100 - T_{m,s})}{(100 - 30)} = \exp\left[-\frac{\pi(0.004).(126.27)}{(1.5/60).(2118)}.(25)\right]$$

Therefore:

$$T_{m,s} = 66.9°C$$

This value is close to the initial value given Hence it is not necessary to repeat the calculation.

Finally, one can determine the rate of heat transfer between oil and the duct wall:

$$Q = \Gamma.C_p.(T_{m,s} - T_{m,e})$$
$$Q = (1.5/60).(2118).(66.9 - 30) = 1953.9\ W$$

13.5.3 Heat transfer coefficients in noncylindrical ducts

Equations (13.44), (13.45) and (13.51), for calculating the change in temperature of the fluid along a heated or cooled pipe can be used for any geometry of the duct by substituting the wetted perimeter value calculated from the geometry.

For noncylindrical ducts the evaluation of the Nusselt number and coefficient of heat transfer, one can use the hydraulic equivalent diameter, D_h (defined in Chapter 7), instead of pipe diameter.

In laminar flow, the values of the Nusselt number varies according to the geometry of the ducts. Table 13.1 provides values of the Nusselt number for various configurations of pipes. In all cases that appear in this table, the Nusselt number is defined by:

$$Nu_D = \frac{h.D_h}{k} \tag{13.56}$$

For turbulent flow, one can use equations (13.54) and (13.55) replacing the hydraulic equivalent diameter for the diameter term, in the evaluation of Reynolds and Nusselt numbers.

Table 13.1 - Values of Nusselt numbers for laminar flow in ducts of different geometries under constant flux and constant temperature conditions (INCROPERA and DEWITT, 2003).

Cross section duct	Condition	
	Constant flux	Constant surface temperature
Circular	4.36	3.66
Rectangular: b/a the 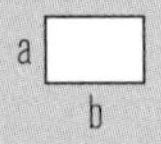		
• b/a = 1	3.61	2.98
• b/a = 2	4.12	3.39
• b/a = 4	5.33	4.44
• b/a = 8	6.49	5.60

13.6 Correlations for Natural Convection

In assessing the correlations for Nusselt number involving coefficient of heat transfer for natural convection, the Reynolds number is replaced by the Grashof number or Rayleigh number.

Unlike in the case of forced convection, in natural convection, the position of the solid surface affects the value of the Nusselt number. This is because in these cases, the profile of fluid velocity and hence the development of thermal and hydrodynamic boundary layers, are affected by the position of the surface. Apart from this, whether the surface is submitted to heating or cooling also affects the development of boundary layers.

Figure 13.10 illustrates three different cases, showing different behaviors of fluid flow in the case of natural convection.

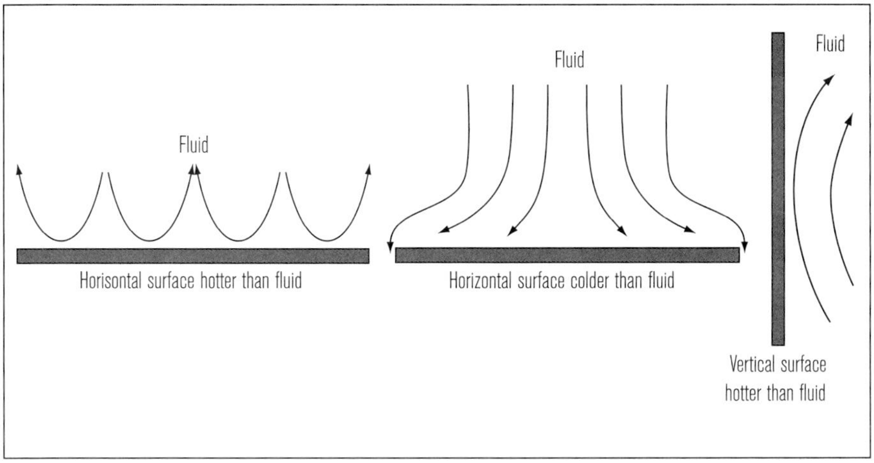

Figure 13.10 - Illustrates three different cases, showing different behaviors of fluid flow in the case of natural convection.

For each of these situations Nusselt number can be evaluated from different correlations. Some of these correlations are presented below for different geometries.

13.6.1 Flat surfaces

For a flat vertical surface, Nusselt number can be estimated using the following correlation (INCROPERA and DEWITT, 2003):

$$Nu_L = \left\{ 0.825 + \frac{0.387.Ra_L^{1/6}}{\left[1 + \left(\frac{0.492}{Pr}\right)^{9/16}\right]^{8/27}} \right\}^2 \qquad (13.57)$$

This correlation is valid for the entire range of values of Rayleigh and Prandtl numbers.

For natural convection, the value of the Rayleigh number serves as a criterion to determine the flow regime. For vertical surfaces, the flow is laminar up to Rayleigh numbers equivalent to 10^9.

This condition applies both to hot as well as cold surfaces relative to the fluid.

In the equation (13.57), the characteristic dimension L is the length of the plate in the vertical direction.

$$L = \frac{A_S}{P} \quad (13.58)$$

For horizontal surfaces, the characteristic dimension is defined by the equation:

where:

- A_S is the surface area of the plate and P its wetted perimeter.

The following are some of the crrelations for determining the Nusselt number ar:

- top surface heated/cold bottom surface (INCROPERA and DEWITT, 2003):

$$Nu_L = 0.54.Ra_L^{1/4} \quad \text{for } 10^4 \leq Ra_L \leq 10^7 \quad (13.59)$$

$$Nu_L = 0.15.Ra_L^{1/3} \quad \text{for } 10^7 \leq Ra_L \leq 10^{11} \quad (13.60)$$

- bottom surface heated/cold top surface:

$$Nu_L = 0.27.Ra_L^{1/4} \quad \text{for } 10^5 \leq Ra_L \leq 10^{10} \quad (13.61)$$

Example

Estimate the heat transfer coefficient for heat transfer to the lateral and upper surfaces of the furnace, as shown in the Figure below. What is the rate of heat loss by convection along these walls and loss by radiation? Compare the results. Discuss what would happen to these rates if the temperature rose to 500°C in the lateral surface . Which of the two rates would be affected by the variation of surface temperature? Explain. Given: Emissivity of wall: 0.7.

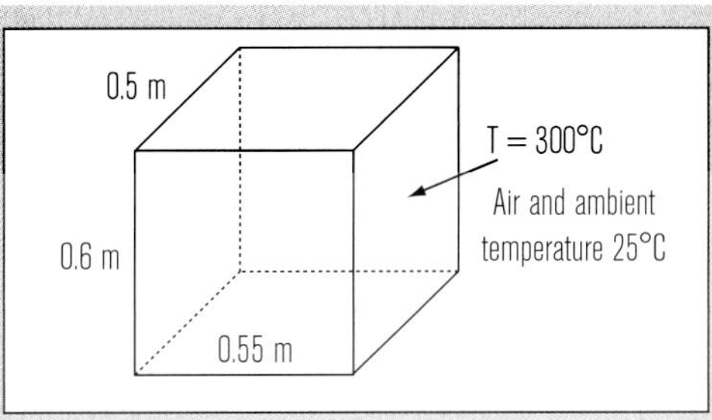

Solution:

To evaluate the rate of heat loss, one has to calculate the heat transfer coefficients by natural convection for two areas. For this Initially, the properties of air temperature in the film has to be determined:

$$T_f = \frac{T_s + T_\infty}{2} = \frac{300 + 25}{2} = 162.5°C = 435.5 \text{ K}$$

$$\rho = 0.8 \text{ kg/m}^3 \qquad \nu = 30.78 \text{ x } 10^{-6} \text{ m}^2/\text{s} \qquad Pr = 0.6871$$

$$k = 36.36 \text{ x } 10^{-3} \text{ W/m.K} \qquad \alpha = 44.8 \text{ x } 10^{-6} \text{ m}^2/\text{s}$$

$$\beta = \frac{1}{T_f} = \frac{1}{435.5} = 0.0023 \text{ K}^{-1}$$

The Rayleigh number is defined by:

$$Ra_L = \frac{g.\beta.(T_s - T_\infty).L^3}{\nu.\alpha}$$

For the lateral surface, the characteristic dimension L is the height of the wall. For the upper surface (horizontal), the characteristic dimension is the relationship between area and perimeter of the surface. Hence:

- Lateral surface (vertical):

$$Ra_L = \frac{g.\beta.(T_s - T_\infty).L^3}{\nu.\alpha} = \frac{(9.8).(1/435.5).(300-25).(0.6)^3}{(30.78 \text{ x } 10^{-6}).(44.8 \text{ x } 10^{-6})} =$$

$$Ra_L = 9.69 \text{ x } 10^8$$

- Upper surface (horizontal):

$$L = \frac{A_{sup}}{P} = \frac{(0.5).(0.55)}{2.(0.5+0.55)} = 0.131 \text{ m}$$

$$Ra_L = \frac{g.\beta.(T_s - T_\infty).L^3}{\nu.\alpha} = \frac{(9.8).(1/435.5).(300-25).(0.131)^3}{(30.78 \text{ x } 10^{-6}).(44.8 \text{ x } 10^{-6})} =$$

$$Ra_L = 1.01 \text{ x } 10^7$$

For side surface (vertical), the correlation for the calculation of the Nusselt number is:

$$Nu_L = \left\{ 0.825 + \frac{0.387.Ra_L^{1/6}}{\left[1+\left(\frac{0.492}{Pr}\right)^{9/16}\right]^{8/27}} \right\}^2$$

Substituting values:

$$Nu_L = 121.11$$

From the Nusselt number, one can calculate the coefficient of heat transfer:

$$Nu_L = \frac{h.L}{k} = 121.11$$

$$h = 7.34 \ W/m^2.K$$

For the upper surface (horizontal) substituting the value of Rayleigh number obtained, above in the following correlation:

$$Nu_L = 0.15.Ra_L^{1/3}$$

Substituting values, we obtain:

$$Nu_L = \frac{h.L}{k} = 32.42$$

$$h = 9.0 \ W/m^2.K$$

The rate of heat loss by convection and radiation can be estimated by:

$$Q = A.h.(T_s - T_\infty) + A.\sigma.\varepsilon.(T_s^4 - T_{viz}^4)$$

For the lateral surface, we get:

$$Q = (0.3).(7.34).(300 - 25) + (0.3).(5.67 \ . \ 10^{-8}).(0.7).[(573)^4 - (298)^4] =$$

$$Q = 605.55 + 1189.67 \ W/m^2 = 1795.22 \ W/m^2$$

For the upper surface, we have:

$$A = (0.55).(0.5) = 0.275 \ m^2$$

$$Q = (0.275).(9.0).(300 - 25) + (0.275).(5.67 \ . \ 10^{-8}).(0.7).[(573)^4 - (298)^4] =$$

$$Q = 680.63 + 1090.53 \ W/m^2 = 1771.16 \ W/m^2$$

If the wall temperature increases,the contributions of both terms convection and radiation, would increase. The radiation term should get affected more, as it grows with the fourth power of the temperatures involved.

13.6.2 Cylinder

For a long cylinder (L/D> 10) in a horizontal position, the Nusselt number can be estimated from the following correlation (INCROPERA and DEWITT, 2003):

$$Nu_D = \left\{ 0.60 + \frac{0.387.Ra_D^{1/6}}{\left[1+\left(\frac{0.559}{Pr}\right)^{9/16}\right]^{8/27}} \right\}^2 \qquad (13.62)$$

$Ra_D \leq 10^{12}$

For vertical cylinders, equation (13.57) can be applied, considering the height L of the cylinder. For this correlation to be valid, the following relationship must be satisfied:

$$\frac{D}{L} \geq \frac{35}{Gr_L^{1/4}} \qquad (13.63)$$

When the above condition is met, the thickness of the boundary layer is much smaller than the diameter of the cylinder.Thus, the curvature of the cylinder has negligible effect on heat transfer.

Example

A-rod of 12.7 mm in diameter is placed horizontally in air at 25°C If the temperature of the bar is 90°C, estimate the rate of heat transfer by natural convection per unit length of the bar. What will the new rate if the bar is placed in contact with air that moves across the piece at a speed of 10 m/s? Compare the results.

Solution

In natural convection, one should determine the Rayleigh number. The physical properties of air are evaluated at the temperature of the film:

$$T_f = \frac{T_{sup} + T_\infty}{2} = \frac{90+25}{2} = 57.5°C = 330.5\ K$$

$\rho = 1.078\ kg/m^3 \quad \nu = 18.4 \times 10^{-6}\ m^2/s \quad Pr = 0.703$

$k = 28.5 \times 10^{-3}\ W/m.K \quad \alpha = 26.0 \times 10^{-6}\ m^2/s \quad \mu = 196 \times 10^{-7}\ Pa.s$

$$\beta = \frac{1}{T_f} = \frac{1}{330.5} = 0.0023\ K^{-1}$$

Substituting the value of the Rayleigh number is:

$$Ra_D = \frac{g.\beta.(T_{sup} - T_\infty).D^3}{\nu.\alpha} = \frac{(9.8).(1/330.5).(90-25).(0.0127)^3}{(18.4\ .\ 10^{-6}).(26.0\ .\ 10^{-6})} = Ra_D = 8252.55$$

Using the following correlation: one can get the Nusselt number and heat transfer coefficient:

$$Nu_D = \left\{0.60 + \frac{0.387.Ra_D^{1/6}}{\left[1 + \left(\frac{0.559}{Pr}\right)^{9/16}\right]^{8/27}}\right\}^2$$

$$Nu_D = 4.17$$

$$h = \frac{Nu_D.k}{D} = \frac{(4.17).(28.5\ .\ 10^{-3})}{(0.0127)} = 9.36\ W/m^2.K$$

For forced convection, the Reynolds number should be evaluated:

$$Re_D = \frac{D.v.\rho}{\mu} = \frac{(0.0127).(10).(1.078)}{(196\ .\ 10^{-7})} = 6985$$

Applying the correlation:

$$Nu_D = 0.3 + \frac{0.62.Re_D^{1/2}.Pr^{1/3}}{\left[1+\left(\frac{0.4}{Pr}\right)^{2/3}\right]^{1/4}}.\left[1+\left(\frac{Re_D}{282{,}000}\right)^{5/8}\right]^{4/5}$$

$$Nu_D = 43.91$$

$$h = 98.54\ W/m^2.K$$

Considering a cylinder of unit length, one can calculate the rate of heat loss by the equation:

$$Q = \pi.D.h.(T_{sup} - T_\infty)$$

For natural convection, we have:

$$Q = \pi.(0.0127).(9.36).(90-25) = 24.27\ W$$

For forced convection, we get:

$$Q = \pi.(0.0127).(98.54).(90-25) = 255.55\ W$$

In the situation analysis, the loss by forced convection is about 10 times greater than that by natural convection.

13.6.3 Sphere

The correlation to estimate the Nusselt number for natural convection around spheres is given by (INCROPERA and DEWITT, 2003):

$$Nu_D = 2 + \frac{0.589.Ra_D^{1/4}}{\left[1 + \left(\frac{0.469}{Pr}\right)^{9/16}\right]^{4/9}} \tag{13.64}$$

$Ra_D \leq 10^{11}$

$Pr \geq 0.7$

13.7 Combined Convection

In all cases discussed so far, forced convection and natural convection were considered individually. This consideration is valid when one process predominates over the other.

A criterion for determining when a particular contribution can be neglected is found through a comparison between the magnitudes of the Reynolds number and Grashof Number. Consider the following situations:

a) $Gr/Re^2 << 1$:

The effects of natural convection are negligible. The value of the Nusselt number can be estimated using only correlations for forced convection.

b) $Gr/Re^2 >> 1$:

The natural convection is predominant. The Nusselt number can be evaluated using only correlations for natural convection.

c) $Gr/Re^2 = 1$:

If the contributions of natural convection and forced convection have the same magnitudes. the Nusselt number should be calculated by incorporating the effects of both.

The situation c is usually identified as combined convection. In this case, Nusselt number can be estimated by combining the effects of natural and forced convection Nusselt number in this case is given by:

$$Nu^a_{Comb} = Nu^a_{Forced} + Nu^a_{Natural} \tag{13.65}$$

where Nu_{Comb}, Nu_{Forced}, $Nu_{Natural}$ are the Nusselt numbers for combined forced and natural, respectively. The exponent *a* is usually a number between 3 and 4.

Equation (13.65) is valid when the effects of natural and forced convection are additive, ie, induce movement of fluid in the same direction or transverse movements.

Where the movement caused by natural and forced convection are in opposite directions, one should use the following relationship:

$$Nu^a_{Comb} = \left| Nu^a_{Forced} - Nu^a_{Natural} \right| \quad (13.66)$$

References

BIRD, R.B.; STEWART, W.E.; LIGHTFOOT, E.N. *Transport phenomena.* New York: John Wiley & Sons, 1960.

GEIGER, G.H.; POIRIER, D.R. *Transport phenomena in metallurgy.* Massachussetts: Addison-Wesley, 1973.

INCROPERA, F.P.; DEWITT, D.P. *Fundamentos da transferência de calor e massa.* 5. ed. Rio de Janeiro: LTC, 2003.

JOO, S.; HAN, J.W.; GUTHRIE, R.I.L. Inclusion behaviour and heat transfer phenomena in steelmaking tundish operations. Part II. Mathematical model for liquid steel in tundishes. *Metallurgical Transactions B,* v. 24B, n. 5, p. 767-77, Oct. 1993.

PEREIRA, R.O.S. et al. Escoamento do aço no molde de lingotamento contínuo de placas. In: CONGRESSO ANUAL DA ABM, 56., 2001, Belo Horizonte, MG. *Anais...* São Paulo: ABM, 2001. p. 1264-73.

Exercises

1 - Consider the duct shown in the figure below. Estimate the temperature of the water coming out of the pipe with internal diameter: 40 mm, outer diameter: 60 mm. Duct length: 10 m. Water flow: 200 L/min. Inlet temperature of water: 25°C.

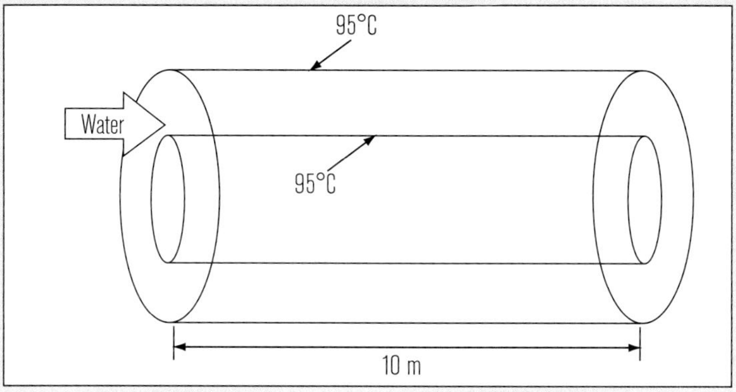

2 - For security reasons, the temperature of the upper surface of a laboratory furnace should not exceed 100°C. Evaluate the thickness of the refractory to ensure that this specification will be met.The upper surface of the furnace has dimensions of 40 x 40 cm and is held in contact with air at 25°C. The maximum operating temperature of the furnace is 800°C and the refractory used has a thermal conductivity of 0.5 W/m.K. The following figure shows a schematic view of the system. Consider one-dimensional heat transfer.

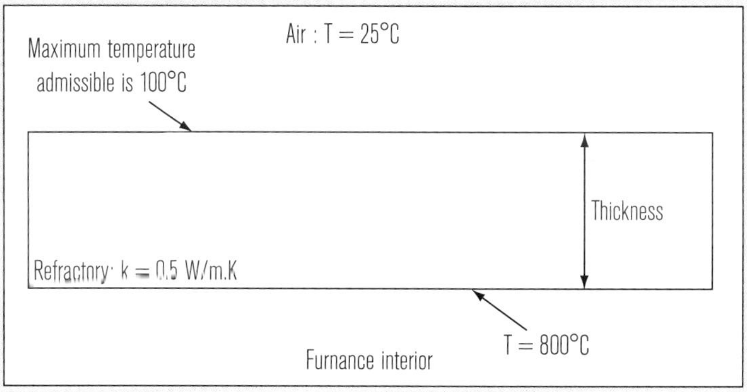

3 - The junction of a thermocouple, which can be considered approximately spherical, is used to measure temperature of air injected into a furnace. Determine the time required for the board to reach a temperature of 472 deg.K, when the air temperature is 473 K. The board is initially at 298 deg. K. To estimate h, consider the properties of air at 400 deg. K. Data:

- Junction: - specific heat: 400 J/kg.K;
- Density: 8500 kg/m^3;
- Thermal conductivity: 20 W/mK;
- Diameter: 0.5 mm;
- Air: speed: 1 m/s.

What would happen this time if the air velocity decreases. Explain your answer.

4 - Water flowing through a tube with a diameter of 25.4 mm and a flow rate of 7 m^3 per hour must be heated from 25 to 80°C for maintaining the surface of the tube at a temperature of 100°C. Determine the length of pipe needed to achieve the above specifications.Analyze what should happen with the length of the tube if its diameter is changed to 38.1 mm.

5 - A bar of steel AISI 304, horizontal 200 mm in diameter, with surface temperature of 1200°C is located in an environment with air at a temperature of 20°C. Evaluate the rate of heat transfer per unit length of the bar due to convection. Is it reasonable to assume that the cooling of the bar in these conditions is isothermal? Justify your answer.

Fundamentals of Heat Transfer by Radiation

Chapter 14

The two mechanisms of heat transfer studied so far, conduction and convection, require a medium through which the heat is transported.

In the third mechanism for heat transport, radiation, heat is transported by electromagnetic waves and, unlike previous mechanisms, does not require a material medium for the purpose. In radiation, heat flow is a function of the difference of the fourth power of surface temperatures involved.

Every type of matter emits or absorbs energy by radiation, at temperatures above 0 deg.K. In gases and solids that are not fully opaque, this emission is volumetric in character. In opaque solids and liquids, the emission of energy by radiation is a surface phenomenon. In this chapter emission or absorption of heat energy by radiation involving the surfaces of bodies will be considered.

In this chapter fundamentals of heat transfer by radiation are discussed.

14.1 Spectrum and Black Body Radiation

The radiation emitted by a body is composed of a continuous spectrum of wavelengths, which indicates the pattern of distribution of energy. This distribution depends on the characteristics of this body and its temperature. Moreover, the intensity of radiation may vary with the direction as well.

The transmitting power of a body at a given temperature can be measured by the following relationship:

$$E = \int_0^\infty E_\lambda . d\lambda \qquad (14.1)$$

where:

- E = power emitted from the body (W/m^2)
- E_λ = power emitted from the body at a given wavelength ($W/m^2.\mu m$);
- λ = wavelength (μm).

In the study of radiation, it is convenient to introduce the concept of black body. This body has the following t general features:

- maximum power emission at a given temperature and at a certain wavelength;
- emission regardless of direction (diffusive emitter);
- absorption of all incident radiation regardless of wavelength.

The power of the blackbody emitter at a given wavelength was determined by Planck (GEIGER and POIRIER, 1973) and is given by the following relationship:

$$E_{\lambda,n} = \frac{C_1}{\lambda^5 . \left[\exp\left(\frac{C_2}{\lambda . T} \right) - 1 \right]} \qquad (14.2)$$

where:

- $E_{\lambda,n}$ = Emitting power of the blackbody at a given wavelength λ;
- C_1 = constant = 3.742×10^8 W.μm^4/m^2;
- C_2 = constant = 1.439×10^4 μm.K;
- T = temperature (K).

Figure 14.1 is a graphical representation of equation (14.2).

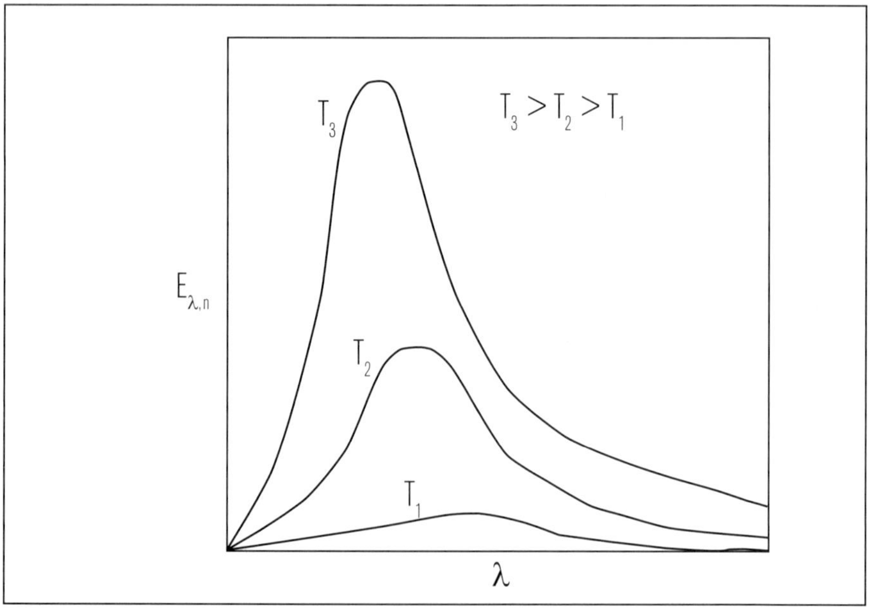

Figure 14.1 - Schematic view of the spectral distribution of emission of a blackbody at different temperatures.

Combining the equations (14.1) and (14.2), one can determine the power emitted from the body in a dark room. The equation obtained is:

$$E_n = \sigma . T^4 \qquad (14.3)$$

where:

- E_n = Emitting power of the black body (W/m^2);
- σ = Stefan-Boltzmann constant (5.67×10^{-8} $W/m^2.K^4$).

14.2 Emissivity

The concept of black body presented in the previous section, is useful to define the behavior of an ideal surface. A real surface has emissive power, that is a fraction of the blackbody emitter. This fraction is called emissivity and is defined by the following equation:

$$\varepsilon_\lambda = \frac{E_\lambda}{E_{\lambda,n}} \qquad (14.4)$$

where:

- ε_λ = emissivity of the blackbody to a wavelength λ.

In the equation (13.4), E_λ and $E_{\lambda,n}$ are evaluated at the same temperature.

One can also define a total emissivity, expressed by:

$$\varepsilon = \frac{E}{E_n} \qquad (14.5)$$

Table 14.1 shows the following ranges of total emissivity for different types of materials. Interestingly, polished metals tend to have lower emissivities than non polished and oxidized.

Table 14.2 presents values of total emissivity for the carbon steel at different temperatures.

Material		T (°C)	ε	T (°C)	ε
Aluminum	Highly polished	227	0.039	580	0.057
	Rusty	200	0.11	600	0.19
Copper	Highly polished	80	0.018	-	-
	Rusty	200	0.57	600	0.57
Iron	Highly polished	180	0.052	227	0.064
	Rusty	20	0.685	-	-
	Cast, polished	200	0.21	-	-
	Molten oxidized	200	0.64	600	0.78
Refractories	Good Radiator	600	0.80-0.85	1000	0.85-0.90
	Bad radiator	600	0.65-0.70	1000	0.75
Water	Thick layer	0	0.95	100	0.963

Table 14.1 - Emissivity bands values for different types of materials (BIRD, STEWART and LIGHTFOOT, 1960).

Table 14.2 - Emissivity of carbon steel as a function of temperature (INGERSLEV and HENEIN, 1997).

Temperature (°C)	Emissivity
316	0.69
482	0.72
649	0.76
816	0.79
982	0.82

For liquid steel, the commonly used emissivity value is 0.28 (ABOUTALEBI, HASAN and GUTHRIE, 1995).

14.3 Absorption, Reflection and Transmission

The flow of energy by radiation incident on a surface is called the total irradiation, G. In the general case, the interaction of energy with the body can generate three fractions, which are seen schematically in Figure 14.2:

- energy absorbed;
- reflected energy;
- transmitted energy.

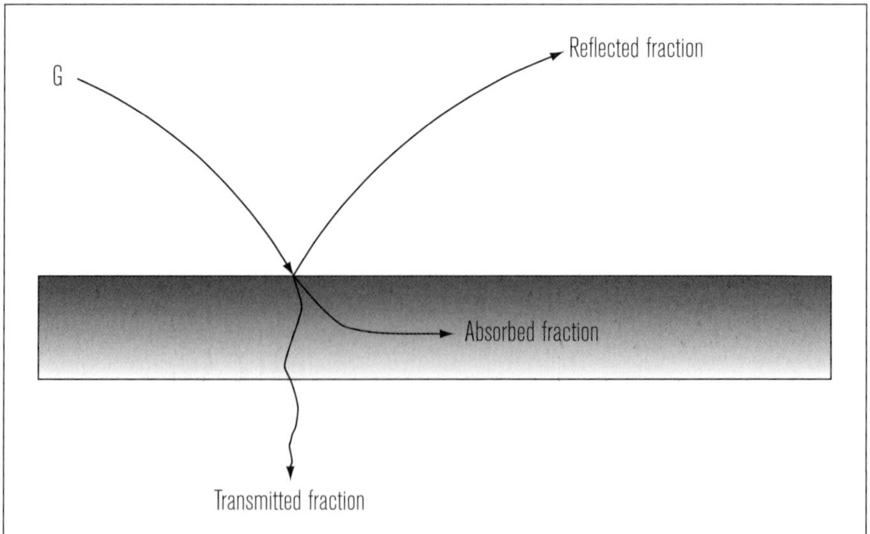

Figure 14.2 - Different fractions of energy from the incident energy by radiation on a body.

Applying the principle of conservation of energy, we have:

$$G = \alpha.G + \rho.G + \tau.G \qquad (14.6)$$

where:

- α = total suface absorbtivity;
- ρ = total surface reflectance;
- τ = total surface transmissivity.

Dividing the equation (14.6) for G, we get:

$$\alpha + \rho + \tau = 1 \qquad (14.7)$$

Each of the terms above represents a number ranging between 0 and 1.

In the case of opaque bodies, the transmissivity is zero, Hence:

$$\alpha + \rho = 1 \qquad (14.8)$$

In the equations above, absorbtivity, reflectivity and transmissivity of the total surface were considered. Similar equations could be written for these properties in a given specific wavelength.

Figure 14.3 illustrates the situation of a body placed inside an isothermal cavity, emitting energy equivalent to that of a blackbody.

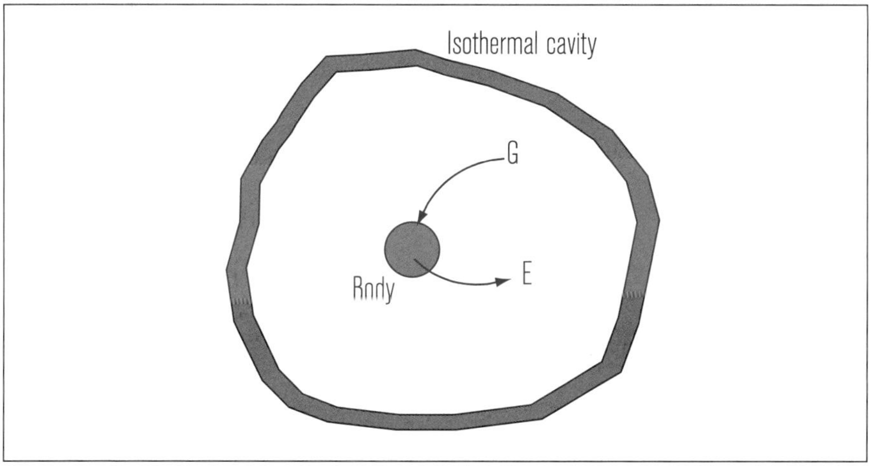

Figure 14.3 - A body placed inside a cavity emitting radiation equivalent to that of a black body.

Considering steady state, one can establish the energy balance for the body, considering only the energy transferred by radiation:

$$[\text{Rate of energy emitted by the body}] - [\text{Rate of energy absorbed by the body}] = 0 \quad (14.9)$$

Defining the above terms:

$$[\text{Rate of energy emitted by the body}] = A_{sup}.E \quad (14.10)$$

$$[\text{Rate of energy absorved by the body}] = A_{sup}.\alpha.E_n \quad (14.11)$$

where A_{sup} is the surface area of the body.

If the body temperature does not vary with time, it is in thermal equilibrium with the walls of the cavity, ie, that they both have the same temperature T. In this case, E and E_n are the emitting power of the body and the cavity at a particular temperature.

Substituting equations (14.10) and (14.11) in (14.9), we obtain:

$$A_{sup}.E = A_{sup}.\alpha.E_n \quad (14.12)$$

Simplifying and rearranging terms, we have:

$$E_n = \frac{E}{\alpha} \quad (14.13)$$

Since α is always less than or equal to 1, equation (14.13) shows that no body can have a emissive power greater than that of a blackbody when both are at the same temperature.

Combining the equations (14.13) and (14.5), we obtain:

$$\alpha = \varepsilon \quad (14.14)$$

Equation (14.14) is the mathematical expression of the law of Kirchhoff. A more general definition is that a body in thermal equilibrium emits the same amount of energy as it absorbs in one direction and at a given wavelength. If it were not this way, a body could send more energy to absorb in a given direction 1 and absorb more than it sent in another direction 2. As a result, the body would be pumping heat from one region to another without work input. This would violate the second law of thermodynamics. A similar analysis can be made to prove it is also independent of the wavelength.

The relation (14.14), despite being valid for a body in thermal equilibrium with a cavity emitting energy equivalent to a black body, is not universally valid as can be seen in the following cases.

Although radiation from a blackbody is a hypothetical concept, it is possible to obtain radiation that are close to that of a blackbody. One way is to use an isothermal enclosure, as shown in Figure 14.4.

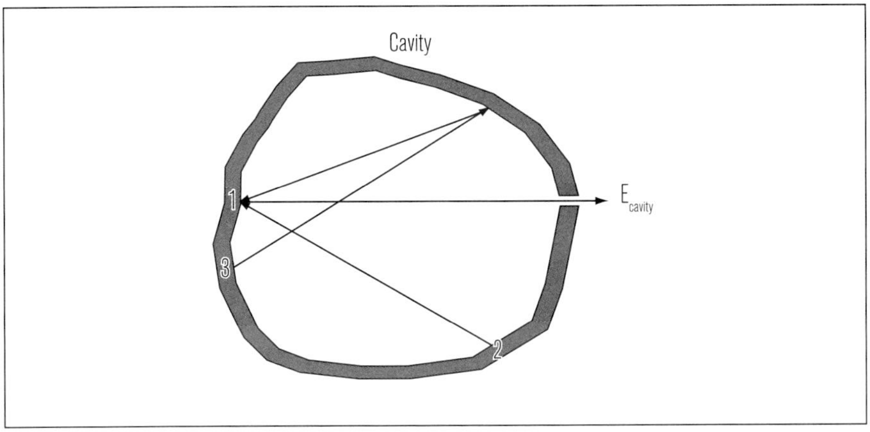

Figure 14.4 - Isothermal enclosure for obtaining radiation equivalent to that of a black body.

The energy that comes out of the area of element 1 as shown in Figure 1 is associated with that element emitted energy combined with the energy that is reflected by it. That is to say, that the energy that strikes it as a result of successive reflections and emissions that occur from other area elements that make up the cavity.

As the entire inner surface of the cavity has the same values of reflectivity and emissivity and is at the same temperature, we can write that the energy coming out of the area element 1 is given by:

$$e_1 = \varepsilon.E_n + \rho.\varepsilon.E_n + \rho^2.\varepsilon.E_n + \rho^3.\varepsilon.E_n + \rho^4.\varepsilon.E_n + ... \quad (14.15)$$

or:

$$e_1 = \varepsilon.E_n.(1 + \rho + \rho^2 + \rho^3 + \rho^4 + ...) \quad (14.16)$$

where ρ = total surface reflectance of individual elements.

Since infinite reflections occur inside the enclosure, the sum of equation (14.16) is composed of endless terms. Because ρ is always less than or equal to 1, the sum of the terms within brackets in equation (13.16) is:

$$(1+\rho+\rho^2+\rho^3+\rho^4+...)=\frac{1}{1-\rho} \quad (14.17)$$

Therefore:

$$e_1=\frac{\varepsilon.E_n}{1-\rho} \quad (14.18)$$

For an opaque body:

$$1-\rho=\alpha \quad (14.19)$$

and according to the Kirchhoff law:

$$\alpha=\varepsilon \quad (14.20)$$

Hence:

$$e_1=E_n \quad (14.21)$$

ie, the energy that comes out of the area element 1 is equivalent to the power of the blackbody emitter.The heated cavity seen in Figure 14.4 is usually used as radiation standard.

14.4 Gray Bodies

An approach very useful in engineering problems is that the absorsbtivity of a surface is independent of the spectral energy distribution of radiation incident on it. A body, that has this feature, is called gray body.

When a gray body at a temperature T_1 exchanges heat by radiation with a body at a temperature T_2, its absorbtivity will not change if the temperature T_2 (and hence the spectral distribution of incident energy) changes. It was also demonstrated in the previous section that if the temperatures T_1 and T_2 are equal (body in thermal equilibrium), the values of emissivity and absorsbtivity are equal.The combination of these two statements suggest that, for a gray body, the emissivity value is also independent of wavelength and is equal to the value of absorbtivity, even if the body is not in thermal equilibrium.

Although no actual body have exactly the behavior of gray body, this is a reasonable approximation in engineering problems and will be considered in developments that will be presented in the next section.

14.5 Exchange of Heat Between Surfaces

When considering the heat exchange by radiation between surfaces, it is necessary to introduce the concept of a view factor, also called the form factor.

14.5.1 View factor

Figure 14.5 shows two surfaces which exchange heat by radiation.

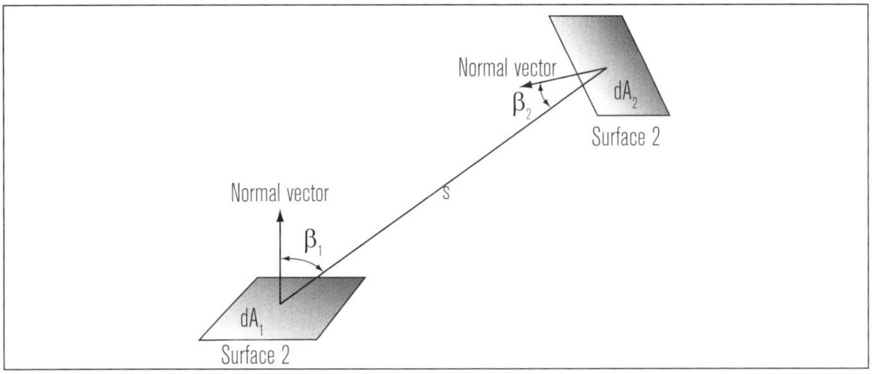

Figure 14.5 - Surfaces exchanging heat by radiation.

The view factor is defined as the fraction of energy leaving surface 1 and reaching surface 2. Assuming that the areas showing diffuse emission and reflection, one can evaluate the view factor of surface 1 to 2 by the following equation:

$$F_{1-2} = \frac{1}{A_1} \int_{A_1} \int_{A_2} \frac{\cos\beta_1 . \cos\beta_2}{\pi s^2} . dA_2 . dA_1 \quad (14.22)$$

and the surface 2 to 1 by:

$$F_{2-1} = \frac{1}{A_2} \int_{A_2} \int_{A_1} \frac{\cos\beta_2 . \cos\beta_1}{\pi s^2} . dA_1 . dA_2 \quad (14.23)$$

Examining the relations (14.22) and (14.23), one can observe:

$$A_1 . F_{1-2} = A_2 . F_{2-1} \quad (14.24)$$

This equation is known as a reciprocal relationship.

When the enclosure is as shown in Figure 14.6, one can also write:

$$\sum_{j=1}^{n} F_{i,j} = 1 \quad (14.25)$$

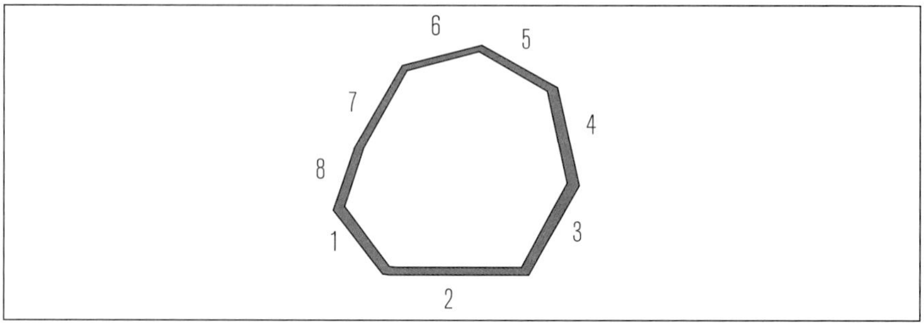

Figure 14.6 - Enclosure with various surfaces exchanging heat by radiation.

For many situations of practical interest, the values of view factors have been determined. The results are usually expressed as equations or graphs. Some of these cases are presented in Figures 14.7 to 14.9 and Tables 14.3 and 14.4.

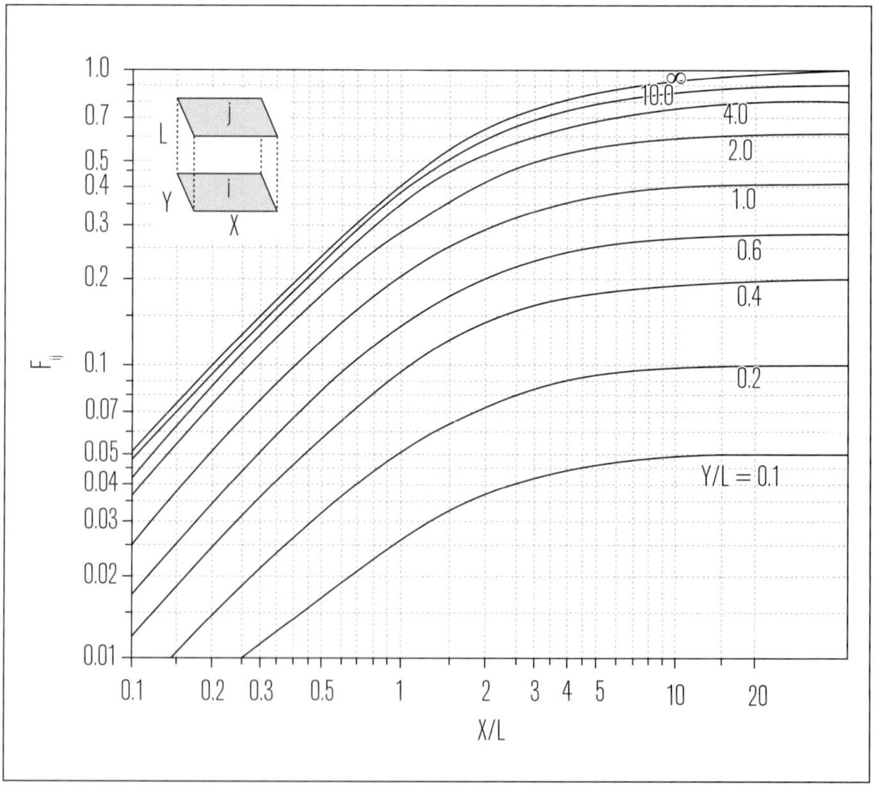

Figure 14.7 - View factors for parallel plates (INCROPERA and DEWITT, 2003).

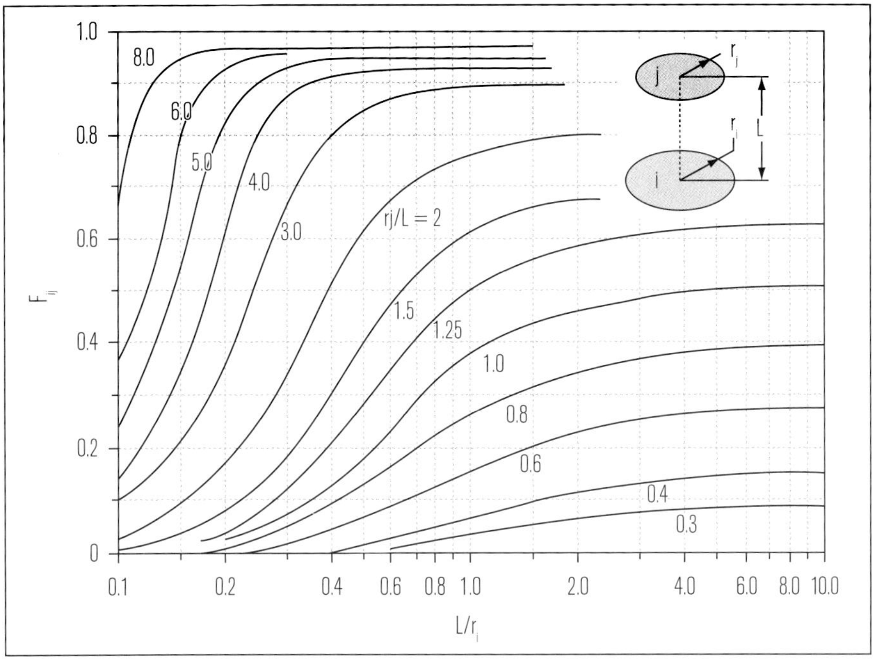

Figure 14.8 - View factors for parallel disks (INCROPERA and DEWITT, 2003).

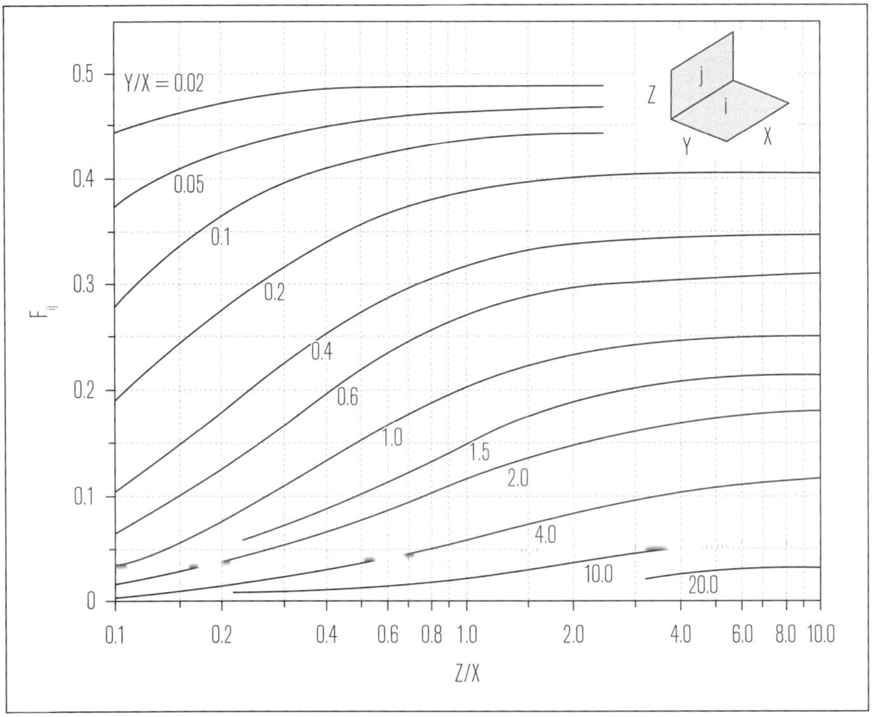

Figure 14.9 - View factors of orthogonal plates (INCROPERA and DEWITT, 2003).

Configuration	Equation
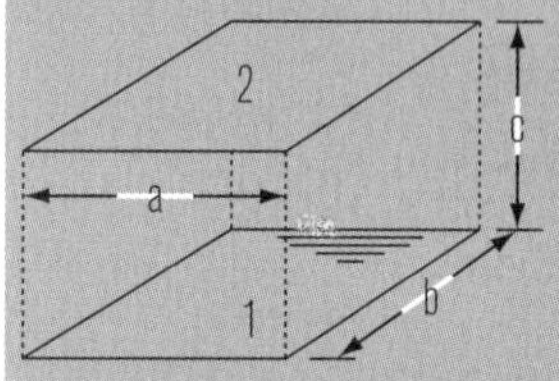	$X = a/c \quad Y = b/c$ $F_{1-2} = \frac{2}{\pi XY}\left\{ \ln\left[\frac{(1+X^2)(1+Y^2)}{1+X^2+Y^2}\right] + X\tan^{-1}X - Y\tan^{-1}Y + X\sqrt{1+Y^2}\tan^{-1}\frac{X}{\sqrt{1+Y^2}} + Y\sqrt{1+X^2}\tan^{-1}\frac{Y}{\sqrt{1+X^2}} \right\}$
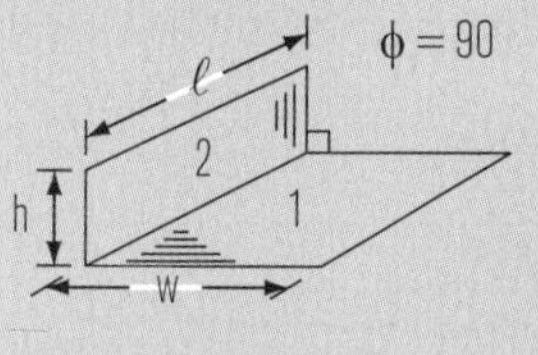	$H = h/\ell \quad W = w/\ell$ $F_{1-2} = \frac{1}{\pi W}\left\{ W\tan^{-1}\frac{1}{W} - \sqrt{H^2+W^2}\tan^{-1}\left(H^2+W^2\right)^{-1/2} + H\tan^{-1}\frac{1}{H} + \frac{1}{4}\ln\left\{ \left[\frac{(1+W^2)(1+H^2)}{1+W^2+H^2}\right] \times \left[\frac{W^2(1+W^2+H^2)}{(1+W^2)(W^2+H^2)}\right]^{W^2} \left[\frac{H^2(1+H^2+W^2)}{(1+H^2)(H^2+W^2)}\right]^{H^2} \right\} \right\}$
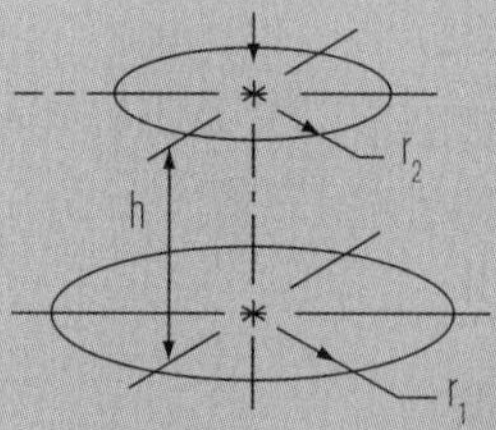	$R_1 = r_1/h, R_2 = r_2/h, \quad X = 1 + \left(1+R_2^2\right)/R_1^2$ $F_{1-2} = \frac{1}{2}\left[X - \sqrt{X^2 - 4\left(R_2/R_1\right)^2}\right]$
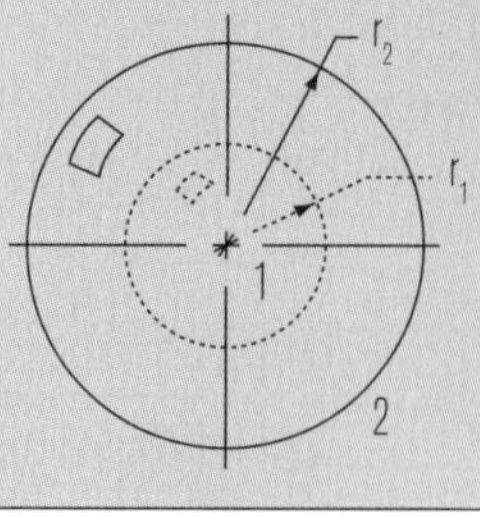	Concentric spheres $F_{1-2} = 1,\ F_{2-1} = \left(r_1/r_2\right)^2,\ F_{2-2} = 1 - \left(r_1/r_2\right)^2$

Table 14.3 - View factors for different configurations of surfaces (LIENHARD IV and LIENHARD V, 2005).

Configuration	Equation
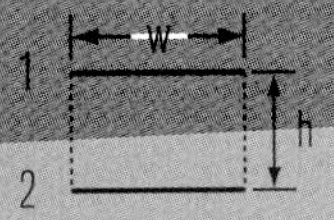	$F_{1-2} = F_{2-1} = \sqrt{1+\left(\frac{h}{w}\right)^2} - \left(\frac{h}{w}\right)$
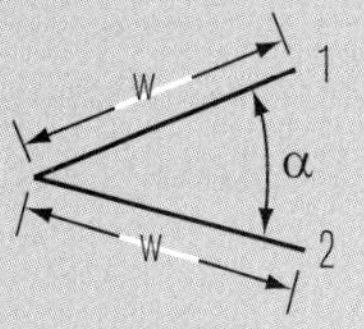	$F_{1-2} = F_{2-1} = 1 - \sin(\alpha/2)$
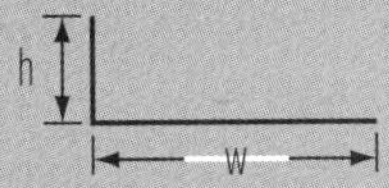	$F_{1-2} = \frac{1}{2}\left(1+\frac{h}{w}-\sqrt{1+\frac{h}{w}^{2}}\right)$
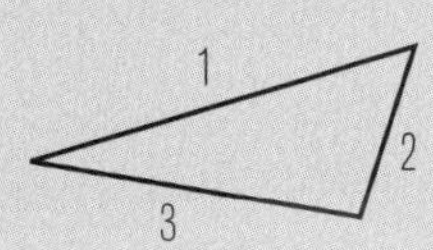	$F_{1-2} = (A_1 + A_2 - A_3)/2A_1$
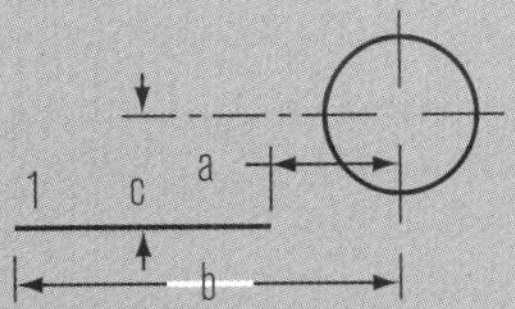	$F_{1-2} = \frac{r}{b-a}\left(\tan^{-1}\frac{b}{c} - \tan^{-1}\frac{a}{c}\right)$
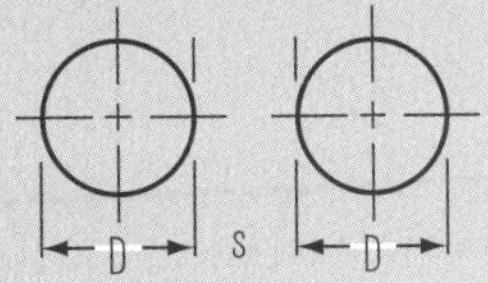	Let X = 1+s/D. Then: $F_{1-2} = F_{1-2} = \frac{1}{\delta}\left(\sqrt{X^2-1}+\sin^{-1}\frac{1}{X}-X\right)$
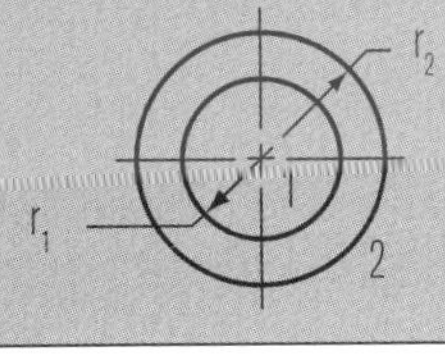	$F_{1-2} = 1$, $F_{2-1} = \frac{r_1}{r_2}$, and $F_{2-2} = 1 - F_{2-1} = 1 - \frac{r_1}{r_2}$

Infinite surfaces in the direction perpendicular to paper plane.

Table 14.4 - View factors for different configurations of surfaces (LIENHARD IV and LIENHARD V, 2005).

14.5.2 Heat exchange between black bodies

The estimate of the rate of heat exchange between two black bodies is facilitated by the fact that a black body can absorb fully the incident energy and has no reflection. Figure 14.10 illustrates two black bodies exchanging heat by radiation.

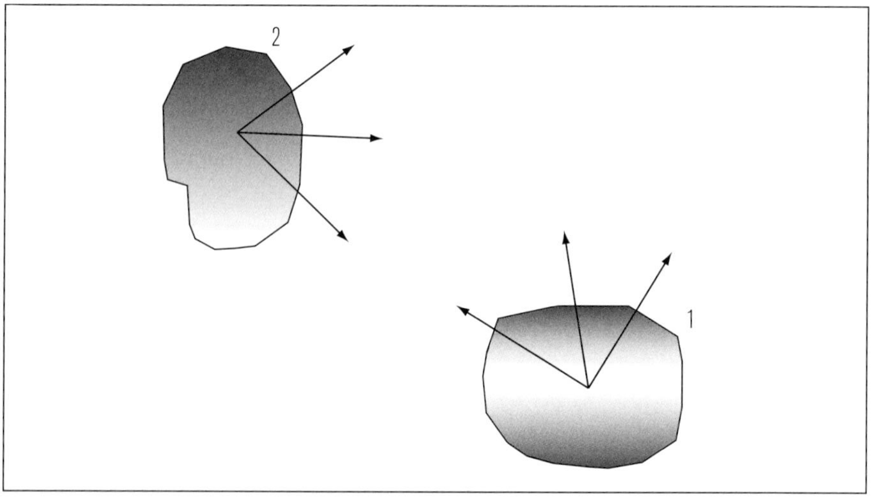

Figure 14.10 - Black surfaces exchanging heat by radiation.

Setting the radiosity, J, and all the energy that leaves a surface per unit area due to radiation, we can write the energy that comes out of the area 1 and reaching 2 is given by:

$$Q_{1\to 2} = A_1.J_1.F_{1\text{-}2} \tag{14.26}$$

For a black body:

$$J_1 = E_n \tag{14.27}$$

Therefore:

$$Q_{1\to 2} = A_1.E_{n,1}.F_{1\text{-}2} \tag{14.28}$$

In turn, the energy that goes from the surface 2 and reaching surface 1 can be expressed as:

$$Q_{2\to 1} = A_2.E_{n,2}.F_{2\text{-}1} \tag{14.29}$$

The heat exchange liquid is then given by:

$$Q_{12} = Q_{1\to 2} - Q_{2\to 1} = A_1.F_{1\text{-}2}.(E_{n,1} - E_{n,2}) \qquad (14.30)$$

To get the equation (14.30), one has to apply the law of reciprocity. As the power of emission of a black body is expressed by:

$$E_n = \sigma.T^4 \qquad (14.31)$$

One can finally write that:

$$Q_{12} = A_1.F_{1\text{-}2}.\sigma(T_1^4 - T_2^4) \qquad (14.32)$$

In an enclosure (as shown in Figure 13.6), consisting of black surfaces, the heat exchange of surface 1 with the other surfaces is expressed by:

$$Q_1 = \sum_{j=2}^{n} A_1.F_{1\text{-}j}.\sigma(T_1^4 - T_j^4) \qquad (14.33)$$

14.5.3 Heat exchange between gray surfaces

The calculation of heat transfer by radiation in respect of gray surfaces is more complex, since these surfaces reflect the incident energy. A convenient way to address this type of problem is to develop an analogy with electrical circuits, as discussed in the next section.

14.5.3.1 Electric analog of gray surfaces

The net rate of heat transfer by radiation from a given area 1 can be expressed by:

$$Q_1 = A_1.(J_1 - G_1) \qquad (14.34)$$

where:

- J_1 represents the radiosity;
- G_1 irradiation of the surface.

Radiosity includes both the energy emitted and reflected by the surface. Thus, one can write:

$$J_1 = E_1 + \rho_1.G_1 \qquad (14.35)$$

The value of Q_1 can also be determined by the difference between the energy emitted and absorbed by the surface:

$$Q_1 = A_1.(E_1 - \alpha_1.G_1) = A_1.\varepsilon_1.(E_{n,1} - G_1) \quad (14.36)$$

Assuming that the surface 1 is opaque and gray, we have:

$$\rho_1 = 1 - \alpha_1 = 1 - \varepsilon_1 \quad (14.37)$$

$$E_1 = \varepsilon_1.E_{n,1} \quad (14.38)$$

Substituting these relations in (14.35), we obtain:

$$J_1 = \varepsilon_1.E_{n,1} + (1 - \varepsilon_1).G_1 \quad (14.39)$$

Rearranging:

$$G_1 = \frac{J_1 - \varepsilon_1.E_{n,1}}{1 - \varepsilon_1} \quad (14.40)$$

Substituting the relation (14.40) in (14.34), we obtain:

$$Q_1 = A_1.\left(J_1 - \frac{J_1 - \varepsilon_1.E_{n,1}}{1 - \varepsilon_1} \right) \quad (14.41)$$

or finally:

$$Q_1 = \frac{E_{n,1} - J_1}{\left(\dfrac{1 - \varepsilon_1}{A_1.\varepsilon_1} \right)} \quad (14.42)$$

A convenient way to interpret the equation (14.42) is through an analogy with Ohm's law of electricity. In this case, $(E_{n,1} - J_1)$ represents the driving force for the occurrence of heat transfer and $(1 - \varepsilon_1)/A_1\varepsilon_1$ is the thermal resistance.

By this analogy, a gray area can be schematically represented in the manner shown in Figure 14.11.

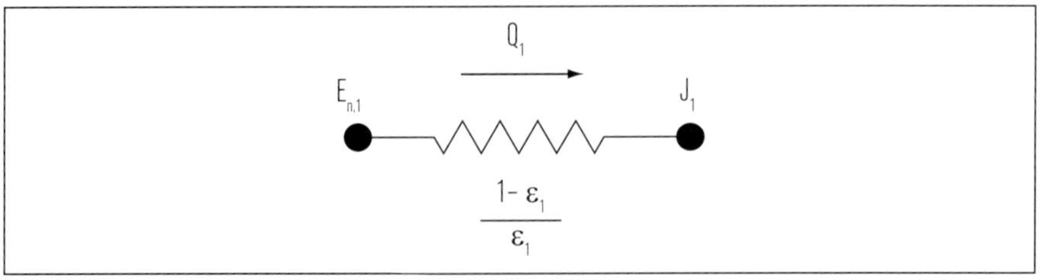

Figure 14.11 - Schematic representation of a gray area.

14.5.3.2 Heat exchange between gray surfaces in an enclosure

The rate of heat exchange by radiation between two gray surfaces can be estimated by the following relationship:

$$Q_{12} = \frac{J_1 - J_2}{\left(\frac{1}{A_1.F_{12}}\right)} \qquad (14.43)$$

In this case, the determination of the value of Q_{12} involves the evaluation of J_1 and J_2.

Some cases of interest in this evaluation are given below.

In an enclosure consisting of *n* surfaces as shown in Figure 14.6, a given area *i* exchanges heat by radiation with other surfaces.Using the analogy with electrical circuit, we can represent this heat exchange as shown in Figure 14.12.

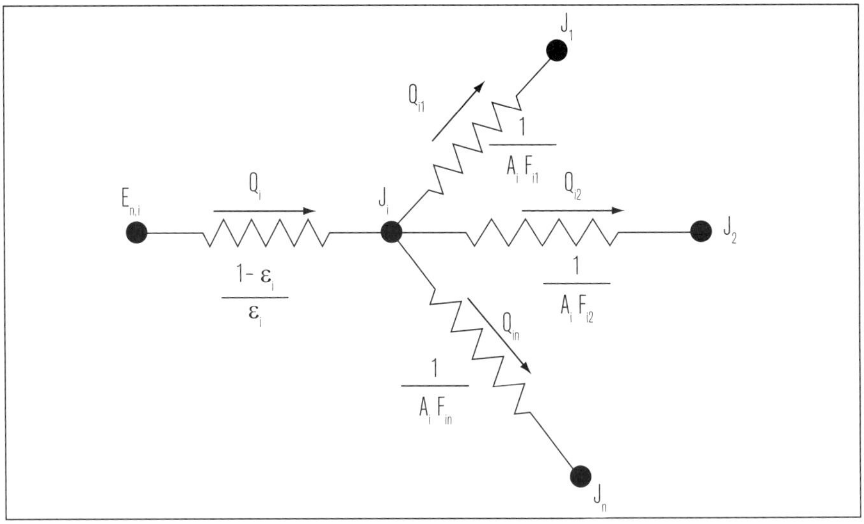

Figure 14.12 - Electrical circuit analog representing the heat exchange by radiation between surfaces in an enclosure.

From the energy balance for the surface *i*, considering only the exchange of heat by radiation, one can write the following relationship:

$$\frac{E_{n,i} - J_i}{\left(\frac{1 - \varepsilon_i}{A_i.\varepsilon_i}\right)} = \sum_{j=1}^{n} \frac{J_i - J_j}{\left(\frac{1}{A_i.F_{ij}}\right)} \qquad (14.44)$$

Equations similar to the expression (14.44) can be written for each of the areas that make up the casing. This way, one gets a system of *n* linear equations, where the unknowns are the values of J_i to J_n. One can solve the system for known values of J, and calculate the rates of heat exchange by radiation between the various surfaces.

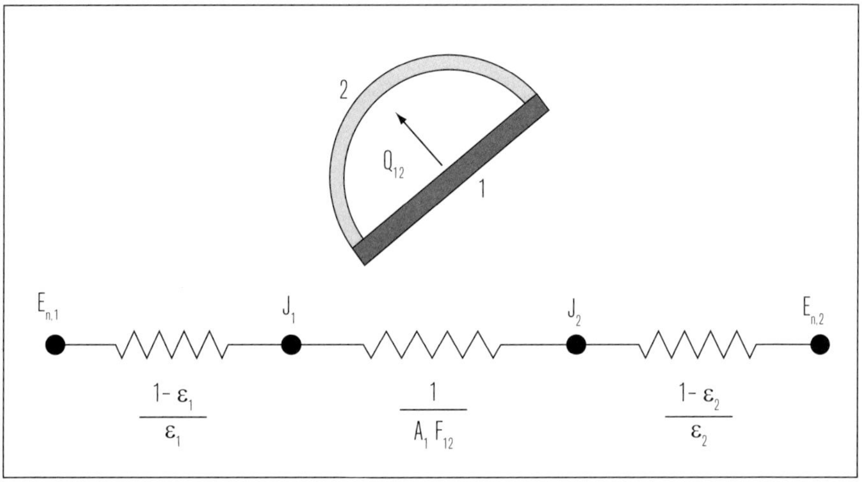

Figure 14.13 - Schematic view of an enclosure with two surfaces and the corresponding electrical circuit for heat exchange by radiation.

Some cases of practical interest involving cases with a smaller number of gray areas. In the case of an enclosure with two surfaces, as shown in Figure 14.13, the rate of heat transfer by radiation from their surfaces can be estimated using the following equation:

$$Q_{12} = \frac{\sigma.(T_1^4 - T_2^4)}{\left(\frac{1-\varepsilon_1}{A_1.\varepsilon_1} + \frac{1}{A_1.F_{12}} + \frac{1-\varepsilon_2}{A_2.\varepsilon_2}\right)} \quad (14.45)$$

A particular case of the situation described above is that of two parallel surfaces and infinite. In this case, one has:

$$A_1 = A_2 \quad (14.46)$$

$$F_{12} = 1 \quad (14.47)$$

With these considerations, the equation (14.45) is simplified to:

$$Q_{12} = \frac{A_1.\sigma.(T_1^4 - T_2^4)}{\left(\frac{1}{\varepsilon_1} + \frac{1}{\varepsilon_2} - 1\right)} \quad (14.48)$$

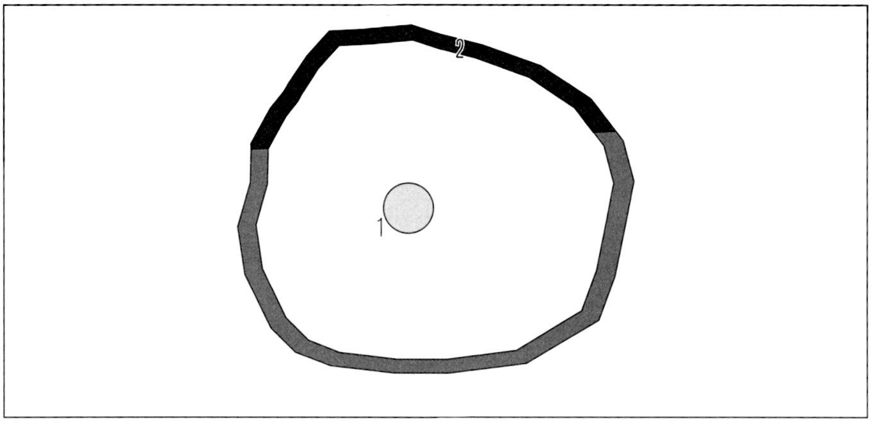

Figure 14.14 - Enclosure with two surfaces, surface 1 being convex and is completely surrounded by surface 2.

In the case shown in Figure 14.14, the following simplifications can be made:

$$A_1 <<< A_2 \tag{14.49}$$

$$F_{12} = 1 \tag{14.50}$$

Equation (14.45) is then simplified to:

$$Q_{12} = A_1.\varepsilon_1.\sigma.(T_1^4 - T_2^4) \tag{14.51}$$

This equation has been widely used in the previous chapters to estimate the rate of heat transfer by radiation between two surfaces.

The analogy with electrical circuits can also be used to introduce the concept of shielding for radiation. This concept is very important and is widely used in the protection of electronic equipment, which are used in environments where there is incidence of high energy radiation. This analysis will be considered the case of two parallel infinite flat surfaces, between which is inserted into a third surface. The two situations are shown in Figure 14.15, with their analog circuits.

For the case without the presence of shielding for radiation, the rate of heat exchange bettween the surfaces 1 and 2 is given by.

$$Q_{12} = \frac{A_1.\sigma.(T_1^4 - T_2^4)}{\left(\frac{1}{\varepsilon_1} + \frac{1}{\varepsilon_2} - 1\right)} \tag{14.48}$$

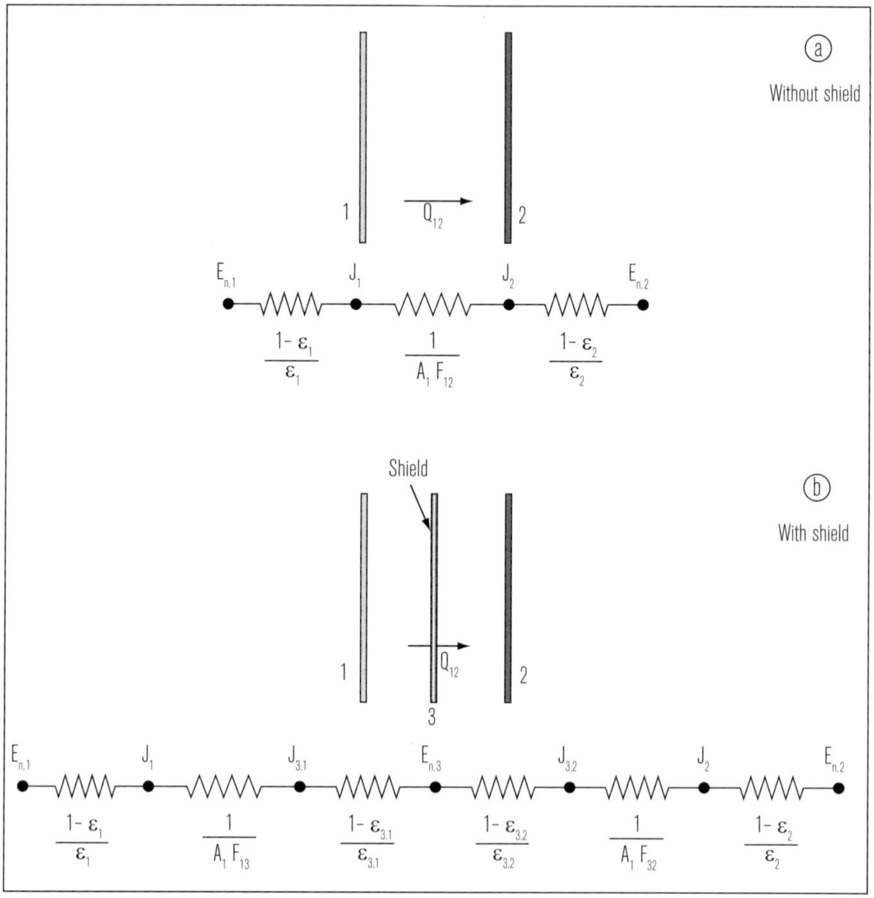

Figure 14.15 - Heat exchange between parallel plates with and without the presence of a shield for radiation.

For case b, the exchange rate of heat exchange by radiation between surfaces 1 and 2 is expressed by:

$$Q_{12} = \frac{A_1.\sigma.(T_1^4 - T_2^4)}{\left(\frac{1}{\varepsilon_1}+\frac{1}{\varepsilon_2}+\frac{2}{\varepsilon_3}-2\right)} \qquad (14.52)$$

To obtain the equation (14.52) it was assumed that the emissivity of two surfaces of the shield have the same value ε_3.

Comparing equations (14.48) and (14.52), it is seen, that the rate of heat transfer in the case a, is always greater than that for b. This reduction in the rate of heat with the presence of shielding is more significant for lower values of ε_3.

The rate of heat transfer can be further reduced if multiple shields are used.

A situation very relevant in metallurgical systems is the reradiating surface.The main feature of this area is the lack of heat transfer liquid. For surfaces which show this behavior it is necessary that they are well insulated and the heat transfer by convection can be neglected.

In the Figure 14.16, an example of reradiating surface with the equivalent electrical circuit.

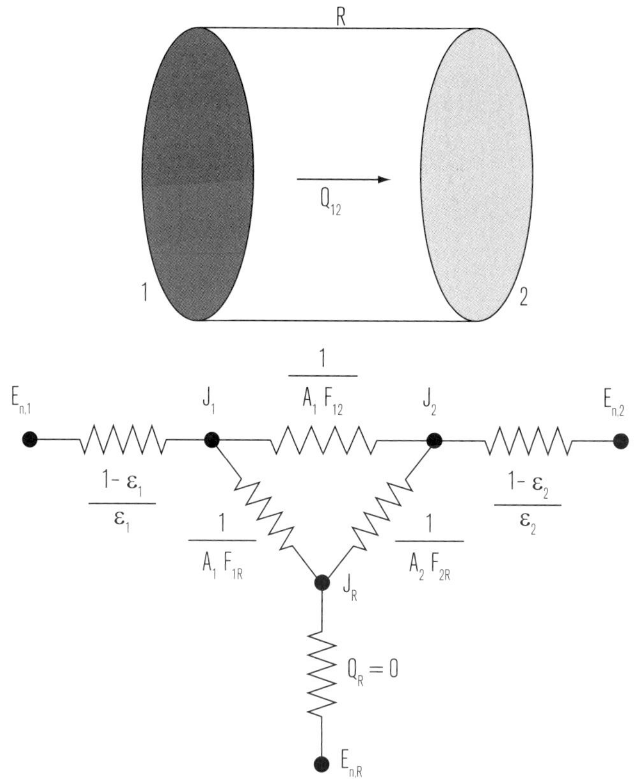

Figure 14.16 - View of an enclosure with three surfaces, one being a reradiating surface, and its corresponding equivalent electrical circuit.

Assuming that the surface reradiating the rate of heat transfer liquid is zero and considering relations valid for electrical circuits, one can propose the following equation to evaluate the rate of heat transfer between the surfaces 1 and 2:

$$Q_{12} = \frac{\sigma.(T_1^4 - T_2^4)}{\left[\frac{1-\varepsilon_1}{A_1.\varepsilon_1} + \frac{1}{A_1.F_{12} + \left(\frac{1}{A_1.F_{1R}} + \frac{1}{A_2.F_{2R}}\right)^{-1}} + \frac{1-\varepsilon_2}{A_2.\varepsilon_2}\right]} \qquad (14.53)$$

Example

Liquid steel to 1873 K is contained in a ladle with geometry that can be considered cylindrical. The internal diameter of the ladle is 2.5 m freeboard is 0.3 m. The neighborhood is at 300 K. Estimate the rate of heat loss by radiation of steel. Consider that the emissivity of the liquid steel is 0.28. What would be the loss of heat to the free board is overlooked?

Solution

A situation analysis here is quite similar to that presented in Figure 14.16, and the areas 1, 2, A, are:

- 1: liquid steel;
- 2: neighborhood;
- R: region near the free board of the ladle.

In this case, the equation (14.53) can be applied. As indicated above, the region 2 represents the neighborhood, which will be considered as a hypothetical surface, approximately like a body for which, $\varepsilon_2 = 1$.

To use the equation (14.53) remains to determine the values of view factors F_{1R} and F_{2R}. For this Table 14.3 can be used resulting in the following parameters ar:

- $r_1 = r_2$ = radius of areas 1 and 2;
- L = freeboard.

With the figures provided, we have:

- $r_1 = r_2 = 1.25$ m;
- L = 0.3 m.

Therefore, we have:

$$R_1 = \frac{r_1}{L} = \frac{1.25}{0.3} = 4.17 \qquad R_2 = \frac{r_2}{L} = \frac{1.25}{0.3} = 4.17$$

$$S = 1 + \frac{1 + R_2^2}{R_1^2} = 2.06$$

$$F_{12} = \frac{1}{2}\left\{ S - \left[S^2 - 4\left(\frac{r_2}{r_1}\right)^2 \right]^{1/2} \right\} = 0.79$$

By the rule of additivity, one can determine F_{1R} and F_{2R} through the list below:

$$F_{1R} = F_{2R} = 1 - F_{12} = 1 - 0.79 = 0.21$$

As $A_1 = A_2$, one can evaluate the rate of heat loss from the steel by the relationship:

$$Q_{12} = \frac{A_1.\sigma.(T_1^4 - T_2^4)}{\left[\dfrac{1-\varepsilon_1}{\varepsilon_1} + \dfrac{1}{F_{12} + \left(\dfrac{1}{F_{1R}} + \dfrac{1}{F_{2R}}\right)^{-1}} + \dfrac{1-\varepsilon_2}{\varepsilon_2}\right]}$$

Replacing data:

$$Q_{12} = \frac{\pi(1{,}25)^2.(5{,}67 \times 10^{-8}).[(1873)^4 - (300)^4]}{\left[\dfrac{1-0{,}28}{0{,}28} + \dfrac{1}{0{,}79 + \left(\dfrac{1}{0{,}21} + \dfrac{1}{0{,}21}\right)^{-1}} + \dfrac{1-1}{1}\right]}$$

$$Q_{12} = 927.981{,}4 \text{ W}$$

If the freeboard is disregarded, the heat loss of the steel was estimated by the ratio:

$$Q_{12} = A_1.\sigma.\varepsilon_1.(T_1^4 - T_2^4) = \pi(1.25)^2.(5.67 \times 10^{-8}).(0.28).[(1873)^4 - (300)^4]$$

$$Q_{12} = 958464.8 \text{ W}$$

Disregarding the free board, the heat loss is slightly higher.

References

ABOUTALEBI, R.; HASAN, M.; GUTHRIE, R.I.L. Coupled turbulent flow, heat, and solute transport in continuous casting processes. *Metallurgical and Materials Transactions B*, v. 26B, n. 4, p. 731-44, Aug. 1995.
BIRD, R.B.; STEWART, W.E.; LIGHTFOOT, E.N. *Transport phenomena.* New York: John Wiley & Sons, 1960.
GEIGER, G.H.; POIRIER, D.R. *Transport phenomena in metallurgy.* Massachussetts: Addison-Wesley, 1973.
INCROPERA, F.P.; DEWITT, D.P. *Fundamentos da transferência de calor e massa.* 5. ed. Rio de Janeiro: LTC, 2003.
INGERSLEV, P.; HENEIN, H. An integral boundary approach for 1 - and 2-D modeling of ingot reheating and cooling. *Iron and Steelmaker*, v. 24, n. 5, p. 75-85, May 1997.
LIENHARD IV, J.H.; LIENHARD V, J.H. *A heat transfer textbook.* Cambridge: Phlogiston, 2005.

Exercises

1 - Repeat the last example given above for the case of the liquid steel, covered with a layer of slag 5 cm thick. The thermal conductivity of the slag and its emissivity are, respectively, 1 W/m.K and 0.8. keeping other parameters constant.

2 - Determine the rate of heat exchange by radiation between the two surfaces seen in the figure below. Consider the two surfaces as black bodies. The surface 1, is at temperature 1600 K and that of 2 is at deg 300 K.

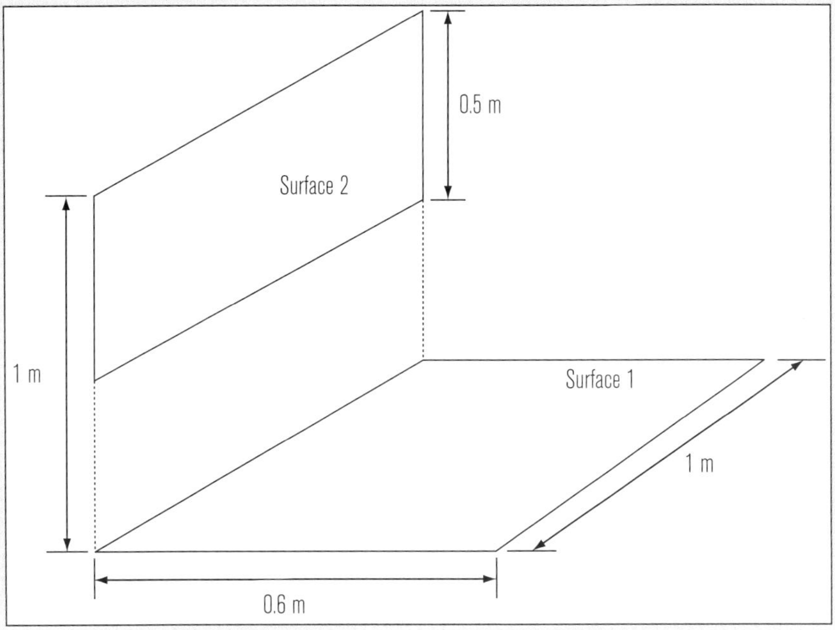

Introduction to the Transport of Species

Chapter 15

The study of transfer of a chemical species from one point to another in a system is motivated by practical considerations, such as:

- achieving uniformity of a chemical species in a product as a result of homogenization;
- increasing reaction rates between species in the refining processes and control of processes for improved productivity and quality.

In general, efficient operation of an industrial process depends on the knowledge of the nature and magnitude of the driving forces based on the fundamentals of transport phenomena with special reference to mass transport, so that objectives in respect of quality of products and productivity are achieved. For example, one can cite desulphurization for achieving low sulfur in an iron or steel melt. Several parameters in the process have to be controlled and for this, apart from thermodynamics of the process, the transport phenomena aspects have to be considered. The same criteria holds good for phosphorous, carbon and control of other elements in the iron or steel melt.

Transfer processes of species are not only important in iron and steel making, but also in the agglomeration processes like pelletizing, sintering in addition to extraction of various metals from ores and other areas like ceramics and polymers. For example, in the drying of pellets, heat transfer promotes evaporation of water and is responsible also for mass transfer of species. The rate of transfer of heat should be controlled in the drying stage through control of flow of hot gases so that there is no sudden evaporation from the pores, causing degradation of the green pellets. As such, in most of the processes like blast furnace, steel making and extraction processes of other metals, all the three aspects of transport phenomena namely heat, mass and momentum transfer play a vital role. For the success of processes measured in terms of productivity and quality interactive knowledge of all the three is essential.

An almost universal feature of metallurgical processes is the role of heterogeneous reactions. Homogeneous reactions also occur, but often coupled with heterogeneous, reactions. This emphasizes the importance of kinetics of reactions. For example, in leaching reactions, the interface area is very important. The controlling step for any reaction has to be investigated. The controlling step may be mass transfer in the boundary layer, chemical reaction at the interface, diffusion control through porous product layer or mixed control involving all the above steps. The case of reduction of iron ores in a reducing gas is an example and there is a vast literature in respect of the mechanism of reduction different ores involving these steps.

Many reactions are termed slag metal reactions, especially in the refining processes. These reactions can be termed as liquid phase reactions and here also the interfacial area playa a very important part. As an example, one can cite the desulfurization of the melt by the top slag. This reaction is given by:

$$[S] + \left(O^{-2}\right) \rightarrow \left(S^{-2}\right) + [O] \qquad (15.1)$$

(where the square brackets and the round brackets represent, respectively, species dissolved in the metal bath and slag). The possible steps in the desulfurization reaction are the following:

- Transfer of reactive species to the reaction interface and in this case slag metal interface. The transference of the species can happen by diffusion;
- Chemical reaction at the interface;
- Removal of reaction products into the liquid phases of metal and slag by mass transfer and diffusion.

So the rate of a reaction like the desulphurization reaction, can be controlled by the transport characteristics of species in the medium which may be mass transfer of the species through the boundary layer, chemical reaction at the at the interface followed by diffusion inside the material. However it is necessary to determine the slowest step in the reaction to find out the controlling step and thereby take measures to improve the process. The speed of removal of the product also plays a crucial part. Composition gradients are usually associated with driving forces for transport by diffusion. These are important in operations such as chemical vapor deposition, nitriding and carburizing of steel. During solidification, the mother liquid gets richer in one of the solute elements and the liquid that solidifies last has a different composition leading to segregation effects, which are injurious to the mechanical properties of materials. Hence during processing, subsequent solidification and mechanical working, strict process control measures have to be are taken in order to minimize defects such as segregation and maximize homogenization so that a quality product results. In the following chapters, phenomenological basis for the transport of species, parameters of transport, steady and unsteady state mass transfer coupled with other phenomena, will be discussed. Finally, applications of the concepts of mass transfer in physical and mathematical modeling of industrial processes will be elaborated.

Fundamentals of Mass Transport

Chapter 16

When chemical species, whether they are elements, compounds or complex ions, move in relation to a coordinate system, one can define the flow of each specie in relation to the same. This is a vector quantity and movement of the medium composed of the species can be expressed by the combination of transport of individual specie.

For example, a flow n_i(kg i/m^2.s) is a measure of the amount of specie i crossing a surface of unit area perpendicular to this vector, per unit time, as can be seen in Figure 16.1.

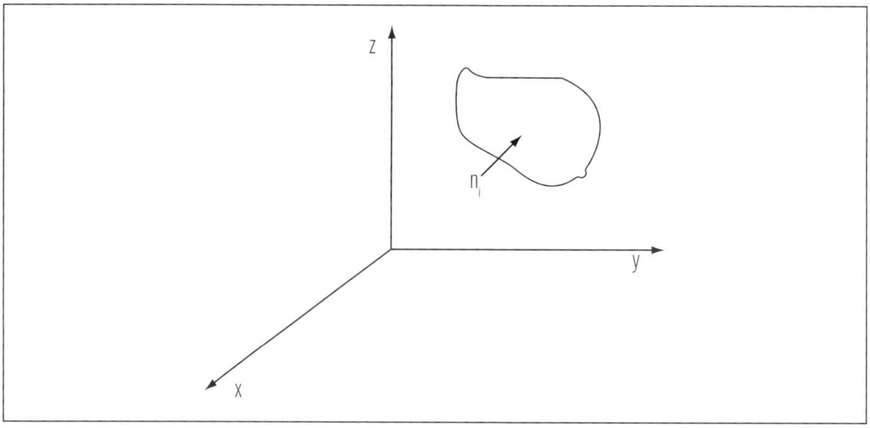

Figure 16.1 - Schematic representation of the vector flow of specie i in cartesian coordinates.

For convenience, this vector can be resolved into its components in relation to the coordinate axes, namely, n_{ix}, n_{iy}, n_{iz}. Also, one can define the absolute velocity of specie i, v_i colinear to the vector flow, which can be estimated from:

$$n_i\left(\text{kg/m}^2.\text{s}\right) = v_i\left(\text{m/s}\right) . \rho_i\left(\text{kg/m}^3\right) \tag{16.1}$$

where:

- ρ_i represents the concentration in terms of mass of specie i

The movement of the medium can be considered as the composite movements of all its species, one can define the total flux (mass) as:

$$n = \sum n_i \tag{16.2}$$

This is the flow perceived by an outside observer, as expressed by the equation of continuity:

$$-\frac{\partial}{\partial t}\rho = \frac{\partial}{\partial x}\rho v_x + \frac{\partial}{\partial y}\rho v_y + \frac{\partial}{\partial z}\rho v_z \quad (16.3)$$

The average velocity (based on mass flow), or the reference velocity, is given by:

$$v = \frac{n}{\rho}(m/s) \quad (16.4)$$

The above expression is consistent with the calculation method used for determining flow. If v is the velocity of the fluid flowing through a control surface, (imaginary or not) perpendicular to this vector and of area dA, then the total mass flow rate is:

$$dM = v\ (m/s).dA\left(m^2\right)\ \rho\left(kg/m^3\right) \quad (16.5)$$

The mass flow given as:

$$n = \frac{dM}{dA} = v\rho \quad (16.6)$$

Different species can have, in general, different values of flow and and velocities. The difference in behavior between them can be characterized by resolving the motion of each in terms of average behavior and a deviation as given below:

$$n_i = \rho_i v + j_i \quad (16.7)$$

ρ_i is the concentration in terms of mass per unit volume of specie i in the medium. Hence, the density of the medium is given by:

$$\rho = \sum \rho_i \quad (16.8)$$

Considering equations (16.4) for average velocity and (16.7):

$$n_i = \frac{\rho_i}{\rho} n + j_i \quad (16.9)$$

In these equations ρ_i v represents the transport of specie i in the medium, moving with velocity v, or by definition, transported by convection. The second term, $j_i(kg/m^2.s)$ is the deviation from this behaviour and is characteristic of the specie and the environment. This is the contribution by diffusion.

Given these definitions, when applied to the summation operator to the expression of flow, extended to all species, that is:

$$\Sigma n_i = (\Sigma \rho_i)\, v + \Sigma j_i \tag{16.10}$$

$$n = \rho v + \Sigma j_i \tag{16.11}$$

Since by definition $n = \rho v$ (Equation 16.4):

$$\Sigma j_i = 0 \tag{16.12}$$

In any situation the evaluation of convective and diffusive contributions depend on the nature of the flow and the phenomenological equations governing the same for the purpose of estimating the diffusive contribution (MILLS, 1999).

One of the other possible methodologies for assessing the transport velocity of a given specie is based on the flow, measured on a molar basis, N_i(moles of i/m^2.s). Adopting the same procedure for the system of equations above:

$$N_i\left(\text{mole/m}^2.\text{s}\right) = v_i\,(\text{m/s}).C_i\left(\text{mole/m}^3\right) \text{ molar flow of specie i} \tag{16.13}$$

$$N = \sum N_i \text{ total molar flow} \tag{16.14}$$

$$C = \sum C_i \text{ total molar concentration} \tag{16.15}$$

$$v^* = \frac{N}{C}[\text{m/s}] \text{ average velocity molar basis} \tag{16.16}$$

$$N_i = C_i v^* + J_i \text{ molar flow of specie i} \tag{16.17}$$

$$\mathbf{N}_i = \frac{C_i}{C}\mathbf{N} + \mathbf{J}_i \text{ or } \mathbf{N}_i = X_i\,\mathbf{N} + \mathbf{J}_i \tag{16.18}$$

$$\Sigma \mathbf{J}_i = 0 \tag{16.10}$$

where:

- C_i(mole of i/m^3) and X_i, are respectively the molar concentration and mole fraction of specie i.

Example

In a given experimental procedure flows of CO and CO_2 at 500°C and 1 atm were measured. The gas phase has 60% CO: $n_{CO} = +1.4$ kg/m²s and $n_{CO_2} = -1.4$ kg/m².s in countercurrent flows. Estimate the diffusive flows using average velocities based on mass flow and molar flow.

In this case, the flows being countercurrent, the total mass flow amounts to:

$$n = n_{CO} + n_{co_2} = +1.4 - 1.4 = 0$$

Hence the mean velocity:

$$v = \frac{n}{\rho} = \frac{0}{\rho} = 0$$

Since $n_{CO} = \rho_{CO}\, v + j_{CO}$ and $n_{CO_2} = \rho_{CO_2}\, v + j_{CO_2}$ the convective contributions are zero and therefore:

$$n_{CO} = j_{CO} = +1.4 \text{ kg/m}^2.\text{s}$$

$$n_{CO_2} = j_{CO_2} = -1.4 \text{ kg/m}^2.\text{s}$$

The flow rate of each specie on a molar basis is given by:

$$N_{CO} = \frac{n_{CO}}{M_{CO}} = \frac{+1.4 \text{ kg/m}^2\text{s}}{0.028 \text{ kg/mole}} = +50 \text{ mole/m}^2.\text{s}$$

$$N_{CO_2} = \frac{n_{CO_2}}{M_{CO_2}} = \frac{-1.4 \text{ kg/m}^2\text{s}}{0.044 \text{ kg/mole}} = -31{,}82 \text{ mole/m}^2.\text{s}$$

Hence the total flow, N = 18.18 mole/m²s.

Assuming that the gases have ideal behavior, the total molar concentration is given by:

$$C = \frac{P}{RT} = \frac{1.013\ 10^5\ (\text{Pa})}{8.31\ (\text{J/K.mole})\ 773\ (\text{K})} = 15.77 \text{ mole/m}^3$$

The average velocity or reference velocity based on molar flux is given by:

$$\dot{v} = \frac{N}{C} = \frac{18.18}{15.77} = 1.15(\text{m/s})$$

The convective contribution is not zero. Hence:

$$N_{CO} = X_{CO}\ N + J_{CO} \text{ or } 50 = 0.6 \times 18.18 + J_{CO}$$

$$N_{CO_2} = X_{CO_2}\ N + J_{CO_2} \text{ or } -31.82 = 0.4 \times 18.18 + J_{CO_2}$$

$$J_{CO} = 39.09 \text{ mole/m}^2\text{s and } J_{CO_2} = -39.09 \text{ mole/m}^2\text{.s}$$

Since there are different options regarding the choice of the reference velocity one may wonder if there can be a choice of convenience. Of course, the flow of a given species cannot be dependent on the chosen reference velocity, as the mathematical framework would also become different. The following example clarifies this point.

Example

For an ideal gas in a binary phase, with reference to the velocity based on molar flows it can be shown that the diffusion coefficients are equal. Consider a gas phase AB, in a capillary container at constant volume, pressure and temperature. Gaseous species A and B can be considered as ideal gases. If there is a concentration gradient between the ends of the capillary molar flows of A and B should be equal but opposite in sign, $N_A = -N_B$. Otherwise in a particular region there would be a certain increase of the number of moles in the gas phase leading to a change in pressure (since A an B are ideal gases and PV = nRT). This situation is not tenable. Also the total molar concentration is constant and is given by:

$$C = \frac{P}{RT} = C_A + C_B \ (\text{mol/m}^3)$$

If $\nabla C_A = -\nabla C_B$ then, the transport equations can be written for both the gases, taking into account that the total molar flow should be zero, and the definition of average velocity based on the molar flow :

$$N_A = C_A\ v^* - D_A\ \nabla C_A$$

$$N_B = C_B\ v^* - D_B\ \nabla C_B$$

where: D_A and D_B represent the diffusion coefficients of species A and B.

Combining the above equations:

$$N_A + N_B = N = (C_A + C_B)v^* - (D_A - D_B)\nabla C_A = 0$$

$$N_A + N_B = N = (C_A + C_B)\frac{N}{C} - (D_A - D_B)\nabla C_A = 0$$

Hence for an ideal gas in a binary phase, with reference to the velocity based on molar flows, the diffusion coefficients are equal:

$$D_A = D_B$$

The terms n(kg/m².s) and N(moles/m².s) have been defined at the very beginning of this chapter. As has been mentioned the reference velocity in the case of gases is based on molar flows. In the case of solids, it is more convenient to consider the mass concentration which would be uniform and constant in any part of the system.

$$\rho = \rho_A + \rho_B = \text{constant} \tag{16.20}$$

In that case the mass flow of A in a given direction is equal (in the module) but is in opposite direction to the mass flow of B, ie $n_A + n_B = 0$. Hence:

$$n_A = (n_A + n_B)\frac{\rho_A}{\rho} - D_A\frac{d\rho_A}{dx} \tag{16.21}$$

$$n_B = (n_A + n_B)\frac{\rho_B}{\rho} - D_B\frac{d\rho_B}{dx} \tag{16.22}$$

Hence:

$$n_A + n_B = (n_A + n_B)\frac{\rho_A + \rho_B}{\rho} - (D_A - D_B)\frac{d\rho_A}{dx} = 0 \tag{16.23}$$

From the above equation it can be concluded that $D_A = D_B$, since $n_A + n_B = 0$.

This is the criteria for describing of transport of a specie in solid or liquid, ie the diffusivity based on the average speed measured in relation to mass flows.

16.1 Phenomenological Basis of Diffusion

The examples above show that the choice of reference velocity can be made arbitrarily, for mathematical convenience. Consequently it is valid for the diffusion term also and does not necessitate any physical justification. However the basis of the reasoning behind phenomenological treatment of diffusion should take into account the real forces responsible for the process which can adequately be represented in mathematical terms. In fact, the term diffusion (ordinary) is the movement of matter at the atomic level, due to thermal vibrations. For example, Figure 16.2 describes the atomic planes in a phase AB. As there is a concentration gradient in the system, the concentration of A decreases from right to left.

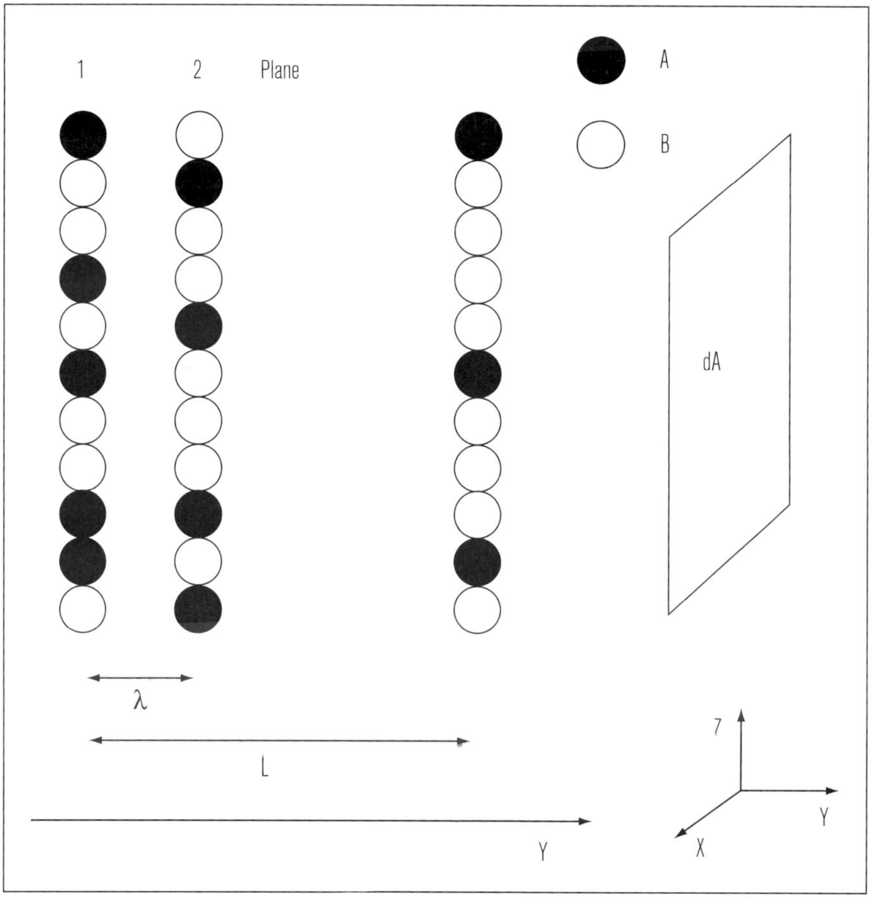

Figure 16.2 - Mechanism of diffusion by jumps (BALLUFFI, ALLEN and CARTER, 2005).

Above absolute zero, atoms of A and B do not hold a fixed position. They vibrate around equilibrium positions of the network. These atoms are capable of jumping from one position to another, with a frequency depending on the temperature, characteristics of the medium, and equal to $\upsilon(s^{-1})$. Considering only the motion of atoms in the direction Y the velocity with which the atoms of A leave a given plane (moving in the direction OY), is given by $\upsilon(s^{-1}).C(atoms/m^2)/2N_0$, where C is the concentration in the plane and N_0 is the Avogadro number. The factor ½ is introduced because the jump could be observed at random in any sense of direction OY. Therefore, the net molar flow, considering different concentrations in the planes 1 and 2, is given by:

$$j_A = \frac{\nu C_1 - \nu C_2}{2N_o} \left(moles/m^2 s\right) \tag{16.24}$$

Depending on the definition of diffusive flux, the relationship between volume concentration and surface concentration amounts to C/λ, where λ, is defined (Figure 15.2) as distance between two atomic planes.

According to Fick´s law:

$$j_A (mole / m^2.s) = -D_A (m^2 / s) \frac{dC(mole/m^3)}{dy(m)} =$$

$$j_A (mole / m^2.s) = -D_A (m^2 / s) \frac{\Delta C(mole/m^3)}{\Delta y(m)}$$

In the atomic level, Ci (i = 1 or 2) is the concentration measured as number of atoms per unit area of the atomic plane (atoms/m²).

To get C (concentration given as moles per unit volume) Ci should be divided by and then by Avogadro´s number, i.e C = Ci /(λ.N) moles/m³.

Molar concentration at points 1 and 2 in Figure 16.2, can be estimated. Considering ΔY = λ, the Fick´s law can be written in terms of the two equations given below:

$$j_A (mole / m^2.s) = -D_A (m^2 / s) \frac{\Delta C(mole/m^3)}{\Delta y(m)} \quad \text{and}$$

$$j_A = -D_A \frac{(C_2 - C_1)/\lambda N_o}{\lambda}$$

But the molar flux is also given by equation (16.24):

$$j_A = \frac{\nu C_1 - \nu C_2}{2N_o}(\text{moles/m}^2\text{s})$$

Comparing the last two equations, one gets $D_A = \nu\lambda^2/2$ (SHEWMON, 1989):

$$j_A = -D_A \frac{dC}{dy} = -D_A \frac{(C_2 - C_1)/\lambda\ N_o}{\lambda} \equiv -\frac{\nu}{2N_o}(C_2 - C_1) \quad (16.25)$$

$$D_A = \nu\lambda^2/2 \quad (16.26)$$

This result, although not directly used for the calculation of diffusion coefficients, clarifies the dependence of the same on thermal vibration, which alters the value of λ (alters the distance between planes or jump frequency or both).

16.2 Equations for Ordinary Diffusion

In general the expression for the driving force of diffusion is written from a phenomenological basis, assuming that each species is affected differently by a given potential gradient. The different effects of the potential gradient, then, account for the deviations from the average behavior of the medium or convection.

For example, there is equilibrium within phase (except in the case of chemical reactions) when the gradients of temperature, pressure and chemical potential (of any species) don't exist. Of the various situations of imbalance the most common is the one where the gradient of chemical potential of specie i, μ_i, is not equal to zero.

$$\nabla\mu_i \neq 0 \quad (16.27)$$

$$\nabla(\mu_i^o + RT \ln \gamma_i X_i) \neq 0 \quad (16.28)$$

where:

- μ_i^o, standard chemical potential,
- X_i, molar fraction of specie, from
- γ_i activity coefficient of specie i.

This equation implies,the relationship, (Figure 16.3) between activity (or chemical potential) and chemical composition (or molar fraction) within a phase (DEHOOF, 2006).

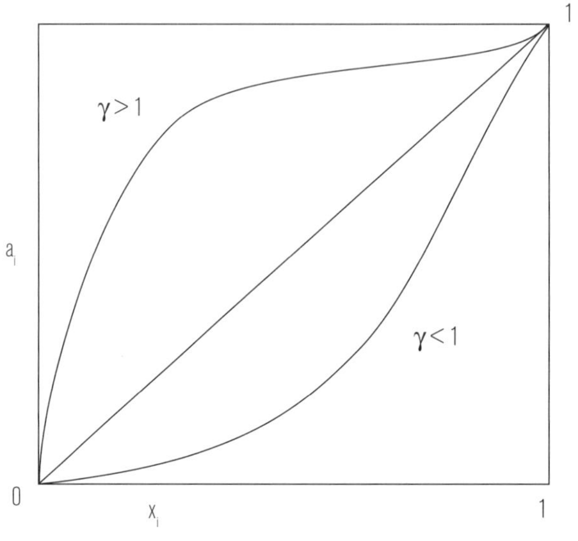

Figure 16.3 - Schematic relationship between activity and mole fraction in an isothermal and isobaric phase; $\gamma > 1$ positive deviation in relation to ideality $\gamma < 1$ negative deviation in relation to ideality.

When $\nabla X_i \neq 0$, It means that inside a phase a gradient of chemical potential is followed by a gradient of composition and vice-versa. The specie tends to get redistributed so that equilibrium is restored, that is to eliminate the chemical potential gradient. That is why either the gradient of chemical potential or the gradient of composition can be considered as the driving force for the diffusion. This is the basic principle behind Fick's law

$$n_i = \rho_i \ v - D_i \ \nabla \rho_i \qquad (16.29)$$

$$N_i = C_i \ v - D_i \ \nabla C_i \qquad (16.30)$$

Equation 16.30 is obtained dividing 16.29 by M_i, atomic mass or molecular weight of the species.

Hence the diffusive flux in relation to the reference velocity based on mass flow, is given by the Fick's diffusion equation.

$$j_i = -D_i \ \nabla \rho_i \ \text{kg/m}^2\text{s} \qquad (16.31)$$

$$j_i = -D_i \ \nabla C_i \ \text{mol/m}^2\text{s} \qquad (16.32)$$

The diffusive mass flux takes place from locations of higher chemical potential to those of lower potential, or from points of higher concentrations to those of lower concentrations, as shown in Figure 16.4. The flow is in the opposite direction to the gradient of concentration, as represented by its negative sign.

D_i is, by definition, the diffusivity or the diffusion coefficient of specie i in the medium. The value of D_i depends on the specie under consideration, temperature, pressure and composition of the medium. Hence D_i in general, has been the subject of several investigations in respect of its experimental evaluation as a function of different variables in the literature. Chapter 17 discusses some methods for its determination.

The multiplicity of definitions of the reference velocity can lead to some confusion. For example, the velocity terms discussed earlier in this chapter (pages 1-5) make clear the differences between the two velocities, given below in equations (16.33) and (16.34) and the respective diffusion coefficients.

$$N_i = C_i \ v - D_i \ \nabla C_i \quad (16.33)$$

$$N_i = C_i \ v^* - D_i \ \nabla C_i \quad (16.34)$$

From the equation (16.34), one can establish an expression for Fick's Law, based on average speed, molar flux and hence diffusion coefficients evaluated through this equation may differ from that calculated by equation (16.33). The choice of the reference velocity that fits in best in a given situation, in mathematical terms should be in a position to ignore or make void the convective contribution, however maintaining the physical basis.

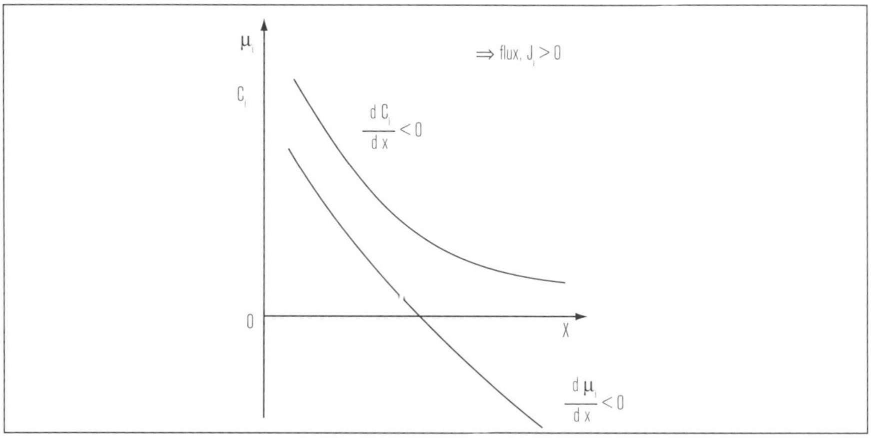

Figure 16.4 - Nature of the diffusive flux vector and the chemical potential gradient.

The Nernst-Einstein equation (KINGERY, BOWEN and UHLMANN, 1976), is defined by: FU = v, where F is the force acting on the specie of interest that results in a velocity v, gives the concept of mobility of a species. U is in the above equation is the mobility, which is a function of the characteristics of the medium and of the specie itself. In the case of ordinary diffusion it is usual to calculate the force from the chemical potential, as given below:

$$\frac{F_i}{N_o} = -\frac{1}{N_o}\frac{d\mu_i}{dx} \ (\text{N/atom}) \tag{16.35}$$

where:

- F_i - force acting on specie i,
- N_0 - Avogadro´s number,
- μ_i - the chemical potential of specie i
- x - the coordinate distance.

For a dilute solution, the chemical potential is given by:

$$\mu_i = \mu_i^0 + RT \ln \gamma_i^0 + RT \ln C_i$$

where γ_i^o is the henrian activity coefficient and the other terms are defined as in equation 16.28.

The derivative of the above equation can be substituted in 16.35. to give:

$$\frac{F_i}{N_o} = -k_B T \frac{1}{C_i}\frac{dC_i}{dx} \ (\text{N/atom}) \tag{16.36}$$

where K_B = R (ideal gas constant)/N_0 (Avogadro`s) is the Boltzmann's constant. The velocity of the specie can be estimated from this force.

$$v_i = -k_B T \frac{1}{C_i}\frac{dC_i}{dx} U_i \tag{16.37}$$

and since the flux is given by $j_i = C_i v_i$ results:

$$j_i = C_i v_i = C_i\{-k_B KT \frac{1}{C_i}\frac{dC_i}{dx} U_i\} \tag{16.38}$$

Hence, $D_i = k_B T U_i$, that is to say that diffusivity is a function of mobility.

Example

Consider steady state transport, of carbon through a thin sheet of iron, of thickness L = 0.1 cm, 800°C, with different gas atmospheres on either side. The surface concentration is equal to 4% and 0% atomic fraction of Carbon. As per the phase diagram (see figure below), one can identify the two phase regions, austenite phase (high carbon) and ferrite (low carbon). The composition of both phases varies along the plate thickness, from 4% to 0%, but exhibits a discontinuity at the point of contact. According to the diagram of equilibrium compositions of ferrite and austenite phases are $C_2 = 2.50\%$ and $C_3 = 0.02\%$ carbon, respectively. The chemical potential varies inside both these phases, but at the point of contact it is equal. It can be seen that lack of thermodynamic equilibrium promotes interphase mass transport through diffusion.

$$\mu_c(800\text{ C, 1 atm phase } \alpha\ 0.02\%) = \mu_c\ (800\text{ C, 1 atm phase } \gamma\ 2.5\%)$$

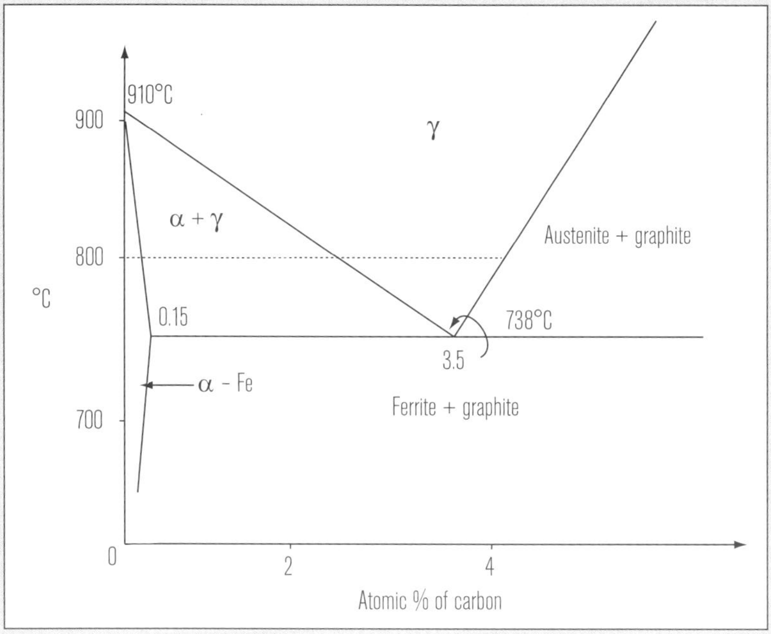

Phase diagram Fe-C (POIRIER and GEIGER, 1994).

Assuming carbon as an impurity and occupies the interstitials (neglecting convection in the medium) and linear gradient of composition within each phase (which is due to the fact that the diffusion coefficients are considered independent of composition $D_c^\alpha = 10^{-6}\text{cm}^2 \times s^{-1}$ and $D_c^\gamma = 10^{-8}\text{cm}^2 \times s^{-1}$). Assuming steady state one can write:

$$N_c^\alpha = N_c^\gamma$$

$$-D_c^\alpha \frac{dC_c^\alpha}{dx} = -D_c^\gamma \frac{dC_c^\gamma}{dx} \text{ (when only diffusion is active)}$$

$$-D_c^\alpha \times \left(\frac{C_3^\alpha - C_4^\alpha}{L-\delta}\right) = -D_c^\gamma \times \left(\frac{C_1^\gamma - C_2^\gamma}{\delta}\right) \text{(for linear profile inside each phase)}$$

where δ is the thickness of the austenite phase. Since $D_c^\alpha = 10^{-6}\text{cm}^2 \cdot s^{-1}$, $C_3 = 0.02\%$, $C_4 = 0.0$, $L = 0.1$ cm and $D_c^\gamma = 10^{-8}\text{cm}^2 \cdot s^{-1}$, $C_1 = 4.0$, $C_3 = 0.02\%$, δ can be easily calculated from the previous equation. In the figure below a schematic diagram giving the variation of chemical potential in the phases are given. There is no discontinuity at the point of contact between the phases, and the nature of chemical potential curve for each phase and that of the curve of composition, justify the form of the phenomenological equation of Fick, namely equation (16.38) (KINGERY, BOWEN and UHLMANN, 1976).

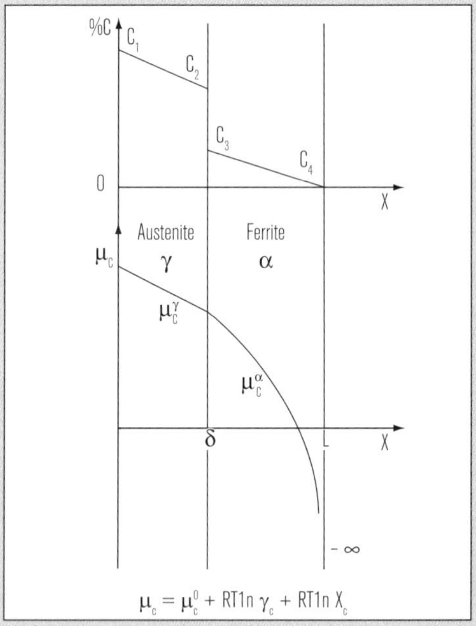

Schematic diagram giving the variation of composition and chemical potential of carbon, $\mu_c = \mu_C^o + RT\ln\gamma_c + RT\ln X_c$ inside the steel sheet.

Example

A classic proposition that characterizes the actual driving force of diffusion is due to Darken and Gurry (1953) and is known as the phenomenon of *up-hill diffusion*. Silicon behaves as a substitutional in the iron matrix while the carbon behaves as interstitial. For this reason variations in terms of concentration of silicon persist for long periods of time as compared to those of carbon. Two pieces of steel, one with initial composition corresponding to 3.8% Si and 0.478% C (left in the figure below) and another with initial composition of 0.44% C, were welded and kept at 1050°C for several days. Both parts become austenitic at this temperature and hence eventually only one phase should be present. The figure below indicates the distribution of solute carbon after 13 days.

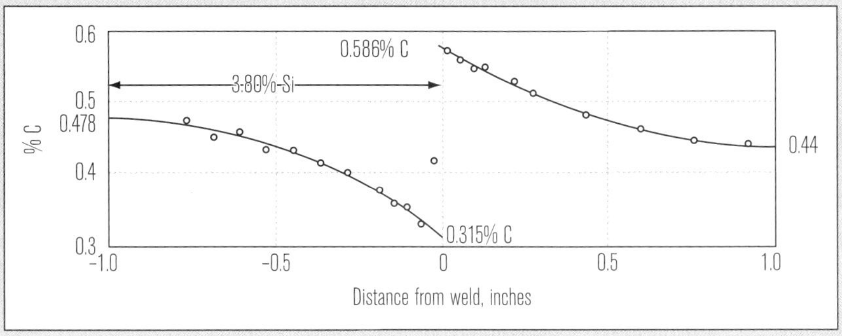

Distribution of carbon as a result of diffusion starting with uniform distribution initially – Carbon migrated from Fe-Si-C alloy to Fe-C alloy (DARKEN and GURRY, 1953).

The thermodynamic behavior of silicon is such that in ferrous alloys, it increases the activity of carbon. Although the pieces in the left has 0.478% C to start with and the one at right has 0.44% C the activity of C in the left number is significantly higher than the activity in the right number.because of presence of Silicon. This trend is maintained for a prolonged period because the diffusion of silicon is comparatively slow, There is a chemical potential gradient because of higher carbon activity in the piece at left (higher value) promoting migration of carbon by diffusion although the right piece has an initial carbon higher than the left thus justifying the phrase *up-hill diffusion*.

16.3 Other Driving Forces of Diffusion

Up to now ordinary diffusion, with single chemical potential gradient has been discussed. In many situations of technological importance, ,ions migrate under the action of external electrical fields. However electrical fields have no effect on complex or neutral substances but have a key role in the operations of electrolysis, electrorefining etc.

A ceramic material can be considered, partially ionic, consisting of anionic and cationic sub-networks, accessible only to their pairs. For example a cation can diffuse in the structure under the action of a chemical potential gradient. If only movement of cations takes, place the resulting structure would show a charge separation, some regions having positive net charge, others with net negative charges. This charge separation would lead to the emergence of an electric field which, which in turn would try to restore the lost balance through the movement of anions. Hence, cations and anions move together causing coupled diffusion, in order to maintain electroneutrality of the structure. Thus an ordinary diffusion can produce an electric field and this leads to the movement of other species. For understanding this phenomenon, one starts with analyzing various driving forces, in a particular system. It can be considered that the various driving forces or potentials are additive. For example, in the case of an electrochemical system there are two driving forces constituting the electrochemical potential.

Electrochemical potential = Chemical potential + Electrical potential

$$\eta_i = \mu_i + z_i F \varphi \tag{16.39}$$

where:

- η_i is the electrochemical potential,
- μ_i the chemical potential,
- z_i is the ionic charge,
- F the Faraday constant,
- φ the electric potential.

So the force acting on the ion can be written on the basis of Nernst Einstein equation (equation 16.35):

$$\frac{F_i}{N_o} = -\frac{1}{N_o}\frac{d(\mu_i + z_i F\varphi)}{dx} \left(\text{N/atom}\right) \tag{16.40}$$

and then, from the concept of mobility (CHIANG, BIRNIE and KINGERY, 1977).

$$v_i = -\frac{U_i}{N_o}\frac{d(\mu_i + z_i F\varphi)}{dx} \tag{16.41}$$

which implies the definition of flow as $j_i = C_i v_i$, and consideration of dilute solution:

$$j_i = C_i v_i = C_i\left[-\frac{U_i}{N_o}\left(RT\frac{1}{C_i}\frac{dC_i}{dx} + z_i F\frac{d\varphi}{dx}\right)\right] \tag{16.42}$$

where $E = \frac{d\varphi}{dx}$ represents, by definition, the electric field. The expression above is usually rewritten as:

$$j_i = C_i v_i = -k_B T U_i\left(\frac{dC_i}{dx} + \frac{z_i F C_i}{RT}\frac{d\varphi}{dx}\right) \tag{16.43}$$

$D_i = k_B\ T\ U_i$ being the diffusion coefficient, by definition. This equation is similar to the equation 16.38 for ordinary chemical diffusion in terms of Boltzmann`s constant and mobility of the species.

For a structure consisting of a cationic sub-lattice of ion 1, with the valency z_1 and an anionic sub lattice of ion 2, with the valency z_2 and neglecting convection one can estimate the rate of transport as:

$$N_1 = -D_1\left(\frac{dC_1}{dx} + \frac{z_1 F C_1}{RT}\frac{d\varphi}{dx}\right) \tag{16.44}$$

$$N_2 = -D_2\left(\frac{dC_2}{dx} + \frac{z_2 F C_2}{RT}\frac{d\varphi}{dx}\right) \tag{16.45}$$

Although the value of the electric field is not known, one can consider that all transport occurs such that electro-neutrality is maintained, ie:

$$C_1\ z_1 + C_2\ z_2 = 0 \tag{16.46}$$

$$C_2 = -\frac{z_1}{z_2}C_1 \tag{16.47}$$

$$\frac{dC_2}{dx} = -\frac{z_1}{z_2}\frac{dC_1}{dx} \tag{16.48}$$

Moreover, in this example, the global electrical charge is zero, which implies:

$$N_1\ z_1 + N_2\ z_2 = 0 \tag{16.49}$$

and then the equations (16.44) and (16.45) can be given as:

$$\mathbf{N}_1\ z_1 = -D_1\left(z_1\frac{dC_1}{dx} + \frac{z_1FC_1z_1}{RT}\frac{d\varphi}{dx}\right) \tag{16.50}$$

$$\mathbf{N}_2\ z_2 = -D_2\left[z_2\left(-\frac{z_1}{z_2}\frac{dC_1}{dx}\right) + \frac{z_2Fz_2}{RT}\left(-\frac{z_1}{z_2}C_1\right)\frac{d\varphi}{dx}\right] \tag{16.51}$$

From the above equations 16.49 to 16.51 one can get (since $N_1\ z_1 + N_2\ z_2 = 0$):

$$-D_1\{z_1\frac{dC_1}{dx} + \frac{z_1FC_1z_1}{RT}\frac{d\phi}{dx}\} +$$
$$-D_2\{z_2(-\frac{z_1}{z_2}\frac{dC_1}{dx}) + \frac{z_2Fz_2}{RT}(-\frac{z_1}{z_2}C_1)\frac{d\phi}{dx}\} = 0$$

Transposing and simplifying:

$$\frac{z_1FC_1}{RT}\frac{d\varphi}{dx} = \frac{D_1 - D_2}{D_1\ z_1 - D_2\ z_2}z_1\frac{dC_1}{dx} \tag{16.52}$$

Substituting this expression in the original transport equation 16.44:

$$\mathbf{N}_1 = -D_1\{\frac{dC_1}{dx} + \frac{z_1FC_1}{RT}\frac{d\phi}{dx}\}$$

and considering the necessary simplifications, one can get:

$$N_1 = -\frac{D_1\ D_2(z_1 - z_2)}{D_1\ z_1 - D_2\ z_2}\frac{dC_1}{dx} \tag{16.53}$$

Where, the diffusion coefficient is:

$$\tilde{D} = \frac{D_1\ D_2(z_1 - z_2)}{D_1\ z_1 - D_2\ z_2} \tag{16.54}$$

Therefore, any species, cations or anions, move complimentary to one another regardless of the relative values of individual diffusion coefficients If, specie 1 has much more diffusing capacity compared to specie 2, that is $D_1 >>> D_2$ then equation 16.54 can be simplified to:

$$\tilde{D} \cong \frac{D_2(z_1 - z_2)}{z_1}$$

In this case the specie that is slow in diffusion controls the process.

According to thermodynamics of irreversible processes the flow of a certain entity (heat, mass, etc.) can be caused by a number of driving forces. It is shown that these are generally related to potential gradients (chemical, electrical etc.).

Neglecting contributions of higher orders it can be written (LASAGA, 1998):

$$J_i = \sum_j L_{ij} X_j \tag{16.55}$$

where:

- J_i represents the flow of entity i;
- X_j the driving force j;
- L_{ij} is the phenomenological coefficient in which the subscripts indicate the importance of this driving force.

This formulation does not explain why a flow of a given entity is caused by some driving force. However it provides a basis for the calculation of the effect of coupling between different phenomena and the relative importance of the driving forces. For example, in a system, where two kinds of irreversible processes are occurring, like mass transport and heat transport it is reasonable to write that each case is a function both driving forces.

$$q = L_{11} X_1 + L_{12} X_2 \text{ (for heat flux)} \tag{16.56}$$

$$j = L_{21} X_1 + L_{22} X_2 \text{ (for mass flow)} \tag{16.57}$$

where X_1 and X_2 represent, respectively, the driving forces for the transport of heat and mass, while L_{ij} denotes a phenomenological coefficient. According to Onsager relation $L_{ij} = L_{ji}$. This factor considers the mutual interference of two processes, L_{12} the amount of heat carried by the species that diffuses and L_{21} the velocity of diffusion due to thermal gradient, or termodifusivity.

Although the mathematical formalisms may vary, one can identify the following driving forces (LASAGA, 1998):

- thermal driving force $\nabla \frac{1}{T}$ (16.58)

t• driving force for ordinary diffusion $-\nabla \frac{\mu}{T}$ (16.59)

• driving force of electro-chemical nature $-\nabla \frac{\varphi}{T}$ (16.60)

• driving force due to chemical reactions occurring in the system $-\frac{1}{T}\Delta G$ (16.61)

As defined before T is the absolute temperature, μ is the chemical potential, ϕ is the electric potential and ΔG is the Gibbs Free Energy.

Comparison with the phenomenological law of Fourier for heat conduction enables one to identify:

$$L_{11} = K\, T^2 \tag{16.62}$$

where, K is the thermal conductivity of the material.

Similarly, Fick's Law yields (LASAGA, 1998):

$$L_{22} = \frac{D}{\frac{\nabla \ln \gamma}{\nabla C} + \frac{1}{C}} \tag{16.63}$$

Note that the phenomenological coefficients are not necessarily constants. For example L_{22} would be a complex function of temperature and composition. It is also observed experimentally that the direct effects (temperature gradient on heat flux, composition gradient on the flow of species) are more pronounced than the cross effects. (COULSON and RICHARDSON, 1993).

Thus, for the case of unidirectional transport of a specie A in a gaseous phase AB, subject to gradients of composition and temperature, one can write:

$$j_A = -D\left(C_T \frac{dX_A}{dz} + \delta \, \frac{1}{T}\frac{dT}{dz} \right) \tag{16.64}$$

where the thermal diffusivity can be set as $D_T = \delta D$. When a temperature gradient is maintained in the gas phase, it establishes a gradient of composition such that the lightest element focuses on the warmer side, while the heavier element gets concentrated in the colder part. This is called the *Soret* effect and the above expression can be applied (COULSON and RICHARDSON, 1993).

Thermal diffusion factor α for some pairs of gases, is defined as:

$$\alpha = \frac{\delta}{X_A(1 - X_A)\,C_T} \tag{16.65}$$

where:

- δ is the phenomelogical coefficient for mass flow due to temperature gradient cross effects;
- X_A is the molar fraction of gas A;
- C_T the total molar concentration.

Values of thermal diffusion factor for some pairs of gases are given Table 16.1.

Table 16.1 - Thermal diffusion factor for some pairs of gases (COULSON and RICHARDSON, 1993).

System		Temperature	Mole fraction	
A	B	°K	X	α
D_2	H_2	288-373	0.48	0.17
He	H_2	273-760	0.50	0.15
N_2	H_2	288-373	0.50	0.34
Air	H_2	258	0.53	0.26
O_2	H_2	90-294	0.50	0.19
CO	H_2	288-373	0.50	0.33
CO_2	H_2	288-456	0.50	0.28
C_3H_8	H_2	230-520	0.50	0.30
N_2	He	287-373	0.50	0.36
		260	0.655	0.37
Ar	He	330	0.90	0.28
		330	0.70	0.31
		330	0.50	0.37
Ne	He	205	0.46	0.31
		330	0.46	0.315
		365	0.46	0.315
O_2	N_2	293	0.50	0.018

The opposite effect of inducing a thermal gradient by maintaining a composition gradient is called the *Dufour effect.*

Besides this, a rigorous mathematical treatment for diffusion is provided by the thermodynamics of irreversible processes, considering various phenomena such as thermoelectric, electrokinetic effects, etc. Thermoelectric phenomena are related to Seebeck-Thomson and Peltier effects etc. (which deal with the generation of a thermocouple emf). The electrokinetic effect deals with the interaction between pressure forces and the electric potential, in relation to the movement of an electrolyte. These are specialized topics and are not covered in this text

References

BALLUFFI, R.W.; ALLEN, S.M.; CARTER, W.G. *Kinetics of materials.* Hoboken: Wiley-Interscience, 2005.
CHIANG, Y.; BIRNIE, D.; KINGERY, W.D. *Physical ceramics*: principles for ceramic science in engineering. New York: John Wiley & Sons, 1977.
COULSON, J.M.; RICHARDSON, J.F. *Chemical engineering.* New York: Pergamon, 1993. v. 1: Fluid flow, heat transfer and mass transfer.
DARKEN, L.S.; GURRY, R.W. *Physical chemistry of metals.* New York: McGraw-Hill, 1953.
DEHOOF, R. *Thermodynamics in material science.* 2. ed. New York: Taylor & Francis, 2006.
KINGERY, W.D.; BOWEN, H.K.; UHLMANN, D.R. *Introduction to ceramics.* 2 ed. New York: John Wiley & Sons, 1976.
LASAGA, A.C. *Kinetic theory in the earth sciences.* New Jersey: Princeton University Press, 1998.
MILLS, A.F. *Basic heat and mass transfer.* New Jersey: Prentice Hall, 1999.
POIRIER, D.H.; GEIGER, G.H. *Transport phenomena in materials processing.* Warrendale: TMS, 1994.
SHEWMON, P. *Diffusion in solids.* 2.ed. New York: John Wiley & Sons, 1989.

Exercises

1 - In an electrorefining cell for a given metal the main reactions are the dissolution, at the vicinity of the impure anode according to the reaction:

$$M \rightarrow M^{+z} + ze$$

and deposition of the same metal, at the purified cathode according to:

$$M^{+z} + ze \rightarrow M.$$

In the ideal cell, there are no parallel reactions, the net effect being cation transport from the anode to the cathode. Consider that in order to maintain the local electroneutrality of the electrolyte, cation has to be transported together with a counter-ion (for example, in a sulphate based solution, SO_4^{-2}). In net terms, the flow of counter-ion must be zero to avoid accumulation in the system. Considering driving forces due to gradients in composition and electric potential, relate the rate of deposition of the metal in the external current.

2 - Estimate the temperature difference that must be maintained at both ends of a capillary 0.3 m long, containing a gas mixture N_2 - H_2, so that the composition at the extremes differ by 5%.

3 - *Up-hill diffusion* is considered to occur in cases where increasing values of composition does not imply increasing values of chemical potential. In *up-hill* diffusion, the diffusive transport of species occurs in the direction of increasing concentration. Hence it becomes necessary that the diffusion coefficient has a negative sign. The following figure relates to a portion of a phase diagram which exhibits a gap of miscibility in solid state. Under the conditions described by the curve of Gibbs free energy shown in the figure, an alloy of composition X decomposes spontaneously into two phases, with composition X_1 and X_2. Give the relationship between gradients of composition and chemical potential of component B, plot of chemical potential and determine the range of composition in which the diffusion coefficient is negative.

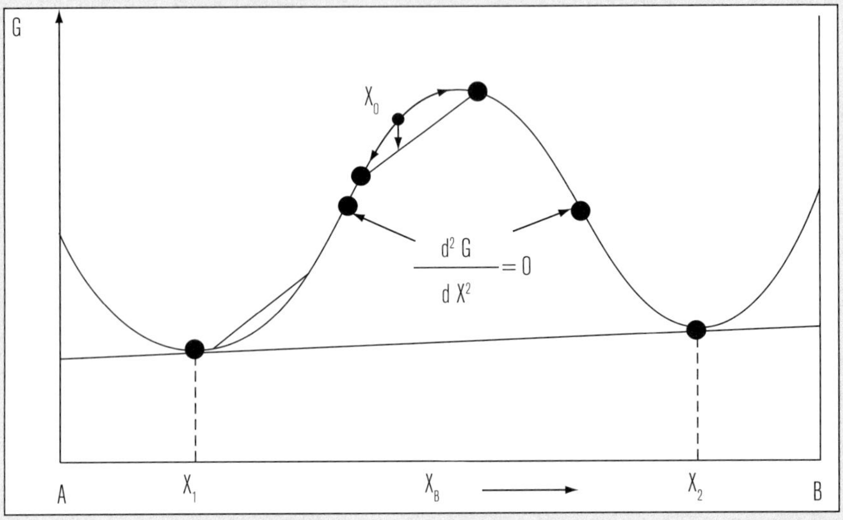

Mass Transport Parameters

Chapter 17

In this chapter, transport parameters like diffusion coefficients and coefficients of mass transport are discussed considering specific situations like:

- Diffusion of an impurity or an interstitial;
- Diffusion through a preferred path, as in the case of a grain boundary;
- Self-diffusion, which does not translate into gradients of composition, is a measure of mobility of one of the species inside the structure. This is important in the studies of recrystallization, grain growth and other areas.

Similarly mass transfer coefficients have been considered for specific applications, depending on the characteristics of the fluid and the object in contact with it.

These aspects including the general equation of transport of species are discussed more elaborately in Chapter 18.

17.1 Diffusion Coefficients

It is possible to determine the diffusion coefficient of a specie experimentally in a particular system. There is lot of data in respect of experimental techniques and actual values of diffusion coefficients in the literature.

Depending on the species and the system, functional relationships can be developed theoretically also. There are correlations between various properties such as viscosity and diffusivity. There are also models for evaluation of diffusion coefficients based on first principles and experimental data as well as semi-empirical methods for determining diffusivity.

Experimental values of diffusivity can have a certain degree of uncertainty. In liquids and gases, it is difficult to separate convection and diffusion. In solids it may depend on the thermo-mechanical history, which determines the distribution of grain sizes and defects. These have an effect on the diffusivity values. Although the units are the same the numerical value may also depend on the choice of the reference velocity.

Aspects of diffusion can be seen both from the macroscopic and microscopic points of view. In the macroscopic treatment, the transport velocity is characterized by a diffusion coefficient, which is determined without much consideration as to what happens in the atomic level. In the atomic or molecular level, diffusivity of defects such as vacancies, interstitials, self or autodiffusion of an atom at its own characteristic crystal lattice are subjects of interest. Diffusion coefficients can also be measured inside a grain, at the grain boundaries, or on the surface of a sample. There are other types of diffusion such as pore diffusion, where a gas diffuses through a porous product layer in a certain process. Diffusion through vey small pores of molecular dimensions can also take place (Knudsen diffusion).

The microscopic treatment enables consideration of simultaneous movement, for example, of a vacancy and a solvent atom. Vacancies are always present and, in this case, the movement of a solvent atom in the crystal structure corresponds to a movement in the opposite direction of a vacancy. This association facilitates the transport in which the movement of solvent through interstitial positions constitutes what is known as self-diffusion. One can also cite the movement of various ionic species in an electrolyte so that electroneutrality is maintained. The development of diffusion models should take into account the peculiarities of each system for satisfactory and reliable determination of diffusivity values.

For the evaluation of influence of temperature on diffusion coefficient Arrhenius equation is widely used.

$$D = D_o e^{-\frac{\Delta E}{RT}} \tag{17.1}$$

where:

- D is the diffusion coefficient;
- D_0 is the pre-exponential factor;
- ΔE is the activation energy of diffusion process;
- R the universal gas constant;
- T the absolute temperature.

For solids, there are theoretical justifications for the use of above equation to evaluate diffusion coefficients as a function of temperature. However in the case of liquids and gases, the validity is quite limited.

17.2 Binary Diffusivity in Gases

As in the case of viscosity and thermal conductivity of binary gas phases, the diffusivity can be estimated by the kinetic theory, which takes into account the motion and collision of non-rigid molecules of gas and includes the energy parameters derived from Lennard-Jones Potential, as explained in the chapter on viscosity (POIRIER and GEIGER, 1994).

$$D(m^2.s^{-1}) = 1.86 \times 10^{-7} \frac{T^{1.5}(1/M_1 + 1/M_2)^{0.5}}{P\, \sigma_{12}^{\ 2}\ \Omega} \qquad (17.2)$$

where:

- T (K) is the absolute temperature;
- M_1 and M_2 (g/mol) are molecular weights of gases in the bianary system;
- P (atm) pressure.

Specific properties of gases important in the kinetic theory of gases such as collision diameter σ_{12} (Å), can be calculated from:

$$\sigma_{12} = (\sigma_1 + \sigma_2)/2 \qquad (17.3)$$

Individual values of collision diameters σ are listed in Table 4.1 (Chapter 4) together with the parameter, collision integral. For a number of gases Ω is a function of $\frac{\varepsilon}{K_B}$. This function is an unique relationship between temperature, and energy parameter ε (minimum energy at equilibrium), as given in Table 17.1. In this case the energy of the pair is calculated as the geometric mean of its components:

$$\frac{\varepsilon}{K_B} = \sqrt{\frac{\varepsilon_1}{K_B}\ \frac{\varepsilon_2}{K_B}} \qquad (17.4)$$

The relationship between collision integral Ω and $\frac{\varepsilon}{K_B}$ can be calculated using numerical regression of data given in Table 17.1 resulting in the following equations:

$$\Omega - 1.444\left(\frac{k_B T}{\varepsilon}\right)^{-0.5116}, \text{ for } \frac{k_B T}{\varepsilon} < 1.30 \qquad (17.5)$$

$$\Omega = 1.0824\left(\frac{k_B T}{\varepsilon}\right)^{-0.1609}, \text{ for } \frac{k_B T}{\varepsilon} > 5.0 \qquad (17.6)$$

$$\log\Omega = 0.2105\left(\log\frac{k_B T}{\varepsilon}\right)^2 - 0.4742\left(\log\frac{k_B T}{\varepsilon}\right) + 0.1553 \qquad (17.7)$$

(for intermediate values).

Besides the above equations proposed by Chapman-Enskog , based on Lennard-Jones potential , several semi-empirical relations have been suggested. For example, the formula of Fuller (POIRIER and GEIGER, 1994; CUSSLER,1997) for the calculation of diffusion coefficient, is given by:

$$D(m^2.s^{-1}) = 10^{-7}\frac{T^{1.75}(1/M_1 + 1/M_2)^{0.5}}{P(V_1^{1/3} + V_2^{1/3})^2} \qquad (17.8)$$

where:

- T(K) represents the absolute temperature;
- M_1 and M_2 (g/mol) are molecular weights, of gases in the bianary system;
- P (atm) pressure, and V1 and V2 are diffusion volumes", (Table 17.2.).

Table 17.1 - Values of Collision Integral for calculating the diffusivity of species (CUSSLER, 1997).

$\frac{K_B T}{\varepsilon}$	Ω	$\frac{K_B T}{\varepsilon}$	Ω	$\frac{K_B T}{\varepsilon}$	Ω	$\frac{K_B T}{\varepsilon}$	Ω
0.30	2.662	1.30	1.273	2.6	0.9878	4.6	0.8568
0.35	2.476	1.35	1.253	2.7	0.9770	4.7	0.8530
0.40	2.318	1.40	1.233	2.8	0.9672	4.8	0.8492
0.45	2.184	1.45	1.215	2.9	0.9576	4.9	0.8456
0.50	2.066	1.50	1.198	3.0	0.9490	5.0	0.8422
0.55	1.966	1.55	1.182	3.1	0.9406	6	0.8124
0.60	1.877	1.60	1.167	3.2	0.9328	7	0.7896
0.65	1.798	1.65	1.153	3.3	0.9256	8	0.7712
0.70	1.729	1.70	1.140	3.4	0.9186	9	0.7556
0.75	1.667	1.75	1.128	3.5	0.9120	10	0.7424
0.80	1.612	1.80	1.116	3.6	0.9058	20	0.6640
0.85	1.562	1.85	1.105	3.7	0.8998	30	0.6232
0.90	1.517	1.90	1.094	3.8	0.8942	40	0.5960
0.95	1.476	1.95	1.084	3.9	0.8888	50	0.5756
1.00	1.439	2.00	1.075	4.0	0.8836	60	0.5596
1.05	1.406	2.1	1.057	4.1	0.8788	70	0.5464
1.10	1.375	2.2	1.041	4.2	0.8740	80	0.5352
1.15	1.346	2.3	1.026	4.3	0.8694	90	0.5256
1.20	1.320	2.4	1.012	4.4	0.8652	100	0.5130
1.25	1.296	2.5	0.9996	4.5	0.8610	200	0.4644
						300	0.4360

Gás	V_i	Gás	V_i
H_2	7.07	CO	18.9
He	2.88	CO_2	26.9
N_2	17.9	N_2O	35.9
O_2	16.6	NH_3	14.9
Ar	20.1	H_2O	12.7
A	16.1	Cl_2	37.7
Kr	22.8	SO_2	41.1

Table 17.2 - Values of diffusion volume for simple gases (CUSSLER, 1997).

Example

Compare the values predicted for D_{co/co_2}, at 1000°C and 2 atm, using Chapman and Fuller equations.

According to Table 4.1 (Chapter 4) Lennard-Jones parameters are given by:

	CO	CO_2
$\frac{\varepsilon}{K_B}$	91.7	195.2
σ	3.69	3.941

For CO-CO_2 pair $\frac{\varepsilon}{K_B}$ is the geometrical mean of individual values,

$$\varepsilon/K_B \; \sqrt{91.7 \times 195.2} = 133.79;$$

- σ is evaluated from the above table;
- $\sigma = (3.69+3.941)/2 = 3.815$ (average value for the two gases).

Hence:

$$\frac{K_B T}{\varepsilon} = 9.515$$

$$\Omega = 1.0824 \text{ x } 9.515^{-0.1609} = 0.7533$$

Substituting in:

$$D(m^2.s^{-1}) = 1.86\times10^{-7}\frac{T^{1.5}(1/M_1+1/M_2)^{0.5}}{P\sigma_{12}{}^2\Omega}$$

$$D(m^2.s^{-1}) = 1.86\ 10^{-7}\frac{1273^{1.5}(1/28+1/44)^{0.5}}{2\times3.815^2\times0.7533} = 9.314\ x\ 10^{-5}$$

From Table 17.2, the diffusion volumes for CO and CO_2 are, 18.9 e 26.9. Substituting this data in Fuller equation:

$$D\ (m^2.s^{-1}) = 10^{-7}\frac{T^{1.75}(1/M_1+1/M_2)^{0.5}}{P(V_1{}^{1/3}+V_2{}^{1/3})^2}$$

$$D(m^2.s^{-1}) = 10^{-7}\frac{1273^{1.75}(1/28+1/44)^{0.5}}{2\ (18.9^{1/3}+26.9^{1/3})^2} = 1.024\times10^{-4}$$

The difference in the values predicted by both the equations is about 10%.

17.3 Diffusivity in Liquids

Unlike gases, semi-theoretical formulation or empirical approach to estimate the diffusivity in liquids is difficult, as the nature of interactions between atoms or molecules, that constitute the liquids, vary widely due to the liberty of movement. For example, liquids can be metals which have cations embedded in a cloud of electrons, or slags which are solutions of anions and cations (ionic theory of slags) or aqueous solutions containing simple or complex ions. Therefore different mechanisms may come to act in the determination of diffusion coefficients in such liquid phases. As such, the situation is complex and no one criteria can be adopted for the evaluation of diffusivity of the species.

However Einstein-Stokes relationship is always remembered as a reference point for determining an initial estimate of diffusivity in liquids. Einstein suggested using the concept of mobility in the case of a solution in which the species of interest is the solute. In this case, the force produced by the chemical potential gradient could be equated with the reaction force which opposes the fluid motion of the solute, as given by Stokes' law, namely:

$$\frac{F_i}{N_o} = -k_BT\frac{1}{C_i}\frac{dC_i}{dx} = 6\pi\eta Rv_i \qquad (17.9)$$

- F_i force acting on the specie;
- N_0 Avogadro number;
- K_B Boltzmann's constant;
- T is absolute temperature;
- C_i concentration of the specie;
- x direction of the movement;
- η viscosity of the medium;
- R atomic radius of the specie;
- V_i velocity of the specie.

From 17.9 the velocity and flux of specie i, can be given as:

$$v_i = -\frac{k_B T}{6\pi\eta R}\frac{1}{C_i}\frac{dC_i}{dx} \tag{17.10}$$

$$j_i = C_i v_i = C_i\left(-\frac{k_B T}{6\pi\eta R}\frac{1}{C_i}\frac{dC_i}{dx_i}\right) = -D_i\frac{dC_i}{dx} \tag{17.11}$$

From 17.10 and 17.11 the D is given in terms of viscosity (Einstein-Stokes equation):

$$D = \frac{k_B T}{6\pi\eta R} \tag{17.12}$$

Two difficulties arise with respect to the use of this equation. The first, is to assign a value to the radius R of the specie that diffuses, as in an aqueous solution where, the ions show variable number of solvation layers. (MESPLEDE and QUEYREL,1996). The second problem is the consideration of the medium, surrounding the solute as being continuous on the basis of Stokes' law. Although this assumption has its limitations, the results of its application are reasonable, as shown in Table 17.3.

Metal	Temperature (°C)	Einstein (nm)	Pauling (nm)*
Na	110 – 630	1.51 – 4.53 Å	1.57 Å
Hg	274 – 364	1.20 – 1.24	1.44
In	448 – 1013	1.36 – 3.06	1.50
Ag	977 – 1397	1.37 – 1.37	1.34
Zn	394 – 837	1.29 – 0.75	1.25
Sn	575 – 956	1.30 – 2.22	1.42

Table 17.3 - Atomic radii determined (PAULING, 1960) and calculated for self diffusion of liquid metals (BIRD, STEWART and LIGHTFOOT, 2002).

Table 17.4 presents experimental values of diffusivity of some elements in liquid iron at different temperatures, together with some activation energy values. As expected, the values are species-specific, varying according to the range of composition and temperature. Some of these can be compared with their counterparts in solid-state diffusion at a later stage.

Silicate slags are liquid oxide solutions with complex structure, because of partial sharing of oxygen atoms of the fundamental units of SiO_4^{4-}. The degree of sharing depends on the temperature and amount of basic oxides present. Diffusivity and many other properties of slags depend on the structure, determined by the presence of basic oxides and temperature.

Table 17.5 presents diffusivity data for some elements at different temperatures in slags in addition to activation energy values.

Table 17.4 - Difusivity of some elements in liquid iron on molar basis (ELLIOT, GLEISER and RAMAKRISHNA, 1963).

Substance	Medium	Concentration	Temperature (°C)	D (m^2/s)	Do (m^2/s)	ΔE (KJ/mole)
C	Fe	0.03% C	1550	7.9×10^{-9}		
C	Fe	2.1% C	1550	7.8×10^{-9}		
C	Fe	3.5% C	1550	6.7×10^{-9}		
Mn	Fe	0 to 4.5% Mn	1725	17×10^{-8}		
N	Fe		1600	3.77×10^{-9}		
P	Fe		1550	4.7×10^{-9}		33.470
S	Fe-C	0.08 to 0.64% S, saturated with C	1390-1560		2.8×10^{-8}	31.400
S	Fe	0.96 ~ 1.25% S	1560-1670		4.9×10^{-8}	35.980
Si	Fe-C	1.5% Si, saturated with C	1400-1600		2.4×10^{-8}	34.300

Table 17.5 - Diffusivity in liquid slags (ELLIOT, GLEISER and RAMAKRISHNA, 1963).

Element	Composition of the slag	Temperature °C	D cm^2/sec	Do cm^2/sec	ΔE kJ/mole
Al^{26} (self diffusion)	43% CaO-10% Al2O3-47% SiO2	1440-1520		4.3×10^4	355.64
Ca	30% CaO-69, 5% P2O5	980	3.0×10^{-6}		
Ca	30% CaO-69, 5% P2O5	1000	4.7×10^{-6}		
Ca	39.4% CaO-21.2% Al2O3-38. 8% SiO2	1300	4.4×10^{-7}		234.30
O^{17} (self diffusion)	40% CaO-20% Al2O3-40% SiO2	1372-1535		1.2×10^6	367.77
P	39.4% CaO-21.2% Al2O3-38.8% SiO2	1300	10×10^{-7}		194.97
P	39.4% CaO-21.2% Al2O3-38.8% SiO2	1400	2×10^{-6}		194.97
P	39.4% CaO-21.2% Al2O3-38.8% SiO2	1450	4.5×10^{-6}		194.97
P	39.4% CaO-21.2% Al2O3-38.8% SiO2	1500	4.4×10^{-6}		194.97
S-1,5% S	50.3% CaO-10.4% Al2O3-39.3% SiO2	1445-1580		1.4	205.02
S-1,5% S	42.5% CaO-9.6% Al2O3-47.9% SiO2	1440	0.8×10^{-6}		
Si (self diffusion)	39% CaO-21 % Al2O3-40% SiO2	1365	4.7×10^{-8}		292.88

17.4 Diffusivity in Solids

Table 17.6 gives the diffusivity values of some elements in iron and its alloys. At least qualitatively, the wide range of variation and temperature dependence can be explained. Hydrogen, relative to iron and other base metals, is an atom of small dimensions, and in fact has the smallest atomic radius among the elements in the periodic table. It occupies interstitial positions. The mechanism of movement between interstitial positions within a crystal lattice is predominant. The diffusion of an interstitial atom does not depend strongly on crystal defects caused by thermal agitation. In fact thermal agitation is present in all solids above zero degree K. The movement of an interstitial from one site to another (Figure 17.1), requires only little distortion resulting in low activation energy.

Therefore, interstitials diffuse rapidly, compared with the diffusion of a substitutional atom. There is little influence of temperature. Furthermore, atoms of comparable size to the atoms of the matrix cannot be interstitial. In this case distortions are required to accommodate them interfering with the organization of the crystal lattice. The movement of such a type requires a ring like rotating cooperative mechanism (Figure 17.2) resulting in diffusion (in bcc metals).

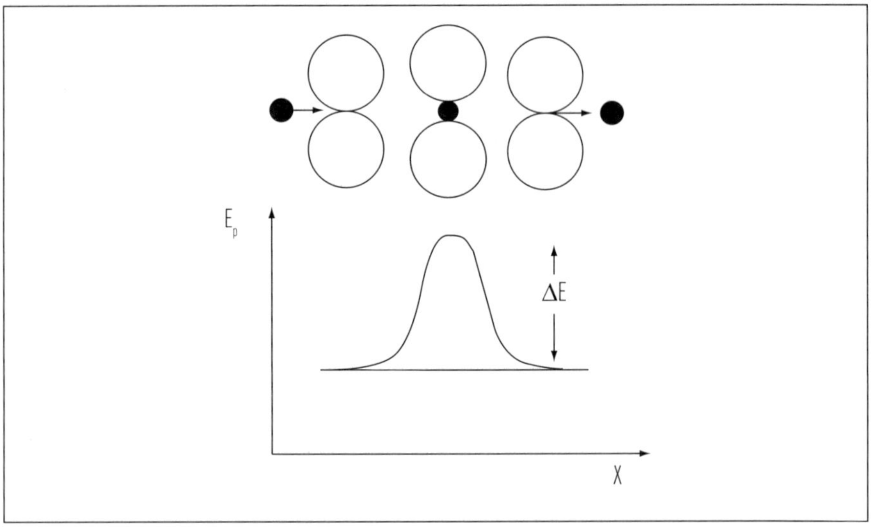

Figura 17.1 - Energy variation relative to the movement of an interstitial.

The second mechanism is the movement of species through the mechanism of vacancy (Figure 17.2). Such a mechanism would be energetically more favorable than in the case of ring like rotating cooperative mechanism It is expected in this case lower diffusivity and higher thermal activation relative to the interstitial movement. The diffusion of manganese in iron can be cited as an example of this type.

Table 17.6 contains data for interstitial and intragranular diffusion (within a grain or single crystal in the absence of diffusion at grain boundaries and surface). Volume (under Type in Table 17.6) indicates cases where both intergranular diffusion (in preferential paths corresponding to the grain boundaries) and intragranular are operative, on the basis of experimental evidence. However it has not been possible to determine these contributions separately. The intragranular diffusion mechanism can operate through interstitial and/or vacancies.

Table 17.6 - Diffusivities of some elements in solid and liquid iiron and its allys (molar basis) (ELLIOT, GLEISER and RAMAKRISHNA, 1963).

Element	Medium	Concentration	Temperature °C	Type	D_o (m^2/s)	ΔE (KJ/mol)
Al	Fe γ		950 ~ 1100		3×10^{-3}	234.000
B	Fe	< 0.05% B	39 ~79	Interstitial	10^{-9}	62.760
B	Fe	0.0003 ~ 0.0035 % B	700 ~ 850	Volume	100	259.400
B	Fe	0 ~ 0.04 %B	950 ~ 1300	Volume	2×10^{-7}	87.860
C	Fe α	0.47 %C	35 ~ 800	Intragranular	2×10^{-6}	84.000
C	Fe α		350 ~ 800	Volume	6.2×10^{-7}	80.330
C	Fe α	0.02 %C	24 ~ 74	Interstitial	3.16×10^{-6}	85.350
C	Fe γ		Stable	Intragranular	7×10^{-6}	133.890
C	Fe γ		900 ~ 1050	Volume	10^{-5}	135.560
C	Fe γ	0.2 ~0.7 %C	1000 ~ 1200		$(0.4 + 0.08\%C)10^{-4}$	131.800
Co	Fe α		700 ~ 850	Volume	4×10^{-5}	225.900
Cr	Fe α	Dilute	Stable	Volume	3	343.000
Cr	Fe γ	Dilute	Stable	Volume	1.8	405.850
H	Fe α	Dilute	15 ~ 900	Volume	0.885×10^{-7}	12.750
H	Fe γ	Dilute	Stable	Volume	1.1×10^{-6}	41.650
Mn	Fe, 0 ~20% Mn		950 ~ 1450		$(0.486 + 0.011 x\%Mn) 10^{-4}$	276.000
N	Fe α	Dilute	0 ~ 600		0.0066×10^{-4}	77.800
N	Fe γ		Stable		3.3×10^{-8}	144.750
Ni	Fe, 0 ~20% Ni		1050 ~ 1450		$(0.344 + 0.012 \%Ni) \times 10^{-4}$	282.500

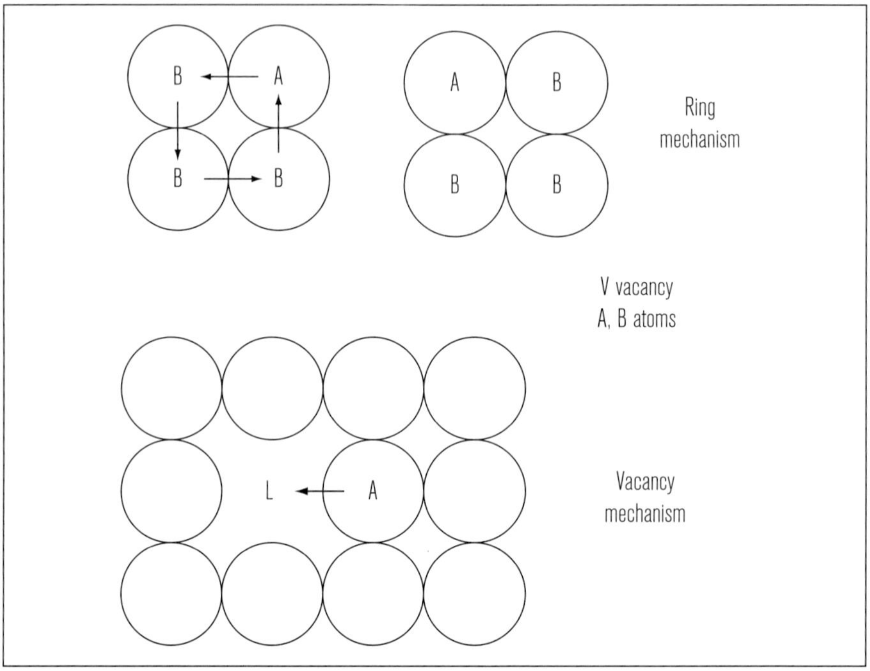

Figure 17.2 - Mechanism of diffusion of noninterstitial atoms.

Diffusion coefficients can be defined or modified to simplify the mathematical treatment in specific situations. For example, the interdiffusion coefficient between two solids A and B can be considered based on the classical experiment of Kirkendall (POIRIER and GEIGER, 1994). In this, two solid pieces, composed of pure A and B mutually soluble, are welded to the top after insertion of a marker (insoluble material) in between. The sample is then subjected to high temperatures in order to activate the diffusion process. Generally, one should expect that the rate of diffusion of A in B (measured by the coefficient D_{AB}) is different from the diffusion of B in A (measured by the coefficient D_{BA}). In the case of $D_{AB} >>> D_{BA}$ in the course of time, a configuration in respect of concentration gradient for species A would result, as represented in Figure 17.3. (Kirkendall effect). This figure implies that the diffusive process brings about a visible movement of convection, which can be detected, from the placement of markers. Hence convection and diffusion cannot be completely separated. The mathematical treatment proposed for this phenomenon starts with the flow equation for each of the two metals, considering diffusion and convective flow with velocity v.

$$N_A = C_A\ v - D_A \frac{dC_A}{dx} \tag{17.13}$$

$$N_B = C_B\ v - D_B \frac{dC_B}{dx} \tag{17.14}$$

Balance of conservation of mass over a control volume corresponding to a slice of the bar, with thickness Δx, and area of cross section dS, (Figure 17.4) assuming absence of chemical reactions, is given by:

Rate of accumulation = Rate of input – Rate of output

$$\frac{d}{dt}(dS\ \Delta x\ C) = dS\ N_x - dS\ N_{x+\Delta x} \tag{17.15}$$

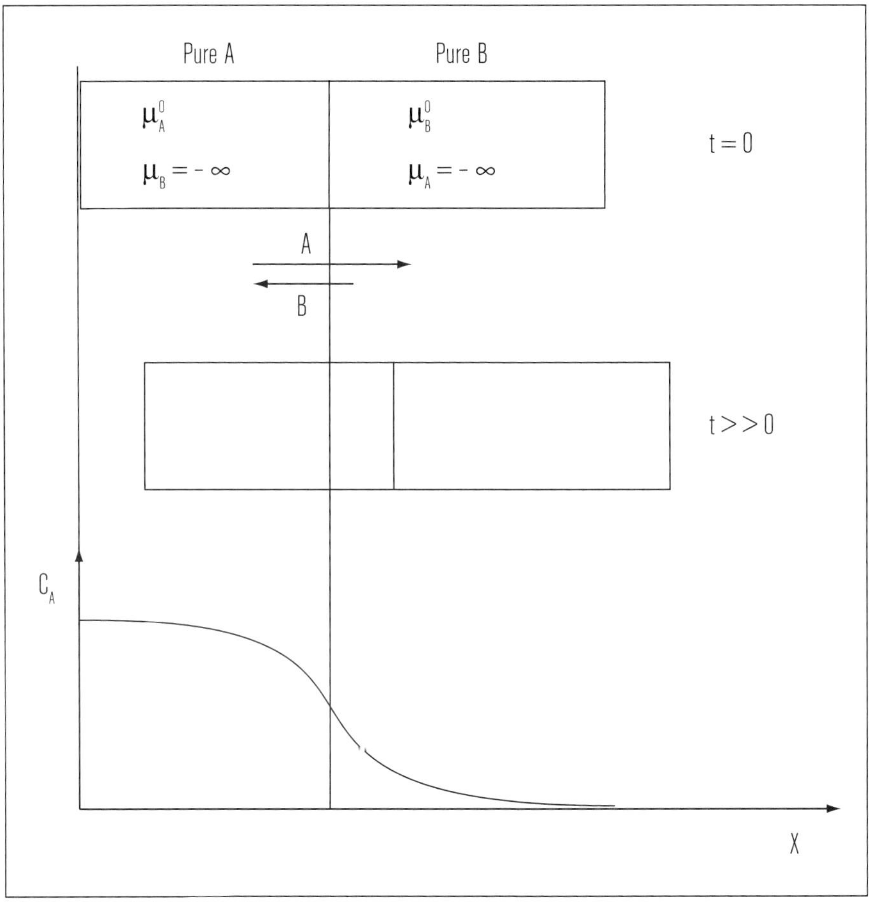

Figure 17.3 - Convective movement induced by diffusion or Kirkendall effect (SHEWMON, 1989; POIRIER and GEIGER, 1994).

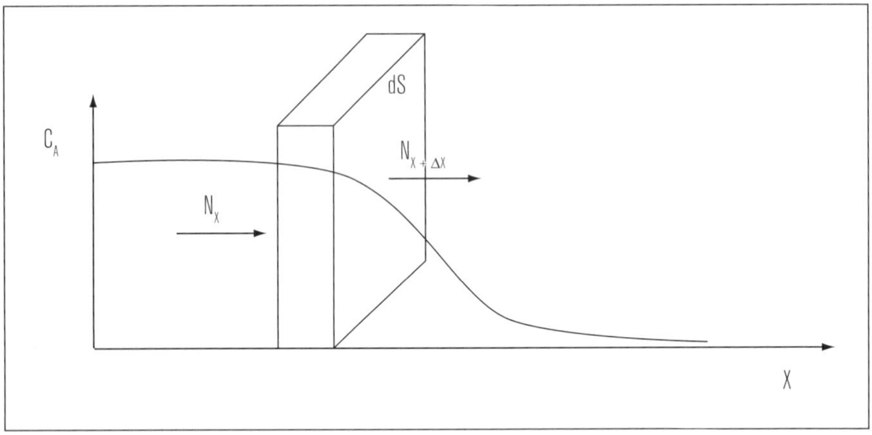

Figure 17.4 - Control volume for balance of conservation of mass.

Taking limits when the thickness tends to zero:

$$\frac{\partial}{\partial t} C = -\frac{\partial}{\partial x} N \tag{17.16}$$

$$\frac{\partial}{\partial t} C = -\frac{\partial}{\partial x}\left(\mathbf{v}C - D\frac{\partial C}{\partial x} \right) \tag{17.17}$$

Considering convection is absent and the diffusion coefficient constant the above equation becomes 2nd law of Fick:

$$\frac{\partial C}{\partial t} = \frac{\partial^2 C}{\partial x^2} \tag{17.18}$$

In cases where convection cannot be ignored, equations should be written for all the species.

$$\frac{\partial}{\partial t} C_A = -\frac{\partial}{\partial x}\left(\mathbf{v}C_A - D_A\frac{\partial C_A}{\partial x} \right) \tag{17.19}$$

$$\frac{\partial}{\partial t} C_B = -\frac{\partial}{\partial x}\left(\mathbf{v}C_B - D_B\frac{\partial C_B}{\partial x} \right) \tag{17.20}$$

Summing up the contributions:

$$\frac{\partial}{\partial t} (C_A + C_B) = -\frac{\partial}{\partial x}\left[v\,(C_A + C_B) - D_A\frac{\partial C_A}{\partial x} - D_B\frac{\partial C_B}{\partial x} \right] \tag{17.21}$$

Volumetric concentrations of A and B remain constant throughout the process. That is to say, that the vacancy concentration is insignificant and hence it can be neglected. Hence the diffusive process need not consider this aspect.

Hence it can be written:

$$C_A + C_B = \text{constant} \quad (17.22)$$

$$\frac{\partial}{\partial t}(C_A + C_B) = 0 \quad (17.23)$$

$$\frac{\partial}{\partial x}(C_A + C_B) = 0 \quad (17.24)$$

This implies:

$$v\,(C_A + C_B) - \frac{\partial C_A}{\partial x}(D_A - D_B) = \text{constant} \quad (17.25)$$

The value of this constant can be evaluated as being equal to zero, assuming a point some distance away from the junction, where convection is negligible and concentration gradient is zero. Then, for any other position:

$$v = \frac{D_A - D_B}{C_A + C_B}\frac{\partial C_A}{\partial x} \quad (17.26)$$

The above equation for v can be introduced in the equation 17.13 for flux of A:

$$N_A = -(X_A D_B + X_B D_A)\frac{dC_A}{dx} \equiv -\tilde{D}\frac{dC_A}{dx} \quad (17.27)$$

In the above equation the definition of interdiffusion coefficient is $\tilde{D} = X_A D_B + X_B D_A$.

One can note that convection is not neglected. However the resulting equation has the same form as the first law of Fick. The interdiffusion coefficient is inherently a function of composition and hence varies along the direction of diffusion.

In the particular case of a dilute solution, for the solute A, the above coefficient can be approximated to $\tilde{D} \cong D_A$, the diffusion coefficient of impurity or solute. As a result Fick's second law, is applicable for the movement of substitutional impurities and analogies related to heat diffusion in solids are also applicable without loss of accuracy. In general, small concentrations of solutes have the power to dramatically change the values of mechanical properties of metals, like the presence of C and N which dissolve interstitially in steels and help to increase the mechanical strength. Therefore, knowing the behaviour and distribution of these is quite relevant.

Table 17.7 and Figure 17.5 present interdiffusion coefficients in some systems of common metals like iron, aluminum and copper. The relatively high activation energy, is consistent with the fact that the alloys are substitutional.

Table 17.7 - Pre-exponential factor and activation energy for cases involving interdiffusion. (Computed fram data in Figure 17.5).

System	Temperature range (K)		D_o (m^2/s)	ΔE (J/K.mole)
C-Fe	1650	1250	2.231E-05	145686
B-Fe	1550	1250	1.712E-07	86860
B-Fe	1100	975	4.371E-04	143544
N-Fe	1000	800	1.805E-07	69503
Ni-Fe	1650	1475	5.615E+02	484176
Cr-Fe	1625	1400	1.590E+04	479554
Si-Fe(3.5% Si)	1575	1100	5.849E-11	44124
Al-Fe	1375	1150	9.345E-08	118133

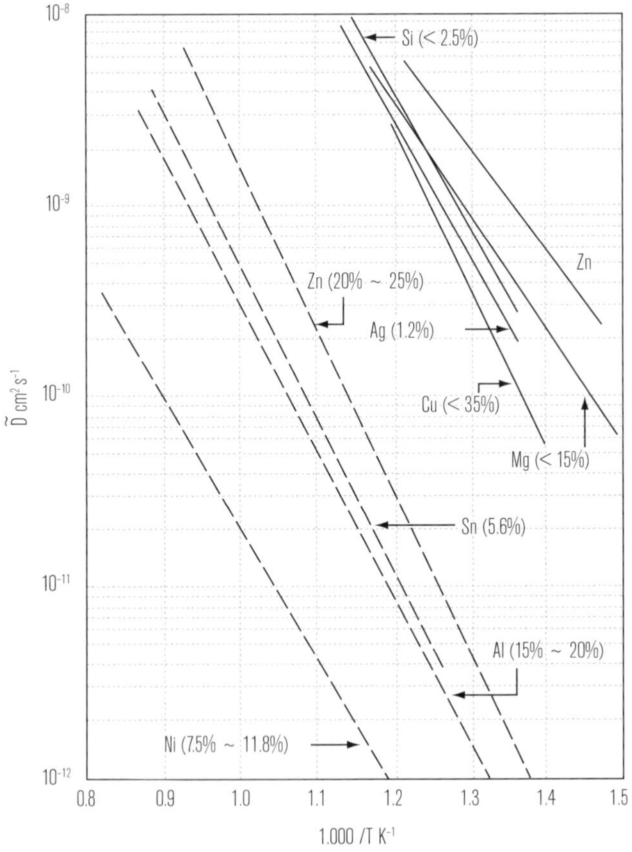

Figure 17.5 - Interdiffusion coefficients for systems of Al and Cu alloys (BIRD, STEWART, and LIGTHFOOT, 2002).

Figure 17.6 presents an idealized scheme of the paths available for diffusion in a polycrystalline structure. In the case of intragranular diffusion of an atom (inside grains) combined with grain boundary diffusion, the differences between these two regions have to be considered in respect of disorder and relative concentration of defects, which are relatively more in the latter. Therefore, the diffusion is rapid in the intergranular regions, but the total flux in each path can be influenced by the area available for flow. The diffusion coefficient resulting from a specific experiment includes, in the general, contributions from both the above mentioned cases. In a polycrystal both the above mechanisms are active. In the case of a single crystal or a very coarse-grained material, the experimentally measured diffusion coefficient tends to be as in the 1st mechanism namely intragranular (neglecting surface contribution).

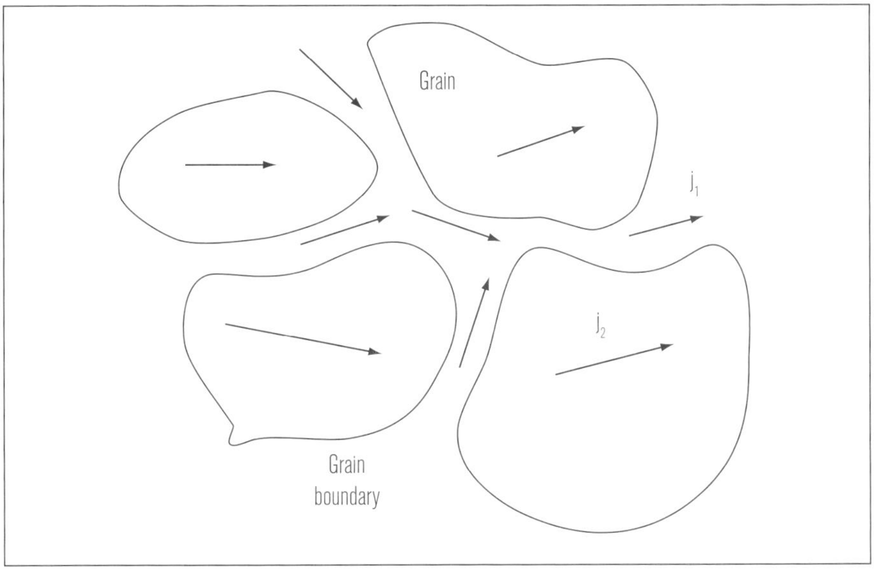

Figure 17.6 - Diffusion paths in a polycrystalline sample, j_1 and j_2 represent diffusive fluxes.

A possible methodology in this case is, (Figure 17.7), to consider the total diffusive flux (j) as the sum of individual streams (j_1 and j_2), for same composition gradient:

$$j\,(A_1 + A_2) = j_1 A_1 + j_2 A_2 \tag{17.28}$$

$$j = -\left(D_1 \frac{A_1}{A_1 + A_2} + D_2 \frac{A_2}{A_1 + A_2}\right) \frac{C_2 - C_1}{L} \tag{17.29}$$

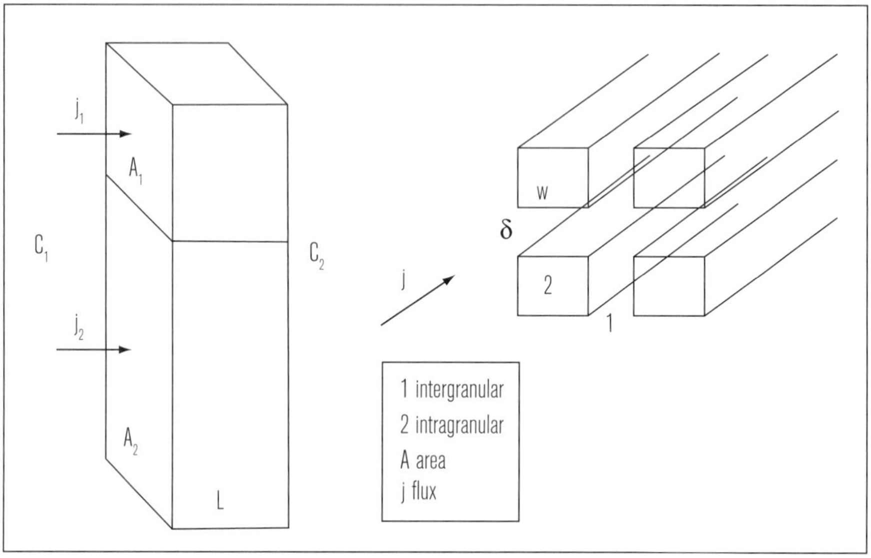

Figure 17. 7 - Difffusion paths considering idealized geometry (WILKINSON, 2000).

In Figure 17.7, j represents the diffusive flux and A is the area through which, this takes place, resulting in an effective diffusion coefficient, (D_{ef}) weighted to reflect the area available for flow in each case namely intergranular (A_1) and intragranular (A_2):

$$D_{ef} = D_1 \frac{A_1}{A_1 + A_2} + D_2 \frac{A_2}{A_1 + A_2} \tag{17.30}$$

Although the intergranular diffusion coefficient is much larger than the intragranular diffusion coefficient (depending on the concentration of defects,which is more in grain boundaries) the opposite occurs, usually in relation to the areas available for flow. If w is the thickness of the intragranular region and δ the thickness of the intergranular region, (Figure 17.7), in the case where $w >> \delta$ which implies a much smaller intergranular area compared to intragranular area:

$$A_1 \cong 4 \frac{1}{2} \delta w \;\; e \;\; A_2 = w^2 \tag{17.31}$$

Hence:

$$D_{ef} = D_1 \frac{2\,\delta}{w} + D_2 \tag{17.32}$$

Example

Pure Nickel has an average grain size of 50 μm, and the diffusion coefficient values for intra and intergranular diffusion are given as:

$$D_2(\text{intra}) = 1.9\times10^{-2} \text{ and } ^{-34175/T} m^2/s$$

$$\delta D_1(\text{inter}) = 3.5\times10^{-15} \text{ and } ^{-13840/T} m^3/s$$

The relative importance of intragranular diffusion can be estimated from:

$$\frac{D_{ef}}{D_2} = 1 + \frac{2\,\delta\,D_1}{w\,D_2}$$

The data are presented in the following figure. At 1250 °K and above the intragranular diffusion predominates; whereas at lower temperatures intergranular diffusion in the grain boundaries becomes important. This is due to the higher thermal activation observed for intragranular (284,00 J/mol) diffusion as compared to intergranular diffusion (115,000 J/mol).

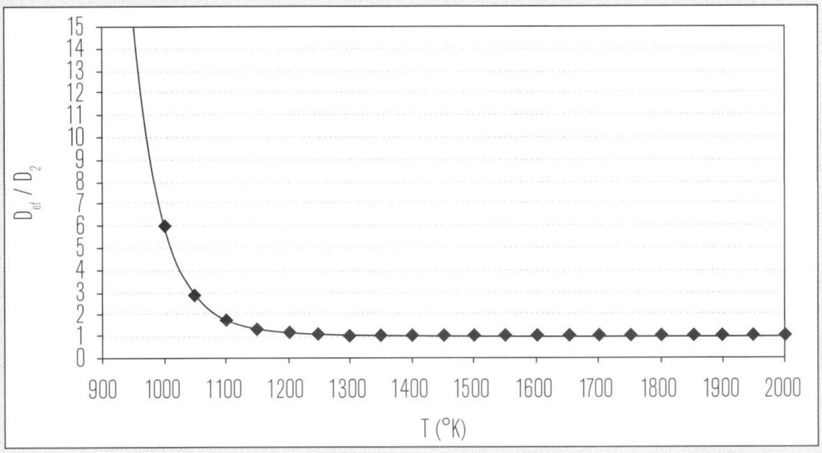

Relation between effective diffusion coefficient and intragranular difffusion coefficient for nickel as a function of temperature (Calculated from data from WILKINSON, 2000).

17.5 Permeability of Gases in Metals

Table 17.8 presents data on diffusivity of some diatomic gases in various metals. Gases dissolve in metals in atomic form and, therefore, the terms refer to the diffusion of atoms of these species. Also as a result, diatomic gases, before being absorbed by the metal, undergo dissociation

$$A_2 \Leftrightarrow 2A; C_A = K\sqrt{P_{A_2}} \tag{17.33}$$

where:

- A_2 represents the diatomic gas and the atomic form;
- K is equilibrium or proportionality constant;
- P_{A2} is the partial pressure of the gas.

The expression, linking the concentration in the metal and the partial pressure of diatomic gas in the gas phase in equilibrium, is known as the Sieverts law. It is strictly valid only for dilute solutions. Otherwise, activity coefficients should be used instead of concentrations. Although the form of dissolution is atomic one can report the amount of "gas" present in the metal in terms of volume of gas that would have been absorbed, for example in the STP. In this case the Sieverts law can be rewritten as.

$$C_A\left[\frac{m^3(\text{CNTP})\text{ of }A_2}{m^3\text{of metal}}\right] = K\sqrt{P_{A_2}\,(\text{atm})} \tag{17.34}$$

The equilibrium constant, K, can be also called solubility. The solubility of gases in metals and alloys depends on temperature and this relationship can be expressed by Arrhenius type of equation for variation of K with temperature.

Table 17.9.gives the solubility of gases in metals and alloys and the activation energy for the process so that the solubility can be calculated at any temperature.

Table 17.8 - Difusivity of some gases (atomic) in metals (ELLIOT, GLEISER and RAMAKRISHNA, 1963; POIRIER and GEIGER, 1994; WILKINSON, 2000).

Gas	Metal	$D_0(m^2/s)$	ΔE (J/K.mole)
H_2	Ni (T<631 K)	6.90×10^{-09}	40300
H_2	Ni (T>631 K)	4.80×10^{-09}	39300
H_2	Cu	1.96×10^{-07}	28800
H_2	Fe α	2.30×10^{-09}	6600
H_2	Al	1.750×10^{-08}	16200
N_2	Fe α	8.00×10^{-07}	76000
O_2	Ag	3.20×10^{-07}	64175

Consider a metal sheet of thickness L, one face being subjected to a pressure P_0(atm) and the other to the pressure P_L(atm) (Figure 17.8). If one assumes that diffusion is relatively slow compared to chemical kinetics, dissociation of gas at the surface is fast and equilibrium is reached on both sides. Then, the metal concentrations, at these faces should be:

$$C^o\left[\frac{m^3(\text{CNTP})\,A_2}{m^3\text{of metal}}\right] = K\sqrt{P_o(\text{atm})} \qquad (17.35)$$

$$C^L\left[\frac{m^3(\text{CNTP})\,A_2}{m^3\text{of metal}}\right] = K\sqrt{P_L(\text{atm})} \qquad (17.36)$$

Since 22.4 STP m3 corresponds to 1 kmole of A_2 (each molecule of A_2 contains two atoms of A) concentrations in terms of k moless A result in:

$$C^o\left(\frac{\text{Kmol A}}{m^3\text{of metal}}\right) = 0.0893\,K\sqrt{P_o(\text{atm})} \qquad (17.37)$$

$$C^L\left(\frac{\text{Kmol A}}{m^3\text{of metal}}\right) = 0.0893\,K\sqrt{P_L(\text{atm})} \qquad (17.38)$$

For steady state and considering constant diffusivity (the cases indicated in the tables are for constant uniform temperature), the concentration profile is linear, as shown in below (Figure 17.8).

Gas	Metal	Ko ($m^3.atm^{-0.5}.m^{-3}$)	ΔE (J/K.mole)
H_2	Ni (T < 631 K)	$1.739 \times 10^{+01}$	17650
H_2	Ni (T > 631 K)	$2.500 \times 10^{+01}$	18650
H_2	Cu	9.490×10^{-02}	43790
H_2	Fe α	$1.261 \times 10^{+02}$	28545
H_2	Al	$2.143 \times 10^{+03}$	112500
N_2	Fe α	5.625×10^{-01}	23580
O_2	Ag	9.063×10^{-01}	30175

Table 17.9 - Solubility of diatomic gases in metals in m^3 (STP)/m^3.atm$^{-1/2}$) (ELLIOT, GLEISER and RAMAKRISHNA, 1963; WILKINSON, 2000).

In the absence of convection, flow can be only diffusive, given by:

$$N_A \cong -D_A\frac{dC_A}{dx} = -D_A\frac{C^L - C^o}{L} \qquad (17.39)$$

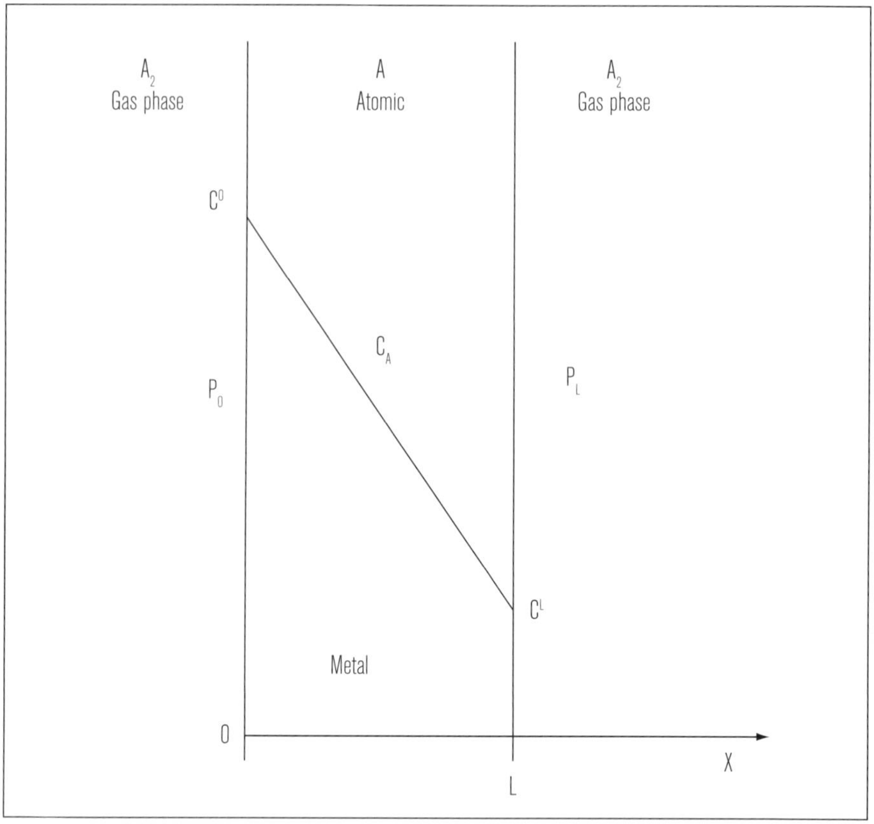

Figure 17.8 - Permeation of a diatomic gas in a metal.

Substituting for concentration terms (Sieverts law):

$$N_A = 0.0893\ D_A \frac{K\sqrt{P_o} - K\sqrt{P_L}}{L} = 0.0893\ K\ D_A \frac{\sqrt{P_o} - \sqrt{P_L}}{L} \qquad (17.40)$$

The above expression gives the flow in terms of kmoles of $A/m^2.s$; if the experimental data is in terms of m^3 of $A^2/m^2.s$, using appropriate conversion units leads to the following equation having a constant K which takes care of all this and D the diffusion coefficient which is characteristic of the metal under consideration:

$$N_A = K\ D_A \frac{\sqrt{P_o} - \sqrt{P_L}}{L}\ m^3 A_2/m^2 s \qquad (17.41)$$

The product $K\ D_A$ is the property characteristic of the gas and the metal in which it diffuses. This parameter enables calculation of the gas flow through a metal wall. This product of diffusivity and equilibrium constant for the dissociation and solution reaction of gas in the metal, is known as permeability P. As already mentioned, the temperature dependence of this parameter can be determined through Arrhenius equation. Data for some metals are given in Table 17.10.

Gas	Metal	P_o ($m^3s^{-1}m^{-1}atm^{-0.5}$)	ΔE (J/K.mole)
H_2	Ni (T<631 K)	1.20×10^{-07}	57950
H_2	Ni (T>631 K)	1.20×10^{-07}	57950
H_2	Cu	1.86×10^{-08}	72590
H_2	Fe α	2.90×10^{-07}	35145
H_2	Al	3.75×10^{-05}	128700
N_2	Fe α	4.50×10^{-07}	99580
O_2	Ag	2.90×10^{-07}	94350

Table 17.10 - Permeability of some gases in metals. (ELLIOT, GLEISER and RAMAKRISHNA, 1963; POIRIER and GEIGER, 1994; WILKINSON, 2000).

Example

Estimate the loss of hydrogen at 25°C, through the walls of a spherical steel vessel. The internal pressure of hydrogen is equal to 2 MPa and, and the external may be neglected. As the wall thickness is small, 0.015 m, compared with the dimensions of the container (3 m diameter), disregard the curvature (ie, it can be assumed unidirectional flow).

Therefore, according, according to Table 16:10 permeability at 25°C can be estimated as $K.D_H = 1.99\ 10^{-13}\ m^3\ s^{-1}\ m^{-1}\ atm^{-1/2}$.

Hence the molar flux of Hydrogen is given by:

$$N_H = 1.99 \times 10^{-13} \frac{\sqrt{P_o}}{0.015} m^3H_2/m^2s$$

Where P_o is internal pressure in atm. Considering that during the course of time the internal pressure lowers one can estimate the initial rate of loss of Hydrogen.

$$N_H = 1.99 \times 10^{-13} \frac{\sqrt{19.74}}{0.015} = 5.9 \times 10^{-11} m^3H_2/m^2s$$

Now assuming that the escape of hydrogen through the walls of the sphere is responsible for the decrease in the number of moles of gas inside of it, and ideal behavior of hydrogen

$$\frac{4\pi R_E^{\ 3}}{3RT}\frac{dP}{dt} = -4\pi R_E^2.KD_H\frac{\sqrt{P}-\sqrt{P_L}}{L}$$

Where R_E is is the internal radius of the sphere, R is gas constant T absolute temperatura. This equation can be integrated over time to get the variation in pressure. For example after 5 days (432,000 s) the pressure can be calculated as follows:

$$\frac{4\pi 1.5^3}{3\times 8.31\times 298}\frac{dP}{dt} = -4\ \pi\, 1.5^2.1.99\times 10^{-13}\frac{\sqrt{P}}{0.015}$$

$$\frac{dP}{dt} = -6.57\times 10^{-8}.\sqrt{P}$$

After transposing the terms, integrating and substituting for the initial pressure one gets:

P = 19.61 atm (there is a pressure drop in 5 days from the initial value of 19,74 atm).

Example

Consider the initial rate of permeation of the case described in the previous example with the one in which a lining of copper, 100 μm is added in order to reduce losses.

This situation is rather exaggerated in the figure below, where the copper layer thickness is drawn out of proportion. The composition profile is linear in both the layers. The concentration of hydrogen at the Cu-Fe interface can be considered in equilibrium with both copper and iron. Equilibrium is assumed in the faces exposed to the gas. It is also reasonable to assume local equilibrium at the interface copper-iron. (2 and 3 in the equation below refer to the interface Cu-Fe and Fe –gas respectively). Hence according to the Sieverts Law.

$$\frac{C^2}{K_{Cu}} = \frac{C^3}{K_{Fe}} = \sqrt{P_{eq}}$$

This relationship does not imply that there should be a gas pocket at the interface. Only the concentrations would be interconnected through the solubility. Moreover, in steadystate, the flow through the copper layer should be equal to the flow in the layer of iron:

$$N_H = -D_H \frac{dC_H}{dx} = -D_H^{Cu} \frac{C^1 - C^0}{L_{Cu}} = -D_H^{Fe} \frac{C^3 - C^2}{L_{Fe}}$$

In this expression, values of diffusivity, concentration and thickness are specific to each layer. According to Sievert's law:

$$N_H = D_H^{Cu} K_{Cu} \frac{\sqrt{P_0} - \sqrt{P_{eq}}}{L_{Cu}} = D_H^{Fe} K_{Fe} \frac{\sqrt{P_{eq}} - \sqrt{P_L}}{L_{Fe}}$$

Under steady state conditions:

$$N_H = \frac{\sqrt{P_0} - \sqrt{P_L}}{\dfrac{L_{Cu}}{D_H^{Cu} K_{Cu}} + \dfrac{L_{Fe}}{D_H^{Fe} K_{Fe}}}$$

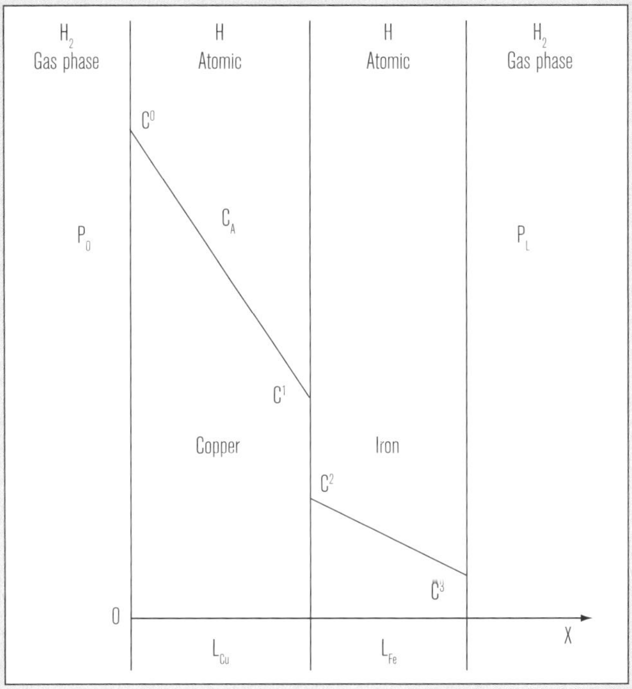

Permeabillity through two layers.

Drawing analogy to an electrical circuit, the numerator of the above equation (concentration difference) is the analogue of potential difference, while the two terms in the denominator are resistances in series. The first is resistance to flow of Hydrogen in the copper layer and the second is the resistance to flow in the iron layer. Permeability values for copper and Iron are given by.

$$K_{Cu}\ D_H^{Cu} = 3.46\times10^{-21}$$

$$K_{Fe} D_H^{Fe} = 1.99\times10^{-13}\ m^3 s^{-1} m^{-1} atm^{-1/2}$$

Hence the resistances are given by:

$$\frac{L_{Cu}}{D_H^{Cu} K_{Cu}} = 2.89\times10^{16} \text{ and } \frac{L_{Fe}}{D_H^{Fe} K_{Fe}} = 7.54\ 10^{10}$$

The flux can be evaluated:

$$N_H = \frac{\sqrt{19.74}}{2.89\times10^{16} + 7.54\times10^{10}} = 1.54\times10^{-16} m^3 H_2/m^2 s$$

It can be concluded that the copper layer offers much more resistance to hydrogen flow and acts as an efficient impermeable medium for that gas reducing the flux by 5 orders of magnitude.

17.6 Diffusion in Pores

Consider, a solid piece (Figure 17.9) whose faces at $x = 0$ and $x = L$ is subject to a chemical potential gradient represented by the concentrations C_A^o and C_A^L respectively. The specie A is present in the liquid or gaseous medium. Most of the solids contain cracks and pores of natural origin or coming from some type of previous treatment. The solid piece can be divided into two regions. One region may be mono or polycrystalline, through which the transport mode can be interstitial,or substitutional, depending on the factors, discussed earlier. There could be locations with pores ,which would be filled by the fluid containing the species if they are open and are interconnected. Diffusion in this situation can be treated similar to the case of intergranular versus intragranular diffusion, identifying the various stages in accordance with the relative importance of transport that occurs within the channels, defined by the morphology of the pores and their distribution. In an extreme case, the solid portion can be considered to be truly impermeable and the regions with the network of pores are available for the transport of the species.

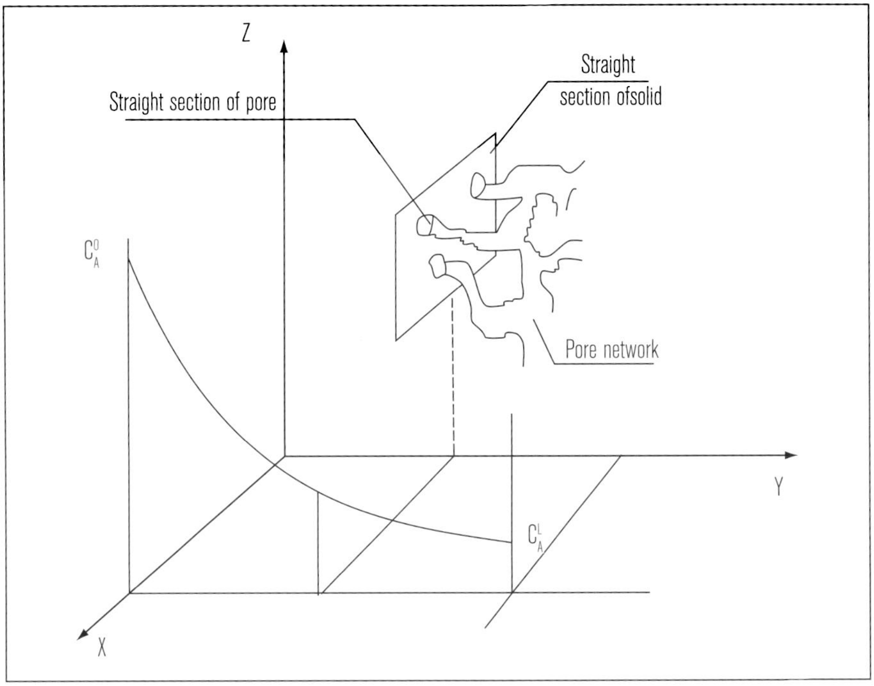

Figure 17.9 - Diffusion through pores of the solid.

The diffusion coefficient of specie A, in a fluid is usually determined by experiments in which the whole area of cross section is available for flow. In general the experimental determination involves measuring the flow through a capillary under the action of a potential gradient. The transport occurs throughout the area of cross section of the capillary and the wall does not interfere with the process. Diffusion coefficients found in literature are typically determined under conditions, in which only the interactions between the constituent atoms or molecules of the medium are involved in the process. However in the case of porous solid this is not the case, since only a part of a given area of cross section of the solid is free for the purpose of diffusion. This means only the cross sectional area of the pores is available for diffusive flow. Hence it is necessary to make a first approximation, considering the porosity ε of the solid, and associated properties so as to define an effective diffusion coefficient (D_{ef}).

$$D_{ef} = \varepsilon D_A \qquad (17.42)$$

According to the same reasoning, the macroscopic capillary diffusion path would be linear While the macroscopic capillary diffusion path is straight and linear, in the case of the fluid containing specie A, it has to pass through the network of pores of the solid which have tortuous paths, unlike in the capillary. Hence a correction factor called the tortuosity factor, τ, has to be introduced for taking into account the lowering of the value of apparent diffusivity in the pores , as given below (THEMELIS, 1995):

$$D_{ef} = \frac{\varepsilon D_A}{\tau} \tag{17.43}$$

Porosity of solids can be measured by routine methods or through the use of nitrogen adsorption-desorption (BET) and others. These give additional data on pore size, pore size distribution and internal surface area of pores. Tortuosity values are generally estimated from the experimental diffusivity data with a model containing this factor. Another aspect that has to be considered is the interaction of the fluid and the species contained in it with the pore walls during diffusion. In the case of a gas, if the number of collisions between molecules is much larger than the number of collisions with the pore walls, (the influence of the walls is minimum) molecular diffusion prevails. The number of collisions between these molecules and the walls are of the same order of magnitude, when the pore size is small or in the case of rarified atmosphere at low pressures (low density of gas). Under these conditions when significant collisions involving the walls take place, one can use Knudsen diffusion coefficient, given by:

$$D_k\left(cm^2.s^{-1}\right) = 9,700\, r\sqrt{\frac{T}{M}} \tag{17.44}$$

where:

- r(cm) is the radius of the pore;
- T(K), is absolute temperature;
- $M(g.mol^{-1})$ molecular weight of the gas.

In this case also the correction factors like in equation 16.43 for tortuosity and porosity of the solid can be incorporated.

$$D_{ef} = \frac{\varepsilon D_K}{\tau} \tag{17.45}$$

A criterion as to which mechanism is prevailing ordinary or molecular diffusion or Knudsen's is based on a comparison between the pore radius and mean free path of gas molecules. This can be estimated from the following equation:

$$\lambda = \frac{1}{\sqrt{2}\ \pi\ d^2\ n} \tag{17.46}$$

where:

- d(cm) is the collision diameter of the molecules;
- n($-/cm^3$) number of molecules for unit volume.

If the pore radius r is of the same order of magnitude as the mean free path then Knudsen diffusion is important. For high pore diameters this is not important and one can use molecular diffusion coefficient.

Example

For illustrating the importance of molecular and Knudsen diffusion, consider a gas phase having CO (35%)-N_2(65%) at 700 K and 0.1 atm pressure. One can estimate the diffusion coefficient using Fuller equation with V_{CO} = 18.9 e V_{N2} = 17.9:

$$D(m^2.s^{-1}) = 10^{-7}\frac{T^{1.75}(1/M_1 + 1/M_2)^{0.5}}{P(V_1^{1/3} + V_2^{1/3})^2} = 10^{-7}\frac{700^{1.75}(1/28 + 1/28)^{0.5}}{0.1(17.9^{1/3} + 18.9^{1/3})^2} = 9.13\times 10^{-4}$$

For ideal gas PV = nRT, hence the number of molecules per unit volume (n) is given by:

$$\frac{PN}{RT} = \frac{1.013\times 10^4\ \times 6.02\times 10^{23}}{8.31\ \times\ 700} = 1.05\times 10^{24}\ (-/m^3)$$

where:

- N is Avogrado number, 6.023×10^{23} (-/mol).

Collision diameter of CO is 3.69×10^{-8} cm;

Collision diameter of Nitrogen is 3.798×10^{-8} cm.

Hence the average is 3.744×10^{-8} cm,

Subsittuting this value for d,

$$\lambda = \frac{1}{\sqrt{2}\ \pi\ d^2\ n} = \frac{1}{\sqrt{2}\ \pi\ (3.744\times 10^{-8})^2\ 1.05\times 10^{18}} = 1.53\times 10^{-4}\ \text{cm}$$

The result is about 150 microns. Hence only pores having this magnitude are important for Knudsen diffusion. For example in the case of a sand mold (SiO2) for foundry the pores have a diameter of around 5×10^{-3} cm. In this case Knudsen diffusion coefficient is given by:

$$D_k(cm^2.s^{-1}) = 9{,}700\ r_e\sqrt{\frac{T}{M}} = 9700 \times 2.5 \times 10^{-3}\sqrt{\frac{700}{28}} = 121\ cm^2.s^{-1}$$

If this value is compared with the value for molecular diffusion one comes to the conclusion that in this case Knudsen diffusion is not important for this pore size. This can be demonstrated in another way considering 2 resistances in parallel involving both the mechanisms:

$$\frac{1}{\tilde{D}} = \frac{1}{D} + \frac{1}{D_K} = \frac{1}{9.13} + \frac{1}{121}$$

This results in $\tilde{D} = 8.48\ cm^2.s^{-1}$. It can also be concluded that the contribution by the second term is insignificant compared to the 1st term.

17.7 Mass Transfer Coefficients

In many situations a given chemical specie gets transported from one phase to the other through an interface between them, where chemical reactions can take place. These fluid phases may be in a state of turbulent flow, in which case the velocity as well as concentration fluctuate as a function of time. This is given schematically in Figure 17.10.

As has been already mentioned, the fluid movement and transport of species also occur simultaneously. In this case, the equations of momentum and mass transport have to be solved simultaneously to get the velocity and concentration profiles. In cases involving heat, mass and momemtum transfer all these three sets of equations relevant to the system have to be solved simultaneously to evaluate the velocity temperature and concentration profiles. In the case of turbulent flow, the same type of equations can be used with time smoothed variables as explained in the earlier chapters. It has also been mentioned that transient problems involving two or two or three dimensions would require in general numerical solutions with the use of computers.

Mass transport, like in the case of heat transfer in turbulent flow, has been treated with the aid of film or transfer coefficients which can be determined experimentally like the heat transfer coefficients as discussed earlier.

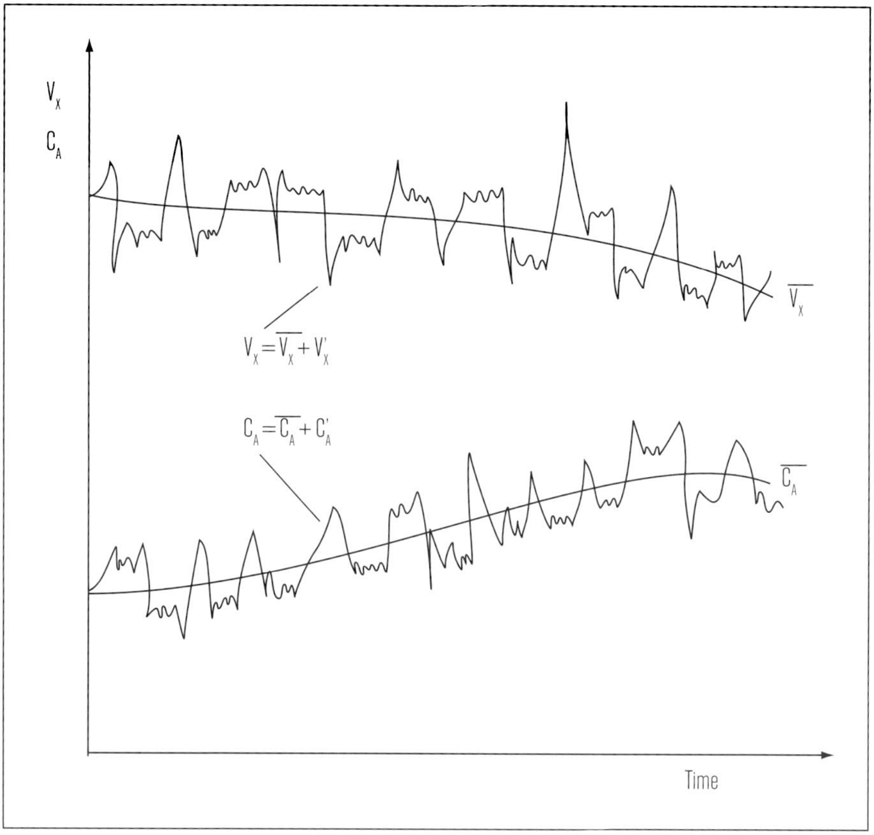

Figure 17.10 - Fluctuations of velocity and concentration at a point as a function of time in turbulent flow.

Transfer coefficients can be calculated. For example, near the solid surface, turbulent flow tends to become laminar. Eddies gradually lose strength, as the flow regime approaches the solid. At the solid/fluid interface it breaks, as suggested in Figure 17.11. Hence the interface mass flow can be calculated from Fick's First Law if concentration profile is known:

$$N_{A,r} = -D_A \frac{\partial C_A}{\partial r} \qquad (17.47)$$

On this basis one can define a parameter namely mass transfer coefficient Km, given by:

$$N_{A,i} = -D_A \frac{\partial C_{A_i}}{\partial r} = K_m \left(\overline{C_A^0} - \overline{C_A^\infty} \right) \qquad (17.48)$$

Where $\overline{C_A^0}$ and $\overline{C_A^\infty}$ are average concentrations of A at the solid surface and inside the fluid respectively.

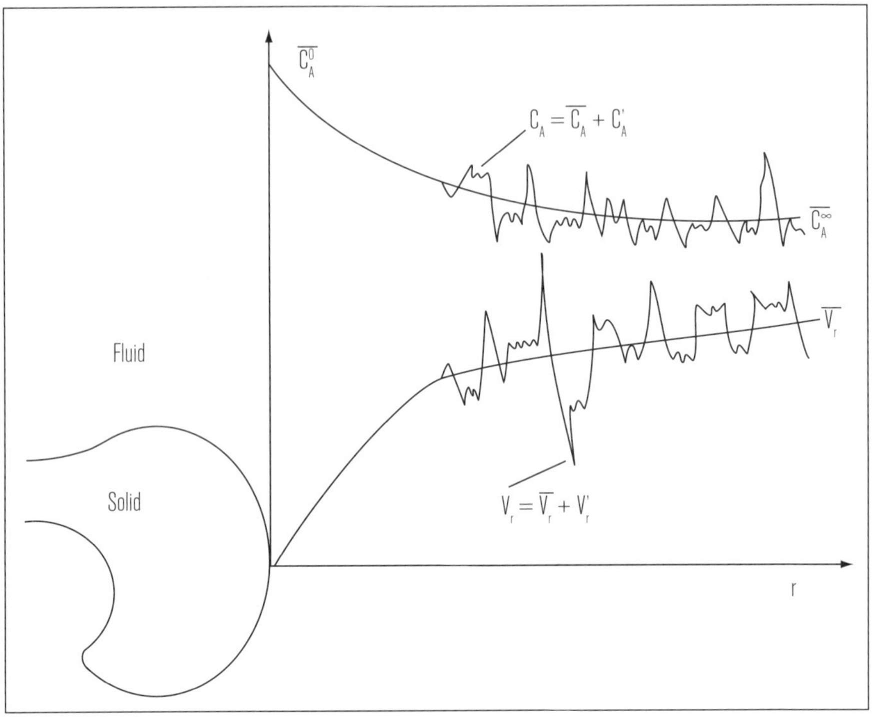

Figure 17.11 - Damping of fluctuations by viscous effects at the proximity of an interface.

Possible techniques for calculating the velocity and concentration profiles are:

1) Direct Numerical Simulation (DNS) This involves solving simultaneous equations of continuity, Navier-Stokes and conservation of the specie A. and others in the system. Eddies of varying sizes, from the magnitude of the system to microscopic dimensions, are characteristic of turbulent flows. These are responsible for turning mechanical energy into heat, ie energy dissipation typical of these processes. Hence even these small eddies have to be simulated to reproduce natural fluctuations in the parameters such as velocity temperature and concentration. For dealing with these small eddies, finer mesh of integration is required and this affects the numerical solution process increasing the computer time. for solution. The cost of CPU time, storage capacity etc. would be prohibitive except for systems of reduced dimensions.

2) Alternatively the treatment could involve applying the average operator:

$$\overline{A} = \frac{1}{t_2 - t_1}\int_{t_1}^{t_2} A(t).dt \qquad (17.49)$$

Here also the equations of continuity, Navier-Stokes and conservation of specie A have to be solved. However this procedure enables the calculation of average values instead of instantaneous values. The original equations contain typical terms of turbulence, such as Reynolds tensors and turbulent viscosity in the Navier Stokes equations as discussed in Chapter 7:

$$\rho\ \overline{v_x^{'} v_y^{'}} = \eta^{\text{turbulent}} . \frac{d\,\overline{v_x}}{dy} \tag{17.50}$$

Or turbulent thermal conductivity term in the energy conservation equation:

$$\rho\, C_p\ \overline{v_x^{'} T^{'}} = K^{\text{turbulent}} . \frac{d\,\overline{T}}{dx} \tag{17.51}$$

and a turbulent diffusivity term for diffusing specie A, in the equation for mass conservation equation for A:

$$\overline{v_x^{'} C_A^{'}} = D_A^{\text{turbulent}} . \frac{d\,\overline{C_x}}{dx} \tag{17.52}$$

The implementation of this method would require application of turbulence models, either macroscopic or microscopic (such as k-ε), to estimate the diffusivities from the turbulent velocity field averages.

Eventually, in the absence of specific knowledge of the concentration profiles, the mass flow could be measured as in the case of heat flow experimentally, enabling to estimate the value of k_m. Applying the concepts of dimensional analysis, as discussed in Chapter 7 and Chapter 13, the experimental data can be summarized in terms of relevant dimensionless groups of the type:

$$f\ (\text{Sh, Re, Sc, geometry}) = 0 \tag{17.53}$$

where:

$$\text{Sh(Sherwood)} = \frac{k_m L}{D_A};\ \text{Re(Reynolds)} = \frac{L \rho V}{\eta};\ \text{Sc(Schmidt)} = \frac{\eta}{\rho\, D_A}$$

- k_m is mass transfer coefficient;
- L characteristic length of the system;
- D_A the diffusion coefficient of specie A in a fluid;
- ρ fluid density;
- η fluid viscosity.

It should be possible to calculate k_m from the geometry of the system, the average properties of the fluid and the average velocity, as it is a function of all these parameters.

The actual distribution of a chemical specie, from the interface between the two phases can be represented, as shown in Figure 17.12. This can be considered a smooth and continuous variation. This profile can be approximated by another consisting of two linear parts. The point of intersection defines the film thickness δ called the boundary layer of concentration.

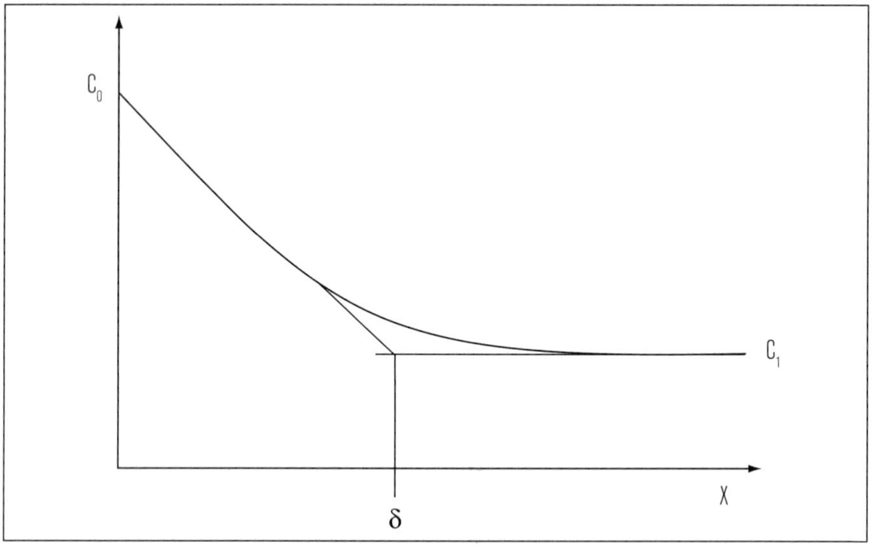

Figure 17.12 - Schematic profile of composition profile of a specie adjacent to the interface.

Assuming steady state and neglecting convection (the thickness of the boundary layer is small and hence no slip condition for the the fluid velocity can be considered) one can write on a molar basis:

$$N = -D\frac{dC}{dx} = \text{constant}$$

Integrating the above equation, for the above concentration limits in the film:

$$N = -D\,\frac{C_1 - C_o}{\delta}$$

the mass transfer coefficient can be defined as

$$k = \frac{N}{C_o - C_1} \text{ and } k = \frac{D}{\delta}$$

From this formulation, it is observed that both the molar flux and the boundary layer thickness are not known. However one can infer the boundary layer thickness, if the mass transfer coefficient can be determined and the diffusion coefficient of the specie is known. Nevertheless, the above definition is useful, as it can give useful information about possible effects of relative velocity, fluid properties such as viscosity, density on the thickness of boundary layer and consequently on the rate of mass transport.

In some specific cases it may be possible to find a value of coefficient of mass transfer from first principles. In most cases these are obtained through experimental work. Dimensional analysis based on Buckingham theorem,has been discussed in Chapters 7 and 13. This technique enables to identify dimensionless groups in a system and these can be correlated through experimental work. For example, one can consider (based on prior knowledge, or subjective analysis of the problem) that the mass transfer coefficient, k_m(m/s), is a function of properties of the fluid such as density, ρ(kg/m^3), viscosity, μ ($kg.m^{-1}.s^{-1}$) relative velocity between the particle and fluid, U($m.s^{-1}$), geometry of the body characterized by, X(m), and finally, the intrinsic diffusivity of the species in a fluid, D($m^2.s^{-1}$). The technique of dimensional analysis is based on Rayleighs method of indiceswhich postulates that a dependent variable like the mass transfer coefficient can be expressed as a function of different physical quantities (independent variables)in the form of a product of power functions:

$$K_m = \rho^a \mu^b X^c U^d D^e \qquad (17.54)$$

The next step is to express the dependent and independent variables above in terms of fundamental units of mass, length, time and temperature.

Denoting each quantity in 17.54 in terms of fundamental units, length (L), mass (M) and Time (T):

$$L\,T^{-1} = (M\,L^{-3})^a\,(M\,L^{-1}\,T^{-1})^b\,L^c\,(L\,T^{-1})^d\,(L^2\,T^{-1})^e \qquad (17.55)$$

Separating the fundamental units with the respective arguments:

$$L\,T^{-1} = L^{-3a-b+c+d+2e}\,M^{a+b}\,T^{-b-d-e} \qquad (17.56)$$

Equating the powers on the basis of dimensional homogeneity, one can write the algebraic equations:

$$1 = -3a - b + c + d + 2e$$

$$0 = a + b$$

$$1 = b + d + e$$

From the above equations the following equivalence between the powers of the variables can be established:

$$b - c - 2e = d - 3a - 1$$

$$b = -a$$

$$b + e = 1 - d$$

Hence, expressing b, c and e in terms of a and d:

$$b = -a$$

$$c = d - 1$$

$$e = 1 - d + a$$

and substituting the same in 17.54:

$$k_m = \rho^a \mu^{-a} X^{d-1} U^d D^{1-d+a} \tag{17.57}$$

Regrouping the above variables one can write:

$$\frac{k_m X}{D} = \left(\frac{\mu}{D \rho}\right)^{d-a} \left(\frac{X V \rho}{\mu}\right)^d \tag{17.58}$$

The above dimensionless groups are used frequently for calculating mass transfer coefficients and are known as:

$$Sh\ (Sherwood) = \frac{k_m X}{D} \tag{17.59}$$

$$Re\ (Reynolds) = \frac{X V \rho}{\mu} \tag{17.60}$$

$$Sc\ (Schmidt) = \frac{\mu}{D \rho} \tag{17.61}$$

In other words Sh = f(Re, Sc), and the exponent values can be obtained after performing experiments with the objective of finding a correlation between the dimensionless numbers graphically or through regression analysis. In some other system other dimensionless numbers can be involved and the same procedure can be adopted defining the dependent and independent variables and carrying out the dimensional analysis.

In certain situations the mass transfer coefficient can be calculated. For example (POIRIER and GEIGER, 1994) Higbie, considers (Figure 17.13) that, packages of uniform concentration of the fluid initially are transported by convection to the interface between the fluid and another phase, under conditions of no slip. These packages can be considered coupled to the interface for a certain time, called the exposure time in which mass transfer occurs between phases. In order to quantify the mass flow, the package of fluid can be considered during this treatment period, as semi-infinite and the mass is transferred only by diffusion. In this case, the concentration gradient within the package (Figure17.13) can be expressed as follows using error function (POIRIER and GEIGER, 1994):

$$\frac{C_A - C_A^0}{C_A^\infty - C_A^0} = \text{erf}\left(\frac{x}{2\sqrt{D_A\, t_e}}\right) \qquad (17.62)$$

where:

- D_A is the diffusion coefficient of species A in the fluid;
- t is time of exposure;
- x distance measured in the fluid and from the interface.

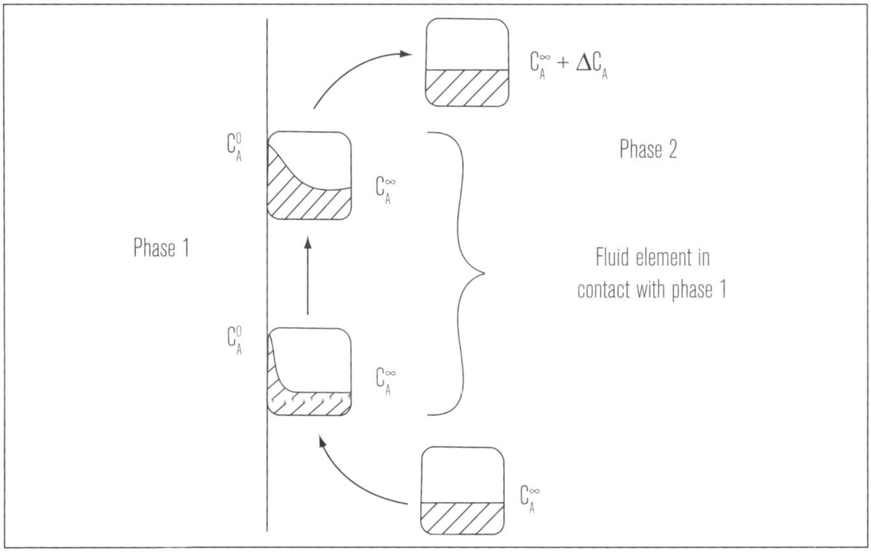

Figure 17.13 - Schematic representation of Higbie.model (POIRIER and GEIGER, 1994).

The error function (erf) equation given above can be solved to determine the concentration gradient (and therefore flow) at position x = 0. Subsequently the flow equation can be integrated over time to get an average value in the range of exposure. Higbie's theory results in an equation for mass transfer coefficient km:

$$k_m = 2\sqrt{\frac{D}{\pi\ t_e}} \tag{17.63}$$

For example if a metal droplet of radius R moves inside a slag with a velocity of V, the time taken to move the distance equivalent to one diameter can be assigned to time required to displace the liquid layer around the sphere or the renewal time. Hence t_e can be estimated as:

$$t_e = \frac{2R}{V} \tag{17.64}$$

Higbie's theory is quite elegant, provided t_e, can be determined with precision, a rather difficult proposition.

Now consider the transfer of a given specie from a spherical particle to a stagnant fluid that surrounds it. One can choose a spherical spherical coordinate system to represent this (Figure 17.14).

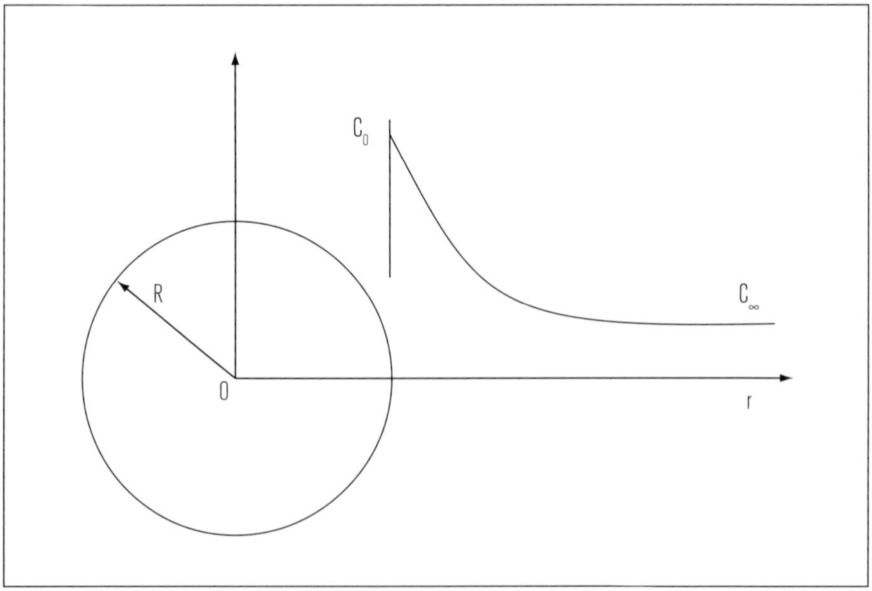

Figure 17.14 - Mass transfer of a species from a sphere to a stagnant fluid.

In the absence of convection and for stagnant fluid and assuming steady state:

$$J = \left(-D\frac{dC}{dr}\right)4\pi\, r^2 \,(mol/s) \quad (17.65)$$

Considering the boundary conditions in Figure 17.14 and separating the variables:

$$J\,\frac{-dr}{r^2} = 4\pi D\, dC \quad (17.66)$$

$$J\left(\frac{1}{\infty}-\frac{1}{R}\right) = 4\pi\, D(C_\infty - C_o) \quad (17.67)$$

The definition of mass transfer coefficient for the situation is:

$$k_m = \frac{J}{4\pi\, R^2(C_o - C_\infty)} \quad (17.68)$$

Comparing equations 17.7 and 17.8 the value of K_m is equal to D/R. Hence:

$$Sh = \frac{k_m\; 2R}{D} = 2 \quad (17.69)$$

One can find a number of correlations in the literature for different systems.There has been extensive research work in this subject. For example small size bubbles moving inside a metal bath can be approximated to rigid spheres. In this case Frossling equation can be used to estimate the mass transfer coefficient for the boundary layer between the bubble and the metal (metal side):

$$Sh = 2 + 0.552\cdot\sqrt{Re}\cdot Sc^{\frac{1}{3}} \quad (17.70)$$

where:

$$Sh = \frac{2\cdot R\cdot k_m}{D_S} \quad (17.71)$$

$$Re = \frac{\rho_L\cdot V\cdot 2R}{\eta_L} \quad (17.72)$$

$$Sc = \frac{\eta_L}{\rho_L\cdot D_S} \quad (17.73)$$

17.8 Analogy with Heat Transfer

The equations of balance of conservation are intrinsically similar for the transport of heat, mass or momentum.This can be taken advantage of for facilitating analytical or numerical solutions, by using dimensionless numbers having different variables and also utilizing the similarities in respect of the differential equations describing the process in terms of heat mass and momentum transfer.

Table17.11 gives various similarities between heat and mass transfer in respect of the important parameters as well as dimensionless groups.

Table 17.11 - Some variables and corresponding dimensionless groups in heat and mass transfer.

Heat transfer	Mass transfer	Observations
$\rho C_p T$	C_A or ρ_A	Measure of transported Quantity (heat and mass)
$K/\rho C_p$	D_A	Difusivity (thermal and mass diffusivity of specie A)
Q	N_A or n_A	Flux (heat or mass)
h	k_m	Heat or mass transfer coefficient
$Pe = U L/\alpha$	$Pem = U L/D$	Inertial force /diffusive force (Pe and Pem are peclet numbers for heat and mass transfer similar to Reynolds number in momentum balance
$Pr = \mu C_p/K$	$Sc = \mu/\rho D$	Ratio of boundary layer thicknesses in both
$Gr = L^3 g \beta \Delta T/v^2$	$Gr = L^3 g \rho \Delta\rho/v^2$	Buoyancy force/diffusive force (natural convection in heat and mass transfer)
$Fo = K t/\rho C_p L^2$	$Fo = D t/L^2$	Penetration distance/Length (characteristic)
$Nu = h L/K$	$Sh = K_m L/D$	Total flux/diffusive flux

In heat transfer, the Nusselt number is expressed in terms of Reynolds, Prandtl or Grashoff etc., whereas in mass transfer the Sherwood number which is similar to Nusselt number is a function of dimensionless parameters, Reynolds Schmidt, Grashoff etc.

$$Nu = f\ (Re, Pr, Gr, Fo...) \tag{17.74}$$

$$Sh = f(Re, Sc, Gr, Fo,\ldots) \tag{17.75}$$

As has been stated in the sections with correlations for heat transfer even mass transfer correlations are valid under certain conditions or flow regimes described by Reynolds number or other dimensionless numbers with other limits for application. For exemple like the validity of Ranz-Marshall equation for spheres as given below:

$$Nu = 2 + 0.66\ Re^{0.5}\ Pr^{0.33} \qquad 1 < Re < 10^4\ ;\ \text{gases and liquids } Pr < 0.7 \qquad (17.76)$$

the Frossling equation for spheres in mass transfer has the following validity.

$$Sh = 2 + 0.52\ Re^{0.5}\ Sc^{0.33} \qquad 1 < Re < 10^4;\ \text{gases e liquids } Pr < 0.7 \qquad (17.77)$$

The above equations are similar and analogical and have the same conditions of validity Like this there are many equations containing dimensionless groups in heat and mass transfer that are analogical. It is useful to keep an idea of what to expect in terms of characteristic values of groups that involve properties of gases, liquids and solids. As mentioned earlier in heat transfer Prandtl number represents the ratio of the diffusivities of momentum and heat. In gases the Prandtl number is close to 1, suggesting that the hydrodynamic boundary layers and temperature are approximately equal. In liquid metals the Prandtl number is much less than 1, implying that heat transmission is easy and therefore thermal boundary layer extends more deeply within the fluid. Hence the thermal boundary layer would be thicker than the corresponding one for momentum.

17.9 Correlations for the Calculation of Mass Transfer Coeficient in Different Applications

The following table gives the correlations for evaluating mass transfer coefficient in situations and proceses that one comes across in the metallurgical operations.

Gas bubbles in a stirred bath	$\frac{k\,d}{D} = 0.13\left[\frac{(P/V)\,d^4}{\rho\,\nu^3}\right]^{1/4}\left(\frac{\nu}{D}\right)^{1/3}$	d = bubble diameter P/V = Power for agitation for unit volume
Gas bubbles in a still liquid bath	$\frac{k\,d}{D} = 0.31\left(\frac{d^3 g\Delta\rho/\rho}{\nu^2}\right)^{1/3}\left(\frac{\nu}{D}\right)^{1/3}$	d= bubble diameter $\Delta\rho$ =Density difference between gas and liquid
Big droplets rising in still liquid bath	$\frac{k\,d}{D} = 0.42\left(\frac{d^3 g\Delta\rho/\rho}{\nu^2}\right)^{1/3}\left(\frac{\nu}{D}\right)^{1/2}$	d = diameter of the droplet > 0.3 cm $\Delta\rho$ = Difference in density between the two phases.
Small droplets rising in still liquid bath	$\frac{k\,d}{D} = 1.13\left(\frac{d\,v_o}{D}\right)^{0.8}$	d = diameter of droplet v_o = velocity of the droplet
Fixed bed	$\frac{k}{v_o} = 1{,}17\left(\frac{d\,v_o}{\nu}\right)^{-0{,}42}\left(\frac{\nu}{D}\right)^{-2/3}$	d= diameter of particle v_o = superficial velocity
Fluidized bed	$\frac{k}{v_o} = 1/\left\{0{,}010 + 0{,}863/\left[\left(\frac{d\,v_o}{\nu}\right)^{0{,}58} - 0{,}483\right]\right\}\left(\frac{\nu}{D}\right)^{0{,}67}$	d= diameter of particle v_o = superficial velocity
Falling liquid film	$\frac{k\,z}{D} = 0{,}69\left(\frac{z\,v_o}{D}\right)^{1/2}$	z = Position along the film v_o = Mean velocity of the film

ν: kinematic viscosity; k: mass transfer coefficient; D: diffusion coefficient; ρ: density; g: acceleration due to gravity.

Table 17.12 - Correlations for calculation of K_m (CUSSLER, 1997).

17.10 Influence of Temperature on Transport Parameters

The influence of temperature on transport parameters is usually discussed separately. This is because the temperature is admittedly one of the variables that affects significantly the rate of transport processes and the chemical reaction rates.

The temperature dependence of the parameters such as diffusion coefficient, viscosity, mass transfer coefficient, thermal conductivity etc. can be quantified using Arrhenius equation. This is important as many are high temperature processes and require estimation of values of these parameters at various temperatures.

$$\alpha = \alpha_0 \cdot e^{\frac{\pm \Delta E}{RT}} \tag{17.78}$$

where:

- α = parameter;
- α_0 = preexponential factor;
- ΔE = activation energy;
- R = universal gas constant;
- T = absolute temperature.

By convention, the activation energy term is preceded by ± so that the the value always remain positive. It is always possible to choose a temperature range for use in Arrhenius equation. Therefore, if the available parameter values of interest are between T_1 and T_2, the activation energy is given by:

$$\frac{\alpha_{T1}}{\alpha_{T2}} = e^{\frac{\pm \Delta E}{R}\left(\frac{1}{T_1} - \frac{1}{T_2}\right)} \tag{17.79}$$

Example

The rates of chemical reactions are in general temperature dependent. Small increases in temperature can lead to large increases in reaction rates. In the case of some reactions, for every 10°C increase in temperature, the chemical reaction rate actually doubles. If the Arrhenius equation is used to quantify this case, the apparent value of activation energy should vary according to the temperature range.

For example if the rate of reaction doubles with an increment of 10°C, say from 25 to 35°C or 1000 to 1010°C, the reaction rates at these two temperatures are related according to the following equation:

$$\frac{K_{T1}}{K_{To}} = e^{\frac{-\Delta E}{RT}\left(\frac{1}{To} - \frac{1}{T1}\right)}$$

This results in the following activation energy values:

To → T1	ΔE
298 K to 308 K	12640 cal/mole
1273 K to 1283 K	224946 cal/mole

Hence the magnitude of activation energy for a particular parameter depends on the temperature also.

Pyrometallurgical processes, usually conducted at temperatures above 800°C, can be separated into two classes, according to a greater or less sensitivity in terms of temperature or in terms of activation energy.

i) Higer values of ΔE in the range of 50 Kcal/mole or more indicate a strong influence of temperature on the variable. In this case, the parameter is said to be strongly influenced by temperature.

ii) Activation energy values ΔE less than 10 kcal/mole indicate weak influence of temperature on the variable. In this case, the parameter is said to be weakly influenced by temperature.

An activation energy of the order of 20,000 cal/mole for a process in aqueous media, conducted at room temperature or close to it, is considered significantly influenced by temperature.

Example

Consider a tube furnace, with 5 cm in diameter, through which a gas (50% H_2, 50% H_2O) flows. The flow rate is 5 L/min. Estimate the mass transfer coefficient of mass transfer between this gas and spherical particles with a diameter of 0.8 cm, temperature of 700 K and 800 K, and 1 atm considering the property values given in the following table.

Physical properties of Hydrogen and water vapor (densities estimated from ideal gás law and viscosities as per procedure described in Chapter 4)

Temperature (K)	800	700
ρ_{H2O} (kg/m³)	0.2746	0.3140
μ_{H2O} (Pa.s)	27.86×10^{-5}	24.25×10^{-5}
ρ_{H2} (kg/m³)	0.0306	0.0408
μ_{H2} (Pa.s)	14.40×10^{-6}	15.89×10^{-6}
$D_{H_2\text{-}H_2O}$(m²/s)	4.96	3.92

The gas velocity is given by:

$$U = \frac{Q}{\text{area}} = \frac{500\ \text{cm}^3 / 60\ \text{s}}{3.1416 \times 5^2 / 4\ \text{cm}^2}$$

Using the data given in Table 16.13 and substituting in Frössling equation it is possible to estimate $K_m^{700K} = 4.94$ cm/s e $K_m^{800K} = 12.4$ cm/s. The activation energy calculated from these at two temperature result in 12.3 Kcal/mol.

The evaluation of temperature influence on parameters based on activation energy is rather subjective. Some processes are influenced strongly and others weakly by temperature. In fact, transport processes are weakly influenced by temperature relative to chemical reactions. Although this is generally true, it cannot be taken for granted for all situations. Each case must be evaluated separately as presented in Table 17.13. The significance of these will be gone thorough in more detail in Chapter 18.

Table 17.13 - Typical values of activation energy in the transport processes (calculated according to data and equations given by ELLIOT, GLEISER, RAMAKRISHNA, 1963; CUSSLER, 1997; POIRIER and GEIGER, 1994).

Species	Medium	Temperature (°C)	ΔE (Kcal/mole)
--	In water Eq. Einstein-Sutherland; $D = \frac{k_B T}{4\pi R \eta}$	20 ~ 40	4.5
Gases A-B	$D_{AB} = \frac{10^{-3} T^{1.75}}{P\left(V_B^{\frac{1}{3}} + V_A^{\frac{1}{3}}\right)^2} \sqrt{\frac{1}{M_A} + \frac{1}{M_B}} \quad \frac{cm^2}{s}$	1173 ~ 1273	3.3
Fe	In liquid Fe-2.5%C	1340 ~1400	15.7
H	In liquid iron	1565~1679	3.3
Al	Sef diffusion in liquid slag 43% CaO, 10% Al_2O_3, 47% SiO_2	1440~1520	85
Na	Self diffusion in liquid slag 34% Na_2O, 66% SiO_2	850~1500	12.4
H	In α iron	15 ~ 900	3.05
Cr	In γ iron	900~1450	69.7
H	In γ iron	900~1400	9.95
Cu-Ni	Interdiffusion, $X_{Cu} D_{Ni} + X_{Ni} D_{Cu}$, 7.5 ~11.8% Níckel	673 ~838	32
H_2-H_2O	Mass transfer coefficient (Frossling equation)	427~527	12.3

References

BIRD, R.B.; STEWART, W.E.; LIGTHFOOT, E.N. *Transport phenomena.* 2. ed. New York: John Wiley & Sons, 2002.
CUSSLER, E.L. *Diffusion mass transfer in fluid systems.* 2. ed. Cambridge: Cambridge University Press, 1997.
ELLIOT, J.F.; GLEISER, M.; RAMAKRISHNA, V. *Thermochemistry for steelmaking.* Massachusetts: Addison-Wesley, 1963. v. 2.
MESPLEDE, J.; QUEYREL, J.L. *Precis de chimie:* solutions aqueuses. [Paris]: Breal, 1996.
PAULING, L. *The nature of the chemical bond.* New York: Cornell University Press, 1960.
POIRIER, D.H.; GEIGER, G.H. *Transport phenomena in materials processing.* Warrendale: TMS, 1994.
SHEWMON, P. *Diffusion in solids.* 2. ed. New York: John Wiley & Sons, 1989.
THEMELIS, N.J. *Transport and chemical rate phenomena.* Switzerland: Gordon and Breach, 1995.
WILKINSON, D.S. *Mass transportation in solids and fluids.* Cambridge: Cambridge University Press, 2000.

Exercises

1 - Diffusivity of manganese in liquid iron is 6.3×10^{-8} m^2/s (up to 1.28% Mn) at 1680°C, and 17×10^{-8} m^2/s at 1725°C. Estimate the activation energy.

2 - The mechanisms of diffusion in solids can be intragranular, intergranular or surface diffusion with activation energies decreasing in that order. Hence it is expected that at a certain critical temperature $0.7 \times T_{Fusion}$, the contribution of one of these become predominant. Which one? Consider the following experimental data

$$D(m^2/s) = 2.3 \times 10^{-9} \exp(-13200/T);$$

$$D(m^2/s) = 0.895 \times 10^{-4} \exp(-22975/T).$$

One of the above is for self diffusion of silver in a polycrystalline matrix (In polycrystalline matrix both intragranular e intergranular diffusion are possible), and the other is for self diffusion in a single crystal. Indicate which value you would use for each case. What is the temperature of transition from which intragranular diffusion becomes dominant? Estimate the intergranular diffusion coefficient at 500°C, , if the intergranular zone thickness is approximately $1/1000^{th}$ of the perimeter of the grain.

3 - In a pyrometallurgical process variations in temperatures in two shifts are given in the figure below. Determine the relation "production in shift A/ production in shift B", assuming that the process is thermally activated with E values of 1, 5, 10, 50 and 100 Kcal/mol.K.

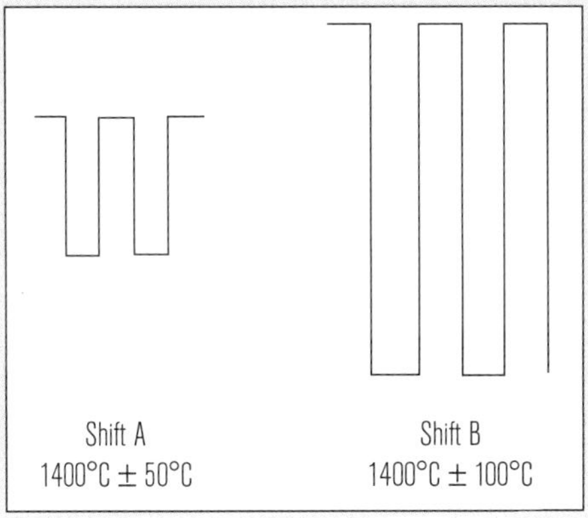

4 - Estimate the initial rate of decarburization of pig iron. Given: carbon content %C = 1,34 + 2,54.10^{-3}T (°C) ; hydrodynamic condition at gas/metal interface (at metal side) is: D/δ = 0.5.10^{-2} cm.s^{-1} (D is carbon diffusion coeficient and the boundary layer thickness). Decarburization is due to an oxygen jet striking the surface at 1550°C. The required rate can be considered as equal to the flux of carbon. The interface concentration can be neglected.

5 - What is time necessary for a graphite sphere immersed in a, pure stagnant iron bath at 1550°C to dissolve completely? Assume that the dissolution rate is equal to the flux of carbon. Also given ρ_{Fe} = 7000 kg.m^{-3}; Dc = 7.8 x $10^{-5}$$cm^2.s^{-1}$ The concentration at the graphite -metal interface is equal to the saturation value namely: %C = 1.34 + 2.54 x 10^{-3} T (°C).

6 - A droplet of liquid iron (6.4 mm diameter) containing initially 10 ppm of H_2 is subjected to a stream of Argon gas (velocity : 3 m.s^{-1}) at 1920 K. Assume that the solubility of H_2 at 1920 K and 1 atm is 31 cm^3/100 g of Fe and that there are no concentration gradients inside the droplet. Estimate the time necessary for removing 90% do H_2, if the rate of removal of H_2 is equal to its transport rate through boundary layer at the gas side.

7 - According to surface renewal theory for a droplet traversing through a fluid the mass transfer coefficient is given by: $K = 2.(D_{AB}/(\pi.t_e))^{0.5}$ where t_e is the time required to the droplet to travel a distance equal to its diameter. Zinc gets removed from lead through a treatment in a fused salt bath containing $PbCl_2$ giving rise to a reaction (Zn (in Pb) + $PbCl_2$ = Pb + $ZnCl_2$), Droplets of Pb-Zn alloy go through the slag layer. Estimate the rate of mass transfer through the boundary layer of the salt assuming: $d_{droplet}$ = 2 x 10^{-3} m;$V_{droplet}$ = 0.1 m/s; D_{PbCl2} = $10^{-9}$$m^2.s^{-1}$; concentration of $PbCl_2$ in the fused salt is 1000 mole.m^{-3}.

8 - Experiments on dissolution of a sample of silver plate (1 cm2 area) in a cyanide solution (0.5 weight %) resulted in dissolution of 4.77 mg of Ag in one hour. Estimate the dissolution rate (moles/cm^2.s). Give the relation between dissolution rate of Ag and transport velocity of CN- . Consider that the process is controlled by transport of CN through the boundary layer. Estimate the boundary layer thickness, if the diffusion coefficient for CN amounts to 1.83 x 10^{-5} cm^2/s.

9 - A iron plate (0.01 cm thickness) is exposed to gaseous atmosphere (having CO) on either side of the plate so as to maintain 1.2% e 0.1% de C on the surfaces, respectively. What is the carbon flux at 1200 C, at steady state if $Dc = 1.625 + 0.747\ \%C - 1.088\ \%C^2 + 1.476\ \%C^3\ cm^2s^{-1}$?

10 - A plastic membrane "A", 0.3 mm thick separates two chambers with pressures of H_2O being 0.05 e 0 atm, respectively. The surface concentrations are in equilibrium with the respective partial pressure of vapours given by C (mole H_2O/cm^3) = 10^{-4}. P_{H2O} (atm). Estimate the flux of H_2O, at steady state. Given diffusion coefficient of H_2O in polymer is $10^{-9}(m^2/s)$.

11 - A small bubble of a desulphurizing gas, e.g. magnesium vapor, travels upward a molten steel bath. Magnesium from the bubble and sulphur from the steel react at the metal/gas interface according to Mg + S = MgS(s). The process is deemed to be controled by mass transfer of sulphur through a boundary layer at the steel side. Thus the process velocity can be estimated calculating the sulphur rate of transport in this region. Estimate K_S (sulphur), at 1873K given $D_s = 3\ 10^{-4}\ cm^2/s$; R = 0.003 m; $\rho_{steel} = 7000\ kg/m^3$; $\eta_{steel} = 0.007$ Pa.s; V = 0,3 m/s.

Steady and Unsteady State Transport of Species

Chapter 18

18.1 Species Conservation Balance

As in the case of transport of momentum or energy balance, conservation equations for chemical species or mass balance can be set up in a contol volume. The balance equation contains terms for rate of accumulation and transfer of species by diffusion and convection as well as generation and consumption of species due to chemical reactions. Like in the earlier chapters of momentum and heat transfer, mass balance can be carried out an infinitesimal control volume, chosen in the respective coordinates (cartesian, cylindrical or spherical) to get differential equations which can be solved by analytical or numerical techniques to get the concentration profiles. This chapter deals with development conservation equations and solutions under various conditions.

18.2 General Equation for Mass Transport

Consider a balance of conservation of the species A, in cartesian coordinate system in a control volume defined by orthogonal edges ΔX, ΔY and ΔZ as given in Figure 18.1. In this infinitesimal control volume a balance of transport terms in all the three directions namely x,y and z can be established.

Rate of accumulation of A within the control volume	=	Net rate of entry into the control volume through diffusion	+	Net rate of entry of the control volume by convection	+	Rate of production/consumption due to chemical reactions in the control volume

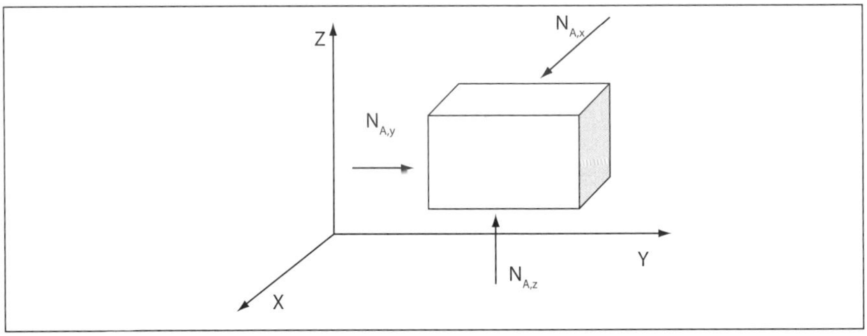

Figure 18.1 - Control Volume for mass balance to get the general equation for mass transport.

The diffusive and convective contributions are included in the transport equations. The various terms have been defined in the previous chapters:

$$N_{A,x} = v_x C_A - D_A \frac{dC_A}{dx} \tag{18.1}$$

$$N_{A,y} = v_y C_A - D_A \frac{dC_A}{dy} \tag{18.2}$$

$$N_{A,z} = v_z C_A - D_A \frac{dC_A}{dz} \tag{18.3}$$

The balance results in the following terms for accumulation, transport generation of species:

$$\frac{d}{dt}\left(\Delta x.\Delta y.\Delta z.C_A\right) = \Delta y.\Delta z.\left(N_{A,x}\left.\right|_{x=x} - N_{A,x}\left.\right|_{x=x+\Delta x}\right) + \Delta x.\Delta z.\left(N_{A,y}\left.\right|_{y=y} - N_{A,y}\left.\right|_{y=y+\Delta y}\right) + \Delta y.\Delta x.\left(N_{A,z}\left.\right|_{z=z} - N_{A,z}\left.\right|_{z=z+\Delta z}\right) + \Delta x.\Delta y.\Delta z.r_A \tag{18.4}$$

where:

- r_A represents the rate of production/consumption within the control volume as a result of a chemical reaction.

Dividing both sides by control volume and taking limits:

$$\frac{\partial C_A}{\partial t} + \frac{\partial N_{A,x}}{\partial x} + \frac{\partial N_{A,y}}{\partial y} + \frac{\partial N_{A,z}}{\partial z} = r_A \tag{18.5}$$

or, for a constant value of D_A:

$$\frac{\partial C_A}{\partial t} + \frac{\partial (v_x C_A)}{\partial x} + \frac{\partial (v_y C_A)}{\partial y} + \frac{\partial (v_z C_A)}{\partial z} - D_A \nabla^2 C_A = r_A \tag{18.6}$$

The same procedure can be followed for other coordinate systems, and the resulting expressions summarized in Table 18.1 (without the species generation or consumption term):

Rectangular	
$$-\frac{\partial C_A}{\partial t} = \frac{\partial N_{A,x}}{\partial x} + \frac{\partial N_{A,y}}{\partial y} + \frac{\partial N_{A,z}}{\partial z}$$	(18.7)
Cylindrical	
$$-\frac{\partial C_A}{\partial t} = \left[\frac{1}{r}\frac{\partial (r\,N_{A\,r})}{\partial r} + \frac{1}{r}\frac{\partial N_{A\,\theta}}{\partial \theta} + \frac{\partial N_{Az}}{\partial z}\right]$$	(18.8)
Spherical	
$$-\frac{\partial C_A}{\partial t} = \left[\frac{1}{r^2}\frac{\partial (r^2\,N_{A\,r})}{\partial r} + \frac{1}{r\sin\theta}\frac{\partial N_{A\,\theta}\sin\theta}{\partial \theta} + \frac{1}{r\sin\theta}\frac{\partial N_{A\,\varphi}}{\partial \varphi}\right]$$	(18.9)

Table 18.1 - Equations of conservation of species in different coordinate systems without considering generation or consumption of the species (POIRIER and GEIGER, 1994).

In the table above, the N_A values have convective terms and hence a knowledge of fluid flow conditions or estimation of velocity field in the system simultaneously (to assess the contribution by convection) is essential. In simple cases, the velocity field can be imposed or forced convection can be considered separately. However momentum and mass transport are interactive. When the changes in compositions are strong enough to induce differences in density this can have an influence on convection. In more complex situations, there can be simultaneous heat, mass and momentum in particular operations with significant interactive effects. In these cases relevant equations have to solved simultaneously to get velocity ,temperature and concentrations profiles. One can cite, as an example, modeling of the blast furnace process. In these cases the situation becomes more complicated as two or three dimensions have to be considered. In such cases, one has to resort to numerical solutions of the equations. For simple processes, the equations with assumptions can be solved through analytical methods. Some examples are discussed below.

Unidimensional problems of transport of species, associated with convection, can be more easily solved if the velocity field is known. Consider the case of a liquid film (BIRD, STEWART and LIGHTFOOT, 2002) of thicknes δ and viscosity η flowing in a vertical plane (Figure 18.2). Assuming laminar unidirectional flow and steady state conditions for an incompressible fluid, in cartesian coordinate system, the velocity profile can be derived adopting the same procedure as given in chapter 5. (section 5.3.2) and is given by:

$$V_z = \frac{\rho\, g\, \delta^2}{2\,\eta}\left[1 - \left(\frac{x}{\delta}\right)^2\right] \tag{18.10}$$

If this film is brought into contact with a gaseous phase containing a soluble species A in the liquid, then it would be possible to analyze different scenarios of transport, depending on the combinations of variables such as diffusivity and contact time. The configuration shown in Figure 18.2, should be valid for the case where the penetration capability of the specie is small. This assumption implies that the viscosity and density of the liquid do not change substantially and hence the velocity profile is not affected. One can develop a mass balance for the component A, assuming that at the gas liquid interface. A has the saturation value of concentration and hence there is diffusion into the liquid as the film drops, it is exposed to this value for more time and hence concentration changes with x and z. A mass balance for specie A in an infinitesimal volume assuming y direction as unity, results in the following equation:

$$\frac{\partial N_{Az}}{\partial z} + \frac{\partial N_{Ax}}{\partial x} = 0 \tag{18.11}$$

where:

$$N_{Az} \cong C_A V_z \tag{18.12}$$

In the z direction the contribution of diffusion can be ignored as A moves in the z direction mainly due to bulk flow. In the x direction A is transported mostly by diffusion.

$$N_{Ax} \cong -D_A \frac{\partial C_A}{\partial x} \tag{18.13}$$

The assumption that in the direction z, transport by convection is much more important than transport by diffusion, is reasonable unless the rate of descent of the liquid is extremely low. On the other hand in the direction x, it is reasonable to ignore convective transport compared to diffusion.

Therefore:

$$\frac{\rho g \delta^2}{2\eta}\left[1-\left(\frac{x}{\delta}\right)^2\right]\frac{\partial C_A}{\partial z} = D_A \frac{\partial^2 C_A}{\partial x^2} \tag{18.14}$$

equation to be solved considering the boundary conditions:

$$Z = 0 \quad C_A = 0 \tag{18.15}$$

$$X = 0 \quad C_A = C_A^o \tag{18.16}$$

$$X = \delta \quad \frac{\partial C_A}{\partial x} = 0 \tag{18.17}$$

Where C_A^o represents the concentration of A in the liquid-gas interface, assumed to be constant throughout the film and equal to the saturation concentration.

A further simplification can be added if it is considered that the region of penetration is restricted to a narrow layer of liquid (relative to the total thickness of the film) where the velocity does not differ substantially from the maximum. Then equation 18.14 can be written as:

$$\frac{\rho g \delta^2}{2 \eta} \frac{\partial C_A}{\partial z} = D_A \frac{\partial^2 C_A}{\partial x^2} \quad (18.18)$$

subject to boundary conditions:

$$Z = 0 \quad C_A = 0 \quad (18.19)$$

$$X = 0 \quad C_A = C_A^o \quad (18.20)$$

$$X = \infty \quad \frac{\partial C_A}{\partial x} = 0 \quad (18.21)$$

Transposing all the constant quantities in equation 18.18 to the right side and considering semi-infinite situation, an error function solution for the concentration profile is possible as given below (Exercise: Solve equation 18.18 with the above boundary conditions and arrive at the following equation).

$$\frac{C_A}{C_A^o} = 1 - \operatorname{erf} \frac{x}{\sqrt{\dfrac{8 \eta D_A z}{\rho g \delta^2}}} \quad (18.22)$$

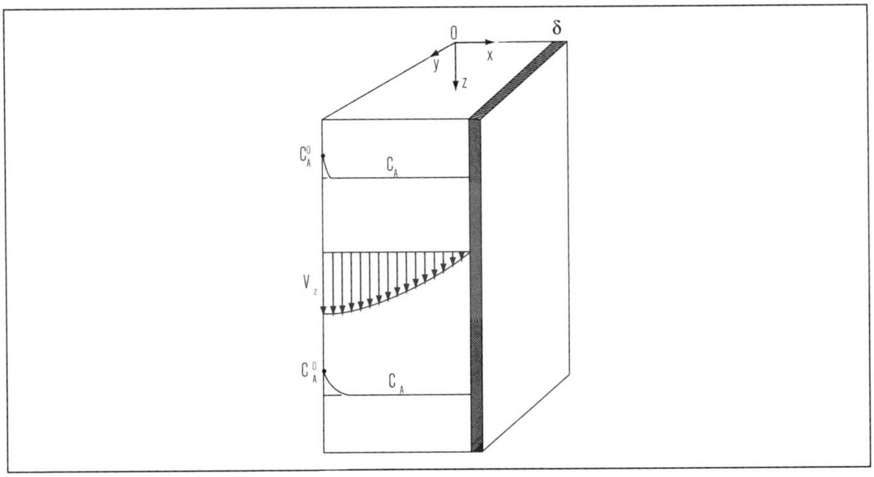

Figure 18.2 - Absorption of a solute in contact with a liquid film downward.

Example

Consider a pipe containing a gas A with a concentration of C_A^o (mol/m^3). This pipe is connected to an environment containing a gas B with concentration C_B^L (mol/m^3), through a duct of length L (m) and radius R (m), as given in the figure below. One needs to pressurize the pipe such that leak through the duct, of gas A is controlled, and the gas in the pipe is not contaminated by the gas in the environment. Since B should not contaminate A, the net flow of B through the duct should point outwards. This means that convection forcing B out of the pipe must be stronger than diffusion driving B into the tube. The critical condition for this is given below:

$$N_B = v^* C_B - D_B \frac{dC_B}{dx} = 0$$

Transposing and integrating considering boundary conditions: (concentrations of gas are given below in the figure)

$$\frac{C_B}{C_B^L} = e^{-\frac{v(L-x)}{D}}$$

This expression can be used to calculate the level of contamination C_B^o depending on the velocity of exhaust gas. For example, for a contamination less than 1 ppm, through a duct of 0.001 x 0.01 (m), where diffusion coefficient in gas phase is D = 10^{-6} m^2/s, one can calculate

$$10^{-6} = e^{-\frac{v(0.01-0)}{10^{-6}}}$$

which implies v = 0.00138 (m/s) and Q = 10^{-3}cm^3/s.

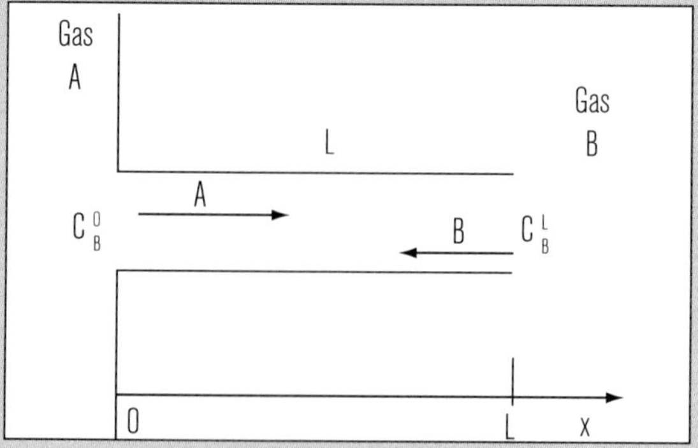

Escape control of gas to prevent contamination.

In some specific situations like the study of combustion of carbon pellet with oxygen gas in an apparatus given in Figure 18.3, at 1000°C and 1 atm, an analysis can be carried out on the basis of diffusion equations coupled with convection (MIDDLEMAN, 1998).

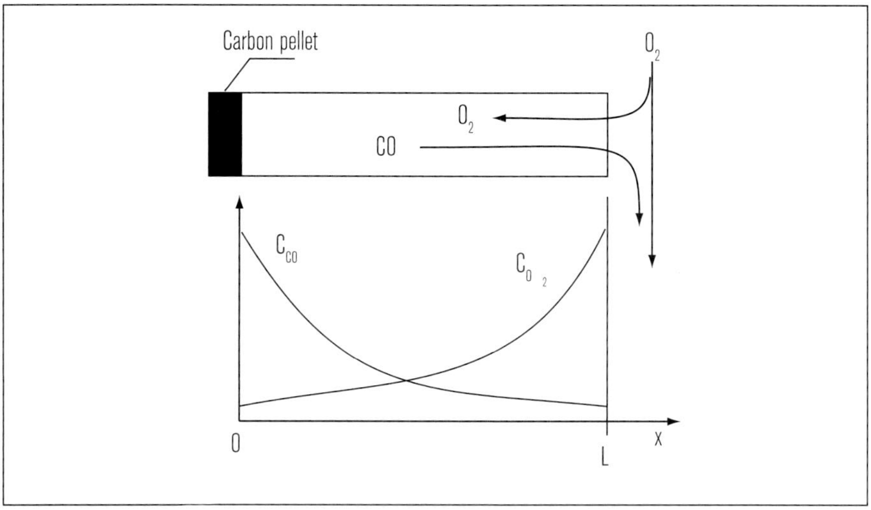

Figure 18.3 - Oxidation of carbon pellet by oxygen at high temperatures.

A thermodynamic analysis indicates that mainly carbon monoxide is produced as a result of the the combustion reaction, $2C(s) + O_2(g) = 2CO(g)$. The gas in the small reaction tube above should be made up of CO, and O_2.

Assuming quasi-stationary state, for which there is no accumulation of reactants and products anywhere in the system:

- the transport rate of O_2 is equal to the rate of consumption by chemical reaction;
- the transport rate of CO is equal to the rate of production by chemical reaction.

Hence according to the stoichiometry of reaction:

$$2C(s) + O_2(g) = 2CO(g)$$

$$-2\,N_{O2} = N_{CO} \tag{18.23}$$

Substituting these in the flux equation for rate of transport of CO:

$$N_{CO} = -D_{CO}\frac{dC_{CO}}{dx} + X_{CO}\left(N_{CO} + N_{O_2}\right) \tag{18.24}$$

$$N_{CO} = -D_{CO}\frac{dC_{CO}}{dx} + X_{CO}\left(N_{CO} - \frac{1}{2}N_{CO}\right) \tag{18.25}$$

$$N_{CO} = -D_{CO}\frac{dC_{CO}}{dx} + X_{CO} \times \frac{1}{2}N_{CO} \tag{18.26}$$

Also assuming ideal behavior of gases, PV = nRT, one can write for the relationship between partial molar concentration, molar fraction and total molar concentration:

$$C_i = x_i . C_T \tag{18.27}$$

$$C_T = \frac{P}{RT} = \frac{1}{0.082 \times 1{,}273} = 9.58 \times 10^{-3} \text{ mole/liter}$$

From these results:

$$N_{CO} = -D_{CO}C_T\frac{dX_{CO}}{dx} + X_{CO} \times \frac{1}{2}N_{CO} \tag{18.28}$$

$$N_{CO} = -\frac{D_{CO} . C_T}{1 - \frac{1}{2}X_{CO}}\frac{dX_{CO}}{dx} \tag{18.29}$$

Boundary conditions for the integration of this equation are derived from the fact that at the interface between carbon pellet and pure carbon monoxide, ie at x = 0 is $X_{CO} = 1$. Similarly at the opposite end of the reaction tube ,it is pure oxygen, ie, at x = L, $X_{CO} = 0$. Hence is possible to get the concentration profiles of CO gas and oxygen in the tube , as given in the Figure 18.3.

Example

A possible arrangement for the determination of diffusivity in a binary gas phase AB is given in the following figure. The apparatus consists of a quartz capillary containing a pool of liquid metal, a gas stream of known composition AB is flowing at other end of the capillary, and the whole system is kept under constant temperature and pressure and is at steady state. Accordingly, it provides a potential for diffusion of species A as a result of the difference in concentration (partial pressure or chemical potential) the interface between the liquid-gas and end of capillary. Similarly there is a concentration gradient established for the component B. The flow of the species can be indirectly measured by a thermo-balance to record the loss of mass by evaporation of the metal in the pool so that the diffusion coefficient can be evaluated.

Supposing manganese is used as the metal in the pool with flow of pure argon at the end of capillary, then the gas phase is Manganese (gas A)-Argon(gas B). The experimental conditions are the following: effective length of capillary, L, equal to 0.02 m, temperature 1600°C, total pressure amounting to 1.013 x 10^5 Pa. At this temperature, the vapour pressure of Manganese above the liquid is 5370 Pa. The weight loss recorded by the thermal balance is 5.84 x 10^{-3} moles of Mn/m^2s^{-1}.

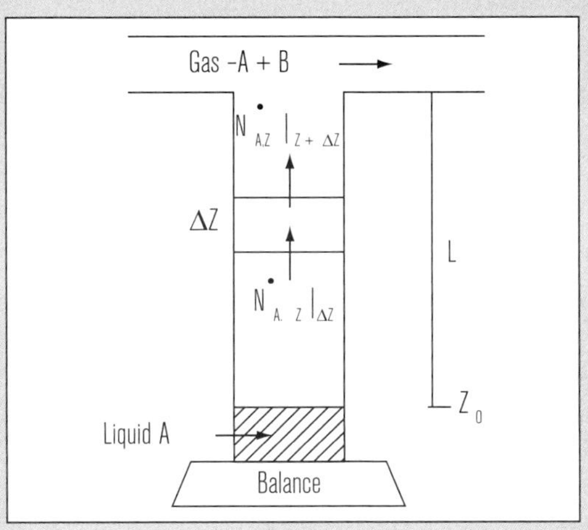

Stefan Apparatus for measurement of diffusivity in binary gases (POIRIER and GEIGER, 1994).

The equation for transport of argon in the gas phase at steady state is given by:

$$N_{Ar} = -D\frac{dC_{Ar}}{dz} + X_{Ar}(N_{Ar} + N_{Mn})$$

Argon is an inert gas, and is nonreactive and insoluble in the metal and hence it can be considered that argon flow is zero.Therefore $N_{Ar} = 0$. Hence in this case the transport equation of Mn only has to be considered

$$N_{Mn} = -D\frac{dC_{Mn}}{dz} + X_{Mn}$$

$$N_{Mn} = -D\,C_T\frac{dX_{Mn}}{dz} + X_{Mn}\,N_{Mn}$$

where:

- C_T represents the total concentration in the gas phase, estimated from $C_T = P_T/RT(\text{moles}/m^3)$;
- X_{Mn} represents the molar fraction of manganese in gas phase;
- D the binary diffusivity;
- N_{Mn} the molar flow of manganese.

In this equation (ideal gas law): P_T is the total pressure, R the gas constant, T the absolute temperature.

Therefore the previous equation can be easily integrated, taking in consideration that the Mole fraction of Mn at z = L, X^L_{Mn} can be considered virtually nil, because at this position there is a stream of pure argon. Also the mole fraction of Mn in the liquid-gas interface, X^o_{Mn} can be estimated as the ratio of the vapor pressure of Mn (assuming local equilibrium at this point) and the total pressure.

$$N_{Mn} - X_{Mn}\,N = -D\,C_T\frac{dX_{Mn}}{dz}$$

$$N_{Mn}\,dz = -D\,C_T\frac{dX_{Mn}}{1-X_{Mn}}$$

Integrating:

$$N_{Mn} = \frac{D\,C_T}{L}\ln\frac{1-X^L_{Mn}}{1-X^0_{Mn}}$$

The only unknown in the above expression is the diffusion coefficient which can be evaluated. As such:

$$C_T = P_T/RT = 1.013 \text{ x } 10^5/(8.31 \text{ x } 1873) = 6.508 \text{ moles}/m^3$$

$$5.84 \text{ x } 10^{-3} \text{ moles of Mn} / m^2 s^{-1.} = D\frac{6.508}{0.02}\ln\frac{1-0}{1-5370/1.013 \text{ x } 10^5}$$

$$D = 13{,}95 \text{ x } 10^{-3} \text{ m2/s.}$$

18.3 Analogies with Heat Transport

There is no perfect physical analogy between heat and mass transport. For example there is no equivalent of radiation in transport of mass. Heat conduction does not involve convection in the medium. Also unlike temperature which is the driving force in heat transfer the composition gradient in mass transport is not necessarily the driving force for diffusion.

However, there are similarities in mathematical forms of differential equations, boundary conditions and some of the correlations. The mathematical methods employed in analytical and numerical methods of resolution, of differential equations are similar in both the fields and are applicable to both. In fact the fields present a significant analogical behaviour.

For example, consider the unidirectional diffusion in a semi-infinite solid in the form of a long bar initially at a uniform temperature. When the temperature is suddenly changed at one end, the other end and the vicinity continues to be at the initial temperature remains so throughout the experiment. Assuming constant properties (thermal conductivity, heat capacity and density) of the material of the bar, the variation of temperature with time, T(x,t), (Figure 18.4), can be described by the differential equation:

$$\frac{\partial T}{\partial t} = \alpha \frac{\partial^2 T}{\partial x^2} \qquad (18.30)$$

and the boundary conditions being,

$$T(x = 0;t) = T_0;\ T(x = \infty;t) = T_i;\ T(x;t = 0) = T_i \qquad (18.31)$$

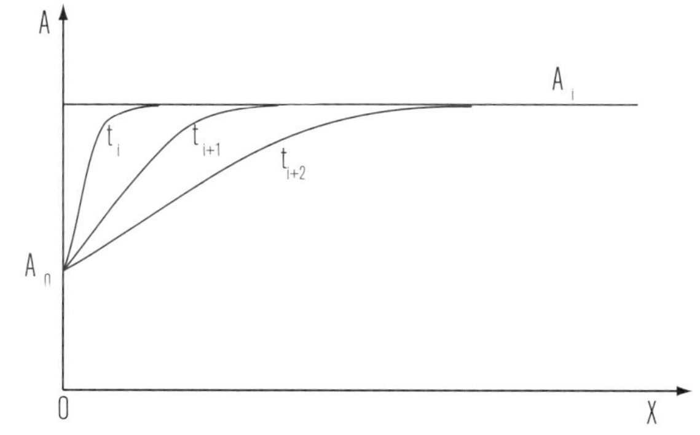

Figure 18.4 - Temperature profile as a function of time and position for semi-infinite solid.

The temperature profiles can be calculated from the error function equation below which is the solution for the differential equation 18.31 (see Chapter 12, section 12.5).

$$\frac{T - T_o}{T_i - T_o} = \text{erf}\left(\frac{X}{2\sqrt{\alpha t}}\right) \tag{18.32}$$

where $\alpha = K/\rho C_p$ is the thermal diffusivity of the material of the bar. The validity of this solution requires that the matrix remains static and that the heat transfer is by diffusion only.

From this expression one can define the distance of penetration δ, which denotes the position of an observer at time t, who can identify any change (for example, at approximately 1%) introduced in the face of the bar, located at z = 0, as:

$$\delta_{1\%} = 4\sqrt{\alpha t}\ \text{m} \tag{18.33}$$

The same equation calculates the instantaneous flow of heat in the face at z = 0 as:

$$q_{z=0} = \frac{K}{\sqrt{\pi \alpha t}}\,(T_o - T_i)\ \text{J/m}^2.\text{s} \tag{18.34}$$

The above equation gives the total quantity of heat exchanged with the environment, in an elapsed time scale equal to t_e:

$$q = 2K\sqrt{\frac{t_e}{\pi \alpha}}\,(T_o - T_i)\ \text{J/m}^2 \tag{18.35}$$

A similar situation can be cited in mass transport. Consider the same bar long enough where there is unidirectional diffusion considering the bar as semi-infinite. The bar has an initial concentration of C_i in respect of an interstitial species like Carbon or Nitrogen (in iron or any other base metal). At one end of the bar this value suddenly changes to C_0 and remains at this value throughout the experiment. Similar to the temperature profiles shown in Figure 18.4 the concentration profiles can be presented for specie A, as a function of concentration and position. Solving the differential equation that describes the variation of concentration, adopting the same procedure as for the heat transport equation, discussed above:

$$\frac{\partial C}{\partial t} = D\frac{\partial^2 C}{\partial x^2} \tag{18.36}$$

This is known as Fick's second law and can be solved with initial and boundary conditions:

$$C(x = 0;t) = C_0; C(x = \infty;t) = C_i; C(x;t = 0) = C_i \qquad (18.37)$$

The solution is similar to the one obtained in heat transport with D substituting for thermal diffusivity and concentration terms for temperatures.

$$\frac{C - C_o}{C_i - C_o} = \text{erf}\left(\frac{X}{2\sqrt{D\,t}}\right) \qquad (18.38)$$

Similar to the case of heat conduction, instantaneous and average mass flow at the interface, penetration, distance etc can be evaluated from this solution knowing the relevant data.

The distance of penetration is given by:

$$\delta_{1\%} = 4\sqrt{D\,t} \quad \text{m} \qquad (18.39)$$

Instantaneous flow of species in the face at z = 0 and the quantity exchanged with the environment are respectively:

$$j_{z=0} = \sqrt{\frac{D}{\pi\,t}}\,(C_o - C_i) \quad \text{mol/m}^2\text{.s} \qquad (18.40)$$

$$m = 2\sqrt{\frac{D\,t_e}{\pi}}\,(C_o - C_i) \quad \text{mol/m}^2 \qquad (18.41)$$

The mathematical and physical analogy between the two problems is quite clear. It should be remembered the conditions are "only the heat or species A moves by diffusion, the matrix remains static".

The exercise of identifying mathematical and physical analogies between transport processes is very important as solutions are facilitated in addition to contributions to innovations and solving new problems.

Example

A gear of low carbon steel is subjected to a surface hardening operation by raising the level of surface carbon through a carburizing treatment. The purpose is to impart to the surface a high hardness after necessary heat treatment,wear resistance etc, and at the same time maintaining the core with low carbon, so as to enable absorption of shock without breakage. Carbon penetrates the steel surface as given in the in the following figure. The flow of carbon is in reality three dimensional and transient and would require numerical solutions. However, as an approximation, one can consider that the depth affected is very small compared with the part dimensions (eg tooth thickness, width of gear). Hence unidirectional mass flow can be considered. Diffusion of carbon can be considered in the semi-infinite medium. Then the distance is given by (for erf = 0.99)

$$\delta_{1\%} = 4\sqrt{D\,t}$$

where D is the diffusion coefficient of carbon in in austenite given by:

$$D = 0.372.e^{-17816/T}\,cm^2 s^{-1}$$

For 1-hour exposure at 1000°C the penetration is given by:

$$\delta_{1\%} = 4\sqrt{3{,}108 \text{ x } 10^{-7} \text{ x } 3{,}600} = 0.134\,cm$$

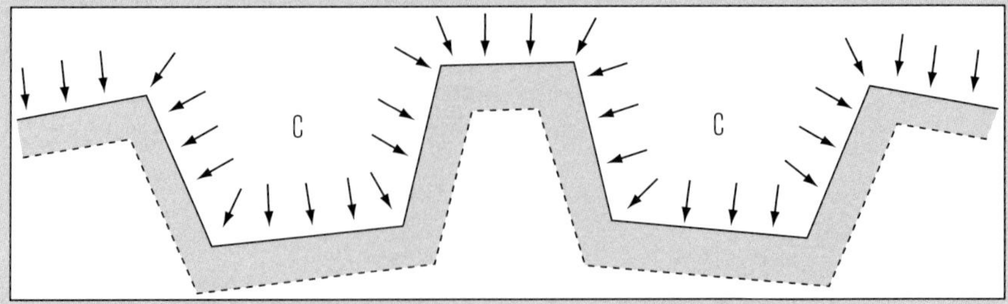

Schematic figure of gear surface and carbon penetration.

This exercise can be extended to other situations like the following. Consider the case of a solid, initially containing specie A, uniformly distributed, with a concentration of C_A^o. It gets suddenly immersed in a fluid, also containing the same species but with a concentration of C_A^∞. As the two phases contain the same species A, a condition of equilibrium can get established (at a constant temperature and pressure) between them (solid and fluid):

$$\mu_A\,(\text{solid}) = \mu_A\,(\text{fluid}) \tag{18.42}$$

where μ_A (solid) and μ_A (fluid) are chemical potentials of component A in the solid phase and fluid phase, respectively.

The partition coefficient at the surface for component A is given by:

$$L_E = \left(\frac{C_A^{\text{fluid}}}{C_A^{\text{solid}}}\right)_{\text{equilibrium}} \tag{18.43}$$

ie, there should be a driving force for mass transfer if the ratio of concentrations, C_A^∞/C_A^0 initially is not according to the equilibrium concentration as shown by equation 18.43. A likely composition profile is given in Figure 18.5.

There are gradients of composition within the two phases. Also note that the composition profile in both the solid and the fluid would be a function of time and position and that there is (unlike the analogous problem of heat transfer) a discontinuity at the solid / fluid interface in terms of concentration of species A.

Like the Biot number in the heat transfer applications, giving the ratio of heat transfer coefficient and thermal conductivity, one can have a similar analogy in the transport of a specie in respect of the ratio between mass transfer coefficient and diffusion coefficient of a particular specie. Biot Number in this case gives an idea regarding the speed of diffusion inside the solid, relative to mass transfer in the fluid at the vicinity of the surface (MILLS,1999):

$$\frac{k_m\,(\text{in fluid}).L}{D_A\,(\text{in solid})} = \text{Biot number for mass transfer} \tag{18.44}$$

For Biot < 0.1. as in heat transfer (temperature gradients inside the solid are insignificant) in this case also the concentration or composition of a specie in the solid can be considered uniform (Figure 18.6).

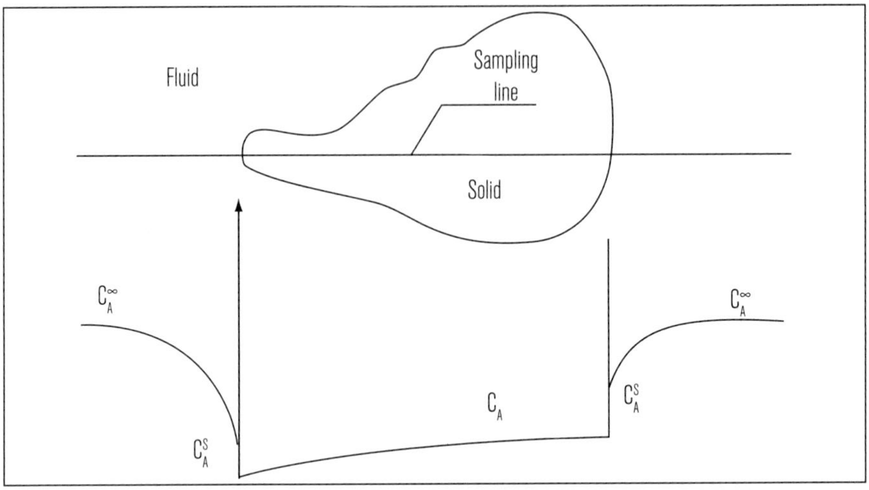

Figure 18.5 - Gradients of concentration for the mass transfer between a solid and a fluid in the transient regime.

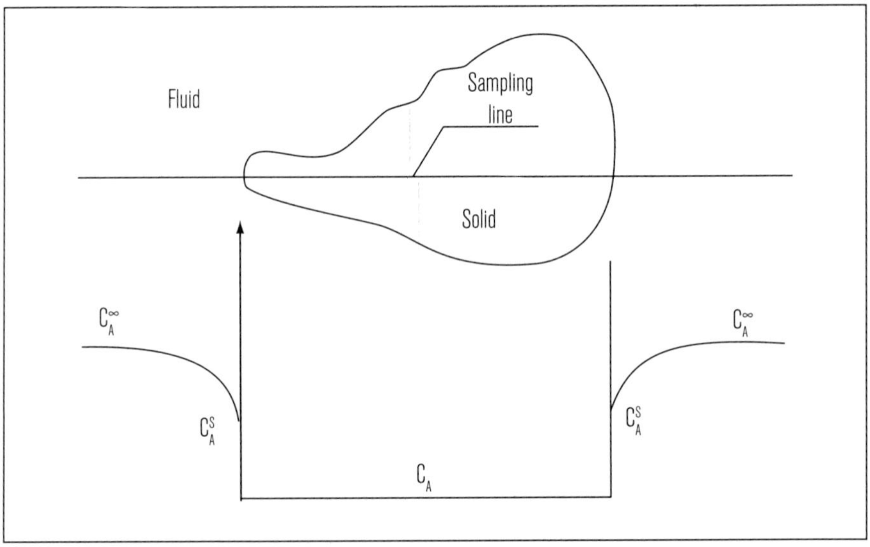

Figure 18.6 - Gradients of concentration for the mass transfer between a solid and a fluid, transient and Newtonian.

A balance of conservation for species A in the particle is given by:

$$\frac{dVC_A}{dt} = \pm S.K_m\ (\overline{C_A^\infty} - C_A^S) \tag{18.45}$$

where V represents the volume of the particle (m^3), S is the surface area of particle (m^2) and the sign ± should be chosen to indicate transport to or from the particle; C_A^S which is the concentration in the fluid-solid interface is unknown. Since the partition coefficient is known for the equilibrium at the interface, $L = \frac{C_A^S}{C_A}$ (note that the local fluid concentration C_A is different from C_A^∞, which makes for the boundary layer), C_A^S in equation 18.45 can be substituted.

$$\frac{dVC_A}{dt} = \pm S.K_m\ (\overline{C_A^\infty} - L.C_A) \tag{18.46}$$

The above equation can be integrated with the initial condition ($t = 0$; $C_A = C_A^o$). In this expression L is the partition coefficient to be determined from the equilibrium constant, established for the surface of the particle.

The treatment corresponds to the analogous one discussed in heat transfer known as Newtonian cooling (or heating) discussed in Chapter 12.

Other analogies from heat transfer can be applied to problems of mass transport in the transient regime.

For example, when a plate of infinite length and thickness 2L, initially at T_i is immersed in a fluid at T_f, a temperature profile (variable over time) develops as given in Figure 18.7. In this case the temperature at the interface of solid and liquid is the same as there is perfect contact between these phases. In a situation of excess fluid T_f also represents the temperature of the solid in thermal equilibrium with the fluid, ie a uniform temperature distribution throughout the system.

This situation has been discussed in Chapter 12, section 12.2. The objective of dealing with it again is to highlight the similarities and differences. In heat transfer dimensionless variables were defined to facilitate solutions:

$$\theta = \frac{T - T_f}{T_i - T_f} \tag{18.47}$$

which transforms the unidirectional unsteady state equation for heat transfer (analogous to Fick's Second Law) as:

$$\frac{\partial T}{\partial t} = (K/\rho\ C_p)\frac{\partial^2 T}{\partial x^2} = \alpha\frac{\partial^2 T}{\partial x^2} \tag{18.48}$$

$$\frac{\partial \theta}{\partial t} = \alpha\frac{\partial^2 \theta}{\partial x^2} \tag{18.49}$$

where $\alpha = K/\rho C_p$ is the thermal diffusivity of the solid, K the thermal conductivity; ρ is the density and C_p the specific heat at constant pressure. The initial condition and boundary conditions are given Figure 18.7:

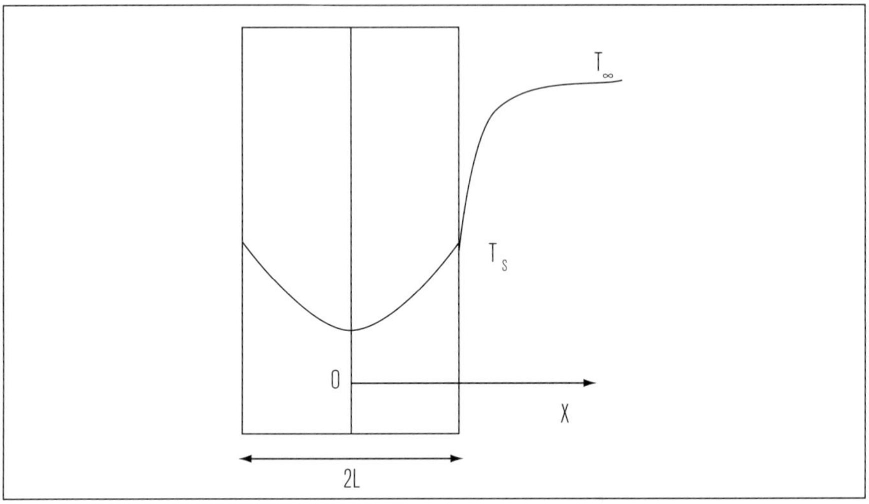

Figure 18.7 - Distribution of temperature in a plate of thickness 2L subject to convective environment.

1) initial condition, uniform distribution of temperature:

$$T(x,t=0) = Ti \tag{18.50}$$

$$\theta\ (x,t=0) = \frac{T_i - T_f}{T_i - T_f} = 1 \tag{18.51}$$

2) boundary condition, symmetry (assumed) with respect to the plane at x = 0:

$$\frac{\partial T}{\partial x}\Big|_{x=0} = 0 \tag{18.52}$$

$$\frac{\partial \theta}{\partial x}\Big|_{x=0} = 0 \tag{18.53}$$

3) boundary condition, representing the condition of no heat accumulation in the interface fluid/plate:

$$-k\frac{\partial T}{\partial x}\Big|_{x=\pm L} = h\ (T_f - T_{x=\pm L}) \tag{18.54}$$

$$-k\frac{\partial \theta}{\partial x}\Big|_{x=\pm L} = -\ h\ \theta_{x=\pm L} \tag{18.55}$$

The introduction of dimensionless variable for temperature θ has the objective to relate the various dimensionless numbers which form part of solution describing the temperature profile (Figure 18.8):

$$\theta = \frac{T - T_f}{T_i - T_f}, \xi = \frac{x}{L}, Fo = \frac{\alpha t}{L^2}, Bi = \frac{h L}{K} \qquad (18.56)$$

As described in Chapter 11, the solution comes (TOSUN, 2002) in the form of an infinite series:

$$\theta = 2\sum_{n=1}^{\infty} \frac{\sin \lambda_n}{\lambda_n + \sin \lambda_n \cos \lambda_n} e^{-\lambda_n^2 Fo} \cos \lambda_n \xi \qquad (18.57)$$

In this expression λ_n is the nth root of the equation:

$$\lambda_n \tan \lambda_n = Bi \qquad (18.58)$$

The first five values of the roots are presented in Table 18.2. For a given Biot number many roots as necessary to ensure convergence. (more details in Chapter 12, Table 12.1).

Bi	λ_1	λ_2	λ_3	λ_4	λ_5
0	0.000	3.142	6.283	9.425	12.566
0.1	0.311	3.173	6.299	9.435	12.574
0.5	0.653	3.292	6.3623	9.477	12.606
1.0	0.860	3.426	6.437	9.529	12.645
2.0	1.077	3.644	6.578	9.630	12.722
10	1.429	4.306	7.228	10.200	13.214

Table 18.2 - Roots of equation 18.58.

It can be shown (TOSUN, 2002) based on the equations above that the heat flow through the plate-fluid interface is given by:

$$q = \frac{4K(T_\infty - T_o)}{L} \sum_{n=1}^{\infty} \frac{\lambda_n \sin^2 \lambda_n}{\lambda_n + \sin \lambda_n \cos \lambda_n} e^{-\lambda_n^2 Fo} \qquad (18.59)$$

The relation between the dimensionless variables is given in the following Figure 18.8.

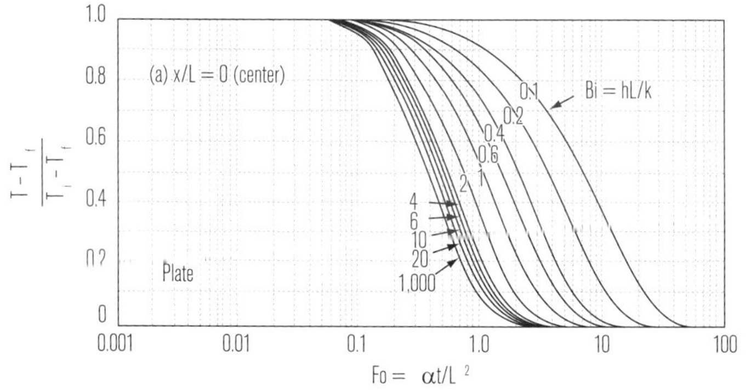

Figure 18.8 Abacus solution of temperature distribution in a plate of thickness 2L subject to convective environment (POIRIER and GEIGER, 1994).

Similar procedure can be adopted for transport of species by defining the appropiate dimensionless variables in mass transfer. Starting from Ficks 2nd law, the equations can be formulated and solved using proper boundary conditions. It is be possible to get analogical relations for mass transfer, like in heat transfer, as given below.

In an analogous situation of transfer of a specie in a plate of thickness 2L, initially at uniform concentration C_A^i, gets immersed in a fluid containing the same specie at a concentration of C_A^∞. A schematic profile of composition that would result is given in Figure 18.9. Unlike the case of heat transfer, treated earlier, there is a discontinuity in the concentration profile at the solid-fluid interface. The composition on the left side of interface (in solid) is not necessarily equal to the composition on the right side of Interface (fluid). There are even distributions in solid and in fluid. However there is a discontinuity of concentration, but not of chemical potential at the interface, which enables to define the partition coefficient:

$$L_E = \left[\frac{C_A\,(\text{solid})}{C_A\,(\text{fluid})}\right]_{\text{equilibrium}} \tag{18.60}$$

In the presence of excess fluid, the solid equilibrium concentration would be given by:

$$L_E = \frac{C_A\,(\text{solid})}{C_A^\infty} \tag{18.61}$$

The profile of concentration in solid can be estimated from the transport equation that considers only the diffusive contribution (ignoring convective contribution) This a fair assumption for impurities and interstitials. Initially, like in the case of heat transfer a dimensionless variable of concentration (analog of temperature) is defined by:

$$\theta = \frac{C_A - L_E\, C_A^\infty}{C_A^i - L_E\, C_A^\infty} \tag{18.62}$$

According to Fick's Second Law, diffusion in solid is governed by:

$$\frac{\partial C_A}{\partial t} = D\frac{\partial^2 C_A}{\partial x^2} \tag{18.63}$$

In terms of 18.62 this is transformed to:

$$\frac{\partial \theta}{\partial t} = D\frac{\partial^2 \theta}{\partial x^2} \tag{18.64}$$

Initial and boundary conditions are given by:

1) initial condition, uniform distribution of composition:

$$C_A(x,t=0) = C_A^i \qquad (18.65)$$

$$\theta\,(x,t=0) = \frac{C_A^i - L_E\,C_A^\infty}{C_A^i - L_E\,C_A^\infty} = 1 \qquad (18.66)$$

2) boundary condition, symmetry (assumed) with respect to the plane at x = 0:

$$\frac{\partial C_A}{\partial x}\Big|_{x=0} = 0 \qquad (18.67)$$

$$\frac{\partial \theta}{\partial x}\Big|_{x=0} = 0 \qquad (18.68)$$

3) boundary condition, for the transfer of species at the fluid –solid interface:

$$\underbrace{-D\frac{\partial C_A}{\partial x}\Big|_{x=\pm L}}_{\text{in solid}} = \underbrace{K_m(C_A^\infty - C_A^S)}_{\text{in fluid}} \qquad (18.69)$$

This last boundary condition is also analogical to the problem of heat transfer discussed in Chapter 12. The heat flux entering the solid is equal to heat flux through the boundary layer. For heat transfer, there is no temperature discontinuity at solid/fluid interface. For mass transfer a discontinuity of the concentration profile at solid/fluid interface is a certainty. However the condition:

$$\underbrace{-D\frac{\partial C_A}{\partial x}\Big|_{x=\pm L}}_{\text{in solid}} = \underbrace{K_m(C_A^\infty - C_A^S)}_{\text{in fluid}} \qquad (18.70)$$

can be rewritten assuming local chemical equilibrium at the interface:

$$-D\frac{\partial C_A}{\partial x}\Big|_{x=\pm L} = K_m\left(C_A^\infty - \frac{C_A\big|_{x=\pm L}}{L_E}\right) = \frac{k_m}{L_E}\left(L_E\,C_A^\infty - C_A\big|_{x=\pm L}\right) \qquad (18.71)$$

$$-D\frac{\partial \theta}{\partial x}\Big|_{x=\pm L} = -\frac{k_m}{L_E}\,\theta_{\,x=\pm L} \qquad (18.72)$$

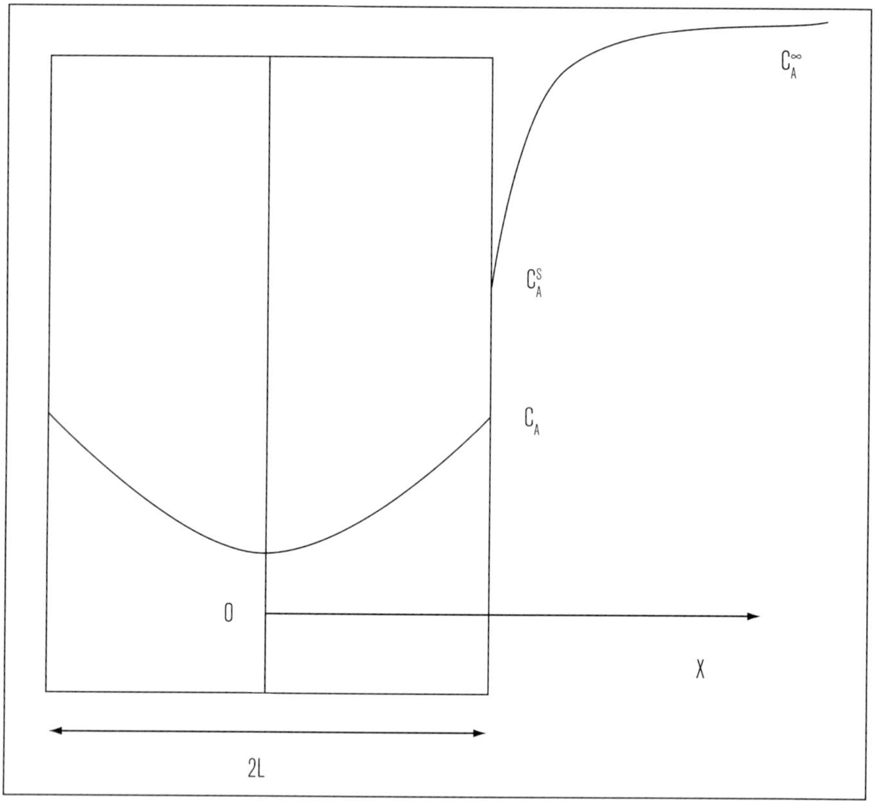

Figure 18.9 - Distribution of specie A between a plate of thickness 2L and a convective medium.

Adopting similar mathematical analogy of heat transfer, the following dimensionless parameters can be identified namely, relative position, Biot number for mass transfer Fourier number for mass transfer and dimensionless concentration respectively:

$$\xi = \frac{x}{L};\ Bi = \frac{k_m\, L}{D\, L_E},\ Fo = \frac{D\, t}{L^2},\ \theta = \frac{C_A - L_E\, C_A^\infty}{C_i - L_E\, C_A^\infty} \tag{18.73}$$

The values of above groups can be substituted in the equations 18.57 and 18.58 (derived for heat transfer situation but valid for the present situation because of analogy and use of dimensionless quantities) to get the following Figures 18.10 and 18.11 for the mass transfer situation giving a relationship between dimensionless concentration, Biot and Fourier (mass transfer) numbers for various geometric positions.

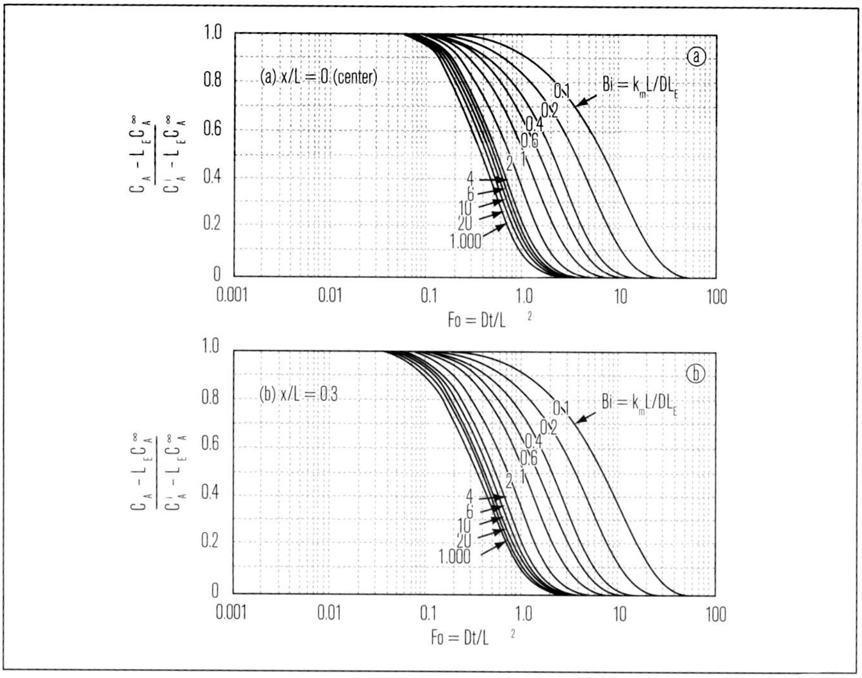

Figure 18.10 - Distribution of specie A within the plate in positions a) center b) 0.3 L from the center.

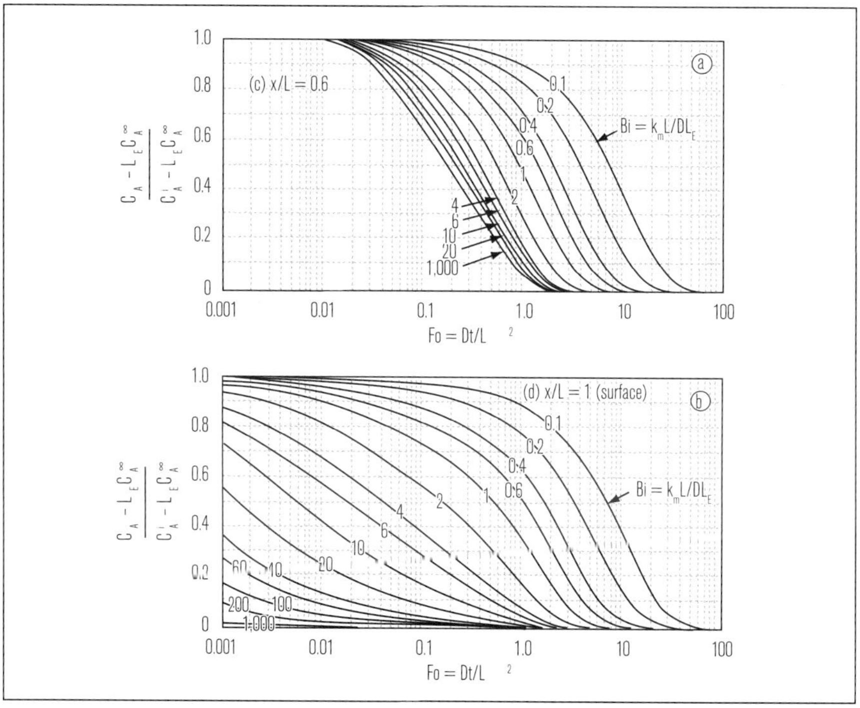

Figure 18.11 - Distribution of the specie A within the plate in positions a) 0.6 L from the center and b) at the surface.

Similar solutions can be worked out for other geometries like a cylinder of infinite length, initially with uniform composition C_A^i subjected to a contact with a fluid with the specie concentration C_A^∞.

The relevant dimensionless numbers defining positions, Biot, Fourier for mass transfer and concentration are similar to the previous case of the plate. However the basic equations are for cylindrical coordinates with similar boundary conditions. The method of analytical solutions is also different using Bessel functions. Values for the dimensionless parameters have to be substituted in equations for cylindrical coordinates discussed in chapter 11 of heat transfer to get graphical solutions like in the case of plate.

$$\xi = \frac{r}{R}; \; Bi = \frac{k_m R}{D L_E}; \; Fo = \frac{D t}{R^2}; \theta = \frac{C_A - L_E C_A^\infty}{C_i - L_E C_A^\infty} \qquad (18.74)$$

Here R is the radius of the cylinder and r the radial distance on a cylindrical coordinate system with axis z being identical with the axis of symmetry of the cylinder.The graphical solution shown in Figure 18.12, comes from analogy with the problem of heat transfer. The resulting solution is in the form of an infinite series described in Chapter 12, section 12.4.

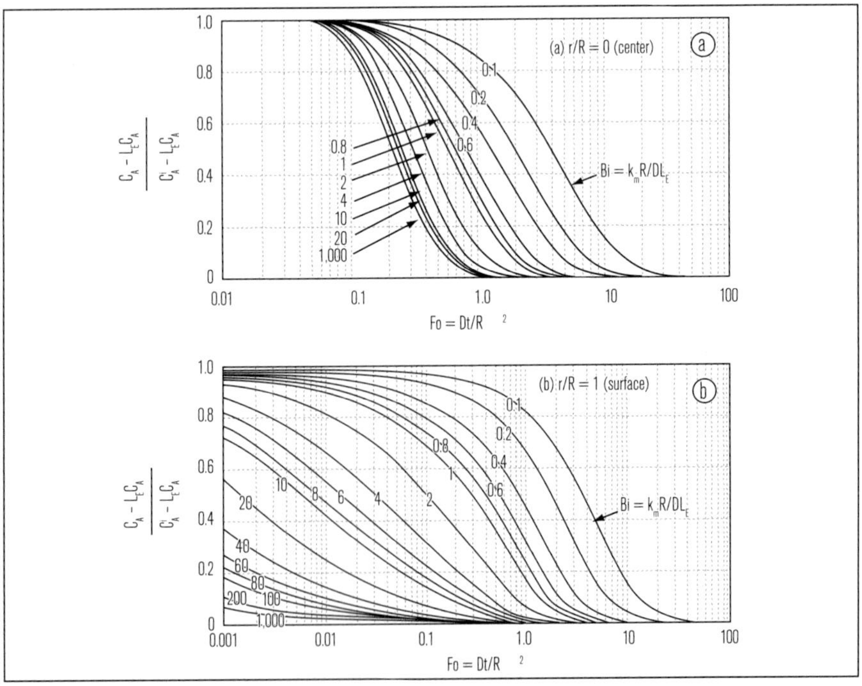

Figure 18.12 - Distribution of the specie A within the cylinder at positions: a) 0.6R from center; and b) at the surface.

The solution in the case of sphere with uniform composition initially, C_A^i, and exposed to a convective environment with a constant concentration, C_A^∞, is given in the Figures 18.13 e 18.14, in terms of dimensionless numbers like in the case of cylinder and plate.

$$\xi = \frac{r}{R};\ Bi = \frac{k_m R}{D L_E};\ Fo = \frac{D t}{R^2};\ \theta = \frac{C_A - L_E C_A^\infty}{C_i - L_E C_A^\infty} \tag{18.75}$$

This type of solution has already been discussed in Chapter 12.

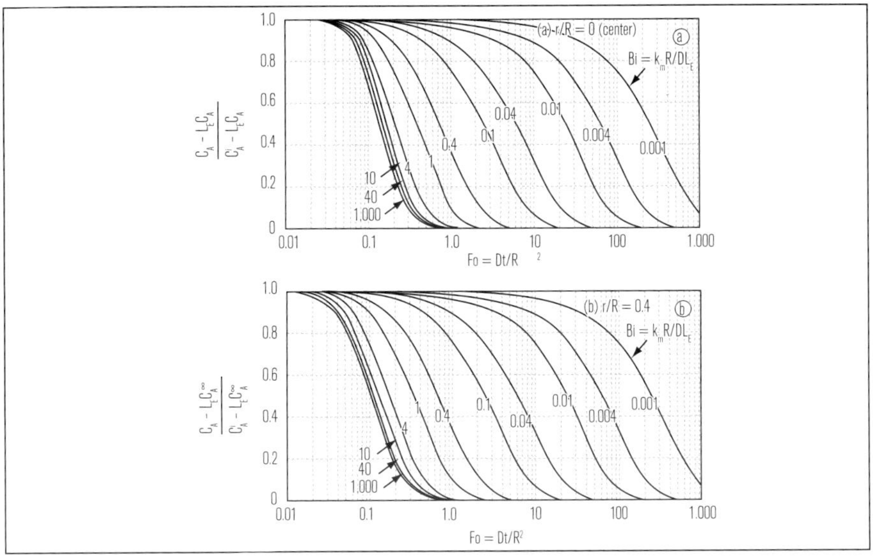

Figure 18.13 - Distribution of specie A in the sphere at positions: a) center; and b) 0.4 R from the center.

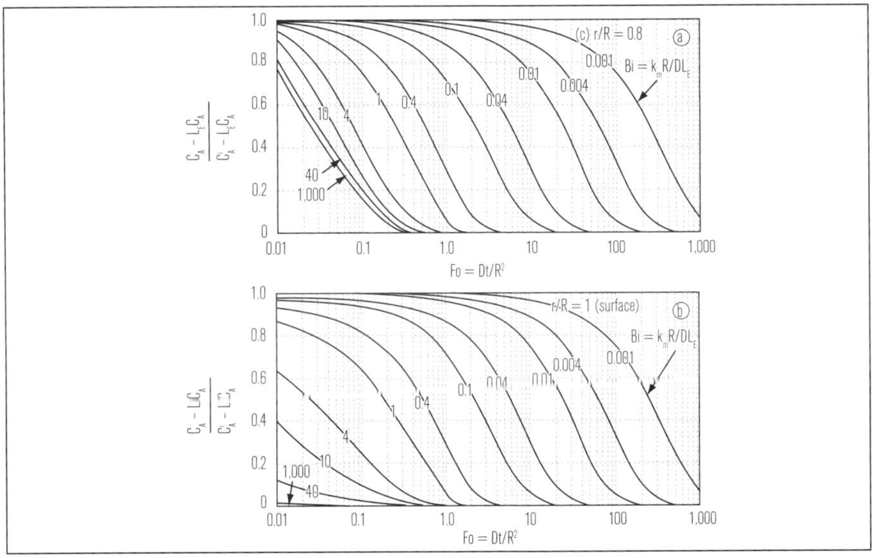

Figure 18.14 - Distribution of specie A in the sphere at positions: a) center; and b) 0.4 R from the center.

18.4 Combination of Solutions

In Chapter 12, two and three-dimensional diffusion of heat in a solid had been discussed (Section 12.6). The same approach can be employed for mass transfer.

Solutions for diffusion of species in a short cylinder can be evaluated as a product solution of an infinite cylinder of radius R and an infinite plate having a thickness of 2L, as shown in Figure 18.15. It can be shown that the product of individual solutions C (x, t) and C (r, t) is the solution for short cylinder. Knowing the Biot number and Fourier number at any location in the cylinder, the dimensionless concentration can be determined. For getting the solution in terms of infinite series the basic equation given below in cylindrical coordinates, can be solved using Bessel functions with incorporation of dimensionless numbers, as explained in the application to the plate using cartesian coordinates.

$$\frac{\partial C}{\partial t} = D\left[\frac{1}{r}\frac{\partial}{\partial r}\left(r\frac{\partial C}{\partial r}\right) + \frac{\partial^2 C}{\partial x^2}\right] \quad (18.76)$$

As given in Figure 18.16, other combinations can be considered. The examples considered in Chapter 12, based on solutions with dimensionless parameters in heat transfer are applicable to mass transport also, as the variables can be expressed in terms of dimensionless quantities. However each case may have some special characteristics, which have to be taken into consideration.

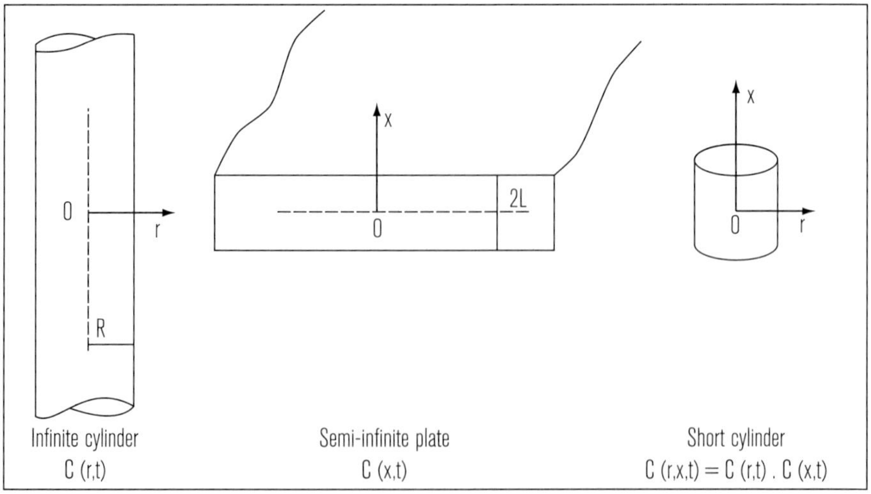

Figure 18.15 - Schematic diagram for building combined solutions.

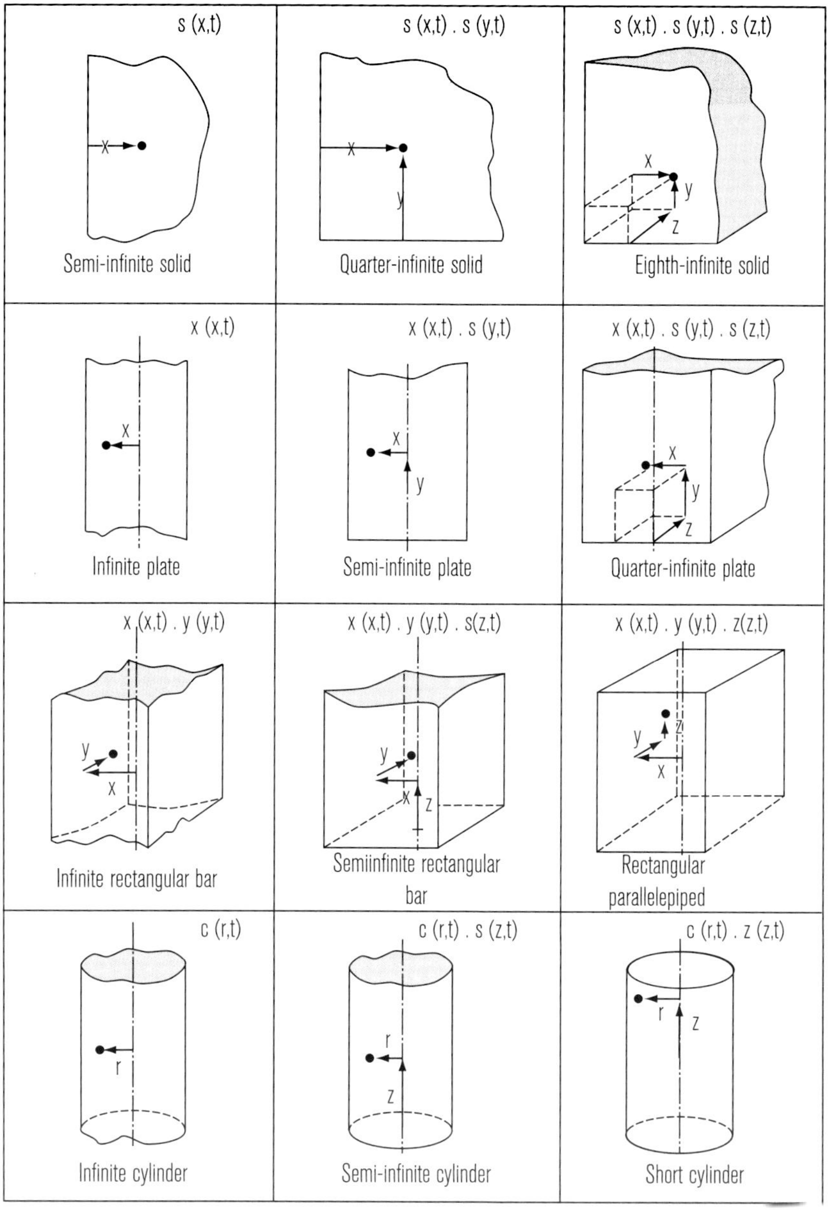

Figure 18.16 - Examples of combined solutions (POIRIER and GEIGER, 1994).

In combined solutions for heat transfer treated in earlier, Chapter 12 the heat transfer by convection between solid body and the fluid can be considered in the same way. However in the analogical cases of transfer of species, treated in previous sections, there are boundary conditions corresponding to the steady concentration in the interface fluid/solid. This condition can be imposed, for example, for contact with a fluid in which the activity of the species is kept constant and one can assume local equilibrium at the interface.

Table 18.3 presents some solutions to the problem of diffusion of species in the solid. These solutions are valid only for the geometry, initial condition specified, under conditions where the surface concentration was kept constant throughout the process. The solution is presented in terms of *average* concentrations.

Diffusion in plate with semi-thickness

I.C.: $C(x,0) = C_i$,

B.C.: $C(L,t) = C_s$,

$$\frac{\partial C}{\partial x}(0,t) = 0$$

Solution:

$$\frac{\bar{C} - C_s}{C_i - C_s} = \frac{8}{\pi^2}\sum_{n=0}^{\infty}\frac{1}{(2n+1)}\exp\left[\frac{(2n+1)^2\,\pi^2}{4}\frac{Dt}{L^2}\right]$$

Diffusion in solid circular with radius R

I.C.: $C(r,0) = C_i$,

B.C.: $C(R,t) = C_s$,

$$\frac{\partial C}{\partial r}(0,t) = 0$$

Solution:

$$\frac{\bar{C} - C_s}{C_i - C_s} = \sum_{n=1}^{\infty}\frac{4}{\xi_n^2}\exp\left[\frac{-\xi_n^2 Dt}{R^2}\right]$$

where: ξn = 2.405, 5.520, 8.654, 11.792, 14.931, for n = 1, 2, 3, 4, 5 etc

Diffusion in sphere of radius R

Same boundary conditions as in the case of cylinder given above

$$\frac{\bar{C} - C_s}{C_i - C_s} = \frac{6}{\pi^2}\sum_{n=1}^{\infty}\frac{1}{n^2}\exp\left[\frac{-n^2\,\pi^2\,Dr}{R^2}\right]$$

Table 18.3 - Analytical solutions for specific boundary conditions in terms of average concentrations for simple geometries (unidimensional unsteady state condition) (POIRIER and GEIGER, 1994)

Example

A steel plate of dimensions 3 m x 1.2 m x 0.1 m, initially showing uniform distribution of hydrogen, is subjected to vacuum for 40 hours, at a temperature. The diffusion coefficient of Hydrogen is $D_H = 10^{-9}$ (m^2s^{-1}). The plate can be considered "infinite", with a thickness of 2L = 0.1 (m). Substituting the values one gets for $Dt/L^2 = 5.8 \times 10^{-2}$ and, entering with this value in the figure below, the dimensionless concentration can be evaluated as $\frac{\overline{C}-C_S}{C_i-C_S} = 0.74$. This implies that about 25% of hydrogen is removed

Which implies that about 25% of hydrogen is removed.

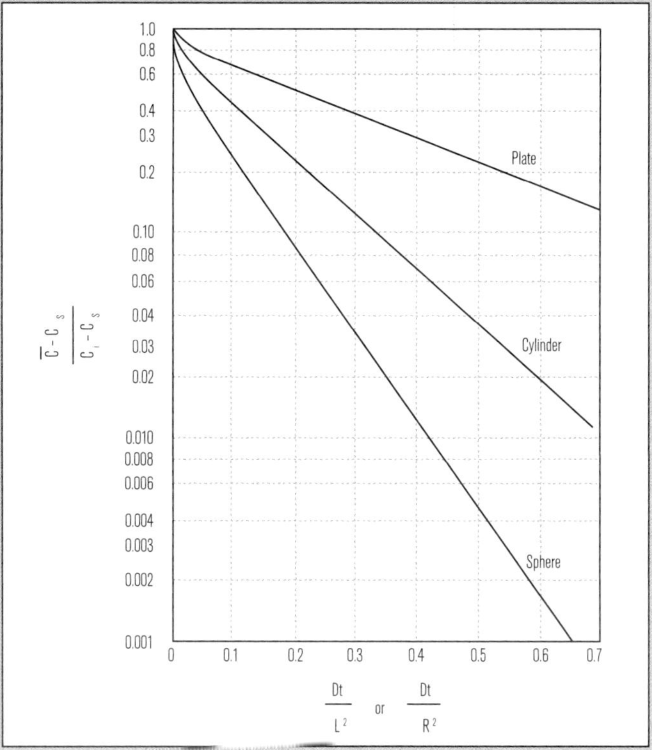

Mean values of concentration as a function of Dt / L^2 for some geometric shapes (POIRIER and GEIGER, 1994).

In many of these examples given, simplifications have been made for enabling analytical solutions. In the case of situations, involving heat, mass and momentum transfer with complicated boundary conditions as well as transient conditions in two or three dimensions simultaneous solution of transport equations for all the three modes and for each species have to be developed and solved through numerical procedures to get temperature, velocity and concentration profiles. In the case turbulent flow the relevant equations have to be considered as discussed before in the chapters of momentum and heat transport.

References

BIRD, R.B.; STEWART, W.E.; LIGTHFOOT, E.N. *Transport phenomena*. 2. ed. New York: John Wiley & Sons, 2002.

MIDDLEMAN, S. *An introduction to mass and heat transfer*. New York: John Wiley & Sons, 1998.

MILLS, A.F. *Basic heat and mass transfer*. New Jersey: Prentice Hall, 1999.

POIRIER, D.H.; GEIGER, G.H. *Transport phenomena in materials processing*. Warrendale: TMS, 1994.

TOSUN, I. *Modelling in transport phenomena*. Amsterdam: Elsevier, 2002.

Exercises

1 - *Chemical Vapor Deposition* (CVD) can be used for deposition of titanium according to the reaction TiI_4 (g) = Ti (s) + $2I_2$ (g) (ΔG^o = 402000 - 118T Joules) at 1 atm and 1500 K. In this process TiI_4 (gas, pure and 1 atm) is passed over the end of a capillary (5 cm long, 5 mm in diameter) at the bottom of which is the substrate on which deposition occurs. Consider the value of diffusion coefficient in the gas phase equal to 0.875 cm^2/s. State the boundary conditions, stoichiometric restrictions (and other necessary conditions as necessary), find the flow of TiI_4 and deposition rate of titanium.

2 - A pure iron plate (2 mm; 1255 K) finds himself subjected to a gaseous atmosphere capable, by reacting $SiCl_4$ (g) = Si + $2Cl_2$ (g) to raise the surface concentration of 3% Si. Estimate the compositions in the center of the plate and the average, after 1 hour of treatment, if $D_{Si\text{-}Fe} = 8 \times 10^{-8}\ cm^2.s^{-1}$.

3 - Consider a thin sheet of steel, with 0.4%C and 2 mm thick, subjected to 950°C in a gaseous atmosphere capable of maintaining the surface concentration of 0.1% C. Determine the maximum amount of time for which there is valid and reliable solution considering semi-infinite solid. For this time, use the equation and calculate the flow of the species through the interface metal/gas and find the total amount of carbon removed, and the average concentration.

4 - A piece of 1020 steel is heated to 1250 K in an atmosphere of CO/CO_2, so that through the reaction 2CO = C + CO_2 the surface concentration is maintained at 1%C. Determine the composition profile after 1 hour, if $Dc = 2 \times 10^{-11}\ m^2.s^{-1}$.

5 - Estimate the % of hydrogen remaining in a piece of steel, cylindrical, massive and long, with a diameter of 0.10 m, after 40 hours of vacuum degassing. Assume $D_H = 1 \times 10^{-11} m^2.s^{-1}$.

6 - Consider that a copper sheet (0.2 mm thick) is in equilibrium with H_2 (4 atm, 1000°C, $D_H = 10^{-6} cm^2 s^{-1}$). After this it is placed in vacuum for 60 s. Estimate the average composition and the composition on the surface if the solubility of H_2 in copper corresponds to 1.44 ppm under 1 atm and 1000°C.

7 - A steel wire (0.01 cm thick) passes through a tube furnace at 1200°C. It contains gas fuel with a potential to raise the surface carbon content of the wire to 0.8% of C. Estimate the average composition of the wire, if the length of the furnace is equal to 1.5 m and wire speed 15 cm/s.

Mass Transport Coupled to Other Phenomena

Chapter 19

19.1 Mass Transport Coupled to Chemical Reactions

A large number of manufacturing processes of metals and other materials involve simultaneous transport of heat, mass, and momentum coupled to chemical reactions. Many processes of extraction and refining of metals like iron making in blast furnace and steel making can be cited as examples. In physical metallurgy, the precipitation of a new phase from an uniform homogeneous matrix initially or the process of heat treatment of metals and alloys, where new phases (with different compositions) form as a function of time, can lead to the establishment of composition gradients (chemical potential) and provoke migration of material species . Hence the kinetics of precipitation reaction and mass transport are clearly linked.

In many processes several chemical reactions can occur with their characteristic mechanisms accompanied by physico-chemical transformations such as evaporation, condensation, melting, adsorption, etc. generating or consuming heat and movement of various species involved. Several stages may be involved with transport of heat and mass within phases or involving multiphases and with complicated mechanisms.

One of the objectives of this discipline namely Transport Phenomena, is to apply the basic principles for the development of mathematical and physical models of a given process in order to quantify, control, and predict the results with changing variables as well as to carry out different simulations in order to arrive at optimum conditions.

Hence a manufacturing process is constituted of many individual steps involving physico-chemical reactions and transformations defined as TF in Figure 19.1, whose mechanisms and kinetics have to be investigated in the laboratory to optimize them in the industrial process. There have been a large number of investigations in respect of these in the literature. The various steps in a reaction can occur in series, parallel or according to a mixed or joint mechanism (Figure 19.1). This division enables each stage to identify a characteristic resistance like in the electric or hydrodynamic analogue. The stages of greater resistance can be considered to be the ones controlling the process. With the identification of this one can focus on how to manipulate the same for attaining greater efficiency.

Schematically a model can be built comprising all the steps, or part of them, chosen on the basis of previous work done on the same and this can be completed when the steps involved (reactions, transformations, mass transport, heat transport etc.) are thoroughly investigated and clarified. However if some steps are not known arbitrary or empirical equations or coefficients can be used to complete the model description.

Models can be built to shed more light on the scientific basis of a process and to test the validity of the assumptions or hypotheses. Models can be used to monitor and control of systems or industrial processes. In the latter case it should be capable of generating a quick response, if a variable is changed and this depends on the way the model is built. The cost and the profit aspect of using the model also should be considered. Depending on the complexity of an industrial process, the processing speeds of a a computer in respect of control of the same would vary and hence it is necessary that the model construction should be capable of quick response for efficient control. Even though the procedure is complex, models not built on robust scientific approach and cannot be flexible for the control of processes. For example it is possible to formulate a purely statistical model, obtained from the analysis of data of a reactor over an extended period of time. However since there is no basic scientific approach, any eventual change in the routine or inclusion of any improvement cannot be taken into account as in this case, the model is not built on the basis of understanding of the process. The development of a model for a given process typically involves:

- assumptions/hypotheses about possible sub-physical-chemical processes that constitute the process and knowledge of thermodynamics and kinetics of the same;
- mathematical description of the sub-processes based on transport phenomena principles and integrating them with the model for the main process.

In this chapter, after a brief review of chemical kinetics, examples, in which coupling of chemical reaction rates, heat and mass transport etc. to develop process models and simple cases with analytical solutions are discussed.

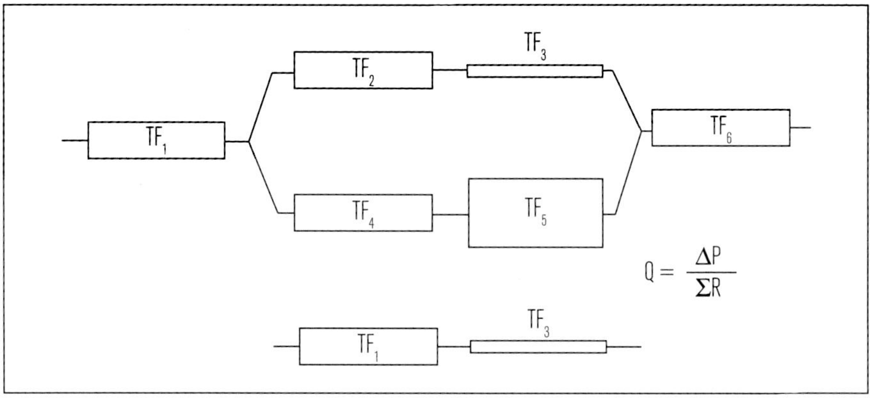

Figure 19.1 - Schematic division of a given process in stages, arranged in series and/or parallel.

19.2 Kinetics of Chemical Reactions

Consider in gas phase, the following reaction:

$$H_2\,(g) + Br_2\,(g) = 2HBr\,(g) \tag{19.1}$$

The reaction, takes place as a result of collisions between molecules of H_2 and Br_2, each with adequate amount of energy to break the bonds in the reactants, H-H, Br-Br respectively and form HBr. If N is the total number of molecules in the sample, and ΔE represents the energy required then, according to Boltzmann distribution, the number of molecules with energy equal to or greater than required, is given by:

$$N = N_o\, e^{\frac{-\Delta E}{k_B T}} \tag{19.2}$$

where:

- K_B is the Boltzmann constant;
- T is the absolute temperature;
- N_0 is the Avogadro's number;
- ΔE is the activation energy.

In other words, above absolute zero, there is always a fraction of molecules with enough energy to take part in the process, but the value of ΔE limits the number of molecules having adequate energy necessary for the reaction. Moreover, the molecules of H_2 and Br_2, should be placed at a favorable angle (orientation), for efficient reaction since the bonds are spatially oriented.

In fact, chemical reactions, such as described above, rarely occur in the way they are written.This should be taken as a general way to present the same, but in reality the processes in the atomic/molecular level take place, through a series of elementary steps (or elementary chemical reactions) which can be identified as the mechanism of the reaction and these reactions can occur in series and or in parallel.

For example, there is experimental evidence that the reaction cited above on the formation of HBr actually takes place, via a mechanism that involves the following elementary steps in series:

- R1: Initiation of the reaction $Br_{2\,(g)} \Rightarrow 2Br_{(g)}$;
- R2: propagation of the reaction $Br_{(g)} + H_{2(g)} \Leftrightarrow HBr_{(g)} + H_{(g)}$;
- R3: propagation of the reaction $H_{(g)} + Br_{2\,(g)} \Rightarrow HBr_{(g)} + Br_{(g)}$;
- R4: End or termination of the reaction $2Br_{(g)} \Rightarrow Br_{2(g)}$.

The terms initiation, propagation and termination of a particular reaction are specific to chemical kinetics. The chemical species in the atomic level H and Br do not appear in the global reaction either as reactants or as products. However they take part in the reaction, as active centers or intermediate species. Therefore, the term initiation of reaction is concerned with the formation of these active centers or the intermediates, and at the end of the reaction they are destroyed by the reaction of termination.

Each of the above mentioned reactions, forms part of the mechanism of formation of HBr from its elements, as given below:

$$Br_{(g)} + H_{2(g)} \Leftrightarrow HBr_{(g)} + H_{(g)} \tag{19.3}$$

This implies that the meeting of a Br atom and a molecule of H_2 to form a molecule of HBr and an atom of H.

In the case of elementary reactions, like the one given above, one can identify the mechanism based on a simplified treatment considering the following generic reaction in gaseous medium:

$$A_{(g)} + B_{(g)} \Leftrightarrow C_{(g)} + D_{(g)} \tag{19.4}$$

For this reaction the symbol ⇔ indicates that the reaction is reversible, that is, it takes place in the forward direction denoted by → and its reverse direction denoted by ←. Combination of both these occurring simultaneously is termed as reversible reaction given by equation 19.4, This combination of the forward component and the reverse component, for three situations are represented schematically in Figure 19.2.

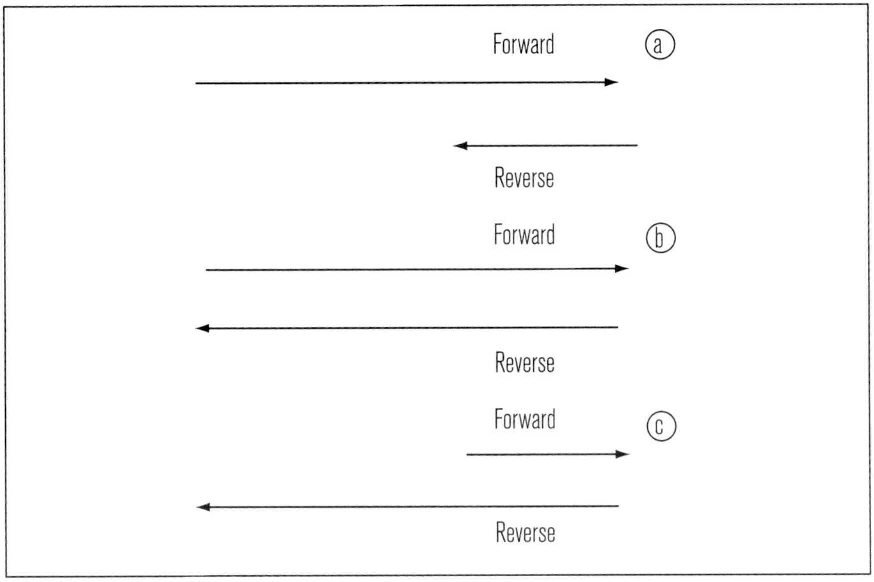

Figure 19.2 - Reversibility during chemical reaction: a) Forward component is larger, there is the production of C and D; b) Forward and reverse are equal, there is equilibrium; c) If reverse component is larger, then there is a production of A and B.

According to the theory of activated complex or the transition state, there is an energetically more favorable way, through which a reaction can occur.This path is described schematically in terms of a reaction coordinate, designated by r (which may be associated with distance). Figure 19.3 illustrates the reaction between an active center and a molecule H_2.

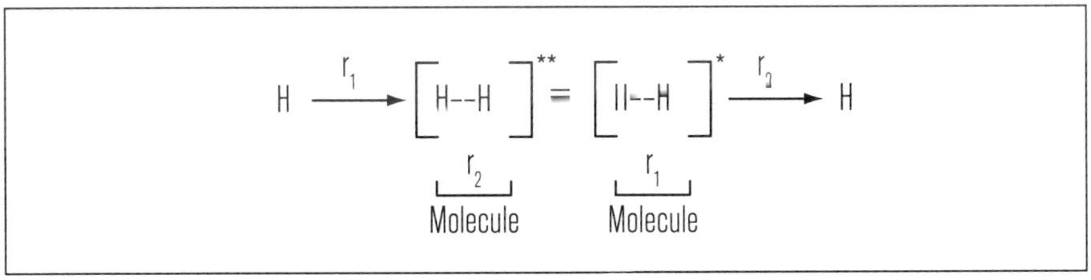

Figure 19.3 - Reaction H-H.

The state of energy of the system represented in Figure 19.3 can be computed though statistical mechanics and the results are in the form of a map in Figure 19.4. In this figure the lines (1), (2) and (3) represent contours of isoenergy, with increasing values towards (1) ⇒ (2) ⇒ (3). There would be a valley, representing different possibilities or statistics (energy) of the existence of molecule H_2*, similar to the molecule H_2**. The surface described by these contours, in this case, should be symmetrical in relation to the AO, where O is the origin (as the reaction itself is the same in both directions). However this cannot be considered a general case.

The dashed line in Figure 19.4 describes the path of processing Reagent R (A, B) into products P (C, D) as shown schematically in Figure 19.5.

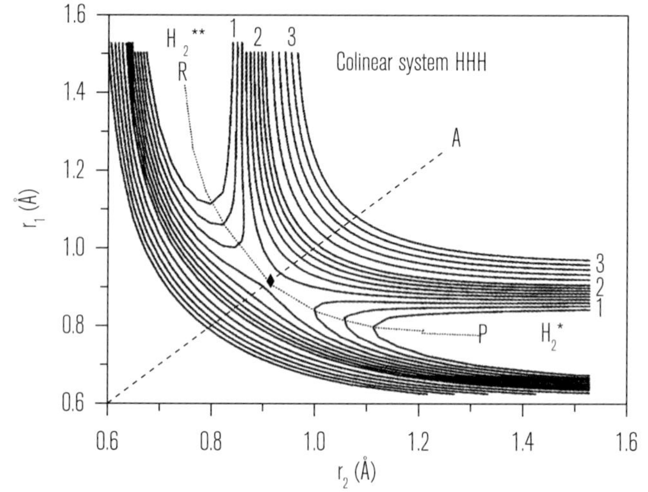

Figure 19.4 - Energy map for unidirectional reaction HHH.

Since there is no unique energetic configuration, either for reagents, or for products, the schematic energy path, as shown in Figure 19.5, should be taken as an average . In this figure, while EA,B and EC,D represent the energies of reactants and products respectively, EZ is defined as the energy of a transition state between reactants and products. Z is not a specific compound, molecule, or specie, but can be defined as a power configuration (RAY, 1993). However, it is customary to refer to EZ as energy of an activated complex. A comprehensive illustration of the energy states is also presented in Figure 19.6. The figure indicates that for the forward reaction, it would be necessary to energize the system (molecules A and B) so that the molecules achieve at least the energy state E_Z or in other words, sufficient energy to overcome an energy barrier, called activation energy symbolized by ΔE. The same considerations apply to the reverse reaction:

$$A + B \Leftarrow C + D \tag{19.5}$$

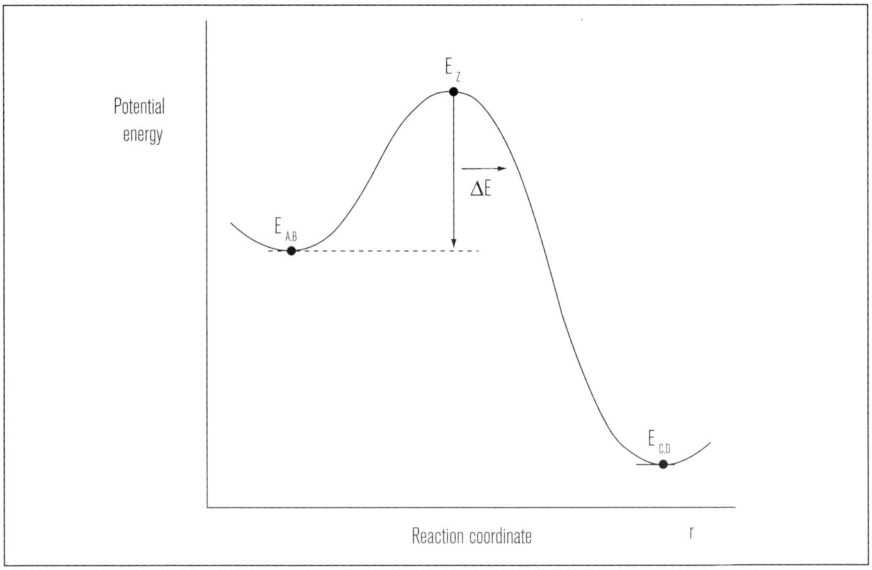

Figure 19.5 - Path of average energy for a given reaction.

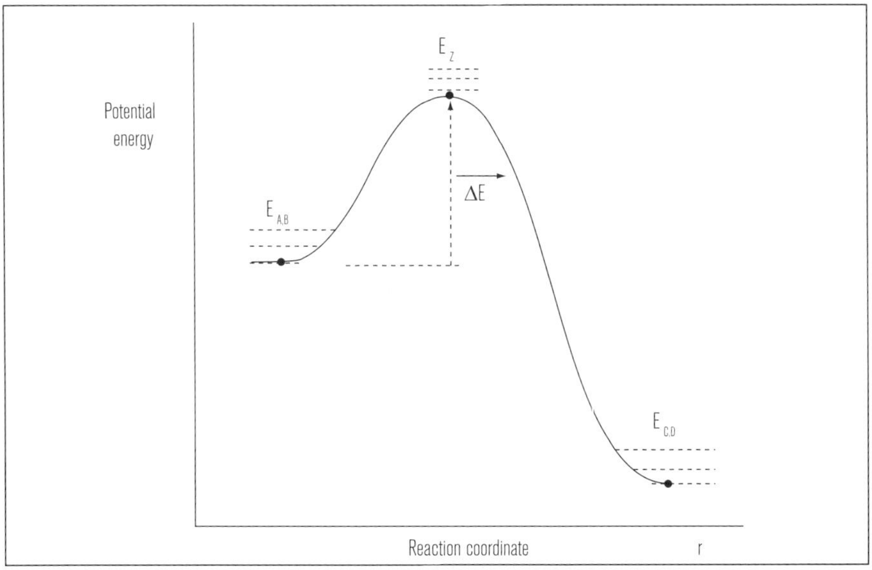

Figure 19.6 - Schematic potential energy curve of reactants, products and active center.

Two assumptions are fundamental to the theory. According to the the first, there exists at any moment, a pseudo-equilibrium between the transition state Z (activated complex) and the reagents. Therefore, for the forward reaction A + B $\Rightarrow$ Z (SOHN and WADSWORTH, 1979).

$$\vec{K}_{eq} = \frac{C_Z}{C_A C_B} \tag{19.6}$$

where:

- C_A, C_B and C_Z, represent the concentrations of species A and B and of the pseudo-species Z;
- $\vec{K}_{eq}$ represents the equilibrium constant.

The Gibbs free energy is given by:

$$\Delta\vec{G}^O = \Delta\vec{H}^O - T\Delta\vec{S}^O \tag{19.7}$$

Hence:

$$\vec{K}_{eq} = e^{\frac{-\Delta\vec{H}^o}{RT}} . e^{\frac{\Delta\vec{S}^o}{R}} \tag{19.8}$$

The values ΔH^o and ΔS^o are heat and entropy of formation of pseudo-species Z but they are not directly measurable.

The other assumption is that the activated complex decomposes in accordance with the universal frequency given by:

$$\upsilon = \frac{k_B T}{h} \tag{19.9}$$

where:

- k_B stands for the Boltzmann constant;
- h for the Planck's constant;
- T for the absolute temperature.

Therefore, the velocity of the forward reaction would be given by:

$$\vec{r} = \upsilon(s) C_Z \left(mol/m^3\right) \tag{19.10}$$

$$\vec{r} = \frac{k_B T}{h}\left(\vec{k}_{eq} C_A C_B\right) \tag{19.11}$$

$$\vec{r} = \frac{k_B T}{h}\left(e^{\frac{-\Delta \vec{H}^o}{RT}} e^{\frac{\Delta \vec{S}^O}{R}} \right)(C_A C_B) \quad (19.12)$$

or finally:

$$\vec{r} = \vec{k} C_A C_B \quad (19.13)$$

The reaction rate is directly proportional to the concentrations of the reactants species. Hence for the following type of reaction:

$$2A \Rightarrow B \text{ or } A + A \Rightarrow B \quad (19.14)$$

one can write:

$$\vec{r} = \vec{k} C_A C_A = \vec{k} C_A^2 \quad (19.15)$$

Other forms of the same equation are:

$$\vec{r} = \frac{k_B T}{h}\left(\vec{k}_{eq} C_A C_B \right) \quad (19.16)$$

$$\vec{r} = \frac{k_B T}{h} e^{\frac{-\Delta \vec{G}^O}{RT}} C_A C_B \quad (19.17)$$

Which indicates, for other parameters being constant, higher the value of ΔG^0, lower is the reaction rate. ΔG^0 can then be associated with a free energy barrier. As $\Delta \vec{G}^O = \Delta \vec{H}^O - T\Delta \vec{S}^O$ this barrier involves an energy component ΔH^0 and an entropy component representing order or disorder, ΔS^0. A given reaction may be energetically favorable (positive value and small ΔH^0) but may have unfavorable configurational aspect indicated by the entropy term (values ΔS^0 negative and small). The condition will result in a reaction with low rate. Thus, the theory takes care of of the orientation of the species during collisions and this also influences the rate of reaction.

Although mathematically, the theory of collisions presents almost identical results like in the case of theory of transition state or activated complex. The first is based on the fact that the forward reaction rate would be proportional to the number of collisions between species A and B (reagents) which again should be proportional to C_A C_B. Not all collisions can be effective, by virtue of the fact that two specific molecules A and B may not possess enough energy to overcome the activation barrier. In this case, the Boltzmann distribution has been used to assess the fraction of molecules with energy greater than the activation energy, ΔE. Finally, even collisions of these energized molecules may not occur at right orientations. This would further reduce the effectiveness of the process. Hence the global rate can be defined as:

$$\vec{r} = \varphi\, C_A C_B e^{\frac{-\Delta E}{RT}} \tag{19.18}$$

where:

- φ would be a measure of the effectiveness of the collisions.

Whatever be the choice in respect of these mechanisms for the following reaction:

$$A + B \Leftrightarrow C + D \tag{19.19}$$

The rate of forward reaction is given:

$$\vec{r} = \vec{k} C_A C_B \tag{19.20}$$

And that of reverse reaction:

$$\overleftarrow{r} = \overleftarrow{k} C_C C_D \tag{19.21}$$

Hence rate of overall reaction is:

$$r = \vec{r} - \overleftarrow{r} = \vec{k} C_A C_B - \overleftarrow{k} C_C C_D \tag{19.22}$$

Which is usually written as:

$$r = k_D C_A C_B - k_R C_C C_D \tag{19.23}$$

where:

- K_D is defined as the rate constante for the forward reaction;
- K_R is the rate constant for the reverse reaction.

Eventually, at equilibrium $\vec{r} = \overleftarrow{r}$:

$$k_{eq} = \frac{\vec{k}}{\overleftarrow{k}} = \frac{C_C C_D}{C_A C_B} \tag{19.24}$$

The above treatment, based on *theory of transition state or activated complex* can be extended to the case of non-ideal solutions. In this case, the equilibrium between the activated complex and the reactants can be written as:

$$k_{eq} == \frac{\gamma_Z C_Z}{\gamma_A C_A \gamma_B C_B} \tag{19.25}$$

where:

- $\gamma_{i=A,B,Z}$ represent the activity coefficients of the reactants and the activated complex, determined as a function of temperature, and composition.

Then the rate of forward reaction can be written as:

$$\vec{r} = \frac{k_B T}{h}\left(\frac{\gamma_A \gamma_B}{\gamma_C}\right) C_A C_B \tag{19.26}$$

Although this equation is more general, in the absence of specific knowledge of the values of activity coefficients, one can express the same in terms of Boltzmann and Plancks constants as before:

$$\vec{k} = \frac{k_B T}{h}\left(\frac{\gamma_A \gamma_B}{\gamma_C}\right) k_{eq} \approx \frac{k_B T}{h} k_{eq} \tag{19.27}$$

Example

As an example of the complexities involved in evaluating the rate of reactions one can consider rate of formation of HBr from H_2 and Br_2 is given by (FOGLER, 2005):

$$r = \frac{k_1[H_2][Br_2]^{1/2}}{k_2 + [HBr]/[Br_2]}$$

As is evident, the reaction rate should be a function of temperature, pressure and composition. The relationship that determines the same is complex, due to the fact that the reaction occurs through a mechanism involving several intermediate steps. It is also difficult to determine the order of the reaction, as it occurs through a mechanism involving several elementary reactions, each one exhibiting its own dependency on the variables mentioned above.

However, some rules have to be remembered , while trying to formulate an expression for the rate of a chemical reaction without prior knowledge and analysis of the mechanisms of these reactions.

At constant temperature, generally the rate of a chemical reaction gradually decreases with time (or extent of reaction) and at equilibrium it is zero. The equation for the rate of an irreversible reaction (or forward or reverse reactions) can be given as:

$$r = kf(C_i) \tag{19.28}$$

where:

- $f(C_i)$ represents a complex function of concentration, whose coefficients can be temperature dependent;
- k is considered as rate constant of the reaction.

The rate of chemical reaction varies with temperature and this dependence is usually expressed through the Arrhenius equation:

$$k = k_O e^{\frac{-\Delta E}{RT}} \tag{19.29}$$

In most cases, we can write:

$$r = k\ \pi\ C_i^{\alpha_i} \tag{19.30}$$

where:

- π represents the multiplication operator;
- C_i the concentration of the reactant i;
- α_i a coefficient to be determined by numerical fit of experimental data in the proposed equation.

α_i is also called order of reaction for component i and can be positive, negative, integer or a fraction.

In practical terms, a metallurgical engineer may not be concerned with the various steps in a particular reaction and details of reaction mechanism. Normally there is uncertainty of industrial data in respect of parameters in a process and for all practical purposes of control, a satisfactory a statistical relationship between process variables and the rate of a reactions like decarurization, desulphrization, dephosphorization can be developed in iron and steel making.

In general chemical reactions are strongly activated by temperature and have high values of activation energy. Higher the temperature higher is the rate of reaction. This is important, since based on this information, ,one can determine if there is a difficulty or hindrance in respect of a particular chemical reaction taking place at high temperatures. The reason could be attributed to a slower mass or heat transfer steps, which could be rate determining. However there are situations where the rate of a reaction apparently decreases with increasing temperature. In these cases a deep analysis of the reaction mechanism has to be carried out to clarify this exception to the general rule. An example is given below.

Example

Consider the reaction of oxidation of nitrogen (II) to nitrogen (IV), according to the reaction:

$$2NO(g) + O_2(g) = 2\ NO_2(g)$$

It is observed experimentally that the reaction rate decreases with increasing temperature, which was explained on the basis of the following reaction mechanism:

- Step I: Formation of intermediate compound:

$$NO(g) + NO(g) = N_2O_2(g)$$

which is considered fast, in conditions of near equilibrium, ie:

$$K_{eq} = \frac{(N_2O_2)}{(NO)^2}$$

- Step II: Formation of NO_2 (slow and irreversible):

$$N_2O_2 + O_2(g) \Rightarrow 2NO_2(g)$$

for which the reaction rate is given by:

$$\frac{d(NO_2)}{dt} = K_2(N_2O_2)\ (O_2)$$

where K_2 is the rate constant for the reaction, determined experimentally.

Combining the two equations above:

$$\frac{d(NO_2)}{dt} = K_{eq}K_2(NO)^2(O_2)$$

Theoretical and experimental data suggest that the above equation is in conformity with the experimental results, namely the reaction rate decreases with increasing temperature. The "discrepancy" is explained as follows. K_2 increases with increasing temperature and at the same time K_{eq} decreases considerably with increasing temperature. Hence the product, K_{eq} K_2 decreases with increasing temperature. Hence there is no contradiction in respect of the general rule that the rate of a chemical reaction is strongly activated by temperature, and increases with the latter.

Example

The value of activation energy obtained from the rates of reaction determined at different temperatures can be considered as apparent activation energy of the process which in reality may be a combination of several other physicochemical steps. For example consider the case where chemical reaction and adsorption are happening together.The adsorption process can be physical or chemical.The physical adsorption is due to van der Waals forces and the amount of energy involved in the process lies in the range of 2~6 kcal/mol. During the chemical adsorption, there is formation of chemical or ionic bonds between the adsorbed species and atoms sat the surface. The energy involved in such processes is about 20 kcal/mol. Just this heat of adsorption is responsible for lowering the activation energy and catalysis of chemical reactions in fluids, as shown below. The reaction in question can be in accordance with the following mechanism:

$A + S \Leftrightarrow A.S$ $\quad$ K_{eq} step in equilibrium

$AS \Rightarrow BS;$ $\quad$ K_1 slow step

$BS \Rightarrow B + S$ $\quad$ K_2 fast step

where:

- S is the active surface site of adsorption;
- A is the reactant;
- B is the product.

The mechanism is, therefore, the one where A is transformed into B by the reaction A ⇔ B through catalysis at surface of a solid, as processing occurs on the surface of the catalyst. If steady state is assumed then the intermediate product BS is produced by the second reaction at the same rate it gets consumed in the third reaction. Hence:

$$K_1(AS) = K_2(BS)$$

From the equilibrium of the first reaction:

$$(AS) = K_{eq}(A).(S)$$

Hence:

$$d(B)/dt = K_1(AS) = K_2(BS) = K_1.K_{eq}(A).(S)$$

The above model indicates that the production rate of B is proportional to the concentration of reagent A and the availability of active sites on the surface of the catalyst. K_1 can be estimated by Arrhenius equation and K_{eq} from Gibbs-Helmholtz equation in the following relation for the rate:

$$d(B)/dt = K_1.K_{eq}(A).(S)$$

The value of apparent activation energy is given by:

$$\text{Apparent } \Delta E = \Delta E_1 + H_{adsorption}$$

where:

- ΔE_1 is the activation energy of the 1st step (slow);
- $H_{adsorption}$ is the enthalpy of adsorption.

In the case of strong adsorption, ie $H_{adsorption}$ is negative and small (relative value) the effect of adsorption would effectively reduce the barrier represented by the activation energy and increase the reaction rate.

Example

A homogeneous chemical reaction takes place between species A and B as follows The relationship between rate of reaction, temperature and concentrations is given in the table.

r(moles/m³.s)	T(K)	CA	CB	ln r	1/T	ln CA	ln CB
0.23	1500	1	0.7	-1.46968	0.000667	0	-0.35667
1.12	1500	2.1	1.5	0.11333	0.000667	0.74194	0.40547
0.75	1500	3.3	0.8	-0.28768	0.000667	1.19392	-0.22314
0.39	1300	2	3.4	-0.94161	0.000769	0.69315	1.22378
0.32	1300	1.8	3	-1.13943	0.000769	0.58779	1.09861
0.7	1400	2	2.2	-0.35667	0.000714	0.69315	0.78846
2	1400	2.7	4.1	0.69315	0.000714	0.99325	1.41099
0.1	1400	0.8	0.9	-2.30259	0.000714	-0.22314	-0.10536

A multiple regression analysis leads to:

$$\ln r = 12.3 - 20019\ 1/T + 0.851 \ln CA + 1.28 \ln CB$$

or

$$r = 220 e^{200191/T} C_A^{0.851} C_B^{1.28}$$

No light is shed on the reaction mechanism. But a workable rate equation is obtained, which can used for all practical purposes.

19.3 Modelling of Processes Involving Chemical Reactions

As has already been mentioned, in general the production processes of metals and other materials are complex operations involving heat and mass transport and chemical reactions. In the following sections development of some of the models where many of the above aspects are present and also different phases like solids and liquids is presented. Simplifications have been made in the description of the various steps that enable analytical solutions. In most cases these models can be adapted in applications one comes across in the industry.

19.3.1 Reaction between a dense solid (nonporous) particle with gas

Existance of a truly nonporous solid is controversial. Usually a solid is made up of dense matter but with pores. A chemical reaction can take place at at the reactin sites (S_{CR}) located either at the solid external surface or inside in the bulk (See figure below). If the rate of reagent transport through the pores network is fast enough then the reagent is capable of penetrating the solid and for the chemical reaction purposes the solid can be considered porous one. Otherwise if the reagent is consumed as soon it gets to the external surface the reaction interface it has no opportunity to penetrate the particle. The reaction is confined to a narrow layer close to the external surface and the particle can be taken as a nonporous one. Leaching of ores at room temperature is an example of the first case (porous solid); and burning of a coal particle can be cited as an example for the second case (nonporous).

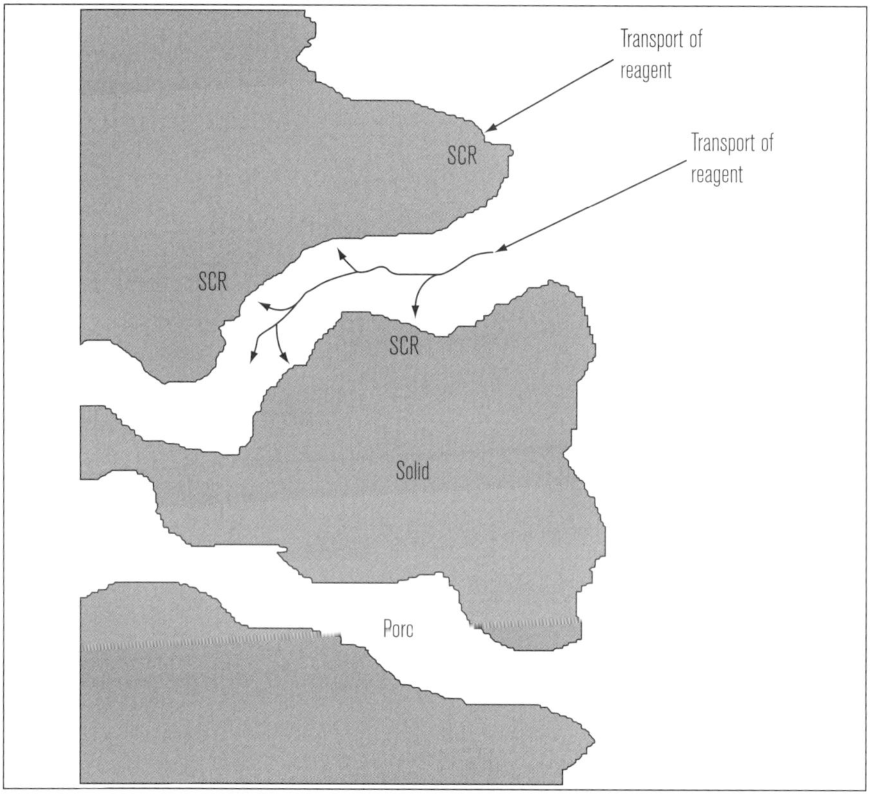

Figure 19.7 - A solid surface seen through a microscope showing surface pores through which a reactant penetrates towards reaction sites.

A vast majority of materials have some degree of porosity through which reactants can diffuse into the solid and products after reaction can diffuse out. At the other extreme, the reagent may be able to penetrate through the network of pores of the solid and reaction should be substantial as the surface area for reaction is increased considerably.

A chemical reaction between a nonporous solid and a gas involving mass transfer and consumption of solid can be given as:

$$M(s) + n\,A(g) \rightarrow n\,B\,(g)\ \Delta G < 0 \tag{19.31}$$

A criterion can be developed to make a distinction between these two extremes. mentioned above. In such a case the, mass transfer through the pores, the diffusivity of the species D_A^{ef} and the distance traveled, L, as a measure of the speed of permeation have to be considered.

$$(D_A^{ef}/L)\ \Delta C_A \tag{19.32}$$

along with another parameter, that describes rate of chemical reaction:

$$K_{RQ}\ DC_A \tag{19.33}$$

At any given instant reaction is observed only on the outer surface of the solid, as the solid is impervious to gases. Figure 19.8, shows a schematic profile of compositions in this system.

Since the reaction, as mentioned before, is restricted to only outer surface area, the rate of permeation is negligible compared to the speed of chemical reaction. In other words, as soon as the reagent reaches the surface there is reaction and the particle is consumed, as permeation is not possible.

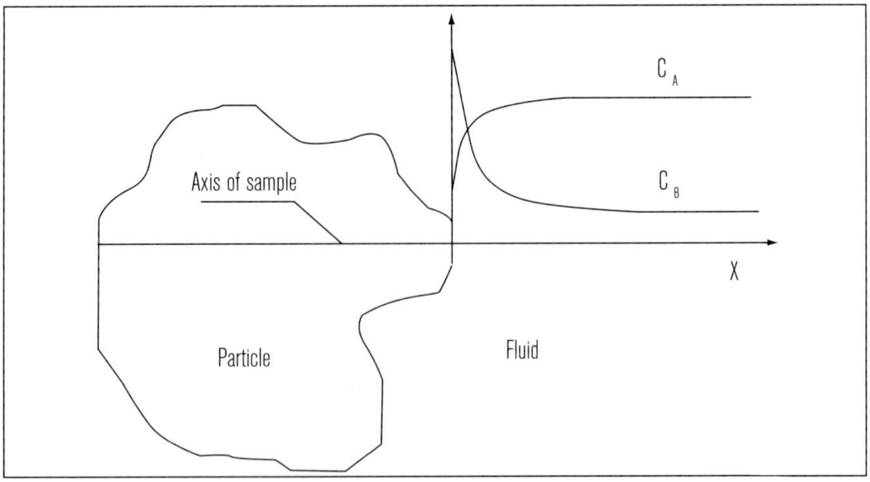

Figure 19.8- Schematic figure for the reaction of impermeable particle with a fluid.

Under these conditions a simplified model can be developed for a spherical solid undergoing reaction only at the outer surface (which implies that the reaction M + nA = nB is restricted to the surface located at r = R), assuming isothermal system and the steps (sub - processes) shown in Figure 19.9, namely:

- mass transport through the boundary layer;
- chemical reaction on the particle surface.

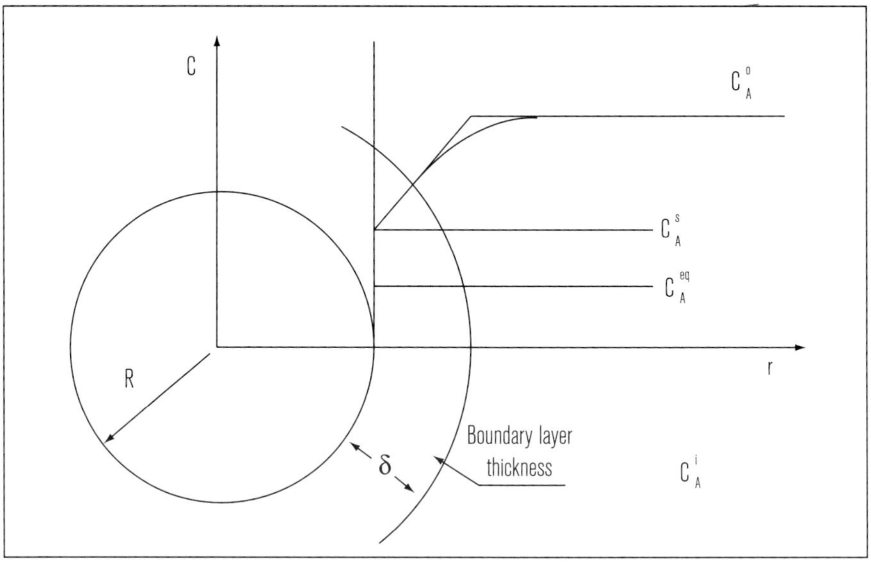

Figure 19.9 - Distribution of species in fluid phase in accordance with the model.

In this case the following steps in the mathematical treatment is adopted.

1) Mass transfer rate of species A, from the bulk to the reaction interface through the boundary layer is given by:

$$V_{TM}\left(\frac{\text{moles of A}}{\text{s}}\right) = \text{Area} \times \text{mass transfer coefficient} \times \text{concentration gradient} \quad (19.34)$$

$$V_{TM}\left(\frac{\text{moles of A}}{\text{s}}\right) = -4\pi R^2\left(m^2\right) \times D_A\left(m^2 \cdot s^{-1}\right) \times \frac{\left(C_A^0 - C_A^i\right)}{\delta}\left(\frac{\text{moles of A}}{m^4}\right) \quad (19.35)$$

The negative sign (-) in the above equation signifies that the flow vector, in this case is in the opposite direction to the positive coordinate axis r.

In these equations, R is the radius of the particle; D_A is the diffusion coefficient of A in gas (to be determined experimentally); C_A^o , C_A^i are the concentrations within the gas and the reaction interface; δ is the thickness of the boundary layer.

The flux V_{TM} does not include convective contribution. Considering the stoichiometry of the reaction M (s) + nA (g) = nB (g) together with the assumption of quasi-stationary state (no accumulation of species in the gas phase) one can write (POIRIER and GAGER, 1994):

Rate of consuption of A(mole/s) = Rate of production of B (mole/s)
Rate of transport of A(mole/s) = Rate of transport of B(mole/s) (19.36)

$$-N_A\left(\frac{\text{moles of A}}{m^2\cdot s}\right) = N_B\left(\frac{\text{moles of B}}{m^2\cdot s}\right) \quad (19.37)$$

$$-D_A\frac{dC_A}{dr} + X_A\left(N_A + N_B\right) = -D_A\frac{dC_A}{dr} \quad (19.38)$$

(diffusive) (convective) (diffusive)

The above equation simplifies to only diffusive contribution.

2) Chemical reaction rate can be evaluated as:

V_{RQ} = Reaction area x Rate constant x Concentration driving force (19.39)

which corresponds to a phenomenological formulation. Assuming first order reaction (mathematical treatment is simpler in this case) and for driving force defined interms of concentration difference at the interface ($C_A^i - C_A^{eq}$), one gets:

$$V_{RQ} = -4\pi R^2\left(m^2\right)\times K_{RQ}\left(m\cdot s^{-1}\right)\times\left(C_A^i - C_A^{eq}\right)\left(\frac{\text{moles of A}}{m^3}\right) \quad (19.40)$$

K_{RQ} is the rate constant for the reaction to be determined experimentally; C_A^{eq} is the equilibrium concentration, which can be calculated from thermodynamic data.

For quasi-stationary state:

Rate of transporte of species A = Rate of consumption of species A (19.41)
V_{TM} V_{RQ}

In equations 19.35 and 19.40, concentration at the interface C_A^i is unknown. However for quasi-steady state conditions the flux value can be expressed in terms of concentration difference and resistances in each step above as in an electrical analogy, Figure 19.10.

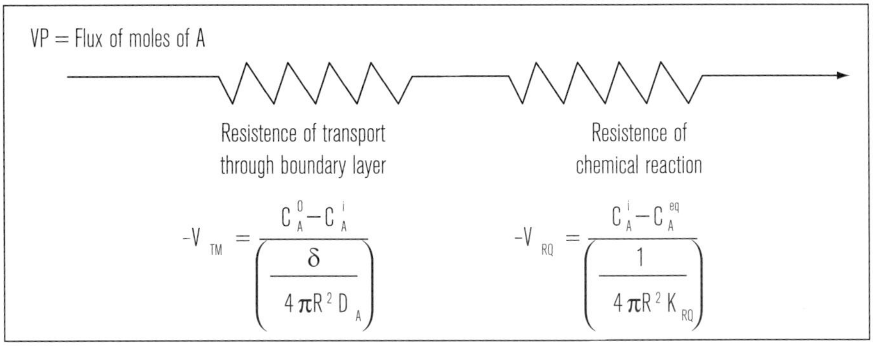

Figure 19.10 - Electrical Analogy of the reaction steps.

Considering resistances as given in Figure 19.10:

$$V_P\left(\frac{\text{moles of A}}{\text{s}}\right) = -\frac{\left(C_A^0 - C_A^{eq}\right)}{\frac{\delta}{4\pi R^2 D_A} + \frac{1}{4\pi R^2 K_{RQ}}} \tag{19.42}$$

$$V_P\left(\frac{\text{moles of A}}{\text{s}}\right) = -\frac{\left(C_A^0 - C_A^{eq}\right)}{R_{TM} + R_{RQ}} \tag{19.43}$$

where R_{TM} and R_{RQ} represent the resistance to mass transport and chemical reaction, respectively.

As with the electrical circuit, the greater resistance controls the flow value and then:

a) For mass transfer through boundary layer controlling the reaction, $R_{TM} >> R_{RQ}$

$$V_p = -\frac{\Delta C}{R_{TM} + R_{RQ}} \approx -\frac{\Delta C}{R_{TM}} \tag{19.44}$$

or

$$V_P\left(\frac{\text{moles of A}}{\text{s}}\right) \cong -\frac{\left(C_A^0 - C_A^{eq}\right)}{\frac{\delta}{4\pi R^2 D_A}} = -4\pi\, R^2 \frac{D_A}{\delta}\left(C_A^0 - C_A^{eq}\right) \tag{19.45}$$

The functional relationship between D_A and T is given by:

$$D_A = D_A^0 \cdot e^{-\frac{\Delta E}{RT}} \tag{19.46}$$

Therefore:

$$V_P\left(\frac{\text{moles of A}}{s}\right) = -4\ \pi\ R^2 \frac{D_A^0 \cdot e^{-\frac{\Delta E}{RT}}}{\delta}\left(C_A^0 - C_A^{eq}\right) \tag{19.47}$$

Experiments at different temperatures would provide the apparent activation energy corresponding to the diffusion process, if the driving force and boundary layer thickness can be maintained (approximately) constant.

In this particular example, it is the gaseous diffusion. This is weakly influenced by temperature. So, if ΔE (of the process, obtained from V_p values at different temperatures) is small (~5000 cal/K.mole) the controlling step for the process is very likely to be mass transfer through the boundary layer.

b) In the case of chemical reaction control, $R_{TM} << R_{RQ}$:

Hence:

$$V_p = -\frac{\Delta C}{R_{TM} + R_{RQ}} \approx -\frac{\Delta C}{R_{RQ}} \tag{19.48}$$

$$V_P\left(\frac{\text{moles of A}}{s}\right) \cong -\frac{\left(C_A^0 - C_A^{eq}\right)}{\dfrac{1}{4\ \pi\ R^2 K_{RQ}}} = -4\ \pi\ R^{\,2} K_{RQ}\left(C_A^0 - C_A^{eq}\right) \tag{19.49}$$

K_{RQ} dependence on temperature is given by:

$$K_{RQ} = K_{RQ}^0 \cdot e^{-\frac{\Delta E}{RT}} \tag{19.50}$$

where, in general (there are exceptions) ΔE, the activation energy, is in the order or greater than 50000 cal/mole. Therefore, in cases of control by chemical reaction it is expected that it is much higher than processes controlled by mass transfer through boundary layer.

A qualitative influence of temperature and degree of agitation on the resistances for controlling steps of a reaction like the one discussed above namely chemical reaction control and mass transfer through boundary layer control or mixed control are given in Figure 19.11.

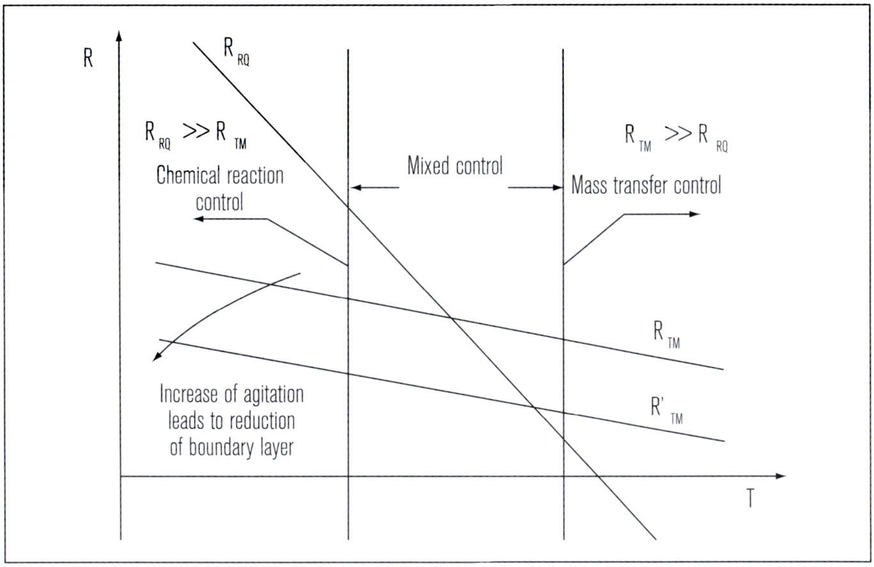

Figure 19.11 - Influence of temperature and agitation on resistances of the controlling steps of the process.

It is observed that chemical reaction control is prominent at relatively low temperatures and mass transfer control through boundary layer is significant at relatively high temperatures.

Changes in rates can be achieved, if changes are made in the controlling resistances. For example, if the temperature of operation is equal to T_1 for which $R_{TM} >> R_{RQ}$, then the rate of the process can be enhanced substantially by increased agitation, thereby decreasing the boundary layer thickness.

Example

Consider carbochlorination TiO_2, contained in briquettes of TiO_2, and graphite, according to the reaction:

$$TiO_2\,(s) + 2C\,(s) + 2\,Cl_2\,(g) = TiCl_4\,(g) + 2\,CO(g)$$

As discussed in the previous section, this reaction is also composed of steps involving chemical reactions and transport of species.

In the case of control by chemical reaction or mass transport the influence of temperature on rate constant (V_P) can be evaluated using Arrhenius equation

$$\ln V_p = \ln V_P^0 - \frac{\Delta E_{apparent}}{RT}$$

Experimental data obtained by Gameiro (1994) are presented in the table and the figure below. As can be seen from the figure, at low temperatures, the values of activation energy are high, suggesting control by chemical reaction and low values of activation energy are obtained at high temperatures suggesting control by mass transfer through boundary layer.

Initial rates of carbochlorination of briquettes TiO_2/graphite with porosity of 30% and containing 20% carbon with chlorine at 1 atm pressure:

T(K)	773	803	823	873	973	1073	1173	1273
r (g TiO_2/min)	0.001	0.005	0.009	0.015	0.021	0.026	0.027	0.030

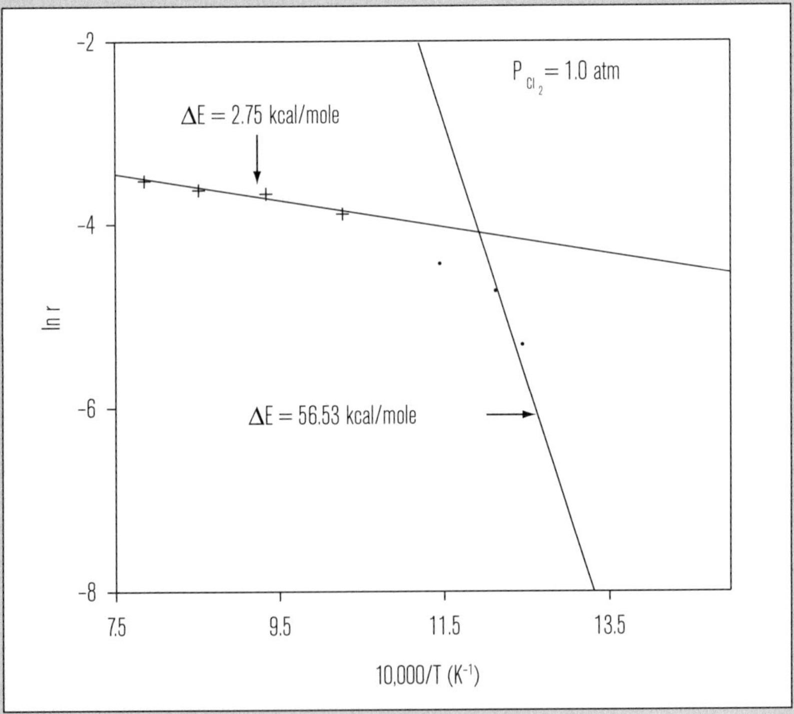

Activation energy for carbochlorination of TiO_2 (GAMEIRO, 1994).

In the gas –solid reactions described above it is usual that the experimental data is reported in terms of fractional reduction or fraction reacted. For example the reaction M (s) + nA (g) = nB (g) implies that the radius of the solid should decrease with time due to reaction with gas.

In fact, on the basis of the variable "molar density of A in the particle" $\rho_M\left(\frac{\text{moles of M}}{m_3}\right)$, one can define the fraction reacted:

$$\alpha = \frac{\text{moles of M at, t = 0-moles of M at, t = t}}{\text{moles of M at, t = 0}} \tag{19.51}$$

$$\alpha = \frac{\frac{4}{3}\pi R_0^3 \cdot \rho_M - \frac{4}{3}\pi R^3 \cdot \rho_M}{\frac{4}{3}\pi R_0^3 \cdot \rho_M} = \frac{R_0^3 - R^3}{R_0^3} \tag{19.52}$$

$$\alpha = 1 - \left(\frac{R}{R_0}\right)^3 \tag{19.53}$$

The quantity of M in the particle radius R (at time t) is given by:

$$\frac{4}{3}\pi R^3 \left(m^3\right) \times \rho_{MO} \tag{19.54}$$

The rate of consumption of M in the particle is equal to:

$$\frac{d}{dt}\left(\frac{4}{3}\pi R^3 \rho_M\right) = 4\pi R^2 \rho_M \frac{dR}{dt}\left(\frac{\text{moles of M}}{s}\right) \tag{19.55}$$

Therefore, assuming the validity of the previously proposed model of coupled diffusion and chemical reaction at the surface (equation 19.42) and considering the stoichiometry of the reaction:

$$4\pi R^2 \rho_M \frac{dR}{dt}\left(\frac{\text{moles of M}}{s}\right) = \frac{1}{n} V_p \left(\frac{\text{moles of A}}{s}\right) \tag{19.56}$$

$$4\pi R^2 \rho_M \frac{dR}{dt} = -\frac{1}{n} \frac{\left(C_A^0 - C_A^{eq}\right)}{\frac{\delta}{4\pi R^2 D_A} + \frac{1}{4\pi R^2 K_{RQ}}} \tag{19.57}$$

This differential equation gives the change in the radius of the particle with time and can be integrated with the initial condition t = 0, R = R_0.

Simplifying, one gets:

$$\frac{R}{R_0} - 1 = -\frac{\left(C_A^0 - C_A^{eq}\right)}{\dfrac{\delta}{D_A} + \dfrac{1}{K_{RQ}}} \times \frac{1}{\rho_M} \quad t \tag{19.58}$$

Considering 19.53 giving the expression for fraction reacted and 19.58 a function g (α), can be defined in terms of fraction reacted.

$$g(\alpha) = \left[(1-\alpha)^{\frac{1}{3}} - 1\right] = -\left[\frac{\left(C_A^0 - C_A^{eq}\right)}{\dfrac{\delta}{D_A} + \dfrac{1}{K_{RQ}}}\right] \times \frac{1}{R_0 \cdot \rho_M} \cdot t \tag{19.59}$$

This predicts a linear variation of g (α) with time for the model to be valid (Figure 19.12) as given by the straight line relationship. The 2nd curve in the graph shows that this model is not suitable for this case:

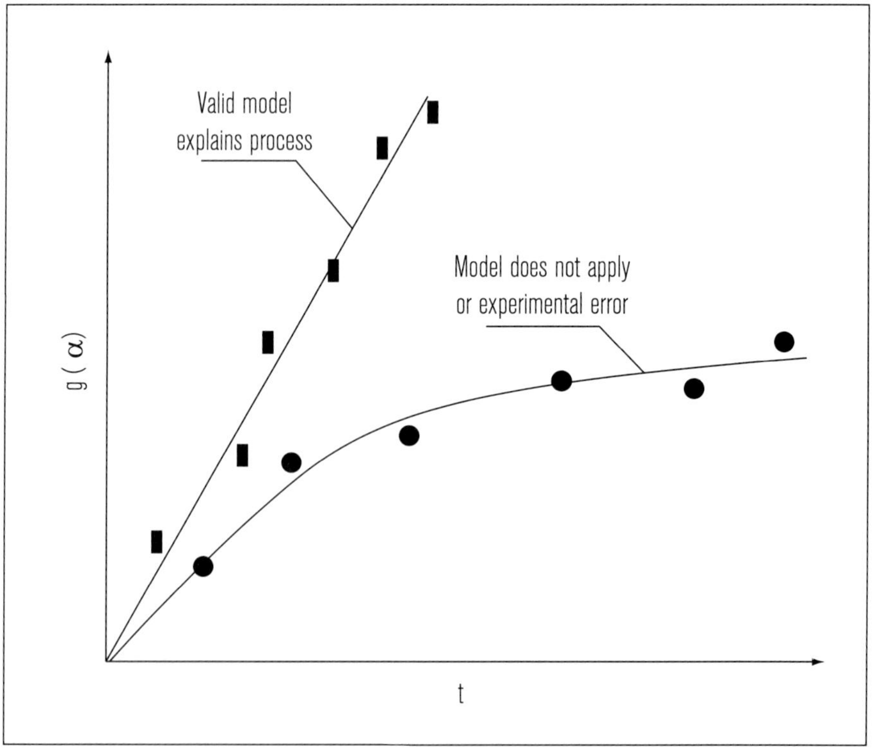

Figure 19.12 - Test of validity of the model for the experimental points (Plot of g (α) as as a function of time.

19.3.2 Reaction of porous particle with gas

In the area of gas solid reactions the other extreme is the reaction of a gas with a porous solid. The gas penetrates the solid and the reaction takes place not only on the surface but also on the walls of the pores (THEMALIS, 1995). A qualitative description of this process is given in Figure 19.13.

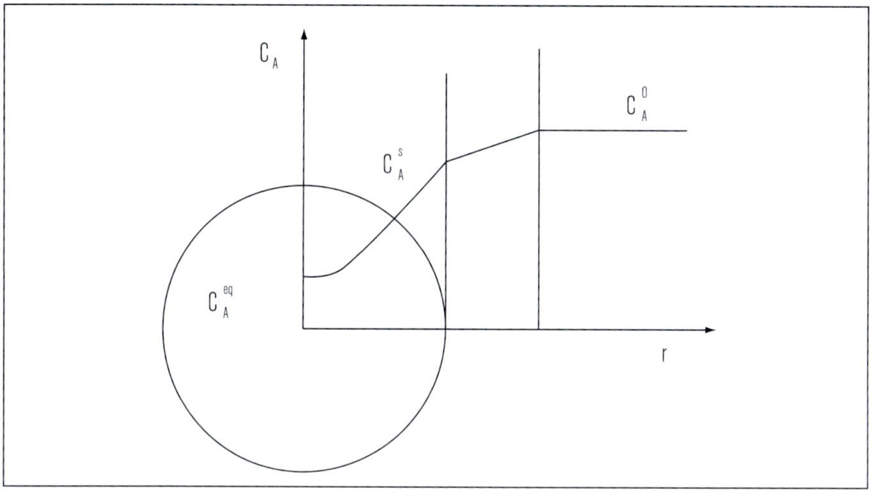

Figure 19.13 - Schematic composition profile for the case of reaction gas with porous solid.

As in the case of nonporous solid discussed before consider the same reaction, M (s) + nA (g) = nB (g), the rate of transport through the boundary layer is given by:

$$V_{TM}\left(\frac{\text{moles of A}}{\text{s}}\right) = -4\pi R^2 \times D_A \times \frac{\left(C_A^0 - C_A^s\right)}{\delta} \tag{19.60}$$

Since the gas diffuses inside the solid in this case a balance of conservation of the species has to be carried out in a control volume corresponding to a spherical shell of infinitesimal thickness located between r and r + Δr considering diffusion through pores and chemical reaction:

$$-4\pi\, r^2\, D_A^{ef} \frac{dC_A}{dr}\Big|_r + 4\pi\, r^2\, D_A^{ef} \frac{dC_A}{dr}\Big|_{r+\Delta r} - 4\pi\, r^2\, \Delta r\, a\, K_{RQ}(C_A - C_A^{eq}) = 4\pi\, r^2\, \Delta r\, \varepsilon \frac{dC_A}{dt} \tag{19.61}$$

In this expression, the first two terms represent the differential for diffusion in the pores, the third for chemical reaction in the walls of the pores. For assessing the rate of chemical reaction one has to know the area taking part in the reaction. Hence the parameter a (m^2/m^3), the surface area of pores per unit volume, has to be incorporated in the term representing the chemical reaction. This parameter can be determined by experimental techniques such as BET or mercury porosimetry. The parameter ε represents the porosity of the particle. The above equation can be written as:

$$\frac{D_A^{ef}}{r^2}\frac{d}{dr}\left(r^2\frac{dC_A}{dr}\right) - a\ K_{RQ}(C_A - C_A^{eq}) = \varepsilon\frac{dC_A}{dt} \tag{19.62}$$

The solid takes part in the reaction inside the pores and also on the surface.

From the condition of symmetry:

$$\left.\frac{dC_A}{dr}\right|_{r=0} = 0 \tag{19.63}$$

It is also necessary to write an expression for the decrease of radius of the particle. All the amount of reagent that crosses the boundary layer, does not penetrate the particle.(1st term in equation 19.64). One part diffuses inside (2nd term) and the other part, reacts at the external surface. Hence there is reduction in the radius with time given by the following equation (the term on the right hand side of 19.64 is based on 19.56):

$$-4\pi R^2 \times D_A \times \frac{\left(C_A^0 - C_A^s\right)}{\delta} - 4\pi R^2 D_A^{ef}\left.\frac{dC_A}{dr}\right|_{r=R} = n\ \rho\ 4\pi R^2\frac{dR}{dt} \tag{19.64}$$

and

$$-4\pi R^2 \times D_A \times \frac{\left(C_A^0 - C_A^s\right)}{\delta} - 4\pi R^2 D_A^{ef}\left.\frac{dC_A}{dr}\right|_{r=R} = (1-\varepsilon)4\pi R^2 a K_{RQ}(C_A^s - C_A^{eq}) \tag{19.65}$$

The simultaneous solution of equations 19.62 to 19.65 is only possible by numerical techniques, and requires some assumption for variation of parameter α with time during the progress of the reaction.

Catalysts are usually made of a porous matrix, which serve as a substrate for a reaction, and that is not consumed. In this case, several simplifications are possible, provided the particle size and specific surface area of the pores do not change.

Example

Consider a solid sphere through which a certain specie A (impurity or interstitial) is able to diffuse. During diffusion A is consumed by a 1st order reaction. The balance of conservation of species performed in a spherical shell, for steady state is given by (FROMENT and BSICHOFF, 1979):

$$N_A\ 4\ \pi\ r^2\ |_r - NA\ 4\ \pi\ r^2\ |_{r+\Delta r} + 4\ \pi\ r^2\ \Delta r\left(-K\ C_A\right) = 0$$

Taking the limit when Δr tends to zero:

$$\frac{d(r^2 N_A)}{dr} + r^2\ K\ C_A = 0$$

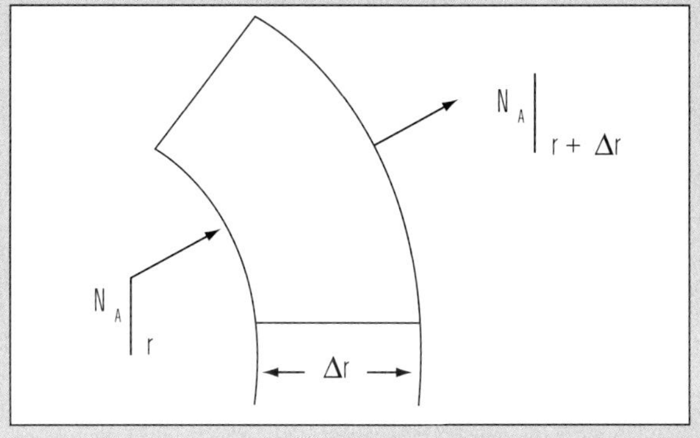

Control Volume for balance of conservation of species in spherical shell.

The introduction of the expression of Fick's second law in spherical coordinates and disregarding any convective effect, gives:

$$\frac{D_A}{r^2}\frac{d}{dr}\left(r^2\frac{dC_A}{dr}\right) - K\ C_A = 0$$

Boundary conditions can be assumed to be:

1) Symmetry in the center of the sphere:

$$r = 0 \qquad \frac{dC_A}{dr} = 0$$

2) Known and constant composition on the surface of the sphere:

$$r = R \qquad C_A = C_A^o$$

The solution in this case is the following (TOSUN, 2002):

$$\frac{C_A}{C_A^o} = \frac{R}{r}\frac{\sinh(\Lambda(r/R))}{\sinh \Lambda}$$

where $\Lambda = \sqrt{\frac{k R^2}{D_A}}$ represents the Thiele module.The significance of the Thiele module can be understood with the help of the following table, which shows the significance of the physico-chemical terms listed in the differential equation describing the flow.

Physical meaning of terms of the conservation equation in the spherical system.

Equation	Physical significance	Parameter
$\frac{D_A}{r^2}\frac{d}{dr}\left(r^2 \frac{dC_A}{dr}\right)$	Diffusion	$\frac{D_A C_A^o}{R^2}$
$K C_A$	Chemical reaction rate	$K C_A^o$

Thiele module can be taken as a measure of the relative intensity of the processes of chemical reaction and diffusion.

19.3.3 Topochemical model for gas solid reactions

Consider a isothermal chemical reaction of the type between a solid and a gas. In this case a gas as product is produced like in the case of reduction of iron oxides where CO gas reacts with iron oxide to give iron:

$$MO\ (s) + nA\ (g) = M\ (s) + nB\ (g) \qquad \Delta G < 0 \qquad (19.66)$$

where the solid MO can be assumed impermeable to gases while solid product, M, is porous, so that the reacting gas can diffuse through porous product layer.

Overall dimensions of the particle do not change throughout the process and the progress of the reaction is schematized in Figure 19.14.

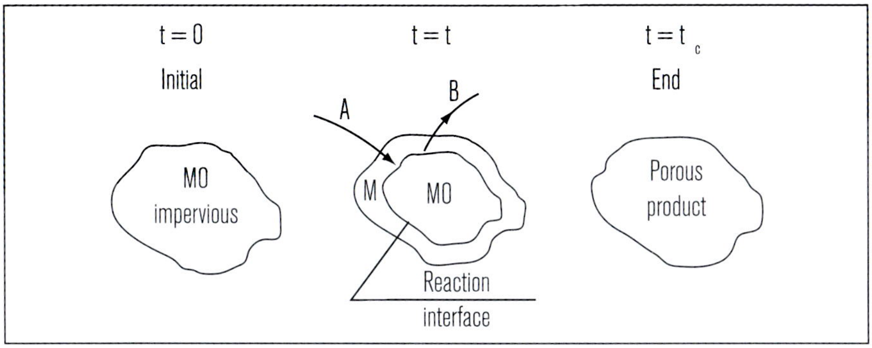

Figure 19.14 - Movement of interface of reaction with time from initial to final according to topochemical model of reaction.

Various steps in this type of gas solid reaction for the reaction of a spherical particle with gas are given below at a certain instant of time t = t as in Figure 19.14 (HABACHI, 1999). Figure 19.15 gives a schematic profile of concentrations for these steps:

1) transport of reagent through the boundary layer in the gas phase surrounding the sphere from the bulk of the gas phase to the surface of the particle;
2) transport of reagent through the porous product layer, from the surface of the particle to the reaction interface at r = R;
3) chemical reaction MO + nA → nB +M at the interface, r = R;
4) transport of product B through the porous product layer;
5) transport of product B through the boundary layer.

Similar to the treatment of the previous model, each of the steps mentioned can be described in terms of rates:

1) Transport through the boundary layer:

$$V_{CL}\left(\frac{\text{moles of A}}{s}\right) = -4\pi R_0^2 \times K_m \times \left(C_A^0 - C_A^S\right) \tag{19.67}$$

where:

- R_0 Is the radius of the particle;
- K_m is a coefficient of mass transfer, to be determined experimentally.

The sign (-) is due to the fact that the species A is getting transported in the negative direction in respect of axis r.

2) Transport through the porous product layer:

The specific flow (per unit area) of a species A can be evaluated from:

$$N_{A,r}\left(\frac{\text{moles of A}}{m^2 \cdot s}\right) = -D_A \frac{dC_A}{dr} + X_A\left(N_A + N_B\right) \tag{19.68}$$

Due to the stoichiometry of the reaction and the assumption of quasi steady state (or no accumulation of species A within the porous structure) one can write $-N_A = N_B$.

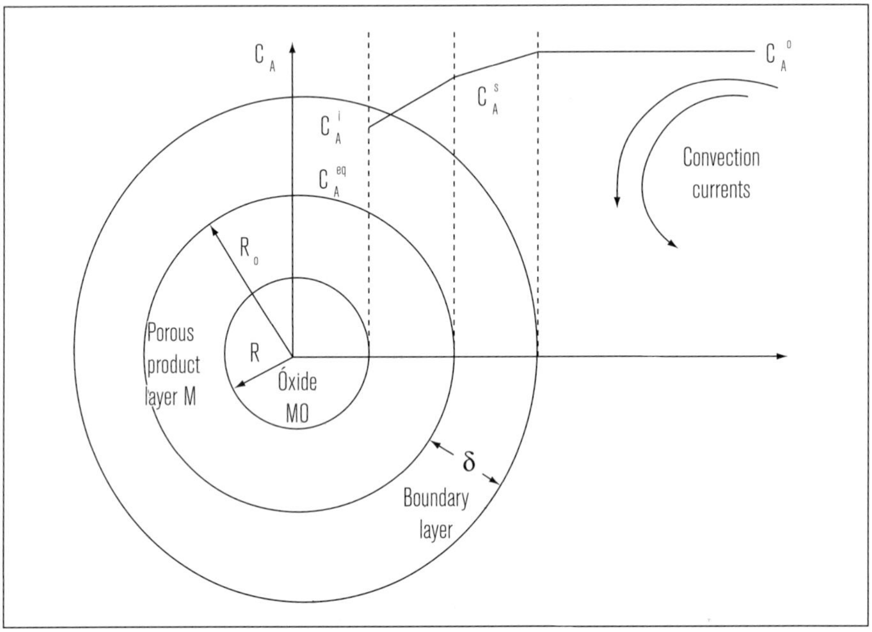

Figure 19.15 - Schematic figure for the reaction model (spherical particle).

Hence:

$$N_{A,r} = -D_A \frac{dC_A}{dr} \quad R \le r \le R_0 \tag{19.69}$$

Then the rate of transport of A through a generic surface located at r = r would be given by:

$$\text{Flow rate (constant)} = 4\pi r^2 \cdot N_{A,r} = 4\pi r^2\left(-D_A \frac{dC_A}{dr}\right) \tag{19.70}$$

The flow rate is constant due to the assumption of quasi-stationary state. The value of this can be evaluated by integrating the above expression with the boundary conditions:

for $r = R_0$ $C = C_A^S$ and for $r = R$ $C = C_A^i$.

Hence:

$$\text{Flow} = -\frac{4\pi D_A \left(C_A^S - C_A^i\right)}{\left(\frac{1}{R} - \frac{1}{R_0}\right)} \qquad R \le r \le R_0 \tag{19.71}$$

Finally, the expression considering the porosity and the tortuosity of the network of pores in the porous product layer (Chapter 17, section 17.6) effective diffusivity $\frac{D_A\,\varepsilon}{\tau}$ should be used in the equation 19.71. Substituting, one gets the flow rate through the porous product layer:

$$V_{PP}\left(\frac{\text{moles of A}}{s}\right) = -4\pi\cdot\frac{D_A\cdot\varepsilon}{\tau}\cdot\frac{\left(C_A^S - C_A^i\right)}{\left(\frac{1}{R} - \frac{1}{R_0}\right)} \tag{19.72}$$

3) For the chemical reaction at the interface between the oxide surface and porous product, assuming first order kinetics, flux can be evaluated:

$$V_{RQ}\left(\frac{\text{moles of A consumed}}{s}\right) = -4\pi R^2 \times K_{RQ} \times \left(C_A^i - C_A^{eq}\right) \tag{19.73}$$

where K_{RQ} must be determined experimentally.

Finally, for each of the steps mentioned above a characteristic resistance can be defined as discussed in previous cases on the basis of electrical analogy:

$$V_P\left(\frac{\text{moles of A}}{s}\right) = -\frac{\left(C_A^o - C_A^{eq}\right)}{\frac{1}{4\pi R_0^2 K_m} + \frac{\tau}{4\pi\,\varepsilon\,\omega D_A}\left(\frac{1}{R} - \frac{1}{R_0}\right) + \frac{1}{4\pi R^2 K_{RQ}}} \tag{19.74}$$

Also, due to the stoichiometry of the reaction and since the oxide in the unreacted core is confined to the range $0 \le r \le R$, one can write:

$$4\pi R^2 \rho_{MO}\frac{dR}{dt}\left(\frac{\text{mols de MO}}{s}\right) = \frac{1}{n}V_P\left(\frac{\text{mols de A}}{s}\right) \tag{19.75}$$

$$4\pi R^2 \rho_{MO} \frac{dR}{dt} = -\frac{1}{n} \frac{\left(C_A^0 - C_A^{eq}\right)}{\frac{1}{4\pi R_0^2\, K_m} + \frac{\tau}{4\pi\, \varepsilon\, D_A}\left(\frac{1}{R} - \frac{1}{R_0}\right) + \frac{1}{4\pi R^2 K_{RQ}}} \tag{19.76}$$

The above equation can be easily integrated with the initial condition: t = 0 and R = R_0 to get the amount reacted at any time interval. As before the equation can be modified in terms of fraction reacted, making it easier to check the validity of the model as was done before.

Example

A spherical sample of ore contains gold particles. For leaching the gold out penetration of the particle by an appropriate reagent is essential. For this case a reaction model based on diffusion control for transport through porous product layer (in this case porous layer of unreacted gangue) can be proposed. If particles of 2 mm show 50% extraction after 1 hour of contact, what would be the the time needed to reach 90% extraction of particles of 1 mm in diameter?

The relevant equation for application in this case,can be obtained by simplifying the equation 19.75 (only diffusion control through porous product layer).

$$4\pi R^2 \rho_{MO} \frac{dR}{dt} = -\frac{1}{n} \frac{\left(C_A^0 - C_A^{eq}\right)}{\frac{\tau}{4\pi\varepsilon D_A}\left(\frac{1}{R} - \frac{1}{R_0}\right)}$$

This equation can be easily integrated to give:

$$\frac{1}{3\,R_o}(R^3 - R_o^3) - \frac{1}{2}(R^2 - R_o^2) = \frac{\varepsilon D_A\left(C_A^0 - C_A^{eq}\right)}{n\,\tau\,\rho} t$$

Dividing both sides of the equation by R_o^2 and considering the definition of fraction reacted, given by equation 19.53, an equation similar to 19.59 can be obtained:

$$g(\alpha) = \frac{-\alpha}{3} - \frac{1}{2}((1-\alpha)^{2/3} - 1) = \frac{\varepsilon D_A\left(C_A^0 - C_A^{eq}\right)}{n\,\tau\,\rho}\frac{t}{R_o^2}$$

The reacted fraction is known and the term $\frac{\varepsilon D_A\left(C_A^0 - C_A^{eq}\right)}{n\,\tau\,\rho}$ can be considered constant, The solution is in the form of variation of fractional conversion with time and the time for a given conversion is proportional to the square of the radius of the particle.

Example

An ore sample is non uniform in size and is made up of particles of 200 microns (33%) 100 microns (33%) and the rest 50 microns average diameter. It is to be leached and the progress of the dissolution will be followed by determining the variation of liquor concentration with time. If the process kinetics can be explained by the topochemical model and the following equation applies for mass transport through boundary layer control:

$$1-(1-\alpha)^{1/3} = \frac{b\,K\,\Delta C}{\rho_M\,R_o}t$$

For an arbitrary value of $\frac{b\,K\,\Delta C}{\rho_M} = 10{,}000$, the following curve for fractional reaction with time can be drawn.

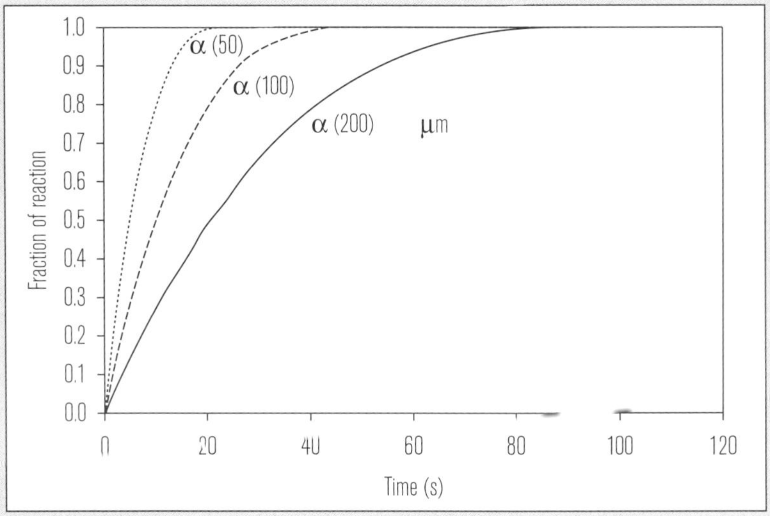

Fraction of reaction as a function of time, for three diameters, under boundary layer control.

It is apparent that smaller the particle faster is the conversion. Also the time of conversion seems to depend on initial radius.

These data of conversion for different diameters can be combined to determine an average degree of conversion as a function of time. If there is no experimental error involved, the same value should result by chemical analysis of the solution. Finally if mass transfer through the boundary layer is rate controlling, then a linear relationship between $1-(1-\alpha)^{1/3}$ and time should result. If it is not so one can mistakenly conclude that the model does not hold good. However from the following plot one cannot reach any definite conclusion as there are 2 straight line portions that form part of the curve. Hence more investigations would be necessary to arrive at a definite conclusion.

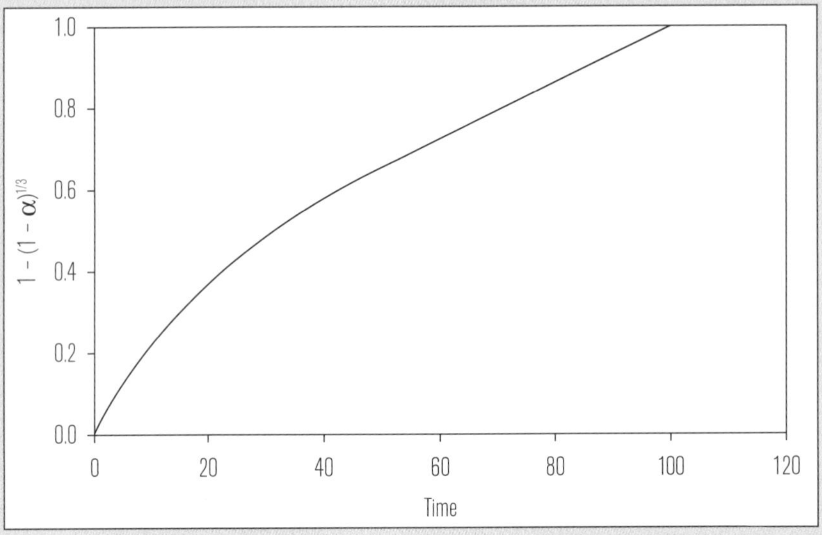

Plot of $g(\alpha) = 1 - (1 - \alpha)^{1/3}$ as a function of time.

19.3.4 Theory of films

The interaction between fluids is very important and common in processing of metals and materials and the the theory of films plays a crucial role in theses interactions. According to this there exists a concentration boundary layer on either side of the interface that separates the two fluids

(BIRD, STEWART and LIGHTFOOT, 2002). Chemical species can cross these boundary layers and take part at the interface in chemical reactions and then the products get transported through the films into the respective phases.

Consider the application of theory of films, to the transfer of a chemical species between the two fluid phases, α and β. For choosing a model for the process initially it is necessary to identify the various steps (MIDDLEMAN, 1998). In the present case:

- Mass transport in the phase α;
- Chemical reaction at the interface $\alpha - \beta$;
- Mass transport in the phase β.

The theory of films can be applied without restriction to cases where one part is in a high state of agitation and hence with a smaller boundary layer thickness relative to the other and the mass transport is restricted to a film adjacent to the interface or concentration boundary layer as shown in Figure 19.16. In the following brackets namely [ ] and parentesis () are used for the phases α and β respectively, the subscripts (i) and (eq) indicate the concentrations of any species at the interface ($\alpha - \beta$) and at equilibrium respectively. For purposes of quantification, one can use phenomenological equations. This does not necessarily devalue the same scientific basis, but rather based on common sense.

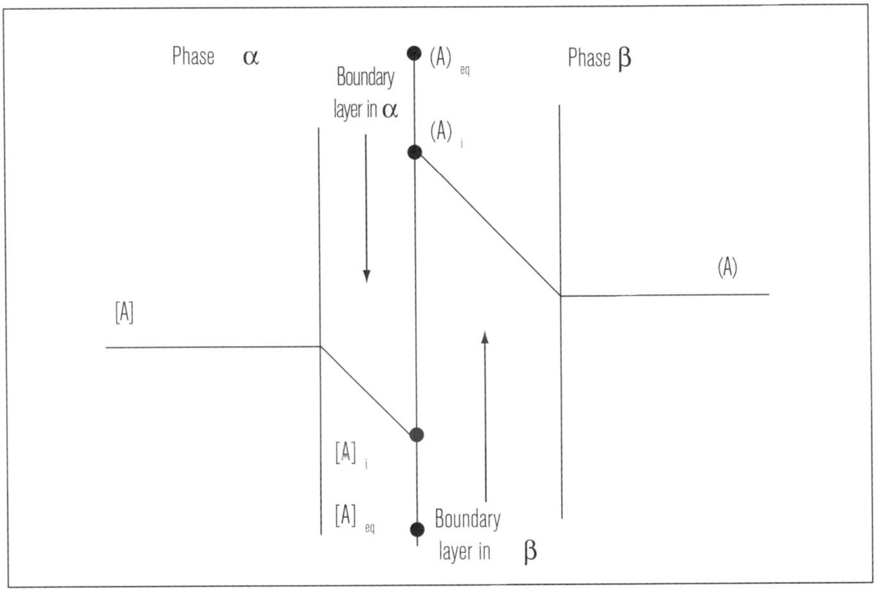

Figure 19.16 - Conceptual and idealized representation of a process of mass transfer between two fluid phases.

Regarding the mass transport in the boundary layer in α, the flow of A is given by:

Flow of A = mass transfer coefficient x interface area x driving force.

$$n_A\left(\frac{\text{moles of A}}{s}\right) = k_\alpha \text{Area}\left([A]-[A]_i\right) \quad (19.77)$$

where:

- K_α is the mass transfer coefficient (m/s);
- Area, is the area of chemical exchange interface m^2.

K_α is generally determined experimentally. As already discussed in the previous chapters, it is a function of physicochemical properties of the medium (diffusion coefficient of the species in the fluid, fluid viscosity, density, temperature, pressure, composition) and the velocity field inside the fluid.

For the interfacial chemical reaction, the same methodology is followed:

Reaction rate = Coefficient × Interface area × Driving force

$$r_A\left(\frac{\text{moles of A}}{s}\right) = k_{RQ}\ \text{Area}\left([A]_i - [A]_{eq}\right) \quad (19.78)$$

The determination of rate constant of a certain chemical reaction can be fairly complex, depending on the mechanism.The value of K_{RQ} has to determined experimentally, as a function of temperature only. The order of the reaction also is evaluated from experiments. The reaction can be considered a 1st order reaction if there is experimental justification for the same. If the order is not equal to one the simplified analytical treatment given below, based on electrical analogy would not apply. However a numerical solution can be easily implemented.

One can simplify the mathematical treatment, even if exact mechanism is not known adopting the following procedure involving a distribution coefficient of a species between the two phases like in the examples discussed earlier.

If $(A)_i$ is the interfacial concentration of A in phase β then the concentration corresponding to equilibrium in phase α will be given by:

$$L = \frac{(A)_i}{[A]_{eq}} \rightarrow\rightarrow [A]_{eq} = \frac{(A)_i}{L} \quad (19.79)$$

Therefore:

$$r_A\left(\frac{\text{moles of A}}{s}\right)=k_{RQ}\text{Area}\left([A]_i-\frac{(A)_i}{L}\right) \tag{19.80}$$

Similarly, for the mass transport in the phase β:

$$n'_A = k_\beta \text{Area}\left((A)_i-(A)\right) \tag{19.81}$$

From the above analysis the following 3 equations result:

$$n_A = k_\alpha \text{Area}\left([A]-[A]_i\right) \tag{19.82}$$

$$r_A = k_{RQ}\text{Area}\left([A]_i-\frac{(A)_i}{L}\right) \tag{19.83}$$

$$n'_A = k_\beta \text{Area}\left((A)_i-(A)\right) \tag{19.84}$$

In any of these three equations, interface concentrations $(A)_i$ and $[A]_i$ represent parameters difficult to measure (as it is difficult to position a sensor just on α side or β side of the interface α/β). Therefore one has to assume quasi-stationary state, whereby there is no accumulation of A. "A gets transported through boundary layer in α , gets consumed by the chemical reaction and then transported through the boundary layer in β:

$$n_A = r_A = n'_A \tag{19.85}$$

With this assumption the interfacial concentrations can be eliminated as the flux is constant resulting in the equation:

$$n_A\left(\frac{\text{moles}}{s}\right)=\frac{\left([A]-\frac{(A)}{L}\right)}{\left(\frac{1}{\text{Area }K_{\alpha}}+\frac{1}{\text{Area }K_{RQ}}+\frac{1}{L\text{ Area }K_\beta}\right)} \tag{19.86}$$

which as has been mentioned earlier has the electrical analogy with circuit of three resistors in series (Figure 19.17).

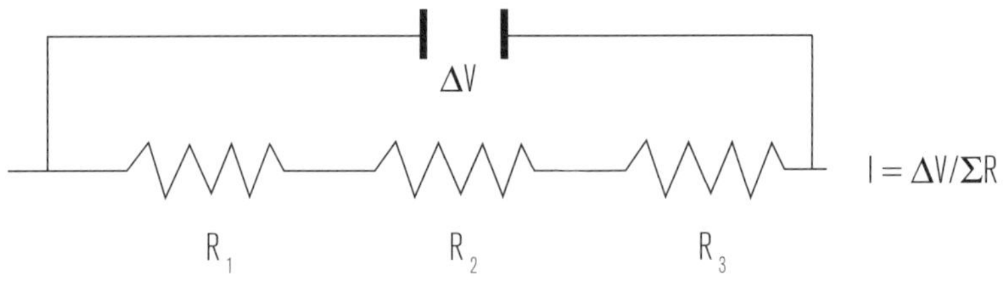

Figure 19.17 - Electrical circuit with resistors in series.

By analogy to define various resistances and control regimes, Table 19.1.

Table 19.1 - Resistances in electrical and mass transfer analogies.

	Electrical analogy	Chemical /Mass transfer analogy
Driving force	Potential difference ΔV	Concentration difference [A] - (A)/L
Flow	Electrons; current I	Specie A
Resistance	R_1	Resistance to transport in the boundary layer (α) $R_{T\alpha} = 1/(K_\alpha \text{ Area})$
Resistance	R_2	Resistance due to chemical reaction $R_{RQ} = 1/(K_{RQ} \text{ Area})$
Resistance	R_3	Resistance to transport in the boundary layer (β) $RTB = 1/(L\, K_\beta \text{ Area})$

The above analogy helps to identify stages in the process, in which, one can change the value of characteristic resistance (by changing the the parameters involved in the definition of that resistance) which would result in substantial changes in the flux.This type of analysis also helps in identifying the stage which controls the process namely the one which offers maximum resistance. For example in a particular case the chemical reaction, a step may have high resistance relative to the others and hence the entire process is controlled by this slower step of high resistance.

$$R_{RQ} > R_{T\alpha} \text{ and } R_{RQ} > R_{T\beta}$$

Other possibilities are shown in Table 19.2:

Order of magnitude of the resistance	Control type
$R_{T\alpha} > R_{RQ}$ and $R_{T\alpha} > R_{T\beta}$	Mass transport in the boundary layer in α
$R_{RQ} > R_{T\alpha}$ and $R_{RQ} > R_{T\beta}$	Chemical reaction
$R_{T\beta} > R_{RQ}$ and $R_{T\beta} > R_{T\alpha}$	Mass transport in the boundary layer in β
$R_{T\alpha} \sim R_{T\beta} > R_{RQ}$	Mixed control, mass transport
$R_{T\alpha} \sim R_{RQ} > R_{T\beta}$	Mixed control, mass transport and reaction
$R_{T\beta} \sim R_{RQ} > R_{T\alpha}$	Mixed control, mass transport and reaction
$R_{T\beta} \sim R_{RQ} \sim R_{T\alpha}$	Mixed control, all steps are important

Table 19.2 - Controlling steps of reactions according to the theory of films in mass transfer.

The above table gives for this particular process the important or dominant steps whose identification can lead to control and manipulation of several operating parameters, that would bring about major changes in the efficiency of the process.This knowledge can be translated into greater stability (smaller fluctuations) and increased the productivity of a certain process. The other parameters that do not appear in the definition of dominant resistance may represent a second choice of action, for the same purpose.

Example

Consider a hearth furnace for refining of metals. It contains a shallow bath of (~0.5 m) metal (steel, hot metal) on a rectangular hearth refractory at the base of the furnace. The heating process to the desired temperature can be achieved by exothermic combustion of gas, oil, etc. Slags of suitable compositions are used to refine the metal generally oxidizing slags for dephosphorization and reducing type of slags for desulphurization. In this type of furnace the chemical exchanges between slag and metal and other reaction of refining take place in the slag metal interface with an area roughly equal to the cross sectional area of the furnace. As a result the process would be very slow and and it would take long time for refining to complete and to get the steel with proper final composition. In fact this was the practice in open hearth furnaces used for steel making upto the middle of last century.

On the other hand in the LD or Basic Oxygen Furnace (general designation for the oxygen converter with oxygen blowing from the top and additional agitation by inert gas injected through the tuyeres at the bottom of the convertor) oxygen jet pierces the metal surface on the top with consequent ejection of metal droplets in the direction of the slag increasing considerably the slag metal interface area of reaction and also leading to the formation of an emulsion with slag which gets agitated with evolution of CO renewing constantly the surface area. This interaction between gas, bubbles, slag and the metal drops and formation of emulsion enhance considerably the speed of refining reactions. Carbon and other elements get rapidly removed by oxidation due to agitation caused by inert gas bubbles in which CO diffuses and the increase of oxidation and refining rates in the ejected metal droplets due to the action of oxygen jet.

Typical values of samples of the emulsion and metal from the bottom of the converter show that the area available for refining is much more in the case of droplets captured in the emulsion than at the bottom of the reactor, as given in the following figure.The reason is clearly the increase in surface area in this process. This effect is amplified as a continuous cycle of drops get ejected inside the slag phase and drop back to the metal bath, which results in a a refining time of about 20 minutes or even less.

On the basis of the film model one can discuss the influence of surface area on reaction or refining rates discussed above between slag and metal. Increasing the interfacial area would lead to a lowering of the resistance in all the three steps namely mass transfer in both the boundary layers and chemical reaction at the interface between the two phases (slag and metal). In fact if one considers that the mass of steel, M_{steel}, of density ρ_{steel} is divided evenly between droplets of radius R, one can write for the flux from equation 19.85 considering the area for diffusion/reaction equivalent to $\frac{3M_{steel}}{R\rho_{steel}}$. Introducing this in the equation one gets:

$$n_A\left(\frac{\text{moles}}{\text{s}}\right) = \frac{3M_{steel}}{R\rho_{steel}} \frac{\left([A] - \frac{(A)}{L}\right)}{\left(\frac{1}{K_\alpha} + \frac{1}{K_{RQ}} + \frac{1}{LK_\beta}\right)} \qquad (19.87)$$

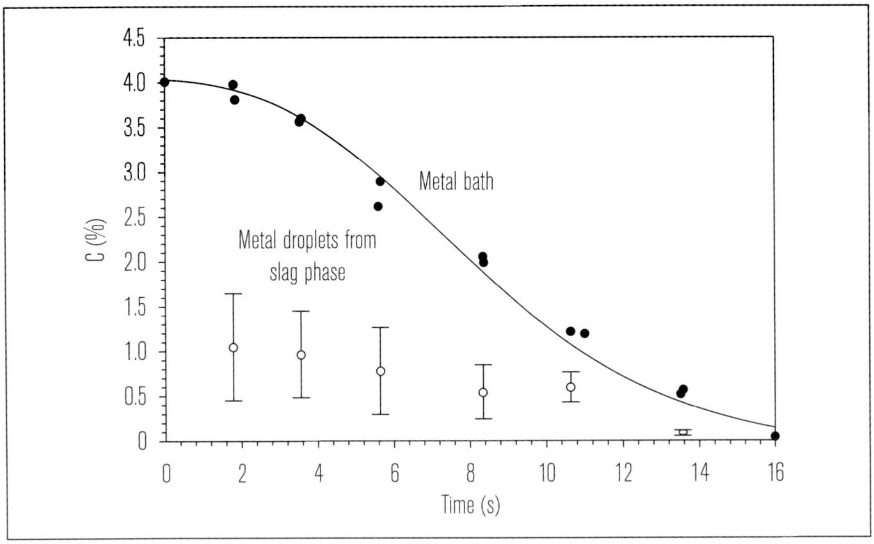

Figure 19.18 - Variation of carbon content during LBE (lance bubbling equilibrium), samples taken from metal drops ejected vs within the metal bath (CICUTTI et al., 1997).

In general, the resistance values in the equation 19.86 are functions of mass transfer coefficients and the reaction rate constant which in turn are determined by variables such as temperature, pressure, composition (interaction between the species that are reacting or being transported to the medium), and hydrodynamic conditions (velocity field in fluid) as well as properties such as viscosity, density etc.

Example

Consider application of film theory to the reaction of desulfurization of hot metal (or steel) with slag. In this case the phase α represents the metal, and the phase β represents the slag. The partition coefficient of sulfur between slag and metal is given by:

$$L_S = \frac{(\%S)_{slag}}{[\%S]_{steel}}$$

This has a high value, 300 ~ 2000, depending on temperature, slag basicity and the oxidation state of metal. There are a number of correlations for calculation of L_S as a function of these parameters in the literature. In this system as can be seen from the figure below the resistances for the steps of mass transport through the boundary layers are not significantly influenced by temperature as that of chemical reaction at the interface. Due to high values of L_S the transport resistance in the slag can be considered not significant whereas the resistance due to chemical reaction at the interface and the transport resistance at the metal side have to be considered. The resistance value for chemical reaction is significant at lower temperatures and decreases considerably at higher temperatures in accordance with the following figure which describes schematically the variation of resistances in this case with temperature

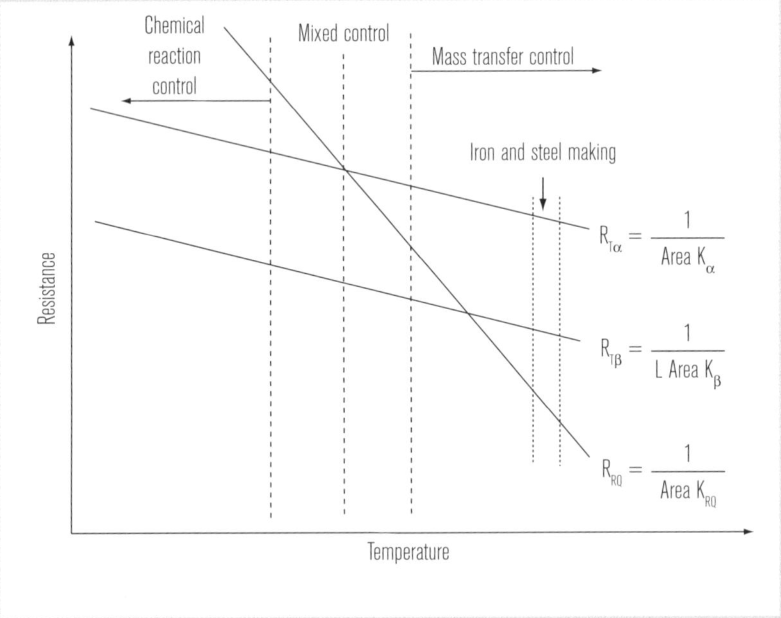

Schematic variation of resistances with temperature in a hypothetical case of desulphurization.

According to the above figure, for identifying possible controls for the process, the dominant values of resistance are mass transport from α (metal) and/or chemical reaction at the interface, ignoring the transport resistance in the slag. At lower temperatures the resistance due to chemical reaction, is quite significant relative to the values for mass transfer. There is a transition temperature range at which the values of resistances for transport from metal to interface and chemical reaction tend to be equal. At the vicinity of this arbitrarily defined temperature range, the region where the resistances are of the same order of magnitude, can be characterized as the ones which contribute to mixed control.

On the right hand side of the mixed control region the resistance for the mass transfer step from metal side becomes significant relative to the resistance value for chemical reaction which suffers strong reduction at higher temperatures leading to a fast reaction, in which the transported reagent to the interface would be consumed immediately at the interface. Under these conditions the resistance to mass transport of species from metal side to interface become relatively more significant.

Similar reasoning would lead us to the conclusion that the range of temperatures (low temperatures) to the left of the mixed control region control range corresponds to chemical reaction control as its resistance value is quite high.

The area between vertical dotted lines define the range of temperatures suitable for operations of refining of iron and steel, since it is located in the control region for mass tranport and this becomes the normal region for any kinetic analysis of the refining process. In this case, the resistance for chemical reaction control and mass transfer term to slag (because of high partition coefficient) can be ignored in the general equation 19.86 and can be written as:

$$n_S = \frac{\left[[S] - \frac{(S)}{L_S}\right]}{\left[\frac{1}{\text{Area } K_{metal}}\right]} = \text{Area } K_{metal}\left[[S] - \frac{(S)}{L_S}\right]$$

19.3.5 Theory of films applied to multiple reactions

The same chemical processes like the ones discussed above, can be analyzed from a different point of view, considering exchange of ionic species between slag and metal during the refining operations. For this purpose a slag can be considered as an ionic liquid containing species such as:

$$SiO_4^{4-};\ Ca^{2+};\ Mg^{2+};\ O^{2-};\ PO_4^{3-};\ S^{2-}$$

even though normally a slag is represented as a molecular mixture of oxides, sulfides and fluorides, like CaO, SiO_2, MgO, P_2O_5, CaS, CaF_2 etc.

Adopting the molecular description the sulfur distribution between metal and slag is given as:

$$[S] \Leftrightarrow (S) \tag{19.88}$$

If one considers ionic species in slag and atomic species in metal the above distribution shuld be really given as an equilibrium between sulfur and oxygen ions in the slag/metal system.

$$\underset{(\text{metal})}{[S]} + \underset{(\text{slag})}{(O^{2-})} = \underset{(\text{slag})}{(S^{2-})} + \underset{(\text{metal})}{[O]} \tag{19.89}$$

The above can be considered as corresponding to two electrochemical reactions of half cells as given below:

$$2e + [S] = (S^{2-}) \tag{19.90}$$

$$(O^{2-}) = [O] + 2e \tag{19.91}$$

It should be kept in mind, that these reactions would occur under conditions of electroneutrality of the slag and metal.

Normally in a refining reactor many chemical reactions take place simultaneously. For example in the oxygen steel making process various reactions due to oxygen saturation of the metal bath are not in equilibrium and oxidation and refining reactions occur simultaneously with different rates.These reactions can be considered on the basis of ionic species and a simultaneous reaction model can be proposed where overall electroneutrality is maintained.

The model assumes that the electrochemical reactions that occur at the slag-metal interface are intrinsically fast, since the temperatures are high and hence chemical reaction is not the controlling step of the process. Each species is transported through boundary layer adjoining the metal side whose resistance is considered, fast chemical reaction at the interface, followed by transport to slag through the boundary layer with negligible resistance as explained before. The composition gradients are given in Figure 19.17. for this situation. For the above conditions the following equation is applicable:

$$\dot{n}_i\left(\frac{\text{mole}}{\text{s}}\right) = \text{Area}\left(\text{m}^2\right)\, K_i\left(\frac{\text{m}}{\text{s}}\right)(C - C_S)\left(\frac{\text{mole}}{\text{m}^3}\right) \tag{19.92}$$

where:

- Area = area of the interface;
- K_i = mass transfer coefficient of the species i;
- C, C_S = Concentrations in the bulk of metal and at the reaction interface respectively.

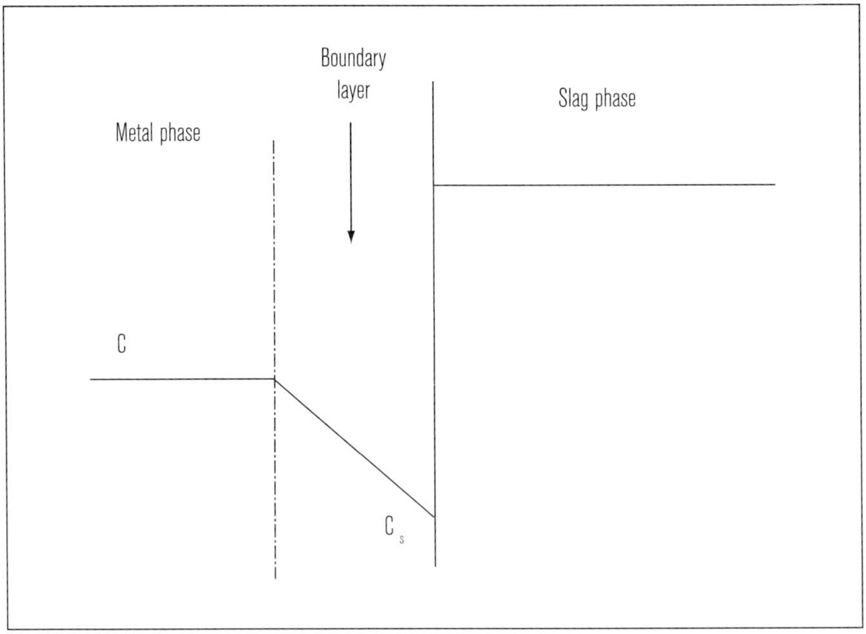

Figure 19.19 - Schematic concentration profile across the slag-metal interface, for model of multiple reactions.

Following is a list of possible reactions:

$$Fe = Fe^{2+} + 2e \quad (19.93)$$

$$2e + O = O^{2-} \quad (19.94)$$

$$Si = Si^{4+} + 4e \quad (19.95)$$

$$P = P^{5+} + 5e \quad (19.96)$$

$$2e + S = S^{2-} \quad (19.97)$$

From the above equations the condition of electroneutrality implies:

$$2\dot{n}_{Fe} + 2\dot{n}_{Mn} + 4\dot{n}_{Si} + 5\dot{n}_{P} - 2\dot{n}_{S} = 0 \quad (19.98)$$

This equation can be used if mass transfer coefficient for each species is known. As explained earlier the value of this coefficient would vary depending on physical-chemical and hydrodynamic conditions. Chemical analysis gives value of concentration within the metal, but the concentration at the interface is not known and has to be estimated through other methods as discussed in earlier sections.

It is assumed that mixing is perfect in the slag phase and only transport resistance on the metal side boundary layer is important and equilibrium prevails at the interface due to fast chemical reaction. Hence the effective equilibrium constants calculated from % compositions and true constants (calculated from thermodynamic data) using expressions for activity coefficients for the reactions listed are given below:

$$(\mathrm{Fe})+(\mathrm{O})=(\mathrm{FeO}) \quad K_{FeO}=\frac{\%FeO}{\%Fe^{int}\ \%O^{int}}=K_{eq;FeO}\cdot\frac{\gamma_{Fe}\ \gamma_{O}}{\gamma_{FeO}} \tag{19.99}$$

$$(\mathrm{Si})+2(\mathrm{O})=(\mathrm{SiO_2})$$

$$K_{SiO2}=\frac{\%SiO_2}{\%Si^{int}\ (\%O^{int})^2}=K_{eq;SiO2}\cdot\frac{\gamma_{Si}\ (\gamma_{O})^2}{\gamma_{SiO2}} \tag{19.100}$$

$$2(\mathrm{P})+5(\mathrm{O})=(\mathrm{P_2O_5})$$

$$K_{P2O5}=\frac{\%P_2O_5}{(\%P^{int})^2\ (\%O^{int})^5}=K_{eq;P2O5}\cdot\frac{(\gamma_{P})^2\ (\gamma_{O})^5}{\gamma_{P2O5}} \tag{19.101}$$

$$(\mathrm{CaO})+(\mathrm{S})=(\mathrm{CaS})+(\mathrm{O})$$

$$K_{CaS}=\frac{\%CaS\ \%O^{int}}{\%CaO\ \%S^{int}}=K_{eq;CaS}\cdot\frac{\gamma_{CaO}\ \gamma_{S}}{\gamma_{CaS}\ \gamma_{O}} \tag{19.102}$$

In these expressions the slag concentrations can be obtained through sampling. However the interface concentrations (defined by superscript (int)) in the above equations are not known and cannot be determined by any sampling procedure. However these can be estimated from the flux equation like (19.91) for each of the species, Fe, Si, P, S, O together with the equation of electroneutrality (19.97).

$$2K_{Si}\left[\%Si-\frac{\%SiO_2}{K_{SiO2}(\%O^{int})^2}\right]+2.5K_P\left[\%P-\sqrt{\frac{\%P_2O_5}{K_{P2O5}(\%O^{int})^5}}\right]$$
$$+K_{Fe}\left(\%Fe-\frac{\%FeO}{K_{FeO}\%O^{int}}\right)-2K_S\left(\%S-\frac{\%CaS\%O^{int}}{K_{CaS}\%CaO}\right)-2K_O\left(\%O-\%O^{int}\right)=0 \quad (19.103)$$

This equation can estimate the interfacial concentration of oxygen since the effective equilibrium constants can be calculated from true equilibrium constants and the activity coefficients, since the composition of the slag is known through sampling and also values of mass transfer coefficients for each species.The following sequence can be used to assess changes in the composition of slag and metal during the course of refining process:

1) calculate the value of the $\%O^{int}$ from the electroneutrality equation;

2) calculate the flux of each species, Fe, Si, P, S, O, etc;

3) recalculate mass and composition of metal and slag from the fluxes;

4) repeat steps 1, 2, and 3.

A similar procedure has been adopted (but more complete in that it includes steps of mass transfer in the boundary layer on the slag side and rate of generation carbon monoxide) by Ohguchi et al. (1994), to evaluate the simultaneous dephosphorization and desulfurization of iron, with slag containing 30% CaO, 20% $CaCl_2$, 20% CaF_2, 30% Fe_2O_3 at 1350°C, in the system and agitation by inert gas, 0.06 (W/kg).

Figures 19.20 and 19.21 give the results of these experiments, for iron containing low and high silicon respectively. Desulfurization is significant in both the cases . In the case of dephosphorization ,this cannot be affirmed, since high silicon pig iron suffers a reversal of phosphorous after about 1,000 seconds probably due to a competition of oxidation reactions of silicon and phosphorus. Even in the case of low silicon pig iron, there is a slight reversal of phosphorus in the middle period of the experiments. Figure 19.20 shows the simulation results using the model. The model can reproduce the effect of reversal of phosphorus, which occurs from the moment when the oxygen consumption for the steps of Si and S transfer reduces the interfacial oxygen activity. Hence $\%O^{int}$ values go below the critical concentration needed to maintain the phosphorus in the slag, ie, by oxidizing the same.

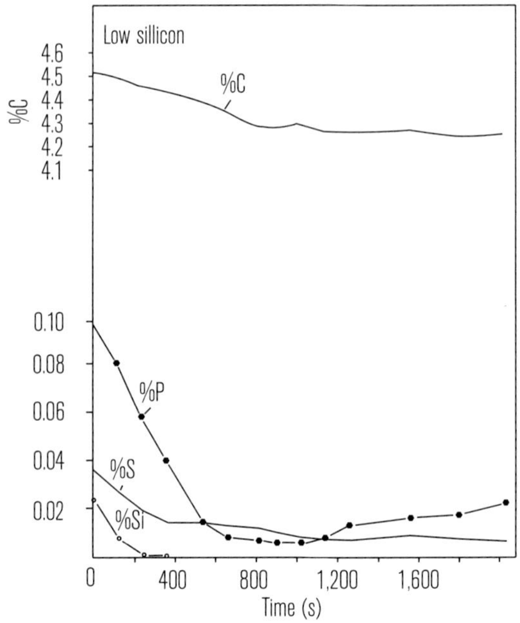

Figure 19.20 - Simultaneous desulphurization and dephosphorization of hot metal (3,020 g) with low silicon and 184 g of slag (OHGUCHI et al., 1984).

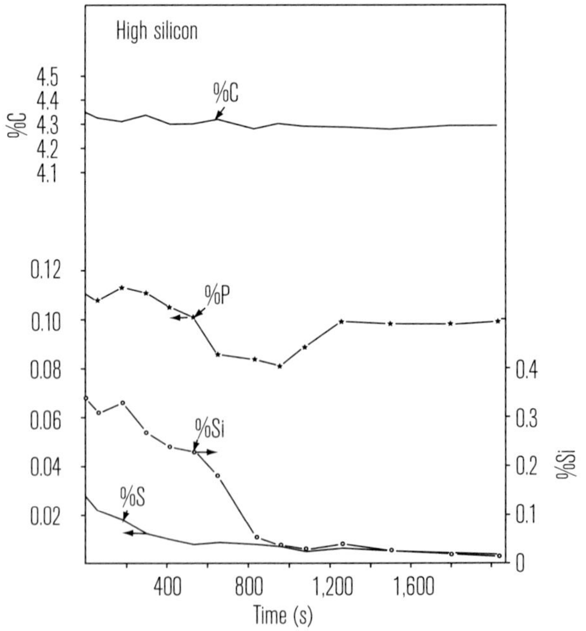

Figure 19.21 - Simultaneous desulphurization and dephosphorizaion of hot metal (3,000 g) with high silicon and 180 g of slag (OHGUCHI et al., 1984).

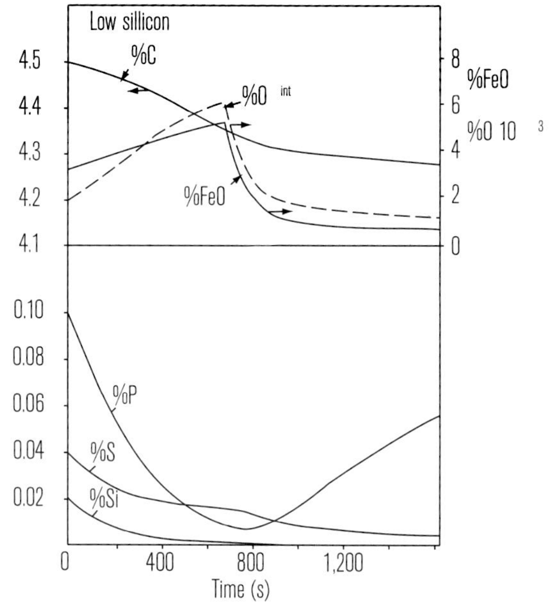

Figure 19.22 - Simulation results for model multiple reactions for low silicon iron, (OHGUCHI et al., 1984).

19.3.6 Model validity test - Example of calcination of lime stone

Turkdogan (1996) conducted experiments on kinetics of calcination of limestone, quantifying the process by thermo-gravimetry and monitoring temperatures inside the limestone particle during calcination by introducing thermocouples, as shown in Figure 19.23.

The outer layer of the particles is CaO, the decomposition product of $CaCO_3$,while the core remains unreacted, with $CaCO_3$ with a uniform temperature distribution equal to T_{DC}.T_{DC} represents the temperature at which the decomposition of carbonate becomes kinetically important, around 900 ~ 1000°C. T_F represents temperature of the furnace where the particle undergoes decomposition. The reaction of carbonate decomposition is highly endothermic and is given by:

$$CaCO_3\ (s) = CaO\ (s) + CO_2\ (g)\ \Delta H_{25C} = 179746\ (J/mol) \qquad (19.104)$$

It may be assumed that heat transport through the product layer could be the controlling step, (ie the higher the rate of heat transfer higher is the intensity of decomposition) quasi- steady state conditions, that is to say "all the energy transferred to the product layer is consumed in the calcination reaction(topochemical) that occurs at the interface $CaO/CaCO_3$. Hence the heat consumption rate should be equal to the rate of heat transport given by:

$$4\ \pi\ R^2\ \rho_{CaO} \frac{dR}{dt}\ .\Delta H = -\frac{4\ \pi\ K_{CaO}\ (T_F - T_{DC})}{\frac{1}{R} - \frac{1}{R_0}} \qquad (19.105)$$

The methodology of development of equations is the same given in the section discussing topochemical reaction (19.3.3). In the above equation ρ_{CaO} represents the molar density of CaO, number of moles per unit volume (mole/m^3), K_{CaO} represents the thermal conductivity of the product of the reaction. The above equation can be simplified,

$$R^2\ \rho_{CaO} \frac{dR}{dt}\ .\Delta H = -\frac{K_{CaO}\ (T_F - T_{DC})}{\frac{1}{R} - \frac{1}{R_0}} \qquad (19.106)$$

and integrated between the limits:

- t = 0, R = R0 initial boundary condition;
- t = t, R = R boundary condition at time t.

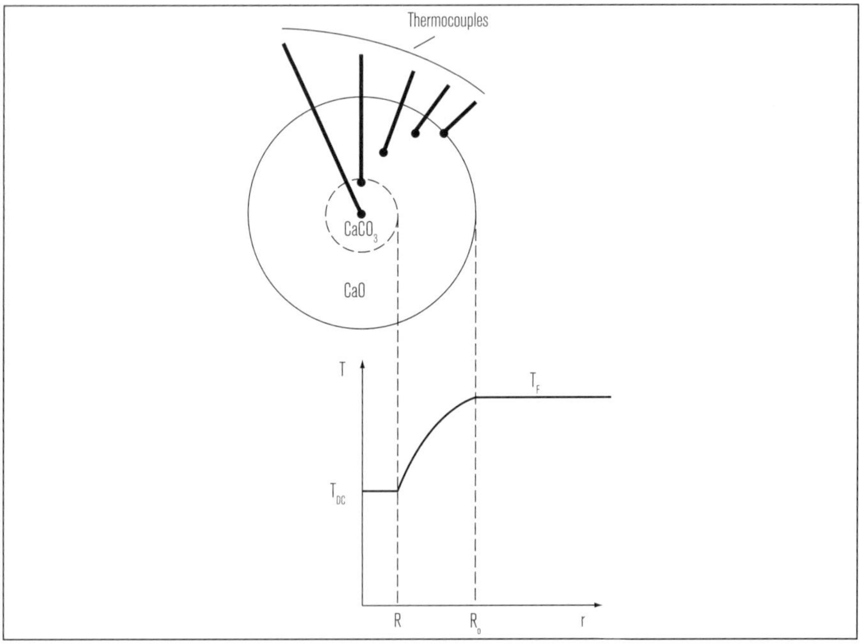

Figure 19.23 - Distribution of temperature in $CaCO_3$ decomposition experiments (TURKDOGAN, 1996).

Integration with the above boundary conditions results in the following equation:

$$\frac{\rho_{CaO}\,.\Delta H\ R_0^2}{K_{CaO}\ (T_F - T_{DC})}\ \frac{1}{3}\left(\frac{R^3}{R_0^3}-1\right)-\frac{1}{2}\left(\frac{R^2}{R_0^2}-1\right)=t \qquad (19.107)$$

As $\alpha = 1-\left(\frac{R}{R_0}\right)^3$ where α represents the fraction reacted, the expression above can be used to define a function g (α) as discussed earlier,

$$g(\alpha)\ = -\frac{\alpha}{3}-\frac{1}{2}[(1-\alpha)^{2/3}-1] = \frac{K_{CaO}\ (T_F - T_{DC})}{\rho_{CaO}\ .\Delta H\ R_0^2}t \qquad (19.108)$$

that is, if the model is valid the plot g (α) *Vs* t should result in a straight line with intercept zero.This would be a test of validity of the model. Experimental verification can be done through monitoring weight of the particle as a function of time (fractional reaction) .The topo-chemical nature of the reaction can be ascertained by observing the cross section of the particle at different times for confirming the shrinking of the core with time.

However in the absence of information from thermocouples (which give evidence of endothermic reaction and the process is controlled by heat transport as discussed earlier) one can suggest the calcination process is controlled by resistance to mass transport of CO_2 through the porous product.layer. In this case the general equation for mixed control (equation 19.76), after eliminating the resistances due to mass transfer through the boundary layer and the chemical reaction reduces to:

$$R^2\rho_{MO}\frac{dR}{dt} = -\frac{\varepsilon\ D_A}{\tau}\frac{\left(C_A^{eq}-C_A^0\right)}{\left(\frac{1}{R}-\frac{1}{R_0}\right)} \qquad (19.109)$$

Considering the definition of fractional reaction as defined by equation $\alpha = 1-\left(\frac{R}{R_0}\right)^3$ and integrating:

$$g\ (\alpha) = -\frac{\alpha}{3}-\frac{1}{2}\left[(1-\alpha)^{2/3}-1\right] = \frac{\varepsilon\ D_A(C_A^{eq}-C_A^0)}{\tau.R_0^2}t \qquad (19.110)$$

This also gives a linear relationship between g (α) and t. Hence the linearity alone cannot be taken as conclusive proof of the validity of the model.The non-linearity would suggest experimental errors also which can be verified by repeating the same. Hence one has to look into the assumptions, that are not valid, experimental errors etc before coming to a conclusion. In other words the same set of experimental data can fit into several models. And hence one has to get maximum experimental evidence before coming to a reasonable conclusion regarding the mechanism.

Example

A simplified model for the removal of carbon and dissolved oxygen during decarburation in steel melt can be developed. The following steps, in series can be considered for the development of the model. The figure below give the mechanism and concentration gradients involved in the process.

1) Transport of carbon and oxygen, from the steel melt to the reaction interface. This transport occurs primarily by convection until the vicinity of the interface metal/gas and then by diffusion to the interface through boundary layer. The steps of transport of carbon and oxygen are in parallel;
2) Reaction between carbon atoms and oxygen atoms forming CO gas molecules and adsorption at the at the interface;
3) Desorption and diffusion of CO gas through the boundary layer to any gaseous environment (argon bubbles etc) and removal by convection.

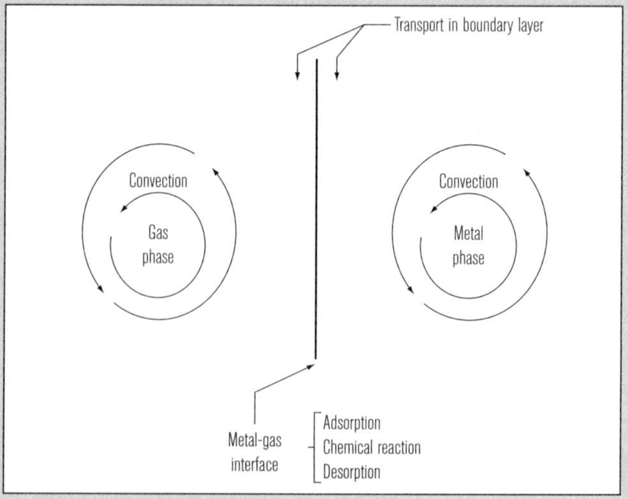

Schematic diagram of a model of decarburization.

Refining processes of iron and steel are given in high temperatures. Thus it is customary to assume that the step corresponding to the chemical reaction is not the limiting step. The transport in the fluid, (gas), would be comparatively fast, so that this step cannot be considered as rate limiting step. Therefore, carbon and oxygen are transported by convection the liquid phase (fast step) and by diffusion through boundary layer to the reaction interface (slow step). The last step should be the rate limiting step determining the time and extent of degassing, as shown in the figure, giving concentration profiles in the boundary layer at the metal side for C and O.

Now consider a typical curve of decarburization, d%C/dt vs t for the oxygen steel making process given in the other figure with geometrical parameters such lance-bath distancer, nozzle hole angle, diameter of the lance holes, cross-sectional area of the reactor, and the rate of blowing, Q (Nm^3/min) being kept constant during the process. The shape of this curve obtained as given in the last figure can then be explained with the above discussion.

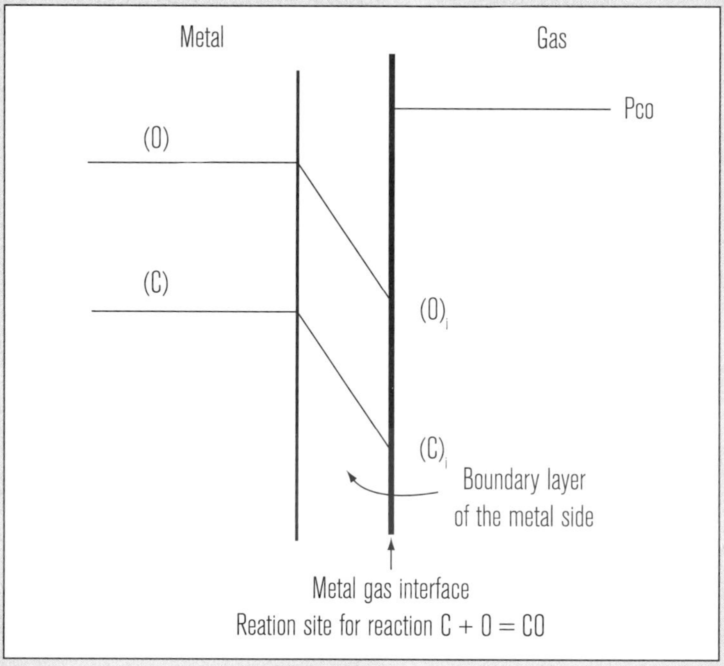

Schematic view of the process of decarburization.

As has been already mentioned, chemical reaction rates are high for processes conducted at high temperatures, and hence this the step (chemical reaction C + O => CO) is not the rate limiting step of the process. Hence the possibilities to control the speed of the process are the steps for the transport of carbon and oxygen through boundary layer. In the early stages of blowing, there would be competition between the oxidation of carbon and those of the metalloids, Si and Mn. As the oxidation of the latter is preferred, the rate of decarburization reaches significant values after the oxidation of Si,Mn etc. The region II corresponds to the period in which there is excess carbon in the bath. Hence the transport of carbon to the reaction site cannot the controlling step.and hence the process is controlled by oxygen transport to the reaction interface. Since the blowing rate of oxygen is constant the rate of decarburization remains constant during this period at levels shown in region II. At a certain point during blowing, characterized by the value of critical carbon value, $C_{critical}$, carbon in the bath can be regarded as comparatively low. Therefore, the transport of carbon is not as effective as like the earlier period with consequent decrease of rate of decarburization. During this period, (region III), the processing speed is controlled by the transport of carbon. The oxygen blown in excess contributes to the over-oxidation of the bath, since not all of it can be consumed.

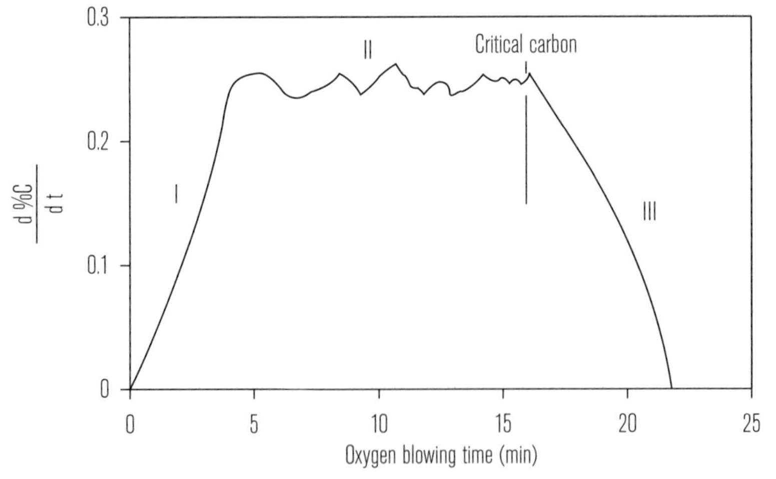

Typical decarburization rates in oxygen steel making process (LD).

Example

A series of manufacturing steps of steel involves chemical exchanges between liquid steel and slag cover. The effectiveness of these depends not only on thermodynamic parameters and kinetic aspects which includes area of contact between the phases; renewal of surfaces undergoing interface reactions. The last aspect is caused by agitation which guarantees the effective contact as well as renewal through mechanical, electromagnetic, pneumatic means etc. In general, this is associated with a stirring energy or energy of agitation. This intensity of agitation due to a gas or stirring energy ε is given by (SZEKELLY, CARLSSON and HELLE, 1989):

$$\varepsilon = \frac{8.31}{22.4\ M} T_S \left[\left(1 - \frac{T_I}{T_S} \right) + \ln \left(1 + \frac{9.81 \rho H}{P_A} \right) \right] Q$$

where:

- ε is the specific energy of agitation (W/t);
- M is the mass of steel (t);
- Q is the gas flow rate (STP L/s);
- T_S is the temperature of steel (K);
- T_I is the temperature of gas being injected (K);
- ρ is the density of steel ($kg.m^{-3}$);
- H is the depth of injection (m);
- P_A is the ambient pressure (Pa).

To quantify the rate of a certain process undergoing agitation one needs the value of coefficient of mass transfer, which is a function of combination of properties of the phases involved, geometry and gas flow. The figure below shows the variation of the apparent mass transfer coefficient for sulfur, K_S as a function of gas flow (or specific energy of agitation). This coefficient is called "apparent" because it is calculated based on cross-sectional area of the ladle and not the effective area of the slag metal interface. The curve is divided into two main regions. Initially one observes a moderate increase in K_S when the flow increases; from a certain value of gas flow K_S increases considerably with flow rate as ε values also should be more pronounced in this region. This change of behavior is attributed to emulsification of slag coverage, with the consequent increase in interfacial area metal-slag resulting in an increase in the apparent value of coefficient of mass transfer.

Riboud and Vasse (1985) developed, by comparing data from industrial- desulphurising facility an equation for calculating the mass transfer coefficient of sulfur between slag and steel in a ladle:

$$K = 500.\left(\frac{D_S \cdot Q}{Area}\right)^{1/2}$$

where:

$D_S = 2.8 \times 10^{-8} \times EXP(-7500/(1.987 \times T)$ is t), ($m^2.s^{-1}$); is the diffusivity of sulfur in steel melt and Q is the gas flow rate at the temperature and pressure of the interface ($m^3.s^{-1}$).

This expression is part of the proposed mathematical model. Due to the high degree of agitation coming from the gas injection there are no concentration gradients within the liquid metal. However there is concentration gradient only in the boundary layer adjacent to the slag (see second figure below). On the other hand in the case of desulphrization involving high values of distribution coefficients of sulfur, the composition gradients at boundary layer on the slag side need not be considered. In such a case a sulfur balance (Accumulation rate = Speed of transport) considering that all sulfur removed from the metal through the boundary layer adjoining the metal, is transferred to the slag, according to the following equation (RIBOUD and VASSE, 1985):

$$-\frac{1000.M_{steel}}{100.M_S}\frac{d(S)}{dt} = \frac{\rho.K_S.Area}{100.M_S}\left[(S) - (S)_i\right]$$

where:

- M_{steel} steel weight (tons);
- M_S atomic weight of sulfur, ($g.mol^{-1}$);
- (S) wt% sulfur in steel;
- $(S)_i$ wt% sulfur at the interface metal/slag;
- t, time (s);
- ρ, density of steel ($kg.m^{-3}$)
- Area, area of the interface metal/slag, m^2;
- K_S, coefficient of mass transfer, ($m.s^{-1}$).

The expression above can easily be combined with that resulting from the assumption of local equilibrium at the interface, ie the distribution of sulfur conforms to the distribution coefficient $L_S = (S)/(S)_i$, and sulfur balance between slag and metal bath, given by:

$$M_{steel}(S)_0 = M_{steel}(S) + M_{slag}(S)$$

An analytical solution can be obtained:

$$\frac{[S]_o - [S]}{[S]_o} = \frac{1}{1+\Lambda} - \frac{EXP[-\beta.(1+\Lambda).t]}{1+\Lambda}$$

Where $(S)_0$ is initial súlfur content in liquid metal and:

$$\Lambda = \frac{M_{steel}}{M_{slag} \cdot L_s} \text{ and } \beta = \frac{\rho . K_S . Area}{1000 . M_{steel}}$$

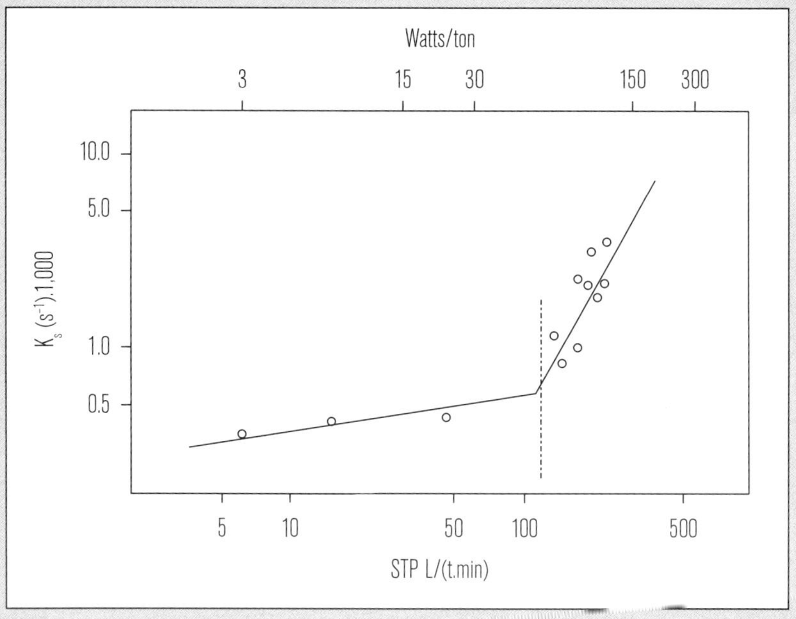

Fxamplo of dopendence of the apparent transfer coefficient of sulfur in the boundary layer metal-slag and gas flow (hence the rate of specific energy contribution).

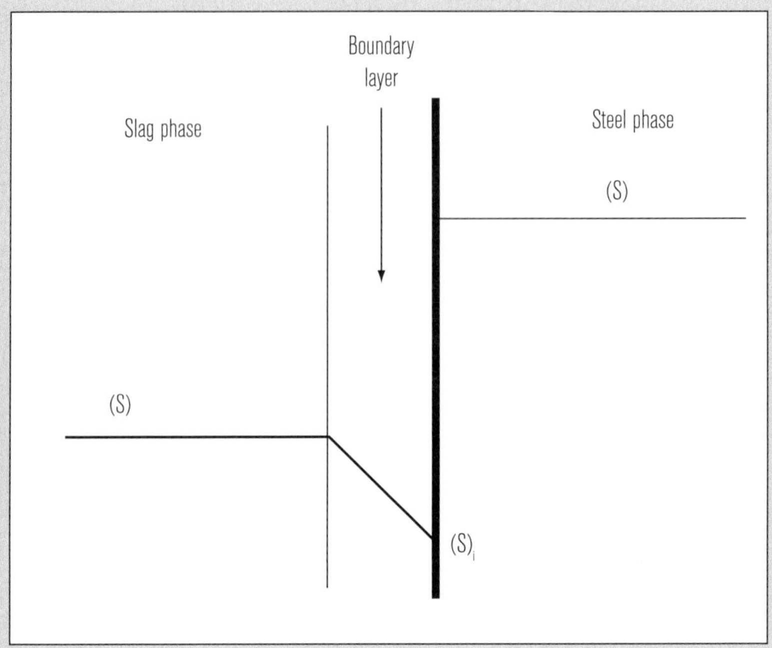

Figure: Profile of concentration of S in the vicinity of the interface metal/top slag (RIBOUD and VASSE, 1985).

Example

For calculating partition coefficient of Sulphur the following data are given: 300 tons of steel, 6 tons of slag composition 10% SiO_2, 35% Al_2O_3, 50 % CaO and 5% MgO, 0.008 STP m^3 gas/s, 1873 K; ladle with a mean diameter of 3 m, 7000 kg steel/m^3. For deoxidation aluminum is used which is the standard practice. Assuming h_{Al} (wt%) equal to 0.03 at equilibrium in the following equation for deoxidation:

$$Al_2O_{3(S)} = 2\,Al_{1\%} + 3\,O_{1\%}$$

$$\log \frac{h_{Al}^2 \cdot h_O^3}{a_{Al_2O_3}} = -\frac{64000}{T} + 20.57$$

From activity data of alumina, given in the ternary diagram below, activity of oxygen h_0 can be calculated from the equation above. Then, one can estimate the value of the partition coefficient of sulfur using the following equation:

$$\log L_s = \frac{(21920 - 54640\ \lambda)}{T} - 43.6\ \lambda - 23.9 - \log h_O + \log f_s$$

where:

- λ, Optical basicity of the slag;
- h_0, activity of oxygen in steel, reference weight%;
- f_s, activity coefficient of sulfur in steel, reference weight%.

Optical basicity can be estimated from:

$\lambda = \sum X_i' \lambda_i$ and $X_i' = \frac{X_i n_O^i}{\Sigma X_i n_O^i}$ where n_O^i is the number of oxygen atoms in the oxide i, X_i and λ_i are the mole fraction and optical basicity of component i of the slag, i = SiO_2, Al_2O_3, CaO, MgO, MnO, FeO, presented in the following table.

Values of optical basicity of some oxides

Species i	SiO_2	Al_2O_3	CaO	MgO
λ_i	0.48	0.61	1.00	0.78

This type of calculation produces curves of iso-values of partition coefficient of sulfur, with which knowing the weights of steel melt and the slag, the thermodynamic retention of sulfur (or sulfur capacity of slag) can be evaluated. The kinetic performance can be gauged by substituting the data in the model proposed by Riboud and Vasse (1985).

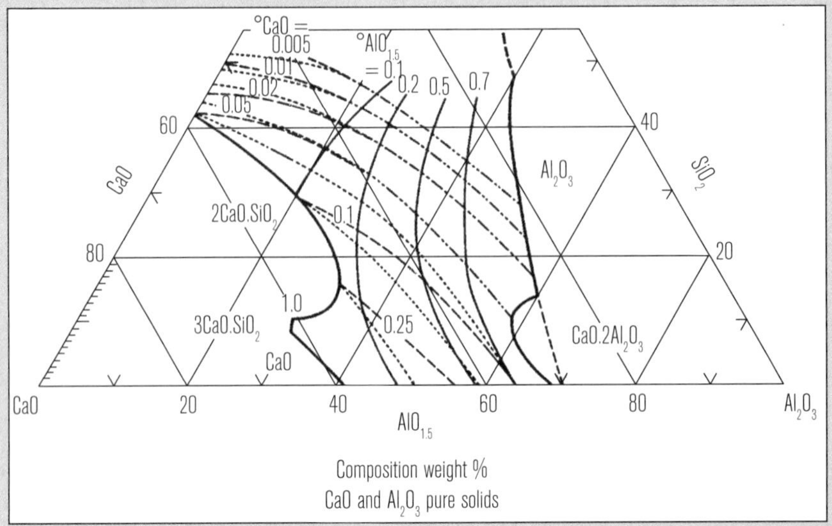

Isoactivity curves of the system $CaO\text{-}SiO_2$, - Al_2O_3 at 1600°C (RIBOUD and VASSE, 1985).

Example

Desulfurization in a Kanbara Reactor (KIRMSE, 2006; SILVA et al., 2009).

A schematic of the kanbara reactor, used for desulfurization of hot metal is given below. A powdered desulfurizing reagent like lime is dispersed throughout the hot metal using a rotating impeller. The desulfurization process depends on the combination of thermodynamic and kinetic parameters. Thermodynamic considerations include the type, composition and relative amount of desulfurizing agent, initial temperature of treatment, mass and composition of hot metal. Kinetic conditions depend mainly on the distribution of the particle size of the desulfurizing agent, intensity of agitation of the bath. Estimate the initial rate of desulfurization.

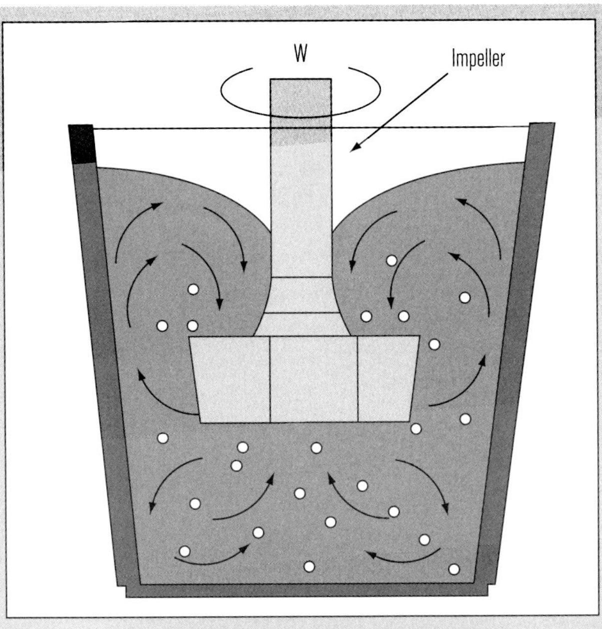

Schematic figure of Kanbara reactor

The following table gives some typical operating parameters of the reactor.

Average data value for desulfurization in Kanbara reactor.

Mass of hot metal	300 tons
Initial sulfur	300 ppm
Impeller rotation speed	100 rpm
Initial temperature	1370°C
Final temperature	1340°C
Total added flux	2470 kg
Specific flux consumption	8.1 kg/ton
Treatment time	12 minutes

As in the case of previous examples cited above on desulfurization, the rate of desulfurization is given by:

$$R = -\frac{1000\,M_L}{100}\frac{d(\%S)}{dt}$$

where:

- R is the rate of desulfurization (kg/s);
- M_L is weight of hot metal (tons);
- [%S] isr the sulfur concentration;
- t for the elapsed time (s).

R is also given bythe following equation considering that Sulfur removal takes place by mass transfer through boundarylayer:

$$R = \frac{k_S \rho_L A}{100}\left[[\%S] - [\%S]_i^P\right] = \frac{k_S \rho_L A}{100}\left[[\%S] - \frac{(\%S)_p}{L_S}\right] \approx \frac{k_S \rho_L A}{100}[\%S]$$

where:

- k_S is the mass transfer coefficient at metal/particle interface (m/s);
- ρ_L is the hot metal density (kg/m^3);
- A is the superficial area of all particles combined (m^2);
- [%S] is the sulfur concentration in hot metal;
- $[\%S]_i^P$ is the sulfur concentration in the hot metal at reagent particle interface;
- $(\%S)_p$ is the sulfur concentration inside the reagent particle;
- L_S is the sulfur partition coefficient.

In the last equation the term $\frac{(\%S)_P}{L_S}$ is quite small in comparison with [%S], since during the initial rate of desulfurization, $(\%S)_P$ is negligible and L_S is bound to be large. Details of aspects of desulfurization have been given in the previous examples. One now can proceed to evaluate the values of each parameter involved.

Considering particles of equal size the total area for mass transfer is given by:

$$A = \frac{3M_{reag}}{\rho_{CaO} R} = \frac{3 \times 2470}{3000 \times 10^{-3}} = 2470 \text{m}^2$$

where, average radius R is equal to 1 mm; and addition of (M_{reag}) amounts to 2470 kg. The reagent has 90% lime/fluorspar (85% CaF_2, SiO_2), with density ρ_{CaO} close to 3000 kg/m^3 ρ_L is the hot metal density can be evaluated from the following figure as 6300 k/m^3.

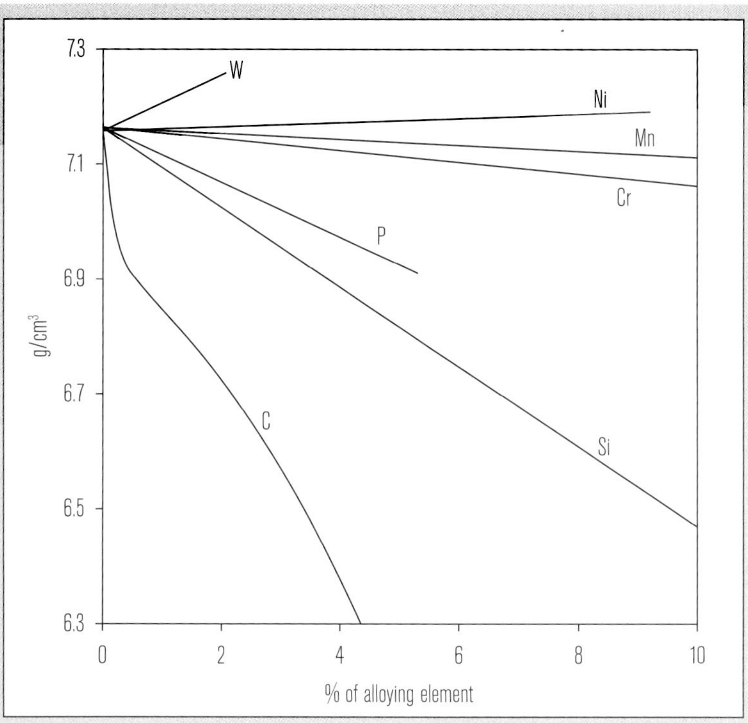

Density of some liquid iron alloys (ELLIOTT, GLEISER and RAMAKRISHNA, 1963).

The mass transfer coefficient is given by (IRONS and CELIK, 1992):

$$k_S = \frac{D_S}{d_P}\left[2 + 0.4\left[\frac{\varepsilon d_P^4 \rho_L^3}{\mu_L^3}\right]^{1/4} \left(\frac{\mu_L}{D_S \rho_L}\right)^{1/3}\right]$$

where:

- K_S is the mass transfer coefficient (m/s);
- D_S is the sulfur diffusivity in hot metal (m^2/s);
- d_p is the particle diameter (m);
- ε is the rate of energy dissipation (W/kg);
- ρ_L is the hot metal density (kg/m^3);
- μ_L is the hot metal viscosity (kg/m.s).

As already seen $\rho_L = 6{,}300$ (kg/m^3) and $d_p = 2 \times 10^{-3}$ (m).

D_S is $2.8 \times 10^{-8}\ e^{-3775/T}$ m^2/s (ELLIOTT, GEIGER and RAMAKRISHNA, , 1963) and hence $D_S = 2.8 \times 10^{-9}$ m2/s at 1650 °K (initial treatment temperature). From the following figure, $\mu_L = 6 \times 10^{-3}$ (kg/m.s).

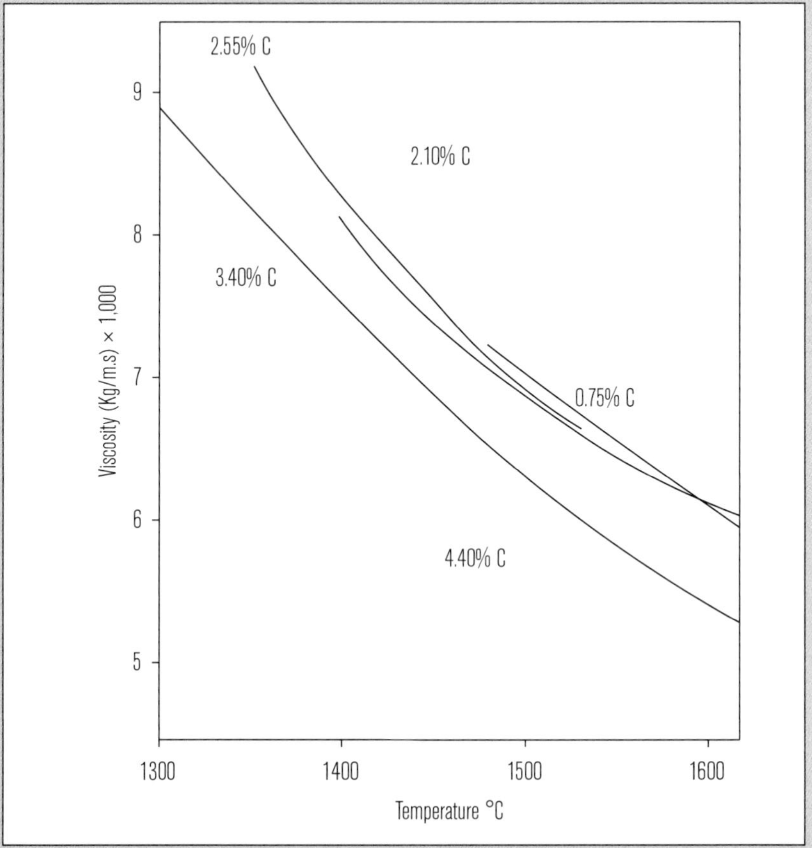

Viscosity of some liquid iron alloys (kg/m.s) (ELLIOTT, GLEISER and RAMAKRISHNA, 1963).

ε, the rate of energy dissipation (W/kg), is a measurement of the amount of energy the impeller provides for agitation. It can be assessed using the relationship between power number and impeller Reynolds number. The impeller Reynolds number is:

$$Re = \frac{\omega D_{imp}^2 \rho_L}{\mu_L} = \frac{(100/60) \times 1.5^2 \times 6300}{6 \times 10^{-3}} \approx 4 \times 10^6$$

where:

- ω is the speed of rotation of the impeller, given in Hertz;
- D_{imp} is the impeller diameter.

For the calculation Reynolds number the value the impeller diameter, taken from an industrial reactor, is 1.5 m, $\rho_L = 6300$ (kg/m³) and $\mu_L = 6 \times 10^{-3}$ (kg/m.s).

Reynolds number calculated above is well in the range of completely turbulent flow (KIRMSE, 2006). From the following figure, the value of power number, for this Reynolds number under turbulent conditions, is around one or slightly more than 1.

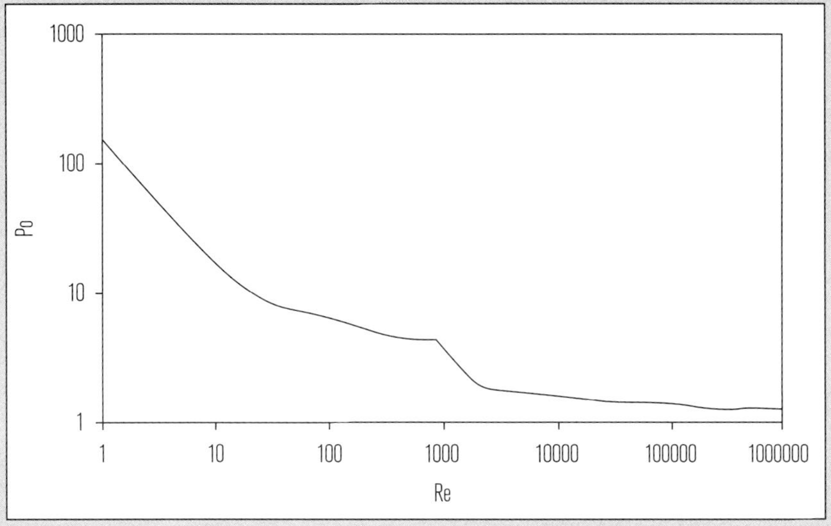

Power Number, P_0, versus Impeller Reynolds Number (KIRMSE, 2006).

Finally, combining the definition of Power Number:

$$Po = \frac{P}{\rho_L \omega^3 D_{imp}^5}$$

and the definition of ε.

$$\varepsilon = \frac{P}{1000\ M_L} = \frac{P_O\ \rho_L\ \omega^3\ D_{imp}^5}{1000 M_L} (W / kg)$$

where:

- P is power provided by the impeller (Watts);
- M_L is hot metal weight (ton);
- P_0 is impeller power number;
- ρ_L is hot metal density (kg/m^3);
- ω is impeller speed of rotation (Hz);
- D_{imp} is impeller diameter (m).

Hence:

$$\varepsilon = \frac{P_O\,\rho_L \omega^3\,D_{imp}^5}{1000\,M_L} = \frac{1 \times 6300 \times (100/60)^3 1.5^5}{1000 \text{ x } 300} = 0.738 (W/kg)$$

Substituting the value in the equation for mass transfer coefficient:

$$k_S = \frac{D_S}{d_P}\left[2 + 0.4\left[\frac{\varepsilon d_P^4 \rho_L^3}{\mu_L^3}\right]^{1/4} \left(\frac{\mu_L}{D_S \rho_L}\right)^{1/3}\right]$$

$$k_S = \frac{2.8 \text{ x } 10^{-9}}{2 \text{ x } 10^{-3}}[2.0 + 0.4\left[\frac{0.738 \times (2 \times 10^{-3})^4 \times 6300^3}{(6 \times 10^{-3})^3}\right]^{1/4}\left(\frac{6 \times 10^{-3}}{2.8 \times 10^{-9} \times 6300}\right)^{1/3} = 2.40 \times 10^{-4} m/s$$

Finally the rate of initial desulfurization, for an initial sulfur concentration of 0.03% can be calculated:

$$-\frac{1000\,M_L}{100}\frac{d[\%S]}{dt} \approx \frac{k_S \rho_L A}{100}[\%S]$$

$$-\frac{d[\%S]}{dt} \approx \frac{k_S \rho_L A}{1000\,M_L}[\%S] = \frac{2.40 \times 10^{-4} \times 6{,}300 \times 2{,}470}{1000 \times 300} \times 0.03 = 3.73 \times 10^{-4} \%S/s$$

This model was applied for analysis of the desulfurization behavior for the industrial Kambara reactor. Data from more than 600 heats have been compared with the theoretical predictions, of sulphur removal as per the procedure given above. The agreement seems to be good, as the ratio of final sulfur/sulfur calculated theroretically is close to one.

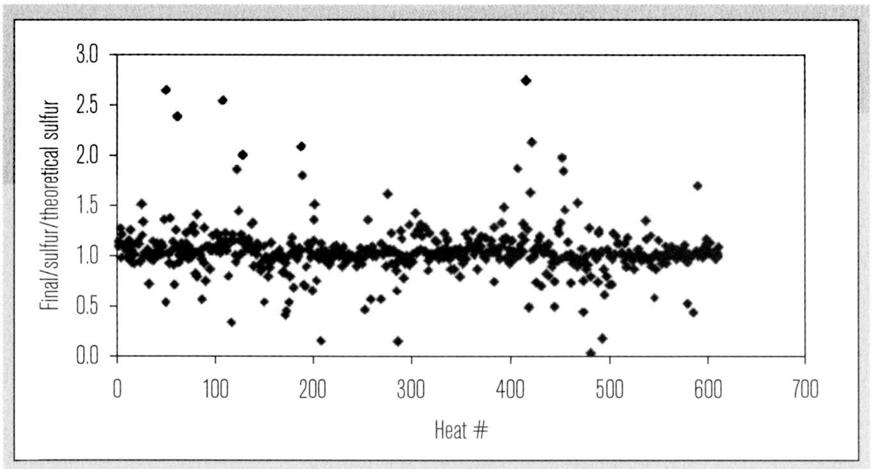

Final sulfur obtained in a kambara reactor compared with that predicted by model.

Using the above model the initial rates of desulfurization was calculated for different impeller velocities and particle diameters of the desulfurizing agent. These parameters seem to have a significant effect on the initial rate of desulfurization, as given by the following figures.

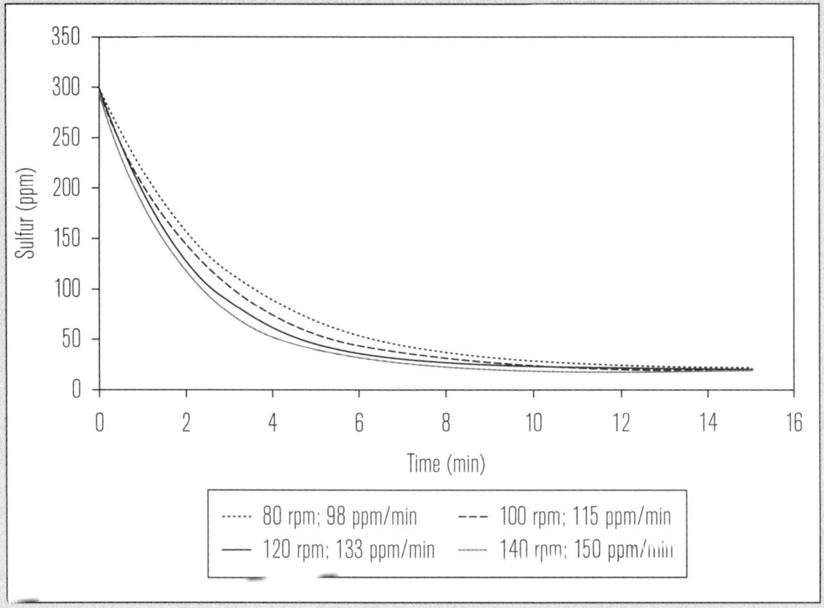

Influence of impeller velocity on the initial desulfurization rate

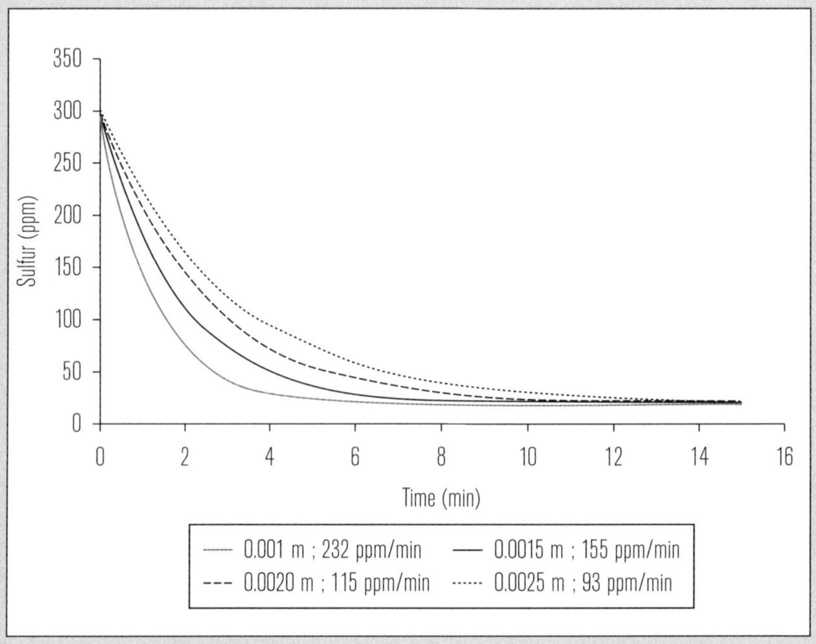

Influence of particle diameter on initial rate of desulfurization in a Kambara reactor.

Example

The rate of decarburization in a metal in contact with a bubble of inert gas takes place through transfer of carbon and oxygen through the boundary layer of metal interface with gas (slow step) followed by rapid interfacial chemical reaction and the rapid transport of the product namely CO by convection into the gas phase as discussed before.

There is literature available for calculating the diameter of the bubbles and mass transfer coefficient. For turbulent conditions in the bath, caused by inert gas bubbling,the following equation gives the average diameter of the bubbles (SANO and MORI, 1980a,b):

$$d_b = 6.9\left(\frac{\sigma}{\rho}\right)^{0.5} \cdot \left(\frac{Q}{\text{Area}}\right)^{0.44} \text{ m}$$

where:

- d_b is the bubble diameter, (m);
- ρ is the density of steel, ($kg.m^{-3}$);
- σ is the interfacial tension gas-bath (N/m);
- Q is the gas flow rate, ($m^3.s^{-1}$);
- Area is the cross-sectional area of flow (m^2).

While the mass transfer coefficient is given by (SZEKELY and EL-KADDAH, 1981).

$$\frac{k \cdot d_b}{D_i} = 2 + 0.4 \left(\frac{\varepsilon \cdot \rho^3 \cdot d_b{}^4}{\eta^3} \right)^{0.25} \cdot \left(\frac{\eta}{\rho \cdot D_i} \right)^{1/3}$$

where:

- D_i stands for diffusivity of element in liquid steel (m^2/s);
- ε for rate of energy dissipation (W/kg);
- η for melt viscosity (Pa.s).

Assessing the rate of dissipation of energy from (SZEKELY, CARLSSON and HELLE, 1989):

$$\varepsilon = \frac{8.31}{22.4\ M} T_S \left[\left(1 - \frac{T_l}{T_S} \right) + \ln \left(1 + \frac{9.81 \rho H}{P_A} \right) \right] Q$$

and from the given data

bath at 1873 K with 100 ppm of oxygen and carbon;

ambient pressure of 0.1 atm;

200 tons of steel in a with 3.5 m depth of metal;

viscosity of steel, 7 mPa.s;

density of steel, 7000 $kg.m^{-3}$;

diffusivity of oxygen in liquid steel $2.5 \times 10^{-9}\ m^2.s^{-1}$;

diffusivity of carbon in liquid steel, $7.9 \times 10^{-9}\ m^2.s^{-1}$;

interfacial tension gas-steel 1.9 N/m;

gas rate of 0.012 STP $m^3.s^{-1}$.

Substituting the values in the above equations,after calculating the bubble diameter, $\varepsilon = 0.134$ W/kg of steel.

And mass transfer coefficients:

$$K_0 = 1.41 \times 10^{-4}\ m.s^{-1} \text{ and } K_C = 3.05 \times 10^{-4}\ m.s^{-1}$$

Therefore, considering the situation in which the bubble has not received any carbon monoxide, in which case the rate of transport would be maximum, and assuming equilibrium at the interface gas/metal, the flow of carbon and oxygen can be estimated using the following equation:

$$n_i = \frac{\pi d_b^2}{4} K_i \frac{1000\, \rho\ \%i}{100\ M_i} \quad moles.s^{-1}$$

where:

- i is the element to be transported, carbon or oxygen;
- % i is the weight % of i;
- M_i is the Atomic weight of element i ($g.mol^{-1}$);
- K_i is the coefficient of mass transfer, ($m.s^{-1}$);
- d_b is the Diameter of bubble (m);
- ρ is the density of steel ($kg.m^{-3}$).

The rate of transport of carbon is calculated as $3.5 \times 10^{-6}\ mol.s^{-1}$.

The rate of oxygen transport is in the order $1 \times 10^{-5}\ mol.s^{-1}$, indicating, therefore, that`controlling step is the transport of oxygen.

19.3.7 The role of interfaces

The processes discussed so far involve heterogeneous reactions, in which the role of interfaces has been important. The following example illustrates catalytic reaction between gaseous components, occurring on a metal surface. A simple model based on adsorption-desorption.is presented.

There is experimental evidence for pick-up of nitrogen in steel melts, when the baths are desulfurized and deoxidized. There can be nitrogen pick up from atmosphere or transfer from the bath to atmosphere depending on the composition of the atmosphere and other conditions. The rate of nitrogen transfer between liquid steel and gaseous atmosphere can be expressed by the following semi-empirical relationship (SHINME, MATSUO and MORISHIGUE, 1988)

$$-\frac{d(N)}{dt} = K_N.(N)^2 \tag{19.111}$$

$$\frac{1}{(N)} - \frac{1}{(N)_0} = K_N.t \tag{19.112}$$

Where (SHINME, MATSUO and MORISHIGUE, 1988):

$$K_N = 60.\frac{A}{V}.\frac{0.974.f_N^2}{\left[1+28.4.(O)+5.34.(S)\right]^2}\ \%^{-1}.\min^{-1} \tag{19.113}$$

where:

- A is the area of gas-metal interface, (cm^2);
- V is the volume of metal, (cm^3);
- f_N is the activity coefficient of nitrogen, reference weight%;
- (O), weight percentage of Oxygen;
- (S), weight percentage of sulfur.

This result is usually explained by the fact that sulfur and oxygen are surface-active elements in liquid steel. Surface-active elements accumulate preferentially in metal-gas interface and diminish interfacial energy and consequently the free energy of the system. So in multicomponent systems the various species can be distributed at the interface in proportions different from within the bulk of the phases. There can be positive adsorption of some spcies when these are present in excess (relative to the concentration within the phase) in the interface, and negative adsorption when there is a deficiency of the same in the interface. In a system of several species some exhibit positive adsorption and others negative adsorption. In a simple analysis, the driving force for adsorption can be considered to be a trend to minimize the free energy of the system. The species which, that cause lowering of interfacial adsorption exhibit positive adsorption and vice versa. According to Gibbs equation:

$$\Gamma = -\frac{1}{RT}\frac{\partial\gamma}{\partial \ln a} \tag{19.114}$$

where:

- Γ(moles/cm^2) and are excess solute at the interface and activity of the same within the liquid respectively.

Adsorption behavior of sulfur and oxygen in liquid iron is shown in Figure 19.24 (SANO, LU and RIBOUD, 1997). Nitrogen adsorption is affected by the positive adsorption of oxygen and/or sulfur at the interface. It is recognized that these elements occupy sites in the metal-gas interface, in which normally the steps of adsorption and desorption of species occur. Occupied sites, *ie* surface area covered, are not accessible to the transfer process (regardless of the direction of the reaction) and hence a reduction in the rate of adsorption of Nitrogen is expected.

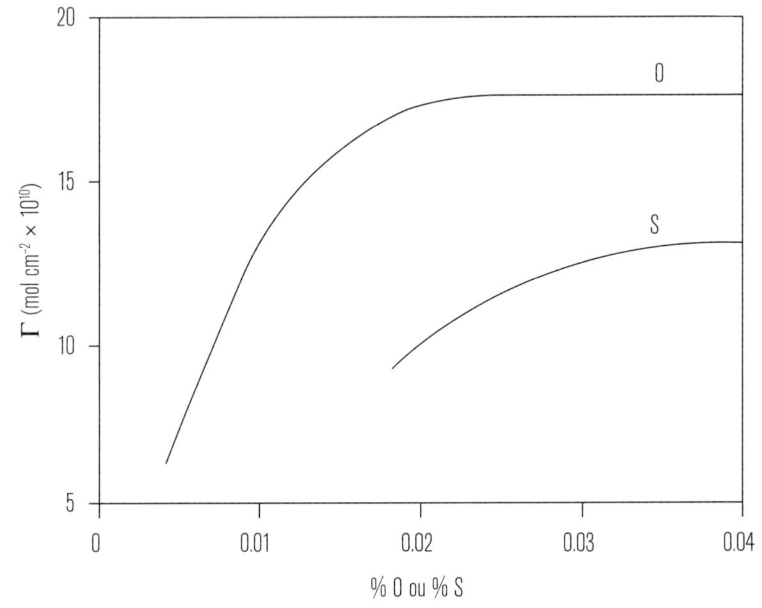

Figure 19.24 - Positive Adsorption of sulfur and oxygen in liquid iron.

Therefore, for Nitrogen adsorption, these active vacant sites would be an important step in the process and the same would be harmed, if the sites are blocked by other elements. Mathematically this effect is expressed by the decrease of KN, when the concentrations of oxygen and sulfur increase. Figure 19.25 shows this effect for increased concentrations of oxygen. As a result, strongly deoxidized steels are especially vulnerable to Nitrogen *pick-up*. A similar effect would not be so strongly felt in the case of hydrogen, because of its small atomic diameter and at the same time giving it a high mobility.

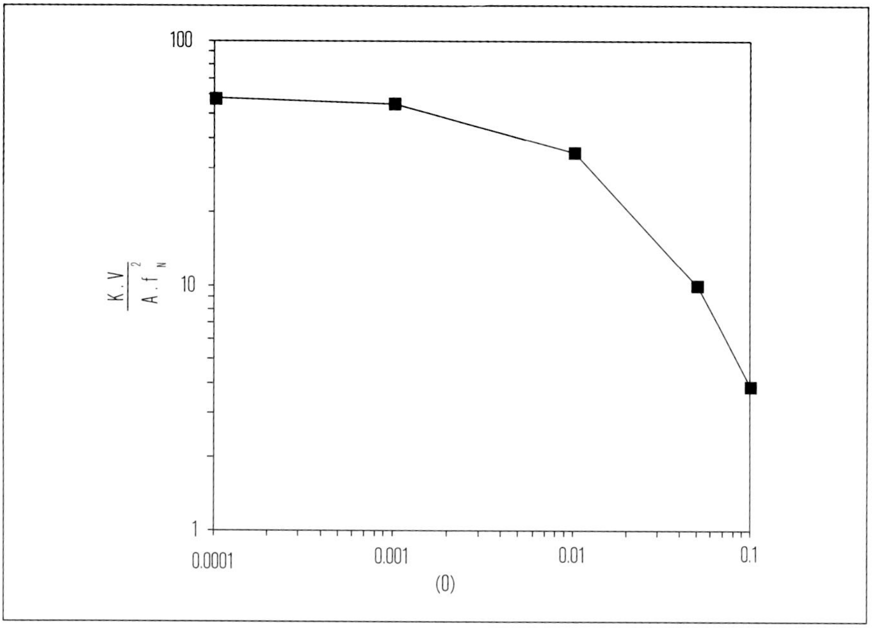

Figure 19.25 - Influence of Oxygen on the kinetics of removal of N_2 (Equation 19.113).

Example

Argon rinsing is common practice during steelmaking but it can give rise to a plume eye exposure of liquid steel to the atmosphere as illustrated in the following Figure. Larger plume eye area results in higher reoxidation rate of molten steel as well as higher Nitrogen pickup and thermal losses. Nitorgen pickup is high especially in deoxidized steel melts during rinsing through the plume eye because of low concentratraion of surface active agnts like Oxygen. One can estimate the Nitrogen pick up rate by the steel melt through the plume eye using the following procedure.

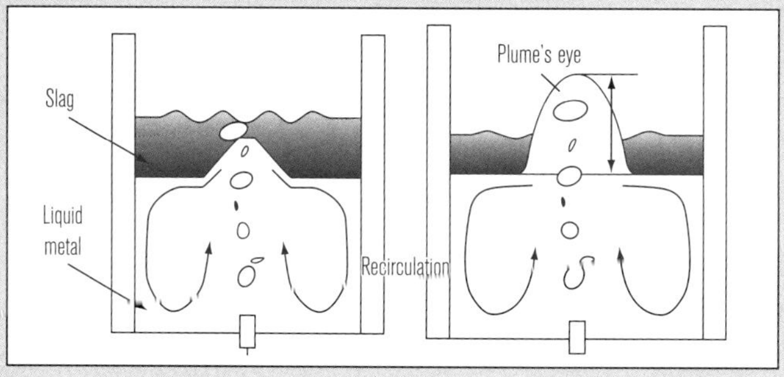

Trajectory of gas bubbles with thick (a) and thin (b) slag layers.

The plume area, according to Iguchi et al. (2004) can be estimated using the following equations (International System of Units):

$$\frac{A_e}{H^2} = 0.60 - 0.652\left(\frac{2g(\rho_m - \rho_s)h}{\rho_m u^2}\right) \text{ for } \left(\frac{2g(\rho - \rho_s)h}{\rho_m u^2}\right) < 0.7$$

$$\log\frac{A_e}{H^2} = -0.222 - 0.741\left(\frac{2g(\rho_m - \rho_s)h}{\rho_m u^2}\right) \text{for} \left(\frac{2g(\rho - \rho_s)h}{\rho_m u^2}\right) < 4$$

where:

- A_e is the area of the eye of the plume;
- H is liquid steel height in the ladle;
- G is the acceleration due to gravity;
- ρ is the density of gas;
- ρ_m is the density of metal;
- ρ_s is the density of the slag;
- h is the thickness of slag layer;
- u, average velocity of steel in the plume.

The liquid velocity(u) in the plume (DEO and BOOM, 1993), can be calculated (Q denotes the gas flow rate, H the level of metal in the ladle and R its radius) assuming gas flow rate of 1.6 m^3/min, radius of the ladle equal to 2.025 m and liquid steel height of 3.65 m:

$$u = 4.78\frac{Q^{1/3}H^{1/4}}{R^{1/4}} = 4.78 \times (1.6/60)^{1/3} \times (3.65)^{1/4} \;/\; (2.025)^{1/4} = 1.66 \text{ m/s}$$

With ρ_m = 7 (ton/m^3), ρ_s = 3.5 (ton/m^3), h = 0.07 (m), H = 2,9 (m) the plume eye area is calculated as follows:

$$\frac{A_e}{H^2} = 0.60 - 0.652\left(\frac{2g(\rho_m - \rho_s)h}{\rho_m u^2}\right) = 0.60 - 0.652\left(\frac{2 \times 9.81\,(7.0 - 3.5) \times 0.7}{7.0 \times 1.66^2}\right) = 0.437$$

A_e = 5.82 m^2 resulting in the eye of the plume radius of R_{op} = 1.36 m.

According to Oeters (1985), the value of diffusion coefficient, at 1600°C, would be D_N= 1.1 × 10^{-8}m^2/ s. Considering the effects of surfactants Sulfur and Oxygen, the mass transfer coefficient can be calculated from the equations (OETERS, 1994):

$$k=\left[2\sqrt{\frac{D_N}{\delta\, t_e}}\right]/(1+178\ \%O+39\ \%S)$$

The exposure time t_e is the ratio between the surface velocity (equal to the the plume velocity) of steel and the distance, traveled (equal to the the radius of the eye) before the flow is deflected into the interior volume. Hence:

$$t_e=\frac{R_{op}}{u}=1.36/1.66=0.82s$$

Therefore mass transfer coefficient can be calculated (for a steel melt with 65 ppm of sulfur):

$$k=\left[2\sqrt{\frac{D}{\pi t_e}}\right]/(1+178\%O+39\%S)=\sqrt{\frac{1.1\times10^{-8}}{3.14\times0.82}}\left(\frac{2}{1+178\times0.0+39\times0.0065}\right)=1\times10^{-4}m/s$$

For the bare steel surface exposed to a Nitrogen containing gaseous environment the flux can be estimated as A_e K C_N which corresponds to an increase in Nitrogen content equal to:

$$\frac{d\,N_{ppm}}{dt}[ppm/s]=A_e\,K\,C_N\frac{10^6\,M_N}{M_{steel}}$$

where:

- M_{steel}(g) is the weight of steel;
- M_N is the atomic weight of Nitrogen(14 g/mole);
- C_N is the Nitrogen gas/metal equilibrium concentration (g atom/m³).

C_N is given by Sievert's law (TURKDOGAN, 1996):

$$\frac{N_{ppm}}{P_{N2}^{1/2}}=k_{eq}\qquad \log k_{eq}=-\tfrac{188}{T}+2.76$$

Hence:

$$C_N[\text{gram atom N/m}^3]=\frac{k_{eq}P_{N2}^{1/2}\rho_m}{M_N}=[456.7\times(0.79)^{1/2}\times7.0]/14=202.96$$

where ρ_m(ton/m^3) represents the density of steel and the partial pressure of Nitrogen is equal to 0.79 atm (79% of air). For 315 tons of liquid steel one gets:

$$\frac{dN_{ppm}}{dt}[ppm/s] = A_e\, K\, C_N \frac{10^6 M_N}{M_{steel}} = \frac{5.82 \times 1.0 \times 10^{-4} \times 203 \times 10^6 \times 14}{315 \times 10^6} = 5.22 \times 10^{-3}$$

19.3.8 Population balance

In many cases, particles and bubbles in a reaction with a fluid medium have a size distribution that may be a function of time and special variables. In the case of a powdery material that dissolves, it is rarely made up of particles of one uniform size. In these applications it may be necessary to account for the contributions from all size classes, on the basis of a population balance. Consider for example the need for adjustments in the composition of the slag in contact with the bath which is almost routinely done. Additions can be made directly to the metal. Whatever be the practice, usually the final slag is the result of a combination between these items of addition, oxidation products of chemical reactions and any residual slag coming from previous operations. For simplifying consider that the additions to the slag are done isothermally and instantaneously.

There may, however be a significant difference between the nominal composition of the slag and the actual composition of the liquid portion, because of the time interval required for heating and dissolution of the additions.The portion of slag that more effectively participates in the steps of mass transfer is the liquid phase.

Then consider injection of spherical particles of CaO and CaF_2 in liquid iron, 30 tons, to reform an initial slag composition 40% CaO, 40% SiO_2, 20% Al_2O_3. In three different cases the slag is present corresponding to 2, 3 and 5 kg/ton of hot metal at 1,400 and 1,500°C. Injection mixture is made up of lime and fluorspar, 95% CaO, 5% CaF_2, with three different particle sizes namely 100, 250 and 500 μm, in three different inert gas flow rates of 15, 20 and 25 m^3/h.

The assumption of isothermal process of dissolution of this mixture can be justified if the powder is deeply injected in the hot metal, which woluld allow for preheating the material before it gets to the surface layer of the slag. Assuming that the dissolution process is controlled by mass transfer through concentration boundary layer around each particle, the following expression can be written for dissolution rate per particle:

$$n_{CaO} = \frac{4\pi r_p^2 k_p \rho_{ts}}{100} \left[(\%CaO)_{sat} - (\%CaO)_{ts} \right] \text{ kg s}^{-1} \qquad (19.115)$$

where:

- r_p is the radius of particle of CaO, (m);
- K_p is the coefficient of mass transfer, (m/s);
- ρ_{ts} is the density of the top slag (kg/m³);
- $\%CaO_{sat}$ is the % of lime in saturated state in the slag;
- $\%CaO_{ts}$ is the % of lime in the top slag.

This flux can be written as a function of variation of the radius of the particle:

$$-\frac{dr_p}{dt} = \frac{k_p \rho_{ts}}{100\rho_p} \left[(\%CaO)_{sat} - (\%CaO)_{ts} \right] \qquad (19.116)$$

This expression gives the rate of change of particle size of CaO ,in terms of driving force ie, the difference between the concentrations of CaO, (instantaneous) and that for saturation. During particle injection, the new particles would join those who are already undergoing dissolution within the bulk of the slag phase. At any given moment, there is a population of particles within the slag having a wide size distribution. For each range of particle size equation 19.116 is applicable. However but it would be necessary to determine the number, in order to determine the total amount of CaO incorporated in a certain time. In systems like this, characterized by the presence of multiple particles of various sizes, a robust analysis involving principles of population balance can be applied. This technique is based on discretization of particle size, defined arbitrarily according to size ranges. From these expressions relating to rate of birth and death in respect of them (particle mobility between the classes) and starting from certain initial size distribution, it is possible to infer the number of particles present in each class, at any instant of time. Computations can be done to estimate the number of particles placed in different classes and to estimate the dissolution rate specific to each class. The overall rate of dissolution, and the total mass dissolved, can be evaluated from these variables.

Details of development of population balance for application to the dissolution of particles during the injection process and technical data for the same are available in the literature (SILVA et al., 2000). With these one can estimate , the values of the parameters involved in the dissolution equation given above. Physicochemical properties such as, slag viscosity, density, % of lime saturation, diffusion coefficient of calcium and oxygen etc as well as and operating parameters like gas flow , immersion depth of lance injection, rate of energy intake etc. are important for this analysis. Some results of this analysis are given below.

Figure 19.26 shows the effect of temperature on the % CaO dissolved in the liquid slag. The rate of dissolution is higher at higher temperatures. However once saturation is reached there is no more dissolution and around this point injection of CaO becomes meaningless. The beneficial effect of temperature increase can be explained by the improvement of transport properties (low viscosity and increase of mass transfer) and mainly by increased driving force of dissolution, as measured by the increase in CaO% saturation.The decrease in particle size, with its implications on the total area of the interface slag/particle is an important factor in the process of dissolution. This effect is more pronounced at lower temperatures as suggested by Figure 19.26. It is observed that particles of larger size (500 μm) injected into slag at low temperatures (1400°C) practically do not dissolve. In this case it is important to increase the proportion of CaF_2 in the mixture injected.

The amount of carry-over slag is important in the efficiency of processes such as desulphurization. In some situations sulfur removed from the metal in the transitional zone of the reactor can be reversed in the permanent reactor, due to the low sulfide capacity of top slag. Hence the reforming of the of carry-over slag is important.

Figure 19.27 reinforces the recommendation that excessive amounts of carry-over slag should be avoided, since they require time to reform. In this sense it seems important for preconditioning of the carryover slag, even before the starting of the injection process and perhaps during pouring.

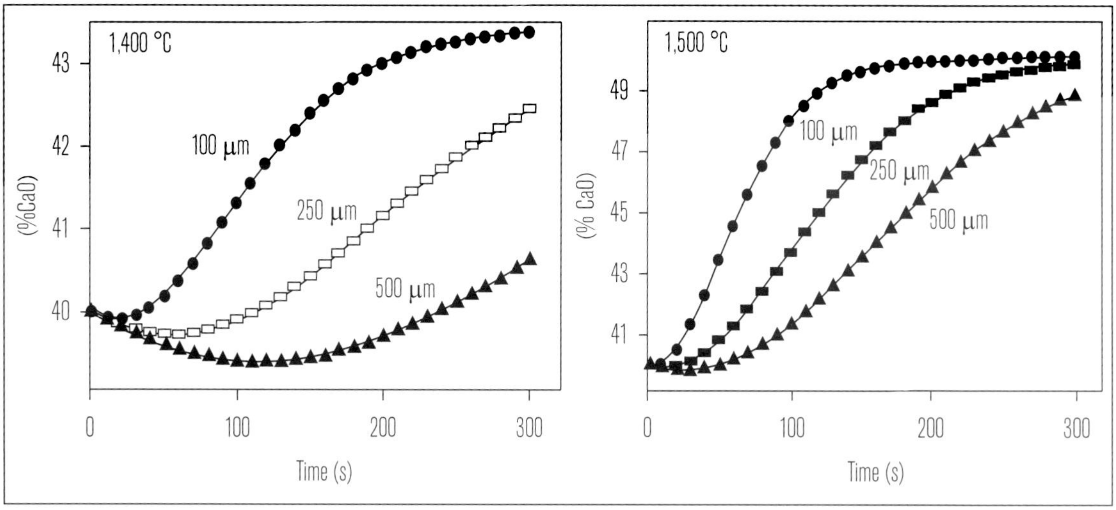

Figure 19.26 - Effect of temperature and particle size on the %CaO in the liquid slag: a) 1400°C; b) 1500°C.

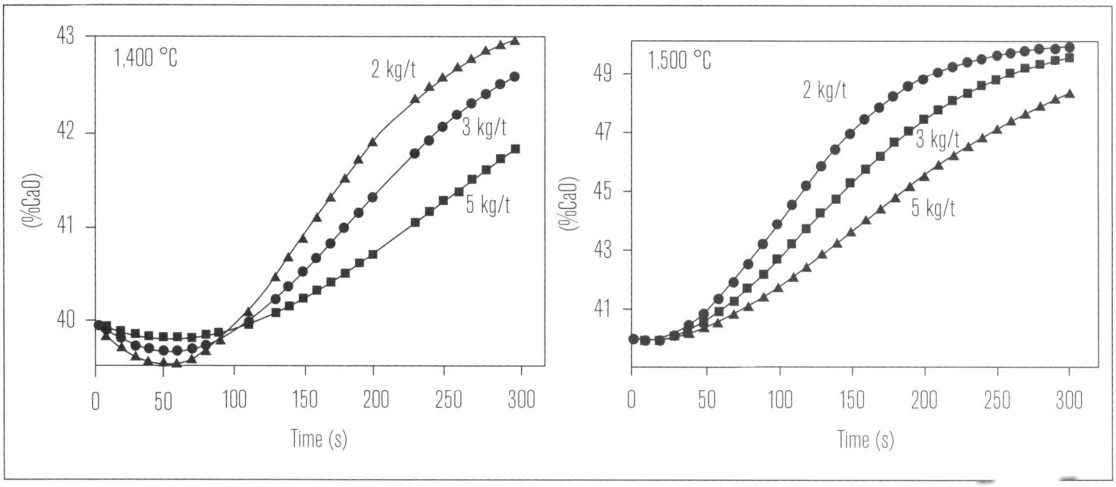

Figure 19.27 - Effect of the amount of carry-over slag on the time required for the reformation of slag: a) 1400°C; b) 1500°C.

19.3.9 Coupled heat and mass transfer

When a small spherical particle of carbon, undergoes combustion in an oxygen atmosphere under ordinary pressure it is reasonable to assume that at high temperature exceeding 1000°C, the reaction product is mainly carbon monoxide.This conclusion comes from analysis of equilibrium thermodynamics of the reaction with excess carbon. The oxygen partial pressure is negligible. Combustion reaction at high temperatures can be given as follows.

$$C_{(s)} + 1/2O_{2(g)} = CO_{(g)} \qquad (19.117)$$

The oxidation occurs on the outside surface of the particle as well as inside the pores, ie, the region's reaction would be a volume that could eventually extend to the center of the particle. However, at elevated temperatures chemical reaction step is not rate controlling step in the process because it is intrinsically fast. Steps of mass tranport like oxygen transport to the particle through boundary layer should be the controlling step. Oxygen is immediately consumed by chemical reaction at the surface and very little reactant gas is available to be transported to the interior of the pores for further reaction This is tantamount to saying that the reaction occurs mostly at the outer surface of the particle with diffusion and interface reaction limited to pores near the surface. Assuming quasi steady state, the rate of oxygen transport to the surface of the particle, is equal to the rate of oxygen consumption:

$$r_{O_2} = n_{O_2} = -4\pi R^2 K_m C_T (X_{O_2} - X_{O_2}^{eq}) \text{ moles of } O_2/s \qquad (19.118)$$

where:

- r_{O2} indicates the rate of oxygen consumption;
- n_{O2} is the rate of transport of oxygen through the boundary layer around the particle;
- R is the radius of the particle;
- K_m is the mass transfer coefficient through the boundary layer;
- $C_T = \frac{P}{RT}$ mols/m³ is the total concentration assuming ideal gas;
- X_{O2} and X_{O2}^{eq} are the mole fractions of oxygen within the gas.

In the case of a single particle X^{eq}_{O2} can be considered unitary and the equilibrium concentration of X^{eq}_{O2} can be considered negligible.

Hence the rate of consumption of carbon, according to the stoichiometry of the reaction is:

$$r_C = -8\,\pi\,R^2\,K_m\,C_T\,X_{O_2} \text{ moles of C/s} \qquad (19.119)$$

that evaluates the rate of decrease of the radius of the particle as:

$$4\,\pi R^2 \rho_C \frac{dR}{dt} = -8\,\pi\,R^2\,K_m\,C_T\,X_{O_2} \qquad (19.120)$$

where:

- ρ_C is the molar density of carbon particle in moles of C/m^3.

The preceding equation could be used to estimate the time for the total consumption of the particle. However some of the parameters involved are temperature dependent and, the reaction being exothermic, it is difficult to ensure that the values remain constant during the process. Assuming quasi-stationary regime of heat transfer, considering heat generated by the oxidation reaction is dissipated by thermal radiation and convection to he surrounding environment (in fact, part of this energy is used to heat the particle) then,

$$h\,(T - T_A) = \varepsilon\,\sigma\,(T^4 - T_A^4) = r_C\,\Delta H_{CO} \qquad (19.121)$$

where:

- h is the heat transfer coefficient through the boundary layer;
- T is the temperature of the particle;
- T_A is the ambient temperature;
- ε is the thermal emissivity of the surface of the particle;
- σ is the Bolztmann's constant;
- ΔH_{CO} is the enthalpy of reaction $C_{(s)} + 1/2O_{2(g)} = CO_{(g)}$.

The following equations have to be simultaneously solved:

$$r_C = -8\,\pi\,R^2\,K_m\,C_T\,X_{O_2} \qquad (19.122)$$

$$h\,(T - T_A) + \varepsilon\,\sigma\,(T^4 - T_A^4) = r_C\,\Delta H_{CO} \qquad (19.123)$$

This can be done by an iterative process involving the temperature that can follow the evolution of the process. The process involves starting with a known particle size and environment temperature; then defines a small time step followed by calculation of rc from equation 19.122; and estimation T from 19.123; again update rc again with 19.122. (MILLS, 1999). The process is repeated until constant values are obtained.

Example

The objective is to estimate the time required to dissolve particles of carbon in liquid iron (steel). These particles are injected into the melt with the aim of increasing the carbon content in the liquid iron.

Assuming that the particle is impermeable and is consumed at the surface reacting with the fluid medium (section 19.3.1) basically there are two steps occurring in series:

1) Chemical reaction of dissolution of carbon, taking place at the particle/ steel melt interface;

2) Transport of carbon to the bulk of the metal.

The chemical reaction taking place is:

$$C \text{ (in graphite, on coke)} = \underline{C} \text{ (dissolved in the steel)}$$

The qualitative composition profile (that would be obtained if one could perform microprobe analysis analysis of Carbon from the interface steel/particle to the bulk of metal) is given in the folowing figure.

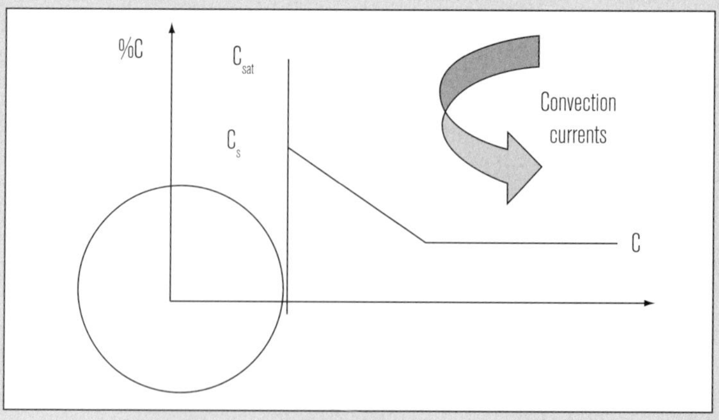

Scheme for the case of impermeable particle reacting with a fluid.

C_{sat} is the saturation value of C in steel melt, depending on the temperature and its composition and can be calculated using the equation,determined from phase diagrams.

$$C_{sat} = 1.34 + 2.54 \times 10^{-3}\ T\ [C]$$

C_S is the actual composition of the steel melt in contact with the particle.

C_{sat} being greater than C_S indicates that there is no local equilibrium at the interface; $Csat \geq C_S \geq C$ is the condition for the dissolution reaction C (particle) = C (steel) which has negative ΔG value. Away from the particle strong convection currents maintain uniform composition of the steel melt equal to C and the value of C varies with time during the course of dissolution. C_S must be greater than C, so that the direction of mass transport is to the bulk of the steel melt.

The rate of dissolution reaction can be given as:

VRQ = (area available for reaction) × Rate constant × driving force ($C_{sat} - C_S$)

$$V_{RQ}\left(\frac{\text{moles of C}}{s}\right) = 4\pi R^2\left(m^2\right) \times K_{RQ}\left(m \cdot s^{-1}\right) \times \left(C_{sat} - C_S\right)\left(\frac{\text{moles of C}}{m^3}\right)$$

This expression K_{RQ} represents the rate constant for the reaction to be determined experimentally.

The second step of the reaction namely transport of carbon from the reaction interface to the bulk through the boundary layer is given by:

V_{TM} = Area x mass transfer coefficient x concentration gradient between surface and bulk.

$$V_{TM}\left(\frac{\text{moles of C}}{s}\right) = 4\pi R^2\left(m^2\right) \times K_m\left(m \cdot s^{-1}\right) \times \left(C_S - C\right)\left(\frac{\text{moles of C}}{m^3}\right)$$

where R is the radius of the particle.

For quasi-stationary state:

$$\underset{V_{TM}}{\text{Rate of transport}} = \underset{V_{RQ}}{\text{Rate of consumption}}$$

Of all the variables mentioned C_S (concentration in the interface) is hard to determine. Considering electrical analogy discussed earlier:

$$V_P\left[\frac{\text{moles of C}}{\text{s}}\right]=\frac{\left(C_{sat}-C\right)}{\dfrac{1}{4\pi R^2K_m}+\dfrac{1}{4\pi R^2K_{RQ}}}$$

In the denominator there are two resistances. One is due to mass transfer in the boundary layer and the other is the resistance of chemical reaction of dissolution of Carbon. As the process takes place in temperatures of around 1600°C, the resistance due to the latter can be ignored and the the controlling step would be mass transport through boundary layer:

$$V_P\left[\frac{\text{moles of C}}{\text{s}}\right]=\frac{\left(C_{sat}-C\right)}{\dfrac{1}{4\pi R^2K_m}}$$

As the carbon flow corresponds to dissolution rate, which is a function of decrease in the radius of the particle:

$$\frac{d}{dt}\left(\rho_C\left[\frac{\text{moles of C}}{m^3}\right]\frac{4}{3}\pi R^3\left[m^3\right]\right)=-\frac{\left(C_{sat}-C\right)}{\dfrac{1}{4\pi R^2K_m}}$$

$$\rho_C\frac{dR}{dt}=-\frac{\left(C_{sat}-C\right)}{\dfrac{1}{K_m}}$$

where ρ_C represents the molar density of Carbon particle. This equation can be integrated to evaluate variation of C with time and variation of radius of the sphere with time. For the dissolution of a single sphere or cylinder one can get the dissolution rate through thermogravimetric methods.

However, the application of the above expression to the actual system should consider several particles with different sizes and would involve detailed calculations based on size distribution of particles. One can assume uniform size distribution (of varying size over time) of particles initially and the number of particles (n_0) remaining unchanged and equal to the initial number (although their respective size are decreasing due to dissolution. Considering that the term $4\pi R^2$ (m^2) as the flow area available for each particle considered, the total flux of carbon to the bulk of the metal due to all the particles is given by:

$$V_{total}\left[\frac{\text{moles of C}}{\text{s}}\right] = n^{\circ} 4\pi R^2 K_m \left(C_{sat} - C\right)$$

which corresponds to a change in composition of steel given by:

$$\frac{M(\text{kg})}{\rho\left(\text{kg}/\text{m}^3\right)} \frac{dC\left(\text{moles}/\text{m}^3\right)}{dt}$$

where M is the mass of steel and ρ the density of the same.

The equation that links the composition of the steel to the amount of carbon particles should be applicable if carbon is not lost by oxidation or other means, ie, if all the carbon from the the particles remain in the steel melt. Hence considering initial carbon (particles and steel) = carbon at time t (particles and steel) one can write for a carbon balance:

$$\frac{M}{\rho} C_0 + n^0 \rho_C \frac{4}{3}\pi R_0^3 = \frac{M}{\rho} C + n^0 \rho_C \frac{4}{3}\pi R^3$$

where Co and Ro are initial concentration of carbon in the steel and the initial radius of particles, respectively. The evolution of the carbon content is given by simultaneous solution of equations given below:

$$\frac{M}{\rho}\frac{dC}{dt} = n^{\circ} 4\pi R^2 K_m \left(C_{sat} - C\right)$$

From Carbon balance transposing the term R can be evaluated:

$$R = \left[\frac{3M}{4n^0 \pi \rho_C \rho}(C_0 - C) + R_0^3\right]^{1/3}$$

From Frössling equation for sphere mass transfer coefficient of the dissolving species in a fluid can be evaluated;

$$Sh = 2 + 0.552 \cdot \sqrt{Re} \cdot Sc^{\frac{1}{3}}$$

Typical property values are:

- η (steel viscosity) = 0.3699 × 10^{-3} exp(4982/T (K) Pa.s pure iron;
- ρ (steel density) = 7015 – 0.883 (T (°C) – 1536) kg/m^3 pure iron;
- D(carbon diffusivity) = 7.8 × 10^{-9}m^2/s in Fe-C melt (up to 2.1% C, 1550°C);
- D(carbon diffusivity) = 6.7 × 10^{-9}m^2/s Fe-C melt (3.5% C, 1550°C);

For the evaluation of mass transfer coefficient using Frossling equation, one needs to know the Reynolds number which requires calculation of relative velocity between the particle and metal. In this system no single value for this variable can be defined. However an average terminal velocity can be calculated considering particle weight buoyant force and frictional effects.

V_T, Terminal velocity, can be estimated by force balance acting on the particle:

$$\frac{4}{3}\pi R^3 g(\rho - \rho_P) - f\left(\frac{1}{2}\rho V_T^2\right)\pi R^2 = 0$$

In this equation f represents the friction factor, and ρ_P the density of the particle.Values of f are normally provided from the Reynolds number, and can be identified three flow regimes.

For laminar flow, $Re = \frac{2R\rho V_T}{\eta} \leq 1$ Stokes equation applies:

$$V_T = \frac{2}{9\eta} R^2 g(\rho - \rho_p)$$

For intermediate *regime,* $1 \leq Re = \frac{2R\rho V_T}{\eta} \leq 1000$ friction factor is given by:

$$f = \frac{18.5}{Re^{0.6}}$$

For turbulant regime $Re = \frac{2R\rho V_T}{\eta} \geq 1000$ f ~ 0.44

Another relationship can be derived between f and Re (through force balance) which is given by:

$$\frac{32}{3}\rho \frac{R^3 g}{\eta^2}(\rho - \rho_P) = f\,Re^2$$

Terrminal velocity can be calculated through any one of these equations depending on the flow regime.

Example

Estimate the time required for the dissolution of a spherical particle of graphite, 3 mm in diameter, through agitation with graphite particles:

For ρ = 7,000 kg/m³, R(graphite) = 0.0015 m; η = 7 x 10^{-3} Pa.s; ρ_P = 2,000 kg/m³ and assuming that the terminal velocity is situated in the middle range of Reynolds number, the above equation:

$$\frac{32}{3}\rho\frac{R^3 g}{\eta^2}(\rho-\rho_P)=f\,Re^2$$ gives 252 257.14 = f Re^2, and from $f=\frac{18,5}{Re^{0.6}}$

results in Re = 898.12; this value is in between 1 < Re <1000.

$Re=\frac{2R\,\rho V_T}{\eta}$ gives V_T = 0.299 m/s. On the other hand one may estimate

$Sc=\frac{\eta}{\rho.D}=\frac{7\times10^{-3}}{7000\times7.8\,10^{-9}}$ = 128.20 and from $Sh=2+0.552\cdot\sqrt{Re}\cdot Sc^{\frac{1}{3}}$ would result Sh = 85.41 and K_m = 2.22 × 10^{-4} m/s.

This value is valid for the initial size of the particle, whose diameter decreases with time. This also can be taken into account by calculating the mass transfer coefficient as a function of time.

In the case of single particle,

$$\rho_C\frac{dR}{dt}=-K_m\left(C_{sat}-C\right)-\approx -K_m C_{sat} \qquad \text{(A)}$$

$\rho_{graphite}$ = 2000 kg/m³, or $\rho_C=\frac{2{,}000\text{ Kg m}^{-3}}{12\times10^{-3}\text{Kg.mole}^{-1}}=166\,666\text{ mole m}^{-3}$

C_{sat} = 5.404% at 1600 °C,

$$C\left(\text{mol.m}^{-3}\right)=\frac{\rho_{steel}\left(\text{kg m}^{-3}\right)\frac{\%C}{100}}{12\times10^{-3}\left(\text{Kg mol}^{-1}\right)}=31\,523\,\text{mole m}^{-3}$$

Substituting the values of molar density of graphite, the mass transfer coefficient and the concentrations C and C_{sat} in equation (A) results in:

dR/dt = -4.2 x 10^{-5} m/s, or R – R_0 = - 4.2 x 10^{-5} t after integration.

For R_0 = 0 the elapsed time is 15.87 s.

Dissolution time for particles with different sizes are given (when a number of particles are involved) for raising the carbon content of the steel melt from 0.02% to 0.8%, considering concentration differences as driving force as given below:

- Saturation % C -% C initial;
- Saturation % C -% C final.

The first case represents a lower limit, the second an upper limit. The actual case is between the two. The time of dissolution as a function of diameter of the particles for the above two cases of driving force are given in the following figure. One can note the significant role of size of the particle on dissolution times.

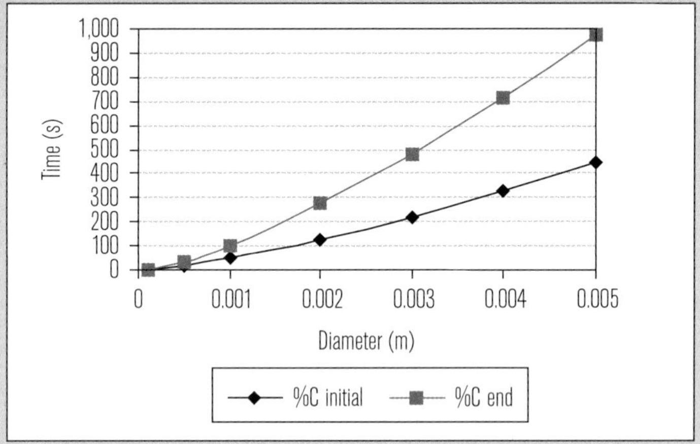

Time of dissolution of graphite particles as a function of size; initial carbon 0.02% C and final carbon 0.8% C.

In this chapter many examples of interphase mass transfer with applications in desulphurization and refining reactions in steel making, gas solid reactions dissolution of carbon as well as its combustion have been dealt with with many simplifying assumptions. Many of the applications given are from published literature. For getting more information about the topics discussed it is recommended that references cited may be consulted. The examples given have the objective of orienting how to proceed about regarding development of simple models from basic principles involving analytical solutions. Complicated models based on simultanenous differential equations can be solved with the help of sophisticated software packages available commercially.

References

BIRD, R.B.; STEWART, W.E.; LIGTHFOOT, E.N. *Transport phenomena*. 2nd ed. New York: John Wiley & Sons, 2002.

CICUTTI, C. et al. Study of slag-metal reactions in an LD-LBE converter. In: EUROPEAN STEELMAKING CONFERENCE, 2., 1997, Taranto, Itália. *Proceedings*... Taranto: [s.n.], 1997.

DEO, B.; BOOM, R. *Fundamentals of steelmaking metallurgy*. New York: Prentice Hall, 1993.

ELLIOTT, J. F.; GLEISER, M.; RAMAKRISHNA, V. *Thermochemistry for steelmaking*. Reading: Addison Wesley, 1963.

FOGLER, H.S. *Elements of chemical reaction engineering*. New Jersey: Prentice Hall, 2005.

FROMENT, G.F.; BSICHOFF, K.B. *Chemical reactor design and analysis*. 2. ed. New York: John Wiley & Sons, 1979.

GAMEIRO, D.H. *A cinética da carbocloração do dióxido de titânio*. 1994. 261 p. Tese (Doutorado em Engenharia Metalúrgica e de Materiais) - Coordenação dos Programas de Pós-Graduação em Engenharia da Universidade Federal do Rio de Janeiro, Rio de Janeiro, 1994.

HABASHI, F. *Kinetics of metallurgical process*. Québec: Laval University Bookstore Zone, 1999.

IRONS, G. A.; CELIK, C. Comparison of iron desulphurisation using calcium carbide and limemagnesium mixtures. *Ironmaking and Steelmaking*, v. 19, n. 2, p. 136-44, 1992

KIRMSE, O.J. *A study of metallurgical characteristics of Kambara reactor through a physical modeling study*. Dissertação (Mestrado) - Universidade Federal de Ouro Preto, Ouro Preto, 2006.

MIDDLEMAN, S. *An introduction to mass and heat transfer*. New York: John Wiley & Sons, 1998.

MILLS, A.F. *Basic heat and mass transfer*. New Jersey: Prentice Hall, 1999.

OETERS, F. *Metallurgy of steemaking*. Dusseldorf: Stahl & Eisen, 1994.

OHGUCHI, S. et al. Simultaneous desulphurization and dephosphorization of molten pig iron. *Ironmaking Steelmaker*, v. 11, n. 4, p. 202-13, 1984.

POIRIER, D. H.; GEIGER, G.H. *Transport phenomena in materials processing*. [S.l.]: TMS, 1994.

RAY, H.S. *Kinetics of metallurgical reactions*. New York: International Science, 1993.

RIBOUD, P.; VASSE, R. Désulfuration de lácier em poche: synthese dês résultats théoriques et industriels. *Revue de Metallurgie*, CIT, v. 82, n. 11, p. 801-10, nov 1985.

SANO, M.; LU, W.K.; RIBOUD, P.V. *Advanced physical chemistry for process metallurgy*. London: Academic Press, 1997.

SANO, M.; MORI, K. Dynamics of bubbles swarms. *Transactions ISIJ*, v. 20, n. 10, p. 668-74, 1980.

SANO, M.; MORI, K. Size of bubbles in energetic gas injection into a liquid metal. *Transactions ISIJ*, v. 20, n. 10, p. 675-681, 1980.

SHINME, K.; MATSUO, T.; MORISHIGUE, M. Acceleration of nitrogen removal in stainless steel under reduced pressure. *Transactions ISIJ*, v. 28, p. 297-304, 1988.

SILVA, C.A. et al. A kinetic model applied to molten pig iron desulfurization by injection of lime based powders. *ISIJ International*, v. 37, n. 1, p. 21-30, 1997.

SILVA, C.A. et al. A kinetic model applied to pig iron desulphurization in the Kanbara reactor. In: INTERNATIONAL CONFERENCE ON MOLTEN SLAGS FLUXES AND SALTS, 8., *Proceedings*... 2009. p. 737-747.

SILVA, I.A. et al. A kinetic study of the dissolution of lime particles in top slags during lime-based powder injection pré-treatment of molten pig iron. In: PROCESS METALLURGY AND STRUCTURAL ENGINEERING OF THE INTERNATIONAL MATERIALS RESEARCH CONGRESS, 10., 2000, Cancun, Mexico. Mexico: Universidad del Mexico, 2000. p. 18-32.

SOHN, Y.H.; WADSWORTH, M.E. *Rate process of extractive metallurgy.* [S.l.]: Plenum Press, 1979.

SZEKELY, J.; CARLSSON, G.; HELLE, L. *Ladle metallurgy.* Berlin: Springer-Verlag, 1989.

SZEKELY, J.; EL-KADDAH, N. A mathematical model for desulfurization kinetics in argon stirred ladle. *Ironmaking and Steelmaking,* v. 6, p. 269-76, 1981.

THEMELIS, N.J. *Transport and chemical rate phenomena.* Switzerland: Gordon and Breach, 1995.

TOSUN, I. *Modeling in transport phenomena.* Amsterdan: Elsevier, 2002.

TURKDOGAN, E.T. *Fundamentals of steelmaking.* London: The Institute of Materials, 1996.

Exercises

1 - A spherical particle of Fe2O3 is reduced with H2 at 900°C. Consider that the reduction process is controlled by diffusion of gases through the porous product layer of metallic iron. Estimate the rate of reduction when the fraction reacted, α, is equal to 0.5 and 0.9, considering d = 2 cm; $D_{H2\text{-}H2O} = 3\ cm^2.s^{-1}$; $X^{eq}_{H2} = 0.62$, $\varepsilon = 0.3$; $\tau = 1.7$. Verify whether this is the correct assumption regarding rate controlling step, if the gas flow is equal to 1 to STP L H_2.min $^{-1}$, in a reactor with 10 cm in diameter.

2 - Argon (2% H_2) is bubbled through a bath of Al at 970 K (solubility of H_2 is equivalent to 10×10^{-6} STP m^3/kg; initial concentration is equivalent to 5×10^{-6} STP m^3/kg). Estimate the flux of H_2, assuming local equilibrium at the interface for the Hydrogen dissolution in the metal.

3 - Assume that the transfer rate of O_2 for the formation of a layer of SiO_2 on the surface of a silicon plate surface, is controlled by diffusion of O_2 gas. Estimate the growth rate of the layer of silica in air at a temperature of 1,300 K.

4 - Values of time (t is the elapsed time in hours) taken for different fractional conversion, α, in a gas solid reaction at three different temperatures are given below:

α	t (hr) at 1,000°C	t (hr) at 950°C	t (hr) at 900°C
0.3	0.2	0.25	0.50
0.4	0.3	0.45	0.90
0.5	0.5	0.70	1.50
0.7	1.15	1.60	3.60
0.9	2.50	2.50	----

Verify possible agreement of these data with the model of unreacted core with variation of size of the particle. Do the same for topochemical reaction model.

5 - Values α for isothermal reduction of iron ore at three different temperatures are:

T°C\t(min)	1	2	3	4	5	6	7
826	0.035	0.0976	0.106	0.13	0.148	0.160	0.178
900	0.05	0.15	0.14	0.17	0.192	0.210	0.230
1020	10	0.18	0.22	0.24	0.255	0.262.	0.270

Construct graphs "α versus t". Consider the various alternative reaction models. Draw graphs "g (α) versus t". Estimate the energy activation.

6 - Wustite is reduced to metallic iron at 700°C with Hydrogen. Calculate the rate of reduction when α becomes equal to 0.4, assuming control by diffusion into the porous product layer. While $\frac{H2O}{H2}_{eq} = 0.39$, $D_{H2-HO} = 7.1 \times 10^{-4}\ m^2.s^{-1}$, $\varepsilon = 0.4$, $d_p = 0.015$ m, $\tau = 2$, $\rho_{wustite} = 4000\ kg/m^3$.

7 - A cube (5 mm) of wustite is reduced by H_2 at 700°C, in a tubular furnace with a diameter of 0.05 m. Assume control by diffusion in the boundary layer and $\frac{H2O}{H2}_{eq} = 0.39$, $D_{H2-HO} = 7.1 \times 10^{-4}\ m^2.s^{-1}$, $\eta = 2.25 \times 10^{-5}$ Pa.s; $Q = 13.3 \times 10^{-6}$ STP m^3/s. Write the expression for the rate of reduction and estimate the total time required for the same.

8 - A spherical particle of NiO (0.01 m of diameter) is reduced to Ni at 700°C by a gas phase. Assume $\frac{H2O}{H2} = 9$, $\frac{H2O}{H2}_{eq} = 100$. Assume control by diffusion through porous product layer and estimate the time required for reduction.

9 -A pyrometallurgical process for the production of steel can be conceptually divided into two steps: mass transport in fluid medium, chemical reaction in the interface. Which would be the rate controlling step? Justify.

10 - A process for manufacturing a solution between two metals, A and B, have the following steps: transport of gas containing species B to the solid-gas interface, interface chemical reaction releasing B; interdiffusion of A and B leading to the formation of the phase. Experiments conducted at various temperatures resulted in high values of activation energy of the process. What is the rate controlling step? Justify.

11 - Consider the reduction of a metal oxide with carbon, occurring in a bed containing particles of of both the materials,represented by the reaction:

$$MO(s) + C(s) = M(s) + CO(g)$$

The above mechanism assumes that the reaction takes place at the points of contact oxide/carbon, which seems inconsistent with the observed reduction rates since the contact surface is small. Since the observed rates are high, it is assumed then that the process occurs via the gas phase through the steps:

1) carbon monoxide reduces the oxide and forms CO_2;
2) carbon dioxide formed in the reduction is carried to the carbon particle;
3) which regenerates CO through the Boudouard reaction.

$$MO(s) + CO(g) = M(s) + CO_2(g)$$

$$CO_2(g) + C(s) = 2CO(g)$$

Assume that the process of reduction is a special case, controlled by the step of transporting CO_2 from the oxide (MO stands for MnO) surface to the carbon surface, where regeneration occurs. Assume that the transport rate is proportional to the flow area between the particles (A), the diffusion coefficient, given by:

$$D = \frac{10^{-3}\, T^{1.75}}{P(v_{CO}^{1/3} + v_{CO2}^{1/3})} \sqrt{\frac{1}{M_{CO}} + \frac{1}{M_{CO2}}}$$

as well as driving force (measured as the difference in equilibrium concentrations between the two points) and to the inverse of the average interspacing between the particles (.δ). Consider that the gases are ideal. Write an expression for the rate of transport for temperatures of 1,500 K and 1,600 K and pressure of 1 atm. Estimate the ratio of transport rates. Estimate the apparent activation energy of the process. Based on the results obtained, investigate whether this process is strongly activated by temperature.

12 - The rate of a process involving mass transport in a fluid medium and interfacial reaction is given by:

$$r = \frac{\Delta C}{\frac{1}{AK_m} + \frac{1}{AK_{RQ}}}$$

where ΔC is the driving force, A the area of interface, K_m and K_{RQ} in the above equation are mass transfer coefficient and rate constant for chemical reaction respectively. Assume:

$$K_m = 4 \times 10^{-6}.e^{-3,000/T},\ K_{RQ} = 30 \times e^{-45,000/T}$$

Find out the ranges of temperature for mass transport control and for reaction control. The critical ratio defining predominance can be taken as equal to 10. What is the controlling step at 1400 K? Determine the range of temperatures at which agitation without alteration of surface area, interferes with the process.

13 - A process of forming a protective layer of SiO_2 on the surface of an M-Si alloy can be decomposed involving two steps: 1 - chemical reaction on the interface 2 - Diffusion of oxygen inside the metal to interface where reaction takes place. It is given $\ln D = -0.72 - 33\,200/T$, where $D(cm^2/s)$ is the diffusion coefficient. The process as a whole is highly activated by temperature. Is the rate controlling step the chemical reaction?Justify.

14 - In a RH desgasser (see figure) the liquid steel with a certain concentration of C(mole carbon/m^3) circulates under vacuum with injection of argon gas in the upleg (1) from the ladle to the vacuum chamber (2). The vacuum enhances the rate of degassing (increase of thermodynamic potential due to reduction of pressure) according to the reaction C + O = CO (g) and the vacuum can reach levels as low as 1 mm Hg Also the metal undergoes considerable agitation in the vacuum chamber. The process is very efficient for the removal of Carbon resulting in low C in the metal. The model of this reactor is assumed to consist of two stages in series. The first is the reaction in the vacuum chamber, where the rate is given by:

$$n_C = -a\,K\,(C_r - C_{eq})$$

Where α is the gas/liquid interfacial area (m^2), C_r is the concentration of carbon in the metal at the vacuum chamber, C_{eq} is the equilibrium concentration, K (m/s) is the rate constant. C_{eq} is in general very low and can be ignored.

The second step is the transfer of steel which did not get refined in the upleg which climbs to the vacuum chamber and, after decarburization, comes back to the ladle through down leg (3). This step can be quantified as:

$$n_C = - Q (C - Cr)$$

The circulation rate of steel melt is Q (m^3/s) and C (mole carbon/m^3) stands for carbon concentration in the metal inside the ladle. Considering V (m^3) as the total volume of steel inside the reactor and quasi steady state conditions the rate of carbon accumulation can be calculated from:

$$V\frac{dC}{dt} = -\frac{\{C - C_{eq}\}}{\frac{1}{aK} + \frac{1}{Q}}$$

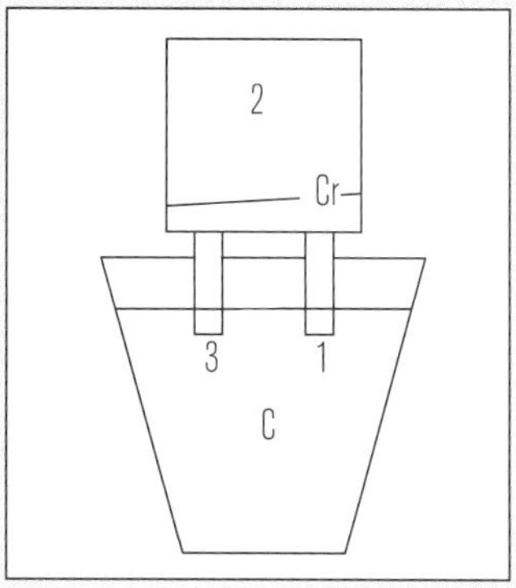

R-H degasser schematic figure.

Hence 1/aK can be taken as the resistance of the reaction occurring in the vacuum chamber, and 1/Q as a resistance to circulation.

1) Derive the preceding equation;
2) Consider V = 45 m^3, Q '= 180 tons/min, and decarburization from 500 ppm to 100 ppm in 15 minutes. Find the total resistance of the process;
3) What is the resistance which controls the process?

15 - The following table presents data (temperature, agitation rate, rate of process) for the cemention reaction Pd^{2+}/Cu. In each temperature, what is the likely controlling step? Justify. Estimate the activation energy in each temperature range. Do the values of activation energy support the controlling steps suggested above?

°C	rpm	Vp	°C	rpm	Vp	°C	rpm	Vp
25	500	3.5	45	900	10.4	65	500	11.5
	1000	3.5		1000	10.3		750	14.3
	1250	3.4		1250	10.1		1000	18

Heat Transfer in Packed Beds

Chapter 20

This chapter deals with the study on thermal exchanges between gas and a beds of particles. The treatment given is specially applicable to granular zone or shaft of the blast furnace for identifying the relevant parameters that are important in heat exchange between gas and solid moving in countercurrent direction. There have been simplified models in recent years (MIWA, 1991; AUSTIN, NOGAMI and YAGI, 1997; TAKATANI, INADA and IJISAWA, 1999; CASTRO, NOGAMI and YAGI, 2002), and the treatment will be presented below is confined to heat transfer only Aspects of mass and momentum transfer have not been considered. There have been however a large number of publications. literature on the subject of heat, mass and momentum transfer in blast furnaces and direct reduction processes during the last few decades (LAHIRI and SESHADRI, 1969). Even books have been published on the subject (OMORI, 1987).

20.1 Heat Transfer in a Counter Current Gas-Solid Bed

The shaft region of the blast furnace as well as direct reduction processes like Midrex etc. are essentially counter current reactors of gas and solid moving in opposite directions (gas has an upward and solid a downward movement), as shown schematically in Figure 20.1.

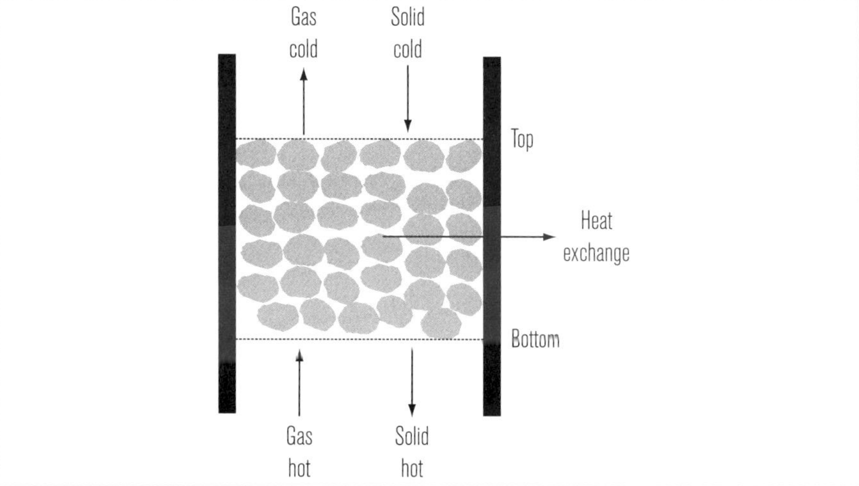

Figure 20.1 - Schematic view of counter current gas-solid. packed bed.

In this type of reactor, the temperature of the gas and the solid vary continuously along the length and depends on the gas flow along the cross section also. (especially in blast furnace).

20.1.1 Mathematical treatment

To develop the mathematical model for heat transfer in countercurrent reactors, some simplifying assumptions have been made:

- The solid particles are isothermal, ie there are no temperature gradients inside the particles;
- The specific heats of gas and solid vary very little over the height of the reactor Hence this variation need not be considered (average values assumed);
- Since the eactor operates at steady state, ie, temperatures, velocities and physical properties of solid and gas do not vary with time;
- The velocities of solid and gas as well as the void fraction of the bed are constant along the reactor. It is also assumed that there is no generation of heat due to chemical reactions in the bed and the system is adiabatic.

For getting the differential equations and solution through them that provide temperature profiles of the solid and gas along the reactor, it is necessary to develop energy balances for these two phases within an infinitismal volume element of the reactor, as shown in Figure 20.2.

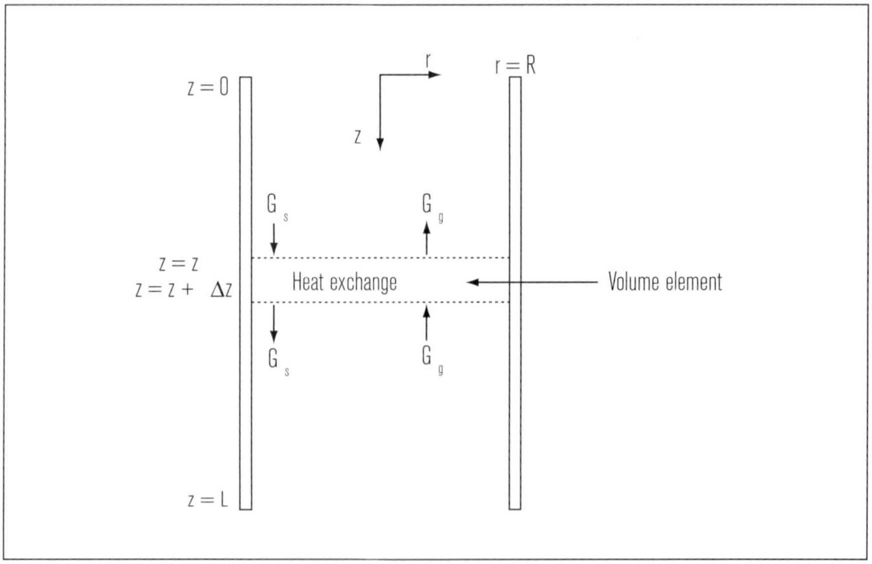

Figure 20.2 - Volume element for heat balance in cylindrical cross section.

The various terms of the energy balance equation in the volume element are:

$$\begin{bmatrix}\text{Rate of energy}\\ \text{input from}\\ \text{gas phase}\end{bmatrix} = \begin{bmatrix}\text{Rate of energy}\\ \text{output from}\\ \text{gas phase at z}\end{bmatrix} + \begin{bmatrix}\text{Rate of energy}\\ \text{transfer from}\\ \text{gas to the solid}\end{bmatrix} \tag{20.1}$$

For gas:

- Rate of energy input from gas phase at z + Δz:

$$\pi R^2 G_g C_p^g \left(T_g\big|_{z+\Delta z} - T_o\right)$$

- Rate of energy output from gas phase at z:

$$\pi R^2 G_g C_p^g \left(T_g\big|_{z} - T_o\right)$$

- Rate of energy transfer from gas phase to solid:

$$\pi R^2 \Delta z\ h_V \left(T_g - T_s\right)$$

where:

- R = radius of the reactor (m);
- G_g = mass flow of gas per unit cross sectional area of the reactor ($kg/m^2.S$);
- C_p = specific heat of gas at constant pressure (J/kg.°C);
- $T_g\big|_{z+\Delta z}$ = gas temperature measured at z Δ + z (°C);
- $T_g\big|_z$ = gas temperature measured at z, (°C);
- T_o = reference temperature (°C);
- h_V = volumetric coefficient of heat transfer ($W/m^3.°C$).

The temperature To is an arbitrary temperature, relative to which energy is evaluated Whatever be its value, it will disappear in the final equation.

The volumetric coefficient of heat transfer is a parameter whose value is prorportional to the rate of heat exchange between gas and solid. However, this coefficient measures the rate of heat transfer per unit reactor volume and not per unit area of contact between gas and solid surface as in the case of heat transfer coefficient h_c.

Replacing portions of the energy balance in (20.1), we obtain:

$$\pi R^2 G_g C_p^g\, T_g\big|_{z+\Delta z} = \pi R^2 G_g C_p^g\, T_g\big|_z + \pi R^2 \Delta z\, h_V \left(T_g - T_s\right) \tag{20.2}$$

Dividing the relation (20.2) by the value of the volume element, $\pi R^2.\Delta z$, and recalling the definition of first order derivative, the following differential equation is obtained for the temperature of the gas:

$$G_g C_p^g \frac{dT_g}{dz} = h_V \left(T_g - T_s\right) \tag{20.3}$$

In the same way the energy balance for solid, can be carried out:

$$\begin{bmatrix}\text{Rate of energy} \\ \text{input from} \\ \text{solid phase}\end{bmatrix} = \begin{bmatrix}\text{Rate of energy} \\ \text{output from} \\ \text{solid phase at z}\end{bmatrix} + \begin{bmatrix}\text{Rate of of energy} \\ \text{transfer from} \\ \text{gas to solid}\end{bmatrix} \tag{20.4}$$

- Rate input of energy in solid from gas at z + Δz:

$$\pi R^2 G_s C_p^s \left(T_s\big|_{z+\Delta z} - T_o\right)$$

- Rate output of energy from solid at z:

$$\pi R^2 G_s C_p^s \left(T_s\big|_z - T_o\right)$$

- Rate of energy transfer from gas to solid:

$$\pi R^2 \Delta z\ h_V \left(T_g - T_s\right)$$

where:

- G_s = mass flow of solid per unit sectional area of the reactor ($kg/m^2.S$);
- C_p^s = specific heat of solid at constant pressure (J/kg.°C);
- $T_s|_{z+\Delta z}$ = temperature of the solid evaluated at z + Δz (°C);
- $T_s|_z$ = temperature of the solid evaluated at z, (°C).

Substituting the terms in the energy balance equation (20.4), one gets:

$$\pi R^2 G_s C_p^s\, T_s\big|_{z+\Delta z} = \pi R^2 G_s C_p^s\, T_s\big|_z + \pi R^2 \Delta z\, h_V \left(T_g - T_s\right) \tag{20.5}$$

Dividing (20.5) by the value of volume element, $\pi R^2.\Delta z$, and recalling the definition of first derivative, the following differential equation for the temperature of the solid can be obtained:

$$G_s C_p^s \frac{dT_s}{dz} = h_V \left(T_g - T_s\right) \tag{20.6}$$

Before solving the system of differential equations (20.3) and (20.6), for obtaining thermal profiles of the solid and gas in the reactor, one can analyze them for other important information. Considering the case when the gas is at at higher temperature than the solid, from the ratio of (20.3) and (20.6), one can get:

$$\frac{G_s C_p^s}{G_g C_p^g} = \frac{\frac{dT_g}{dz}}{\frac{dT_s}{dz}} \tag{20.7}$$

The left side of the above expression is can be defined as parameter α:

$$\frac{G_s C_p^s}{G_g C_p^g} = \alpha \tag{20.8}$$

where $G_s G_p^s$ and $G_g G_p^g$ can be regarded as heat flow of solid and gas, respectively, α is the ratio is the ratio of heat flow of solid and gas. Rearranging the expression (20.7), one gets:

$$\frac{dT_g}{dz} = \alpha \frac{dT_s}{dz} \tag{20.9}$$

For this expression, it can be seen that α is the proportionality factor relating the temperature variation of gas and solid along the length of reactor. Therefore, this factor is associated with the the nature of thermal profile of gas and solid in the reactor.

Suppose the solid temperature is T_s at $z = 0$, and the gas is at a temperature T_g^0 at $z = H$:

a) For $\alpha < 1$, according to equation (20.9), changes in gas temperature along the reactor (dT_{go}/dz) are smaller than those for the solid, as shown by the thermal profile given in Figure 20.2 . The concavity of the curves in the downward direction can be attributed to this.

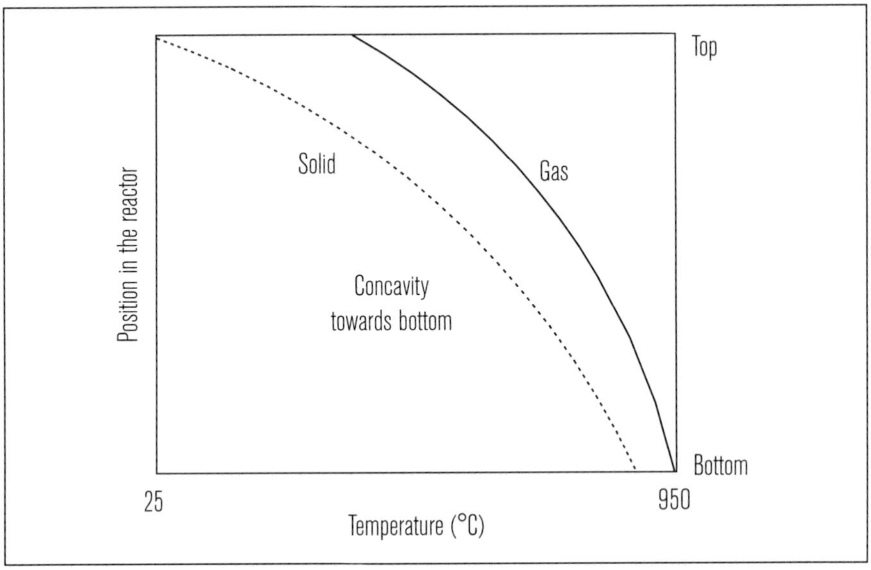

Figure 20.3 - Schematic temperature profiles for solid and gas: $\alpha < 1$.

b) For $\alpha = 1$. according to equation (20.9), the change in the temperatures of gas and solid is equal, resulting in a thermal profile given in Figure 20.4 namely two parallel lines.

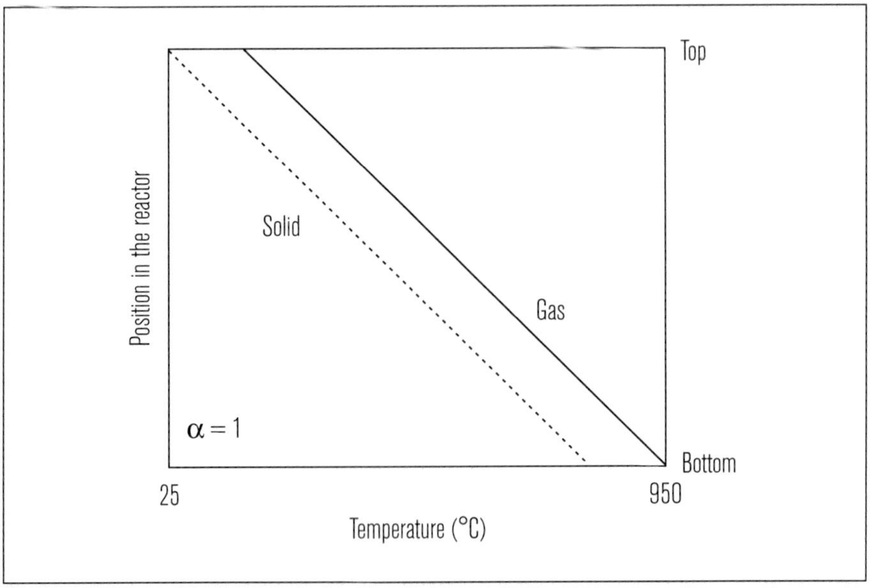

Figure 20.4 - Schematic temperature profiles for solid and gas: $\alpha = 1$.

c) $\alpha > 1$ For this case, variations in gas temperature are higher than those of solid, resulting in the upward concavity of the curves of thermal gradient as shown in Figure (20.5).

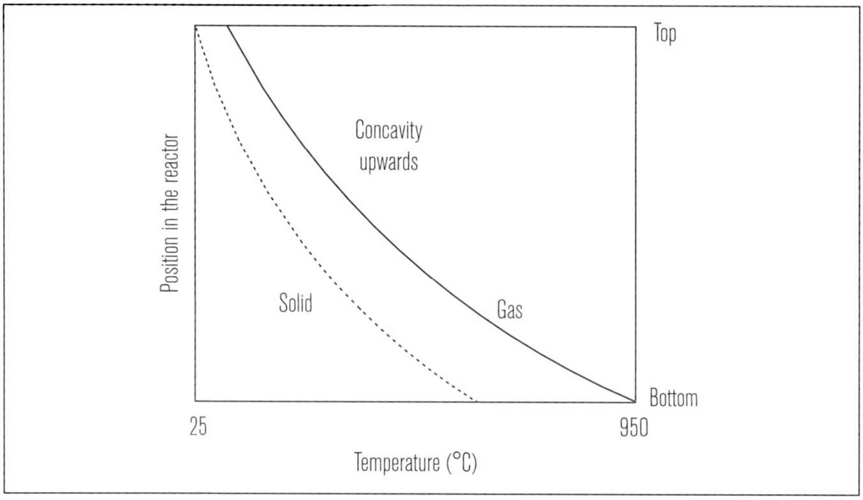

Figure 20.5 - Schematic temperature profiles for solid and gas for $\alpha > 1$.

It is worth noting that the parameter α determines the curvature of the thermal profile, r in respect of transfer of heat between gas and solid. However it does not give any notion about the heat exchange, which would require definition of another parameter.

Rarranging equations (20.3) and (20.6):

$$\frac{dT_g}{dz} = \frac{h_V}{G_g C_p^g}\left(T_g - T_s\right) \tag{20.10}$$

$$\frac{dT_s}{dz} = \frac{h_V}{G_s C_p^s}\left(T_g - T_s\right) \tag{20.11}$$

The terms containing the heat transfer coefficients in the above equations can be defined as ϕ_g and ϕ_s, respectively as given below:

$$\phi_g = \frac{h_V}{G_g C_p^g} \tag{20.12}$$

$$\phi_s = \frac{h_V}{G_s C_p^s} \tag{20.13}$$

An increase in the value of ϕ_s, causes an increase in the rate of heating of the solid, for a given temperature difference between solid and gas. The temperature profiles are determined by the values of α, ϕ_s and ϕ_g, of which only two are independent variables since:

$$\alpha = \frac{\phi_g}{\phi_s} \tag{20.14}$$

Therefore, setting the values of two of the above parameters, it is possible to determine the temperature profiles of the solid and gas in the reactor.

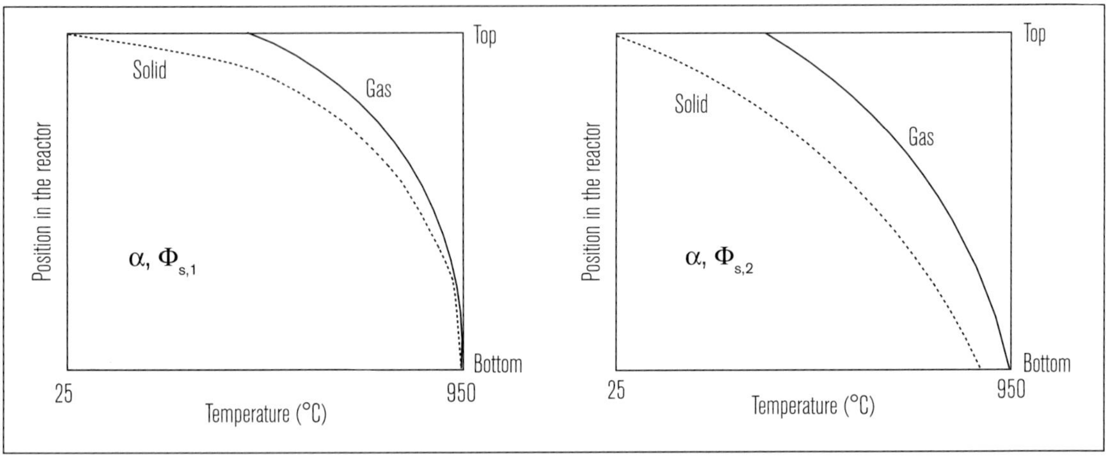

Figure 20.6 - Schematic temperature profiles for solid and gas: $\phi_{s,1} > \phi_{s2}$.

The effect of two different values of the parameter ϕ_s (or ϕ_g) and for a fixed value of α is given in Figure 19.6. From the temperature profiles, it can be noted that a higher ϕ_s value means an increased rate of heating of the solid in the region nearer to z = 0 and this results in an isothermal region near z = H.

The differential equations (20.3) and (20.6), can be resolved for boundary conditions given below to get solid and gas temperature profiles in the reactor:

$$z = 0 \qquad T_s = T_s^o$$

$$z = H \qquad T_g = T_g^o$$

The equations for calculating the temperature of the solid and gas respectively are given below in dimensionless form:

$$\frac{T_s - T_s^o}{T_g^o - T_s^o} = \frac{\exp\left[(\alpha - 1)\phi_s z\right] - 1}{\alpha \exp\left[(\alpha - 1)\phi_s H\right] - 1} \tag{20.15}$$

$$\frac{T_g - T_s^o}{T_g^o - T_s^o} = \frac{\alpha \exp\left[(\alpha - 1)\phi_s z\right] - 1}{\alpha \exp\left[(\alpha - 1)\phi_s H\right] - 1} \quad (20.16)$$

The expressions are valid for the factor α not equal to 1 (one).When it is equal to one, the dimensionless temperature terms for solid and gas are given by:

$$\frac{T_g - T_s^o}{T_g^o - T_s^o} = \frac{\phi_s z + 1}{\phi_s H + 1} \quad (20.17)$$

$$\frac{T_s - T_s^o}{T_g^o - T_s^o} = \frac{\phi_s z}{\phi_s H + 1} \quad (20.18)$$

Example

Using the expressions for thermal profile obtained above, determine the thermal profiles for heating of sinter and coke in counter current reactor of 10 m length and the hot gases have the temperature of 950°C at the bottom and solids at the top at 25°C.

Data:

Gas:

- Flow rate G_g = 0.97 kg/m^2.s;
- Specific heat C_p = 1129.7 J/kg.°C.

Coke:

- Flow rate G_s = 0.19 kg/m^2.s;
- Specific heat C_p^s = 1464.4 J/kg.°C;
- h_V = 251 W/m^3.°C.

Sinter:

- Flow rate G_s = 0.69 kg/m^2.s;
- Specific heat C_p^s = 962.3 J/Kg.°C;
- h_V = 2510 W/m^3.°C.

Solution

For calculating the temperature profile of the coke, the values of the following parameters are evaluated from the given data:

$$\alpha = \frac{G_s C_p^s}{G_g C_p^g} = 0.25$$

$$\Phi_S = \frac{h_V}{G_s C_p^s} = 0.90 (m^{-1})$$

As $\alpha \neq 1$, equation (20.15), can be used to get the results given in the following table.

Temperature Profile for counter current of heating of coke

Position z (m)	Temperature of coke (°C)
0	25.0
1	480.0
2	712.0
3	829.0
4	889.0
5	920.0
6	935.0
7	943.0
8	947.0
9	949.0
10	950.0

In the case of sinter similarly:

$$\alpha = \frac{G_s C_p^s}{G_g C_p^s} = 0.61$$

$$\phi_S = \frac{h_V}{G_s C_p^s} = 3.78 \ (m^{-1})$$

Again, $\alpha \neq 1$, we use the equation (20.15). The results are given in the table below:

Temperature Profile for counter current heating of sinter

Position z (m)	Sinter temperature (°C)
0	25.0
1	741.0
2	903.0
3	939.0
4	948.0
5	949.0
6	950.0
7	950.0
8	950.0
9	950.0
10	950.0

Thermal profiles of coke and sinter are given below in the figure. It can be seen that coke gets heated at a slower rate due to its lower volumetric heat transfer coefficient compared to that of sinter.

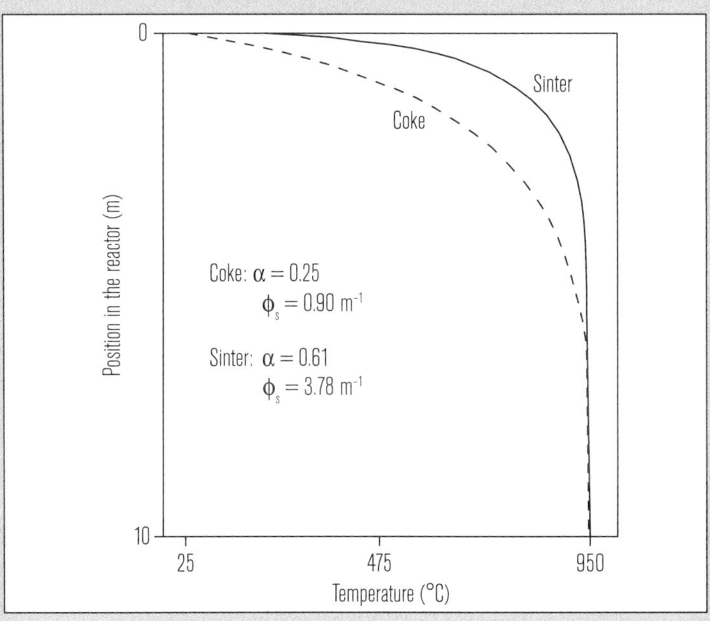

20.1.2 Volumetric heat transfer coefficient

The importance of the parameter ϕ_s in the rate of heating the solid with gas in a countercurrent gas as discussed above,is due to the volumetric heat ransfer coefficient. One can see from equation (20.19), that ϕ_s is directly proportional to the volumetric coefficient of heat transfer, h_V. Therefore, it would be interesting to analyze the effect of some parameters on the value of h_V and correlations available for determining the same.

Similar to the normal heat transfer coefficient based on area (Chapter 13), the volumetric coefficient of heat transfer also can be experimentally determined. The results in this case are usually given in the form of mathematical correlations expressing h_V as a function of variables such as gas velocity , particle diameter and gas temperature etc.

One of the commonly used correlations, for determining the volumetric heat transfer coefficient was developed by Kitaev, Yaroshenko and Suchkov (1967), is given below:

$$h_V = A_S \frac{V_o^{0.9} T_g^{0.3} M'}{d^{0.75}} \ (\mathrm{kJ / m^3 h °C}) \tag{20.19}$$

where:

- A_s = coefficient that depends on the solid material. Values of A_s given for some materials are: iron ore = 670, Sinter = 630, limestone = 695, Coke = 710;
- V_0 = Superficial velocity in the empty reactor cross section measured under standard condition: 0°C and 1 atm pressure;
- T_g = gas temperature (°C);
- d = particle diameter (m);
- M' = coefficient that is a function of bed porosity.

For arriving at this correlation Kitaev, Yaroshenko and Suchkov (1967) have reviewed a number of other expressions and performed experiments using industrial materials, like sinter,iron ore etc varying the velocity of gas, temperature, particle size etc.

The value of the volumetric coefficient of heat transfer increases as the gas velocity increases, other terms in the equation being constant. Hence increase in gas velocity would accelerate heat exchange between gas and solid in the reactor.

Another parameter that influences the value of h_V is the average diameter of the particles. From equation (20.19), it is obvious that the volumetric heat transfer coefficient increases with decrease in particle size, ie, the rate of heat exchange between gas and solid increases as the particles become smaller.

Increase in the gas temperature results in a higher value of h_V. Normally, the knowledge of average gas temperature along the reactor is used to determine the value of volumetric heat transfer coefficient h_V.

The value of parameter A_s, depends on the solid material.It can be noted that the values are practically the same for sinter, iron ore, limestone and coke.

According to Kitaev, Yaroshenko and Suchkov (1967), the parameter M' for a bed with void fraction equal to 0.2, is around 0.50. However, M' values for the beds vary with the type of material and the void fraction of the bed. The correlation suggested by Kitaev, Yaroshenko and Suchkov (1967) fits well the data provided by Heidemann (1968), as given below:

- Coke: M' = 0.425;
- Sinter: M' = 1.16.

The variation of M 'is associated with the difference in the void fraction of beds of these two types of materials.

Example

Calculate the volumetric coefficient of heat transfer to the sinter and coke under the following conditions:

- Gas flow: $Q = 22.35\ Nm^3/s$;
- Cross section of the reactor: $S_o = 29.8\ m^2$;
- Average diameter of coke: d = 0.06 m;
- Mean diameter of the sinter: d = 0.01 m.

Average gas temperature inside the reactor is 950°C.

Solution

For determining the superficial velocity of gas in standard conditions, the flow rate of gas is divided by the cross section of the reactor:

$$v_o = \frac{Q}{S_o} = \frac{22.35}{29.80} = 0.75 \text{ m/s}$$

Using the relation (20.19), one gets:

For coke:

- $A_S = 710$;
- $d = 0.06$ m;
- $M' = 0.425$.

$$h_V = A_S \frac{V_o^{0.9} T_g^{0.3} M'}{d^{0.75}} = 710 \frac{0.75^{0.9} 950^{0.30} 0.425}{0.06^{0.75}} = 15054 \text{ kJ/hm}^3\,°\text{C}$$

$$\text{or } h_V \cong 4.18 \text{ W/m}^3\,°\text{C}$$

For the sinter:

- $A_S = 630$;
- $d = 0.01$ m;
- $M' = 1.16$.

$$h_v = A_s \frac{V_o^{0.9} T_g^{0.3} M'}{d^{0.75}} = 630 \frac{0.75^{0.9} 950^{0.30} 1.16}{0.01^{0.75}} = 139000 \text{ kJ/hm}^3\,°\text{C}$$

$$\text{or hv} \cong 38.6 \text{ W / m}^3\,°\text{C}$$

The volumetric heat transfer coefficient of sinter is about nine times greater than that for coke. This difference is mostly due primarily to the different particle diameters of coke and sinter.

20.2 Application to Blast Furnace

Figure 20.7 shows a thermal profile of the blast furnace, obtained through experimental results with the use of sensors etc. This profile does not match the ones outlined in the previous examples.

This difference is due primarily to the assumptions made in the above mathematical treatment of heat transfer in beds. These are not fully valid in the case of the blast furnace, as mass transfer aspects involving chemical reactions such as reduction of iron ores by CO, gasification of coke, decomposition of lime stone etc and the heat liberated or absorbed by these reactions, have not been considered. Also the gas that comes from the softening and melting zone undergoes change in temperature and composition in the shaft region. The heat transfer rates across the cross section varies due to channeling causing different flow rates in different locations. Hence the complete model in the shaft region of the blast furnace should be atleast bidimensional considering equations of heat, mass and momentum transfer. These equations have to be solved simultaneously for getting the temperature profiles of solid, gas, fractional reduction profiles of iron oxides and concentration profiles of the gases along the length of blast furnace shaft region.

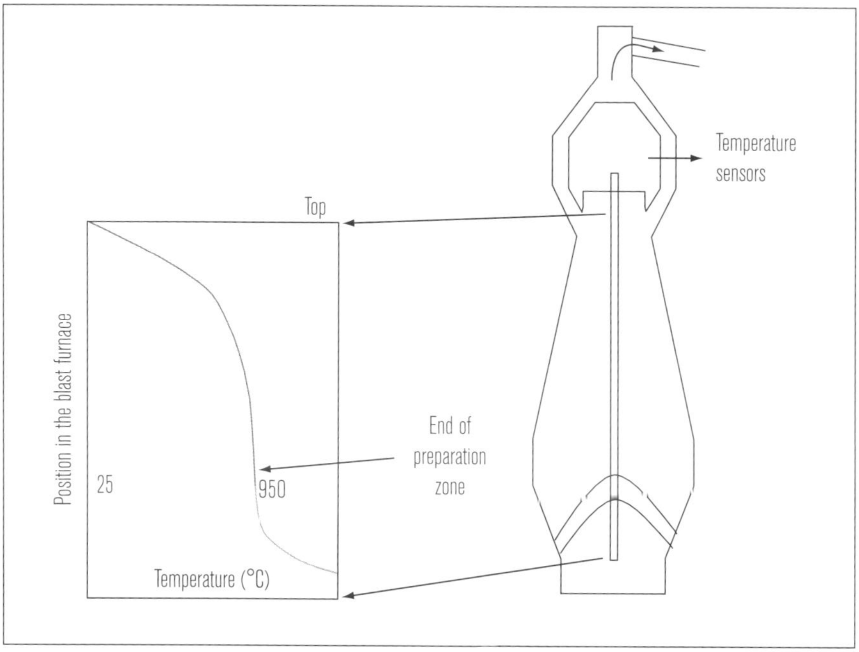

Figure 20.7 - Temperature profile in blast furnace-7, Seraing, Belgium (HEIDEMANN, 1968).

Among the chemical reactions, the gasification reaction of coke carbon , given below , is important, as it is endothermic and the heat absorption is significant in the lower shaft region of the blast furnace:

$$CO_{2_{(g)}} + C_{(s)} = 2CO_{(g)}$$

This rate of this endothermic reaction is quite high above a certain critical temperature. Hence it influences greatly the thermal profile of the blast furnace. For coke this temperature is about 950°C, which is exactly the temperature of the reserve zone or isothermal region, shown in Figure 20.7. For charcoal, this critical temperature is around 800°C.

In thermochemical studies of blast furnace process, it is usual to divide the furnace into two parts:

- Upper zone where temperatures are less than or equal to critical temperature cited above depending on the type of fuel;
- Lower zone where temperatures higher than the critical temperatures prevail.

One can now analyze the thermal profile given in the Figure 20.7, considering blast furnace with two zones or two independent reactors as above. The following comments are relevant:

- The upper region or zone of preparation has a behavior similar to that of a reactor with α values less than one, as seen in Figure 20.3;
- The lower zone of development or active chemical reactions with α value greater than than one, with a thermal profile similar to the Figure (20.5).

In the preparation zone , the main reactions that occur are the reduction of iron oxides to wustite. The enthalpy of these reactions are not significant like the gasification reaction and hence they do not have a significant effect on the thermal profile. Hence the mathematical treatment of thermal exchanges in the counter current reactor given before is applicable to this region.

$$\frac{G_s C_p^s}{G_g C_p^g} = \alpha \qquad (20.8)$$

On the other hand ,the lower high temperature zone (development zone) behaves like a reactor with α value greater than 1 (one). This can be attributed to the gasification reaction involving large absorption of heat.

Increase in the specific heat of solid, C_p^{S}, causes an increase in α value. The effect is similar to that of an endothermic reaction on the thermal profile of a reactor. The significance of the comparison is just like for an increase in the specific heat of solid, it is necessary to provide more heat to raise the temperature of the solid to the desired level. Similarly, the endothermic reaction also has to be compensated in terms of heat. In the thermal reserve zone most of the indirect reduction of iron oxides to wustite stage takes place and hence the thermal profile in this zone of preparation has a decisive influence on the process. Therefore, it is necessary to investigate the influence of various operating variables that affect this zone. With this objective, a heat transfer study with a bed of two different solids (metallic charge and solid fuel) has been given in the following.

20.2.1 Countercurrent packed beds with two different solids

As mentioned earlier the solid fuel and the metallic charge present inside the furnace, have different behaviors in relation to the thermal exchanges. However there can be a single treatment involving both these materials in respect of development of energy balances in a manner similar to the one adopted in section (20.1). With this methodology, one can obtain expressions for the calculation of the temperature profile in the zone of preparation of the blast furnace.

The energy balance is carried out in the same control volume element given in Figure (20.2).

a) Energy balance for the solid fuel:

- Rate of output of energy at z + Δ z:

$$\pi R^2 G_c C_p^c \left(T_{C|z+\Delta Z} - T_0 \right)$$

- Rate of energy input of energy in z:

$$\pi R^2 G_c C_p^c \left(T_{C|z} - T_0 \right)$$

- Rate of energy transfer from the gas to fuel:

$$\pi R^2 \Delta Z \, x_c h_v^c \left(T_g - T_c \right)$$

where:

- G_c = mass flow of fuel per unit area of the reactor (kg/m²$.s);
- C_p^c = specific heat of fuel (J/kg.°C)
- T_c = temperature of the fuel (°C);
- x_c = fractional volume of fuel in the burden;
- h_v^c = volumetric heat transfer coefficient gas to fuel (W/m³).

Combining the various terms of energy balance:

$$\pi R^2 G_c C_p^c T_s\Big|_{z+\Delta z} = \pi R^2 G_c C_p^c T_s\Big|_z + \pi R^2 \Delta z\, x_C h_p^c \left(T_g - T_c\right) \quad (20.20)$$

Dividing the above equation by the volume element $\pi R^2 \Delta z$ and using the definition of first derivative:

$$G_c C_p^c \frac{dT_c}{dz} = x_C h_v^c \left(T_g - T_c\right) \quad (20.21)$$

b) Energy balance for the metallic charge:

- Rate of energy output at z + Δz:

$$\pi R^2 G_s C_p^s \left(T_{s|z+\Delta Z} - T_0\right)$$

- Rate of input of energy at z:

$$\pi R^2 G_s C_p^s \left(T_{s|z} - T_0\right)$$

- Rate of energy transfer from gas to metallic charge:

$$\pi R^2 \Delta Z\, x_s h_v^s \left(T_g - T_s\right)$$

where:

- G_s = mass flow of metal loading per unit area of the reactor (kg/m².s);
- C_p^s = specific heat of metallic charge (J/kg.°C);
- T_s = temperature of the metallic charge (°C);
- x_s = the volume of the metallic charge in the blast furnace;
- h_v^s = volumetric heat transfer coefficient for gas to the metallic charge (W/m³.°C);

The terms in the balance equation reduce to:

$$G_s C_p^s \frac{dT_s}{dz} = x_S h_v^s \left(T_g - T_s\right) \quad (20.22)$$

c) Energy balance for the gas:

- Rated output of energy at z + Δz:

$$\pi R^2 G_g C_p^g \left(T_{g|z+\Delta Z} - T_0\right)$$

- Rate of energy input at z:

$$\pi\ R^2 G_g C_p^g \left(T_{g|z} - T_0\right)$$

- Rate of energy transfer from gas to solid:

$$\pi\ R^2 \Delta Z\ x_c h_v^c \left(T_g - T_c\right) + \pi\ R^2 \Delta Z\ x_s h_v^s \left(T_g - T_c\right)$$

Like before the balance gives:

$$G_g C_p^g \frac{dT_g}{dz} = x_c h_v^c \left(T_g - T_c\right) + x_s h_v^s \left(T_g - T_s\right) \qquad (20.23)$$

Equations (20.21) (20.22) and (20.23) can be solved simultaneously for evaluation of thermal profiles of gas and solid in zone of preparation, in the following form:

$$\frac{dT_c}{dz} = \frac{x_c h_v^c \left(T_g - T_c\right)}{G_c C_p^c} \qquad (20.24)$$

$$\frac{dT_s}{dz} = \frac{x_s h_v^s \left(T_g - T_s\right)}{G_s C_p^s} \qquad (20.25)$$

$$\frac{dT_g}{dz} = \frac{x_c h_v^c \left(T_g - T_c\right)}{G_c C_p^c} + \frac{x_s h_v^s \left(T_g - T_s\right)}{G_s C_p^s} \qquad (20.26)$$

Some of the terms in the above equations can be substituted by the following parameters:

$$\phi_c = \frac{x_c h_v^c}{G_c C_p^c} \qquad (20.27)$$

$$\phi_s = \frac{x_s h_v^s}{G_s C_p^s} \qquad (20.28)$$

$$\alpha_c = \frac{G_c C_p^c}{G_g C_p^g} \qquad (20.29)$$

$$\alpha_s = \frac{G_S C_p^s}{G_g C_p^g} \qquad (20.30)$$

The solution to the thermal profiles can then be given in terms of the above parameters. The following solution for the thermal profiles of the gas, metallic charge and the fuel resulted after mathematical solution of the equations.

$$T_g = C_1 + C_2 \exp(\lambda_2 z) + C_3 \exp(\lambda_3 z) \tag{20.31}$$

$$T_s = C_1 + C_2 \frac{\phi_s}{\phi_s + \lambda_2} \exp(\lambda_2 z) + C_3 \frac{\phi_s}{\phi_s + \lambda_3} \exp(\lambda_3 z) \tag{20.32}$$

$$T_C = C_1 + C_2 \frac{\phi_c}{\phi_c + \lambda_2} \exp(\lambda_2 z) + C_3 \frac{\phi_c}{\phi_c + \lambda_3} \exp(\lambda_3 z) \tag{20.33}$$

In the above expressions, the values of λ_2 and λ_3 are given by:

$$\lambda_2 = \frac{-k + \sqrt{k^2 - 4k'}}{2} \tag{20.34}$$

$$\lambda_3 = \frac{-k - \sqrt{k^2 - 4k'}}{2} \tag{20.35}$$

$$k = \phi_s (1 - \alpha_s) + \phi_c (1 - \alpha_c) \tag{20.36}$$

$$k' = \phi_s \phi_c (1 - \alpha_c - \alpha_s) \tag{20.37}$$

The constants C_1, C_2 and C_3 can be determined using the boundary conditions. In the case of preparation zone:

$$z = H \quad T_g = T_g^0$$

$$z = 0 \quad T_c = T_c^0$$

$$z = 0 \quad T_s = T_s^0$$

where:

- $T_c^0 T_s^0$ are charging temperatures of coke and metallic burden respectively;
- T_g^0 is the emperature of the gas at the entry of zone of preparation.

$$\Delta = \frac{\phi_c}{\phi_c + \lambda_3} \left[\frac{\phi_c}{\phi_c + \lambda_2} - \exp(\lambda_2 H) \right] + \frac{\phi_s}{\phi_s + \lambda_3} \left[\exp(\lambda_2 H) - \frac{\phi_c}{\phi_c + \lambda_2} \right] + \exp(\lambda_3 H) \left(\frac{\phi_c}{\phi_c + \lambda_2} - \frac{\phi_s}{\phi_s + \lambda_2} \right) \tag{20.38}$$

Using the boundary conditions, one can get the following relations for the calculation of these constants.

$$\Delta C_1 = \frac{\phi_c}{\phi_c+\lambda_3}\left[T_g^o\frac{\phi_s}{\phi_s+\lambda_2} - T_s^o\exp(\lambda_2 H)\right] + \frac{\phi_s}{\phi_s+\lambda_3}\left[T_c^o\exp(\lambda_2 H) - T_g^o\frac{\phi_c}{\phi_c+\lambda_2}\right] + \tag{20.39}$$

$$+\exp(\lambda_3 H)\left(T_s^o\frac{\phi_c}{\phi_c+\lambda_2} - T_c^o\frac{\phi_s}{\phi_s+\lambda_2}\right)$$

$$\Delta C_2 = \frac{\phi_c}{\phi_c+\lambda_3}\left(T_s^o - T_g^o\right) + \frac{\phi_s}{\phi_s+\lambda_3}\left(T_s^o - T_c^o\right) + \exp(\lambda_3 H)\left(T_c^o - T_s^o\right) \tag{20.40}$$

$$\Delta C_3 = \frac{\phi_s}{\phi_s+\lambda_2}\left(T_c^o - T_g^o\right) + \exp(\lambda_2 H)\left(T_s^o - T_c^o\right) + \frac{\phi_c}{\phi_c+\lambda_2}\left(T_g^o - T_s^o\right) \tag{20.41}$$

$$C_1 = \frac{\Delta C_1}{\Delta} \tag{20.42}$$

$$C_2 = \frac{\Delta C_2}{\Delta} \tag{20.43}$$

$$C_3 = \frac{\Delta C_3}{\Delta} \tag{20.44}$$

To determine the temperature profiles it is necessary to go through the following steps:

- Evaluation of the parameters ϕ_s, ϕ_c, α_s and α_c;
- Determination of λ_2 and λ_3 with equations (20.34) and (20.35);
- Calculation of C_1, C_2 and C_3 using equations (20.42) (20.43) and (20.44);
- Evaluation of temperature T_g, T_s and T_c for various values of z, using equations (20.31) (20.32) and (20.33).

All these procedures become simple and fast using a computer for the calculations.

Example

Calculate temperature profiles of gas, metallic charge and coke in the preparation zone of a blast furnace. The following date are given:

Gas:

- Flow volume: $Q = 22.35\ Nm^3/s$;
- Average density: $\rho_g = 1.3\ kg/m^3$;
- Specific heat $C_p = 1129.7\ J/kg.°C$.

Sinter:

- Mass flow rate: Q_S = 20.72 kg/s;
- Specific heat C_p^c = 962 J/kg.°C;
- Volume fraction in the burden x_s = 0.6;
- Diameter d = 0.01 m.

Coke:

- Mass flow rate: Q_c = 5.83 kg/s;
- Specific heat C_p^c = 1464 J/kg.°C;
- Volume fraction in the burden x_c = 0.4;
- Diameter d = 0.06 m.

Cross section of shaft region = S_o = 29.8 m^2

- Height H = 5 m.

The temperature of the gas is 950°C and the temperature burden (sinter and coke) is 25°C.

Solution

For the determination of volumetric heat transfer coefficient, the gas velocity is given by:

$$V_o = \frac{Q}{S_o} = \frac{22.35}{29.8} = 0.75 \text{ m/s}$$

Assuming an average gas temperature of around 900°C and using the relation (20.25), heat transfer coefficient for coke can be calculated:

$$h_v^c = A_S \frac{V_o^{0.9} T_g^{0.3} M'}{d^{0.75}} = 710 \times \frac{0.75^{0.9} \times 900^{0.3} \times 0.425}{0.06^{0.75}}$$

$$h_v^c = 14{,}436.5 \text{ kJ/hm}^3 \text{ or } h_v^C = 4.01 \text{ kW/m}^3.°C$$

Sinter:

$$h_v^s = A_S \frac{V_o^{0.9} T_g^{0.3} M'}{d^{0.75}} = 670 \times \frac{0.75^{0.9} \times 900^{0.3} \times 1.16}{0.01^{0.75}}$$

$$h_v^s = 136.762.4 \text{ kJ/hm}^3.°C \text{ or } h_v^s = 37.99 \text{ kW/m}^3.°C$$

The mass flow per unit area of gas, coke and sinter are given by:

$$G_g = \frac{Q\,\rho_g}{S_o} = \frac{22.35 \times 1.3}{29.8} = 0.975\ \text{kg/m}^2\text{s}$$

$$G_c = \frac{Q_c}{S_o} = \frac{5.83}{29.8} = 0.1956\ \text{kg/m}^2\text{s}$$

$$G_s = \frac{Q_s}{S_o} = \frac{20.72}{29.8} = 0.6953\ \text{kg/m}^2\text{s}$$

With these values, we can calculate the parameters that determine the thermal behavior of the zone of preparation:

$$\varphi_c = \frac{x_c h_v^c}{G_c C_p^c} = 5.746\ \text{m}^{-1}$$

$$\varphi_s = \frac{x_s h_v^s}{G_s C_p^s} = 34.067\ \text{m}^{-1}$$

$$\alpha_c = \frac{G_c C_p^c}{G_g C_p^g} = 0.2602$$

$$\alpha_s = \frac{G_s C_p^s}{G_g C_p^g} = 0.6075$$

Using the equations developed earlier, it is possible to evaluate:

$$k = \varphi_s(1-\alpha_s) + \varphi_c(1-\alpha_c) = 34.067(1-0.6075) + 5.746(1-0.2601) = 17.623\ \text{m}^{-1}$$

$$k' = \varphi_s\varphi_c(1-\alpha_c-\alpha_s) = 5.746 \times 34.067 \times (1-0.6075-0.2601) = 25.9172\ \text{m}^{-2}$$

$$\lambda_2 = \frac{-k+\sqrt{k^2-4k'}}{2} = \frac{-17.623+\sqrt{17.623^2-4\times 25.9172}}{2} = -1.6195\ \text{m}^{-1}$$

$$\lambda_3 = \frac{-k+\sqrt{k^2-4k'}}{2} = \frac{-17.623-\sqrt{17.623^2-4\times 25.9172}}{2} = -16.0035\ \text{m}^{-1}$$

With the aid of a computer, one can determine the values of C_1, C_2 and C_3 and the thermal profiles in the zone of preparation. The results are given in the table below:

$C_1 = 950.2143°C$; $C_2 = -704.1135°C$ and $C_3 = -9998.6024°C$

	Temperature (°C)		
Height (m)	Gas	Sinter	Coke
0.00	147.50	25.00	25.00
0.25	478.72	453.68	297.20
0.50	636.87	621.20	513.96
0.75	741.22	730.78	659.19
1.00	810.80	803.84	756.09
1.25	857.22	852.58	820.72
1.50	888.18	885.08	863.83
1.75	908.83	905.77	892.59
2.00	922.61	921.23	911.78
2.25	931.80	930.88	924.57
2.50	937.93	937.32	933.11
2.75	942.02	941.61	938.80
3.00	944.75	944.48	942.60
3.25	946.57	946.39	945.14
3.50	947.78	947.66	946.83
3.75	948.59	948.51	947.96
4.00	949.13	949.08	948.71
4.25	949.49	949.47	949.21
4.50	949.73	949.70	949.54
4.75	949.89	949.88	949.77
5.00	950.00	949.99	949.92

The above data have been plotted in the following figure showing the temperature profiles of gas, sinter and coke in the preparation zone.

The following observations in respect of thermal profiles given in the figure are relevant:

- The heating rate of solids is high for the conditions given;
- The temperature of top gas is around 150°C, close to the actual values obtained in blast furnaces;

- Sinter gets heated up faster than coke. At 0.75 m below the top of the burden, the temperature difference between gas and sinter is around 10°C, while the difference between coke and gas is around 80°C;
- There is an isothermal region, where temperatures of gas and solids are practically the same around 950°C. This region corresponds to the thermal reserve zone of coke blast furnace.

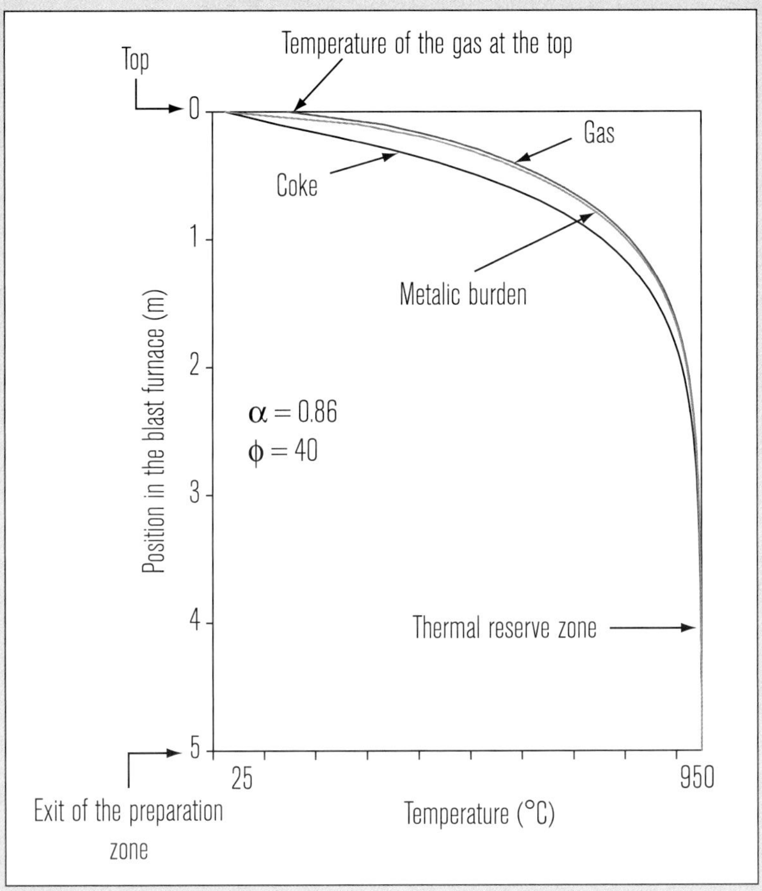

From the equations developed, one can analyze the effect of some operating variables on the thermal profile of the zone of preparation of the blast furnace.

20.2.2 Influence of operational parameters on the thermal profile of the blast furnace

The gas flow rate is an important parameter in the determination of thermal profile of the zone of of preparation of the blast furnace, since it affects the values of α_c and α_s, the volumetric coefficient of heat transfer, and the values of ϕ_c and ϕ_s.

To examine its effect, consider case of a coke oven gas with flow rates of 30 and 20 Nm^3/s. In the preparation zone, other parameters for the solid charge like coke and sinter are the same as in the previous example.

a) Gas flow rate: 30 Nm^3/s.

Using the same equation takes the form of previously, we have:

$$V_o = \frac{Q}{S_o} = \frac{30}{29.8} = 1.007 \text{ m/s}$$

For an average gas temperature exceeding 900°C, the volumetric coefficient of heat transfer is given by:

Coke:

$$h_v^c = A_S \frac{V_o^{0.9} T_g^{0.3} M'}{d^{0.75}} = 710 \times \frac{1.007^{0.9} \times 900^{0.3} \times 0.425}{0.06^{0.75}}$$

$$h_v^c = 19{,}311.3 \text{ kJ/m}^3.°\text{C or } h^C_v = 5.36 \text{ kW/m}^3.°\text{C}$$

Sinter:

$$h_v^s = A_S \frac{V_o^{0.9} T_g^{0.3} M'}{d^{0,75}} = 670 \times \frac{1.007^{0.9} \times 900^{0.3} \times 1.16}{0.01^{0.75}}$$

$$h_v^s = 178{,}295.3 \text{ kJ/m}^3.°\text{C or } h_v^s = 49.53 \text{ kW/m}^3.°\text{C}$$

The mass flow rate of gas equals:

$$G_g = \frac{Q\, \rho_g}{S_o} = \frac{30 \times 1.3}{29.8} = 1.309 \text{ kg/m}^2\text{s}$$

The parameters for the thermal behavior in the preparation zone has the following values:

$$\varphi_c = \frac{x_c h_v^c}{G_c C_p^c} = 7.49 \text{ m}^{-1}$$

$$\varphi_s = \frac{x_s h_v^s}{G_s C_p^s} = 44.4 \text{ m}^{-1}$$

$$\alpha_c = \frac{G_c C_p^c}{G_g C_p^g} = 0.194$$

$$\alpha_s = \frac{G_s C_p^s}{G_g C_p^g} = 0.453$$

With these values and with the aid of a computer programme, one can evaluate the thermal profiles. The results are shown in Table 20.1 and Figure 20.8.

Table 20.1 - Temperature profiles for gas flow of 30 Nm^3/s.

	Temperature °C		
Height (m)	Gas	Sinter	Coke
0.00	352.17	25.00	25.00
0.25	823.69	808.81	627.79
0.75	907.73	905.11	846.96
1.00	937.13	935.65	917.06
1.25	945.88	945.41	939.47
1.50	948.68	948.53	946.63
1.75	949.58	949.53	948.92
2.00	949.87	949.85	949.65
2.25	949.96	949.95	949.89
2.50	949.99	949.98	949.96
2.75	950.00	950.00	949.99
3.00	950.00	950.00	950.00
3.25	950.00	950.00	950.00
3.50	950.00	950.00	950.00
4.00	950.00	950.00	950.00
4.25	950.00	950.00	950.00
4.50	950.00	950.00	950.00
4.75	950.00	950.00	950.00
5.00	950.00	950.00	950.00

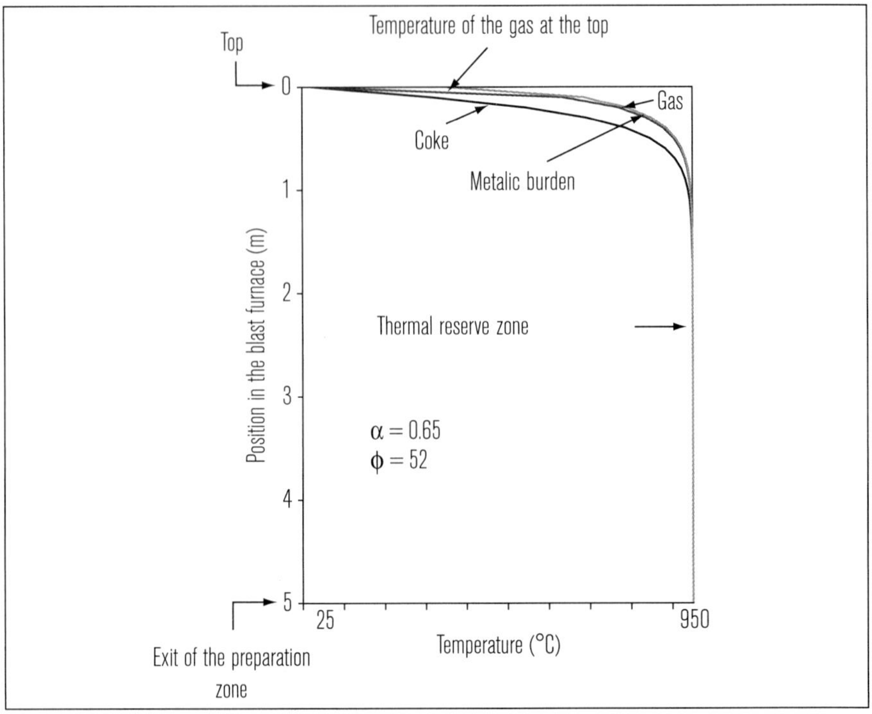

Figure 20.8 - Temperature profiles in the zone of preparation for a gas flow rate of 30 Nm³/s.

b) Gas flow rate of 20 Nm^3/s

Repeating the previous calculations, the values obtained are:

- Gas velocity: $V_o = 0.671$ m/s.

Volumetric heat transfer coefficient:

- Coke: $h^c_v = 3.72$ $kW/m^3.°C$;
- Sinter: $h^s_v = 34.37$ $kW/m^3.°C$.

Gas flow rate: $G_g = 0.873$ $kg/m^2.s$.

Parameters of heat transfer:

- $\phi_c = 5.20$ m^{-1};
- $\phi = 30.8$ m_s^{-1};
- $\alpha_c = 0.291$;
- $\alpha_s = 0.769$.

With these values, the results can be seen in Table 20.2 and Figure 20.9.

Height (m)	Temperature °C		
	Gas	Sinter	Coke
0.00	58.21	25.00	25.00
0.25	175.55	163.41	105.47
0.50	259.23	248.84	193.90
0.75	334.65	325.20	275.05
1.00	403.35	394.74	349.03
1.25	465.97	458.12	416.46
1.50	523.04	515.89	477.92
1.75	575.06	568.54	533.93
2.00	622.47	616.52	584.99
2.25	665.68	660.25	631.51
2.50	705.06	700.12	673.92
2.75	740.96	736.45	712.57
3.00	773.67	769.57	747.80
3.25	803.48	799.75	779.91
3.50	830.66	827.25	809.17
3.75	855.43	852.32	835.84
4.00	878.00	875.17	860.15
4.25	893.58	896.00	882.31
4.50	917.33	914.98	902.50
4.75	934.42	932.28	920.91
5.00	950.00	948.05	937.68

Table 20.2 - Temperature profiles in the area in preparation for a flow rate of 20 Nm^3/s.

A comparison of the results (for gas flows of 30 and 20 Nm^3/s, in addition to the example discussed earlier, for flow of 22.35 Nm^3/s) is given below:

Gas flow (Nm^3/s)	Top gas temperature(°C)
20.00	58.21
22.35	147.50
30.00	352.17

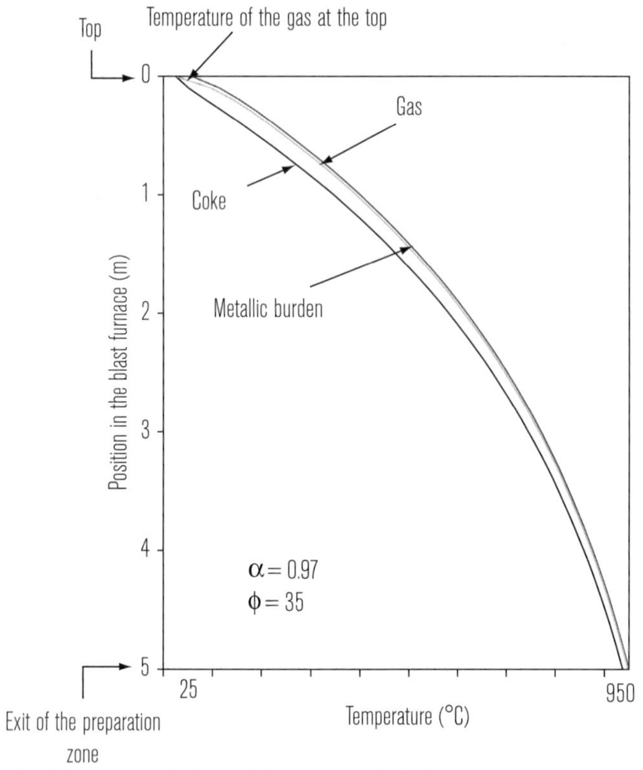

Figure 20.9 - Temperature profiles in the zone of preparation for a gas flow rate of 20 Nm^3/s.

From the above data comparing the 3 different gas flow rates ,the following observations can be made:

- As the gas flow rate increases, (other parameters kept constant), there is elevation of top gas temperature;
- When the gas flow is decreased, the rates of heating of sinter and coke are decreased. For a flow rate of 30 Nm^3/s, the isothermal region (or thermal reserve zone) forms about two meters below the level of charge. When the gas flow decreases to 20 Nm^3/s, there is no thermal reserve zone.

These results are interesting, when one remembers that in the actual blast furnace, the gas distribution is not uniform throughout its cross section. The bed porosity also varies across the cross section and the height. Hence the gas velocity varies in the radial direction of the furnace. As a result, the heating profiles of the metallic charge and coke in the zone of preparation vary according to position.

Assuming that the cylindrical portion of the zone of preparation is composed of three coaxial cylinders or zones of independent counter current reactors with different velocities in each , thus simulating qualitatively the gas flow in the shaft of the blast furnace , the following observations can be made:

- Central zone: In this region, the gas velocity is higher. The thermal behavior is similar to the case analyzed above,for the case of gas flow rate of 30 Nm^3/s.The rate of heating of metallic burden is high and the thermal reserve zone appears a few meters below the level of charge.The temperature of top gas is high;
- Peripheral zone (along the wall of the furnace): Here the gas velocity is low.The thermal profiles would be similar to the case with a flow rate of 20 Nm^3/s, the rate of heating of the solid is reduced and there is no thermal reserve zone.The temperature of top gas in this region is low;
- Intermediate zone: In this region the thermal behavior would be similar to the situation where the gas flow rate equal to 22.35 Nm^3/s. The heating rate of solids is less than that of the central cylinder but larger than the peripheral. The thermal reserve zone appears at lower levels.

Figure (20.10) shows the temperature profile of the top gas as a function of radial position of a blast furnace (central driving). These are based on temperature measurements with sensors carried out by Capelani and Schneider (1982), in the blast furnace of Fos (Solmer). The values obtained confirm these observations: the central region has higher temperatures, while in the periphery the gas temperatures are relatively lower.

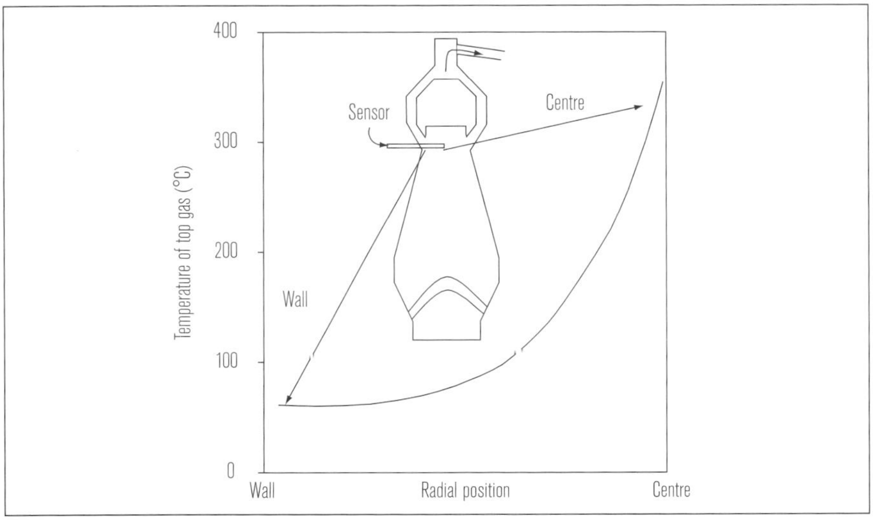

Figure 20.10 - Distribution of radial gas temperature in the top of the blast furnace Fos (Solmer) determined by Capelani and Schneider (1982).

Besides the gas flow rate, many other parameters influence the heat exchange process in the preparation zone such as like α_s, α_c, ϕ_c and ϕ_s, whose effect on thermal profiles have been discussed earlier. Some other parameters can be defined as given below:

$$\alpha_T = \frac{G_c C_p^c + G_s C_p^s}{G_g C_p^g} = \alpha_c + \alpha_s \tag{20.45}$$

$$\phi_T = \phi_c + \phi_s \tag{20.46}$$

The equation 20.45 defines the ratio of heat flow of solids (metallic charge and solid fuel) and heat flow of gas. The effect of this parameter (20.45) on the thermal profile is identical to that α discussed earlier in respect of concavity curves for thermal profile.The factor α_T is associated with the heating rate of solids. Higher its value higher, is the heating rate.

The influence of particle size of solids can also be analyzed, keeping all other parameters constant. A reduction in the average size of particles of charge and fuel does not affect the value of α_T, but increases ϕ_T, thus increasing the volumetric heat transfer coefficient. Hence the rate of heating of solid favors formation of a thermal reserve zone (in the case of $\alpha_T < 1$).

The temperature of top gas normally indicates the efficiency of thermal exchanges in the zone of preparation Hence it is important to check the effect of parameters ϕ_T and α_T on the value of this temperature.

To investigate the effect of α_T the earlier model developed for thermal exchanges was used keeping the value of ϕ_T, constant. The top gas temperatures for various values of α_T have been computed and the results are given in Figure (20.11). In the calculations It was considered that the gas enters the preparation zone at 950°C, and the burden enters the top at 25°C The height of preparation zone is 5 m. From the figure it can be observed, that the top gas temperature increases linearly with decreasing value of α_T. That is to say, that for a given thermal efficiency exchange (given by the value of ϕ_T), when more gas is used to heat a given amount of solid, higher is the top gas temperature.

Figure (20.12) also shows the effect of α_T on the temperature of top gas, obtained through experimental measurements by Itaya et al. (1982), in the blast furnace 2 of Mizushimo. Plant. One can observe the similarity between the Figure 20.11 and 20.12 in respect of influence of α_T factor.

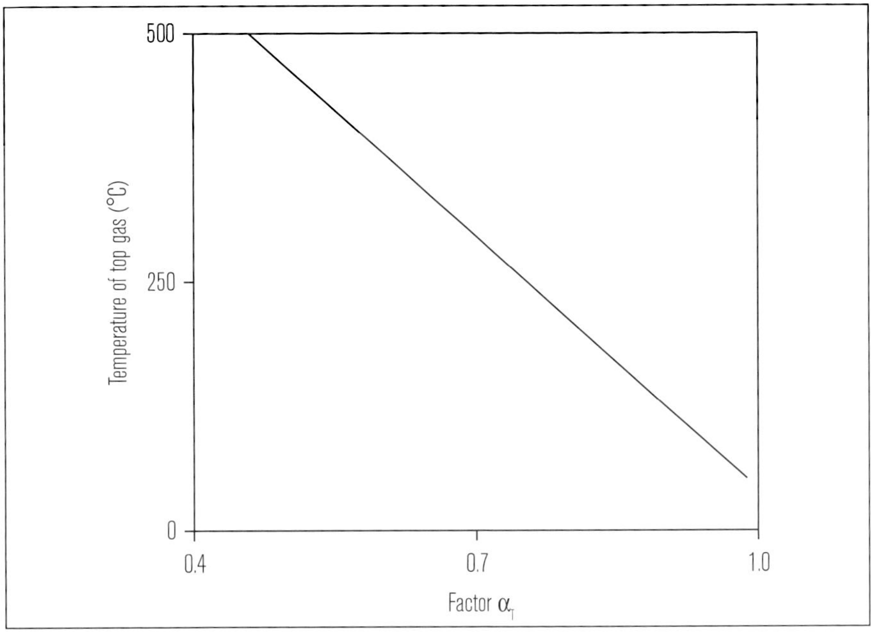

Figure 20.11 - Influence of factor α_T, on the temperature of top gas, calculated by the thermal exchange model.

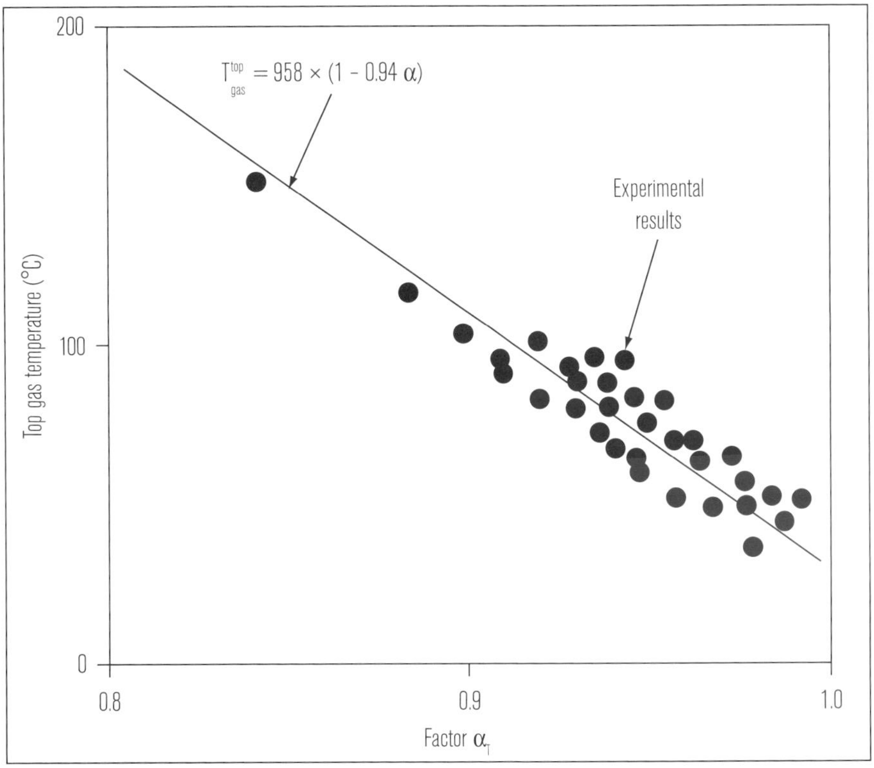

Figure 20.12 - Influence of factor α_T, on the temperature of the top.gas measured in blast furnace 2 of the Mizushima plant (ITAYA et al., 1982).

On the basis of results obtained, Itaya et al. (1982) proposed the following relationship for calculating the temperature of top gas:

$$T_g^T = 958(1 - 0.94\alpha_T) \qquad (20.47)$$

where:

- T^T_g is the temperature of top gas (°C).

Using the model of thermal exchanges, the correlation is obtained is:

$$T_g^T = 944.3(1 - 0.97\alpha_T) \qquad (20.48)$$

Some aspects of blast conditions through the tuyers affect the value of α_T factor The enrichment of oxygen in air decreases the gas volume going up and hence α_T value rises.The increase in humidity of the blast increases the volume of gas and reduces α_T. The injection of auxiliary fuel such as fuel oil, tar etc. increases the gas volume and decreases the flow rate of solid fuel causing a reduction in the value of α_T.

The influence of the value of ϕ_T on top gas temperature has been carried out the same methodology of thermal exchange profile of gas-solid system for the zone of preparation for the following values of variables and parameters.

α_T factor = 0.85, T^g_o = 950°C, $T_c^o = T_s^o$ = 25°C, H = 5 m

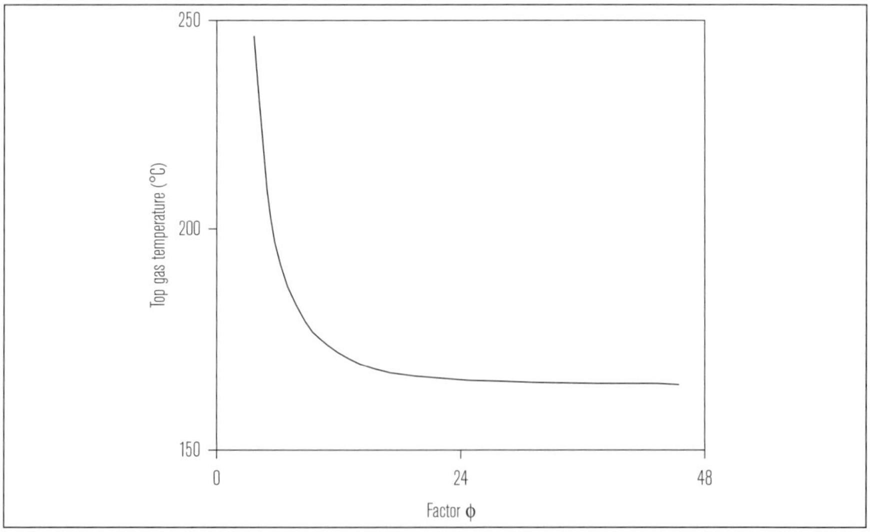

Figure 20.13 - Influence factor ϕ_T on top gas temperature using the model of heat exchange.

The results of variations of ϕ_T through changes in proportionally values of ϕ_s and ϕ_c such that $\phi_s/\phi_c = 5$. The results are shown in Figure (20.13). It can be seen that the value of ϕ_T affects the temperature of the top gas upto a certain point (around $\phi_T = 24$), beyond which the variations in ϕ_T do not cause further changes in the temperature of the top gas. This point varies depending on the value of factor α_T.

This fact can be explained. ϕ_T measures actually the efficiency of heat exchange. Higher the ϕ_T, more efficient is the heat exchange. For low values of ϕ_T heat exchange between the gas and the burden takes place at a slow rate. Hence when the gas passes through the zone of preparation, heat transfer is not efficient As a result, top gas has a higher temperature. With increased values of ϕ_T heat exchange is faster and becomes more efficient. Hence the gas leaving the top has a lower temperature and the burden leaving the preparation zone is at higher temperature relatively.

From a certain value of ϕ_T, solids in the lower region of the preparation zone reach thermal equilibrium with gas (solid and gas at the same temperature), implying the presence of thermal reserve zone. Thereafter, increasing the value of ϕ_T does not change the temperature of top gas, but shifts the position of the thermal reserve zone nearer to the the top of furnace that is to say to higher levels.

Hence, the value of φ_T for a given value of α_T, (less than one), can be used as an indicator for the presence of the thermal reserve zone in the blast furnace. However, it must be remembered that the factor φ_T, given by equation (20.46) denotes an average over the entire cross section of the blast furnace. As in the blast furnace the gas flow is not uniform, the values of φ_T and α_T, may vary from point to point. There can be regions where the thermal reserve zone is present a few meters below the top and there may be others where there are is no thermal reserve zones.

Finally, it is important to mention that the γ factor, defined as the ratio between the calculated temperature of the top gas (assuming perfect thermal exchanges in the zone of preparation) and the measured value, is also an indicator of the efficiency of heat exchange.The value of the temperature of top gas, calculated assuming that the gas and solids reach the thermal equilibrium at the exit of the zone of preparation, represents also the minimum temperature of the top gas. If the measured value exceeds the calculated value the indication is that that somewhere in the furnace, the calculated value, the indication is that that somewhere in the furnace, at the exit of the preparation zone, heat exchange is not perfect and hence the solids do not reach the same temperature as the gas.

20.3 Thermal Profile in Relation to Reduction Reactions in the Preparation Zone

The extent of the reduction reactions of iron oxides in the preparation zone, has been estimated in terms of ω factor (NICOLLE et al., 1980) This factor has a significant influence in determining the coke consumption in the blast furnace. A number of factors affect the efficiency of reduction reactions, the important being the thermal profile in the area of preparation zone and its and length. Both these two parameters are influenced by α_T, and ϕ_T.

20.3.1 Effect of thermal profile

Temperature affects the rates of reduction reactions of iron oxides. The rate constant is a function of temperature. The driving force for the reaction, is the difference between the concentrations of gas in the furnace and the gas concentration in equilibrium with the oxide being reduced.The rate constant increases with increasing temperature. However the driving force varies with the type of reduction reaction

- Reduction of magnetite to wustite: For a fixed concentration of gas, the driving force of the reaction increases with increasing temperature;
- Reduction of wustite to iron: For a given gas concentration, increasing temperature lowers the driving force for this reduction step.

Thus, the temperature profile in the preparation zone is crucial in respect of chemical reactions of reduction.

In the case where the heating of the metallic burden is fast, the reserve zone is formed just below the level of charge, and the temperature levels in the zone of preparation becomes high.In this situation, the reduction of iron oxides to the wustite is favored. However, the production of metallic iron becomes difficult, because the driving force gets diminished for the occurrence of this reaction.

When the heating of the metallic charge is slower, the tendency for the disappearance of the thermal reserve zone increases and the temperature levels in the zone of preparation are lower. In this case, the reduction to wustite stage is more difficult, than reduction of the same to metallic iron.

If the metallic burden has poor reducibility, it is difficult to produce metallic iron in the preparation zone. However in this case it is desirable to have rapid heating of the burden with the formation of thermal reserve zone just below the level of charge, to be able to achieve maximum reduction to the wustite stage.

When the metallic burden has high reducibility, reduction to metallic iron is possible in the preparation zone.

Hence there should be a optimum thermal profile with not too high temperatures so as not to compromise the driving force for the production of iron and not too low, so as not to diminish the the value of rate constant for the reduction reaction.

In the Figure (20.14), obtained by Nicolle et al. (1980), shows the factor ω as a function of α_T factor for metallic burdens of different reducibilities. φ_T values for the curves are not given Values of α_T, alone cannot determine the thermal profile in the preparation zone. When this value diminishes, gas flow increases , (as implied in the figure,) and it also increases the value of φ_T. Figure (20.14) shows that as α_T, values get less, values of φ_T increase, resulting in the appearance of the thermal reserve zone near to the top. On the contrary, increase in the values of α_T, leads to lower values of φ_T resulting in a trend of diminishing or disappearance of the thermal reserve zone (see Figures 20.8. and 20.9).

Figure (20.14) also highlights the role of the reducibility of the metallic burden. If the reducibility is high, there is an optimum value of α_T (and ϕ_T) that maximizes the production of metallic iron, including the possibility of obtaining ω less than zero. On the right side of this point, the thermal levels are low leading to lower values of rate constants leading to lower reduction. On the left side of this point, temperature levels are higher and the driving force for the production of metallic iron is reduced. The figure also implies that that a highly reducible metallic charge reducible is very important to obtain low values of ω factor.

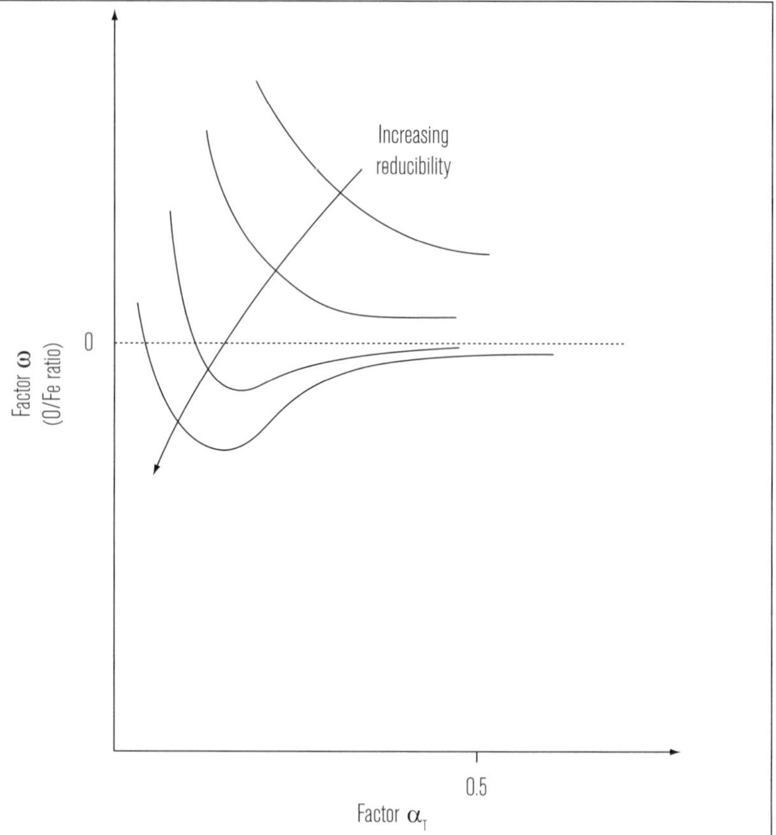

Figure 20.14 - Effect of factor α_T on the thermal profile and deviation from ideality factor (ω), in respect of reduction according Nicolle et al. (1980).

20.3.2 Effect of the length of the preparation zone

In addition to the thermal profile, the length of the zone of preparation, affects the residence time of the metallic charge in this region, influences the degree of reduction of iron oxides and, consequently, the factor ω.

The length of the preparation zone depends on the level of the softening and melting zone in the lower part of the furnace. If this is shifted up to higher levels, then the length of the preparation zone gets diminished. Figure (20.15) shows the effect of α_T on the position of softening and melting. zone. It appears that an increase in the value of α_T through a reduction in gas flow, reduces the length of this zone.That is to say that increase in values of α_T causes an increase in the length of preparation zone and increase the residence time of metallic burden in this zone to get a high degree of reduction. Hence it is possible to have appropriate values α_T (and ϕ_T) to maximize the extent of the reduction reactions of the oxides of iron in the zone of preparation.by having the right thermal profile and length of this zone.

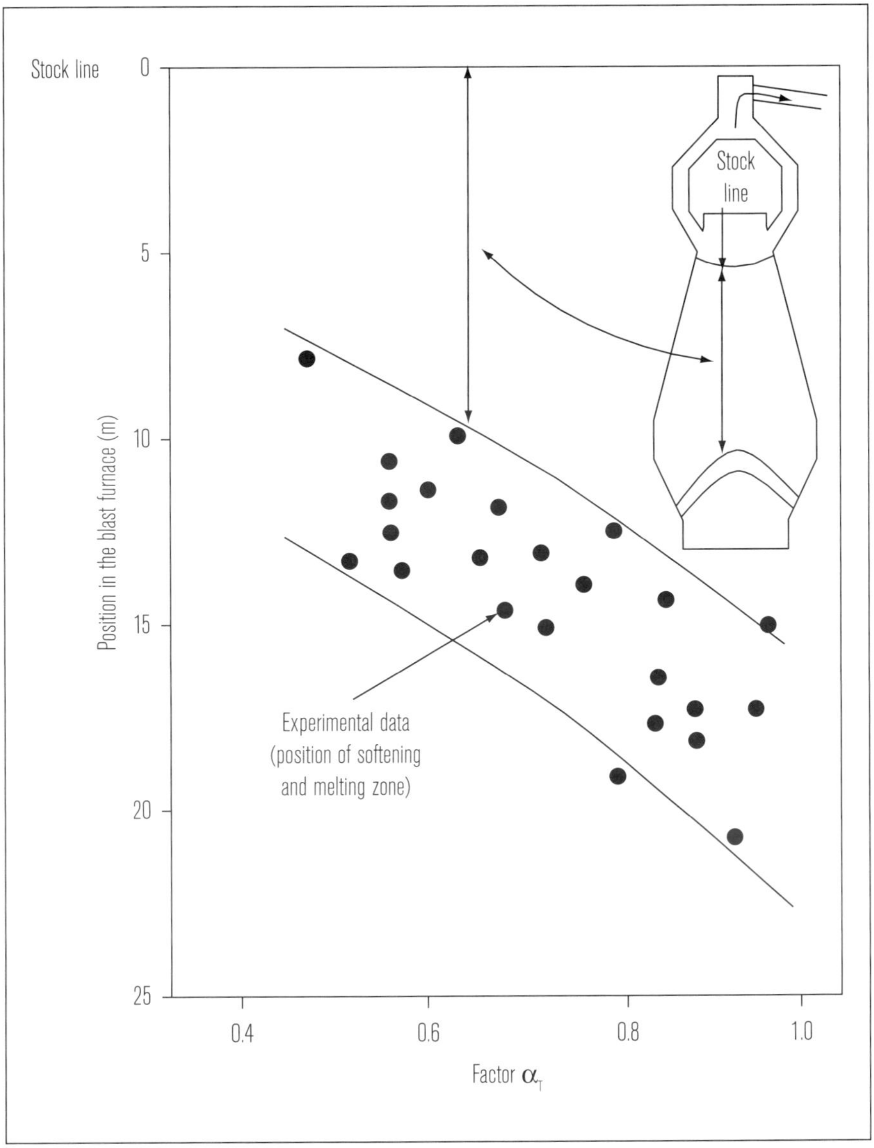

Figure 20.15 - Effect of factor α_T on the level of the zone of softening and melting, according to Ohno, Kondo and Fukushima (1983).

Figure (20.16), obtained by Kajikawa et al. (1983), shows the effect of factor α_T on the degree of utilization of gas. The following observations can be made:

- there is a optimum value of α_T characteristic for maximum utilization of reducing gas for obtaining high grade of reduction and this value varies with the reducibility of metallic burden;
- the reducibility of metallic charge has an important effect on the degree of utilization of the reducing gas in the zone of preparation;
- thermal profile in the area of preparation (for each point in the figure) is not well defined, that is corresponding ϕ_T values are not given.

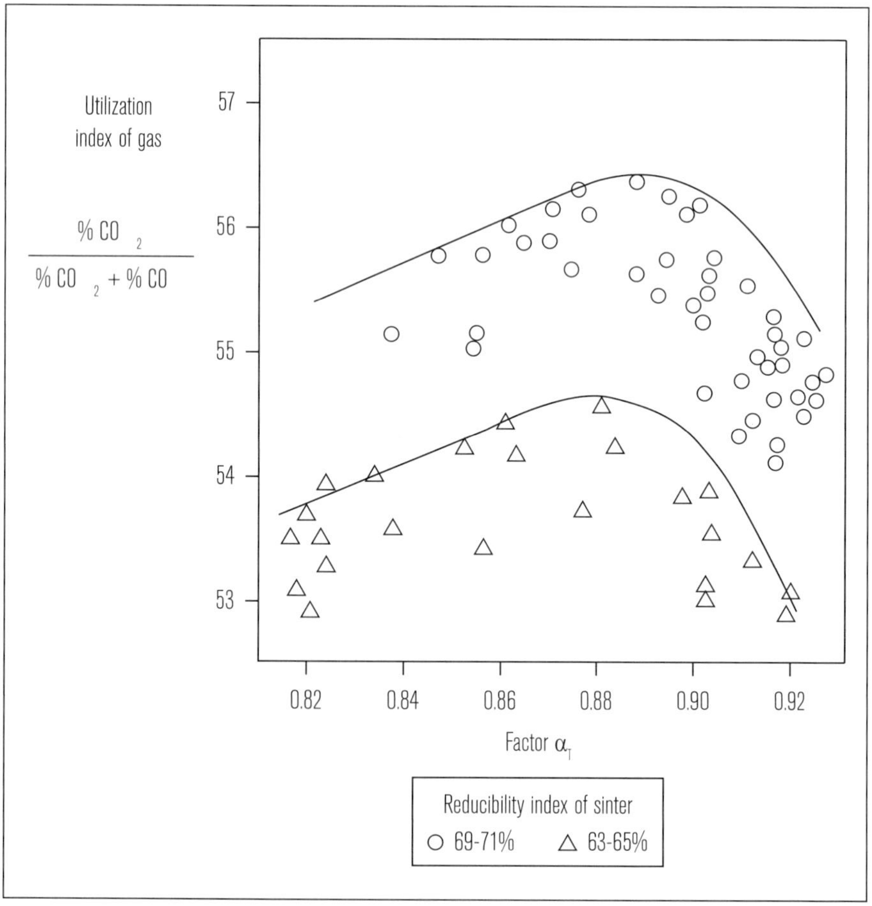

Figure 20.16 - Effect of α_T on the degree of utilization of gas in the blast furnace of Fukuyama 3 (KAJIKAWA et al., 1983).

In a blast furnace with no uniform gas flow the values of α_T vary over the cross section. Hence the extent of reduction a in the preparation zone varies from point to point. As a result, the value of the factor ω will be different for points located along the wall and the center Actually ω can be calculated by a thermo-chemical model and it would represent an average value for the degree of oxidation of metallic burden leaving the zone of preparation.

Finally, some comments can be made on the operation of blast furnace for low fuel consumption.

As coke consumption reduces, there is a reduction in the volume of gas flowing through the furnace per unit mass of iron produced, and this is reflected in the thermal profile of the zone of preparation.

In terms of the parameters that influence the thermal exchanges in the blast furnace, α_T (tending to 1) causes a decrease of ϕ_T. Therefore, these two factors would act resulting in the disappearance of the thermal reserve zone and lowering of temperature levels in the zone preparation. The resulting thermal profiles would be similar to those shown in Figure (20.9)

As shown in Figure (20.14), for these types of thermal profiles, obtaining low values of the factor ω and carbon consumption is associated with the use of a metallic burden with high reducibility. so that there is efficient reduction in the preparation zone with low thermal levels (relatively low temperature).

This fact highlights the importance of the reducibility of the metallic burden in the reduction of coke consumption in blast furnaces.

It is important to remember that an increase in the value of α_T causes a reduction in the level of the zone of softening and melting. In order to maintain the permeability of the furnace, it is necessary to use coke and metallic burden with high resistance to degradation (during reduction) so that fines produced do not have adverse influence on the permeability ensuring efficient gas solid contact. This is essential both for heat transfer and for reduction of oxides.

References

AUSTIN, P.R.; NOGAMI, H.; YAGI, J.I. A mathematical model of four phase motion and heat transfer in the blast furnace. *ISIJ International*, v. 37, n. 5, p. 458-67, 1997.

CAPELANI, R.; SCHNEIDER, M. Moyens de mesures et investigacion particulières pour un meilleur réglage du haut-forneau. *Rèvue de Métallurgie* - CIT, n. 3, p. 201-13, mars 1982.

CASTRO, J.A.; NOGAMI, H.; YAGI, J.I. Numerical investigation of simultaneous injection of pulverized coal and natural gas injection with oxygen enrichment to the blast furnace. *ISIJ International*, v. 42, n. 11, p. 102-11, 2002.

HEIDEMANN, M. C*ontribution a l'etude des échanges de chaleur entre un gaz et un lit granule heterogene:* application a la cuve de haut-forneau. Nancy: Institut de Recherches de la Sidérurgie Française, 1968. 88 p.

ITAYA, H. et al. Modèle mathématique simplifiée pour l'évaluation de la répartition des températures dans un haut-fourneau. *Rèvue de Métallurgie - CIT*, v. 79, p. 443-50, mai 1982.

KAJIKAWA, S. et al. Low fuel operation at Fukuyama no. 3 blast furnace, NKK. *Iron & Steelmaker*, v. 10, n. 9, p. 34-43, Sept. 1983.

KITAEV, B.I.; YAROSHENKO, Y.G.; SUCHKOV, V.D. *Heat exchange in shaft furnaces*. Oxford: Pergamon, 1967. 301 p.

LAHIRI A.K.; SESHADRI, V. An analysis of blast furnace process based on dynamic model. *Journal of the Iron and Steel Institute*, p. 293-307, Mar. 1969.

MIWA, E.K. *Modelo global do alto-forno baseado nos princípios dos fenômenos de transporte.* 1991. 302 p.Tese (Doutorado em Engenharia Metalúrgica e de Minas) – Escola de Engenharia da Universidade Federal de Minas Gerais, 1991.

NICOLLE, R. et al. La reduction dasn le haut-fourneau et la recherché des tres faibles au mille de coke. *Rèvue de Métallurgie* - CIT, n. 10, p. 769-79, oct. 1980.

OHNO, Y.; KONDO, K.; FUKUSHIMA, T. Modèle mathématique de la zone de ramollissement-fusion du haut-fourneau et son application à l'éxploitation. *Rèvue de Métallurgie - CIT*, n. 10, p. 809-25, oct. 1983.

OMORI, Y. *Blast furnace:* mathematical models. London: Elsevier Applied Science, 1987.

TAKATANI, K.; INADA, T.; IJISAWA, Y. Three-dimensional dynamic simulator for blast furnace. *ISIJ International*, v. 39, n. 1, p. 15-22, 1999.

Exercises

1 - Calculate the volumetric heat transfer coefficient of coke and sinter under the following conditions:

- flow of dry air: 3863.5 Nm^3/min;
- nitrogen in top gas: 55%;
- average diameter of the blast furnace: 11.0 m;
- average temperature of the gas in the shaft region 900°C;
- sinter diameter: 18 mm;
- coke diameter: 52 mm.

2 - Calculate the heat-flow of gas, sinter and coke for the following conditions:

- monthly production of pig iron: 140 tons;
- hours of operation in a month: 676.5 h;
- sinter consumption: 1650 kg/ton of iron;
- coke consumption: 450 kg/ton of iron;
- flow of dry air: 3863.5 Nm^3/min;
- analysis of the top gas: N_2 = 55%, CO = 22%, CO_2 = 21%, H_2 = 2%.

3 - Estimate the volume fraction of the blast furnace shaft occupied by sinter from the data provided below:

- sinter consumption: 1650 kg/ton of iron;
- coke consumption: 450 kg/ton of iron;
- bulk density of sinter: 1500 kg/m^3;
- bulk density of coke, 480 kg/m^3.

4 - Using the results of the previous problems, calculate the parameters of thermal exchanges: α_c, α_s, α_T, ϕ_c, ϕ_s, ϕ_T.

5 - With the values of parameters in the problem 4, calculate the temperature profiles of gas, coke and sinter along the zone of preparation of the blast furnace. Consider also the following values:

- height of the zone of preparation: 15 m;
- gas temperature at the entrance of the preparation: zone T_g = 950°C;
- temperatures of sinter and coke burden: $T_s = T_c$ = 25°C.

Some Applications in Process Modeling

Chapter 21

In the earlier chapters, mostly fundamentals of transport phenomena and some applications in metallurgical engineering, have been discussed. In this chapter, as in case of Chapter 20 on blast furnace, application of these principles in steel making, refining and physical modeling of metallurgical processes are dealt with.

21.1 Physical and Mathematical Modeling

Consider mixing of a liquid metal, due to bubbling of inert gas through the tuyere or porous plug located at the bottom of the ladle. In such cases the time required to eliminate the gradients in respect of composition and temperature can be determined as a function of variables related to the characteristics of gas and liquid metal such as density, viscosity, flow rate of gas, volume of metal in the ladle as well as other important aspects such as the geometry of injection devices (type, number and position of injectors) and that of the ladle as shown in Figure 21.1.

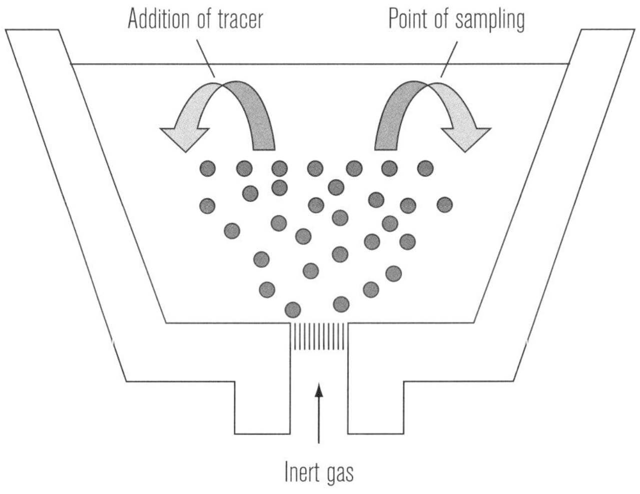

Figure 21.1 - Inert gas injection for mixing in ladle.

A normally adopted procedure is the addition of a tracer (by definition a substance unique and distinguishable from the liquid metal, and when added to it, does not affect the distribution of the parameter that is to be evaluated) in the form of a pulse to the metal all at once at a single time. The metal can be sampled in another position during time intervals to evaluate the tracer concentration ias a function of time, resulting in a curve, as shown in Figure 21.2. Here, C and Co are the instantaneous and infinite time concentrations of added tracer respectively.

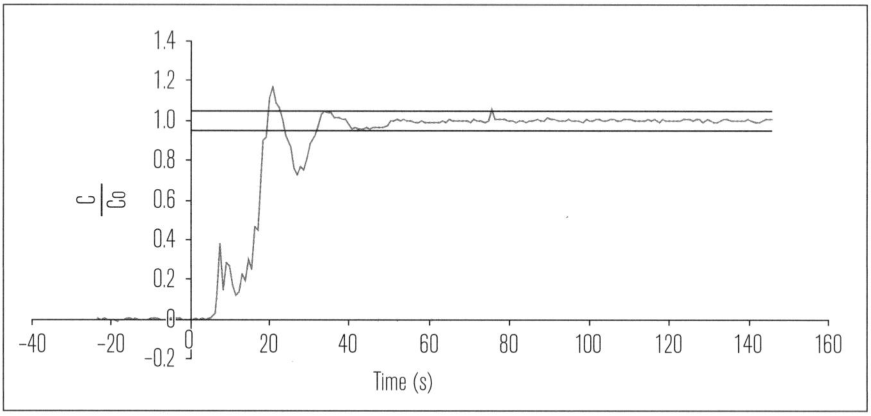

Figure 21.2 - Typical variation of concentration of tracer during mixing.

The nominal concentration Co indicated in the graph is achieved in infinite time. Hence a convention is adopted for determining the mixing time τ, by setting $\tau_{0.95}$ that is the time required for the concentration to reach 95% of the nominal concentration, based on the amount of tracer added initially. This also takes care of fluctuations in composition to stabilize within the range of 5% error from the nominal concentration.

This type of experimental technique given above for assessing mixing time or time for homogenization of the bath and the influence of various parameters on the same is difficult to carry out during the regular plant practice because of the following factors.

- Interruption in routine operations in the plant.
- Decreased productivity because of this interruption and losses as a result of the same and risk of accidents etc in any procedural change.
- Introduction of new materials like tracers etc. (or radioactive isotopes) is risky and involve wastage of materials and man hours. Hence in general the industrial practices are seldom interfered with or interrupted to avoid losses and compromise on quality of the end product.

The above reasons are mainly responsible for the development of physical modeling techniques, whereby an industrial process is simulated in the laboratory with the objective of studying the influence of different variables on the process parameters and arriving at optimum values for maximum productivity and quality with minimum cost of experimentation. These models in general are built to scale, adopting similarity principles between the model and prototype and using materials at ambient temperatures like water in the place of the liquid metal, oil or kerosene to simulate slag and gases for multiphase simulations.

For example the experiment for evaluating the mixing time for a liquid metal in a ladle given above can be simulated, through the addition of KCl as a tracer to water (instead of liquid metal) in a ladle built to a scale. Addition of this tracer can be done at a convenient location and the conductivity of the solution can be monitored at the other location since the presence of K and Cl ions change the conductivity of the medium. Conductivity is a function of concentration and can be monitored by sensors placed at various points. The ladle models are normally constructed of transparent acrylic material to facilitate viewing of interaction between gas phase and the liquid phases (bubbling, the way bubbles interact) during agitation of the bath. Similar concentration profiles with time are obtained in this case also like that shown in Figure 21.2. The effect of variables such as rate of injection of gas, bubble diameter, type and position of nozzles etc on mixing time can be studied and the results can be carried over to the prototype. This practically eliminates experimentation in the industrial reactor and associated problems and disadvantages.

Szekelly, Carlsson and Helle (1989) investigated the mixing phenomena especially in processes with gas injection generating bubbles in the melt. According to them the driving force for mixing should be mainly due to the energy dissipated because of expansion of the bubbles during their rise inside the melt and the contribution of kinetic energy of gas being injected in the melt is relatively insignificant. This energy dissipation is equivalent to the work done during expansion of the bubble, which is subjected to successively lower pressures, from the point of injection to the surface of the bath progressively. Considering isothermal system and ideal gas behavior, the work done is given by:

$$dW = PdV = P\left(\frac{-nRT}{P^2}dP\right) = -nRT\frac{dP}{P} \qquad (21.1)$$

where:

- W = work due to gas expansion, (J);
- P = pressure acting on the bubble (Pa);
- T = bath temperature (K);
- n = number of moles of gas in the bubble;
- R = universal gas constant (J/K.mol).

Integrating the above equation considering pressures at the injection point and the surface of the ladle:

$$W = -nRT\ln\left(\frac{P_A}{P_A + \rho_L gL}\right) = nRT\ln\left(1 + \frac{\rho_L gL}{P_A}\right) \qquad (21.2)$$

where:

- $P_{A,}$ = the pressure at the surface of the bath (Pa);
- ρ_L = the density of the bath, (kg/m³);
- L = the depth of injection, (m);
- g = the acceleration of gravity, (m/s²);
- $P_A + \rho_L gL$ = the pressure at the point of injection, (Pa).

The work of expansion leads to the energy dissipated into the liquid (power) and this is the energy that is important in mixing:

$$\dot{W} = \dot{n}RT\ln\left(1 + \frac{\rho_L gL}{P_A}\right) \qquad (21.3)$$

where:

- $\dot{W}$ = the energy or power dissipated inside the melt;
- $\dot{n}$ = the gas flow rate (moles/s).

At this point it is possible to compare the energy dissipated as per equation 21.3 considered to be the primary contributor for the energy of the mixing process in the liquid metal based on typical industrial values, with kinetic energy of the gas leaving the nozzle ($1/2v^2$), namely: $\frac{1}{2} Q(m^3/s)\, \rho(kg/m^3)\, [Q(m^3/s)/\pi Ro^2(m^2)]^2$.

In this expression Q and ρ are respectively gas flow rate and density of gas and Ro represents the radius of the nozzle. This comparison is given as an exercise at the end of this chapter.

If the mass of material getting agitated in the ladle is M(kg), the rate of energy dissipation ε is given by:

$$\varepsilon = \frac{\dot{n}RT}{M_L}\ln\left(1+\frac{\rho_L gL}{P_A}\right) \qquad (21.4)$$

Figure 21.3 shows a comparison of results from industrial data and physical modeling, which suggests a relationship between mixing time and rate of energy dissipation given below:

$$\tau = a.\varepsilon^{-b}$$

τ; a; b are mixing time and characteristic constants (functions of geometry and configuration of vessel), respectively.

Szekely, Carlsson and Helle (1989) considered that the above relationship is valid for all types of reactors some of which are given in Figure 21.3. However it is more prudent that a relationship like this has to be determined experimentally for a specific class of reactor. In general the values of "a" and "b" are to be taken as functions of geometry and number of nozzles geometry of vessel, etc.

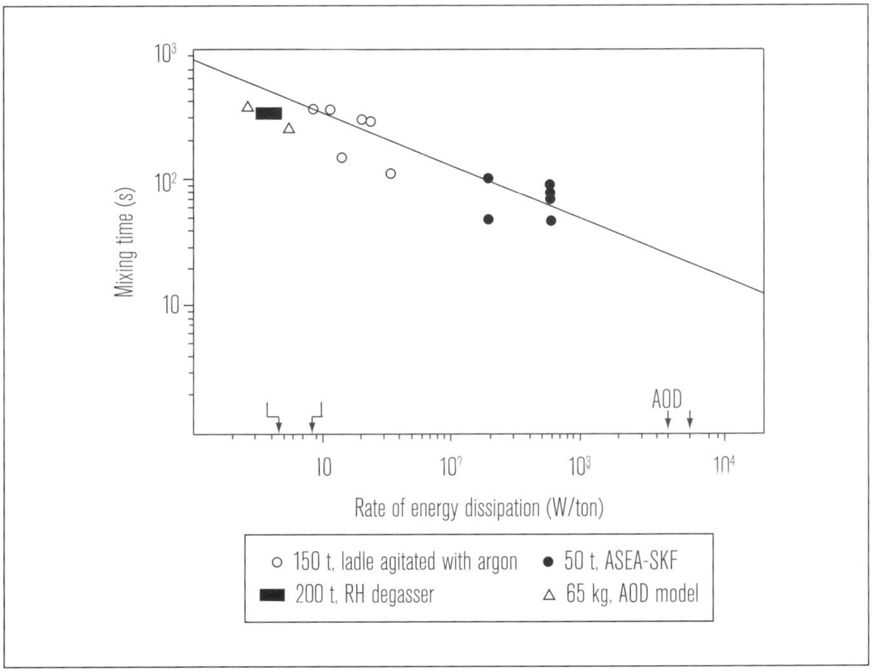

Figure 21. 3 - Relation between mixing time and dissipation energy for different type of reactors (SZEKELY, CARLSSON and HELLE, 1989).

The gas from the injection system may not have sufficient time to reach thermal equilibrium with the metal bath due low residence time in the bath. In this case the gas injected undergoes expansion and this energy should be added as a contribution due to expansion by isobaric heating effect near the injector element, ie $\dot{W} = \dot{n}\ R(T-T_{inj})$ where T_{inj} is the temperature of injection. Then the expression for dissipation of energy becomes:

$$\varepsilon = \frac{\dot{n}\,RT}{M_L}\left[\left(1-\frac{T_I}{T}\right)+\ln\left(1+\frac{\rho_L gL}{P_A}\right)\right]\ (W/Kg) \qquad (21.5)$$

21.2 Criteria of Similarity

According to general theory of models the results obtained in the laboratory model of an industrial unit are applicable to the industrial practice when there is complete similarity between the model (operated in the laboratory) and prototype (industrial reactor). That is to say all similarity criteria such as geometric, dynamic as well as mechanical, thermal and chemical are satisfied. In such cases the correlation between dimensionless numbers describing the above similarity principles in the model can be transported to the prototype and important parameters for optimization of the process can be determined.

For example, consider fluid motion within a reactor, as depicted in Figure 21.4.

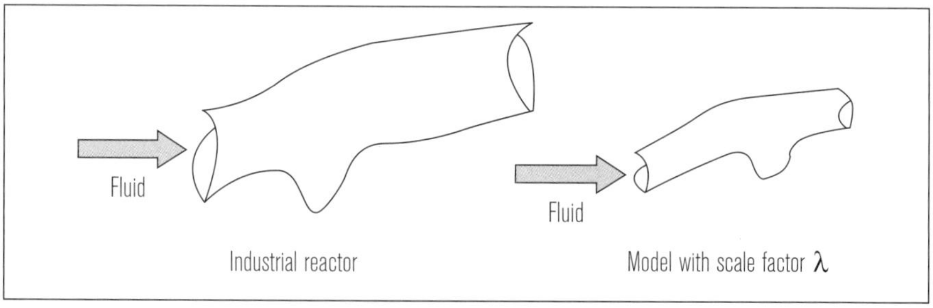

Figure 21.4 - Schematic diagram of industrial reactor and laboratory model built adopting a geometric scale factor.

For the study of the flow a model can be built in the laboratory using a geometric scale factor λ:

$$\lambda = \frac{\text{Characteristc length in the model}}{\text{Characteristc length in the prototype}}$$

The fluid flow characteristics can be analysed for determining other similarity criteria.

Assuming that the fluid is Newtonian and incompressible, the following equations are applicable to both model and prototype.

a) Equation of continuity, $\nabla \cdot V = 0$ (21.6)

b) Navier-Stokes Equation, $\rho \frac{DV}{Dt} = -\nabla P + \eta \nabla^2 V + \rho g$ (21.7)

where:

- $\nabla = \frac{\partial}{\partial x}\vec{i} + \frac{\partial}{\partial y}\vec{j} + \frac{\partial}{\partial z}\vec{k}$, is the gradient operator;
- $\nabla^2 = \frac{\partial^2}{\partial x^2} + \frac{\partial^2}{\partial y^2} + \frac{\partial^2}{\partial z^2}$, is the Laplace operator;
- $\frac{D}{Dt} = \frac{\partial}{\partial t} + \frac{\partial}{\partial x}V_x + \frac{\partial}{\partial y}V_y + \frac{\partial}{\partial z}V_z$, is the substantive operator derived over the movement or Stokes operator.

One can define the following dimensionless variables in terms of characteristic values of velocity (u), length (L), pressure (P_0) and substitute in the above equations:

$$x^* = \frac{x}{L} \;;\; y^* = \frac{y}{L} \;;\; z^* = \frac{z}{L} \;;\; V^* = \frac{V_i}{V} \;;\; P^* = \frac{P - P_0}{\rho V^2} \;;\; t^* = \frac{t}{\left(\frac{L}{V}\right)} \tag{21.8}$$

$$\nabla^* \equiv L\nabla \equiv L\left(\frac{\partial}{\partial x}\vec{i} + \frac{\partial}{\partial y}\vec{j} + \frac{\partial}{\partial z}\vec{k}\right) \equiv \left(\frac{\partial}{\partial x^*}\vec{i} + \frac{\partial}{\partial y^*}\vec{j} + \frac{\partial}{\partial z^*}\vec{k}\right) \tag{21.9}$$

$$\nabla^{*2} = L^2\nabla^2 = \frac{\partial^2}{\partial x^{*2}} + \frac{\partial^2}{\partial y^{*2}} + \frac{\partial^2}{\partial z^{*2}} \tag{21.10}$$

$$\frac{\partial}{\partial t^*} = \left(\frac{L}{V}\right)\frac{\partial}{\partial t} \tag{21.11}$$

On this basis the equations of continuity and Navier Stokes, can be written as:

a) Equation of continuity

$$\nabla^* V^* = 0 \tag{21.12}$$

b) Navier Stokes equation:

$$\frac{DV^*}{Dt^*} = -\nabla^* P^* + \frac{1}{\left(\frac{LV\rho}{\eta}\right)} \times \nabla^{*2} V^* + \frac{1}{\left(\frac{V^2}{Lg}\right)} \tag{21.13}$$

These equations are valid for both the model and the prototype and represent a family of equations describing the velocity fields. A specific equation of this group may be needed for solving a particular problem for which initial conditions and boundary conditions are known. The solutions for the model and the industrial unit (prototype) are the same if the coefficients of the terms of the differential equations are the same. Hence the velocity fields obtained solving the above equations in the case of the model should be applicable to the prototype if:

$$\left(Re = \frac{LV\rho}{\eta}\right)_{Protype} = \left(Re = \frac{LV\rho}{\eta}\right)_{Model}$$

$$\left(F_r = \frac{V^2}{Lg}\right)_{Prototype} = \left(F_r = \frac{V^2}{Lg}\right)_{Model} \tag{21.14}$$

A similar procedure can be adopted in heat and mass transfer considering the respective basic equations of heat balance and mass balance in the case of model and prototype to arrive at dimensionless relationships like in the above case. This method, as such involves scrutiny of the differential equations of heat mass and momentum balance to identify the dimensionless parameters like Reynolds number etc. which can be used to simulate the results of model with the prototype to get new information.

In some cases the equations are not known but the researcher knows from experience or intuitively the relevant variables in the process. As already described in detail in earlier chapters criteria of similarity can also be examined by:

- Rayleighs method of indices;
- Buckingham theorem.

Using the above procedure , the relevant dimensionless numbers in a certain process can be identified and used in physical modelling to determine the parametric relationships between the model and prototype.

In the literature one comes across a large number of dimensionless numbers applicable in various fields involving heat mass and momentum transfer. For example Table 21.1 (SZEKELY, CARLSSON and HELLE, 1989) presents some groups that are of importance in the physical modeling of a tundish for the continuous casting process.

Group	Definition	Application
Reynolds	$\frac{\rho LV}{\eta}$	Fluid Flow
Froude	$\frac{\mu^2}{gL}$	Wave and surface behaviour, jet flow
Modified Froude	$\frac{\rho_g V^2}{\left[\left(\rho_L - \rho_g\right) gL\right]}$	Behavior of gas/liquid systems
Peclet	$\frac{LV}{\alpha}, \frac{LV}{D}$	Forced convection, mass transfer
Weber	$\frac{\rho LV^2}{\sigma}$	Bubble formation, atomization of liquid jets
Morton	$\frac{g\eta L^4}{\rho L\sigma^3}$	Speed of bubbles in liquids
Schmidt	$\frac{\eta}{\rho D}$	Mass transfer
Prandtl	$\frac{\eta C_p}{k}$	Free and forced convection

Table 21.1 - Relevant dimensional groups in the physical modeling of tundish in the continuous casting process.

Example

Consider physical modeling of a reactor which simulates liquid steel processing. It uses water at 25°C (in the model) with the objective of studying the velocity field of steel melt under laminar flow conditions. In this case the first step is to identify the important dimensionless groups and the compatible scale factor:

The dimensionless groups are Reynolds and Froude (as they are related to fluid flow and according to the analysis given earlier) are given by:

$$Re = \frac{L\rho V}{\eta} \text{ and } Fr = \frac{V^2}{Lg}$$

If V_m and V_p are velocities, and similarly the corresponding densities and viscosities of the liquids in the model and prorotype, then according to similarity principles:

$$\frac{L_m \rho_m V_m}{\eta_m} = \frac{L_p \rho_p V_p}{\eta_p}$$

According to Reynolds criterion, the scale factor would be:

$$\frac{V_m}{V_p} = \frac{\frac{\eta_m}{\rho_m}}{\frac{\eta_p}{\rho_p}} \times \frac{1}{\frac{L_m}{L_p}} = \frac{\frac{\eta_m}{\rho_m}}{\frac{\eta_p}{\rho_p}} \cong \frac{1}{\lambda}$$

One gets th above result , as the properties values can be taken as approximately, for steel melt and water respectively:

$$\rho_p = 7000 \text{ kg/m}^3 \qquad \eta_p \cong 7 \times 10^{-3} \text{ Pa.s}$$

$$\rho_m = 1000 \text{ kg/m}^3 \qquad \eta_m \cong 1 \times 10^{-3} \text{ Pa.s}$$

This justifies the scale factor given by the above equation, based on Reynolds criterion.

Similarly, Froude criterion is given by:

$$\frac{V_m^2}{L_m g_m} = \frac{V_p^2}{L_{mp} g_p} \Rightarrow \frac{V_m^2}{L_m} = \frac{V_p^2}{L_{mp}}$$

Hence:

$$V_m = V_p \, \lambda^{1/2}$$

This calculation suggests that the similarity in this case can be achieved only if $\lambda = 1$, with the construction of a full-scale model for satisfying both Reynolds and Froude criteria simultaneously.

Considering the dimensionless numbers for the tundish in the continuous casting process for model and prototype given in Table 21.2, it can be seen that for $\lambda = 1$ it is impossible to meet all the criteria, if one wants to study various aspects of the problem (the formation of drops, motion of bubbles, etc.) in addition to fluid movement with the model in the laboratory. The values for the last 3 items in the tables vary widely in the model in relation to the prototype.

Group	Steel System scale	Model-scale
Reynolds	1	1.1
Froude	1	1.0
Weber	1	3.1
Morton	1	44.0
Modified Froude	1	7.0

Table 21.2 - Calculated scale values of various dimensionless groups for steel and water systems. The values were normalized by taking the steel system as reference.

From the above one can conclude that rarely, it would be possible to construct and operate a model compatible in every aspect with the prototype. Hence it is implied that the model may be applied in specific cases depending upon the physicochemical phenomena to be studied and may have to sacrifice some other criteria of similarity dictated by another dimensionless quantity.

Dimensionless numbers are associated with the ratio between the different variables that determine the behavior of the system. For example (but not exclusively) the ratio of forces:

$$Re = \frac{\text{Inertial force}}{\text{Viscous force}} = \frac{\rho L^2 V^2}{L^2 \eta \frac{V}{L}} = \frac{\rho L V}{\eta} \qquad (21.15)$$

$$Fr = \frac{\text{Inertial force}}{\text{Gravitational force}} = \frac{\rho L^2 u^2}{L^3 \rho g} = \frac{u^2}{Lg} \qquad (21.10)$$

These criteria can be used, in some cases, to determine the relevant dimensionless numbers involved in a system.

Example

Consider the flow pattern of steel melt within the LD vessel. Impact and shear forces transmitted by the oxygen jet at the surface, produce an oscillating cavity at the liquid metal surface. As a result, there is ejection of metal droplets, which in the supernatant "phase", participate in the formation of an emulsion of slag, gas and liquid metal phases. This also may contain other components not dissolved in the metal or slag phase. The emulsion formation is important as it increases in the interfacial area of metal/slag and this can also happen due to atomization of the metal caused by the oxygen jet. The rate of refining increases in respect of these droplets due to increase in the surface area. The figure below compares the progress of decarburization determined from the chemical analysis of samples of droplets in the emulsion and the metal samples collected at the bottom of the reactor. The droplets from the emulsion taken at the same time, have a much lower carbon content relative to that of bulk metal.

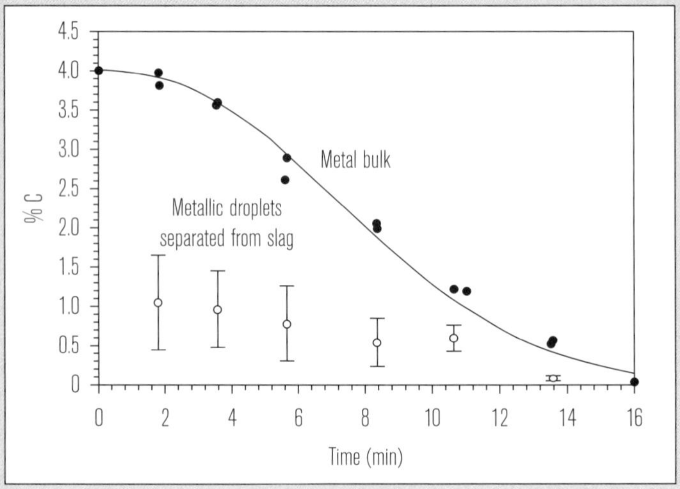

Decarburisation of metal droplets and liquid metal in the LD converter.

The formation of emulsions are due to the following interactive forces between metal slag and bath:

a) Force of inertia of the gas, as well as impact at the gas liquid interface is responsible for the recirculation of metal (STREETER, 1958):

$$\rho \cdot d^2_{lance} \cdot \mu^2_{gas}$$

where ρ, d_{lance}, μ_{gas} are density of gas, internal diameter and gas velocity at the lance nozzle, respectively.

b) Weight of the fluid to be mixed, acts as measure of stability of the metal pool:

$$D^2_{Reactor} \cdot H_{Metal} \cdot \rho_{Metal} \cdot g$$

In the above $D_{reactor}$, H_{metal}, ρ_{metal}, are internal diameter of the vessel; liquid metal height, density of metal bath and acceleration due to gravity , respectively.

The mechanism of formation of emulsions due to these forces is given in the figure below.

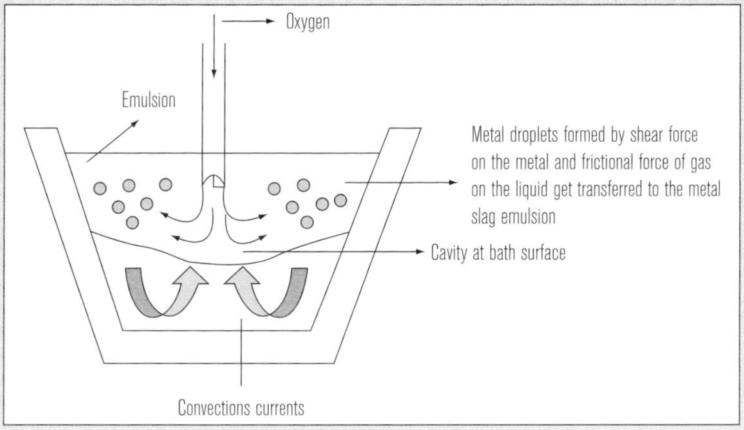

Formation of emulsion in the LD vessel.

In the case of a model simulating an industial process, the behavior of the liquid in the model should be similar to the behavior of the liquid in the prototype. That is to say that the proportion of these forces defined by equations given under (a) and (b) above, should be same in the model and prototype Hypothetically any model can be built on this basis to study the process iin the laboratory and determine the best possible combination of variables for maximum efficiency.

$$\left[\frac{\rho\, d_{lance}\, u^2_{gas}}{D^2_{reactor} H_{liquid}\, \rho_{liquid}\, g}\right]_{model} = \left[\frac{\rho\, d_{lance}\, u^2_{gas}}{D^2_{reactor} H_{liquid}\, \rho_{liquid}\, g}\right]_{prototype}$$

One can also consider that formation of emulsion is also controlled by the ratio between inertial forces and gas/metal interfacial forces. In other words Weber number $\frac{\rho L u^2}{\sigma}$ is an addtional similarity criterion (σ is the interfacial tension):

$$\left[\frac{\rho H_{liquid}\, u^2_{gas}}{\sigma}\right]_{model} = \left[\frac{\rho H_{liquid}\, u^2_{gas}}{\sigma}\right]_{prototype}$$

Example

In the case of simulation of fluid behaviour in a ladle, when an inert gas is injected through a porous plug at the bottom of a ladle to agitate the metal, criteria of similarity would be the dimensionless Froude and Reynolds numbers. The latter must be calculated taking into account the turbulent viscosity, since the flow regime is turbulent. Estimation of the turbulent viscosity η_{uurb} can be can be done using the following equation (SAHAI and GUTHRIE, 1982):

$$\eta_{turb} = 5.5 \times 10^{-3} \rho_L H \left(\frac{gQ}{D}\right)^{1/3}$$

where: ρ_L is the density of the metal; H is the level of the liquid in the ladle; Q is the gas flow rate; D is the diameter of the ladle; g is the gravitational acceleration.

One can write the Navier - Stokes equations in dimensionless form for the turbulent flow regime, with velocities (u_L)model and (u_L) prototype in the model and prototype, in the same way it was done for the laminar regime to get the dynamic similarity criteria between themin terms of turbulent Reynolds and Froude numbers. Hence:

$$Re_{turb} = \frac{\rho_l H u_l}{\eta_{turb}} \quad \text{or} \quad Re_{turb} \cong \frac{\rho_l H u_l}{\rho_l H \left(\frac{Qg}{D}\right)^{\frac{1}{3}}} = \frac{u_l}{\left(\frac{Qg}{D}\right)^{\frac{1}{3}}}$$

and according to Reynolds similarity criterion:

$$\left[\frac{u_L}{Q^{1/3}}\right]_{model} = \left[\frac{u_L}{Q^{1/3}}\right]_{prototype} \frac{1}{\left(\frac{D_{model}}{D_{prototype}}\right)^{1/3}}$$

The scale factor is defined by equation (21.5), and hence:

$$\lambda = \frac{D_{model}}{D_{prototype}}$$

When Froude number is the similarity criterion, it can be written:

$$\frac{(u_L)_{model}}{(u_L)_{prototype}} = \sqrt{\lambda}$$

Substituting this in the Reynolds criterion given above one gets:

$$\left(\frac{Q_m}{Q_p}\right)^{1/3} = \frac{(u_L)_{model}}{(u_L)_{prototype}}\left(\frac{D_{model}}{D_{prototype}}\right)^{1/3} = \lambda^{\frac{1}{3}}\ddot{e}^{1/2}$$

Hence one can now estimate the flow rate to be used in the model compatible with that in the prototype:

$$Q_{mod\,el} = \lambda^{5/2} Q_{prototype}$$

Example

Consider simulation of alloy additions to liquid steel during pouring, using a scale model $\lambda = 1/5$, in a ladle with an average diameter of 4 m. Since the alloy has a lower density compared to steel, in order to increase the recovery of alloying element, additions are made near the point of impact, to take advantage of convection currents generated, as can be seen in the following figure.

In the case of simultaneous occurrence of various phenomena, such as fluid motion, projections of metal drops, emulsification, frictional forces due to additions and the process of their dissolution etc. analysis of the process can be done in stages. For example consider the similarity criterion for the movement of the fluid interface in terms of dimensionless Reynolds, Froude and Weber numbers on the basis of data given for each parameter like velocity, viscosity, interfacial tension etc. for this case, substituted in the equations below:

$$Re = \frac{L\rho u}{\eta} = \frac{4 \times 7000 \times 10}{7 \times 10^{-3}} \cong 4 \times 10^7$$

$$Fr = \frac{u^2}{Lg} = \frac{10^2}{4 \times 9.8} \cong 2.5$$

$$We = \frac{L\rho u^2}{\sigma} = \frac{7000 \text{ x } 4 \text{ x } 10^2}{1.63} \cong 42 \text{ x } 10^6$$

Therefore based on the above results, one can conclude that the various forces have the following important relationship:

Inertial force/gravitational force >> viscous forces

Inertial force/gravitational force >> interfacial forces

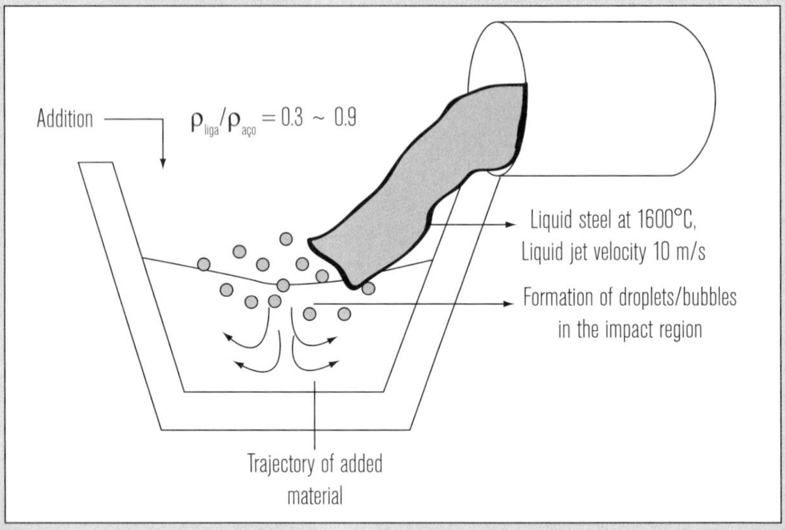

Alloy additions to steel melt in ladle during teeming.

The values calculated for the dimensionless numbers above show that forces of inertia and gravity are the most important with regard to fluid motion. In case it is not possible to satisfy all the criteria simultaneously (Fr, Re, We), one can adopt the Froude criterion as it represents the ratio of the forces relevant to the movement for the case above.

As the geometric similarity implies the relationship between lengths as the scale factor:

$$\lambda = \frac{L_m}{L_p}$$

one can write relationship between areas and volumes as:

$$\frac{A_m}{A_p} = \left(\frac{L_m}{L_p}\right)^2 = \lambda^2 \text{ and } \frac{V_m}{V_p} = \left(\frac{L_m}{L_p}\right)^3 = \lambda^3$$

Also as the Froude criterion provides the relationship between the velocities:

$$\frac{u_m}{u_p} = \lambda^{1/2}$$

the relation between flows would be given by:

$$\frac{Q_m}{Q_p} = \left(\frac{u_m A_m}{u_p A_p}\right) = \lambda^{5/2}$$

This also leads to time scale defined by:

$$\frac{t_m}{t_p} = \frac{\frac{L_m}{u_m}}{\frac{L_p}{u_p}} \approx \frac{L_m}{L_m} \text{ x } \frac{1}{\frac{u_m}{u_p}} = \sqrt{\lambda}$$

These relations show how the model could be operated to ensure similarity of the movement of liquid.

The dispersion of addition inside the steel melt can be described by the balance of forces acting on the particles. The variation of the velocity of the particle with time can be written in terms of buoyant force given by the difference between the first two terms in the right hand side added to the frictional force (the third term in the right hand side), as has been discussed in the earlier chapters (here r is the radius of the addition; u the density of addition; the addition velocity; g the gravitational acceleration; ρ the liquid density ; u_r the slip velocity between the melt and the addition and f the friction drag factor):

$$\frac{4}{3}\pi r^3 \rho_s \frac{du}{dt} = \frac{4}{3}\pi r^3 \rho_s g - \frac{4}{3}\pi r^3 \rho g + f\left(\frac{1}{2}\rho u_r |u_r|\right)\pi r^2$$

This equation can be rewritten in order to neglect the vector form in the balance equation only for determining the criteria of similarity:

$$\frac{4}{3}\pi r^3 \rho_s \frac{du}{dt} = \frac{4}{3}\pi r^3 (\rho_S - \rho) g \pm \frac{1}{2} f \rho u_r^2 \pi r^2$$

or

$$\frac{4}{3}\frac{\rho_s}{\rho}\frac{r}{u_r^2}\frac{du}{dt} \approx \frac{4}{3}\frac{r g}{u_r^2}\left(\frac{\rho_S}{\rho} - 1\right) + \frac{f}{2}$$

In the dimensionless form the above equation suggests that, a similar trajectory for addition in the model and the prototype can be obtained if the coefficients of the equation above applicable to model and prototype are identical as also other equation parameters, ie:

$$f_{model} = f_{prototype} \text{ and } \frac{\rho_s}{\rho} = \gamma_{model} = \gamma_{prototype}$$

Here the friction factor can be found as a function of the Reynolds number based on addition radius and slip velocity:

$$Re = \frac{\rho 2 r u_r}{\eta}$$

The relationships are:

$$f = 24/Re \text{ for } Re < 1;\ f = \frac{18.5}{Re^{0.6}} \text{ for } 1 < Re < 1000;\ f = 0.44 \text{ for } Re > 1000$$

If the above restriction hold one can write the additional similarity criterion:

$$\left(\frac{r}{u_r^2}\right)_{model} = \left(\frac{r}{u_r^2}\right)_{prototype}$$

Example

A certain steel company intends to reduce the losses of aluminum ($\rho = 2700\ kg/m^3$) by dispersing them through injection in the steel melt in the form of spherical pellets (2 cm in diameter, 20 m/s) against a static bath ($\eta = 7 \times 10^{-3}$ Pa.s, $\rho = 7000\ kg/m^3$) The penetration depth of the addition should be maximum in order to increase the degree of recovery, but without inducing premature erosion of the refractory bottom of the ladle. In a preliminary study of the process, it is suggested that a simulation model in the scale 1:3, using water ($\eta = 1 \times 10^{-3}$ Pa.s, $\rho = 1000\ kg/m^{-3}$) as the working fluid and use of resin pellets simulating aluminium pellets. Determine the diameter, density, and speed of entry of the resin particle so that there is complete similarity condition between the model and prototype. Compare the friction factors in both the cases.

Solution

As the steel bath is not static it is necessary to take into account the criteria of similarity in respect of the movement of fluid. For the trajectory of the particle one can write:

a) $\gamma_{model} = \gamma_{prototype}$ or $\frac{\rho_{resin}}{1000} = \frac{2700}{7000}$

b) $\left(\frac{r}{u_r^2}\right)_{model} = \left(\frac{r}{u_r^2}\right)_{prototype}$ or $\left(\frac{r}{u_r^2}\right)_{model} = \frac{0.1}{20^2}$

The scale factor used (1:3) of the ladle model can be extended to enable the estimation of resin size.

$$\frac{r_{model}}{r_{prototype}} = \frac{1}{3} \rightarrow \frac{r_{model}}{0.01} = \frac{1}{3} \rightarrow r_{model} = \frac{0.01}{3}$$

which implies:

$$u_r = 20\,x\sqrt{\frac{1}{3}}$$

All the above considerations have to be taken into account to evaluate the relation friction factor vs. Reynolds number both in the case of model and prototype and these have to be equal. If this restriction cannot be met, appropriate changes have to be made and the calculations have to be redone adopting some other scale factor for the addition. The limit for this flexibility is the ratio between the diameter of the addition and the diameter of the ladle which is in general negligible.

Example

Simulation of flotation of non-metallic inclusions in liquid steel can be done through physical modelling. For this conditions of similarity between the trajectories of particles used for simulating non-metallic inclusions in the model and the prototype have to be compatible. According to Sahai and Emi (1996), the choice of size and density of particles in the model simulating the inclusions in the steel melt in respect of inclusion removal, sintering and coalescence of particles occurring in the actual practice are vital to the success of the results simulation.

The balance of forces acting on a spherical particle as discussed earlier is given by:

$$\frac{4}{3}\pi r^3 \rho_{inc} \frac{du}{dt} = \frac{4}{3}\pi r^3 (\rho - \rho_{inc}) g + \frac{1}{2} f \rho u_r^2 \pi r^2$$

where, u is the absolute velocity of the inclusion; u_r is the relative velocity inclusion/fluid; ρ and ρ_{inc} are densities of metal an inclusion; f is the friction factor.

In dimensionless terms, the preceding equation can be rewritten as:

$$\frac{4}{3}\frac{r}{u_r^2}\frac{\rho_{inc}}{\rho}\frac{du}{dt} = \frac{4}{3}\frac{r g \left(1 - \frac{\rho_{inc}}{\rho}\right)}{u_r^2} + \frac{1}{2} f$$

Similar trajectories of the inclusions in the model and the prototype will be obtained if the corresponding coefficients of the previous equation are equal. Thus:

$$\left[\frac{r g \left(1 - \frac{\rho_{inc}}{\rho}\right)}{u_r^2}\right] = \left[\frac{r g \left(1 - \frac{\rho_{inc}}{\rho}\right)}{u_r^2}\right]_{prototype}$$

or,

$$\frac{r_{model}}{r_{prototype}} = \left(\frac{u_r^{model}}{u_r^{prototype}}\right)^2 \left[\frac{\left(1 - \frac{\rho_{inc}}{\rho}\right)_{prototype}}{\left(1 - \frac{\rho_{inc}}{\rho}\right)_{model}}\right]$$

There are indications that the slip velocity, u_r, would be the same or has the magnitude of the terminal velocity of the particle in liquid steel. Therefore, if the particles obey the Stokes law, ie, that the velocities of the rising particles through the liquid metal, are given by:

$$u_t = \frac{2\rho r^2 g}{9\eta}\left(1 - \frac{\rho_{inc}}{\rho}\right)$$

Also, as shown in the following table, if water is chosen as the working fluid for the model, the kinematic viscosity of the liquids will be nearly equal, $(\eta/\rho)_{water} = (\eta/\rho)_{steel}$.

Properties of fluids int prototype and model.

Property	Value
η_{water}	1.14×10^{-3} kg/m.s
η_{steel}	6.7×10^{-3} kg/m.s
ρ_{water}	1000 kg/m³
ρ_{steel}	7000 kg/m³

Considering that the kinematic viscosities are the same in model and prototype and with the use of the above equation for terminal velocity, the earlier equation changes to:

$$\frac{r_{model}}{r_{prototype}} = \left(\frac{r_{model}}{r_{prototype}}\right)^4 \times \frac{\left(1-\frac{\rho_{inc}}{\rho}\right)^2_{model}}{\left(1-\frac{\rho_{inc}}{\rho}\right)^2_{prototype}} \times \frac{\left(1-\frac{\rho_{inc}}{\rho}\right)_{prototype}}{\left(1-\frac{\rho_{inc}}{\rho}\right)_{model}}$$

Simplifying the above equation, one gets:

$$\frac{r_{model}}{r_{prototype}} = \left[\frac{\left(1-\frac{\rho_{inc}}{\rho}\right)_{model}}{\left(1-\frac{\rho_{inc}}{\rho}\right)_{prototype}}\right]^{-\frac{1}{3}}$$

This equation shows the relationship between size and density of the particles in the prototype and model, so that there can be similarity between their trajectories in the industrial reactor and physical model. If particles of polyethylene are used ($\rho_{inc} = 920$ kg/m³) to simulate inclusions of alumina ($\rho_{inc} = 3000$ kg/m³) the above equation gives:

$$r_{model} = 1.92\ r_{prototype}$$

Generally, the range of sizes of interest of the inclusions is 20 to 150 microns. A non-metallic inclusion of diameter equal to 100 μm in the prototype corresponds to a particle of 0.192 mm in diameter with a density of 920 kg/m³ in the model.

One can also consider for a range of sizes of inclusions rising in the liquid steel with different terminal velocities defined by Stokes Law. Hence the rising velocity of the particle in the model and prototype:

$$u_{model} = \left[\frac{2\,r^2 g\left(\rho - \rho_{inc}\right)}{9\,\eta}\right]_{model} \quad \text{and} \quad u_{prototype} = \left[\frac{2 r^2 g\left(\rho - \rho_{inc}\right)}{9\eta}\right]_{prototype}$$

If the model is operated in accordance with the similarity criterion of Froude, as in the case of tundish in the continuous casting process, it was shown that:

$$\frac{U_{model}}{U_{prototype}} = \sqrt{\lambda}$$

where U is the liquid velocity and λ the scale factor.

There is similarity between the behaviour of inclusion in the model and prototype, ie, liquid and inclusion trajectories will be similar if:

$$\left(\frac{U}{u}\right)_{model} = \left(\frac{U}{u}\right)_{prototype}$$

This implies on:

$$\frac{u_{model}}{u_{prototype}} = \frac{U_{model}}{U_{prototype}} = \sqrt{\lambda}$$

Combining this last equation with that giving inclusion velocity leads to:

$$\frac{u_{model}}{u_{prototype}} = \frac{r^2_{model}}{r^2_{prototype}} \frac{(\Delta\rho)_{model}}{(\Delta\rho)_{prototype}} \frac{\eta_{prototype}}{\eta_{model}} = \frac{U_{model}}{U_{prototype}} = \sqrt{\lambda}$$

$$\frac{r^2_{model}}{r^2_{prototype}} \frac{(\Delta\rho)_{model}}{(\Delta\rho)_{prototype}} \frac{\eta_{prototype}}{\eta_{model}} = \sqrt{\lambda}$$

$$\frac{r^2_{model}}{r^2_{prototype}} \frac{\left(1-\frac{\rho_{inc}}{\rho}\right)_{model}}{\left(1-\frac{\rho_{inc}}{\rho}\right)_{prototype}} \frac{\left(\frac{\eta}{\rho}\right)_{prototype}}{\left(\frac{\eta}{\rho}\right)_{model}} = \sqrt{\lambda}$$

This condition assures that the trajectories of particles in the model and the prototype are similar. Since the values of kinematic viscosity (η/ρ) of the liquid in the model and the prototype can be considered almost the same (water and steel). Hence:

$$\frac{r_{model}}{r_{prototype}} = \left[\frac{\left(1-\frac{\rho_{inc}}{\rho}\right)_{prototype}}{\left(1-\frac{\rho_{inc}}{\rho}\right)_{model}}\right]^{1/2} \lambda^{\frac{1}{4}}$$

This equation correlates the radius or size of the inclusion in the model and prototype, so that the trajectories of these are similar for both model and prototype, considering also differences in density between the solid and fluid, in these reactors respectively.

21.3 Ideal Reactors

Flow behaviour of fluids within reactors where chemical reactions take place are usually studied in detail by chemical engineers. The metallurgist also should have an idea about the concepts involved as these are very important for the understanding the flow behaviour of liquid metal at high temperatures like in the tundish of continuous casting and others. In these reactors determination of parameters such as residence time distribution of various fluid elements in the interior of a reactor that operates under continuous flow regime is of some importance.

Consider an experiment in which a tracer (substance, which dispersion can be monitored, but does not interfere in the velocity field pre-existing) is injected into the feed point of a reactor, and in the form of a pulse (at once) Figure 21.5.

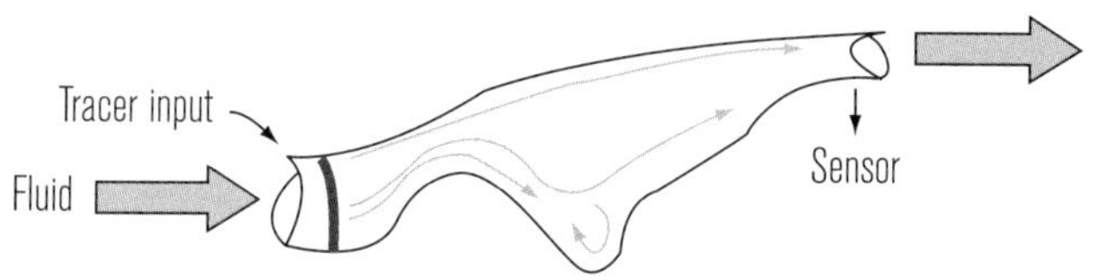

Figure 21.5 - Schematic representation of a continuous flow reactor.

A sensor located at the point of discharge to assess the dispersion of the tracer measures the concentration as a function of time as given in Figure 21.6. Normally, the various fractions of the tracer may not stay inside the reactor for the same time frame, that is, one observes a residence time distribution measured from the concentration curve. In fact, if E(t) represents the Residence Time Distribution Function (RTD), then by definition, the quantity (E(t). dt) representing the fraction of fluid elements with residence time between t and t + dt is given by:

$$E(t)\,.\,dt = \frac{Q\left(\frac{m^3}{s}\right) \times C_T(t)\left(\frac{mols}{m^3}\right) \times dt(s)}{M_0(mols)} \tag{21.17}$$

where:

- Q is the flow rate;
- C_T is the concentration of tracer in the effluent;
- t is the elapsed time;
- M_0 is the amount of tracer added.

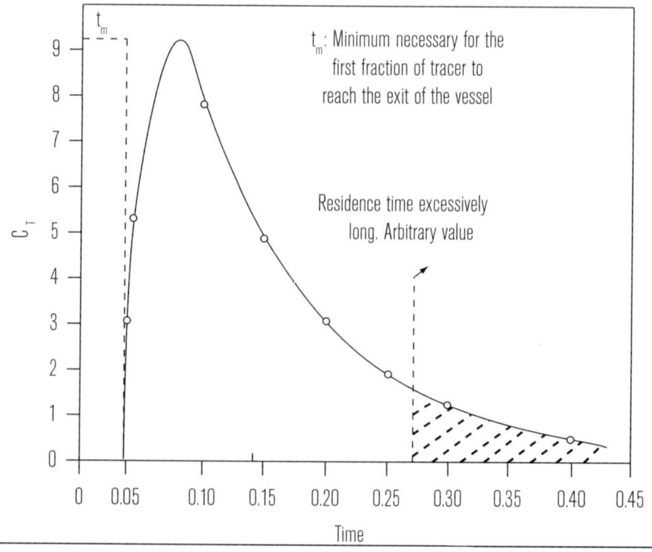

Figure 21.6 - Schematic C vs t curve for the tracer addition (pulse).

The function E(t) is useful to define various parameters:

1) Mean residence time,

$$\tau = \int_{t=0}^{t=\infty} t.\left[E(t).dt\right] \tag{21.18}$$

2) Fraction of fluid with residence times between t_1 and t_2,

$$\int_{t_1}^{t_2} .E(t).dt \tag{21.19}$$

3) Fraction of fluid that remains inside the reactor for a period of time more than t_2,

$$\int_{t_2}^{\infty} E(t).dt \tag{21.20}$$

4) Cumulative function F (t), fraction of fluid elements with residence time less than t,

$$F(t) = \int_{0}^{t} E(t).dt \tag{21.21}$$

and others of importance, such as dispersion variance

It is possible to show that the Mean Residence Time is equal to the Nominal Residence Time, defined as the time required to change a working volume of the reactor, ie

$$T = \frac{V_R\left(m^3\right)}{Q\left(\frac{m^3}{s}\right)} = \tau \tag{21.22}$$

where V_R is the working volume.

The nominal residence time also means the residence time if all the fluid elements remain inside the reactor for the same amount of time that is for the same time interval. This type of flow is known as Piston Flow.

Example

Consider the curve C versus t constructed using the data in the following table for addition of pulse of tracer to a continuous reactor, with a flow rate 1.7 m³/min, in continuous operation.

A mass balance applied to the reactor, and considering inert tracer and nonreactive walls of the reactor gives:

$$Mo \equiv \int_{t=0}^{t=\infty} Q\, C_T dt$$

Mo is the amount of tracer added and can be calculated from the area under the "C_T versus t" curve

Data for C_T versus t diagram.

t(min)	0	5	6	7	8	9	9.5	10	11	12	14	20	25	30	35
Moles/m³	0	0	4	8	13	15	14.5	12	8	6	5	4	2	1	0.1

Mo can be evaluated, by numerical integration, using the following figure:

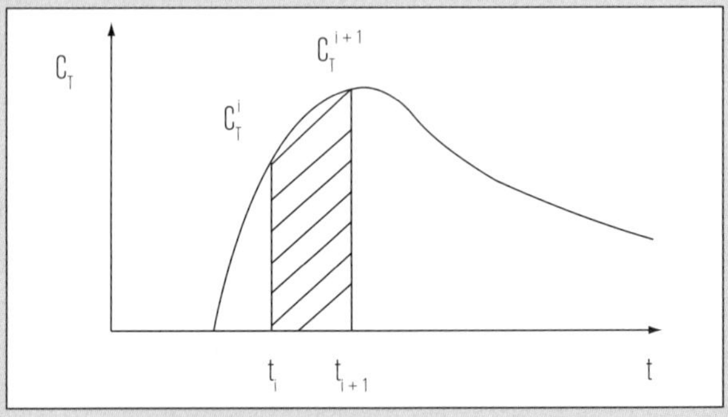

Variables for numerical integration in the C-T diagram.

$$\text{Infinitesimal area} = \frac{C_T^i + C_T^{i+1}}{2}\left(t_{i+1} - t_i\right)$$

$$Mo = Q\int_{t=0}^{t=\infty} C_T\ dt \cong Q\sum \frac{C_T^i + C_T^{i+1}}{2}\left(t_{i+1} - t_i\right)$$

The mean residence time τ is given by:

$$\tau = \int_{t=0}^{t=\infty} t \cdot E(t) dt = \int_{t=0}^{t=\infty} t \cdot \left(\frac{Q \cdot C_T}{M_0} \right) dt = \frac{Q}{M_0} \int_{t=0}^{t=\infty} t \cdot C_T \, dt$$

$$\tau = \frac{Q}{Mo} \Sigma \left(\frac{t_i + t_{i+1}}{2} \right) . \left(\frac{C_T^i + C_T^{i+1}}{2} \right) . (t_{i+1} - t_i)$$

The validity of the following relation given earlier can be checked.

$$T = \frac{V_R}{Q} = \tau$$

From the above one can estimate the working volume of the reactor.

As in the study of ideal gases, it is customary to postulate the existence of idealized flow reactors. These have specific behavior, with which the behavior of real reactors can be compared.

A piston flow reactor is presented schematically in Figure 21.7.

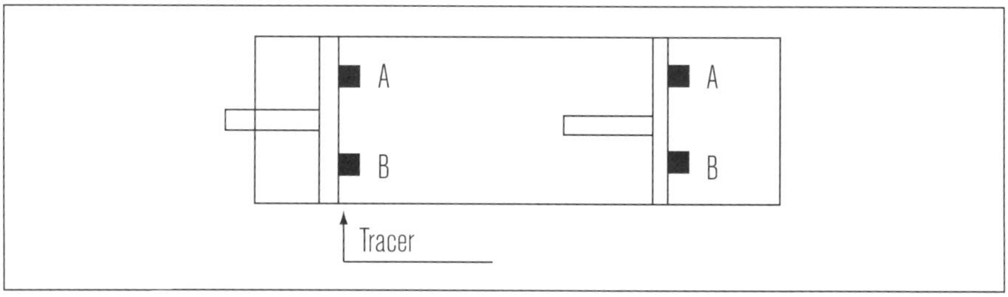

Figure 21.7 - Schematic diagram of a piston flow reactor.

The mixing is absent in radial and longitudinal direction in the fluid, and the residence time of fluid elements is unique. So for this reactor an experiment of adding a tracer in pulse would result in a response curve in the form of a pulse as shown in Figure 21.8.

The perfect mixing or mix reactor represents a case of other extreme. In this there are no gradients of quantities under consideration within the working volume. However the value of the quantity can vary over time, (Figure 21.9). This implies that the concentration of the species of interest inside the reactor, at any point is equal to the effluent concentration.

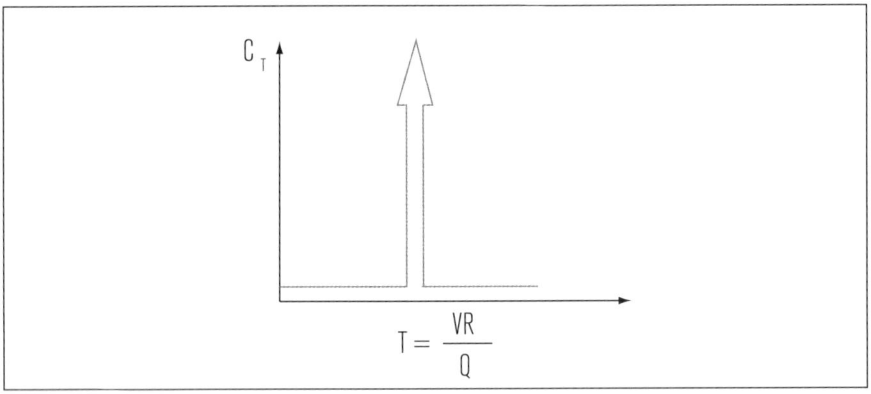

Figure 21.8 - Dispersion curve in a piston flow reactor.

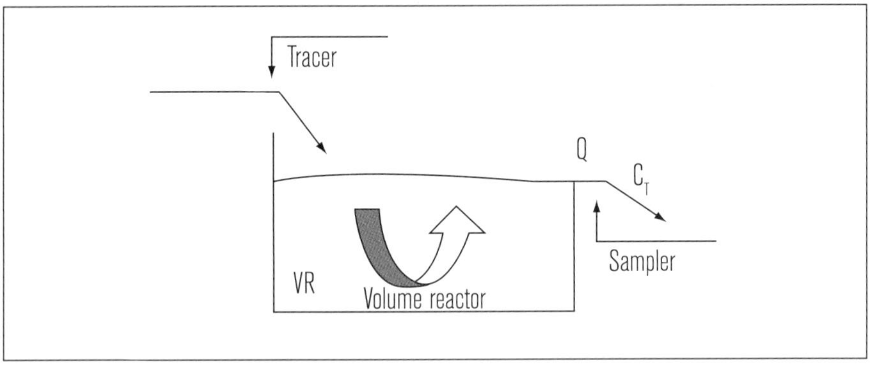

Figure 21.9 - Schematic representation of perfect mix reactor.

Example

Consider first-order reaction occurring in reactors (piston flow, perfect mixing) with the following characteristics:

$$V_R = 1000 \text{ L}, Q = 10 \text{ lpm}, C_E = 1 \text{ mole/L}; K_{RQ} = 6.7 \times 10^{-3} \text{ min}^{-1}$$

operating on a permanent basis. Estimate the degree of conversion in each of them.

a) Piston flow reactor:

For this a balance of conservation applied on a control volume corresponding to an infinitesimal slice of the reactor, located at distance x from the entrance and a thickness Δx is:

Accumulation Rate = Input rate - Output rate - Rate of consumption

$$\frac{d}{dt}\{Area \cdot \Delta x \cdot C\} = Q \cdot C\big|_{x=x} - Q \cdot C\big|_{x=x+\Delta x} - Area \cdot \Delta x \cdot K_{RQ} \cdot C$$

For steady state conditions:

$$0 = QC\big|_{x} - QC\big|_{x+\Delta x} - Area \cdot \Delta x \cdot K_{RQ} \cdot C$$

$$0 = -Q\left\{\frac{C\big|_{x+\Delta x} - C\big|_{x}}{\Delta x}\right\} - Area \cdot K_{RQ} \cdot C$$

Using the definition of derivative:

$$\frac{dC}{dx} = -\frac{Area}{Q} \cdot K_{RQ} \cdot C$$

This differential equation provides the variation of concentration along the reactor. Multiplying the numerator and denominator of the right hand side of the above equation by L, the first term in the right side becomes V_R/Q which is the residence time. Hence:

$$\frac{dC}{dx} = -\frac{\tau}{L} \cdot K_{RQ} \cdot C$$

This equation that can be integrated with the boundary conditions:

$x = 0$ and $C = C_{E;}$ $x = L$ and $C = C_S$

Substituting the data given:

$V_R = 1000$ L, $Q = 10$ lpm, $C_E = 1$ mole/L; $K_{RQ} = 6.7 \times 10^{-3}$ min $^{-1}$

$$\ln\frac{C_E}{C_S} = -\tau \cdot K_{RQ} \quad \text{and} \quad \frac{C_S}{1} = e^{-\frac{1000}{10}6.7\times10^{-3}} \cong 0.511$$

b) Perfect mix reactor:

For the perfect mix reactor also,

Accumulation Rate = Entry feed - Output rate - Rate of Consumption

The respective terms are:

$$\frac{d}{dt}\{V_R \cdot C_S\} = Q \cdot C_E - Q \cdot C_S - V_R \cdot K_{RQ} \cdot C_S$$

Considering steady state and transposing the terms,

$$\frac{C_S}{C_E} = \frac{Q}{Q + V_R \cdot K_{RQ}} = \frac{10}{10 + 1000 \times 6.7 \times 10^{-3}} \cong 0.598$$

Hence the piston flow reactor has a higher conversion efficiency as the value of C_S for the piston flow is lower than that for the perfect mix.

Now consider an experiment of adding tracer in pulse, in a perfect mix reactor operating in steady state (Figure 21.10). A balance of conservation of mass of tracer applied to the working volume VR, and valid for $t \geq 0$ (t equal to zero is the moment of addition of tracer) is written as:

Accumulation Rate = Input rate - Output rate - Rate of Consumption

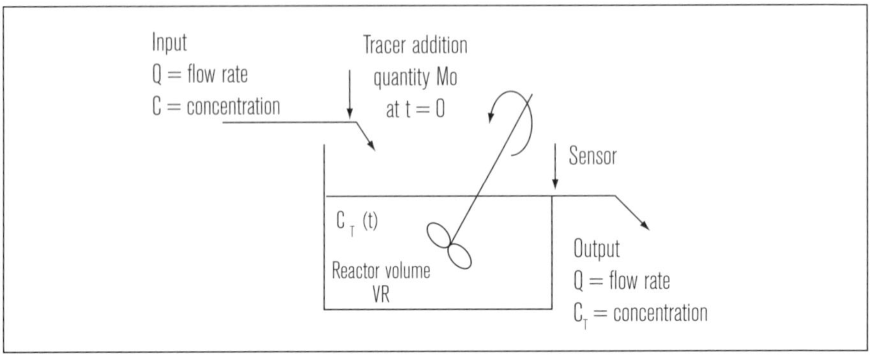

Figure 21.10 - Addition of tracer in the form of pulse in a perfect mix reactor.

The term accumulation here signifies the temporal variation of tracer in the control volume. Over a time interval the tracer gets diluted and escapes with the effluent and this is given by:

$$\frac{d}{dt}\left(V_R \cdot C_T\right) \tag{21.23}$$

The term corresponding to the input of the tracer is zero, because it is administered only in the form of pulse at the beginning of the experiment. Similarly the term for the chemical reaction can be eliminated since, the tracer is not reactive.

Considering uniformity of composition within the reactor, the output term is equivalent to Q .CT, The following equation can be integrated for the initial condition t = 0, C = C_T^o. Hence:

$$V_R \frac{dC_T}{dt} = -Q \cdot C_T \tag{21.24}$$

or

$$\frac{C_T}{C_T^0} = e^{-\frac{t}{\left(\frac{V_R}{Q}\right)}} \equiv e^{-\frac{t}{\tau}} \tag{21.25}$$

where:

$$C_T^o = \frac{M_o}{V_R} \tag{21.26}$$

represents the nominal initial composition (at the moment immediately after addition) in the reactor.

This also implies that the distribution curve residence time is described by: $E(t) = \frac{1}{\tau}.e^{-\frac{t}{\tau}}$ as given in Figure 21.11.

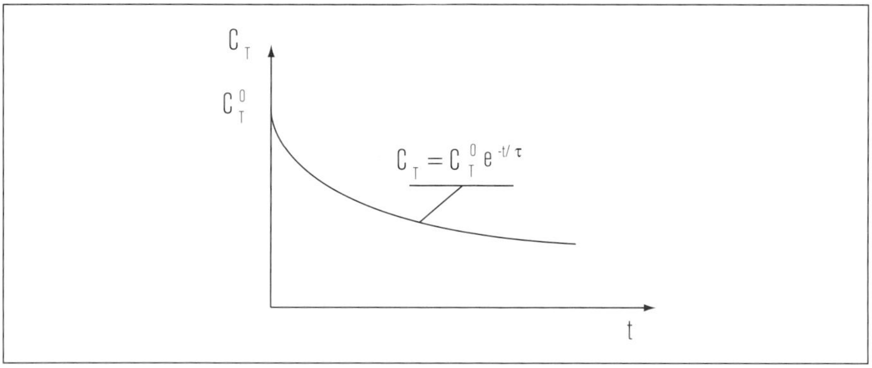

Figure 21.11 - Dispersion Curve, for perfect mix reactor.

21.4 Combination Model of Reactors

Consider a continuous reactor operating in steady state with flow and working volume defined by Q and V_R as before for the evaluation of residence time as given below:

$$T = \frac{VR}{Q} = \tau \tag{21.27}$$

Reactors with different internal geometric configurations can be compared in respect of RTD curves as given in Figure 21.12. It can be observed that average time does not change but only the degree of dispersion for the same. Whatever be the case this parameter can be defined as "minimum time of residence of fluid elements"; or rather minimum time required for the first quantities of tracer to be identified in the effluents, (t_{min}) Maximum value of t_{min} is equal to the nominal residence time for piston flow and the minimum value is zero in the case of perfect mix reactor or in the case of a actual reactor presenting a short circuit between input and output.

Actual reactors can be considered as combination of several idealized reactors. For example, if the value of t_{min} is close to the nominal residence time of the reactor then it approaches the ideal behaviour of piston flow, (Table 21.3.) A parameter defined as FPF, that is "the Fraction of reactor that behaves as a Piston Flow reactor" is given below for estimating this:

$$FFP = \frac{t_{min}}{T} \tag{21.28}$$

Table 21.3 - Minimum residence times in different reactors (from Figure 21.12).

Reactor	Perfect mix	Actual(1)	Actual(2)	Piston flow
t_{min}	0	$t^{(1)}_{min}$	$t^{(2)}_{min}$	T
$\frac{t_{min}}{T}$	0	$\frac{t^{(1)}_{min}}{T}$	$\frac{t^{(2)}_{min}}{T}$	1

Profiles of some of the distribution curves indicate that relatively large elements of volume of fluid remain in the reactor relatively for long period of time. This means a long time is required to expel a contaminant or chemical from the reactor. Such behavior may be associated with a sector in the reactor where there are currents that cause recirculation with little contact with the rest of the fluid as shown in Figure 21.13. There is interaction between the material contained in the portion undergoing recirculation and the rest of the fluid, but this interaction is not effective. For all practical purposes, the material contained in the recirculation regions are gradually and slowly released to the mainstream, creating a sort of tail that extends long. That is why this fraction of the reactor is called Dead Volume. The criterion for determining this portion of the reactor is, arbitrary and could involve defining a time period associated with it far too extensive. According to one of the propositions one can estimate a parameter, FDV "Fraction of reactor that behaves as Dead Volume", which can be defined as the fraction of elements that have residence time greater than twice the nominal time of residence:

$$FVM = \int_{2T}^{\infty} E(t).dt \tag{21.29}$$

This model, explained in Figure 21.14, is completed assuming that the remaining fraction corresponds to FPM, "a Fraction of the volume that behaves like a Perfect Mix reactor", ie FPM + FDV + FPF = 1.

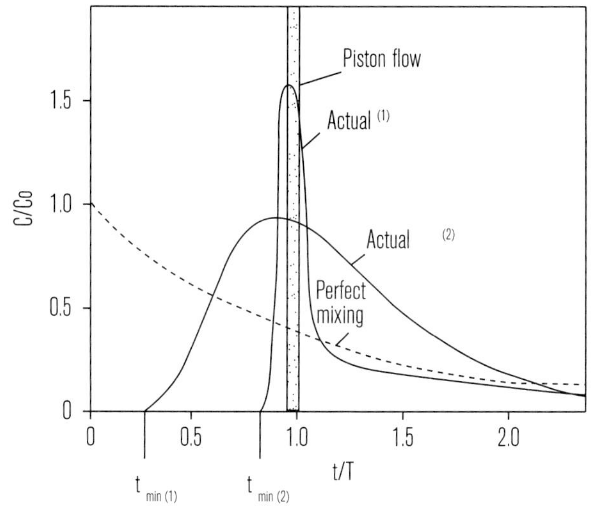

Figure 21.12 - RTD curves for different geometric configurations, for fixed flow and reactor volume.

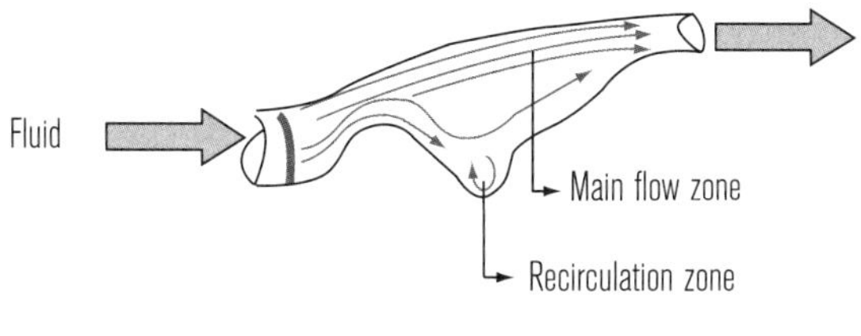

Figure 21.13 - Region of recirculation and dead volume.

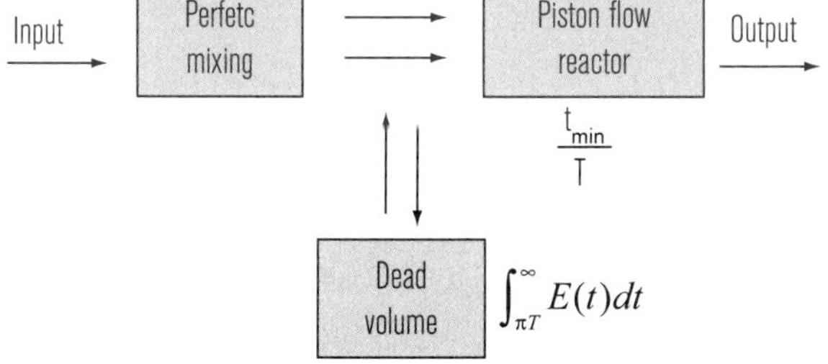

Figure 21.14 - A model of combination of reactors.

Example

Consider a tundish of the continuous casting unit for which the efficiency of flotation of inclusions is given by (WOLF, 2003):

$$r = 1 - e^{\frac{-u_{inc}\,t}{H}}$$

where r is the inclusion removal rate during steady state casting; t is the time available for flotation of inclusions inside tundish, u_{inc} is the terminal velocity of rise of inclusion in steel, and H is the liquid level in the tundish.

Comparing different reactors of same volume and flow rate, T, the nominal residence time would be the same for all inclusions. However:

- The perfect mix reactor would present the worst performance in respect of possibility of flotation of inclusions, since part of them immediately would enter the mold as soon as they enter the inside of the distributor.
- The inclusions in the piston flow have the same time t for flotation that is the maximum amount of time available for the purpose.

Hence the effective time available for inclusion separation should not be taken as the nominal residence time $T = V_R/Q$ but rather a fraction of it, which is t_{min}:

$$r = 1 - e^{\frac{-u_{inc}\,t_{min}}{H}}$$

Consider sequential casting of steels of different compositions, C_1 (first ladle) and C_2 (second ladle) in the continuous casting process. The above procedure leads to the production of intermediate material composition, or intermix as given in the following figure.

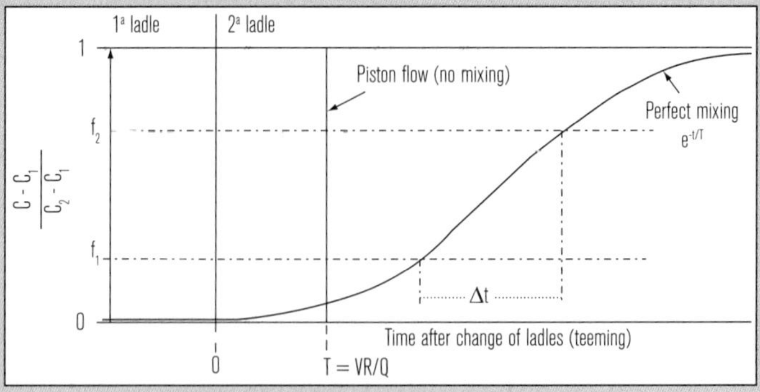

Production of "intermix" in continuous casting (SESHADRI, SILVA, and SILVA, 2005).

The range of dimensionless compositions of steel melt designated by regions f_1 and f_2 are considered non-standard and hence not acceptable. This has to be rejected as scrap. The specific values of f_1 and f_2 depend on the tolerance limit imposed by the customer. The figure presents the case of two extreme behaviours. In the case of piston flow there is no mixing in the reactor as a result of transition. The curve presents a discontinuity after the nominal residence time and the effluent concentration abruptly changes from C_1 to C_2. If this point is detected, until reaching the same, there is no production of intermix. The other extreme case is that corresponding to perfect mixing reactor. As soon as the second ladle begins to teem, there is a change in the composition of the effluent. The curve of transition can be calculated from the flow and volume of the reactor. Depending on the time interval between the compositions which represent the range, not acceptable, one can be set a range corresponding to this transition namely Δt, and the amount discarded (out of range; Rejected = $Q.\Delta t$) can be estimated. Hence to minimize this rejection, one should maximize the piston flow fraction. This type of reasoning can be repeated for other curves of transition, considering specific characteristics of the reactors under investigation.

21.5 Determination of the Rate of Circulation of Liquid in the RH Reactor

In Figure 20.15 a schematic diagram of RH reactor for degassing steel is given. It essentially contains a vacuum chamber for degassing in which the steel melt circulates due to the action of inert gas injection in the up leg.The liquid descends to the ladle through the down leg. There have been several improvements in recent years like injection of oxygen in the vacuum chamber to accelerate decarburization and to reduce carbon content of the steel using the combination of high vacuum and high oxygen potential.One of the important parameters of the process is the circulation rate which is a measure of rate of movement of liquid metal through the up leg into the vacuum chamber and to the ladle through the down leg. Circulation rate of steel influences the rate of renewal of steel surface exposed to vacuum in the chamber and consequently the rate of degassing. The circulation rate is expressed terms of tons of steel per minute and this parameter can be determined by physical modeling techniques. For this purpose , a water model of the industrial process can be constructed based on geometric and dynamic similarities as explained before. The circulation rate of water can be measured as a function of gas flow rate. To determine the circulation rate in the physical model the following procedure can be adopted.

The rate of movement of liquid is measured by injection of a tracer (a solution of Potassium Chloride (KCl)) of known concentration in the form of pulse, at the junction between the upleg and the vacuum chamber. The tracer alters the conductivity value of the water. The measurement of conductivity is carried out in the down leg continuously with a conductivity meter.

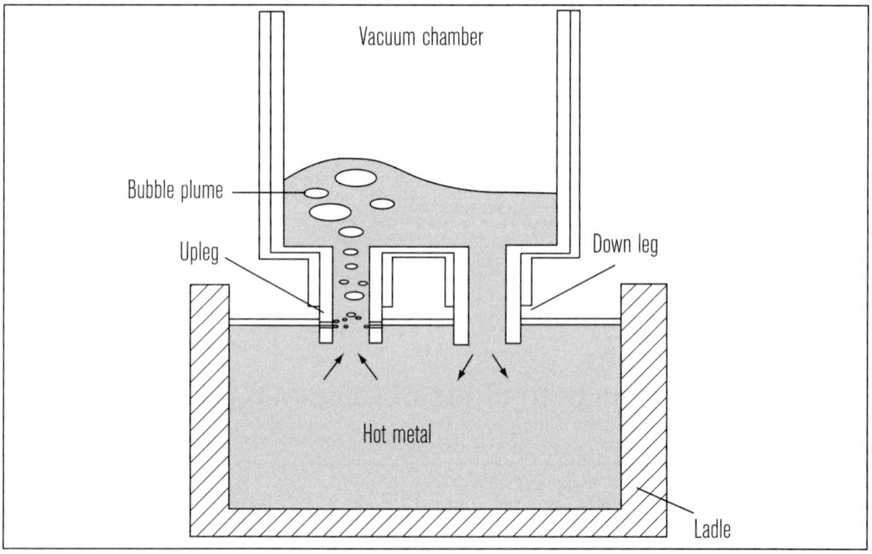

Figure 21.15 - Schematic diagram of RH reactor for secondary refining of steel.

The change in conductivity of the solution, caused by the injection of the tracer is continuously monitored with a A/D acquisition board. A prior calibration, gives the variation of conductivity with the tracer concentration. With this procedure typical curve of concentration versus time, results as shown in Figure 21.16.

In the figure the area under the curve A1, is due to first round of the tracer, KCl, at the down leg. Therefore it represents the total mass of tracer injected. In fact, visual observation of aniline addition at the same point shows that the indicator is dispersed almost instantaneously in the liquid inside the vacuum chamber, and then replaced with clean water, previously contained in the ladle. Hence, between t_1 and t_2, the initial concentration Co, reaches its maximum - Cmax - and goes back to the original value Co. Therefore, the total mass of tracer can be related to the area A1 under the curve.:

$$M = \int_{t1}^{t2} Q \times C \times dt \qquad (21.30)$$

where:

- M is the mass of tracer;
- Q is the liquid circulation rate; C is the tracer concentration.

The increase in concentration due to the addition of M (g KCl) is given by:

$$\Delta C = \frac{M}{V_{H_2O}} \qquad (21.31)$$

where V_{H2O} is the volume of water in the reactor

Hence a mass balance gives:

$$M = Q \times \int_{t1}^{t2} C \times dt = Q \times A_1 = \Delta C \times V_{H_2O} \qquad (21.32)$$

$$Q = \frac{\Delta C \times V_{H_2O}}{A_1} \qquad (21.33)$$

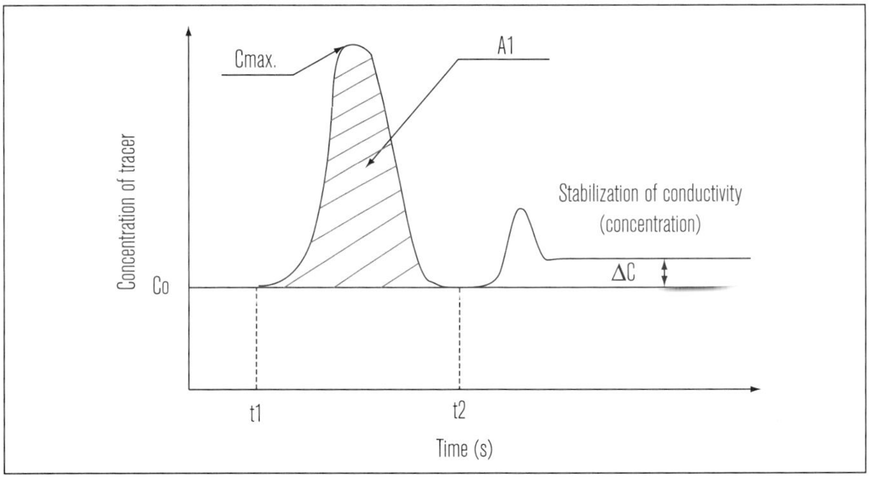

Figure 21.16 - Typical dispersion curve for tracer injected into the vacuum chamber of RH degasser physical model.

An example of results obtained by this technique is shown in Figure 20.17. The experiments were conducted in a model built in the laboratory (1:5 scale) simulating a reactor of 300 tons capacity with nominal circulation rate of 180 ton/min. (SESHADRI et al., 2003, 2004, 2006).

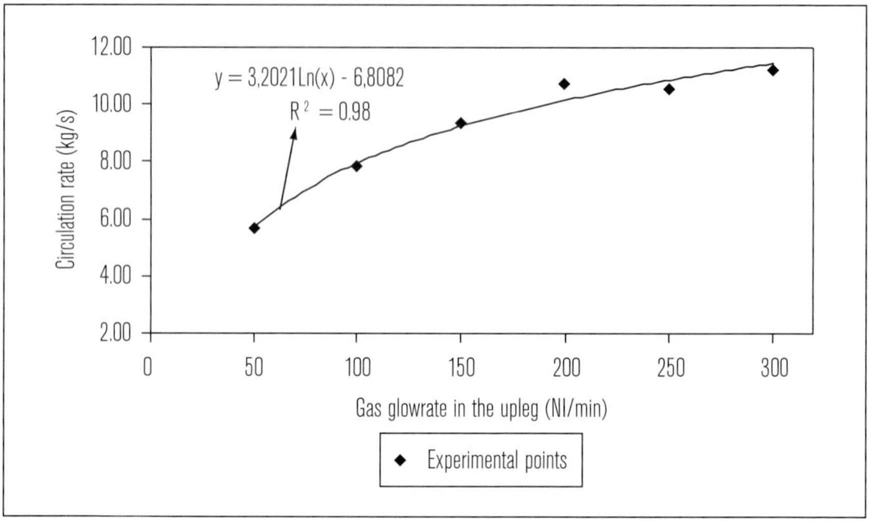

Figure 21.17 - Rate of circulation in RH physical model 1:5 scale, industrial reactor of 300 tones.

21.6 Determination of Degassing Rates

Reactions of degassing have to be considered in the removal of gases dissolved in metals. In the case of steel melt the reactions that are important are:

$$H = \frac{1}{2} H_{2(g)} \tag{21.34}$$

$$N = \frac{1}{2} N_{2(g)} \tag{21.35}$$

$$C + O = CO_{(g)} \tag{21.36}$$

The degassing reactions are favored by the application of vacuum and inert gas injection (with gas bubbles into which the dissolved gases diffuse) Increased contact area of metal/gas, by increasing the bubble area (large number of smaller bubbles) in addition to high values of activity of dissolved elements in the melt contribute to increased degassing rates.

The technique of adsorption/desorption of CO_2 in NaOH solution can be used to simulate the degassing of liquid steel. It assumes that the adsorption-desorption equilibrium of CO_2 in caustic solutions, which involves the species $H_2O(l)$, H^+, OH^-, $H_2O_3(aq)$, HCO_3^-, CO_3^{2-}, $CO_2(aq)$ and

N_a^+ is quickly reached when the solution in question comes into contact with a gas stream containing $CO_{2(g)}$. The distribution of species in equilibrium can be determined from the degree of acidity or pH of the solution. For example, for a given pH value and a known amount of NaOH added to an also known volume of water (so that N_a^+ concentration is also known) one can consider the following:

a) Equilibrium:

$$H_2O_{(l)} = H^+ + OH^- \qquad \log\ K1 = -3.9788 - 2985.3/T \tag{21.37}$$

$$H_2CO_3aq = HCO_3^- + H^+ \qquad \log K2 = -4.6526 - 506.42/T \tag{21.38}$$

$$HCO_3^- = CO_3^= + H^+ \qquad \log K3 = -7.5153 - 838.49/T \tag{21.39}$$

$$CO_2(aq) + OH^- = HCO_3^- \qquad \log K4 = -0.6738 + 2478.9/T \tag{21.40}$$

b) For electro neutrality:

$$H^+ + Na^+ = HCO_3^- + 2\ CO_3^= + OH^- \tag{21.41}$$

A pH meter is immersed in the caustic soda solution used for simulation of the liquid metal so that Hydrogen ion concentration can be monitored continuously.

$$\left(H^+\right) = 10^{-pH} \tag{21.42}$$

Hydroxyl ion concentration is given by:

$$\left(OH^-\right) = K1\ /\ \left(H^+\right) \tag{21.43}$$

and similarly the concentrations of all the other ionic species can be evaluated.

Figure 20.10 shows data from equilibrium calculation for (Na^+) = 0.02 m According to the same, in a very limited range of pH (in this case between 6 and 7), the reaction would be predominantly desorption - adsorption of CO_2.

$$CO_{2(g)} = CO_{2aq} \tag{21.44}$$

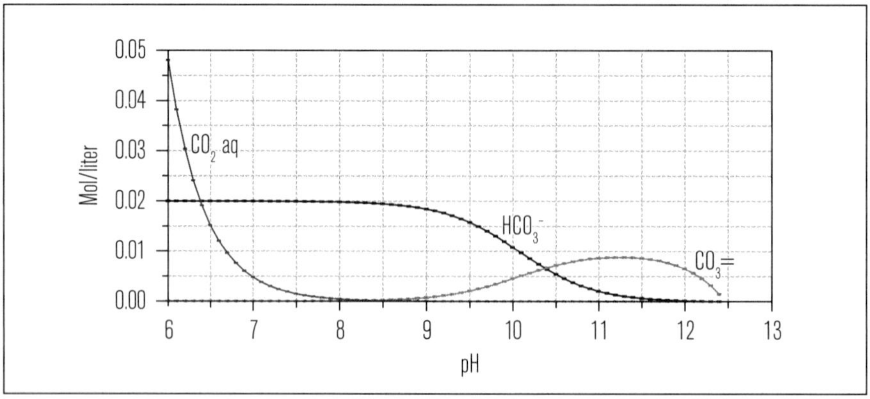

Figure 21.18 - Equilibrium between ionic species in the system, (NaOH) = 0.02 m solution and CO_2.

Since the concentrations of other species remain virtually unchanged, the procedure adopted consists of first, saturating aqueous alkaline solution with CO_2 by insufflation followed by desorption by air insufflation.

The variation (increase) of pH during the procedure is monitored through a A/D data acquisition board and then related, via the above calculations, to the variation of concentration of CO_2.

A macroscopic kinetic model can be proposed, assuming perfect mixing and mass transport control at the interface:

$$\mathrm{Ln}\left(\frac{C}{Co}\right) = -\left(\frac{K \times A}{V}\right) \times t \tag{21.45}$$

where:

- C is the Concentration of CO_2 (mol/l);
- Co is the initial concentration of CO_2 (mol/l);
- A is the area of the gas liquid interface (m^2);
- V is the liquid volume (l);
- t is the operation time (s);
- K is the coefficient of mass transport at the interface liquid - gas.

Then, the slope of log C versus t, provides the value of the kinetic parameter $\frac{K\,A}{2.303\,V}$ as illustrated in Figures 21.19 and 21.20 for the specific case of a 1:5 scale model of a ladle of 80 tons capacity, with inert gas injection through porous plugs.(SESHADRI, SILVA and SILVA, 2005). As it is suggested the total gas flow rate is more important to degasification than gas distribution between nozzles.

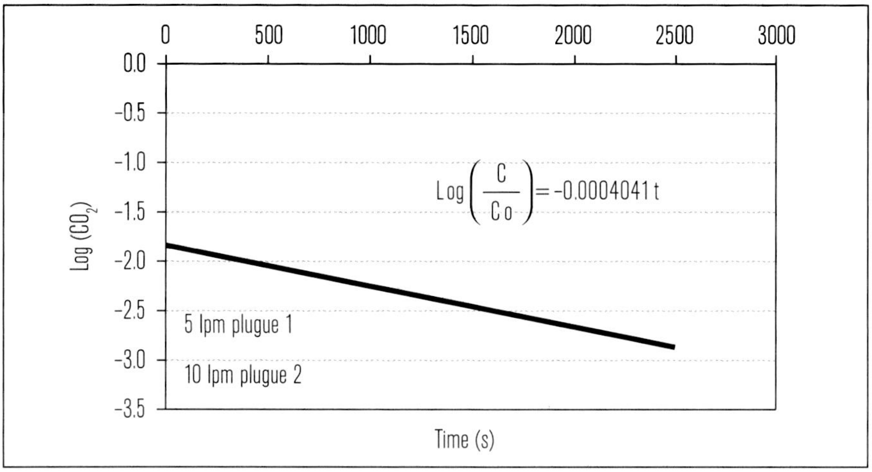

Figure 21.19 - Ttypical curve of variation of CO_2 concentration with time in adsorption/ desorption experiment in NaOH solution.

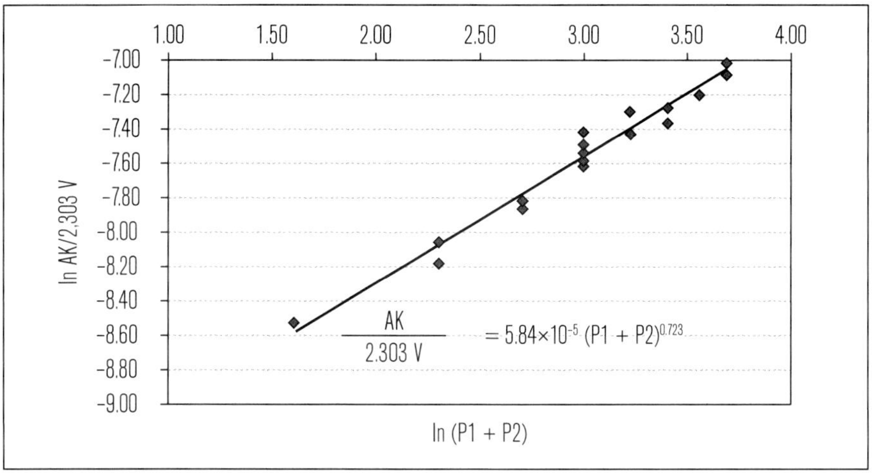

Figure 21.20 - Influence of injection of inert gas, as determined in 1:5 model of ladle, 80-ton rated capacity, P1 and P2 are flow rates of gas through the porous plugs. (SESHADRI et al., 2006).

21.7 Rate of Transfer of a Solute Between Two Liquid Phases

In physical modeling techniques, it is usual to choose water and kerosene or oil to represent bulk or liquid metal phase and slag phase respectively. These are used to simulate the partition of elements like sulfur between the slag and metal in the laboratory using aphyzical model. Many investigations involving top slag, metal and gas systems can be simulated by this technique.

Kerosene dissolves iodine to the extent of approximately 5.7 g/100 ml. On the other hand the solubility of the same in pure water is negligible (about 0.335 g/of water at 25°C). In order to make a significant amount of iodine transferred from kerosene to an aqueous solution an appropriate amount of potassium iodide can be added to the water. After the decomposition reaction the solubility increases considerably.

$$KI = K^{+} + I^{-} \qquad (21.46)$$

$$I_2\left(\text{kerosene}\right) + I^{-} = I_3^{-}\ \left(\text{aqueous}\right) \qquad (21.47)$$

Considering the above fact instead of water, an aqueous solution of potassium iodide can be chosen for simulating liquid steel in a 80 tons ladle with gas injection (25 lpm) through nozzles, located at the bottom. Kerosene saturated with iodine can be used to represent the slag phase Iodine gets transferred from kerosene (slag) into the water (metal) and the transfer rate can be used to estimate the variables influencing this process of mass transfer including the geometrical and operational parameters of a reactor. The amount of iodine in aqueous solution can be determined by titration with sodium thiosulfate (0.006 mol/L in this example). Figure 21.21 shows the required amount of sodium thiosulfate solution for titration of 100 mL of aqueous solution.

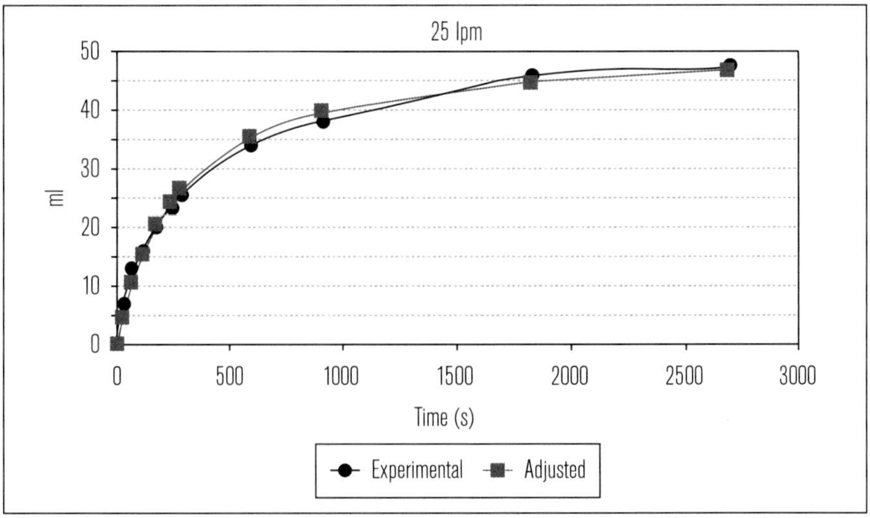

Figure 21.21 - Volume of solution of sodium thiosulfate (0.006 mol/L) needed to titrate 100 ml of aqueous solution; in the physical model of ladle of 80 ton capacity and scale 1:5, while injecting 25 lpm of gas (SESHADRI et al., 2006).

A macroscopic model for mass transfer between the kerosene and the aqueous solution can be written as:

$$V \frac{dC}{dt} = AK\,(C_{eq} - C) \tag{21.48}$$

In the above equation, V is the volume of aqueous solution, C is the concentration of iodine in aqueous solution, A is the area of interface kerosene/water solution, K is mass transfer parameter; C_{eq} is the iodine equilibrium concentration.

This model assumes perfect mixing in the bulk of water and kerosene, except in a region corresponding to the boundary layer.

This equation may be integrated considering the following boundary conditions:

- $t = 0$ $C = 0$, (there is no iodine (I_3^-) in the original aqueous solution)
- $t = \infty$ $C = C_{eq}$ (equilibrium condition obtained after a long period).

$$\frac{C_{eq} - C}{C_{eq}} = e^{-\frac{AK}{V}t} \tag{21.49}$$

After substituting in the above equation the experimental data, (Figure 21.21) the kinetic parameter $\frac{AK}{V}$ can be evaluated by taking the derivative of the equation 21.49:

$$\frac{dC}{dt} / C_{eq} \Big|_{t=0} = AK\ / V \tag{21.50}$$

In most cases, the exact value of A is not known due to slag emulsification. Hence it is usually estimated on the basis of area of cross section of the reactor. As such the kinetics is characterized by the factor $\frac{AK}{V}$. Higher its value , higher is the reaction rate.

The relations obtained can be adjusted by regression in order to identify the value of the kinetic parameter, as given in Figure 21.22.

Figure 21.22 - Kinetic Parameter describing the mass transfer between kerosene (slag) and aqueous solution (steel) for the physical model of a ladle of 80 tons , 1:5 scale, 80 tons capacity, and. Q being the gas flow rate in the model in lpm (SESHADRI et al., 2006).

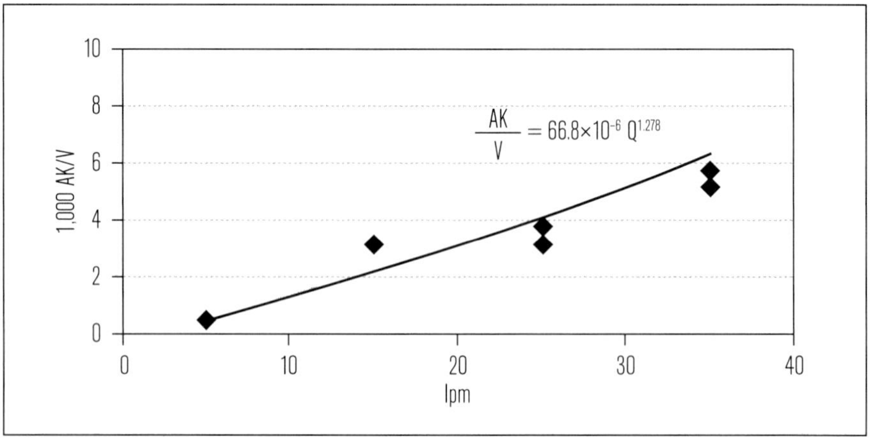

There is still some controversy about the value of the exponent in Q (inert gas flow rate). It varies from ¼ for low gas flow rates to 2.1 for high gas flow rates. This change of behavior is attributed to emulsification of metal/slag phases in the case of higher flow rates. In these experiments, due to low viscosity of kerosene and density very close to that of water, emulsification is significant, especially in the high flow rate region.

21.8 Dissolution of Alloys in Liquid Steel

Any metal or alloy added to a steel melt for the purpose of changing the composition or for deoxidizing etc., has initially an internal temperature gradient at the surface. A solid steel shell forms around it initially (ENGH, 1992). This shell prevents the mass transport between the addition and the liquid steel, and impedes dissolution in the initial period. However the shell starts dissolving, when its temperature reaches steel melt temperature. From that instant , there is regular dissolution. It is possible that an intermetallic compound can form between iron-alloy interface which may have a higher or lower temperature of fusion relative to the steel melt.

Additions of Class I can be defined as those with lower melting temperature compared to the temperature of liquid steel. They are easily dissolved and assimilated in the melt as they are usually in the liquid form after the steel shell melts back. In this situation the mixing time, counted after remelting of the outer shell, would be in the order of 1 ~ 2 minutes, as verified from experimental observations in the ladle model. However there could be situations where in the final stages of the steel shell cycle the additions do not reach melting temperatures, which would require additional period of heating. Figure 21:23 indicates the time periods required for incorporating additions of Class I.

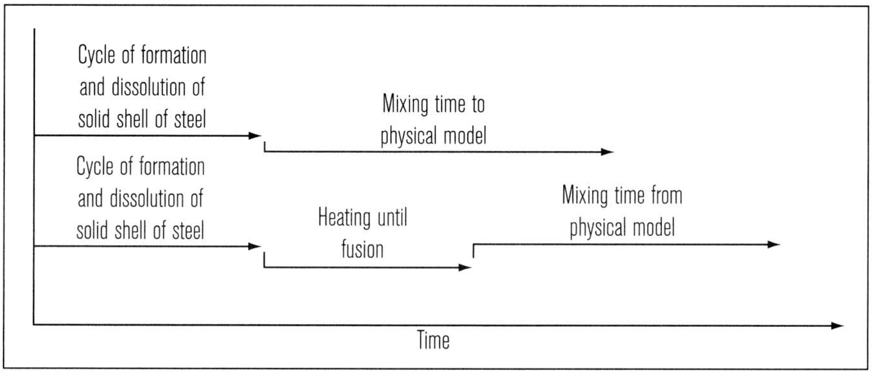

Figure 21.23 - Time periods needed to incorporate additions - Class I, $T_{fusion} < T_{steel}$.

Additions of Class II are classified as those who have melting temperatures above that of the bath. The incorporation of these occurs by dissolution or mass diffusion after the disappearance of the shell, as given in Figure 21.24. Here again, after the period of dissolution, the mixing time determined experimentally for ensuring homogeneity, should be added to get the total time needed for incorporation of the addition.

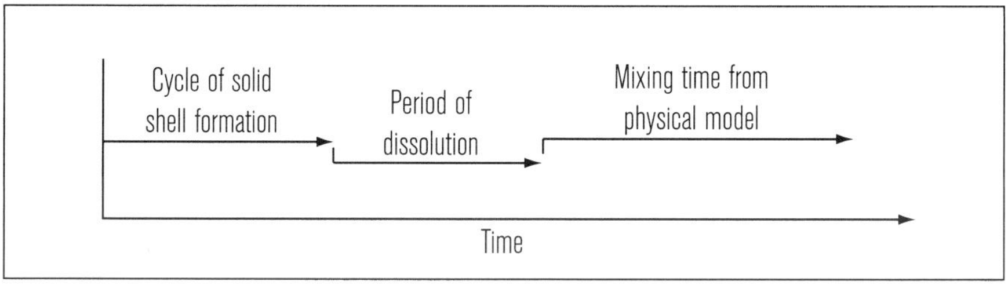

Figure 21.24 - Periods of time needed to incorporate additions Class II, $T_{fusion} > T_{steel}$.

Accurate determination of the cycle period of the appearance and disappearance of the shell of the added material can be ascertained through a mathematical model based on thermophysical properties of steel as well as the material added and variation of these values with temperature as it changes from room temperature to the temperature of the steel melt during addition.

The temperature profile in the particle added (spherical) during cycle of the formation and disappearance of steel shell is, schematically, shown in Figure 21:25.

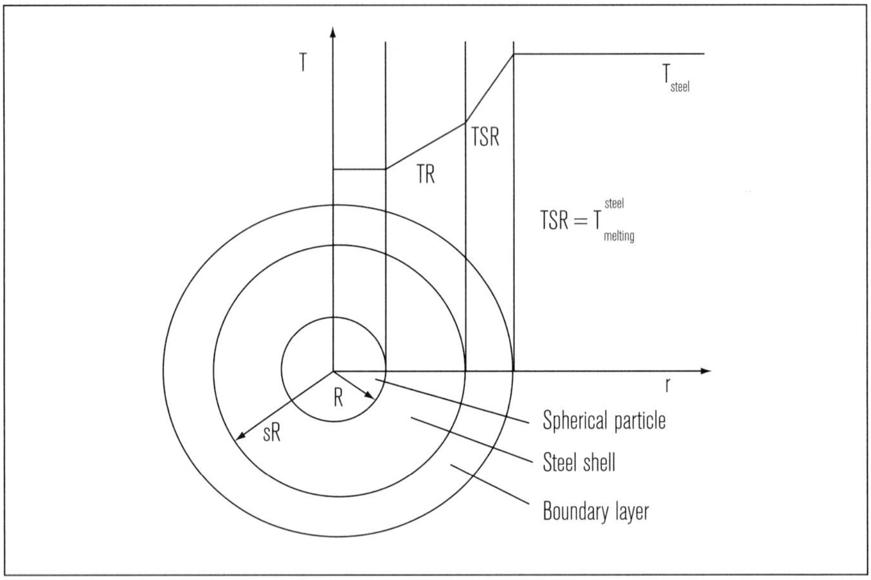

Figure 21.25 - Temperature profile of the spherical particle added with steel shell formation at time t.

The spherical particle of radius R, develops a solidified shell at a given moment changing the radius, equivalent to r = sR. The temperature distribution inside the particle is considered uniform, having a value TR (which varies with time). The temperature of the interface shell/liquid steel, TSR, is the melting temperature of steel, 1810 K. There are temperature gradients in the shell (TSR toTR) and around the shell (T_{steel} to TSR). Assuming a spherical symmetry, the flow of heat through the spherical surface at r = R, at the interface addition/shell, is given by:

$$\text{Flux} = 4\,\pi\, r^2 \left(-K \frac{dT}{dr}\right)\Big|_{r=R} \tag{21.51}$$

A similar equation can be written heat flow for r = sR. Alternatively, assuming quasi- steady state conditions, as discussed in the earlier chapters in heat transfer,one can write:

$$\text{Flux 1} = -\frac{4\,\pi\, K^S(TSR - TR)}{\frac{1}{R} - \frac{1}{sR}}(J/s) \tag{21.52}$$

where K^s (40 W/mK) is the thermal conductivity of solid steel. This quantity of heat is used to heat the core of the particle added. Hence:

$$\frac{4}{3}\pi R^3 \rho C_p \frac{dTR}{dt} = \frac{4\pi K(TSR - TR)}{\frac{1}{R} - \frac{1}{sR}} \tag{21.53}$$

where ρ and C_p represent density and specific heat of the added material

The heat flow to the shell interface from the liquid can be evaluated knowing the heat transfer coefficient though the boundary layer, h:

$$F2 = -4\pi (sR)^2 h(T_{steel} - TSR) \tag{21.54}$$

The heat transfer coefficient can be estimated from the following equations:

$$Nu = 2 + 0.6 \cdot \sqrt{Re} \cdot Pr^{\frac{1}{3}} \tag{21.55}$$

where:

$$Nu = \frac{2\ sR\ h}{K^L} \qquad Re = \frac{\rho \cdot U_r \cdot 2sR}{\eta} \quad Pr = \frac{C_p^L \eta}{K^L} \tag{21.56}$$

K^L and C_p^L represent the thermal conductivity and specific heat of liquid steel, respectively while η is the viscosity for the same.

Typical values of the properties of pure liquid iron are:

- $\eta = 0.3699 \times 10^{-3} \times e^{4982/T(K)}$ Pa.s;
- $\rho = 7015 - 0.883\ (T(°C) - 1536)$ kg/m^3;
- K^L =40 W/m.K;
- C_p^L = 824 J/kg.K.

The value of the relative velocity (U_r) is estimated assuming its maximum value, as the terminal velocity in liquid steel:

$$Re = \frac{2sR\rho U}{\eta} \leq 1 \qquad U = \frac{2}{9\eta}(sR)^2 g \Delta\rho \tag{21.57}$$

$$1 < R_e < 1{,}000 \qquad U = \sqrt{\frac{8gsR\Delta\rho}{3\rho f}} \quad f = \frac{18.5}{Re^{0.6}} \tag{21.58}$$

$$R_e > 1{,}000 \qquad U = \sqrt{\frac{8gsR\Delta\rho}{3\rho f}} \quad f = 0.44 \tag{21.59}$$

In these equations, as discussed in the earlier chapters, f is the friction factor. $\Delta\rho$ is the difference (in module) of densities between addition material together with steel shell and steel melt and can be estimated using the following relation:

$$\Delta\rho = \frac{\rho + (s^3 - 1)\,\rho^S_{steel}}{s^3} - \rho^L_{steel} \qquad (21.60)$$

In the above equation the indices S and L indicate solid and liquid, respectively. And s is defined as the ratio between the radius of the shell (sR) and the radius of the addition (R).

Then the growth of the shell is controlled by the difference between the heat flows (Figure 20.26) . The shelI remelts if more heat arrives (Flux 2) than that crossing sR/R boundary(Flux 1)).

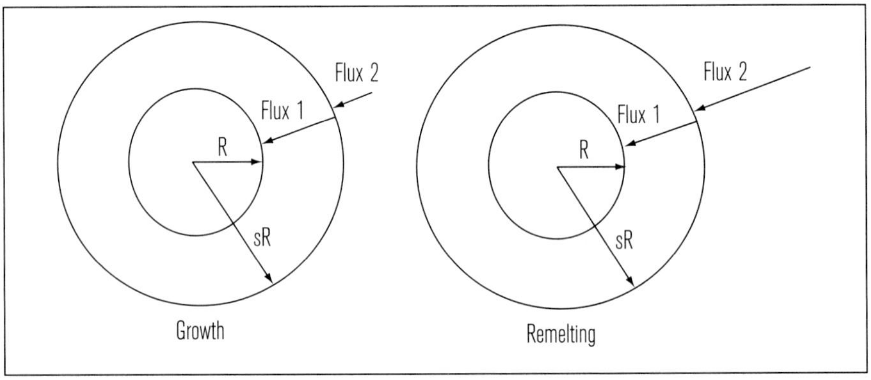

Figure 21.26 - Growth and remelting of the shell in terms of heat flows.

An energy balance considering heat flows that cause melting or solidification at the interface at r = sR, is given by:

$$4\pi R^3 \rho^S_{steel}\, \Delta H_{melting}\, s^2 \frac{ds}{dt} = \text{Flux } 2 - \text{Flux } 1 \qquad (21.61)$$

where:

- $\Delta H_{melting}$ (250 KJ/kg) is the latent heat of melting for steel.

A numerical procedure (Figure 21.27) can be implemented by adopting integration over an infinitesimal interval of time, Δt and initial condition (to avoid singularity)

$t = 0 \qquad\qquad s = 0.001$

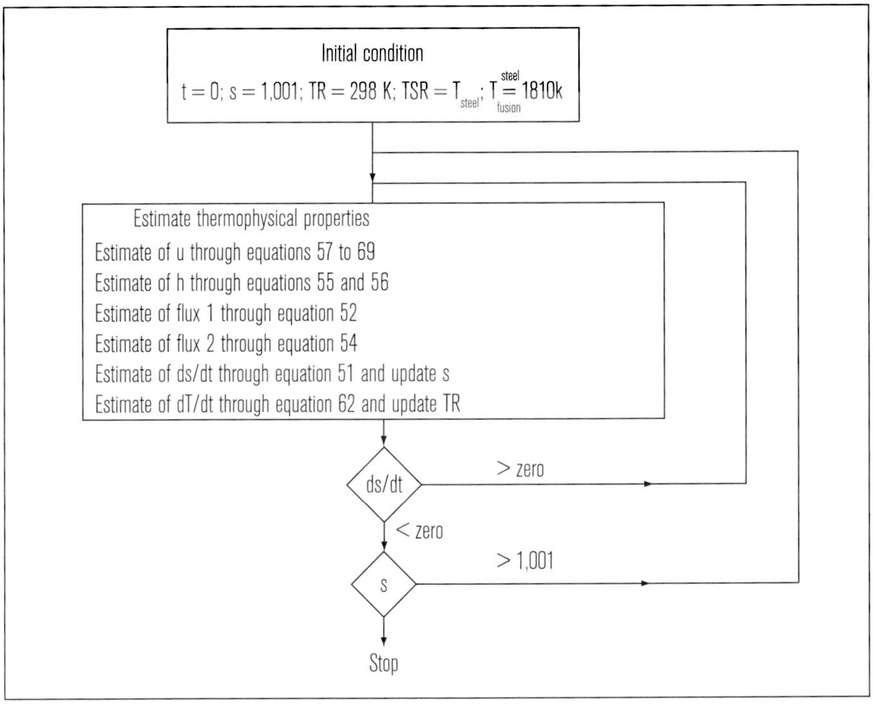

Figure 21.27 - Flow Sheet for evaluating the cycle of formation and remelting of steel shell during addition of an alloy to steel melt. (equation numbers are given without prefix 21).

If the addition belongs to Class I, $T_{melting} < T_{steel}$ there could be a situation, during the progress of remelting of the steel shell, in which the melting temperature is not reached. In this case, additional heating for melting may be necessary and this can be estimated assuming a temperature profile, as outlined in Figure 21.28.

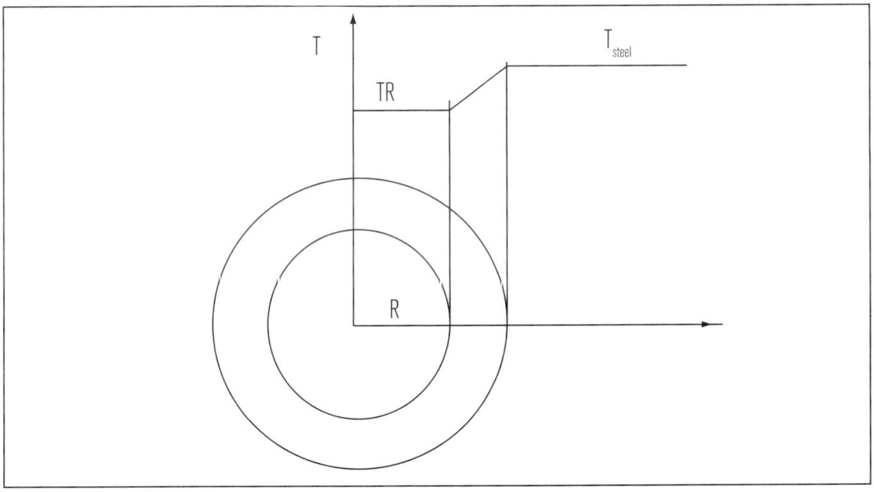

Figure 21.28 - Temperature profile after remelting of steel shell on the surface of the material added.

A balance of energy at the surface of the particle after the melting of the shell gives:

$$\frac{4}{3}\pi R^3 \rho C_p \frac{dTR}{dt} = 4\pi R^2 h (T_{steel} - TR) \qquad (21.62)$$

This equation can be integrated with the initial condition TR = TR° (core temperature at the final stage of shell) to give:

$$t_o = -\frac{R\rho C_p}{3h} \ln \frac{T_{steel} - T_{melting}}{T_{steel} - TR^o} \qquad (21.63)$$

where $T_{melting}$ indicates the melting temperature of addition. This additional time must be added to the time required for melting.

$$\frac{4}{3}\pi R^3 \rho \Delta H^{addition}_{melting} = 4\pi R^2 h (T_{steel} - T_{melting}) t_1 \qquad (21.64)$$

T the value of h can be obtained from the above equations, making s = 1.

Finally, it is to be observed that addition of Class II, those for which $T_{melting} > T_{steel}$ are incorporated by dissolution, diffusion of matter through a boundary layer of liquid steel around them. The dissolution takes place after remelting of the steel shell. A schematic diagram of the composition profile in the boundary layer is given in Figure 21.29.

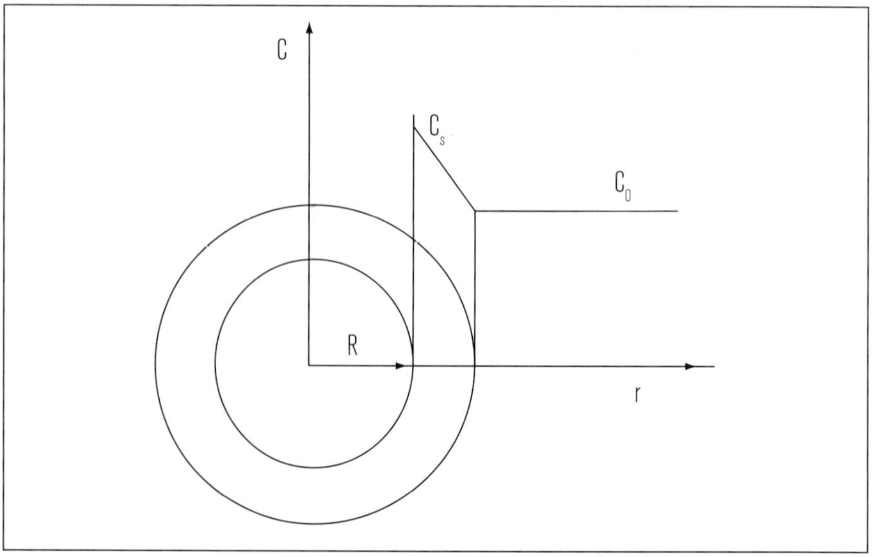

Figure 21.29 - Composition profile for incorporation of addition and dissolution by diffusion through boundary layer.

C_s and C_0 are the interface(between the surface of alloy and steel melt) and bulk concentrations (mol/m³) in the system as given in the figure. For simplifying local equilibrium can be assumed at the interface, in which case the value of C_s would be the saturation value of the material in the steel melt given by the respective phase diagram Fe-X (alloying element).

The radius of the added alloying element X varies as a function of time:

$$\frac{d}{dt}\left(\frac{4}{3}\pi R^3 \rho_X\right) = -4\pi R^2 k_m (C_S - C_O) \qquad (21.65)$$

where:

- ρ_x is the concentration of X in the solid addition, in mol/m³,
- K_m is the mass transfer coefficient through the boundary layer at the interface.

Hence:

$$\frac{dR}{dt} = -\frac{k_m}{\rho_X}(C_S - C_O) \qquad (21.66)$$

C_0 is the concentration of the element dissolving in the steel melt. This varies with time and can be estimated by considering the amount of addition and particle size distribution. A numerical procedure for arriving at the time necessary for the total dissolution of all particles can be formulated. However the worst case scenario in terms of dissolution rate can occur at the end of the process, when the concentration of the element dissolving is close to the nominal concentration of the alloy, C_0^o. In this case the driving force for the dissolution reaches its lowest value. The upper limit for the period of dissolution can be estimated as:

$$\frac{dR}{dt} = -\frac{k_m}{\rho_X}(C_S - C_O^O) \qquad (21.67)$$

Integrating this equation:

$$t = \frac{R_O \rho_X}{k_m (C_S - C_O^O)} \qquad (21.68)$$

where:

- C_0^o and R_0 represent the nominal alloy concentration after complete dissolution and the initial radius of the material added respectively.

K_m is to be evaluated for the conditions prevailing during additions in the liquid steel, considering temperatures varying from the steel shell temperature to the melt temperature. That is for some value of temperature in between T_{steel} and $T_{melting}^{steel}$.

Similar to the calculation of the heat transfer coefficient the mass transfer coefficient also can be evaluated from the following equations.

$$Sh = 2 + 0.552 \cdot \sqrt{Re} \cdot Sc^{\frac{1}{3}} \tag{21.69}$$

where:

$$Sh = \frac{2R\,k_m}{D} \qquad Re = \frac{\rho \cdot U_r \cdot 2R}{\eta} \qquad Sc = \frac{\eta}{\rho D} \tag{21.70}$$

and D is the diffusion coefficient of the alloying element in the liquid steel.

Values of D can be estimated from the equation:

$$D = \frac{k_B T}{2\pi\eta d}\left(\frac{55.85 + m_2}{2\,m_{Fe}}\right)^{1/2} \tag{21.71}$$

where:

- $k_B = 1.38 \times 10^{-23}$ (J/K) Boltzmann's constant;
- d(m) is the diameter of the atomic species (2) that diffuses (Figure 21.71);
- m_2(g/mole) is the atomic mass of species (2) that diffuses;
- T is the absolute temperature.

Table 21.4 - Some values of atomic diameter, $\overset{o}{A}$ for calculation of diffusion coefficient of different metals.

Element	Diameter	Element	Diameter	Element	Diameter	Element	Diameter
Li	3.04	Ca	3.94	Co	2.5	Cd	2.98
Be	2.24	Ti	2.90	Ni	2.5	Sn	3.02
In	3.72	V	2.62	Cu	2.56	Ba	4.34
Mg	3.20	Cr	2.50	Zn	2.66	Yes.	2.86
Al	2.86	Mn	2.74	Nb	2.863	W	2.74
P	2.20	Fe	2.48	Mo	2.72	Pt	2.78

It is apparent that the time required for formation and remelting of the shell around the particles and the process of dissolution are all controlled by simultaneous heat and mass transfer at the interface of liquid/solid. The parameters controlling these phenomena, namely k_m and h can be evaluated assuming that the bath is well stirred, in which the relative velocity between solid and steel would be approximately equal to the terminal velocity. In the absence of agitation the conditions of heat and mass transfer are affected with significant reductions in the values of k_m and h. Consequently the dissolution rates become smaller and the assimilation of added material gets delayed. An estimate of this effect can be obtained by making u = 0 in the calculations carried out earlier.

In order to evaluate the effect of of different levels of agitation the following equation can be used (SZEKELLY and EL-KADDAH, 1981).

$$\frac{k_m 2R}{D} = 2 + 0.4 \left\{\frac{\varepsilon (2R)^4}{\nu^3}\right\}^{1/4} Sc^{1/3} \tag{21.72}$$

where:

- $\nu = \frac{\eta}{\rho}$ is the kinematic viscosity,
- $\varepsilon (m^2 s^{-3})$ the rate of energy dissipation per unit mass of bath.

The corresponding equation for evaluating the heat transfer coefficient is:

$$\frac{h\ 2R}{K} = 2 + 0.4 \left\{\frac{\varepsilon (2R)^4}{\nu^3}\right\}^{1/4} Pr^{1/3} \tag{21.73}$$

Figures 21.30, 21.31 and 21.32 provide estimates of the cycle of shell formation and melting obtained according to the methodology described earlier. The simulation conditions are given in Table 21.5.

Table 21.5 - Thermophysical properties for the calculation of shell thermal conductivity (k), specific heat (Cp), density (ρ); (LS) = Liquid steel; SS = Solid steel; AD = addition.

R_0 (m)		k^{LS} W/m.K	40	C_p^{LS} J/K.kg	824	ρ^{LS} kg/m^3	6950.541	k^{SS} W/mK	40
C_p^{SS} J/K.kg	715	ρ^{SS} kg/m^3	8000	ρ^{AD} kg/m^3	5000	C_p^{AD} J/K.kg	500	η Pas	0.005214

As can be seen from the Figure 21.30 the solidified portion reaches an additional thickness of about 20% of the original radius. At the end of the cycle of shell formation and melting the core temperature reaches the temperature of melting of steel (Figure 21.31). The cycle time varies almost linearly with the radius of the addition material and this is important only for additions of greater size (Figure 21.32).

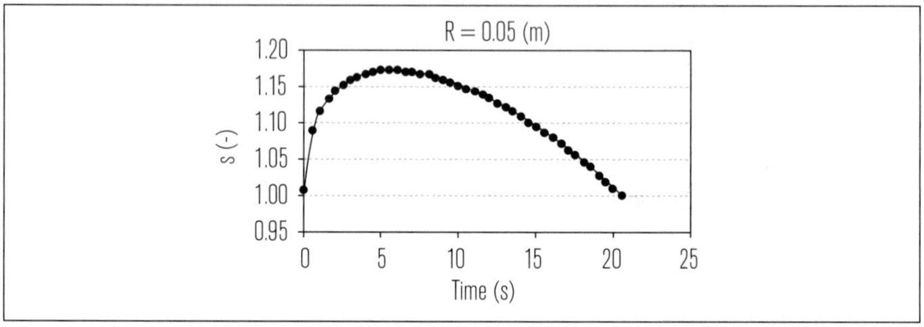

Figure 21.30 - Cycle of formation and melting of shell for a spherical particle with a radius of 0.05 m.

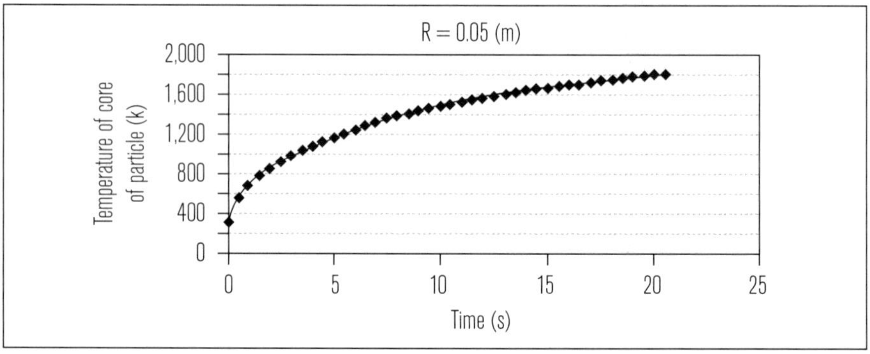

Figure 21.31 - Variation of core temperature with time for an added particle with a radius of 0.05 m.

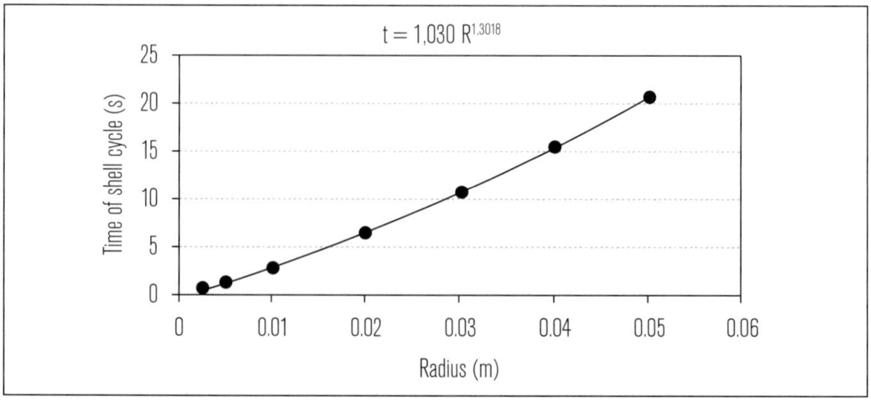

Figure 21.32 - Cycle of formation/fusion of shell as a function of radius of material added.

A comparison between cycle times for the shell formation and melting is given in Table 21.6. The heat transfer coefficient has been estimated assuming:

1) melt/particle slip velocity is the terminal velocity, V_t;

2) particle is in a static bath (Nu = 2);

3) Estimation of rate of energy dissipation using equation 21.72.

The dissipation rate is be estimated for a ladle with 80 tons of steel, with bath height of 2 meters, injection of 300 lpm (stp) and ambient pressure of 0.5 atm.ure of 0.5 atm. Table 21.6 givens the cycle times of shell, t(s), according to three methods of calculating the heat transfer coefficient.

Radius (m)	t (Vt)	t (Nu = 2)	t (dissipation energy)
0.0025	0.41	2.268	0.199
0.005	1.059	9.044	0.431
0.01	2.621	36.161	0.933
0.02	6,382	144.633	2.015
0.03	1.717	325.418	3.142
0.04	15.461	578.517	4.289
0.05	20.54	903.931	5.455

Table 21.6 - Cycle times of of shell, t(s), according to three methods of calculating the heat transfer coefficient.

The data above was plotted in the form of a graph and is given in Figure 21.33. The cycle time is higher for bath without stirring or agitation (Nu = 2). For the other 2 cases the behavior is similar and the cycle times are shorter.

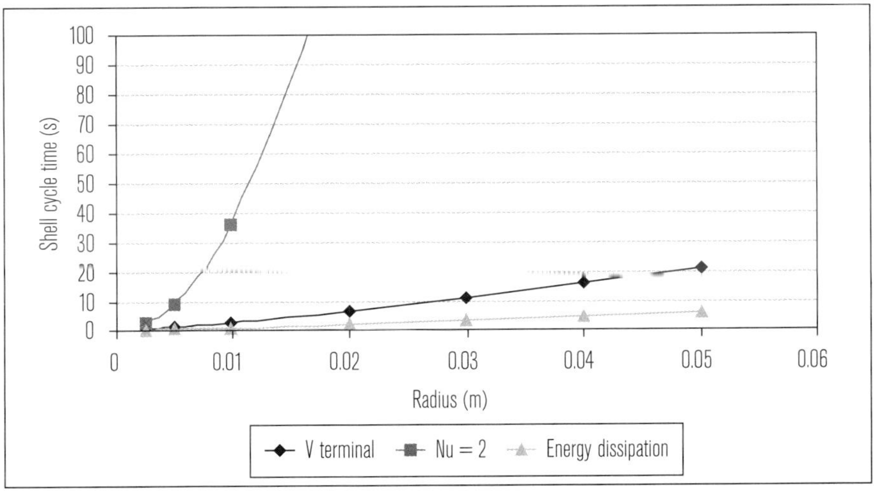

Figure 21.33 – Cycle time for formation and melting of shell for the 3 cases as per Table 21.6 as a function of radius of the particle.

Time of dissolution of Niobium, a Class II addition material, can be evaluated, using the above methodology. The data for Niobium is given in Table 21.7. The time of dissolution can be was evaluated on the basis of mass transfer coefficient , calculated adopting the methodology used for the calculation of heat transfer coefficient in the case of cycle time (Figure 21.33). The results are given in Figure 21.34. The first part of the figure gives the variation in radius of the addition material on the basis of $R = s\ R_0$, ie, the dissolution process ends when s = 0. The variation of s with time is given on the basis of terminal velocity and dissipation energy (the methods give different values of mass transfer coefficient which are responsible for the differences in the rate of dissolution). The next figure gives the dissolution rate for static conditions of bath (Nu = 2). One can see that the rate of dissolution is much slower in this case:

Table 21.7 - Parameters for Niobium for calculating the time of dissolution in a ladle containing liquid steel.

R_0 (m)	0.0025	Csat mole/m^3	22600	Co mole/m^3	2200
ρ kg/m^3	8.570	Visc Pás	0.005214	D m^2/s	2.482 x 10^{-9}
steel kg	80.000	Pa atm	0.5	Level m	2
ε m^2/s^3	0.0565	Atomic wt.	92.91	Gas STP m^3/min	0.3

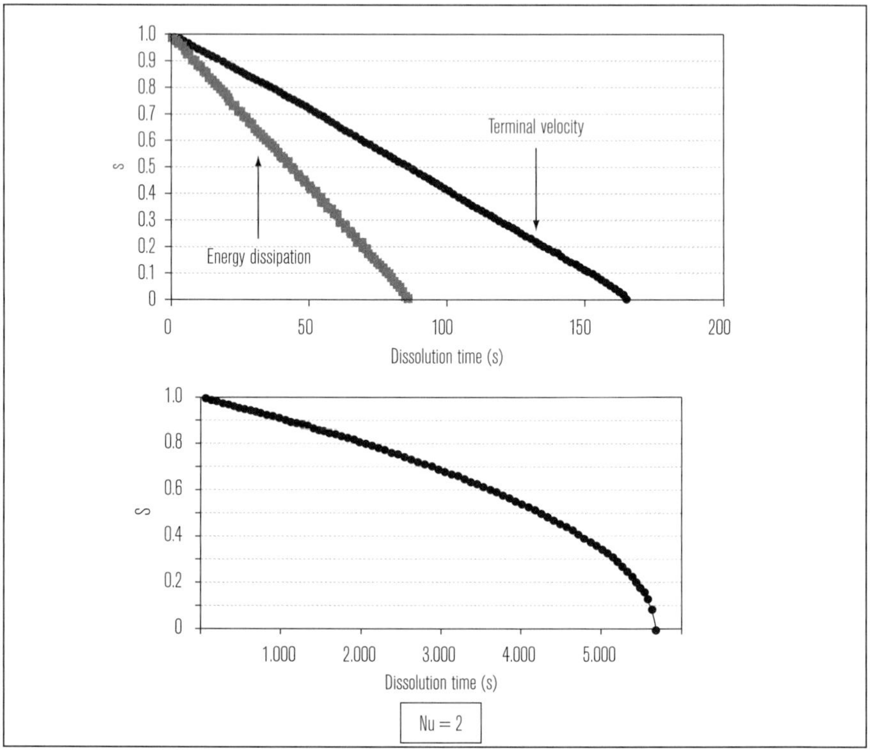

Figure 21.34 - Dissolution times of niobium, according to the three models for calculating the mass transfer coefficient.

Example

Niobium is one of the several elements that can be introduced into the molten steel melt to modify its mechanical properties. Niobium addition, even in small quantities, results in a decrease in the grain size of steel and an increase in the mechanical strength (see figures given in the following pages). Decrease in grain size is the only mechanism for conciliating an increase in mechanical strength with an increase in toughness of steels, This is the reason for the great interest in steels microalloyed with niobium.

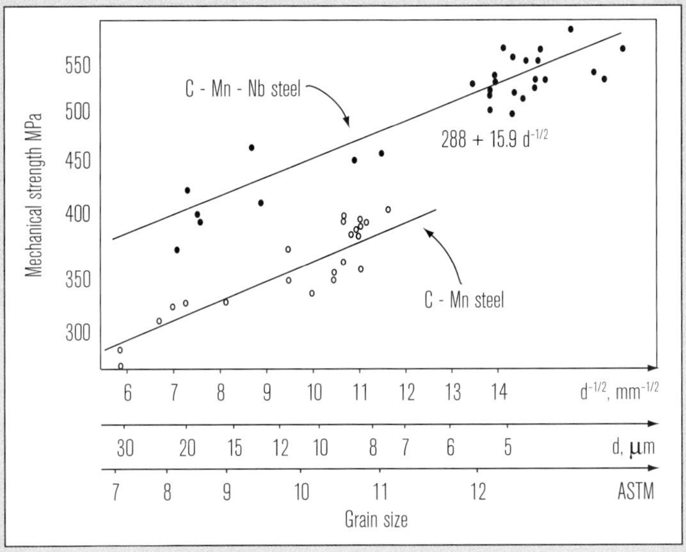

Effect of niobium addition on grain size and mechanical strength of steel.

Additions to liquid steel can be made in various ways and there is always a concern about the consistency and reproducibility of the alloying element recovery (measured as the ratio between the amount of alloying element effectively incorporated and the amount added to the steel). Ferro-niobium additions made on the liquid metal surface, in secondary refining ladles can result in erratic yields, if measures are not taken for preventing its interaction with the slag and atmosphere at the top of the reactor. The use of bath agitation by bubbling with argon accomplishes well this task, as it leaves the surface free for iron alloy additions, resulting in recovery rates exceeding 95%. When a greater control of these additions is required, injection processes of particulates or wires containing the desired element (Figure below) can be used. These processes are recommended, as they promote better contact between the addition and the bath and thereby improve the efficiency of the process.

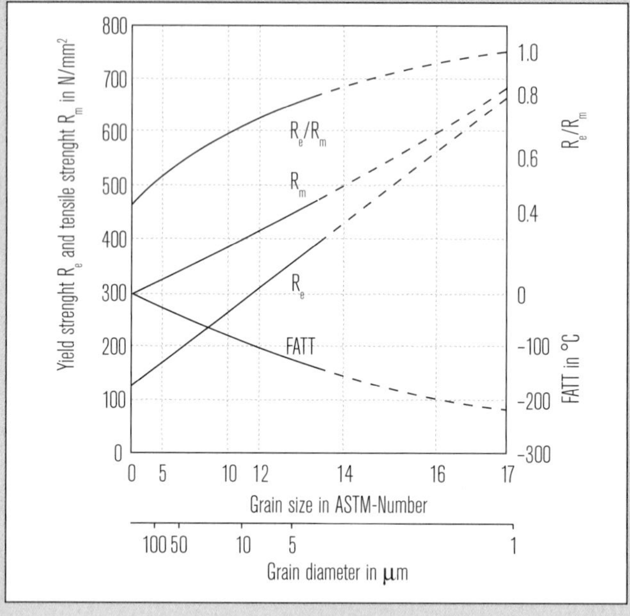

Effect of grain size on mechanical strength and toughness of microalloyed steel.

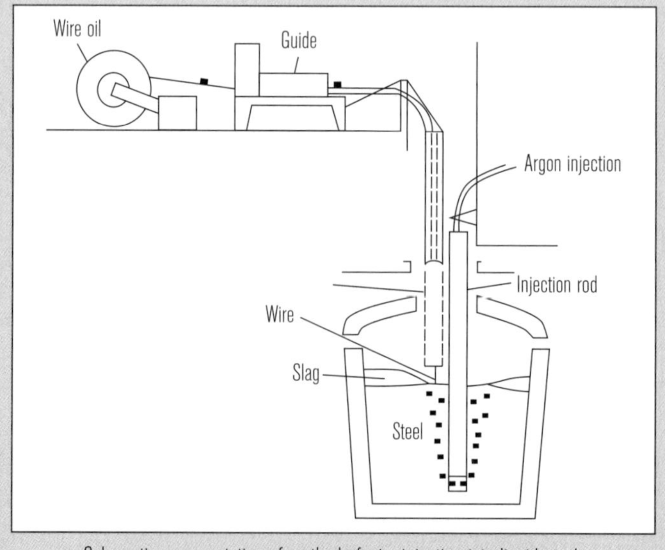

Schematic representation of method of wire injection into liquid steel.

In the wire injection process, wires are unwound from a coil and injected into steel under constant downward vertical speed. The speed value must be chosen to ensure maximum immersion depth (to ensure that the addition released by the wire attains largest possible residence time inside bath), but avoiding hitting the bottom of the ladle, which could lead to mechanical instability. Regardless of these aspects, the alloying elements contained in the wire injected in steel begin to be effectively incorporated only after the period of remelting of the "frozen" steel shell formed around the wire. Niobium injection using wires involves an additional component namely the master alloy. The master alloy (Fe-Nb, 66.5% by weight) is an intermediate compound characterized by fragility, not appropriate for wire forming. Hence a powdered material (<2 mm) is used, jacketed by a low carbon steel cylindrical tube of approximately 13 mm diameter.

Evaluation of the time required for incorporation of a powdered addition, enclosed into a steel cylindrical shell, requires knowing the time period for forming and remelting of the steel layer around the wire, and considering the newly released particles assimilation process.

The formation of shell on the jacket surface can be described considering a cylindrical coordinates system, centered on the wire's longitudinal axis, and symmetry, as denoted by the following equations:

$$\frac{1}{r}\frac{\partial}{\partial r}\left(k\, r\frac{\partial T}{\partial r}\right) = \rho\, u\, C_p \frac{\partial T}{\partial z} \qquad \text{Stationary observer}$$

$$\frac{1}{r}\frac{\partial}{\partial r}\left(k\, r\frac{\partial T}{\partial r}\right) = \rho C_p \frac{\partial T}{\partial t} \qquad \text{Observer fixed on the wire}$$

Acceptable boundary conditions include perfect contact between the particles and the jacket inner surface, and between the jacket outer surface and the steel layer (shell) solidified around it, in other words, equal temperatures at these points. Because of symmetry, the temperature is minimum at the center of the wire, and a convective heat transport condition is applicable at the interface between the solidified layer and the liquid steel, at moving position sR ($s \geq 1$). T_m is the average temperature of the steel, and T_s is the shell surface temperature. This condition is given by:

$$q = -h\, 2\pi(sR)dL\,(T_m - T_s)$$

The convective heat transfer coefficient can be estimated from the following:

$$Nu = 0.029\,Re^{0.80}\,Pr^{0.33} \qquad \text{For liquidmetals } Pr < 1$$

$$\frac{hL}{k} = 0.029\left(\frac{L\rho u}{\eta}\right)^{0.80}\left(\frac{C_p\,\eta}{k}\right)^{0.33}$$

In this expression, L represents an expected value for wire penetration into the metallic bath. Specific values for the liquid steel properties, used for this estimate, would be 40 w/m.K, 7,000 kg/m^3, 6.5 mPas, 824 J/K.kg, for thermal conductivity, density, viscosity and specific heat, respectively. Note that the heat transfer coefficient increases as the wire speed increases, which helps in reducing the life period of existence of the solidified shell. On the other hand, the point of detachment of the addition would tend to be deeper. These contradictory tendencies are not analyzed in this example. However the assumed speed value amounts to 1.5 m/s.

The position of the moving interface between the solidified shell and the liquid metal is defined by the thermal balance that relates the quantity of heat extracted by the shell , the quantity of heat provided by the surrounding metal and the quantity of heat due to transformation of phase at r = sR, (as given in the figure below) The thermal balance is given by.

$$-2\pi(sR)dL\,K^S\frac{dT}{dr} + h\,2\pi(sR)dL\,(T_m - T_s) = -\Delta H\,2\pi(sR)dL\frac{d\,sR}{dt}\rho^S$$

where:

- K^S, ρ^S indicate properties of solid steel, thermal conductivity and density respectively,
- ΔH enthalpy of melting.

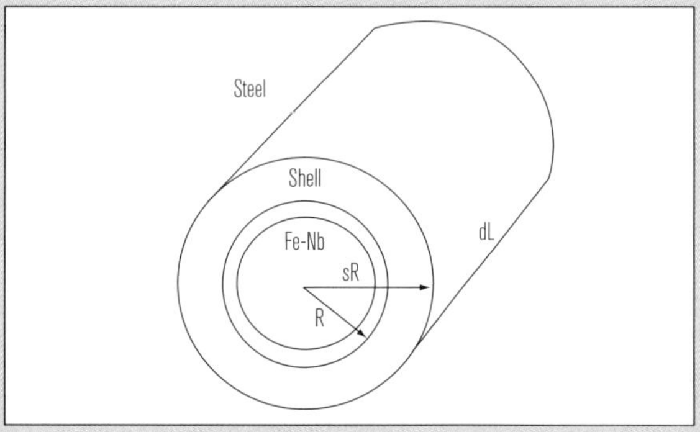

Cross-section of wire for iron-niobium addition; typical values are 13 mm external radius (R), and 0.4 mm wall thickness.

An approximate mathematical solution for this problem can be achieved considering a temperature profile as depicted in the figure given below.

First, the heat flux (per unit of length of wire) passing through the shell, at a given time can be estimated as:

$$\text{Flux}\,1 = -\frac{2\pi K^S(TSR - TR)}{\ln s}(J/m.s)$$

This quantity of heat is used to heat the iron-niobium core. The resulting temperature increase is given by:

$$\pi R^2 \rho C_p \frac{dTR}{dt} = \frac{2\pi K^S(TSR - TR)}{\ln s}$$

In this equation, ρ is the iron-niobium density (approximately equal to 8100 kg/m^3) and C_p is its specific heat. In the absence of specific data, it can be estimated as 510 J/Kg.K, by using the Kopp-Neumann equation. Using Kopp-Neumann law, the specific heat C_p per mass unit (in J•kg^{-1}•K^{-1}) for alloys with N components can be calculated with the following equation:

$$C_p = \sum_{i=1}^{i=N} C_p^i X_i$$

where C_p^i and X_i represent the specific heat and the weight fraction of each component in the alloy added

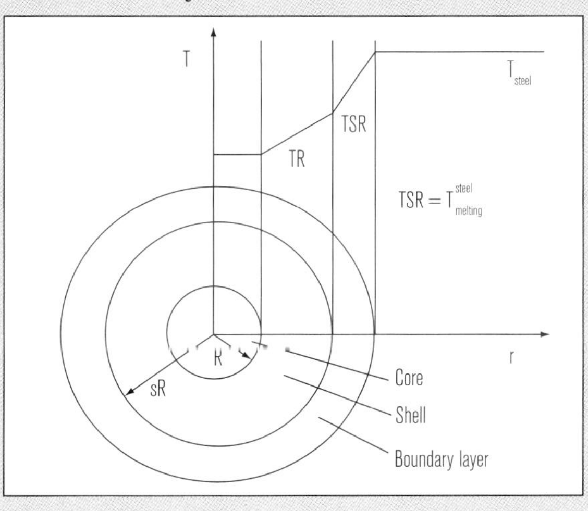

Temperature profile assumed for semi-analytical calculation of the shell cycle.

Moreover, the quantity of heat passing through the boundary layer around the shell, towards the core, is given by:

$$\text{Flux 2} = -h\, 2\pi(sR)(T_{steel} - TSR)\ J/m.s$$

which would allow estimating the instantaneous shell radius variation as:

$$2\pi sR \frac{d\, sR}{dt} \rho^S \Delta H = -(\text{Flux 1} - \text{Flux 2})$$

Here, the enthalpy of steel melting ΔH = 267000 J/kg, and the specific mass of solid steel, ρ^s = 7300 kg/m^3.

The following figure shows the variation of s parameter as well as of the core temperature as a function of time (estimated from the scheme of equations given above). It can be noted that the duration of the solidified shell is approximately 2 seconds. This suggests that the iron-niobium particles would be released from an approximate depth of 3 m.

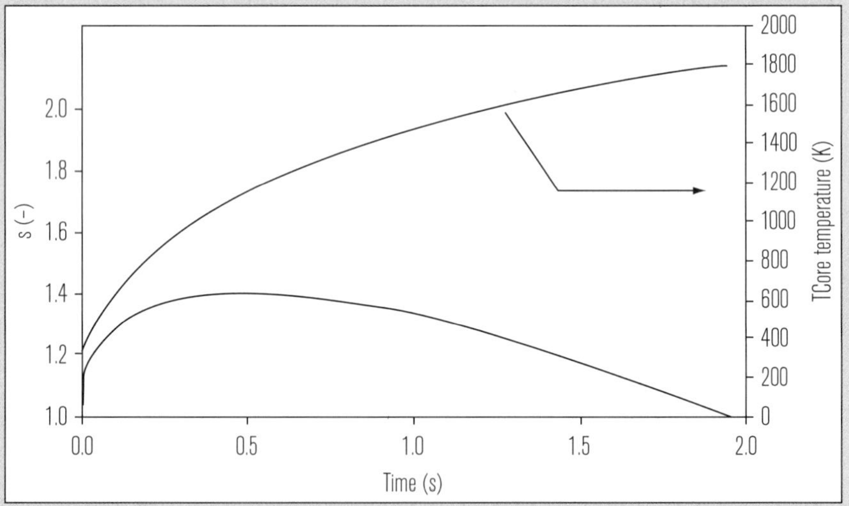

Shell thickness and wire core temperature, as a function of time. Wire diameter of 13 mm; superheat of 80 K; wire speed of 1.5 m/s.

Once released from the wire, the iron-niobium particles can interact with the liquid steel. A "typical" steel manufacturing temperature of 1600°C has been considered. The reported melting range (liquid/solid) of an iron-niobium alloy (66% by weight) is 1580°C ~ 1630°C. It is actually safer to assume that the final stage of incorporation occurs by diffusion (provided that fusion is not completely achieved). In this case, assuming the controlling step as mass transfer at a boundary layer around each particle, it can be written:

$$\pi D^2 \frac{dD/2}{dt} \rho_{Fe-Nb} = -K_m \pi D^2 \Delta C_{Nb}$$

In this expression, D represents the average diameter of the particles, around 2 mm, and ρ_{Fe-Nb} represents the niobium molar density in the alloy. The alloy density is approximately 8100 kg/m^3 (at 66%Nb, 99.9 g/mol) and implies an approximate molar density of 57982 mol/m^3. ΔC_{Nb} is the difference between niobium concentrations at particle/bath interface, and inside bath (away from the particle), it is the driving force for the dissolution process. The first concentration is not known with precision, as it depends on the relative dissolution rates of iron and niobium. As an estimate, one can assume local equilibrium at the liquid steel/iron-niobium particle interface. According to the iron-niobium phase diagram, at 1,600°C, addition of increasing amounts of niobium to the pure liquid iron leads to the formation of a second solid phase at 37.5% by weight of niobium. Hence the concentration at saturation point amounts to C_{Nb}= 8100 x 10^3 g/m^3 x 0.375/92.9 g/mol (in the absence of data on the specific mass of this solution). Moreover, as niobium is used for microalloying, participating in small percentages in the steel, the second concentration may be neglected in comparison with the first.

Finally, the mass transfer coefficient has to be evaluated. Steel and the addition densities are similar and, inside a static bath, the iron-niobium particles would tend to sediment at the bottom of the ladle. This does not occur due to recirculation movements deliberately induced by inert gas or electromagnetic agitation. The magnitude of the mass transfer coefficient K_m depends on the relative velocity between the particle and the fluid. Assuming an intermediate turbulence condition, the terminal fall velocity of the addition into steel here assumed as corresponding to the relative velocity between the addition and the steel is given by:

$$V^2 = \frac{4}{3}\frac{Dg}{f}\frac{\Delta\rho}{\rho} = \frac{4}{3}\frac{0.002 \times 9.81}{f}\frac{8100-7000}{7000} = 4.11 \times 10^{-3}/f$$

Where f is the friction factor

$$f = \frac{18.5}{Re^{0.6}} = \frac{18.5}{(\rho DV / \eta)^{0.6}} = \frac{18.5}{(7000 \times 0.002 \times V / 0.0065)^{0.6}} = \frac{0.185}{u^{0.6}}$$

Then, V = 0.0659 m/s which allows estimating the Reynolds number that validates the calculations:

$$Re = \frac{\rho DV}{\eta} = \frac{7000 \times 0.002 \times 0.0659}{0.0065} = 142$$

In this case, K_m is estimated from the following relation:

$$Sh = 1 + 0.744\ Re^{0.48}\ Sc^{0.333} \qquad Sc > 200;\ 100 < Re < 200$$

$$\frac{K_m D}{D_{Nb-Fe}} = 1 + 0.724 \left(\frac{\rho VD}{\eta} \right)^{0.48} \left(\frac{\eta}{\rho D_{Nb-Fe}} \right)^{0.333}$$

requiring a known value of diffusion coefficient for niobium in liquid iron obtained from the following relation:

$$D_{Nb-Fe} = \frac{KT}{2\pi\ \eta\ d_{atom}} \left(\frac{m_{Fe} + m_{Nb}}{2m_{Nb}} \right)$$

where K is the Boltzmann constant ($1.38\ 10^{-23}\ m^2\ kg\ s^{-2}\ K^{-1}$, T is the temperature (1873 K), η is the steel viscosity ($6.5\ 10^{-3}$ Pa.s), m_{Fe} is the atomic weight of iron (55.85 g/mol), m_{Nb} is the atomic weight of Niobium (92.9 g/mol), and d_{atom} is its atomic diameter of Niobium ($2.86\ 10^{-10}$ m), resulting in D_{Nb-Fe} $3.54\ 10^{-9}\ m/s^2$. It can be noted that the mass transfer coefficient varies throughout the process, as it is a function of particle diameter (decreasing with time), and because the relative velocity between the particle and the steel varies. For these preliminary calculations, an average diameter value of 2 mm is considered, with which the mass transfer coefficient can be evaluated:

$$\frac{K_m D}{D_{Nb-Fe}} = 1 + 0.724 \times 142^{0.48} \left(\frac{0.0065}{7000 \times 3.5410^{-9}} \right)^{0.333} = 52.3$$

giving the value for mass transfer coefficient around 9.26×10^{-5} m/s.

With these values, the average speed of dissolution can calculated as:

$$K_m \pi D^2 \Delta C_{Nb} = 9.26 \times 10^{-5}\ 3.1416 \times 0.002^2 \times 32687 = 3.80 \times 10^{-5}\ mol/s \text{ or } 28\ mg/cm^2.s$$

After appropriate simplifications, the equation can also be integrated to yield the total time of dissolution;

$$t = \frac{D\,\rho_{Fe-Nb}}{2K_m\,\Delta C_{Nb}} = 19.17\,s$$

which implies a total time of assimilation (shell formation/remelting plus dissolution) close to 21 s.

Recent experimental work (CRUZ et al., 2010a,b) shows that dissolution rates of FeNb are considerably higher than those predicted from diffusional dissolution models. As a matter of fact, FeNb starts melting at temperatures below the normal temperature range of molten steel refining (around 1600°C). Hence the final step in the dissolution of this alloy in molten steel could be simple fusion of the same followed by dispersion to the surrounding steel bath. Hence the rate of dissolution should be faster , compared to the process of diffusion controlled dissolution.

The following figure presents experimental data on the incorporation of Niobium in steel as a function of time by sampling time to time. First the low carbon steel melt was deoxidized, and Ferroniobium particles with dimensions of 7 mm or 15 mm in diameter were added to the bath at 1575°C or 1625°C. It can be observed that the concentration of Nb in the metallic bath reaches the desired values between 0.05% and 0.06% within around 15 s.

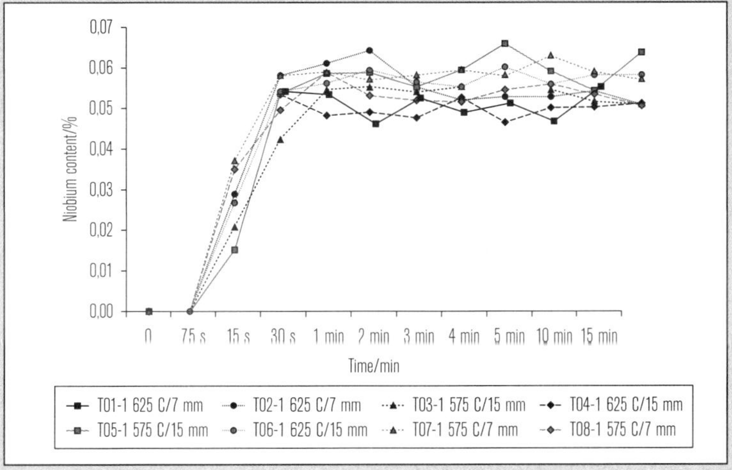

Evolution of the concentration of Niobium in molten steel, for temperatures of 1575°C and 1625°C and particle diameters of FeNb 7 mm and 15 mm (CRUZ et al., 2010a,b).

When the solid particles of FeNb are added to the steel melt at around 1600 deg C. The particles get heated due to heat transfer accompanied by a certain drop in the temperature of the melt near those locations. This is followed by the formation of a solid steel shell around the Ferroniobium particle. This shell attains the maximum thickness and thereafter, when the particle attains the temperature of steel melt, it begins to remelt. After complete remelting of the shell, the Ferroniobiun particle comes into contact with the steel melt. Then the dissolution starts through a combined mechanism of melting and dispersion. This mechanism of incorporation of the alloy by stages is represented in the following figure.

Stage I (A $\rightarrow$ B): Formation of a solidified steel shell surrounding the FeNb particle Stage (B $\rightarrow$ C) Remelting of the solid shell in the liquid steel,

Stage (C $\rightarrow$ D) Fusion of core particle FeNb (this would be followed by dispersion liquid/liquid).

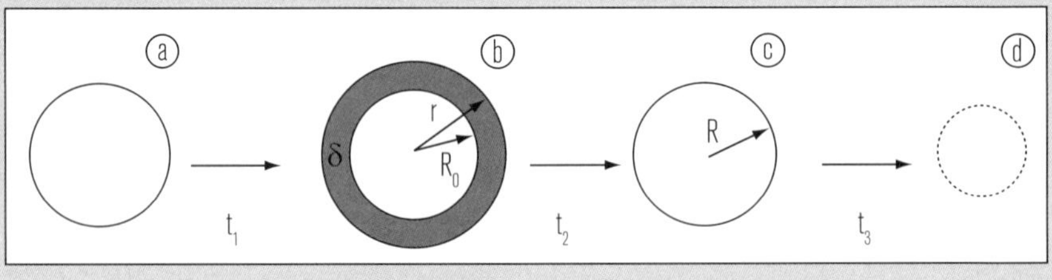

Stages in the process of dissolution of a spherical particle of FeNb in liquid steel.

The interval of time required for remelting of the shell can be estimated by an energy balance considering that first the heat is transferred from liquid steel to the solid- liquid interface This is then used for the fusion of the solidified layer The energy balance for this stage can be written as follows:

1) Heat transferred from the melt to the particle = area x heat transfer coefficient x temperature difference:

$$4\pi r^2 h \Delta\theta \text{ (Joule/s)}$$

Where, h can be substituted from the Nusselt number

$$Nu = \frac{h\ 2r}{k} \text{ or } \frac{Nu\ k}{2r} = h$$

Hence,

$$4\pi r^2 \frac{Nu\,k}{2r} \Delta\theta \text{ (Joule / s)}$$

This quantity of heat energy is necessary for heating the shell up to the steel melt temperature. The rate of this heat transfer can be given as a function of r, the outer radius , which includes the shell thickness:

$$\left\{4\pi r^2 \frac{dr}{dt}\right\} \rho_{steel} \left[L_{fusion}^{steel} + c_p^{steel} \Delta\theta\right] \text{ (Joule / s)}$$

After separation of variables and integration with limits, defining the shell thickness, one can get the time of dissolution of the shell:

$$t_2 = \int_{R_O}^{R_o+\delta} \frac{2\rho_{steel}\left[L_{fusion}^{steel} + c_p^{steel}\Delta\theta\right]}{k\,Nu} r\,dr$$

In the above equation Ro, δ, ρ_{steel}, L_{fusion}^{steel}, c_p^{steel}, $\Delta\theta$, k, Nu and r represent respectively initial radius of the particle, thickness of the shell, density of solid steel, latent heat of fusion of steel, specific heat of liquid steel, superheat of liquid steel in terms of temperature, thermal conductivity of liquid steel, Nusselt number and the radius of the particle with the shell.

In the case of the experimental results described above for the dissolution, if natural convection is the controlling step, the correlation between Nusselt number in terms of Grashoff and and Prandtl numbers is given by:

$$Nu = 2 + 0.6\,Gr^{1/4}\,Pr_f^{1/3};\; Gr = \frac{(2r)^3 \rho_{steel,\,liquid}\,\Delta\rho\, g}{\mu_{steel}^2};\; Pr = \frac{c_p^{steel}\,\mu_{steel}}{k}$$

Nusselt number as well as the time for shell dissolution can be evaluated for typical values given below: $\rho_{steel\ liquid}$ = 7000 kg/m^3; ρ_{steel} = 7,600 kg/m^3; c_p^{steel} = 596 J/kg.K; k = 33 W/m.K, μ_{steel} = 6.5 miliPa.s; L_{fusion}^{steel} = 267000 J/kg. The buoyancy factor in the Grashoff equation can be evaluated considering thermal expansion of liquid steel , where is coefficient of thermal expansion equal to 1.25 x 10^{-4} K^{-1}, for liquid steel; $\Delta\theta$ can be assumed to be around 50°C.

The duration of the first stage namely the formation of solidified shell is taken as approximately equal to the time for necessary for remelting of the same. Hence:

$$t_1 = t_2 = \int_{R_o}^{R_o+\delta} \frac{2\rho_{steel}\left[L_{fusion}^{steel} + c_p^{steel}\Delta\theta\right]}{k\ Nu}\, rdr$$

If the controlling step in the last stage is heat transfer, the time required for melting and dispersion of Ferronobium particle is given by:

$$t_3 = \int_{0}^{R_o} \frac{2\rho_{FeNb}\left[L_{fusion}^{FeNb} + c_p^{FeNb}\Delta\theta\right]}{k\ Nu}\, rdr$$

In this equation, ρ_{FeNb} is the density of Ferroniobium, c_p^{FeNb} is the specific heat of the liquid phase of the alloy , L_{fusion}^{FeNb} is the latent heat of fusion, k is the thermal conductivity of the liquid phase and c_p^{FeNb}, L_{fusion}^{FeNb} and k are respectively specific heat, heat of fusion and thermal conductivity. The Nusselt number can be evaluated using the following correlation.

$$Nu = 2 + 0.6\,Gr^{1/4}\ Pr_f^{1/3};\ Gr = \frac{(2R)^3 \rho_{FeNb,\,liquid}\ \Delta\rho\ g}{\mu_{FeNb}^2};\ Pr = \frac{c_p^{FeNb}\ \mu_{FeNb}}{k}$$

In the above correlation, $\Delta\rho = \rho_{FeNb} - \rho_{steel}$ and it can be assumed that the viscosity value of liquid Ferroniobium (in the absence of reliable data) is close that of the steel melt, μ_{steel}.

With the above procedure one can calculate the total time of dissolution of Ferroniobium in steel melt adopting this mechanism.

The following figure compares the results of the model based on the above mechanism with that of a model ,which assumes that the last stage is diffusion controlled (SISMANIS, 1987). The times predicted by the former model for the dissolution process of Ferroniobium are closer to the experimental results. . For example, according to this model the time of incorporation is around 25 seconds for particles of 1.5 cm whereas the diffusion controlled model predicts , it would be more than 60 seconds.

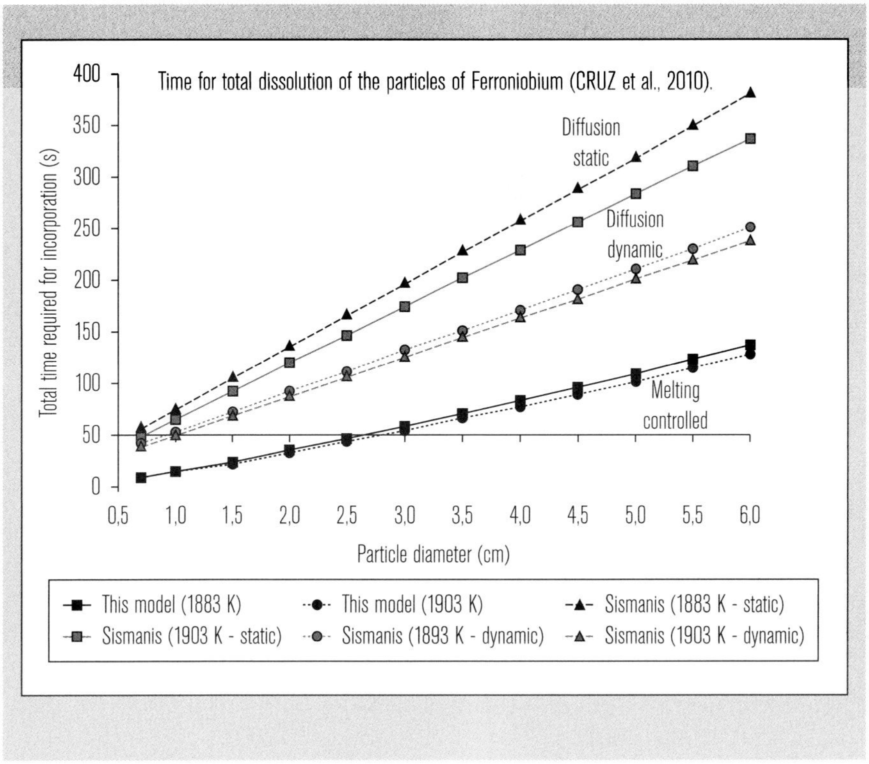

References

CRUZ, E.B. et al. Dissolução de ferronióbio em aços líquidos e principais recomendações práticas para sua adição. In: SEMINÁRIO DE ACIARIA - INTERNACIONAL, 41., 2010, Resende, RJ. *Anais*... São Paulo: ABM, 2010a. p. 188-200.

CRUZ, E.B. Dissolution of ferroniobium in liquid stel and best practicew to incrase niobium recovery during ladle refining. In: BAOSTELL BAC 2010. *Proceedings*... [S.l: s.n.], 2010b. p. B-34-39.

ENGH, T.A. *Principles of metal refining*. Oxford: Oxford Science, 1992.

POIRIER, D.H.; GEIGER, G.H. *Transport phenomena in materials processing*. Warrendale: TMS, 1994.

SAHAI, Y.; EMI, T. Criteria for water modeling of melt flow and inclusion removal in continuous casting tundish. *ISIJ International*, v. 36, n. 9, p. 1166-1173, 1996.

SAHAI, Y.; GUTHRIE, R.I.L. Hydrodynamics of gas stirred melts: Part II. Axisymmetric flows. *Metallurgical Transactions B*, v. 13, n. 2, p. 203-211, 1982.

SESHADRI, V. et al. Decarburization rates in RH-KTB degasser of CST steel plant through physical modeling study. *Ironmaking and Steelmaking*, v. 33, n. 1, p. 34-38, 2006.

SESHADRI, V.; SILVA, C.A.; SILVA, I.A. Physical Modeling simulations of Refining Processes in Brazilian Steel Industry. *Scandinavian Journal of Metallurgy*, v. 34, p. 340-352, 2005.

SESHADRI, V. et al. Operational improvements for RH-KTB degasser of Companhia Siderúrgica de Tubarão (CST) steel plant through a physical modeling study. In: Metallurgical and materials processing. Principles and technologies. v. 2: High temperature metals production. *TMS*, 2003. p. 701-713.

SESHADRI, V. et al. Fluid flow and mixing characteristics in RH degasser of Companhia Siderurgica de Tubarão and influence of bottom gas injection and nozzle blockage through physical modeling study. *Ironmaking and steelmaking*, v. 31, n. 1, p. 37-42, 2004.

SISMANIS, P. G. *The dissolution of niobium and zirconium in liquid steel.* 1987. 274 p. Thesis (Department of Mining and Metallurgicall Engineering) – McGill University, Montreal, 1987.

STREETER, V.L. *Fluid mechanics.* New York: McGraw Hill, 1958.

SZEKELLY, J.; EL-KADDAH, N. A mathematical model for desulphurization kinetics in argon stirred ladle. *Ironmaking and Steelmaking*, v. 8, p. 269, 1981.

SZEKELLY, J.; CARLSSON, G.; HELLE, L. *Ladle metallurgy, materials research and engineering.* New York: Springer, 1989.

WOLF, M.M. Slab caster Tundish configuration and operation: a review. In: HEASLIP, L.J. *Continuous casting.* Warrendale: Iron and Steel Society, 2003. v. 10, p. 22-35.

Exercises

1 - A physical model of LD, is to be constructed using water and air as working fluids. One of the criteria of similarity on the interaction between the jet of oxygen and the metal bath is the dimensionless modified Weber number.

$$We = \frac{\rho_g u^2}{(\rho g \sigma)^{1/2}}$$

which is the ratio of inertial force of the oxygen jet (causing the cavity on the metal surface and the ejection of metal drops) and forces involving the weight of the metal and interfacial tension (responsible for the integrity of the bath). Another dimensionless number that is important is the momentum number which is a measure of transfer of momentum from the jet to the metal bath. Greater the distance between the oxygen lance and the bath less is the force due to inertia of the jet, acting on the liquid metal This is quantified by the momentum number (h_a is the lance to bath distance).

$$M = \frac{\rho_g u^2 d^2}{\rho g h_a^3}$$

Consider typical values of properties of gas and liquid in the industrial system (oxygen/steel) and the model (air/water), presented in the table and find the flow rate of air in the model as a function of the scale factor.

Typical properties of densities and interfacial tension of the fluids in model and prototype.

	ρ_g(kg/m³)	σ (N/m)	ρ (kg/m³)
Prototype	1	1.8	7000
Model	1	0.070	1000

2 - The dimensions of a tundish are (in meters) 1.5 (depth) x 0.8 (width) x 4.5 (length) and the same is used for casting rate of 5 tons per minute. The first one third part of the tundish behaves as a perfect mix reactor in the vertical direction. There is no longitudinal or vertical mixing in the other portions of the reactor (the remaining of the reactor is of piston flow). What is the proportion of inclusions 45 μm diameter that is dragged to the mold? Inclusion is silica, for which $\rho_{SiO_2} = 2.32$ g/cm^3 and consider for steel $\eta = 7 \times 10^{-3}$ Pa.s.

3 - Consider a bubble of calcium, 3 cm in diameter, initially at 50 cm depth in a bath of steel at 1550°C. Estimate the pressure of calcium in the bubble and the number of moles contained in it. Estimate the velocity of rise of the bubble and the time required for the same to rise up to 45 cm depth. Estimate the mass transfer coefficient for the boundary layer around the bubble, assuming D_{Ca} (diffusion coefficient of calcium in metal) is equal to 4×10^{-4}cm^2/s. The bubble can be regarded as a rigid sphere. Consider the relationship between partial pressure and the concentration of calcium dissolved in the metal bath which is Ca (mol/cm^3) = 3.5×10^{-4} P_{Ca} (atm). Assume local equilibrium at the gas-metal interface and find the concentration of calcium in this interface. Assuming that the calcium concentration is negligible in the metallic phase, estimate the flow of calcium due to the composition gradient. Estimate the amount of dissolved of calcium when the bubble rises from 50 cm to 45 cm deep.

4 - To study the pressure drop in the bed of a pig iron blast furnace, a model in scale 1:20, is to be constructed on the basis of similarity principles. A typical unit has the following characteristics.

1) mean diameter of the reactor equal to 9.3 (m);
2) gas flow rate equal to 7,300 (Nm^3/min);
3) gas composition 20%CO, 24% CO_2, 54 % N_2;
4) average temperature of 1,073 K, average pressure of 1.5 atm;
5) void fraction of the bed equal to 0.45 and shape factor of particles equal to 1;
6) an average diameter of coke and ore equal to 5(cm) and 1.5 (cm), respectively.

Estimate the gas flow through the bed using Ergun equation.What are the laminar and turbulent contributions? Estimate their relative importance. Simplify the Ergun equation in terms of dimensionless numbers like Reynolds, $\frac{\rho D_p u}{6\eta(1-w)}$ Euler and a third one involving void fraction and shape factor.

The pressure drop in packed bed operation like in the blast furnace is mainly caused by the gas flow through the voids in the bed and the friction offered by solid surfaces during flow. However there is pressure drop contribution due to the friction offered by the walls of the reactor. In the following figure (POIRIER and GEIGER, 1994) it is shown that the ratio of the pressure drop measurement (including pressure drop for the bed and that due to walls) and the theoretical loss (estimated by the equation of Ergun) is a function of Reynolds number and the ratio of diameter of the reactor and diameter of the particles. When the fraction of the Y axis is closer to 1 the wall effect is minimum (also in cases where the container diameter is very large compared to the particle diameter, more than 20 times).

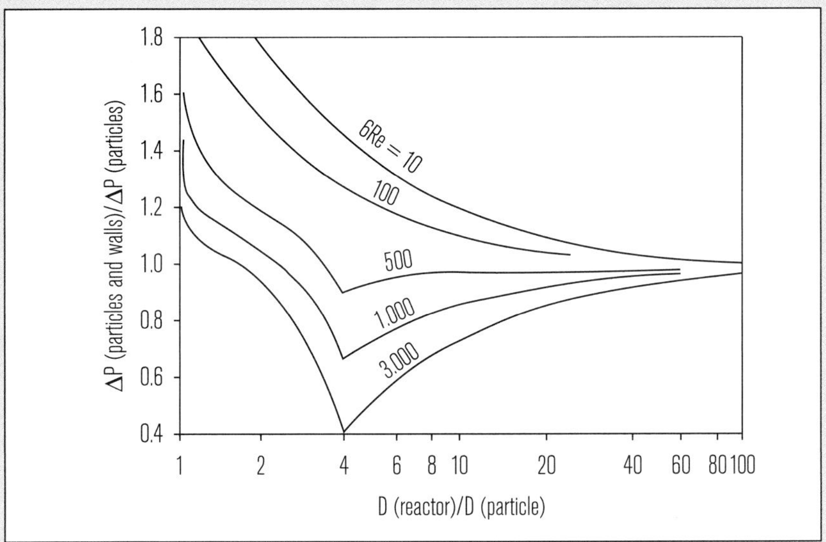

Verify the validity of wall effect being negligible in an actual blast furnace and translate this condition to the model (find the order of magnitude of the particles to be used in the model). Provide an example of the calculations for determining the operating conditions (flowrate of air and particle diameter) in the model. Note that the particles may not obey the ratio 1:20 used to scaling the reactor and that there are plenty of diameter choices of convenience. Also the particles should be as large as possible, to facilitate handling, visualization in the bed and to minimize surface effects that arise when the size is too small.

5 - It is proposed to model the flow of Al (900°C) inside a reactor, utilizing water (25°C) to simulate the flow of metal. Determine the scale factor to be employed. The relevant properties are:

Water:

- $\rho = 1000\ kg/m^3$;
- $\eta = 10^{-3}$ Pa.s.

Aluminum:

- $\rho = 2.385 - 0.28\ (T - T_f)\ kg/m^3$;
- $\eta = 0.1492\ 10^{-3}\ e^{1984/T}$ Pa.s;
- $T_f = 946$ K.

6 - A continuous fluidized bed reactor is used for drying of spherical and uniform particles. The fluidized bed behaves like a perfect mix reactor, and the drying process is controlled by heat transfer through the boundary layer in the gas particle interface. Consider the following parameters:

- h = coefficient of heat transfer, $J/m^2.C^{-1}.s^{-1}$;
- T_g = the gas temperature (K);
- T_p = the particle temperature (K);
- R = radius of particle (m);
- H = Enthalpy involved in heating and vaporization (J/kg water);
- ρ = density of the particle (kg/m^3);
- X = weight fraction of water, (kg water/kg of particle);
- M = is the mass of particles in the fluidized bed (kg);
- G = mass flow of particles $Kg.s^{-1}$.

Find the time necessary for total drying of the particle, and the expression that relates the mass fraction of water, X, with time.

Find the mean residence time of particles in the reactor, and the expression for the distribution of residence times.

Estimate the average fraction of moisture, if the average residence time is equal to three times the time required for complete drying of a particle (some particles shortcircuit, even if the mean residence time is large and leave the reactor in moist condition).

7 - It is intended to use a combination of model reactors to characterize the performance of a leaching tank operating in continuous regime. Assume that the dissolution is controlled by mass transport through the boundary layer, and topochemical in nature Find a expression relating residence time and dissolution. What reactor would be more efficient, piston flow or perfect mix? Assume that the following are known:

- QL and QS = Volumetric flow rates of liquid and solids respectively;
- VR = working volume of the reactor;
- Ro = the particle radius;
- ρ = the molar density of the metal to be leached in the particle;
- C_{eq} and C = the equilibrium concentration and concentration of the metal in the liquor in the reactor respectively;
- K = is the mass transfer coefficient in the boundary layer.

8 - Liquid metal is stirred by a gas injected through porous plug located at the bottom of a ladle. Experiments suggest that there is a functional relationship between the mixing time, τ and energy density, ε which is of the type:

$$\tau = a.\ \varepsilon^{-b}$$

It is also known that the mixing energy is proportional to the flow rate of gas and it would take triple the flow in order to reduce the time by half. What is the relationship between flow and the consumption of gas, needed to reduce the mixing time by 80% of the original?

9 - Addition of oxidants or alloying elements in steel by injecting in the form of wires is a common practice. In this process, the wire is injected vertically with constant speed, so that penetration is maximum (and decreasing losses by contact with the slag or atmosphere). This way high recoveries can be obtained. Care must be taken so that the wire does not reach the bottom of the reactor, causing deflection of the wire (operational instabilities) and premature wear of the coating. To study such a system with the help of a cold model a preliminary analysis has to be made to consider the effectiveness of the operation to establish an appropriate combination of time of residence of the wire inside the metal (L/u) and the time required for the assimilation of the addition, $\left(t = \frac{\rho Cp}{Sh}\right)$. Like in the earlier cases the data given are:

- L = level of liquid in the steel ladle;
- u = speed of wire injection;
- ρ = density of addition;
- Cp = specific heat of the addition;
- S = ratio of surface area/volume of the wire;
- h = is the heat transfer coefficient between wire and fluid.

Consider that the heat transfer coefficient also depends on the injection speed of the wire, and is given by:

$$Nu = a\, Re^{0.8}\, Pr^{0.33}$$

Find a relationship that would be useful to determine the speed of the wire in the model (water at 60°C) in terms of speed in the industrial system (Steel liquid at 1600°C, aluminum at 25°C).

10 - Make a comparison between kinetic energy of gas being injected in a steel ladle and the energy due to isothermic gas expansion. Consider that the metal density is 7000 kg/m^3, steel level at 2.5 m; 300 tons of steel, gás flow rate of 0.5 m^3/min (stp) and gas nozzle of 1 mm diameter. (five of them).

Index